PROTEIN AND PEPTIDE FOLDING, MISFOLDING, AND NON-FOLDING

WILEY SERIES IN PROTEIN AND PEPTIDE SCIENCE

VLADIMIR N. UVERSKY, Series Editor

Metalloproteomics • Eugene A. Permyakov

Instrumental Analysis of Intrinsically Disordered Proteins: Assessing Structure and Conformation • Vladimir Uversky and Sonia Longhi

Protein Misfolding Diseases: Current and Emerging Principles and Therapies • Marina Ramirez-Alvarado, Jeffery W. Kelly, Christopher M. Dobson

Calcium Binding Proteins • Eugene A. Permyakov and Robert H. Kretsinger

Protein Chaperones and Protection from Neurodegenerative Diseases • Stephan Witt

Transmembrane Dynamics of Lipids • Philippe Devaux and Andreas Herrmann

Flexible Viruses: Structural Disorder in Viral Proteins • Vladimir Uversky and Sonia Longhi

Protein and Peptide Folding, Misfolding, and Non-Folding • Reinhard Schweitzer-Stenner

PROTEIN AND PEPTIDE FOLDING, MISFOLDING, AND NON-FOLDING

Edited by
REINHARD SCHWEITZER-STENNER

A JOHN WILEY & SONS, INC., PUBLICATION

Copyright © 2012 by John Wiley & Sons, Inc. All rights reserved

Published by John Wiley & Sons, Inc., Hoboken, New Jersey
Published simultaneously in Canada

No part of this publication may be reproduced, stored in a retrieval system, or transmitted in any form or by any means, electronic, mechanical, photocopying, recording, scanning, or otherwise, except as permitted under Section 107 or 108 of the 1976 United States Copyright Act, without either the prior written permission of the Publisher, or authorization through payment of the appropriate per-copy fee to the Copyright Clearance Center, Inc., 222 Rosewood Drive, Danvers, MA 01923, (978) 750-8400, fax (978) 750-4470, or on the web at www.copyright.com. Requests to the Publisher for permission should be addressed to the Permissions Department, John Wiley & Sons, Inc., 111 River Street, Hoboken, NJ 07030, (201) 748-6011, fax (201) 748-6008, or online at http://www.wiley.com/go/permissions.

Limit of Liability/Disclaimer of Warranty: While the publisher and author have used their best efforts in preparing this book, they make no representations or warranties with respect to the accuracy or completeness of the contents of this book and specifically disclaim any implied warranties of merchantability or fitness for a particular purpose. No warranty may be created or extended by sales representatives or written sales materials. The advice and strategies contained herein may not be suitable for your situation. You should consult with a professional where appropriate. Neither the publisher nor author shall be liable for any loss of profit or any other commercial damages, including but not limited to special, incidental, consequential, or other damages.

For general information on our other products and services or for technical support, please contact our Customer Care Department within the United States at (800) 762-2974, outside the United States at (317) 572-3993 or fax (317) 572-4002.

Wiley also publishes its books in a variety of electronic formats. Some content that appears in print may not be available in electronic formats. For more information about Wiley products, visit our web site at www.wiley.com.

Library of Congress Cataloging-in-Publication Data:

Protein and peptide folding, misfolding, and non-folding / edited by Reinhard Schweitzer-Stenner.
  p. cm. – (Wiley series in protein and peptide science ; 13)
  Includes bibliographical references and index.
  ISBN 978-0-470-59169-7
  1. Protein folding. 2. Peptides. I. Schweitzer-Stenner, Reinhard.
  QP551.P75 2012
  572'.633–dc23
                                              2011044305

Printed in the United States of America

10 9 8 7 6 5 4 3 2 1

CONTENTS

Introduction to the *Wiley Series on Protein and Peptide Science*   xiii

Preface   xv

Contributors   xix

INTRODUCTION   1

1   **Why Are We Interested in the Unfolded Peptides and Proteins?**   3
Vladimir N. Uversky and A. Keith Dunker

    1.1   Introduction, 3
    1.2   Why Study IDPs?, 4
    1.3   *Lesson 1:* Disorderedness Is Encoded in the Amino Acid Sequence and Can Be Predicted, 5
    1.4   *Lesson 2:* Disordered Proteins Are Highly Abundant in Nature, 7
    1.5   *Lesson 3:* Disordered Proteins Are Globally Heterogeneous, 9
    1.6   *Lesson 4:* Hydrodynamic Dimensions of Natively Unfolded Proteins Are Charge Dependent, 14
    1.7   *Lesson 5:* Polymer Physics Explains Hydrodynamic Behavior of Disordered Proteins, 16
    1.8   *Lesson 6:* Natively Unfolded Proteins Are Pliable and Very Sensitive to Their Environment, 18
    1.9   *Lesson 7:* When Bound, Natively Unfolded Proteins Can Gain Unusual Structures, 20

1.10 *Lesson 8:* IDPs Can Form Disordered or Fuzzy Complexes, 25
1.11 *Lesson 9:* Intrinsic Disorder Is Crucial for Recognition, Regulation, and Signaling, 25
1.12 *Lesson 10:* Protein Posttranslational Modifications Occur at Disordered Regions, 28
1.13 *Lesson 11:* Disordered Regions Are Primary Targets for AS, 30
1.14 *Lesson 12:* Disordered Proteins Are Tightly Regulated in the Living Cells, 31
1.15 *Lesson 13:* Natively Unfolded Proteins Are Frequently Associated with Human Diseases, 33
1.16 *Lesson 14:* Natively Unfolded Proteins Are Attractive Drug Targets, 35
1.17 *Lesson 15:* Bright Future of Fuzzy Proteins, 38
Acknowledgments, 39
References, 40

I CONFORMATIONAL ANALYSIS OF UNFOLDED STATES 55

2 Exploring the Energy Landscape of Small Peptides and Proteins by Molecular Dynamics Simulations 57
Gerhard Stock, Abhinav Jain, Laura Riccardi, and Phuong H. Nguyen

2.1 Introduction: Free Energy Landscapes and How to Construct Them, 57
2.2 Dihedral Angle PCA Allows Us to Separate Internal and Global Motion, 61
2.3 Dimensionality of the Free Energy Landscape, 62
2.4 Characterization of the Free Energy Landscape: States, Barriers, and Transitions, 65
2.5 Low-Dimensional Simulation of Biomolecular Dynamics to Catch Slow and Rare Processes, 67
2.6 PCA by Parts: The Folding Pathways of Villin Headpiece, 69
2.7 The Energy Landscape of Aggregating $A\beta$-Peptides, 73
2.8 Concluding Remarks, 74
Acknowledgments, 75
References, 75

3 Local Backbone Preferences and Nearest-Neighbor Effects in the Unfolded and Native States 79
Joe DeBartolo, Abhishek Jha, Karl F. Freed, and Tobin R. Sosnick

3.1 Introduction, 79
3.2 Early Days: Random Coil—Theory and Experiment, 80
3.3 Denatured Proteins as Self-Avoiding Random Coils, 82
3.4 Modeling the Unfolded State, 82

- 3.5 NN Effects in Protein Structure Prediction, 86
- 3.6 Utilizing Folding Pathways for Structure Prediction, 87
- 3.7 Native State Modeling, 88
- 3.8 Secondary-Structure Propensities: Native Backbones in Unfolded Proteins, 92
- 3.9 Conclusions, 92
  Acknowledgments, 93
  References, 94

4 Short-Distance FRET Applied to the Polypeptide Chain 99
Maik H. Jacob and Werner M. Nau

- 4.1 A Short Timeline of Resonance Energy Transfer Applied to the Polypeptide Chain, 99
- 4.2 A Short Theory of FRET Applied to the Polypeptide Chain, 101
- 4.3 DBO and Dbo, 105
- 4.4 Short-Distance FRET Applied to the Structured Polypeptide Chain, 107
- 4.5 Short-Distance FRET to Monitor Chain-Structural Transitions upon Phosphorylation, 116
- 4.6 Short-Distance FRET Applied to the Structureless Chain, 120
- 4.7 The Future of Short-Distance FRET, 125
  Acknowledgments, 125
  Dedication, 126
  References, 126

5 Solvation and Electrostatics as Determinants of Local Structural Order in Unfolded Peptides and Proteins 131
Franc Avbelj

- 5.1 Local Structural Order in Unfolded Peptides and Proteins, 131
- 5.2 ESM, 134
- 5.3 The ESM and Strand-Coil Transition Model, 137
- 5.4 The ESM and Backbone Conformational Preferences, 138
- 5.5 The Nearest-Neighbor Effect, 141
- 5.6 The ESM and Cooperative Local Structures—Fluctuating β-Strands, 141
- 5.7 The ESM and β-Sheet Preferences in Native Proteins— Significance of Unfolded State, 144
- 5.8 The ESM and Secondary Chemical Shifts of Polypeptides, 145
- 5.9 Role of Backbone Solvation in Determining Hydrogen Exchange Rates of Unfolded Polypeptides, 148
- 5.10 Other Theoretical Models of Unfolded Polypeptides, 148
  Acknowledgments, 149
  References, 149

6 Experimental and Computational Studies of Polyproline II Propensity 159
W. Austin Elam, Travis P. Schrank, and Vincent J. Hilser

- 6.1 Introduction, 159
- 6.2 Experimental Measurement of PII Propensities, 161
- 6.3 Computational Studies of Denatured State Conformational Propensities, 168
- 6.4 A Steric Model Reveals Common PII Propensity of the Peptide Backbone, 172
- 6.5 Correlation of PII Propensity to Amino Acid Properties, 175
- 6.6 Summary, 180
  - Acknowledgments, 180
  - References, 180

7 Mapping Conformational Dynamics in Unfolded Polypeptide Chains Using Short Model Peptides by NMR Spectroscopy 187
Daniel Mathieu, Karin Rybka, Jürgen Graf, and Harald Schwalbe

- 7.1 Introduction, 187
- 7.2 General Aspects of NMR Spectroscopy, 189
- 7.3 NMR Parameters and Their Measurement, 191
- 7.4 Translating NMR Parameters to Structural Information, 202
- 7.5 Conclusions, 213
  - Acknowledgments, 215
  - References, 215

8 Secondary Structure and Dynamics of a Family of Disordered Proteins 221
Pranesh Narayanaswami and Gary W. Daughdrill

- 8.1 Introduction, 221
- 8.2 Materials and Methods, 223
- 8.3 Results and Discussion, 226
  - Acknowledgments, 235
  - References, 235

II DISORDERED PEPTIDES AND MOLECULAR RECOGNITION 239

9 Binding Promiscuity of Unfolded Peptides 241
Christopher J. Oldfield, Bin Xue, A. Keith Dunker, and Vladimir N. Uversky

- 9.1 Protein–Protein Interaction Networks, 241
- 9.2 Role of Intrinsic Disorder in PPI Networks, 242
- 9.3 Transient Structural Elements in Protein-Based Recognition, 243

9.4 Chameleons and Adaptors: Binding Promiscuity of Unfolded Peptides, 256
9.5 Principles of Using the Unfolded Protein Regions for Binding, 262
9.6 Conclusions, 266
Acknowledgments, 266
References, 266

10 Intrinsic Flexibility of Nucleic Acid Chaperone Proteins from Pathogenic RNA Viruses 279
Roland Ivanyi-Nagy, Zuzanna Makowska, and Jean-Luc Darlix

10.1 Introduction, 279
10.2 Retroviruses and Retroviral Nucleocapsid Proteins, 280
10.3 Core Proteins in the *Flaviviridae* Family of Viruses, 288
10.4 Coronavirus Nucleocapsid Protein, 290
10.5 Hantavirus Nucleocapsid Protein, 291
Acknowledgments, 293
References, 293

III AGGREGATION OF DISORDERED PEPTIDES 307

11 Self-Assembling Alanine-Rich Peptides of Biomedical and Biotechnological Relevance 309
Thomas J. Measey and Reinhard Schweitzer-Stenner

11.1 Biomolecular Self-Assembly, 309
11.2 Misfolding and Human Disease, 310
11.3 Exploitation of Peptide Self-Assembly for Biotechnological Applications, 326
11.4 Concluding Remarks, 340
Acknowledgments, 340
References, 340

12 Structural Elements Regulating Interactions in the Early Stages of Fibrillogenesis: A Human Calcitonin Model System 351
Rosa Maria Vitale, Giuseppina Andreotti, Pietro Amodeo, and Andrea Motta

12.1 Stating the Problem, 351
12.2 Aggregation Models: The State of The Art, 354
12.3 Human Calcitonin hCT as a Model System for Self-Assembly, 356
12.4 The "Prefibrillar" State of hCT, 358
12.5 How Many Molecules for the Critical Nucleus?, 361
12.6 Modeling Prefibrillar Aggregates, 366
12.7 hCT Helical Oligomers, 366

12.8 The Role of Aromatic Residues in the Early Stages of Amyloid Formation, 372
12.9 The Folding of hCT before Aggregation, 373
12.10 Model Explains the Differences in Aggregation Properties between hCT and sCT, 374
12.11 hCT Fibril Maturation, 375
12.12 α-Helix →β-Sheet Conformational Transition and hCT Fibrillation, 377
12.13 Concluding Remarks, 378
Acknowledgments, 378
References, 379

13 Solution NMR Studies of Aβ Monomers and Oligomers 389
Chunyu Wang

13.1 Introduction, 389
13.2 Overexpression and Purification of Recombinant Aβ, 390
13.3 Aβ Monomers, 393
13.4 Aβ Oligomers and Monomer–Oligomer Interaction, 403
13.5 Conclusion, 406
References, 406

14 Thermodynamic and Kinetic Models for Aggregation of Intrinsically Disordered Proteins 413
Scott L. Crick and Rohit V. Pappu

14.1 Introduction, 413
14.2 Thermodynamics of Protein Aggregation—the Phase Diagram Approach, 415
14.3 Thermodynamics of IDP Aggregation (Phase Separation)— MPM Description, 420
14.4 Kinetics of Homogeneous Nucleation and Elongation Using MPMs, 425
14.5 Concepts from Colloidal Science, 427
14.6 Conclusions, 433
Acknowledgments, 433
References, 434

15 Modifiers of Protein Aggregation—From Nonspecific to Specific Interactions 441
Michal Levy-Sakin, Roni Scherzer-Attali, and Ehud Gazit

15.1 Introduction, 441
15.2 Nonspecific Modifiers, 442
15.3 Specific Modifiers, 454
Acknowledgments, 465
References, 466

16 Computational Studies of Folding and Assembly of Amyloidogenic Proteins   **479**
J. Srinivasa Rao, Brigita Urbanc, and Luis Cruz

- 16.1 Introduction, 479
- 16.2 Amyloids, 480
- 16.3 Computer Simulations, 485
- 16.4 Summary, 514
  - References, 515

INDEX   **529**

INTRODUCTION TO THE *WILEY SERIES ON PROTEIN AND PEPTIDE SCIENCE*

Proteins and peptides are the major functional components of the living cell. They are involved in all aspects of the maintenance of life. Their structural and functional repertoires are endless. They may act alone or in conjunction with other proteins, peptides, nucleic acids, membranes, small molecules, and ions during various stages of life. Dysfunction of proteins and peptides may result in the development of various pathological conditions and diseases. Therefore, the protein/peptide structure–function relationship is a key scientific problem lying at the junction point of modern biochemistry, biophysics, genetics, physiology, molecular and cellular biology, proteomics, and medicine.

The *Wiley Series on Protein and Peptide Science* is designed to supply a complementary perspective from current publications by focusing each volume on a specific protein- or peptide-associated question and endowing it with the broadest possible context and outlook. The volumes in this series should be considered required reading for biochemists, biophysicists, molecular biologists, geneticists, cell biologists, and physiologists as well as those specialists in drug design and development, proteomics, and molecular medicine, with an interest in proteins and peptides. I hope that each reader will find in the volumes within this book series interesting and useful information.

First and foremost I would like to acknowledge the assistance of Anita Lekhwani of John Wiley & Sons, Inc. throughout this project. She has guided me through countless difficulties in the preparation of this book series and her enthusiasm, input, suggestions, and efforts were indispensable in bringing the *Wiley Series on Protein and Peptide Science* into existence. I would like to take this opportunity to thank everybody whose contribution in one way or another has helped and supported this project. Finally, a special thank you goes to my wife, sons, and mother for their constant support, invaluable assistance, and continuous encouragement.

VLADIMIR UVERSKY
September 9, 2008

PREFACE

The unfolded state of peptides and proteins has attracted a considerable interest over the last 10–15 years for a variety of reasons. First, the discovery of the existence of so-called intrinsically disordered proteins and (IDPs) peptides indicated that in contrast to a central dogma of modern biochemistry proteins do not necessarily have to adopt a well-defined secondary and tertiary structure in order to perform biological functions [1]. IDPs are known to play biologically relevant roles, acting as inhibitors, scavengers, and even facilitating DNA/RNA–protein interaction [2–5]. Some IDPs such as α-synuclein, τ-protein, and β-amyloid are involved in neurodegenerative diseases, for example, Parkinson's and Alzheimer's because of their propensity for self-aggregation and fibril formation [6–9]. Moreover, experimental and theoretical evidence has been provided for the notion that the unfolded state is structurally less disordered as predicted by the statistical (or random) coil model, which is built on the assumption that all amino acid residues besides proline can sample the entire sterically accessible region of the Ramachandran plot [10]. In this book, all these issue are addressed in detail by the contributing authors.

The introductory chapter describes the reasons why research on IDPs and peptides has so significantly expanded over the last years. The authors briefly describe the difference between structured (folded) and disordered (unstructured, unfolded) protcins and list the different functions such proteins and peptides can perform. Their chapter is subdivided into "lessons," which show that (1) disorder is encoded in the amino acid sequence; (2) IDPs are highly abundant in nature and globally heterogeneous; (3) their hydrodynamic properties are charge dependent and describable in terms of polymer physics concepts; (4) IDPs are very pliable and therefore convert into (partially) folded systems upon binding to specific targets (other proteins, peptides, or membrane surfaces) and/or can become involved in multiple processes such as recognition, regulation, and signaling; and (5) IDP-type segments of proteins and peptides can be involved in posttranslational modifications, such as phosphorylation or methylation (to name only a few), and in alternative splicing. The authors also emphasize the role of IDPs in human diseases and their role as drug targets.

Part I of the book deals with the conformational analysis of unfolded peptides and proteins. Chapter 2 describes recent efforts to explore the energy landscape of small peptides and proteins with molecular dynamics simulations. It delineates how a principal component analysis can be employed to obtain an accurate, artifact-free energy landscape. While the results suggest that the landscape of unfolded systems is very complex, they are at variance with the classical random coil model in that they suggest the existence of a countable number of metastable states. Chapter 3 describes research on coil libraries of a large set of proteins, the results of which have led the authors to conclude that conformational ensembles sampled in the unfolded state depend on the amino acid sequence and that the 20 natural amino acids exhibit different conformational preferences, which are modified by their nearest neighbors. In their chapter the authors provide evidence for the notion that these conformational preferences bias the folding pathway. They refer to a suite of web-based applications that graphically display the individual conformational preferences of amino acids and the nearest-neighbor effects. The authors of Chapter 4 have pioneered the use of very special donor–acceptor pairs for measuring the end-to-end distance by fluorescence resonance energy transfer. In their chapter they describe the basic aspects of their method and provide several examples to demonstrate its applicability. Chapter 5 summarizes recent work on how solvation and electrostatic interactions determine the backbone conformation of unfolded peptides and proteins. One very important conclusion for nuclear magnetic resonance (NMR) spectroscopists is that chemical shift values cannot be used as indicators of conformations in the same way for solvated and non-solvated residues. In Chapter 6 the authors investigate different polyproline II (PPII) propensity scales of amino acid residues reported in the literature. Their results led them to the conclusion that PPII propensities do not correlate, for example, with secondary-structure propensities. This notion will certainly ignite some debate in the future. In Chapter 7, the authors show how different types of NMR experiments can be used to determine a set of J-coupling constants that depend differently on the dihedral angles of residues. These coupling constants can be used to determine the conformational ensembles of unfolded peptides. Part I of the book concludes with Chapter 8, which reports the results of NMR measurements on an intrinsically unstructured linker domain in a subunit of a Replication Protein A.

Chapters 9 and 10 comprise Part II, on "molecular recognition." Chapter 9 focuses on so-called hub-proteins, which are capable of binding to a variety of different partners. The authors discuss several models designed to rationalize the mediation of protein–protein interactions by intrinsic disorder. The authors of Chapter 10 describe properties of RNA chaperons concerning interactions with nucleic acid that take place during the replication of widespread pathogenic RNA viruses such as retroviruses and flaviviruses. They provide evidence for how mutations of the chaperone can yield replication-defective viral particles, which they relate to unfolding or misfolding.

Part III of the book is dedicated to peptide and protein aggregation. Chapter 11 presents an overview of the aggregation properties of alanine-based polypeptides. Chapter 12 reviews research of human calcitonin, a 32-residue polypeptide synthesized and secreted by the C cells of the thyroid and involved in calcium regulation and bone dynamics. The authors focus on characterizing the early stage of self-aggregation at which metastable aggregates are formed. Chapter 14 is a theoretical contribution. The authors use concepts from colloid and polymer physics to obtain phase diagrams for IDPs, which encompass the state of self-aggregation. Chapter 15 is on modifiers of peptide and protein aggregation, namely salts, ionic liquids, and osmolytes. Chapters 13 and 16 deal with the classical self-aggregating IDP, the β-amyloid peptide $A\beta_{1-41(42)}$. The author of Chapter 13 reviews recent advances in NMR spectroscopy for investigating the early phase of $A\beta$ self-aggregation. This involves the determination of equilibrium and rater constants. Finally, in Chapter 16 the authors show how $A\beta$ self-aggregation can be explored computationally by molecular dynamics simulation techniques. Their review reports the results of simulations aimed at elucidating the influence of inhibitors on the aggregation process.

Altogether, this book provides the interested reader with a rather broad, though certainly still incomplete, overview of the wealth of experimental and theoretical techniques that are currently used to explore IDPs. Moreover, it highlights some most recent discoveries and theories that will certainly stimulate discussions.

I like to thank Dr. Vladimir Uversky for inviting me to serve as an editor of this book. I am indebted to Anita Lekhwani, Senior Commissioning Editor at Wiley-Blackwell, and Stephanie Sakson, Project Manager at Toppan Best-set Premedia, for their valuable assistance at various stages of the editing and publication process. Finally, I gratefully acknowledge the assistance of my son David Stenner, who has helped me to produce the subject index of this book.

REINHARD SCHWEITZER-STENNER

Note: Color versions of selected figures are available at ftp://ftp.wiley.com/public/sci_tech_med/protein_peptide.

REFERENCES

1 Uversky, V. N. (2002) What does it mean to be natively unfolded? *Eur J Biochem 269*, 2–12.
2 Uversky, V. N. (2008) Natively unfolded proteins. In: Creamer, T. P., ed., *Unfolded proteins. From denaturated to intrinsically disordered*, Nova, Hauppauge, NY.
3 Li, X., Romero, P., Rani, M., Dunker, A. K., and Obradovic, Z. (1999) Predicting protein disorders for N, C, and internal regions, *Genome Informatics 10*, 30.
4 Romero, P., Obradovic, Z., Li, X., Garner, E. C., Brown, C. J., and Dunker, A. K. (2001) Sequence complexity of disordered proteins, *Proteins 42*, 38–48.

5. Dunker, A. K., Lawson, J. D., Brown, C. J., Williams, R. M., Romero, P., Oh, J. S., Oldfield, C. J., Campen, A. M., Ratliff, C. M., Hipps, K. W., Ausio, J., Nissen, M. S., Reeves, R., Kang, C., Kissinger, C. R., Bailey, R. W., Griswold, M. D., Chiu, W., Garner, E. C., and Obradovic, Z. (2001) Intrinsically disordered protein., *J Mol Graphics and Modelling 19*, 26–59.
6. Dobson, C. M. (1999) Protein misfolding, evolution and disease, *Trends Biochem Sci 24*, 329–332.
7. Hamley, I. W. (2007) Peptide fibrillization, *Angewandte Chemie-International Edition 46*, 8128–8147.
8. Mukrasch, M. D., Markwick, P., Biernat, J.,]von Bergen, M., Bernado, P., Greisinger, C., Mandelkow, E., Zweckstetter, M., and Blackledge, M. (2007) Highly populated turn conformations in natively unfolded tau protein identified from residual dipolar couplings and molecular simulation, *J Am Chem Soc 129*, 5235–5243.
9. Bernado., P., Bertoncini, C. W., Griesinger, C., Zweckstetter, M., and Blackledge, M. (2005) Defining long-range order and local disorder in native α-synuclein using residual dipolar couplings, *J Am Chem Soc 127*, 17968–17969.
10. Flory, P. J. (1969) *Statistical mechanics of chain molecules*, Wiley & Sons, New York.

CONTRIBUTORS

Pietro Amodeo, Istituto di Chimica Biomolecolare de Consiglio Nazionale delle Ricerche, Compensario Olivetti, Pozzuoli, Italy.

Giuseppina Andreotti, Istituto di Chimica Biomolecolare de Consiglio Nazionale delle Ricerche, Compensario Olivetti, Pozzuoli, Italy.

Franc Avbelj, National Institute of Chemistry, Ljubljana SI 1115, Slowenia.

Scott L. Crick, Department of Biomedical Engineering, Washington University, St. Louis, MO 63130, USA.

Luis Cruz, Department of Physics, Drexel University, Philadelphia, PA 19104, USA.

Jean-Luc Darlix, Unitè de Virologie Humaine (412), Ecole Normale Supéieure de Lyon et Institut National de la Santé et de la Recherche Médicale, IRF 128, 46 allé d'Italie, 69364 Lyon, France.

Joe DeBartolo, Department of Biochemistry and Molecular Biology, University of Chicago, Chicago, IL 60637, USA.

Gary Daughdrill, Department of Cell Biology, Microbiology, and Molecular Biology and the Center for Biomolecular Identification and Targeted Therapeutics, University of South Florida, 3720 Spectrum Blvd., Tampa, FL 33612, USA.

A. Keith Dunker, Center for Computational Biology and Bioinformatics, Institute for Intrinsically Disordered Protein Research, Department of Biochemistry and

Molecular Biology, Indiana University School of Medicine, Indianapolis, IN 46202, USA.

W. Austin Elam, T.C. Jenkins Department of Biophysics, Johns Hopkins University, Baltimore, MD 21218, USA.

Karl F. Freed, Department of Chemistry, University of Chicago, Chicago, IL 60637, USA.

Ehud Gazit, Department of Molecular Microbiology and Biotechnology, George S. Wise Faculty of Life Sciences, Tel-Aviv University, Tel-Aviv 69978, Israel.

Jürgen Graf, Institut für Organische Chemie, Ruprecht Karls-University, Im Neuenhainer Feld 270, D-69120 Heidelberg, Germany.

Vincent J. Hilser, T.C. Jenkins Department of Biophysics, Johns Hopkins University, Baltimore, MD 21218, USA.

Roland Ivanyi-Nagy, Molecular Parasitology Group, The Weatherall Institute of Molecular Medicine, University of Oxford, Oxford, OX3 9DS, United Kingdom.

Maik H. Jacob, School of Engineering and Science, Jacobs University Bremen, Campus Ring 1, D-28759 Bremen, Germany.

Abhishek Jha, Agios Pharmaceuticals, Cambridge, MA 02139, USA.

Abhinav Jain, Biomolecular Dynamics, Institute of Physics, Albert Ludwigs University, 79104 Freiburg, Germany.

Michal Levy-Sakin, Department of Molecular Microbiology and Biotechnology, George S. Wise Faculty of Life Sciences, Tel-Aviv University, Tel-Aviv 69978, Israel.

Zuzanna Makowska, Department of Biomedicine, University of Basel, CH-4031 Basel, Switzerland.

Daniel Mathieu, Institut für Organische und Biologische Chemie, Johann Wolfgang Goethe Universität, 60439 Frankfurt, Germany.

Thomas J. Measey, Department of Chemistry, University of Pennsylvania, Philadelphia, PA 19104, USA.

Andrea Motta, Istituto di Chimica Biomolecolare de Consiglio Nazionale delle Ricerche, Compensario Olivetti, Pozzuoli, Italy.

Pranesh Narayanaswami, Department of Chemistry, PO Box 644630, Washington State University, Pullman, WA 99164-4630, USA.

Werner Nau, School of Engineering and Science, Jacobs University Bremen, Campus Ring 1, D-28759 Bremen, Germany.

Phuong H. Nguyen, Institut für Physikalische und Theoretische Chemie, Johann Wolfgang Goethe Universität, 60439 Frankfurt, Germany.

Christopher J. Oldfield, Center for Computational Biology and Bioinformatics, Institute for Intrinsically Disordered Protein Research, Department of Biochemistry and Molecular Biology, Indiana University School of Medicine, Indianapolis, IN 46202, USA.

Rohit V. Pappu, Department of Biomedical Engineering, Washington University, St. Louis, MO 63130, USA.

J. Srinivasa Rao, Department of Physics, Drexel University, Philadelphia, PA 19104, USA.

Laura Riccardi, Biomolecular Dynamics, Institute of Physics, Albert Ludwigs University, 79104 Freiburg, Germany.

Karin Rybka, Institut für Organische und Biologische Chemie, Johann Wolfgang Goethe Universität, 60439 Frankfurt, Germany.

Roni Scherzer-Attali, Department of Molecular Microbiology and Biotechnology, George S. Wise Faculty of Life Sciences, Tel-Aviv University, Tel-Aviv 69978, Israel.

Travis P. Schrank, Department of Human Biological Chemistry and Genetics, University of Texas Medical Branch at Galveston, Galveston, TX 77555, USA.

Harald Schwalbe, Institut für Organische und Biologische Chemie, Johann Wolfgang Goethe Universität, 60439 Frankfurt, Germany.

Reinhard Schweitzer-Stenner, Department of Chemistry, Drexel University, Philadelphia, PA 19104, USA.

Tobin R. Sosnick, Department of Biochemistry and Molecular Biology, University of Chicago, Chicago, IL 60637, USA.

Gerhard Stock, Biomolecular Dynamics, Institute of Physics, Albert Ludwigs University, 79104 Freiburg, Germany.

Brigita Urbanc, Department of Physics, Drexel University, Philadelphia, PA 19104, USA.

Vladimir N. Uversky, Department of Molecular Medicine, University of South Florida, Tampa, FL 33612, USA.

Rosa Maria Vitale, Istituto di Chimica Biomolecolare de Consiglio Nazionale delle Ricerche, Compensario Olivetti, Pozzuoli, Italy.

Chunyu Wang, Biology Department Rm 2229, Center for Biotechnology and Interdisciplinary Studies Rensselaer Polytechnic Institute, 110 8th Street, Troy, NY 12180-3590, USA.

Bin Xue, Center for Computational Biology and Bioinformatics, Institute for Intrinsically Disordered Protein Research, Department of Biochemistry and Molecular Biology, Indiana University School of Medicine, Indianapolis, IN 46202, USA.

INTRODUCTION

1

WHY ARE WE INTERESTED IN THE UNFOLDED PEPTIDES AND PROTEINS?

Vladimir N. Uversky and A. Keith Dunker

1.1. INTRODUCTION

In addition to transmembrane, globular, and fibrous proteins, it is becoming increasingly recognized that the protein universe includes intrinsically disordered proteins (IDPs) and proteins with intrinsically disordered regions (IDRs). These IDPs and IDRs are biologically active and yet fail to form specific three-dimensional (3-D) structures, existing instead as collapsed or extended dynamically mobile conformational ensembles [1–7]. These floppy proteins and regions are known as pliable, rheomorphic [8], flexible [9], mobile [10], partially folded [11], natively denatured [12], natively unfolded [3, 13], natively disordered [6], intrinsically unstructured [2, 5], intrinsically denatured [12], intrinsically unfolded [13], intrinsically disordered [4], vulnerable [14], chameleon [15], malleable [16], four-dimensional (4D) [17], protein-clouds [18], and dancing proteins [19], among several other terms. The variability of terms used to describe such proteins and regions is a simple reflection of their highly dynamic nature and the lack of the unique 3-D structure. None of these terms or their combinations is completely appropriate, as the majority of them have been borrowed from the fields such as protein folding or crystallography, which

---

Protein and Peptide Folding, Misfolding, and Non-Folding, First Edition. Edited by Reinhard Schweitzer-Stenner.
© 2012 John Wiley & Sons, Inc. Published 2012 by John Wiley & Sons, Inc.

are not directly related to the biologically active proteins that normally exist as structural ensembles.

Since these proteins are highly abundant in any given proteome [20], the role of disorder in determining protein functionality in organisms can no longer be ignored. Native biologically active proteins were conceptualized as parts of the "protein trinity" [21] or the "protein quartet" [22], models where functional protein might exist in one of the several conformations—ordered, collapsed–disordered (molten globule-like), partially collapsed–disordered (pre-molten globule-like), or extended–disordered (coil-like)—and protein function might be derived from any one of these states and/or from the transitions between them. Disordered proteins are typically involved in regulation, signaling, and control pathways [23–25], which complement the functional repertoire of ordered proteins, which have evolved mainly to carry out efficient catalysis [26].

1.2. WHY STUDY IDPS?

Ordered globular proteins are characterized by rigid 3-D structures. The presence of such rigid structures implies that the Ramachandran angles and backbone atoms of each residue undergo non-isotropic small-amplitude motions relative to their local neighborhood and are characterized by the equilibrium positions defined by their time-averaged values. The atom fluctuations are caused by two factors, random thermal motion and small cooperative conformational changes of the local sequence neighborhood, and are known to be correlated with local residue packing [27]. Contrarily to this very static behavior, intrinsically disordered or natively unfolded proteins exist as dynamic ensembles in which atom positions and backbone Ramachandran angles vary significantly over time with no specific equilibrium values and typically involve non-cooperative conformational changes [6].

The kindred of proteins and protein domains, which have been shown *in vitro* to have little or no ordered structure under physiological conditions, is rapidly amplifying. In fact, over the past decade there has been an exponential increase in the amount of studies dedicated to intrinsically disordered or natively unfolded proteins, starting from a few papers in the early 1990s, and ending with about 300 papers in 2011. A special database, DisProt, was created to keep information about these proteins [28]. There are currently more than 620 proteins in this database.

The growing interest in this class of proteins is determined by several factors. The first issue is the structure–function relationship. Although the importance of protein dynamics for protein functions was recognized long ago, the existence of biologically active but extremely flexible proteins questioned the cornerstone paradigm of a protein science according to which a rigid well-folded 3-D structure is required for protein function. In a recent review, Professor Livesay emphasized, "In the same way that static photos of a dance recital certainly fail to reflect the completeness and grandeur of the performance, discrete structural snapshots lack sufficient information to completely describe protein dynamics" [19]. Since the structure and function of IDPs represent a "shape-shifting dance," new ways of analyzing protein structure analysis and investigating protein functionality are necessary.

The lack of rigid globular structure under physiological conditions was proposed to provide IDPs with a considerable functional advantage, as their large plasticity allows them to interact efficiently with several different targets [2–4, 7]. Furthermore, a disorder/order transition induced in disordered proteins during the binding to specific targets *in vivo* provides a unique mechanism for the decoupling of binding specificity and affinity and might represent a simple tool for the regulation of numerous cellular processes, including transcription, translation, and cell cycle control [2–4, 7, 29]. Evolutionary persistence of the IDPs represents additional confirmation of their importance and raises intriguing questions on the role of protein disorder in biological processes.

Second, biomedical aspects are also of great importance. It has been established that deposition of some natively unfolded proteins is related to the development of several neurodegenerative disorders [30, 31]. Examples include Alzheimer's disease (deposition of amyloid-β, tau-protein, α-synuclein fragment NAC); Niemann–Pick disease type C, sub-acute sclerosing panencephalitis, argyrophilic grain disease, myotonic dystrophy, and motor neuron disease with neurofibrillary tangles (accumulation of tau-protein in the form of neurofibrillary tangles); Down's syndrome (nonfilamentous amyloid-β deposits); Parkinson's disease, dementia with Lewy bodies (LBs), LB variant of Alzheimer's disease, multiple system atrophy, and Hallervorden–Spatz disease (deposition of α-synuclein in the form of LBs and Lewy neurites [LNs]).

Finally, IDPs represent an attractive subject for the biophysical characterization of unfolded polypeptide chain under physiological conditions. A large variety of biophysical and biochemical methods have been applied to the structural description of these proteins. This includes proton nuclear magnetic resonance (^{1}H-NMR), heteronuclear NMR, circular dichroism (CD), optical rotatory dispersion (ORD), Fourier transform infrared spectroscopy (FTIR), intrinsic and extrinsic fluorescence, small angle X-ray scattering (SAXS), small angle neutron scattering (SANS), dynamic and static light scattering, gel electrophoresis, gel filtration, sedimentation, viscometry, scanning calorimetry, proteolytic mapping, epitope mapping, and electron microscopy (summarized in Ref. [32]). The special term "natively unfolded" has been used since it was introduced in 1994 to describe the behavior of tau-protein [12]. Although large amounts of experimental data have been accumulated and several disordered proteins have been rather well characterized, no systematic analysis of structural data for the kindred of IDPs has yet been carried out. This lack of methodical inspection of the conformational behavior of IDPs led frequently to confusion.

1.3. *LESSON 1:* DISORDEREDNESS IS ENCODED IN THE AMINO ACID SEQUENCE AND CAN BE PREDICTED

Similar to ordered proteins, the correct folding of which into relatively rigid biologically active conformations is determined by their amino acid sequences, the lack of rigid structure in IDPs is also encoded in the specific features of their amino acid sequences. Some of these proteins have been discovered due to their unusual amino acid sequence compositions. The absence of regular structure in these proteins has been explained by the specific features of their amino acid sequences including the presence of numerous uncompensated charged groups (often negative), that is, a high net charge at neutral pH, arising from the extreme pI values in such proteins [13, 33, 34], and a low hydrophobic amino acid residue content [33, 34].

The analysis of charge and hydropathy has been shown to be sufficient to distinguish structured and some disordered proteins [3]. In fact, by comparing 275 natively folded and 91 natively unfolded proteins (i.e., proteins that at physiological conditions have been reported to have the NMR chemical shifts of a random coil, and/or lack significant ordered secondary structure, as determined by CD or FTIR, and/or show hydrodynamic dimensions close to those typical of an unfolded polypeptide chain), it has been shown that the combination of low mean hydrophobicity and relatively high net charge represents an important prerequisite for the absence of a compact structure in proteins under physiological conditions [3]. This observation was used to develop a charge–hydropathy (CH) plot method for distinguishing ordered and extended disordered proteins based only on their net charges and hydropathies [3]. According to this approach, natively unfolded proteins are specifically localized within a specific region of CH phase space and are separated from compact ordered proteins by a linear boundary [3]. From the physical viewpoint, such a combination of low hydrophobicity and high net charge as a prerequisite for intrinsic unfoldedness makes perfect sense: a high net charge leads to charge–charge repulsion, and low hydrophobicity means less driving force for protein compaction. In other words, these features are characteristic of IDPs with coil-like (or close to coil-like) structures.

A more detailed analysis was carried out to gain additional information on the compositional difference between ordered and disordered proteins. Comparison of a non-redundant set of ordered proteins with several data sets of disorder (where proteins were grouped based on different techniques, such as X-ray crystallography, NMR, and CD, used to identify disorder) revealed that disordered regions share at least some common sequence features over many proteins [1, 35]. In fact, the disordered proteins/regions were shown to be significantly depleted in bulky hydrophobic (Ile, Leu, and Val) and aromatic amino acid residues (Trp, Tyr, and Phe), which would normally form the hydrophobic core of a folded globular protein, and also possess low content of Cys and Asn residues. The depletion of disordered protein in Cys is also crucial as this amino acid residue is known to have a significant contribution to the protein conformation stability via the disulfide bond formation or being involved in coordination of different prosthetic groups. These depleted residues, Trp, Tyr, Phe, Ile, Leu, Val, Cys, and Asn, were proposed to be called order-promoting amino acids. On the other hand, IDPs were shown to be substantially enriched in Ala, polar, disorder-promoting, amino acids, namely Arg, Gly, Gln, Ser, Glu, and Lys, and also in the hydrophobic but structure-braking Pro [4, 26, 36–38]. Note that these biases in the amino acid compositions of disordered proteins are also consistent with the low overall hydrophobicity and high net charge characteristic of the natively unfolded proteins.

In addition to amino acid composition, the disordered segments have also been compared with the ordered ones by various attributes such as hydropathy, net charge, flexibility index, helix propensities, strand propensities, and compositions for groups of amino acids such as Trp + Tyr + Phe (aromaticity). As a result, 265 property-based attribute scales [36] and more than 6000 composition-based attributes (e.g., all possible combinations having one to four amino acids in the group) have been compared [39]. It has been established that 10 of these attributes, including 14 Å contact number, hydropathy, flexibility, β-sheet propensity, coordination number, content of major disorder-promoting residues (Arg + Ser + Pro + Glu), bulkiness, content of major order-promoting residues (Cys + Trp + Tyr + Phe), volume, and net charge, provide fairly good discrimination between order and disorder [4]. Later, 517 amino acid scales (including a variety of hydrophobicity scales, different measures of side chain bulkiness, polarity, volume, compositional attributes, the frequency of each single amino acid, and so on) were analyzed to construct a new amino acid attribute, for example, a novel amino acid scale that discriminates between order and disorder [40]. This scale outperformed the other 517 amino acid scales for the discrimination of order and disorder and provided a new ranking for the tendencies of the amino acid residue to promote order or disorder (from order-promoting to disorder-promoting): Trp, Phe, Tyr, Ile Met, Leu, Val, Asn, Cys, Thr, Ala, Gly, Arg, Asp, His, Gln, Lys, Ser, Glu, and Pro [40].

The fact that the sequences of ordered and disordered proteins are noticeably different raised three important conclusions: (1) IDPs clearly constitute a separate entity inside the protein kingdom; (2) these proteins can be reliably predicted using various computational tools [41]; (3) since peculiarities of amino acid sequence determine protein structure, structurally, these proteins should be very different from ordered globular proteins.

1.4. *LESSON 2:* DISORDERED PROTEINS ARE HIGHLY ABUNDANT IN NATURE

Intrinsic disorder in proteins is a common phenomenon. Based on the assumption that the absence of rigid structure is encoded in the specific features of the amino acid sequence, several predictors of naturally disordered regions (PONDRs) have been developed [1, 4, 42, 43]. Using these predictors, IDPs were indicated to be widespread. In one experiment, more than 15,000 out of 91,000 proteins in the then-current Swiss Protein database were identified as having long regions of sequence that shared the distinguishing sequence attributes of known IDRs [42]. In a second experiment, the commonness of intrinsic disorder was estimated by predicting disorder for whole genomes, including both known and putative protein sequences. Such predictions have been published for 31 genomes that span the three kingdoms. The percentage of sequences in each genome with segments predicted to have ≥40 consecutive disordered residues was used to gain an overview of proteomic disorder.

For so many consecutive predictions of disorder, the false-positive error rate was estimated from ordered proteins to be less than 0.5% of the segments of 40 and less than 6% of the fully ordered proteins [4, 43]. The eukaryotes exhibited more disorder by these measures than either the prokaryotes or the archaea, with *Caenorhabditis elegans*; *Arabidopsis thaliana*; *Saccharomyces cerevisiae*; and *Drosophila melanogaster* predicted to have 52–67% of their proteins with such long predicted regions of disorder, and bacteria and archaea predicted to have 16–45% and 26–51% of their proteins with such long disorder regions, respectively [43]. The increased prediction of disorder in eukaryotes compared with the other kingdoms has been suggested to be a consequence of the increased need for cell signaling and regulation [4, 43].

To understand the level of abundance of intrinsic disorder in the preeminent source of protein structural information, the Protein Data Bank (PDB), the amino acid sequences of proteins whose structures are determined by X-ray crystallography were compared with the corresponding sequences from the Swiss-Prot database [44]. The analyzed data set included 16,370 structures, which represent 18,101 PDB chains and 5434 different proteins from 910 different organisms (2793 eukaryotic, 2109 bacterial, 288 viral, and 244 archaeal). The analysis revealed that the complete sequences of only approximately 7% of proteins were observed in the corresponding PDB structures, and only approximately 25% of the total data set had >95% of their lengths observed in the corresponding PDB structures [44]. This clearly showed that the vast majority of PDB proteins were shorter than their corresponding Swiss-Prot sequences and/or contained numerous residues, which were not observed in maps of electron density.

According to their appearance in corresponding PDB entries, the residues in the Swiss-Prot sequences were grouped into four general categories: "Observed" (residues in structured regions); "Not observed" (residues regions with missing electron density, potentially disordered); "Uncharacterized" (residues that were not in the PDB sequence but were present in the Swiss-Prot sequence); and "Ambiguous" (residues of a single PDB chain associated with multiple PDB structures, which are observed in some 3-D structures but not in others; these residues occur due to the high redundancy of PDB) [44]. Next, the amino acid compositions and disorder propensities of residues in these four categories were analyzed by four different disorder predictors (PONDR® VL-XT, VL3-BA, VSL1P, and IUPred). "Not observed," "Ambiguous," and "Uncharacterized" regions were shown to possess the amino acid compositions typical for IDPs. The vast majority of residues in the "Observed" data set were predicted to be ordered, whereas the "Not observed" regions were mostly disordered. The "Uncharacterized" regions possessed some tendency toward order. Disorder predictions for the short "Ambiguous" regions were ambiguous, whereas long "Ambiguous" regions (>70 amino acid residues) were mostly predicted to be ordered, suggesting that they are likely to be wobbly domains [44]. The major conclusion of this study was really surprising since completely ordered proteins were not highly abundant in PDB and many PDB sequences had disordered regions. In fact, ~10% of the PDB proteins contained long disordered or ambiguous regions (with length $L > 30$ amino acids), and ~40% of the PDB proteins possessed short disordered or ambiguous regions (with $10 \leq L < 30$ amino acids long) [44].

1.5. *LESSON 3:* DISORDERED PROTEINS ARE GLOBALLY HETEROGENEOUS

Typically, structural classification of IDPs is based on the definitions and terms elaborated for conformations observed in the unfolding/refolding pathways of typical globular proteins, which are known to exist in at least four different conformations, ordered, molten globule, pre-molten globule, and unfolded states [45–50]. The structural properties of the *molten globule* are well known and have been systematized in a number of reviews [e.g., Ref. [47]]. The protein molecule in this intermediate state has a globular structure typical of ordered globular proteins, preserves high secondary-structure content, and possesses the native-like folding pattern, but has no (or has only a trace of) rigid cooperatively melted tertiary structure. Molten globules are more accessible to proteolysis and are characterized by high affinity to the hydrophobic fluorescence probes (such as 8-anilinonaphthalene-1-sulfonate [ANS]). The averaged value for the increase in the hydrodynamic radius of a protein molecule in the molten globule state compared with that of ordered state is 15–20%, which corresponds to a volume increase of ~50%. The protein molecule in the premolten globule has no rigid tertiary structure and does not possess globular structure but is characterized by a considerable secondary structure. The protein molecule in the pre-molten globule state is considerably less compact than in the molten globule or folded states, but it is still more compact than the random coil. It can effectively interact with the hydrophobic fluorescent probe ANS, although weaker than the molten globule does.

Based on their structural properties, IDPs are separated in two different groups. Members of the first group, despite their flexibility, are rather compact and possess a well-developed secondary structure; that is, they show properties typical of the molten globule [47]. Proteins from the other group behave almost as a random coil [51] or as pre-molten globules [48]. The proteins from this second group constitute a class of natively unfolded (or extended intrinsically disordered) proteins, which are extremely flexible, essentially non-compact (extended), and have little or no ordered secondary structure under physiological conditions. The general conformational properties of intrinsically unfolded proteins are summarized below. Here we will mostly focus on the structural characteristics, which make natively unfolded proteins exceptional among others. These are low compactness, absence of globularity, low secondary-structure content, and high flexibility.

1.5.1. Compactness

The most unambiguous characteristic of the conformational state of a globular protein remains the hydrodynamic dimensions of the macromolecule. It was noted long ago that hydrodynamic techniques may help to recognize when a globular

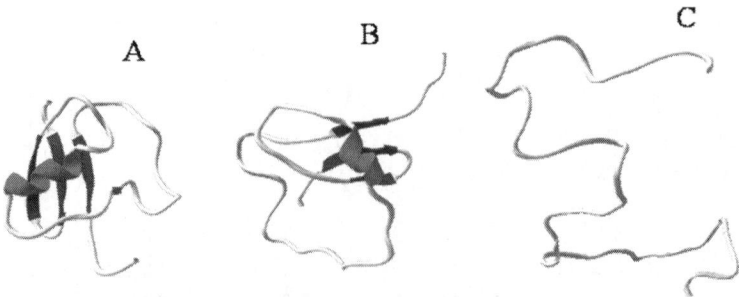

Figure 1.1. Illustration of the hydrodynamic volume variation for a polypeptide chain of 100 amino acids in different intrinsically disordered configurations: (A) collapsed (molten globule-like) disorder; (B) extended (pre-molten globule-like) disorder; (C) extended (coil-like) disorder.

protein has lost all of its non-covalent structure, that is, when it became unfolded [51]. This is because an essential increase in the hydrodynamic volume is associated with the unfolding of a protein molecule. Therefore, conformational states of a globular protein, ordered, molten globule, pre-molten globule, and unfolded, may be easily discriminated by the degree of compactness of the polypeptide chain. Similarly, different types of IDPs (molten globule-like, pre-molten globule-like, and coil-like) may be distinguished by their degree of compactness (Fig. 1.1). Furthermore, ordered and unfolded conformations of globular proteins possess very different molecular mass dependencies of their hydrodynamic radii, R_S [22, 51, 52]. Comparison of the $\log(R_S)$ versus $\log(M)$ curves for natively unfolded proteins with the same dependencies for the ordered, molten globular, pre-molten globular, and urea- or GdmCl-unfolded globular proteins revealed that these $\log(R_S)$ versus $\log(M)$ dependencies for different conformations of globular proteins are described by straight lines [22, 53]. This analysis revealed that according to their $\log(R_S)$ versus $\log(M)$ dependencies, natively unfolded proteins may be divided in two groups, with one of these groups behaving as random coils in poor solvent, and with protein in the second group being close, with respect to their hydrodynamic characteristics, to pre-molten globules (see Fig. 1.2).

This is a very important observation, which may definitely help in understanding the physical nature of the natively unfolded proteins. In fact, it is well established that the behavior of unfolded proteins obeys the theoretical and empirical rules that apply to linear random coils [51]. Particularly, it is known that the hydrodynamic dimensions of random coils depend essentially on the quality of solvent. A poor solvent encourages the attraction of macromolecular segments and hence a chain has to squeeze. On the other hand, in a good solvent, repulsive forces act primarily between the segments and the macromolecule conforms to a loose fluctuating coil [49, 51]. Water is a poor solvent, whereas solutions of urea and GdmCl are rather good solvents, with GdmCl being closer to the ideal one. This difference in solvent quality may account for the observed divergence in $\log(R_S)$ versus $\log(M)$ dependen-

LESSON 3

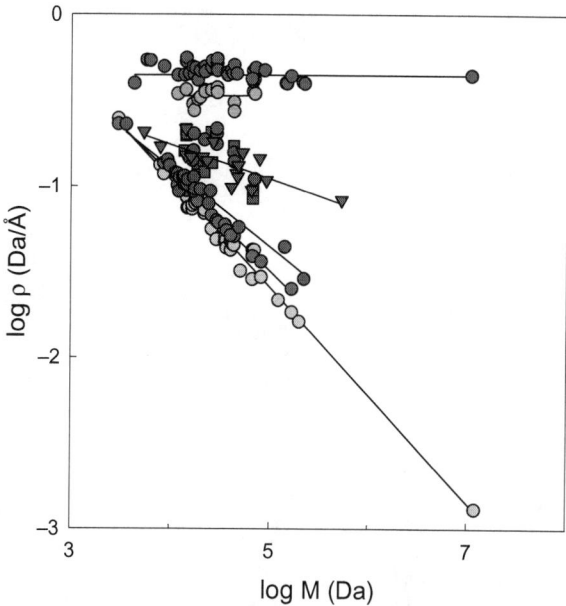

Figure 1.2. Variation of the density of protein molecules, ρ, with protein molecular weight, M, for ordered (red circles), molten globule (green circles), pre-molten globules (dark yellow symbols, where intermediates accumulated during the unfolding by urea or GdmCl are shown by circles); proteins with intact disulate bridges in 8 M urea or 6 M GdmCl are shown as squares; native pre-molten globules are shown as reversed triangles); native coils (blue circles); proteins without cross-links or with reduced cross-links unfolded in 8 M urea (pink circles); proteins without cross-links or with reduced cross-links unfolded in 6 M GdmCl (turquoise circles). The solid lines represent the best fit of the data. See color insert.

cies for the coil-like part of IDPs. The existence of a well-defined difference between the $\log(R_S)$ versus $\log(M)$ dependencies for globular proteins unfolded by urea and GdmCl should also be noted in this respect (see Fig. 1.2).

1.5.2. Globularity

Another very important structural parameter is the degree of globularization, which reflects the presence or absence of a tightly packed core in the protein molecule. In fact, it has been shown that the globular proteins in pre-molten globule-like conformations are characterized by low (coil-like) intramolecular packing density [46, 49]. This information could be extracted from the analysis of SAXS data (Kratky plot), whose shape is sensitive to the conformational state of the scattering protein molecules [54, 55]. It has been shown that a scattering curve in the Kratky coordinates has a characteristic maximum when the globular protein is in the native state or in the molten globule state (i.e., has a globular structure): if a protein is completely

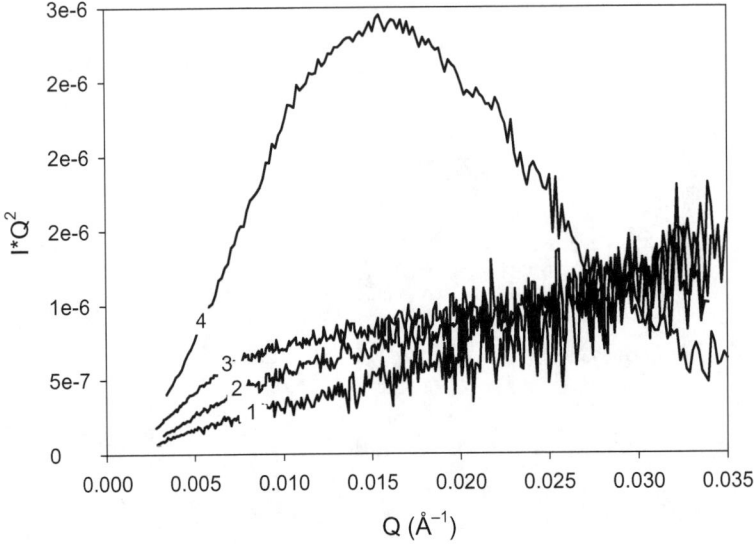

Figure 1.3. Kratky plots of SAXS data for natively unfolded α-synuclein [1], prothymosin α [2], and caldesmon 636–771 fragment [3]. The Kratky plot of a typical ordered protein SNase is shown for comparison [4].

unfolded or a pre-molten globule conformation (has no globular structure), such a maximum will be absent on the respective scattering curve. Analysis of several natively unfolded proteins by this technique revealed that representatives of both classes of natively unfolded proteins (coil-like and pre-molten globule-like) are characterized by the absence of a rigid globular core (see Fig. 1.3).

1.5.3. Secondary Structure

Far-ultraviolet CD spectra of a number of natively unfolded proteins (such as α-synuclein, prothymosin α, phosphodiesterase γ-subunit, caldesmon 636–771 fragment, and many others) possess distinctive spectral features with characteristic deep minima in the vicinity of 200 nm, and relatively low ellipticity at 220 nm (Fig. 1.4). This characteristic shape of far-UV CD spectra is a useful criterion for the selection of natively unfolded proteins. Figure 1.5 represents a "double wavelength" plot, $[\theta]_{222}$ versus $[\theta]_{200}$ dependence, which may be used to assort natively unfolded proteins into two non-overlapping groups. Approximately half of the studied proteins was characterized by far-UV CD spectra characteristic of almost completely unfolded polypeptide chains, with $[\theta]_{200} = -(18,900 \pm 2800)$ degree·cm²/dmol and $[\theta]_{222} = -(1700 \pm 700)$ degree·cm²/dmol, whereas the other half possessed spectra typical of the pre-molten globule state of globular proteins, being consistent with the existence of some residual secondary structure (with $[\theta]_{200} = -(10,700 \pm 1300)$ degree·cm²/dmol and $[\theta]_{222} = -(3900 \pm 1100)$ degree·cm²/dmol) [22, 53]. Some of

LESSON 3

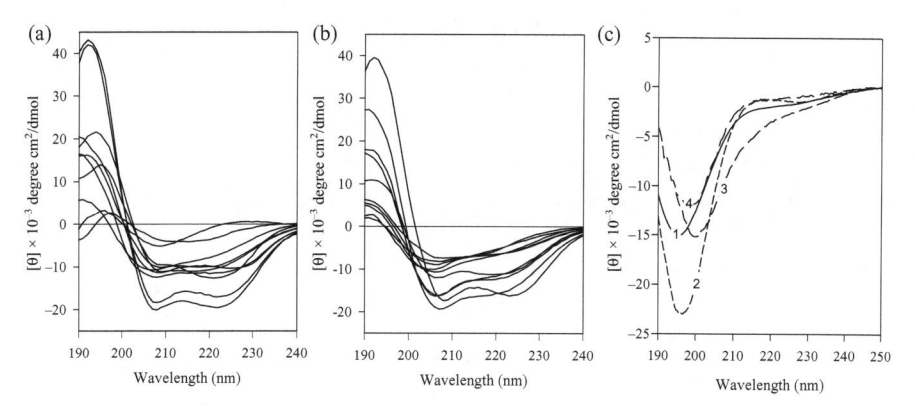

Figure 1.4. Far-UV CD spectra of nine ordered proteins: α-lactalbumin (bovine), α-lactalbumin (human), β-lactamase, ribonuclease A, retinol binding protein, apo-form, carbonic anhydrase B, phosphoglycerate kinase, cytochrome c, apomyoglobin (a), their molten globule folding intermediate states (b), and extended IDPs, α-synuclein [1], prothymosin α [2], caldesmon 636–771 fragment [3], and phosphodiesterase γ-subunit [4] (c).

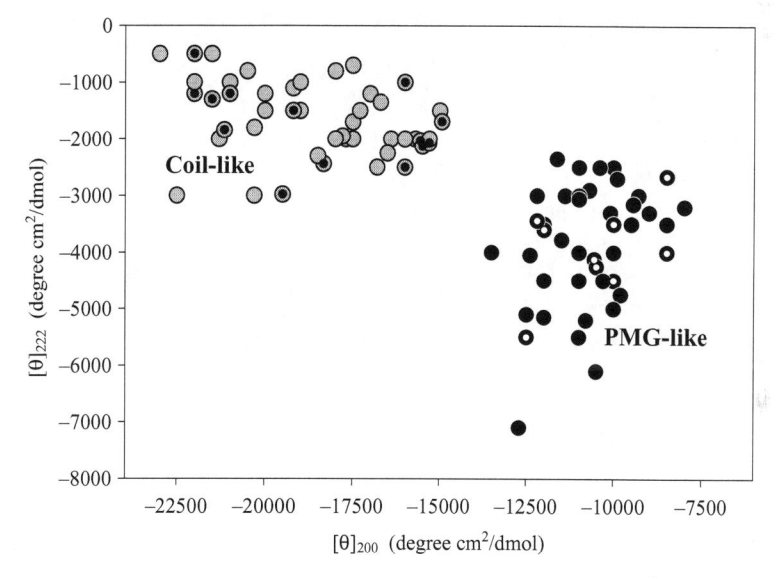

Figure 1.5. Analysis of far-UV CD spectra in terms of double wavelength plot, $[\theta]_{222}$ versus $[\theta]_{200}$, allows the natively unfolded proteins division on coil-like (gray circles) and pre-molten globule-like subclasses (black circles). Intrinsic pre-molten globules (PMG) and intrinsic coils for which the hydrodynamic parameters were measured are marked by white-dotted and black-dotted symbols, respectively.

the natively unfolded proteins were simultaneously characterized by CD and hydrodynamic methods. Intrinsic pre-molten globules and intrinsic coils studied by both techniques are indicated in Figure 1.5 as white-dotted and black-dotted symbols, respectively. These data are consistent with the important conclusion that more compact polypeptides (with pre-molten globule-like hydrodynamic characteristics) possess larger amounts of ordered secondary structure than less compact coil-like natively unfolded proteins. Thus, the simultaneous application of CD and hydrodynamic techniques leaves no doubt that natively unfolded proteins should be subdivided into two structurally distinct groups: intrinsic coils and intrinsic pre-molten globules [22, 53].

The analysis of these spectra yielded a low content of ordered secondary structure (α-helices and β-sheets).

This lack of ordered secondary structure is also confirmed by the FTIR spectroscopy of secondary-structure composition of natively unfolded proteins, such as tau-protein [12], α-synuclein [13, 56], β- and γ-synucleins [57], α_s-casein [58], cAMP-dependent protein kinase inhibitor [59], nucleoporins (Nups) [60, 61], and many others. Importantly, even the caldesmon 636–771 fragment, which was shown to have hydrodynamic properties typical of the pre-molten globule, possesses far-UV CD characteristic of essentially distorted polypeptide chains. Thus, the low overall content of ordered secondary structure could be considered a general property of extended IDPs.

1.5.4. High Flexibility

The fact that IDPs are characterized by an increased intramolecular flexibility may be easily derived from a large amount of NMR studies (summarized in Refs. [2–4, 6, 32, 62–66]). Furthermore, recent advances in NMR technology (especially the use of heteronuclear multidimensional approach) have even opened the way to detailed structural and dynamic description of these proteins [62–66]. Increased flexibility of natively unfolded proteins is indirectly confirmed by their extremely high sensitivity to protease degradation *in vitro* [4, 6, 32]. The functional implication of large intramolecular flexibility has also been recognized. Particularly, it has been noted that the intrinsic lack of structure and high flexibility may be considered a major functional advantage of a protein, which in this way attains the ability to bind several different targets. Precise control over the thermodynamics of the binding process may also be achieved in this way.

1.6. *LESSON 4:* HYDRODYNAMIC DIMENSIONS OF NATIVELY UNFOLDED PROTEINS ARE CHARGE DEPENDENT

Since extended intrinsically disordered (or natively unfolded) proteins are characterized by the low proportion of hydrophobic residues combined with the high content of charged amino acids, they represent an ideal system for the analysis of the role of charge interactions for the conformational properties of unfolded proteins, and for testing the quantitative descriptions and predictions of polymer theory for the influence of charged amino acids on chain dimensions [67]. Recently, using single-molecule Förster resonance energy transfer (FRET), long-range distance distributions and dynamics of several natively unfolded proteins were analyzed. These proteins with very different properties were chosen to sample the range of sequence compositions found in natural proteins [67], as represented by the CH-plot discussed above [3]: the globular cold shock protein Csp*Tm*, which is stably folded even in the absence of ligands; the N-terminal domain of HIV-1 integrase, which folds only upon binding of Zn^{2+} ions and is otherwise denatured; and human prothymosin α, one of the natively unfolded proteins with the largest fraction of charged amino acids identified so far, which does not assume a well-defined folded structure under any known conditions and does not contain regular secondary structure, but plays crucial roles in different biological processes including cell proliferation, transcriptional regulation, and apoptosis. These three proteins were specifically labeled with a donor (Alexa Fluor 488, Molecular Probes, Eugene, OR) and an acceptor (Alexa Fluor 594, Molecular Probes) fluorophore, and investigated with confocal single-molecule fluorescence spectroscopy under a variety of conditions to evaluate the efficiency of energy transfer for individual molecules freely diffusing through the focal spot of the laser beam [67]. This analysis allows finding subpopulations of proteins characterized by different energy transfer efficiencies and distinguishing changes in the conformational properties within one of these subpopulations from a change in their relative abundances. The analysis revealed very peculiar and characteristic responses of the unfolded states of these three proteins to the increasing GdmCl concentration. The unfolded state of the heat shock protein Csp*Tm*, which was detectable by FRET even in the absence of GdmCl, continuously expanded with the increase in the denaturant concentration, whereas the GdmCl dependence on the energy transfer efficiency for the intrinsically disordered HIV-1 integrase and human prothymosin α possessed a more complex behavior, indicating the presence of partial collapse of the unfolded polypeptide chain at low denaturant concentrations followed by the denaturant-induced expansion similar to Csp*Tm* at higher GdmCl concentrations [67]. This conclusion was based on the presence of the remarkable "rollover" of the efficiency of the energy transfer below approximately 0.5 M GdmCl, which was absent when the uncharged denaturant urea was used for the analogous experiments. Importantly, the amplitude of the mentioned rollover of the efficiency of the energy transfer was proportional to the protein charge density. Based on these observations, it has been concluded that natively unfolded proteins can exhibit a prominent expansion at low ionic strengths, which correlate with their net charge and that these pronounced effects of charges on the dimensions of unfolded proteins might have important implications for their cellular functions [67].

This conclusion was further supported by a comprehensive analysis of phenylalanine-glycine (FG) repeat-containing Nups [68], which contain large intrinsically disordered domains with multiple phenylalanine–glycine repeats (FG domains) and form the nuclear pore complex gating the nucleocytoplasmic transport in eukaryotes. A systematic analysis of the hydrodynamic dimensions revealed that under the physiological conditions *in vitro* the FG domains of nucleoporins adopt distinct categories of intrinsically disordered structures, such as molten globule, pre-molten globule, relaxed coil, extended coil (as in urea), or very extended coil (as in GdmCl) [68]. Furthermore, the category of intrinsically disordered structure in a given FG domain was related to its amino acid composition, namely to the content of charged residues, where more charged FG domains possessed larger hydrodynamic dimensions. FG nucleporins with higher charge densities were shown to be more dynamic than the collapsed-coil FG domains. Furthermore, these relaxed- or extended-coil FG domains were shown to repel other FG domains, whereas the collapsed-coil FG domains were able to bind each other to form oligomers, clearly suggesting that there is a functional need in cells to have some FG domains aggregate and other FG domains repel, therefore providing a molecular basis for two different gating mechanisms operating at the nuclear pore complex at distinct locations, one acting as a hydrogel and the other as an entropic brush [68]. Therefore, the abundance and peculiarities of the charged residues distribution within the protein sequences might determine the physical and biological properties of natively unfolded proteins.

1.7. *LESSON 5:* POLYMER PHYSICS EXPLAINS HYDRODYNAMIC BEHAVIOR OF DISORDERED PROTEINS

Each protein is believed to be a unique entity that has a quite unmatched primary sequence, which governs its 3-D structure (or luck thereof) and ensures specific biological functions. Therefore, understanding the effect of sequence variance on the biological performance presents a difficult challenge. However, natural polypeptides have originated as random copolymers of amino acids, which were adjusted or "selected" over evolution based on their functional capacity [69, 70]. Despite their differences in primary amino acid sequences, protein molecules in a number of conformational states behave as polymer homologues, suggesting that the volume interactions can be considered a major driving force responsible for the formation of equilibrium structures or structural ensembles [71]. For example, ordered globular proteins and molten globules (both as folding intermediates of globular proteins and as examples of collapsed IDPs) exhibit key properties of polymer globules, where the fluctuations of the molecular density are expected to be much less than the molecular density itself. Extended IDPs (both intrinsic coils and intrinsic pre-molten globules) and ordered proteins in the pre-molten globule intermediate state possess properties of squeezed coils since water is a poor solvent for a polypeptide. In fact, even high concentrations of strong denaturants (e.g., urea and GdmCl) are very likely to be poor solvents for protein chains, resulting in the preservation of an extensive residual structure even under these harsh denaturing conditions [71].

Based on these and related observations, and given the fact that IDPs (especially their extended forms) are characterized by significant amino acid composition biases, the behavior of low-complexity polypeptides (e.g., homopolypeptide and block copolypeptides) can mimic reasonably well the overall polymeric behavior of more complex systems (e.g., IDPs and IDRs). Intriguingly, recent studies revealed that polar homopolypeptides without hydrophobic groups (e.g., polyglutamine or glycine–serine block copolypeptides), used to model IDPs, prefer collapsed ensembles in aqueous media [72–77]. Furthermore, even polyglycine tends to form heterogeneous ensembles of collapsed structures in water [77]. These results clearly suggest that water is a poor solvent for polypeptide backbone alone and for the IDPs containing long tracts of polar amino acid residues.

A systematic analysis of conformational behavior of protamines, arginine-rich IDPs involved in the condensation of chromatin during spermatogenesis, and protamine-like peptides by a combination of molecular simulations and fluorescence experiments revealed that there is a charge-driven coil-to-globule transition in these highly charged polypeptides, where the net charge per residue serves as the discriminating order parameter (Rohit Pappu, personal communication). Overall, the increase in the size of a polypeptide chain with increasing net charge per residue can be attributed to the increase in the intramolecular electrostatic repulsions between similarly charged side chains and the favorable solvation of these moieties. Pappu also pointed out that there are at least three different classes of globule-forming polar/charged IDPs. The first class is composed of polar tracts, which collapse due to the water being a poor solvent for a backbone and non-charged side chains. The second class is represented by weak polyelectrolytes and weak polyampholytes, which have low per residue net charge and low fractions of positively and/or negatively charged residues. These IDPs form collapsed structures since the driving force responsible for the collapse is not overcome by the intramolecular electrostatic repulsion between the charged side chains and by their favorable free energies of solvation. Furthermore, if such IDPs possess a polyampholytic nature, their globular state is additionally stabilized by electrostatic interactions between the oppositely charged side chains. Finally, IDPs from the third class are strong polyampholytes characterized by high fractions of positively and/or negatively charged residues but low per residue net charges. Such IDPs can form collapsed structures stabilized mostly by multiple electrostatic interactions between solvated side chains of opposite sign (Rohit Pappu, personal communication).

Clearly, IDPs with very high net charges are expected to be more extended and behave more similar to random coils (i.e., similar to conformations adopted by proteins in the denaturant GdmCl). The validity of this hypothesis was recently illustrated via the analysis of the set of Nups containing long natively unfolded domains with phenylalanine–glycine repeats (FG domains). These Nups constitute a gate of the nuclear pore complex (NPC), where the FG domains form a malleable network of disordered polypeptides, which selects and size-discriminates against diffusing macromolecules [68]. In this study, most Nup FG domains were shown to adopt collapsed molten-globular configurations and were characterized by a low content of charged amino acids. Others adopted more extended, coil-like conformations, were structurally more dynamic, and were characterized by a high content of charged amino acids. Many Nups contained both types of structures in a biphasic distribution along their polypeptide chain. For example, the N-terminus of Nsp1 (AA 1–172; Nsp1n) had a low charged-AA content of 2% and adopted molten globular configurations, whereas the remainder of the Nsp1 FG domain (AA 173–603; Nsp1m) had a charged AA content of 36% and adopted extended-coil configurations [68].

1.8. *LESSON 6:* NATIVELY UNFOLDED PROTEINS ARE PLIABLE AND VERY SENSITIVE TO THEIR ENVIRONMENT

Although the mechanisms of function of any biological subject (including proteins) cannot be understood without appropriate consideration of this subject's environment, IDPs represent an extreme case of environmental pliability. Amino acid biases characteristic of IDPs determine their structural variability and lack of a rigid well-folded structure. This structural plasticity is necessary for the unique functional repertoire of IDPs, which is complementary to the catalytic activities of ordered proteins. Amino acid biases also drive atypical responses of IDPs to changes in their environment. In general, the conformational behavior of IDPs is characterized by the low cooperativity (or the complete lack thereof) of the denaturant-induced unfolding, lack of the measurable excess heat absorption peak(s) characteristic of the melting of ordered proteins, "turned out" response to heat and changes in pH, and the ability to gain structure in the presence of various binding partners [78].

1.8.1. Effects of Strong Denaturants

IDPs, being highly dynamic, are characterized by low conformational stability, which is reflected in low steepness of the transition curves describing their unfolding induced by strong denaturants or even in the complete lack of the sigmodal shape of the unfolding curves. This is in strict contrast to the solvent-induced unfolding of ordered globular proteins, which is known to be a highly cooperative process. In fact, we can find here an extreme case of the cooperative transition, which is an all-or-none transition where a cooperative unit includes the whole molecule; that is, no intermediate states can be observed in the transition region. Based on the analysis of the unfolding transitions in ordered globular proteins, it has been concluded that the steepness of urea- or GdmCl-induced unfolding curves depends strongly on whether a given protein has a rigid tertiary structure (i.e., it is ordered) or is already denatured and exists as a molten globule [79, 80]. Although the denaturant-induced unfolding of a native molten globule can be described by a shallow sigmoidal curve (e.g., see Ref. [81]), urea- or GdmCl-induced structural changes in native pre-molten globules or native coils are non-cooperative and typically seen as monotonous featureless changes in the studied parameters. This is due to the low content of the residual structure in these species.

1.8.2. Temperature Effects

The analysis of the temperature effects on structural properties of several extended IDPs revealed that native coils and native pre-molten globules possess so-called "turned out" response to heat [78]. Figure 1.6A depicts the temperature-dependence of $[\theta]_{222}$ for α-synuclein, caldesmon 636–771 fragment, and phosphodiesterase γ-subunit and clearly shows that these three proteins partially fold as the temperature is increased. Therefore, temperature-induced changes in the far-UV CD spectrum of several extended IDPs were interpreted in terms of the temperature-induced forma-

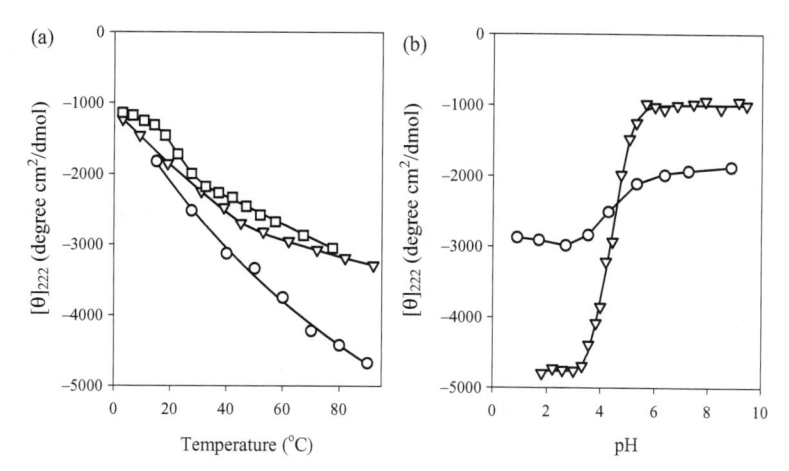

Figure 1.6. Effect of environmental factors on conformational properties of natively unfolded proteins. (a) Temperature-induced changes in far-UV CD spectrum ($[\theta]_{222}$ versus temperature dependence) measured for α-synuclein (triangles), phosphodiesterase γ-subunit (squares), and caldesmon 636–771 fragment (circles). (b) pH-induced structure formation ($[\theta]_{222}$ versus pH dependence) in the natively unfolded α-synuclein (circles) and prothymosin α (triangles).

tion of secondary structure [56, 82, 83]. These heating-induced structural changes in extended IDPs were completely reversible and consistent with the partial. The effects of elevated temperatures may be attributed to the increased strength of the hydrophobic interaction at higher temperatures, leading to a stronger hydrophobic attraction, which is the major driving force for folding.

1.8.3. Structure-Promoting Effects of Extreme pH

Extended IDPs are also characterized by the "turned out" response to changes in pH [56, 84–87], where a decrease (or increase) in pH induced partial folding of extended IDPs due to a decrease in their high net charge present at neutral pH, thereby decreasing charge–charge intramolecular repulsion and permitting hydrophobicity-driven collapse to the partially folded conformation [78]. Figure 1.6B illustrates this unusual conformational behavior of extended IDPs representing the pH dependence of $[\theta]_{222}$ for α-synuclein and prothymosin α. These data are consistent with the conclusion that changes in pH can induce partial folding of natively unfolded proteins.

1.8.4. Protein Chameleons

In general, the pliable nature of the IDPs allows them to be highly adjustable and to morph their structure. For example, α-synuclein was shown to adopt a series of different conformations depending on its environment, being able to stay substantially unfolded, or fold into an amyloidogenic partially folded conformation, or into α-helical or β-structured species, both monomeric and oligomeric. Furthermore, it was shown to form several morphologically different types of aggregates, including oligomers (spheres, doughnuts, and worms), amorphous aggregates, and amyloid-like fibrils. Based on this astonishing conformational behavior and structural plasticity, the concept of a protein chameleon was introduced [15].

1.9. LESSON 7: WHEN BOUND, NATIVELY UNFOLDED PROTEINS CAN GAIN UNUSUAL STRUCTURES

Due to their lack of rigid structure, combined with the high level of intrinsic dynamics and almost unrestricted flexibility at various structure levels in the non-bound state, and due to their unique capability to adjust to the structure of the binding partner, IDPs are characterized by a diverse range of binding modes, creating a multitude of unusual complexes, many of which are not attainable by ordered proteins [88]. Some of these complexes are relatively static, resemble complexes of ordered proteins, and therefore are suitable for structure determination by X-ray crystallography. Among these static complexes are molecular recognition features (MoRFs), wrappers, chameleons, penetrators, huggers, intertwined strings, long cylindrical containers, connectors, armature, tweezers and forceps, grabbers, tentacles, pullers, and stackers or β-arcs [88]. These binding modes are shown in Figure 1.7 and briefly described below.

1.9.1. MoRFs

MoRFs are short, interaction-prone IDP segments that undergo disorder-to-order transitions upon binding and are abundantly involved in molecular recognition [89–91]. MoRFs, being identified as short structured fragments of disordered proteins involved in interaction with globular partners, were structurally classified according to their structures in the bound state: α-MoRFs form α-helices, β-MoRFs form β-strands, and ι-MoRFs form structures without a regular pattern of backbone hydrogen bonds [89, 91]. MoRF typically constitutes one contiguous segment fitted into a grove at the surface of the ordered partner.

1.9.2. Flexible Wrappers

Some IDPs wrap around their ordered binding partners. Complexes of this type are polyvalent ordered complexes where several ordered segments of a disordered protein bind to disjoint and spatially distant binding sites on the surface of the globular protein. Typically, the ordered segments of such wrapping IDPs are connected by the disordered regions. Secondary-structure elements of wrappers almost do not possess intramolecular interactions, forming very intensive intermolecular contacts with the binding partner [92–94]. Many proteins interacting with DNA or RNA are flexible wrappers. For example, numerous transcription factors, regulatory proteins, and other proteins that interact with DNA contain multiple zinc finger motifs. The zinc finger motifs act as independently folded globular domains that are separated by flexible linker regions. Zinc finger domains are disordered in the absence of zinc. Proteins often contain multiple zinc fingers connected by flexible linkers and wrap around the DNA in a spiral manner. The zinc finger-containing proteins typically interact with the major groove along the double helix of DNA. In the bound state, the zinc fingers are arranged around the DNA strand in such a way that the α-helix of each finger contacts the DNA, forming an almost continuous stretch of α-helices around the DNA molecule [95].

1.9.3. Penetrators

In complexes of some IDPs with other proteins or RNA, significant parts of IDPs penetrate deep inside the structure of their binding partners. For example, in the crystal structure of the 30S ribosome, many ribosomal proteins, in addition to globular domains, contained extended internal loops or long N- or C-terminal extensions that were not seen in structures of the isolated proteins but which were associated intimately with the RNA inside the ribosome [96]. The most illustrative example of this penetrating mode is S12, which had a globular domain at the interface side and a long N-terminal extension that threaded its way through the 30S subunit to emerge on the back side to interact with proteins S8 and S17 [96].

1.9.4. Huggers

Typically, monomers constituting the oligomers formed as a result of folding coupled to binding are highly intertwined [97–99], kind of hugging each other. Typically, complexes of this kind are binary. However, there are several examples of group huggers, where monomers clasp more than one partner.

1.9.5. Intertwined Strings

Coiled coils represent a common structural motif in proteins, where up to seven long α-helices intertwined together similar to strings of a rope [100, 101]. This motif is formed by approximately 3–5% of all amino acids in proteins [102], with the most common members of this family being dimers and trimers. Individual α-helices in coiled coil are wrapped around each other into a left-handed helix to form a supercoil. In addition to the left-handed coiled coils, there are right-handed coiled coils [103, 104]. Coiled coils represent a relatively simple but tightly packed structure. Importantly, partners involved in the coiled-coil formation are typically disordered in the non-bound form.

1.9.6. Long Cylindrical Containers

Multichain coiled coils can assemble into long hollow cylinders containing a continuous axial pore with binding capacities for several hydrophobic compounds [105].

22

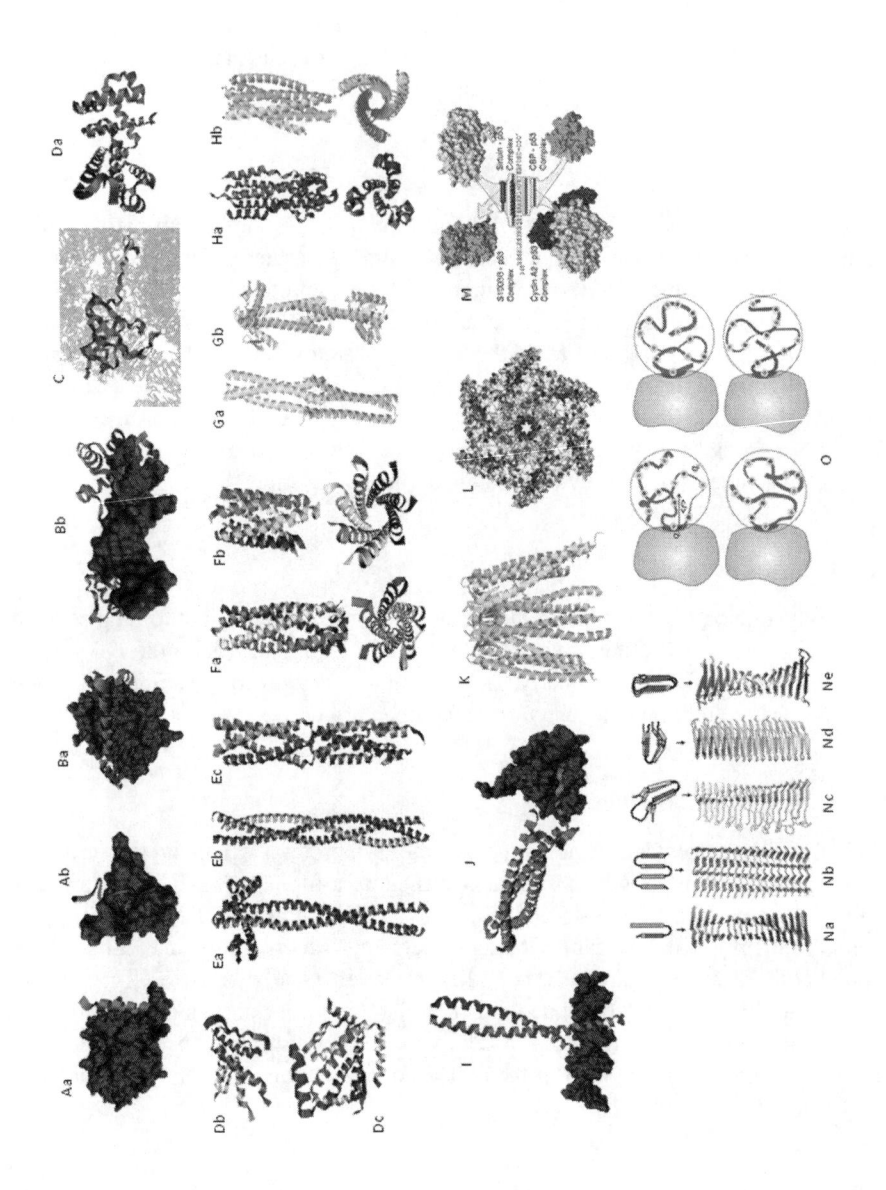

Figure 1.7. A portrait gallery of disorder-based complexes. Illustrative examples of various interaction modes of intrinsically disordered proteins are shown. (A) **MoRFs:** (Aa) α-MoRF, a complex between the botulinum neurotoxin (red helix) and its receptor (a blue cloud) (PDB ID: 2NM1); (Ab) i-MoRF, a complex between an 18-mer cognate peptide derived from the α1 subunit of the nicotinic acetylcholine receptor from *Torpedo californica* (red helix) and α-cobratoxin (a blue cloud) (PDB ID: 1LXH). (B) **Wrappers:** (Ba) rat PP1 (blue cloud) complexed with mouse inhibitor-2 (red helices) (PDB ID: 2O8A); (Bb) a complex between the paired domain from the *Drosophila* paired (prd) protein and DNA (PDB ID: 1PDN). (C) **Penetrator:** Ribosomal protein S12 embedded into the rRNA (PDB ID: 1N34). (D) **Huggers:** (Da) *E. coli trp* repressor dimer (PDB ID: 1ZT9); (Db) tetramerization domain of p53 (PDB ID: 1PES); (Dc) tetramerization domain of p73 (PDB ID: 2WQI). (E) **Intertwined strings:** (Ea) dimeric coiled coil, a basic coiled-coil protein from *Eubacterium eligens* ATCC 27750 (PDB ID: 3HNW); (Eb) trimeric coiled coil, salmonella trimeric autotransporter adhesin, SadA (PDB ID: 2WPQ); (Ec) tetrameric coiled coil, the virion-associated protein P3 from Caulimovirus (PDB ID: 2O1J). (F) **Long cylindrical containers:** (Fa) pentameric coiled coil, side and top views of the assembly oligomeric matrix protein (PDB ID: 1FBM); (Fb) side and top views of the seven-helix coiled coil, engineered version of the GCN4 leucine zipper (PDB ID: 2HY6). (G) **Connectors:** (Ga), human heat shock factor binding protein 1 (PDB ID: 3C19); (Gb) the bacterial cell division protein ZapA from *Pseudomonas aeruginosa* (PDB ID: 1W2E). (H) **Armature:** (Ha) side and top views of the envelope glycoprotein GP2 from Ebola virus (PDB ID: 2EBO); (Hb) side and top views of a complex between the N- and C-terminal peptides derived from the membrane fusion protein of the Visna (PDB ID: 1JEK). (I) **Tweezers or forceps:** A complex between c-Jun, c-Fos, and DNA. Proteins are shown as red helices, whereas DNA is shown as a blue cloud (PDB ID: 1FOS). (J) **Grabbers:** Structure of the complex between βPIX coiled coil (red helices) and Shank PDZ (blue cloud) (PDB ID: 3L4F). (K) **Tentacles:** Structure of the hexameric molecular chaperone prefoldin from the archaeum *Methanobacterium thermoautotrophicum* (PDB ID: 1FXK). (L) **Pullers:** Structure of the ClpB chaperone from *Thermus thermophilus* (PDB ID: 1QVR). (M) **Chameleons:** The C-terminal fragment of p53 gains different types of secondary structure in complexes with four different binding partners, cyclinA (PDB ID: 1H26), sirtuin (PDB ID: 1MA3), CBP bromo domain (PDB ID: 1JSP), and s100ββ (PDB ID: 1DT7). (N) **Stackers or β-arcs:** (Na) stack of β-arches, β-amyloid; (Nb), superpleated β-structure (Sup35p, Ure2P, α-synuclein); (Nc) stack of β-solenoids (prion); (Nd) stack of β-arch dimers (insulin); (Ne) β-solenoids. Modified from Ref. [116]. (O) **Dynamic complexes:** Schematic representation of the polyelectrostatic model of Sic1-Cdc4 interaction. Schematic of an IDP (ribbon) interacting with a folded receptor (gray shape) through several distinct binding motifs and an ensemble of conformations (indicated by four representations of the interaction). The IDP possesses positive and negative charges (depicted as blue and red circles, respectively), giving rise to a net charge q_l, while the binding site in the receptor (light blue) has a charge q_r. The effective distance $\langle r \rangle$ is between the binding site and the centre of mass of the IDP. Reproduced from Ref. [119]. See color insert.

1.9.7. Connectors and Armature

Being formed, coiled coils can be used in the subsequent formation of higher order oligomers, where segments of coiled coils are used to bring oligomeric partners together [106–108]. Coiled coil can serve as an armature, around which a more complex structure is built [109, 110].

1.9.8. Tweezers and Forceps

Many transcription factors form coiled-coil dimers that interact with DNA. Here, the coiled-coil dimer grips the major grove of DNA in a forceps-like manner [111].

1.9.9. Grabbers

In several instances, the ends of the coiled coil form an extensive β-sheet interaction with binding partners [112].

1.9.10. Tentacles

In its crystal structure, the hexameric molecular chaperone prefoldin resembles a jellyfish with a body consisting of a double β-barrel assembly, from which six long tentacle-like coiled coils are protruding. The distal regions of the coiled coils contain hydrophobic patches, which are utilized in the multivalent binding of non-native proteins [113].

1.9.11. Pullers

The *Thermus thermophilus* chaperone ClpB is a two-tiered hexameric ring with a set of 85-Å-long and mobile coiled coils that are located on the outside of the hexamer and act as mechanical pullers [114]. Here, the concerted motions of these coiled coils cause adjacent subunits to move in opposite directions, generating the mechanical force required to pull aggregates apart [114].

1.9.12. Chameleons

One of the most unique features of IDPs is their ability to gain, in a template-dependent manner, very different structures in the bond form. This capability is illustrated by the C-terminal binding region of p53, the same short segment of which binds to four unrelated partners adopting different conformations (an α-helix, a β-sheet, and two differently laid irregular structures) when bound to the different partners [7, 115].

1.9.13. Stackers or β-arcs

The unifying model of the amyloid fibrils is "β-arcades," which are the columnar structures produced by in-register stacking of "β-arcs," strand-turn-strand motifs in which the two β-strands interact via their side chains, not via the polypeptide backbone as in a conventional β-hairpin [116].

1.10. *LESSON 8:* IDPS CAN FORM DISORDERED OR FUZZY COMPLEXES

In addition to the static complexes considered above, where bound partners have fixed structures, some IDPs do not fold even in their bound state, forming so-called disordered, dynamic, or fuzzy complexes with ordered proteins [82, 117–121], other disordered proteins [122–124], or biological membranes [125, 126]. In complexes of some of these IDPs with their binding partners, the disordered regions flanking the interaction interface but not the interface itself remain disordered. Such mode of interaction was recently described as "the flanking fuzziness" in contrast to "the random fuzziness" when the disordered protein remains entirely disordered in the bound state [127, 128]. It is also expected that a similar binding mode can be utilized by disordered proteins while interacting with nucleic acids and other biological macromolecules [88].

Physically, binding is considered as joining objects together, and suggests spatial and temporal fixation of bound partners. The formation of protein complexes with specific binding partners is expected to bring some fixation (at least at the binding site). Therefore, the mentioned disordered complexes, where interaction of a disordered protein with the binding partners is not accompanied by a disorder-to-order transition within the interaction interface clearly, cannot be described by the classical binding paradigm. The contradiction can be resolved assuming that the ordered binding partner or disordered protein contains multiple low-affinity binding sites. The existence of several similar binding sites combined with a highly flexible and dynamic structure of disordered protein creates a unique situation where any binding site of disordered protein can interact with any binding site of its partner with almost equal probability, in a staccato manner. The low affinity of each individual contact implies that each of them is not stable and can be readily broken. Therefore, such disordered or fuzzy complex can be envisioned as a highly dynamic ensemble in which a disordered protein does not present a single binding site to its partner but resembles a "binding cloud," in which multiple identical binding sites are dynamically distributed in a diffuse manner (see Fig. 1.7O). In other words, in this staccato-type interaction mode, a disordered protein rapidly changes multiple binding sites while probing the binding site(s) of its partner [88]. An additional factor that can help holding dynamic complex together could be a weak long-range attraction between protein molecules [129]. This long-range attraction is universal for all protein solutions and has a range several times that of the diameter of the protein molecule, much greater than the range of the screened electrostatic repulsion [129].

1.11. *LESSON 9:* INTRINSIC DISORDER IS CRUCIAL FOR RECOGNITION, REGULATION, AND SIGNALING

The functional importance of being disordered has been intensively analyzed [2, 4, 6, 21, 23–25, 130–136]. The majority of IDPs undergo a disorder-to-order transition upon functioning [2, 4, 6, 21, 23–25, 130–137]. When disordered regions bind to signaling partners, the free energy required to bring about the disorder-to-order transition takes away from the interfacial, contact free energy, with the net result that a highly specific interaction can be combined with a low net free energy of association [4, 130]. High specificity coupled with low affinity seems to be a useful pair of properties for a signaling interaction so that the signaling interaction is reversible. In addition, a disordered protein can readily bind to multiple partners by changing shape to associate with different targets [4, 138, 139]. In addition to decoupled specificity and strength of binding, disorder has several clear advantages for functions in signaling, regulation, and control [1, 2, 4, 140–143]:

1. Increased speed of interaction due to greater capture radius and the ability to spatially search through interaction space.
2. Strengthened encounter complex allows for less stringent spatial orientation requirements.
3. Efficient regulation via rapid degradation.
4. Increased interaction (surface) area per residue.
5. A single disordered region may bind to several structurally diverse partners.
6. Many distinct (structured) proteins may bind to a single disordered region.
7. Intrinsic disorder provides the ability to overcome steric restrictions, enabling larger interaction surfaces in protein–protein and protein–ligand complexes than those obtained with rigid partners.
8. Unstructured regions fold to specific bound conformations, which can be very different, according to the template provided by structured partners.
9. Efficient regulation via posttranslational modification (PTM); that is, phosphorylation, methylation, ubiquitination, sumoylation, and so on.
10. Ease of regulation/redirection and production of otherwise diverse forms by alternative splicing (AS).
11. The possibility of overlapping binding sites due to extended linear conformations.
12. Diverse evolutionary rates with some IDPs being highly conserved and other IDPs possessing high evolutionary rates. The latter can evolve into sophisticated and complex interaction centers (scaffolds) that can be easily tailored to the needs of divergent organisms,
13. Flexibility that allows masking (or not) of interaction sites or that allows interaction between bound partners.
14. Binding fuzziness where different binding mechanisms (e.g., via stabilizing the binding-competent secondary-structure elements within the contacting region, or by establishing the long-range electrostatic interactions, or being involved in transient physical contacts with the partner, or even without any apparent ordering) can be employed to accommodate peculiarities of interaction with various partners.

This clearly suggests that there is a new two-pathway protein structure–function paradigm, with sequence-to-structure-to-function for enzymes and membrane transport proteins, and sequence-to-disordered ensemble-to-function for proteins and protein regions involved in signaling, regulation, and control [4, 21, 22, 53, 132, 133].

Among the various functions found for disordered regions, even superficial analysis of natively unfolded proteins revealed that many of them undergo disorder-to-order transitions when stabilized by binding with specific targets [3]. In fact, for the majority of proteins described in a previous study, the existence of ligand-induced folding was established. Examples include induced structure formation upon binding with DNA (or RNA) for protamines, Max protein, high-mobility group proteins HMG-14 and HMG-17, osteonectine, Ser-Asp (SD) repeat protein D (SdrD protein), chromatogranins A and B, Δ131Δ fragment of SNase, and histone H1. Other examples include the folding of cytochrome c in the presence of heme, the folding of osteocalcine induced by cations, secondary-structure formation in parathyroid hormone-related protein induced by membrane association, structure formation in glucocorticoid receptor brought about by association with trimethylamine N-oxide, folding of histidine-rich protein II induced by heme; and structure formation and compaction of prothymosin α mediated by zinc [3]. Therefore, among the major functions of these unstructured, IDPs are nucleic acid binding, metal ion binding, heme binding, and interaction with membrane bilayers [3].

More than 150 proteins have been identified in early studies as containing functional disordered regions, or being completely disordered, yet performing vital cellular roles [4, 132]. Twenty-eight separate functions were assigned for these disordered regions, including molecular recognition via binding to other proteins, or to nucleic acids [132, 133]. An alternative view is that functional disorder fits into at least six broad classes based on their mode of action [5, 144].

A computational study was carried out for the evaluation of a correlation between the functional annotations in the Swiss-Prot database and the predicted intrinsic disorder [145–147]. The approach is based on the hypothesis that if a function described by a given keyword relies on intrinsic disorder, then the keyword-associated protein would be expected to have a greater level of predicted disorder than the protein randomly chosen from the Swiss-Prot. To test this hypothesis, functional keywords associated with 20 or more proteins in Swiss-Prot were found and corresponding keyword-associated data sets of proteins were assembled. For each keyword-associated set, 1000 length-matching and number-matching sets of random proteins were drawn from Swiss-Prot. Order–disorder predictions were carried out for the keyword-associated sets and for the matching random sets. If a function described by a given keyword were carried out by a long region of disordered protein, one would expect the keyword-associated set to have a greater amount of predicted disorder compared with the matching random sets. The keyword-associated set would be expected to have a lower amount of predicted disorder compared with the random sets if the keyword-associated function were carried out by a structured protein. Given the predictions for the function-associated and matching random sets, it is possible to calculate the P-values, where a P-value > 0.95 was used to define a disorder-associated function and a P-value < 0.05 was used to define an order-associated function. Intermediate P-values are ambiguous [145–147]. The application of this approach revealed that out of 710 Swiss-Prot keywords, 310 functional keywords were associated with ordered proteins, 238 functional keywords were attributed to disordered proteins, and the remaining 162 keywords yield ambiguity in the likely function–structure associations [145–147].

Interestingly, most of the structured protein-associated keywords were related to enzyme activities, which is a result in accordance with the previous discussion indicating that structure formation is for enzyme catalysis rather than just for molecular recognition. As for the disordered protein-associated keywords, most were related to signaling and regulation, again in accordance with the arguments given above that the thermodynamics of disorder-to-order upon binding are favorable to binding reversibility and thus to signaling.

Based on the analysis of then-known data the Protein Trinity paradigm was formulated [4, 21]. According to this model, functional intracellular proteins (or their functional regions) can exist in any of the three thermodynamic states: ordered, molten globule, and random coil. Function can arise from any of the three conformations and transitions between them. "In this view, not just the ordered state, but any of the three states can be the native state of a protein" [4]. Data presented in this review show that natively unfolded proteins, which were originally considered as random coils, are not uniform and can be grouped into two structurally different subclasses, which were designated as intrinsic coils and intrinsic pre-molten globules. These observations bring a new player, the native pre-molten globule, to the protein function field [22]. Therefore, it has been suggested that the Protein Trinity should be extended to the Protein Quartet model, with function arising from four specific conformations (ordered forms, molten globules, pre-molten globules, and random coils) and transitions between any two of the states [22].

1.12. LESSON 10: PROTEIN POSTTRANSLATIONAL MODIFICATIONS OCCUR AT DISORDERED REGIONS

In a study of the functions associated with more than 100 long disordered regions, many were found to contain sites of protein PTM [132, 133]. These PTMs included phosphorylation, acetylation, fatty acylation, methylation, glycosylation, ubiquitination, and ADP-ribosylation, suggesting the possibility that protein modifications commonly occur in regions of disorder. A particular advantage of disorder for regulatory and signaling regions is that changes, such as protein modification, lead to large-scale disorder-to-order structural transitions: such large-scale structural changes are not subtle and therefore could be an advantage for signaling and regulation as compared with the much smaller changes that would be expected from the decoration of an ordered protein structure.

Protein phosphorylation and dephosphorylation are crucial for signaling. Indeed, about one-third of eukaryotic proteins are phosphorylated [148]. Many sites of protein phosphorylation were found to be in regions structurally characterized as intrinsically disordered [132, 133]. This conclusion was based on several lines of evidence, such as a very small number of PDB structures for both the unphosphorylated and phosphorylated forms of the same protein [149, 150]; the fact that the residues of the phosphorylation site often have extended, irregular conformation consistent with disordered structure [150]; the fact that the segments containing phosphorylation sites not only lack secondary structure but are held in place by side chain burial and also by backbone hydrogen bonds to the surrounding kinase side chains [151–156]; the fact that regions flanking the sites of phosphorylation are enriched in the disorder-promoting amino acids [150]; the fact that the sequence complexity distribution of the residues flanking phosphorylation sites matches almost exactly the complexity distribution obtained for IDPs [150]; and the fact there is a high correspondence between the prediction of disorder and the occurrence of phosphorylation [150].

Ubiquitination, the reversible modification of proteins by the covalent attachment of ubiquitin, is implicated in the regulation of a variety of cellular processes and is involved in many diseases. Recently, 141 new ubiquitination sites were identified using a combination of liquid chromatography, mass spectrometry, and mutant yeast strains [157]. The detailed analysis of the sequence biases and structural preferences around known ubiquitination sites indicated that the properties of these sites were similar to those of IDRs. In agreement with this computational study, structural information about the ubiquitination sites is sparse. In fact, despite the large size of PDB, only 7% of currently known ubiquitination sites in yeast could be confidently mapped to protein structures. The analysis of 3-D structures of 32 homologous protein chains (with 15 of them being 100% identical with query proteins) containing 28 ubiquitination sites revealed that 10 ubiquitination sites were in crystal or interchain/intrachain contacts, and therefore the assignment of these sites to a specific structural element should be made with caution. Of the 18 sites that could be confidently assigned to ordered regions, 11 were located within coils (two of which were close to the observed disordered regions), four within helices, and three within strands. The majority of the sites within coils and helices were surface exposed and had high B-factor values indicating high flexibility [157]. It has also been pointed out that along with the lack of structural information for the majority of experimentally detected ubiquitination sites, there were several examples of ubiquitination sites located in the experimentally confirmed disordered regions [157]. Based on these observations it has been concluded that the involvement of flexible and disordered protein regions into various aspects of ubiquitination process provides a strong support for the functional importance of such regions.

In addition to protease digestion, ubiquitination, and phosphorylation, several other types of PTMs, such as acetylation, fatty acid acylation, and methylation, have also been observed to occur in regions of intrinsic disorder [132, 133, 147]. From these findings, it is tempting to suggest that sites of protein modification in eukaryotic cells universally or at least very commonly exhibit a preference for IDRs. For all of the examples discussed earlier, the modifying enzyme has to bind to and modify similar sites in a wide variety of proteins. If all the regions flanking these sites are disordered before binding to the modifying enzyme, it is easy to understand how a single enzyme could bind to and modify a wide variety of protein targets.

1.13. *LESSON 11:* DISORDERED REGIONS ARE PRIMARY TARGETS FOR AS

Alternative splicing is a process by which two or more mature mRNAs are produced from a single precursor pre-mRNA by the inclusion and omission of different segments [158, 159]. The "exons" are joined to form the mRNA and the "introns" are left out [160]. So far AS has been commonly observed only in multicellular eukaryotes [161]. For humans and other mammals, multiple proteins are often produced from a single gene since 40–60% the genes yield proteins via the AS mechanism [162–164]. It was hypothesized that AS very likely provides an important mechanism for enhancing protein diversity in multicellular eukaryotes [165]. AS has affects on a diversity of protein functions such as protein–protein interactions, ligand binding, and enzymatic activity [166–168]. Therefore, it comes as no surprise that abnormal AS has been associated with numerous human diseases, including myotonic dystrophy [169], Axoospermia [170], Alzheimer's disease [171], Parkinson's disease [172, 173], and cancer [174].

Removal of a piece of sequence from a structured protein would often lead to dysfunctional protein folding, most often causing loss of function (sometimes, however, the AS isoform of structured proteins can maintain function, albeit typically with a reduction in activity). Why, then, is the AS phenomenon so common in nature? The analysis of the effect of AS on structured proteins revealed that AS-induced alterations are generally small in size, are usually located on the protein surface, and are most often located in coil regions [175]. Given the small sizes and locations of the changes resulting from AS, the different splice variants were predicted to fold into the same overall structures, with only slight structural perturbations that could be functionally important [175, 176].

The structural implications given above are interesting, but only a small fraction of AS events have been mapped to structured proteins. Since 40–60% of mammalian (human) genes are estimated to undergo AS, and since there are several thousand mammalian proteins in PDB [44], we would expect to find several thousand examples to study. So far, however, despite exhaustive searches of PDB, only 20 examples have been reported [175]. Based on the failure to find a significant number of examples of AS that map to regions of structure, it was hypothesized that the protein folding problems discussed earlier would be solved for different isoforms if the alternatively spliced regions of mRNA were to code for regions of IDP. If AS were to map to ID regions, both multiple and long splice variants would be allowed because structural perturbation would not be a problem.

To test this hypothesis, a collection of human proteins with structurally characterized regions of order and disorder was built and an exhaustive search on AS for all of these proteins was performed. This generated a set of 46 human proteins with 75 alternatively spliced segments, all of which were located in structurally characterized regions [177]. Importantly, of these 75 alternatively spliced regions of RNA, 43 (57%) coded for entirely disordered protein regions, 18 (24%) coded for both ordered and disordered protein regions (with the splice boundaries very often in, or very near to, the disordered regions), and just 14 (19%) coded for fully structured regions [177]. Next, to increase the number of examples, all the Swiss-Prot proteins labeled as having AS isoforms were identified. This approach generated a set of 558 proteins with 1266 regions that are absent from one isoform due to AS. Disorder/order propensities of these AS proteins and regions were predicted together with the disorder/order propensities of the 46 structurally characterized proteins and for their 75 regions that were affected by AS. This analysis revealed a perfect correlation between predictions and observations of disorder in the 46 structurally characterized proteins. For the 1266 regions from Swiss-Prot, the predicted abundance of disorder closely matched the corresponding predictions for the 75 with known structure. These data strongly suggest that AS occurs mostly in regions of RNA that code for disordered protein [177].

These findings have crucial functional implications. Since disorder plays various roles in protein functions and in protein–protein interaction networks, modification of such functions could be readily accomplished by AS within disordered regions. Thus, a linkage between AS and signaling by disordered regions provides a novel and plausible mechanism for understanding the origins of cell differentiation, which ultimately gives rise to multicellular organisms in nature [177].

1.14. LESSON 12: DISORDERED PROTEINS ARE TIGHTLY REGULATED IN THE LIVING CELLS

It is clear now that the IDPs are real, abundant, diversified, and vital. Functions of IDPs are complementary to the catalytic activities of ordered proteins [2–4, 21–26, 29, 53, 132, 133, 135, 145–147, 178, 179]. Many disorder-related functions (e.g., signaling, control, regulation, and recognition) are incompatible with well-defined, stable 3-D structures [2–5, 21, 22, 24–26, 29, 53, 135, 145–147, 179, 180]. Intrinsic disorder is assumed to provide several functional advantages for its carriers, including increased interaction surface area, structural plasticity to interact with several targets, high specificity for given partners combined with high k_{on} and k_{off} rates that enable rapid association with the partner without an excessive binding strength, and the ability to fold upon binding and accessible PTM sites.

The highly dynamic nature of IDPs is a visual illustration of the chaos. However, the evolutionary persistence of these highly dynamic proteins, their unique functionality, and involvement in all the major cellular processes evidence that this chaos is tightly controlled [181]. To answer the question of how these proteins are governed and regulated inside the cell, Gsponer et al. conducted a detailed study focused on the intricate mechanisms of the IDP regulation [182]. To this end, all the *S. cerevisiae* proteins were grouped into three classes using one of the available disorder predictors, DisoPred2 [183]: (1) 1971 highly ordered proteins containing 0–10% of the predicted disorder; (2) 2711 moderately disordered proteins with 10–30% predicted disordered residues; and (3) 2020 highly disordered proteins containing 30–100% of the predicted disorder. Then, the correlations between intrinsic disorder and the various regulation steps of protein synthesis and degradation were evaluated.

To examine the transcription of genes encoding IDPs and ordered proteins, the transcriptional rates and the degradation rates of the corresponding transcripts were compared [182]. This analysis revealed that the transcriptional rates of mRNAs encoding IDPs and ordered proteins were comparable. However, the IDP-encoding transcripts were generally less abundant than transcripts encoding ordered proteins due to the increased decay rates of the former.

Tight regulation of IDP abundance was also observed at the protein level. IDPs were shown to be less abundant than ordered proteins due to the lower rate of protein synthesis and shorter protein half-lives. As the abundance and half-life in a cell of certain proteins can be further modulated via their PTMs such as phosphorylation [184], the experimentally determined yeast kinase–substrate network was analyzed next. IDPs were shown to be substrates of twice as many kinases as were ordered proteins. Furthermore, the vast majority of kinases whose substrates were IDPs were either regulated in a cell-cycle dependent manner, or activated upon exposure to particular stimuli or stress [182]. Therefore, PTMs may not only serve as an important mechanism for the fine-tuning of the IDP functions, but possibly they are necessary to tune the IDP availability under different cellular conditions.

In addition to *S. cerevisiae*, similar regulation trends were also found in *Schizosaccharomyces pombe* and *Homo sapiens* [182]. Based on these observations it has been concluded that both unicellular and multicellular organisms appear to use similar mechanisms to regulate the availability of IDPs. Overall, the study by Gsponer et al. clearly demonstrated that there is an evolutionarily conserved tight control of synthesis and clearance of most IDPs. This tight control is directly related to the major roles of IDPs in signaling, where it is crucial to be available in appropriate amounts and not to be present longer than needed [182]. It has been also pointed out that although the abundance of many IDPs is under strict control, some IDPs could be present in cells in large amounts and/or for long periods either due to specific PTMs or via interaction with other factors, which could promote changes in cellular localization of IDPs or protect them from the degradation machinery [4, 147, 150, 184, 185]. Overall, it has been concluded that the chaos seemingly introduced into the protein world by the discovery of IDPs is under tight control [181].

In an independent study, a global-scale relationship between the predicted fraction of protein disorder and RNA and protein expression in *E. coli* was analyzed [186]. It has been shown that the fraction of protein disorder were positively correlated with both measured RNA expression levels of *Escherichia coli* genes in three different growth media (lysogeny broth (LB) rich medium and nitrogen- and carbon-free (N-C-) minimal media supplemented with glycerol as a carbon source and either ammonium chloride or arginine as a nitrogen source) and predicted abundance levels of *E. coli* proteins. When a subset of 216 *E. coli* proteins that are known to be essential for the survival and growth of this bacterium was analyzed, the correlation between protein disorder and expression level became even more evident. In fact, essential proteins had on average a much higher fraction of disorder (0.36), had a higher number of proteins classified as completely disordered (19% vs. 2% for *E. coli* proteome), and were expressed at a higher level in all three media than an average *E. coli* gene [186]. To better understand the function–disorder relationship for highly expressed *E. coli* proteins, a literature search was carried out for a group of proteins that had high levels of predicted intrinsic disorder. This study revealed that the disorder predictions matched well with the experimentally elucidated regions of protein flexibility and disorder [186]. A direct link between protein disorder and protein level in *E. coli* cells could be because disordered proteins may carry out essential control and regulation functions that are needed to respond to various environmental conditions. Another possibility is that IDPs might undergo more rapid degradation compared with structured proteins, which cells can counter by increasing the mRNA levels of the corresponding genes. In this case, higher synthesis and degradation rates could make the levels of these proteins very sensitive to the environment, with slight changes in either production or degradation leading to significant shifts in protein levels [186].

More evidence for the tight control of IDPs inside the cell came from the analysis of cellular regulation of so-called vulnerable proteins [14]. The integrity of soluble protein functional structures is maintained in part by a precise network of hydrogen bonds linking the backbone amide and carbonyl groups. In a well-ordered protein, hydrogen bonds are shielded from water attack, preventing backbone hydration and the total or partial denaturation of the soluble structure under physiological conditions [187, 188]. Since soluble protein structures may be more or less vulnerable to water attack depending on their packing quality, a structural attribute, protein vulnerability, was introduced as the ratio of solvent-exposed backbone hydrogen bonds, which represent local weaknesses of the structure, to the overall number of hydrogen bonds [14]. It has also been pointed out that structural vulnerability can be related to protein intrinsic disorder as the inability of a particular protein fold to protect intramolecular hydrogen bonds from water attack may result in backbone hydration leading to local or global unfolding. Since binding of a partner can help to exclude water molecules from the microenvironment of the preformed bonds, a vulnerable soluble structure gains extra protection of its backbone hydrogen bonds through the complex formation [187]. To understand the role of structure vulnerability in transcriptome organization, the relationship between the structural vulnerability of a protein and the extent of co-expression of genes encoding its binding partners was analyzed [14]. This study revealed that structural vulnerability can be considered a determinant of transcriptome organization across tissues and temporal phases [14]. Finally, by interrelating vulnerability, disorder propensity, and co-expression patterns, the role of protein intrinsic disorder in transcriptome organization was confirmed since the correlation between the extent of intrinsic disorder of the most disordered domain in an interacting pair and the expression correlation of the two genes encoding the respective interacting domains was evident [14].

1.15. *LESSON 13:* NATIVELY UNFOLDED PROTEINS ARE FREQUENTLY ASSOCIATED WITH HUMAN DISEASES

Because IDPs play crucial roles in numerous biological processes, it was not too surprising to find that some of them are involved in human diseases. For example, a number of human diseases originate from the deposition of stable, ordered, filamentous protein aggregates, commonly referred to as amyloid fibrils. In each of these abnormal protein deposition-mediated pathological states, a specific protein or protein fragment changes from its natural soluble form into insoluble fibrils, which accumulate in a variety of organs and tissues [189–195]. Approximately 20 different proteins are known so far to be involved in protein deposition diseases. These proteins are unrelated in terms of sequence or starting structure. Several IDPs are found in this list of 20 proteins, and they are associated with the development of several neurodegenerative diseases [195, 196]. An incomplete list of disorders associated with IDPs includes Alzheimer's disease (deposition of amyloid-β, tau-protein, α-synuclein fragment NAC [197–200]), Niemann–Pick disease type C, subacute sclerosing panencephalitis, argyrophilic grain disease, myotonic dystrophy, and motor neuron disease with neurofibrillary tangles (accumulation of tau-protein in the form of neurofibrillary tangles [199]); Down's syndrome (nonfilamentous amyloid-β deposits [201]); Parkinson's disease, dementia with LB, diffuse LB disease, LB variant of Alzheimer's disease, multiple system atrophy, and Hallervorden–Spatz disease (deposition of α-synuclein in the form of LB, or Lewy neuritis [202]); prion diseases (deposition of PrP^{SC} [203]); and a family of polyQ diseases, a group of neurodegenerative disorders caused by expansion of GAC trinucleotide repeats coding for polyQ in the gene products [204]. Furthermore, most mutations in rigid globular proteins associated with accelerated fibrillation and protein deposition diseases have been shown to destabilize the native structure, increasing the steady-state concentration of partially folded (disordered) conformers [189–195].

The disorders just mentioned are called conformational diseases, as they are characterized by conformational changes, misfolding, and aggregation of an underlying protein. However, there is another side to this coin: protein functionality. In fact, many of the proteins associated with the conformational disorders are also involved in recognition, regulation, and cell signaling. For example, functions ascribed to α-synuclein, a protein involved in several neurodegenerative disorders, include binding fatty acids and metal ions; regulation of certain enzymes, transporters, and neurotransmitter vesicles; and regulation of neuronal survival (reviewed in [202]). Overall, about 50 proteins and ligands interact and/or co-localize with this protein. Furthermore, α-synuclein has remarkable structural plasticity and adopts a series of different monomeric, oligomeric, and insoluble conformations (reviewed in Ref. [15]). The choice between these conformations is determined by the peculiarities of the protein environment, assuming that α-synuclein has an exceptional ability to fold in a template-dependent manner. Based on these observations, we hypothesize that the development of the conformational diseases may originate from the misidentification, misregulation, and mis-signaling, accompanied by misfolding. In other words, mutations and/or changes in the environment may result in protein confusion, for which the identity of the protein becomes lost, thus reducing its capability to recognize proper binding partners and leading to the formation of nonfunctional and deadly aggregates.

Recent analysis of so-called polyglutamine diseases gives support to this hypothesis [205]. Polyglytamine diseases are a specific group of hereditary neurodegeneration caused by expansion of CAG triplet repeats in an exon of disease genes that leads to the production of a disease protein containing an expanded polyglutamine, polyQ, stretch. Nine neurodegenerative disorders, including Kennedy's disease, Huntington's diseases, spinocerebellar atrophy -1, -2, -3, -6, -7, and -17, and dentatorubral pallidoluysian atrophy belong to this class of diseases [206–209]. In most polyQ diseases, expansion to over 40 repeats leads to their onset [209]. It has been emphasized that molecular processes such as unfolded protein response, protein transport, synaptic transmission, and transcription are implicated in the pathology of polyQ diseases [205]. Importantly, more than 20 transcription-related factors have been reported to interact with pathological polyQ proteins. Furthermore, these interactions were shown to repress the transcription, leading eventually to neuronal dysfunction and death (reviewed in [205]). These results suggest that polyQ diseases represent kind of transcriptional disorder [205], supporting our misidentification hypothesis for at least some of the conformational disorders.

Disorder is also very common in cancer-associated proteins. In 2002, a study found that 79% of cancer-associated proteins and 66% of cell-signaling proteins contain predicted regions of disorder of 30 residues or longer [23]. In contrast, only 13% of a set of proteins with well-defined ordered structures contained such long regions of predicted disorder. Here, cancer-associated proteins were defined as those human proteins in Swiss-Prot containing the keyword "oncogene" (this included anti- and proto-oncogenes) or containing the word "tumor" in the description field. In experimental studies, the presence of disorder has been directly observed in several cancer-associated proteins, including p53 [210], p57^{kip2} [211], Bcl-X$_L$ and Bcl-2 [212], c-Fos [213], and most recently, a thyroid cancer-associated protein, TC-1 [214].

1.16. LESSON 14: NATIVELY UNFOLDED PROTEINS ARE ATTRACTIVE DRUG TARGETS

Since many proteins associated with various human diseases are either completely disordered or contain long disordered regions [215, 216], and since some of this disease-related IDPs/regions are involved in recognition, regulation, and signaling, these proteins/regions clearly represent novel potential drug targets. It is recognized now that the possibility of interrupting the action of disease-associated proteins (including through modulation of protein–protein interactions) presents an extremely attractive objective for the development of new drugs. The rational design of enzyme inhibitors depends on the classical view of protein function, which states that 3-D structure is an obligatory prerequisite for function. While generally applicable to many enzymatic domains, this view has persisted to influence views concerning all protein functions despite numerous examples to the contrary. Due to failure to recognize the important role of disorder in protein function, current and evolving methods of drug discovery suffer from an overly rigid view of protein function. This is most apparent in the observation that the vast majority of currently available drugs target the active site of enzymes, presumably since these are the only proteins for which the order-function paradigm is generally applicable.

Disordered proteins often bind their partners with contiguous residues of relatively short length, which become ordered upon binding [90, 217, 218]. Targeting small molecules to the disordered regions of proteins should enable the development of more effective drug discovery techniques. Drugs targeting these regions will likely function through inducing the disordered region to form an ordered structure dissimilar to its structure in complex with its binding partner, thereby preventing binding. The principles of small molecule binding to disordered regions have not been well studied, but sequence-specific, small-molecule binding to short peptides has been observed [219]. An interesting twist of this disorder-based approach for drug discovery is that using disordered regions as drug targets can be described as inducing order to *prevent* function.

In agreement with above-mentioned concepts, small molecules, "Nutlins," have been recently discovered that inhibit the p53–Mdm2 interaction by mimicking the helix in p53 that binds to Mdm2 [220, 221]. The tumor suppressor protein p53 is at the center of a large signaling network. It regulates expression of genes involved in numerous cellular processes, including cell cycle progression, apoptosis induction, DNA repair, as well as others involved in responding to cellular stress [222]. When p53 function is lost, either directly through mutation or indirectly through several other mechanisms, the cell often undergoes cancerous transformation [223, 224]. Cancers showing mutations in p53 are found in the colon, lung, esophagus, breast, liver, brain, reticuloendothelial tissues, and hemopoietic tissues [223].

p53 is regulated by several different mechanisms including inhibition of its activity by binding to E3 ubiquitin ligase Mdm2, which binds to a short stretch of p53, residues 13–29. This region of p53 is within the transactivation domain; thus p53 cannot activate or inhibit other genes when Mdm2 is bound. Mdm2 ubiquitinates p53 and thus targets it for destruction. Mdm2 also contains a nuclear export signal that causes p53 to be transported out of the nucleus. Although X-ray crystallographic studies of the p53–Mdm2 complex reveal that the Mdm2 binding region of p53 forms a helical structure that binds into a deep groove on the surface of Mdm2 [225], NMR studies show that the unbound N-terminal region of p53 lacks a fixed structure, although it does possess an amphipathic helix that forms a secondary structure part of the time [210] and therefore represents an illustrative example of the α-MoRF concept. This amphipathic helix seen in the unbound state is the same helix that binds to Mdm2. A close examination of the interface between the proteins reveals that Phe^{19}, Trp^{23}, and Leu^{26} of p53 are the major contributors to the interaction, with the side chains of these three amino acids pointing down into a crevice on the Mdm2 surface.

Because of the apparent simplicity of the interface, as well as the importance of the p53-Mdm2 interaction, this protein–protein interaction has been investigated as a possible drug target by many researchers. Several successful peptide inhibitors of the interaction have been created [226–229]. These peptides were all derived from the region of p53 that binds to Mdm2. Additionally all successful peptide inhibitors contained the three crucial residues involved in the interaction, Phe^{19}, Trp^{23}, and Leu^{26} [221].

Several small molecules were recently found to block the p53–Mdm2 interaction [221, 230–232]. While some of these were natural products, others were from a class of *cis*-imidazolines called "Nutlins." These latter molecules increased the level of p53 in cancer cell lines. This drastically decreased the viability of these cells, causing most of them to undergo apoptosis. When one of the Nutlin compounds was given orally to mice, researchers saw a 90% inhibition of tumor growth compared with the control. The structure of Nutlin-2 was shown to mimic the crucial residues of p53, with two bromophenyl groups fitting into Mdm2 in the same pockets as Trp^{23} and Leu^{26}, and an ethyl–ether side chain filling the spot normally taken by Phe^{19} [230–232].

This research demonstrates that finding small molecules to target regions of proteins normally bound by disordered proteins is a feasible approach. It is anticipated that by studying this drug interaction, it will be possible to identify regions of other proteins that can be mimicked by small molecules. Remarkably, the disorder prediction for p53 using PONDR VL-XT software showed a sharp downward spike indicating predicted ordered region near the N-terminus of the protein. Furthermore, the α-MoRF identifier was able to recognize the region of p53 that binds to Mdm2 as a region of molecular recognition [115].

This successful Nutlin story marks the potential beginning of a new era, *the signaling–modulation era*, in targeting drugs to protein–protein interactions. Importantly, this druggable p53–Mdm2 interaction involves a disorder-to-order transition. Principles of such a transition are generally understood and therefore can use to find similar drug targets [233]. In addition to Nutlins, seven types of promising drug molecules that act by blocking protein–protein interactions have been described [234, 235]. While protein disorder is not mentioned in any of the papers describing how a small molecule can block protein–protein interactions, the disorder-based analysis revealed that four of these interactions involve one structured partner and one disordered partner, with 3 of the 4 disordered segments becoming helix upon binding. Therefore, the p53–Mdm2 complex is not the only member of this class currently known to be blocked by a small drug-like molecule. We fully expect many more examples to appear shortly, and for some of these examples to lead to useful drug molecules. Since p53–Mdm2-like interactions are likely to be very common, they clearly define a cornucopia of new drug targets that would operate by blocking disorder-based protein–protein interactions.

For these examples, the drug molecules mimic a critical region of the disordered partner, which folds upon binding, and compete with this region for its binding site on the structured partner. We argue that these druggable sites are likely to operate by the coupled binding and folding mechanism and utilize interaction sites that are small and compact enough to be easily mimicked by small molecules. We have developed methods for predicting such binding sites in disordered regions [236] and have elaborated the bioinformatics tools to identify which disordered binding regions can be easily mimicked by small molecules [233].

A complementary approach to finding small molecules inhibiting disorder-based protein–protein interactions has been developed in the laboratory of Prof. Steven Metallo (see, e.g., [237]). Deregulation of the c-Myc transcription factor is involved in many types of cancer, making this oncoprotein an attractive target for drug discovery. In order to bind DNA, regulated target gene expression, and function in most biological contexts, c-Myc must dimerize with its obligate heterodimerization partner, Max, which lacks a transactivation segment. c-Myc, which is intrinsically disordered as a monomer, undergoes coupled binding and folding of its basic-helix-loop-helix-leucine zipper domain (bHLHZip) upon heterodimerization with its partner protein Max. One approach to c-Myc inhibition has been to disrupt this dimeric complex. In a search for effective inhibitors of the c-Myc–Max interactions, a series of small molecules was analyzed to find seven Myc inhibitors, which were shown to bind to one of three discrete sites within the 85-residue bHLHZip domain of c-Myc. These binding sites are composed of short contiguous stretches of amino acids that can selectively and independently bind small molecules. Inhibitor binding induces only local conformational changes, preserves the overall disorder of c-Myc, and inhibits dimerization with Max. Furthermore, binding of inhibitors to c-Myc was shown to occur simultaneously and independently on the three independent sites. Based on these observations it has been concluded that a rational and generic approach to the inhibition of protein–protein interactions involving IDPs may therefore be possible through the targeting of intrinsically disordered sequence [237].

Ideally, a drug that targets a given protein–protein interaction should be tissue specific. Although some proteins are unique for a given tissue, many more proteins have very wide distribution, being present in several tissues and organs. How can one develop tissue-specific drugs targeting such abundant proteins? Often, tissue specificity for many of the abundant proteins is achieved via the AS of the corresponding pre-mRNAs, which generates two or more protein isoforms from a single gene. Estimates indicate that between 35% and 60% of human genes yield protein isoforms by means of alternatively spliced mRNA [162]. The added protein diversity from AS is thought to be important for tissue-specific signaling and regulatory networks in multicellular organisms. Recently, it has been established that the regions of AS in proteins are enriched in intrinsic disorder [177]. Since disorder is frequently utilized in protein binding regions, having AS of pre-mRNA coupled to regions of protein disorder was proposed to lead to tissue-specific signaling and regulatory diversity [177]. Therefore, associating AS with protein disorder enables the time- and tissue-specific modulation of protein function. These findings open up a unique opportunity to develop tissue-specific drugs modulating the function of a given IDP/IDR (with a unique profile of disorder distribution) in a target tissue and not affecting the functionality of this same protein (with different disorder distribution profile) in other tissues.

1.17. *LESSON 15:* BRIGHT FUTURE OF FUZZY PROTEINS

The last 10–15 years witnessed a real revolution in our understanding of the protein structure–function relationships. The fact that there is an entire class of polypeptides that do not have rigid structures but possess crucial biological function was underappreciated and ignored for a long time despite numerous examples scattered in the literature. The work that started as an attempt to understand what is special about several natively unfolded proteins characterized in our group produced a real explosion of interest in non-folded proteins with biological functions. A new field was created and a lot of intriguing information was produced related to the IDPs in general and to the natively unfolded proteins in particular. This chapter discussed some general concepts related to the natively unfolded proteins and represents some of our data to support these concepts.

Based on the data summarized in this chapter, natively unfolded proteins are characterized by low overall hydrophobicity and high net charge. They possess large hydrodynamic volumes and low contents of ordered secondary structure. They are characterized by the absence of a tightly packed core. They are very flexible but may adopt relatively rigid conformations in the presence of natural ligands. In comparison with ordered globular proteins, natively unfolded polypeptides possess "turn out" responses to changes in the environment, as their structural complexities increase at high temperatures or at extreme pHs.

An intriguing property of natively unfolded proteins is their ability to undergo disorder-to-order transition upon function. The degree of these structural rearrangements varies over a very wide range, from coil to pre-molten globule transitions to formation of rigid ordered structures.

Multiple roles of natively unfolded proteins in pathogenesis of human diseases should not be ignored. Because of the numerous structural adjustments, perturbations, interactions, and functions ascribed to natively unfolded proteins, it is reasonable to suggest that they are potentially prone to misfold. In such cases, the development of different diseases associated with natively unfolded proteins (e.g., various synucleinopathies) may originate from the misregulation, missignaling, and misidentification of a corresponding protein, accompanied by or resulting from its misfolding. In fact, mutations and/or changes in the environment may reduce the capability of a natively unfolded protein to recognize proper binding partners, thus leading to the formation of nonfunctional and deadly aggregates.

More than a decade of intensive studies on IDPs revealed a number of unique features related to their structural properties, abundance, distribution, functional repertoire, regulation, involvement in disease pathogenesis, and so on. However, the amount of data produced so far is just a small tip of a humongous iceberg. IDPs continue to bring discoveries on a regular basis. More discoveries and breakthroughs are expected in future due to advances in novel experimental and computational tools for studies focused on IDPs. Modern protein science is at a turning point.

ACKNOWLEDGMENTS

This work would be impossible without numerous collaborators whose enthusiasm and help drove the studies on natively unfolded proteins for many years. In no particular order the list of people contributed to this project at its different stages includes: Eugene Permyakov, Joel Gillespie, Vyacheslav Abramov, Anthony Fink, Larissa Munishkina, Pierre Souillac, Sebastian Doniach, Ian Millett, Daniel Denning, Kevin Lee, Alexander Sigalov, Michael Rexach, Sergei Permyakov, Keith Oberg, Stefan Winter, Jie Li, Oxana Galzitskaya, Leonhard Kittler, Gunter Lober, Olga

Tcherkasskaya, Seung-Jae Lee, Min Zhu, Amy Manning-Bog, Alison McCormack, Donato Di Monte, Kiowa Bower, I-Hsuan Liu, Gregory Cole, John Goers, Ghiam Yamin, Charles Glaser, Elisa Cooper, Jeffrey Cohlberg, Mark Hokenson, Sudha Rajamani, Joseph Zbilut, Alessandro Giuliani, Alfredo Colosimo, Julie Mitchell, Mauro Colafranceschi, Norbert Marwan, Charles Webber, Jr., Christopher Oldfield, Yugong Cheng, Marc Cortese, Celeste Brown, Mikhail Karymov, Yuri Lyubchenko, Pedro Romero, Lilia Iakoucheva, Zoran Obradovic, Véronique Receveur-Bréchot, Jean-Marie Bourhis, Bruno Canard, Sonia Longhi, Chad McAllister, Yoshiko Kawano, Alexander Lushnikov, Andrew Mikheikin, Andrey Vartapetyan, Predrag Radivojac, Slobodan Vucetic, Timothy R. O'Connor, Jag Bhalla, Geoffrey Storchan, Caitlin MacCarthy, Maureen Harrington, Tanguy LeGall, Patrizia Polverino de Laureto, Laura Tosatto, Erica Frare, Oriano Marin, Angelo Fontana, Megan Sickmeier, Justin Hamilton, Vladimir Vacic, Agnes Tantos, Beata Szabo, Peter Tompa, Jake Chen, King Pan Ng, Gary Potikyan, Rupert Savene, Christopher Denny, Vinay Singh, Yue Zhou, Joseph Marsh, Julie Forman-Kay, Jingwen Liu, Zongchao Jia, Millie Georgiadis, Amrita Mohan, Ann Roman, Jiangang Liu, Narayanan Perumal, Eric Su, Fei Ji, Niels Klitgord, Michael Cusick, Marc Vidal, Chad Haynes, Saima Zaidi, Ya Yin Fang, and Jessica Chen. We are grateful to all former and current colleagues for their priceless contributions, assistance, and support. This work was supported in part by grants R01 LM007688-01A1 (to A. K. D and V. N. U.) and GM071714-01A2 (to A. K. D and V. N. U.) from the National Institutes of Health and the Program of the Russian Academy of Sciences for Molecular and Cellular Biology (to V. N. U.). We gratefully acknowledge the support of the Indiana University—Purdue University Indianapolis Signature Centers Initiative.

REFERENCES

1. Dunker, A. K., Garner, E., Guilliot, S., Romero, P., Albrecht, K., Hart, J., Obradovic, Z., Kissinger, C., and Villafranca, J. E. (1998) Protein disorder and the evolution of molecular recognition: Theory, predictions and observations, *Pac Symp Biocomput*, 473–484.
2. Wright, P. E. and Dyson, H. J. (1999) Intrinsically unstructured proteins: Re-assessing the protein structure-function paradigm, *J Mol Biol 293*, 321–331.
3. Uversky, V. N., Gillespie, J. R., and Fink, A. L. (2000) Why are "natively unfolded" proteins unstructured under physiologic conditions? *Proteins 41*, 415–427.
4. Dunker, A. K., Lawson, J. D., Brown, C. J., Williams, R. M., Romero, P., Oh, J. S., Oldfield, C. J., Campen, A. M., Ratliff, C. M., Hipps, K. W., Ausio, J., Nissen, M. S., Reeves, R., Kang, C., Kissinger, C. R., Bailey, R. W., Griswold, M. D., Chiu, W., Garner, E. C., and Obradovic, Z. (2001) Intrinsically disordered protein, *J Mol Graph Model 19*, 26–59.
5. Tompa, P. (2002) Intrinsically unstructured proteins, *Trends Biochem Sci 27*, 527–533.
6. Daughdrill, G. W., Pielak, G. J., Uversky, V. N., Cortese, M. S., and Dunker, A. K. (2005) Natively disordered proteins. In: Buchner, J. and Kiefhaber, T. (eds.), *Handbook of protein folding*. Wiley-VCH, Verlag GmbH & Co. KGaA, Weinheim, Germany, pp. 271–353.

REFERENCES

7. Uversky, V. N. and Dunker, A. K. (2010) Understanding protein non-folding, *Biochim Biophys Acta 1804*, 1231–1264.

8. Holt, C. and Sawyer, L. (1993) Caseins as rheomorphic proteins: Interpretation of primary and secondary structures of the as1-, b-, and k-caseins, *J Chem Soc Faraday Trans 89*, 2683–2692.

9. Pullen, R. A., Jenkins, J. A., Tickle, I. J., Wood, S. P., and Blundell, T. L. (1975) The relation of polypeptide hormone structure and flexibility to receptor binding: The relevance of X-ray studies on insulins, glucagon and human placental lactogen, *Mol Cell Biochem 8*, 5–20.

10. Cary, P. D., Moss, T., and Bradbury, E. M. (1978) High-resolution proton-magnetic-resonance studies of chromatin core particles, *Eur J Biochem 89*, 475–482.

11. Linderstrom-Lang, K. and Schellman, J. A. (1959) Protein structure and enzyme activity. In: Boyer, P. D., Lardy, H., and Myrback, K. (eds.), *The enzymes*, 2nd ed., Academic Press, New York, pp. 443–510.

12. Schweers, O., Schonbrunn-Hanebeck, E., Marx, A., and Mandelkow, E. (1994) Structural studies of tau protein and Alzheimer paired helical filaments show no evidence for beta-structure, *J Biol Chem 269*, 24290–24297.

13. Weinreb, P. H., Zhen, W., Poon, A. W., Conway, K. A., and Lansbury, P. T., Jr (1996) NACP, a protein implicated in Alzheimer's disease and learning, is natively unfolded, *Biochemistry 35*, 13709–13715.

14. Chen, J., Liang, H., and Fernandez, A. (2008) Protein structure protection commits gene expression patterns, *Genome Biol 9*, R107.

15. Uversky, V. N. (2003) A protein-chameleon: Conformational plasticity of alpha-synuclein, a disordered protein involved in neurodegenerative disorders, *J Biomol Struct Dyn 21*, 211–234.

16. Fuxreiter, M., Tompa, P., Simon, I., Uversky, V. N., Hansen, J. C., and Asturias, F. J. (2008) Malleable machines take shape in eukaryotic transcriptional regulation, *Nat Chem Biol 4*, 728–737.

17. Tsvetkov, P., Asher, G., Paz, A., Reuven, N., Sussman, J. L., Silman, I., and Shaul, Y. (2008) Operational definition of intrinsically unstructured protein sequences based on susceptibility to the 20S proteasome, *Proteins 70*, 1357–1366.

18. Dunker, A. K. and Uversky V. N. (2010) Drugs for "protein clouds": targeting intrinsically disordered transcription factors, *Curr Opin Pharmacol 10*(6), 782–788.

19. Livesay, D. R. (2010) Protein dynamics: Dancing on an ever-changing free energy stage, *Curr Opin Pharmacol 10*, 706–708.

20. Uversky, V. N. (2010) The mysterious unfoldome: Structureless, underappreciated, yet vital part of any given proteome, *J Biomed Biotechnol 2010*, 568068.

21. Dunker, A. K. and Obradovic, Z. (2001) The protein trinity—Linking function and disorder, *Nat Biotechnol 19*, 805–806.

22. Uversky, V. N. (2002) Natively unfolded proteins: A point where biology waits for physics, *Protein Sci 11*, 739–756.

23. Iakoucheva, L. M., Brown, C. J., Lawson, J. D., Obradovic, Z., and Dunker, A. K. (2002) Intrinsic disorder in cell-signaling and cancer-associated proteins, *J Mol Biol 323*, 573–584.

24. Dunker, A. K., Cortese, M. S., Romero, P., Iakoucheva, L. M., and Uversky, V. N. (2005) Flexible nets: The roles of intrinsic disorder in protein interaction networks, *FEBS J* 272, 5129–5148.
25. Uversky, V. N., Oldfield, C. J., and Dunker, A. K. (2005) Showing your ID: Intrinsic disorder as an ID for recognition, regulation and cell signaling, *J Mol Recognit* 18, 343–384.
26. Radivojac, P., Iakoucheva, L. M., Oldfield, C. J., Obradovic, Z., Uversky, V. N., and Dunker, A. K. (2007) Intrinsic disorder and functional proteomics, *Biophys J* 92, 1439–1456.
27. Halle, B. (2002) Flexibility and packing in proteins, *Proc Natl Acad Sci U S A* 99, 1274–1279.
28. Sickmeier, M., Hamilton, J. A., LeGall, T., Vacic, V., Cortese, M. S., Tantos, A., Szabo, B., Tompa, P., Chen, J., Uversky, V. N., Obradovic, Z., and Dunker, A. K. (2007) DisProt: The database of disordered proteins, *Nucleic Acids Res* 35, D786–D793.
29. Dunker, A. K., Oldfield, C. J., Meng, J., Romero, P., Yang, J. Y., Chen, J. W., Vacic, V., Obradovic, Z., and Uversky, V. N. (2008) The unfoldomics decade: An update on intrinsically disordered proteins, *BMC Genomics* 9, Suppl 2, S1.
30. Uversky, V. N. (2009) Intrinsic disorder in proteins associated with neurodegenerative diseases, *Front Biosci* 14, 5188–5238.
31. Uversky, V. N. (2008) Alpha-synuclein misfolding and neurodegenerative diseases, *Curr Protein Pept Sci* 9, 507–540.
32. Receveur-Brechot, V., Bourhis, J. M., Uversky, V. N., Canard, B., and Longhi, S. (2006) Assessing protein disorder and induced folding, *Proteins* 62, 24–45.
33. Hemmings, H. C., Jr, Nairn, A. C., Aswad, D. W., and Greengard, P. (1984) DARPP-32, a dopamine- and adenosine 3':5'-monophosphate-regulated phosphoprotein enriched in dopamine-innervated brain regions. II. Purification and characterization of the phosphoprotein from bovine caudate nucleus, *J Neurosci* 4, 99–110.
34. Gast, K., Damaschun, H., Eckert, K., Schulze-Forster, K., Maurer, H. R., Muller-Frohne, M., Zirwer, D., Czarnecki, J., and Damaschun, G. (1995) Prothymosin alpha: A biologically active protein with random coil conformation, *Biochemistry* 34, 13211–13218.
35. Garner, E., Cannon, P., Romero, P., Obradovic, Z., and Dunker, A. K. (1998) Predicting disordered regions from amino acid sequence: Common themes despite differing structural characterization, *Genome Inform Ser Workshop Genome Inform* 9, 201–213.
36. Williams, R. M., Obradovi, Z., Mathura, V., Braun, W., Garner, E. C., Young, J., Takayama, S., Brown, C. J., and Dunker, A. K. (2001) The protein non-folding problem: Amino acid determinants of intrinsic order and disorder, *Pac Symp Biocomput*, 89–100.
37. Romero, P., Obradovic, Z., Li, X., Garner, E. C., Brown, C. J., and Dunker, A. K. (2001) Sequence complexity of disordered protein, *Proteins* 42, 38–48.
38. Vacic, V., Uversky, V. N., Dunker, A. K., and Lonardi, S. (2007) Composition profiler: A tool for discovery and visualization of amino acid composition differences, *BMC Bioinformatics* 8, 211.
39. Li, X., Obradovic, Z., Brown, C. J., Garner, E. C., and Dunker, A. K. (2000) Comparing predictors of disordered protein, *Genome Inform Ser Workshop Genome Inform* 11, 172–184.
40. Campen, A., Williams, R. M., Brown, C. J., Meng, J., Uversky, V. N., and Dunker, A. K. (2008) TOP-IDP-scale: A new amino acid scale measuring propensity for intrinsic disorder, *Protein Pept Lett* 15, 956–963.

REFERENCES

41 He, B., Wang, K., Liu, Y., Xue, B., Uversky, V. N., and Dunker, A. K. (2009) Predicting intrinsic disorder in proteins: An overview, *Cell Res 19*, 929–949.

42 Romero, P., Obradovic, Z., Kissinger, C. R., Villafranca, J. E., Garner, E., Guilliot, S., and Dunker, A. K. (1998) Thousands of proteins likely to have long disordered regions, *Pac Symp Biocomput*, 437–448.

43 Dunker, A. K., Obradovic, Z., Romero, P., Garner, E. C., and Brown, C. J. (2000) Intrinsic protein disorder in complete genomes, *Genome Inform Ser Workshop Genome Inform 11*, 161–171.

44 Le Gall, T., Romero, P. R., Cortese, M. S., Uversky, V. N., and Dunker, A. K. (2007) Intrinsic disorder in the Protein Data Bank, *J Biomol Struct Dyn 24*, 325–342.

45 Uversky, V. N. and Ptitsyn, O. B. (1994) "Partly folded" state, a new equilibrium state of protein molecules: Four-state guanidinium chloride-induced unfolding of beta-lactamase at low temperature, *Biochemistry 33*, 2782–2791.

46 Uversky, V. N. and Ptitsyn, O. B. (1996) Further evidence on the equilibrium "pre-molten globule state": Four-state guanidinium chloride-induced unfolding of carbonic anhydrase B at low temperature, *J Mol Biol 255*, 215–228.

47 Ptitsyn, O. B. (1995) Molten globule and protein folding, *Adv Protein Chem 47*, 83–229.

48 Uversky, V. N. (2003) Protein folding revisited. A polypeptide chain at the folding-misfolding-nonfolding cross-roads: Which way to go? *Cell Mol Life Sci 60*, 1852–1871.

49 Tcherkasskaya, O. and Uversky, V. N. (2001) Denatured collapsed states in protein folding: Example of apomyoglobin, *Proteins 44*, 244–254.

50 Ptitsyn, O. B. (1994) Kinetic and equilibrium intermediates in protein folding, *Protein Eng 7*, 593–596.

51 Tanford, C. (1968) Protein denaturation, *Adv Protein Chem 23*, 121–282.

52 Uversky, V. N. (1993) Use of fast protein size-exclusion liquid chromatography to study the unfolding of proteins which denature through the molten globule, *Biochemistry 32*, 13288–13298.

53 Uversky, V. N. (2002) What does it mean to be natively unfolded? *Eur J Biochem 269*, 2–12.

54 Glatter, O. and Kratky, O. (1982) *Small angle X-ray scattering*, Academic Press, London.

55 Semisotnov, G. V., Kihara, H., Kotova, N. V., Kimura, K., Amemiya, Y., Wakabayashi, K., Serdyuk, I. N., Timchenko, A. A., Chiba, K., Nikaido, K., Ikura, T., and Kuwajima, K. (1996) Protein globularization during folding. A study by synchrotron small-angle X-ray scattering, *J Mol Biol 262*, 559–574.

56 Uversky, V. N., Li, J., and Fink, A. L. (2001) Evidence for a partially folded intermediate in alpha-synuclein fibril formation, *J Biol Chem 276*, 10737–10744.

57 Uversky, V. N., Li, J., Souillac, P., Millett, I. S., Doniach, S., Jakes, R., Goedert, M., and Fink, A. L. (2002) Biophysical properties of the synucleins and their propensities to fibrillate: Inhibition of alpha-synuclein assembly by beta- and gamma-synucleins, *J Biol Chem 277*, 11970–11978.

58 Bhattacharyya, J. and Das, K. P. (1999) Molecular chaperone-like properties of an unfolded protein, alpha(s)-casein, *J Biol Chem 274*, 15505–15509.

59 Thomas, J., Van Patten, S. M., Howard, P., Day, K. H., Mitchell, R. D., Sosnick, T., Trewhella, J., Walsh, D. A., and Maurer, R. A. (1991) Expression in *Escherichia coli*

and characterization of the heat-stable inhibitor of the cAMP-dependent protein kinase, *J Biol Chem 266*, 10906–10911.

60 Denning, D. P., Uversky, V., Patel, S. S., Fink, A. L., and Rexach, M. (2002) The *Saccharomyces cerevisiae* nucleoporin Nup2p is a natively unfolded protein, *J Biol Chem 277*, 33447–33455.

61 Denning, D. P., Patel, S. S., Uversky, V., Fink, A. L., and Rexach, M. (2003) Disorder in the nuclear pore complex: The FG repeat regions of nucleoporins are natively unfolded, *Proc Natl Acad Sci U S A 100*, 2450–2455.

62 Dyson, H. J. and Wright, P. E. (2001) Nuclear magnetic resonance methods for elucidation of structure and dynamics in disordered states, *Methods Enzymol 339*, 258–270.

63 Dyson, H. J. and Wright, P. E. (2002) Insights into the structure and dynamics of unfolded proteins from nuclear magnetic resonance, *Adv Protein Chem 62*, 311–340.

64 Dyson, H. J. and Wright, P. E. (2004) Unfolded proteins and protein folding studied by NMR, *Chem Rev 104*, 3607–3622.

65 Eliezer, D. (2007) Characterizing residual structure in disordered protein States using nuclear magnetic resonance, *Methods Mol Biol 350*, 49–67.

66 Eliezer, D. (2009) Biophysical characterization of intrinsically disordered proteins, *Curr Opin Struct Biol 19*, 23–30.

67 Muller-Spath, S., Soranno, A., Hirschfeld, V., Hofmann, H., Ruegger, S., Reymond, L., Nettels, D., and Schuler, B. (2010) From the Cover: Charge interactions can dominate the dimensions of intrinsically disordered proteins, *Proc Natl Acad Sci U S A 107*, 14609–14614.

68 Yamada, J., Phillips, J. L., Patel, S., Goldfien, G., Calestagne-Morelli, A., Huang, H., Reza, R., Acheson, J., Krishnan, V. V., Newsam, S., Gopinathan, A., Lau, E. Y., Colvin, M. E., Uversky, V. N., and Rexach, M. F. (2010) A bimodal distribution of two distinct categories of intrinsically disordered structures with separate functions in FG nucleoporins, *Mol Cell Proteomics 9*, 2205–2224.

69 Lau, K. F. and Dill, K. A. (1990) Theory for protein mutability and biogenesis, *Proc Natl Acad Sci U S A 87*, 638–642.

70 Ptitsyn, O. (1995) Molten globule and protein folding, *Adv Protein Chem 47*, 83–229.

71 Tcherkasskaya, O. and Uversky, V. N. (2003) Polymeric aspects of protein folding: A brief overview, *Protein Pept Lett 10*, 239–245.

72 Vitalis, A. and Pappu, R. V. (2009) ABSINTH: A new continuum solvation model for simulations of polypeptides in aqueous solutions, *J Comput Chem 30*, 673–699.

73 Vitalis, A., Wang, X., and Pappu, R. V. (2007) Quantitative characterization of intrinsic disorder in polyglutamine: Insights from analysis based on polymer theories, *Biophys J 93*, 1923–1937.

74 Vitalis, A., Wang, X., and Pappu, R. V. (2008) Atomistic simulations of the effects of polyglutamine chain length and solvent quality on conformational equilibria and spontaneous homodimerization, *J Mol Biol 384*, 279–297.

75 Wang, X., Vitalis, A., Wyczalkowski, M. A., and Pappu, R. V. (2006) Characterizing the conformational ensemble of monomeric polyglutamine, *Proteins 63*, 297–311.

76 Williamson, T. E., Vitalis, A., Crick, S. L., and Pappu, R. V. (2010) Modulation of polyglutamine conformations and dimer formation by the N-terminus of huntingtin, *J Mol Biol 396*, 1295–1309.

REFERENCES

77 Tran, H. T., Mao, A., and Pappu, R. V. (2008) Role of backbone–solvent interactions in determining conformational equilibria of intrinsically disordered proteins, *J Am Chem Soc 130*, 7380–7392.

78 Uversky, V. N. (2009) Intrinsically disordered proteins and their environment: Effects of strong denaturants, temperature, pH, counter ions, membranes, binding partners, osmolytes, and macromolecular crowding, *Protein J 28*, 305–325.

79 Ptitsyn, O. B. and Uversky, V. N. (1994) The molten globule is a third thermodynamical state of protein molecules, *FEBS Lett 341*, 15–18.

80 Uversky, V. N. and Ptitsyn, O. B. (1996) All-or-none solvent-induced transitions between native, molten globule and unfolded states in globular proteins, *Fold Des 1*, 117–122.

81 Neyroz, P., Zambelli, B., and Ciurli, S. (2006) Intrinsically disordered structure of *Bacillus pasteurii* UreG as revealed by steady-state and time-resolved fluorescence spectroscopy, *Biochemistry 45*, 8918–8930.

82 Permyakov, S. E., Millett, I. S., Doniach, S., Permyakov, E. A., and Uversky, V. N. (2003) Natively unfolded C-terminal domain of caldesmon remains substantially unstructured after the effective binding to calmodulin, *Proteins 53*, 855–862.

83 Uversky, V. N., Permyakov, S. E., Zagranichny, V. E., Rodionov, I. L., Fink, A. L., Cherskaya, A. M., Wasserman, L. A., and Permyakov, E. A. (2002) Effect of zinc and temperature on the conformation of the gamma subunit of retinal phosphodiesterase: A natively unfolded protein, *J Proteome Res 1*, 149–159.

84 Uversky, V. N., Gillespie, J. R., Millett, I. S., Khodyakova, A. V., Vasiliev, A. M., Chernovskaya, T. V., Vasilenko, R. N., Kozlovskaya, G. D., Dolgikh, D. A., Fink, A. L., Doniach, S., and Abramov, V. M. (1999) Natively unfolded human prothymosin alpha adopts partially folded collapsed conformation at acidic pH, *Biochemistry 38*, 15009–15016.

85 Konno, T., Tanaka, N., Kataoka, M., Takano, E., and Maki, M. (1997) A circular dichroism study of preferential hydration and alcohol effects on a denatured protein, pig calpastatin domain I, *Biochim Biophys Acta 1342*, 73–82.

86 Lynn, A., Chandra, S., Malhotra, P., and Chauhan, V. S. (1999) Heme binding and polymerization by *Plasmodium falciparum* histidine rich protein II: Influence of pH on activity and conformation, *FEBS Lett 459*, 267–271.

87 Johansson, J., Gudmundsson, G. H., Rottenberg, M. E., Berndt, K. D., and Agerberth, B. (1998) Conformation-dependent antibacterial activity of the naturally occurring human peptide LL-37, *J Biol Chem 273*, 3718–3724.

88 Uversky, V. N. (2011) Multitude of binding modes attainable by intrinsically disordered proteins: A portrait gallery of disorder-based complexes, *Chem Soc Rev 40*, 1623–1634.

89 Mohan, A. (2006) MoRFs: A dataset of molecular recognition features. *Master's thesis*, Indiana University, School of Informatics, p. 59.

90 Oldfield, C. J., Cheng, Y., Cortese, M. S., Romero, P., Uversky, V. N., and Dunker, A. K. (2005) Coupled folding and binding with alpha-helix-forming molecular recognition elements, *Biochemistry 44*, 12454–12470.

91 Vacic, V., Oldfield, C. J., Mohan, A., Radivojac, P., Cortese, M. S., Uversky, V. N., and Dunker, A. K. (2007) Characterization of molecular recognition features, MoRFs, and their binding partners, *J Proteome Res 6*, 2351–2366.

92 Galea, C. A., Nourse, A., Wang, Y., Sivakolundu, S. G., Heller, W. T., and Kriwacki, R. W. (2008) Role of intrinsic flexibility in signal transduction mediated by the cell cycle regulator, p27 Kip1, *J Mol Biol 376*, 827–838.

93 Galea, C. A., Wang, Y., Sivakolundu, S. G., and Kriwacki, R. W. (2008) Regulation of cell division by intrinsically unstructured proteins: Intrinsic flexibility, modularity, and signaling conduits, *Biochemistry 47*, 7598–7609.

94 Graham, T. A., Weaver, C., Mao, F., Kimelman, D., and Xu, W. (2000) Crystal structure of a beta-catenin/Tcf complex, *Cell 103*, 885–896.

95 Luscombe, N. M., Austin, S. E., Berman, H. M., and Thornton, J. M. (2000) An overview of the structures of protein-DNA complexes, *Genome Biol 1*, REVIEWS001.

96 Brodersen, D. E., Clemons, W. M., Jr, Carter, A. P., Wimberly, B. T., and Ramakrishnan, V. (2002) Crystal structure of the 30 S ribosomal subunit from *Thermus thermophilus*: Structure of the proteins and their interactions with 16 S RNA, *J Mol Biol 316*, 725–768.

97 Teschke, C. M. and King, J. (1992) Folding and assembly of oligomeric proteins in *Escherichia coli*, *Curr Opin Biotechnol 3*, 468–473.

98 Xu, D., Tsai, C. J., and Nussinov, R. (1998) Mechanism and evolution of protein dimerization, *Protein Sci 7*, 533–544.

99 Gunasekaran, K., Tsai, C. J., and Nussinov, R. (2004) Analysis of ordered and disordered protein complexes reveals structural features discriminating between stable and unstable monomers, *J Mol Biol 341*, 1327–1341.

100 Mason, J. M. and Arndt, K. M. (2004) Coiled coil domains: Stability, specificity, and biological implications, *Chembiochem 5*, 170–176.

101 Liu, J., Zheng, Q., Deng, Y., Cheng, C. S., Kallenbach, N. R., and Lu, M. (2006) A seven-helix coiled coil, *Proc Natl Acad Sci U S A 103*, 15457–15462.

102 Wolf, E., Kim, P. S., and Berger, B. (1997) MultiCoil: A program for predicting two- and three-stranded coiled coils, *Protein Sci 6*, 1179–1189.

103 Harbury, P. B., Plecs, J. J., Tidor, B., Alber, T., and Kim, P. S. (1998) High-resolution protein design with backbone freedom, *Science 282*, 1462–1467.

104 Stetefeld, J., Jenny, M., Schulthess, T., Landwehr, R., Engel, J., and Kammerer, R. A. (2000) Crystal structure of a naturally occurring parallel right-handed coiled coil tetramer, *Nat Struct Biol 7*, 772–776.

105 Ozbek, S., Engel, J., and Stetefeld, J. (2002) Storage function of cartilage oligomeric matrix protein: The crystal structure of the coiled-coil domain in complex with vitamin D(3), *EMBO J 21*, 5960–5968.

106 Low, H. H., Moncrieffe, M. C., and Lowe, J. (2004) The crystal structure of ZapA and its modulation of FtsZ polymerisation, *J Mol Biol 341*, 839–852.

107 Zhao, X., Ghaffari, S., Lodish, H., Malashkevich, V. N., and Kim, P. S. (2002) Structure of the Bcr-Abl oncoprotein oligomerization domain, *Nat Struct Biol 9*, 117–120.

108 Liu, X., Xu, L., Liu, Y., Tong, X., Zhu, G., Zhang, X. C., Li, X., and Rao, Z. (2009) Crystal structure of the hexamer of human heat shock factor binding protein 1, *Proteins 75*, 1–11.

109 Malashkevich, V. N., Schneider, B. J., McNally, M. L., Milhollen, M. A., Pang, J. X., and Kim, P. S. (1999) Core structure of the envelope glycoprotein GP2 from Ebola virus at 1.9-A resolution, *Proc Natl Acad Sci U S A 96*, 2662–2667.

110 Malashkevich, V. N., Singh, M., and Kim, P. S. (2001) The trimer-of-hairpins motif in membrane fusion: Visna virus, *Proc Natl Acad Sci U S A 98*, 8502–8506.

111 Glover, J. N. and Harrison, S. C. (1995) Crystal structure of the heterodimeric bZIP transcription factor c-Fos-c-Jun bound to DNA, *Nature 373*, 257–261.

REFERENCES

112 Im, Y. J., Kang, G. B., Lee, J. H., Park, K. R., Song, H. E., Kim, E., Song, W. K., Park, D., and Eom, S. H. (2010) Structural basis for asymmetric association of the betaPIX coiled coil and shank PDZ, *J Mol Biol 397*, 457–466.

113 Siegert, R., Leroux, M. R., Scheufler, C., Hartl, F. U., and Moarefi, I. (2000) Structure of the molecular chaperone prefoldin: Unique interaction of multiple coiled coil tentacles with unfolded proteins, *Cell 103*, 621–632.

114 Lee, S., Sowa, M. E., Watanabe, Y. H., Sigler, P. B., Chiu, W., Yoshida, M., and Tsai, F. T. (2003) The structure of ClpB: A molecular chaperone that rescues proteins from an aggregated state, *Cell 115*, 229–240.

115 Oldfield, C. J., Meng, J., Yang, J. Y., Yang, M. Q., Uversky, V. N., and Dunker, A. K. (2008) Flexible nets: Disorder and induced fit in the associations of p53 and 14-3-3 with their partners, *BMC Genomics 9*, Suppl 1, S1.

116 Kajava, A. V., Baxa, U., and Steven, A. C. (2010) Beta arcades: Recurring motifs in naturally occurring and disease-related amyloid fibrils, *FASEB J 24*, 1311–1319.

117 Borg, M., Mittag, T., Pawson, T., Tyers, M., Forman-Kay, J. D., and Chan, H. S. (2007) Polyelectrostatic interactions of disordered ligands suggest a physical basis for ultrasensitivity, *Proc Natl Acad Sci U S A 104*, 9650–9655.

118 Mittag, T., Orlicky, S., Choy, W. Y., Tang, X., Lin, H., Sicheri, F., Kay, L. E., Tyers, M., and Forman-Kay, J. D. (2008) Dynamic equilibrium engagement of a polyvalent ligand with a single-site receptor, *Proc Natl Acad Sci U S A 105*, 17772–17777.

119 Mittag, T., Kay, L. E., and Forman-Kay, J. D. (2010) Protein dynamics and conformational disorder in molecular recognition, *J Mol Recognit 23*, 105–116.

120 Mittag, T., Marsh, J., Grishaev, A., Orlicky, S., Lin, H., Sicheri, F., Tyers, M., and Forman-Kay, J. D. (2010) Structure/function implications in a dynamic complex of the intrinsically disordered Sic1 with the Cdc4 subunit of an SCF ubiquitin ligase, *Structure 18*, 494–506.

121 Sigalov, A. B., Kim, W. M., Saline, M., and Stern, L. J. (2008) The intrinsically disordered cytoplasmic domain of the T cell receptor zeta chain binds to the nef protein of simian immunodeficiency virus without a disorder-to-order transition, *Biochemistry 47*, 12942–12944.

122 Sigalov, A., Aivazian, D., and Stern, L. (2004) Homooligomerization of the cytoplasmic domain of the T cell receptor zeta chain and of other proteins containing the immunoreceptor tyrosine-based activation motif, *Biochemistry 43*, 2049–2061.

123 Sigalov, A. B., Zhuravleva, A. V., and Orekhov, V. Y. (2007) Binding of intrinsically disordered proteins is not necessarily accompanied by a structural transition to a folded form, *Biochimie 89*, 419–421.

124 Pometun, M. S., Chekmenev, E. Y., and Wittebort, R. J. (2004) Quantitative observation of backbone disorder in native elastin, *J Biol Chem 279*, 7982–7987.

125 Sigalov, A. B. and Hendricks, G. M. (2009) Membrane binding mode of intrinsically disordered cytoplasmic domains of T cell receptor signaling subunits depends on lipid composition, *Biochem Biophys Res Commun 389*, 388–393.

126 Sigalov, A. B., Aivazian, D. A., Uversky, V. N., and Stern, L. J. (2006) Lipid-binding activity of intrinsically unstructured cytoplasmic domains of multichain immune recognition receptor signaling subunits, *Biochemistry 45*, 15731–15739.

127 Tompa, P. and Fuxreiter, M. (2008) Fuzzy complexes: Polymorphism and structural disorder in protein–protein interactions, *Trends Biochem Sci 33*, 2–8.

128. Hazy, E. and Tompa, P. (2009) Limitations of induced folding in molecular recognition by intrinsically disordered proteins, *Chemphyschem 10*, 1415–1419.
129. Liu, Y., Fratini, E., Baglioni, P., Chen, W. R., and Chen, S. H. (2005) Effective long-range attraction between protein molecules in solutions studied by small angle neutron scattering, *Phys Rev Lett 95*, 118102.
130. Schulz, G. E. (1979) Nucleotide binding proteins. In: Balaban, M. (ed.), *Molecular mechanism of biological recognition*. Elsevier/North-Holland Biomedical Press, New York, pp. 79–94.
131. Pontius, B. W. (1993) Close encounters: Why unstructured, polymeric domains can increase rates of specific macromolecular association, *Trends Biochem Sci 18*, 181–186.
132. Dunker, A. K., Brown, C. J., Lawson, J. D., Iakoucheva, L. M., and Obradovic, Z. (2002) Intrinsic disorder and protein function, *Biochemistry 41*, 6573–6582.
133. Dunker, A. K., Brown, C. J., and Obradovic, Z. (2002) Identification and functions of usefully disordered proteins, *Adv Protein Chem 62*, 25–49.
134. Dyson, H. J. and Wright, P. E. (2002) Coupling of folding and binding for unstructured proteins, *Curr Opin Struct Biol 12*, 54–60.
135. Dyson, H. J. and Wright, P. E. (2005) Intrinsically unstructured proteins and their functions, *Nat Rev Mol Cell Biol 6*, 197–208.
136. Plaxco, K. W. and Gross, M. (1997) Cell biology. The importance of being unfolded, *Nature 386*, 657, 659.
137. Spolar, R. S. and Record, M. T., Jr (1994) Coupling of local folding to site-specific binding of proteins to DNA, *Science 263*, 777–784.
138. Karush, F. (1950) Heterogeneity of the binding sites of bovine serum albumin, *J Am Chem Soc 72*, 2705–2713.
139. Kriwacki, R. W., Hengst, L., Tennant, L., Reed, S. I., and Wright, P. E. (1996) Structural studies of p21$^{Waf1/Cip1/Sdi1}$ in the free and Cdk2-bound state: Conformational disorder mediates binding diversity, *Proc Natl Acad Sci U S A 93*, 11504–11509.
140. Dunker, A. K., Obradovic, Z., Romero, P., Kissinger, C., and Villafranca, E. (1997) On the importance of being disordered, *PDB Newsletter 81*, 3–5.
141. Romero, P., Obradovic, Z., Li, X., Garner, E. C., Brown, C. J., and Dunker, A. K. (2001) Sequence complexity of disordered protein, *Protein Struct Funct Genet 42*, 38–49.
142. Brown, C. J., Takayama, S., Campen, A. M., Vise, P., Marshall, T. W., Oldfield, C. J., Williams, C. J., and Dunker, A. K. (2002) Evolutionary rate heterogeneity in proteins with long disordered regions, *J Mol Evol 55*, 104–110.
143. Cortese, M. S., Uversky, V. N., and Dunker, A. K. (2008) Intrinsic disorder in scaffold proteins: Getting more from less, *Prog Biophys Mol Biol 98*, 85–106.
144. Tompa, P. and Csermely, P. (2004) The role of structural disorder in the function of RNA and protein chaperones, *FASEB J 18*, 1169–1175.
145. Xie, H., Vucetic, S., Iakoucheva, L. M., Oldfield, C. J., Dunker, A. K., Uversky, V. N., and Obradovic, Z. (2007) Functional anthology of intrinsic disorder. 1. Biological processes and functions of proteins with long disordered regions, *J Proteome Res 6*, 1882–1898.
146. Vucetic, S., Xie, H., Iakoucheva, L. M., Oldfield, C. J., Dunker, A. K., Obradovic, Z., and Uversky, V. N. (2007) Functional anthology of intrinsic disorder. 2. Cellular components, domains, technical terms, developmental processes, and coding sequence diversities correlated with long disordered regions, *J Proteome Res 6*, 1899–1916.

147 Xie, H., Vucetic, S., Iakoucheva, L. M., Oldfield, C. J., Dunker, A. K., Obradovic, Z., and Uversky, V. N. (2007) Functional anthology of intrinsic disorder. 3. Ligands, post-translational modifications, and diseases associated with intrinsically disordered proteins, *J Proteome Res 6*, 1917–1932.

148 Marks, F. (1996) *Protein phosphorylation*, Wiley, VCH Weinheim, New York, Basel, Cambridge, Tokyo.

149 Johnson, L. N. and Lewis, R. J. (2001) Structural basis for control by phosphorylation, *Chem Rev 101*, 2209–2242.

150 Iakoucheva, L. M., Radivojac, P., Brown, C. J., O'Connor, T. R., Sikes, J. G., Obradovic, Z., and Dunker, A. K. (2004) The importance of intrinsic disorder for protein phosphorylation, *Nucleic Acids Res 32*, 1037–1049.

151 Bossemeyer, D., Engh, R. A., Kinzel, V., Ponstingl, H., and Huber, R. (1993) Phosphotransferase and substrate binding mechanism of the cAMP-dependent protein kinase catalytic subunit from porcine heart as deduced from the 2.0 A structure of the complex with Mn2+ adenylyl imidodiphosphate and inhibitor peptide PKI(5-24), *EMBO J 12*, 849–859.

152 Narayana, N., Cox, S., Shaltiel, S., Taylor, S. S., and Xuong, N. (1997) Crystal structure of a polyhistidine-tagged recombinant catalytic subunit of cAMP-dependent protein kinase complexed with the peptide inhibitor PKI(5-24) and adenosine, *Biochemistry 36*, 4438–4448.

153 Lowe, E. D., Noble, M. E., Skamnaki, V. T., Oikonomakos, N. G., Owen, D. J., and Johnson, L. N. (1997) The crystal structure of a phosphorylase kinase peptide substrate complex: Kinase substrate recognition, *Embo J 16*, 6646–6658.

154 ter Haar, E., Coll, J. T., Austen, D. A., Hsiao, H. M., Swenson, L., and Jain, J. (2001) Structure of GSK3beta reveals a primed phosphorylation mechanism, *Nat Struct Biol 8*, 593–596.

155 Hubbard, S. R. (1997) Crystal structure of the activated insulin receptor tyrosine kinase in complex with peptide substrate and ATP analog, *Embo J 16*, 5572–5581.

156 McDonald, I. K. and Thornton, J. M. (1994) Satisfying hydrogen bonding potential in proteins, *J Mol Biol 238*, 777–793.

157 Radivojac, P., Vacic, V., Haynes, C., Cocklin, R. R., Mohan, A., Heyen, J. W., Goebl, M. G., and Iakoucheva, L. M. (2010) Identification, analysis, and prediction of protein ubiquitination sites, *Proteins 78*, 365–380.

158 Sambrook, J. (1977) Adenovirus amazes at Cold Spring Harbor, *Nature 268*, 101–104.

159 Black, D. L. (2003) Mechanisms of alternative pre-messenger RNA splicing, *Annu Rev Biochem 72*, 291–336.

160 Gilbert, W. (1978) Why genes in pieces? *Nature 271*, 501.

161 Ast, G. (2004) How did alternative splicing evolve? *Nat Rev Genet 5*, 773–782.

162 Stamm, S., Ben-Ari, S., Rafalska, I., Tang, Y., Zhang, Z., Toiber, D., Thanaraj, T. A., and Soreq, H. (2005) Function of alternative splicing, *Gene 344*, 1–20.

163 Brett, D., Hanke, J., Lehmann, G., Haase, S., Delbruck, S., Krueger, S., Reich, J., and Bork, P. (2000) EST comparison indicates 38% of human mRNAs contain possible alternative splice forms, *FEBS Lett 474*, 83–86.

164 Johnson, J. M., Castle, J., Garrett-Engele, P., Kan, Z., Loerch, P. M., Armour, C. D., Santos, R., Schadt, E. E., Stoughton, R., and Shoemaker, D. D. (2003) Genome-wide

survey of human alternative pre-mRNA splicing with exon junction microarrays, *Science 302*, 2141–2144.
165 Graveley, B. R. (2001) Alternative splicing: Increasing diversity in the proteomic world, *Trends Genet 17*, 100–107.
166 Minneman, K. P. (2001) Splice variants of G protein-coupled receptors, *Mol Interv 1*, 108–116.
167 Thai, T. H. and Kearney, J. F. (2004) Distinct and opposite activities of human terminal deoxynucleotidyltransferase splice variants, *J Immunol 173*, 4009–4019.
168 Scheper, W., Zwart, R., and Baas, F. (2004) Alternative splicing in the N-terminus of Alzheimer's presenilin 1, *Neurogenetics 5*, 223–227.
169 Roberts, R., Timchenko, N. A., Miller, J. W., Reddy, S., Caskey, C. T., Swanson, M. S., and Timchenko, L. T. (1997) Altered phosphorylation and intracellular distribution of a (CUG)n triplet repeat RNA-binding protein in patients with myotonic dystrophy and in myotonin protein kinase knockout mice, *Proc Natl Acad Sci U S A 94*, 13221–13226.
170 Ma, K., Inglis, J. D., Sharkey, A., Bickmore, W. A., Hill, R. E., Prosser, E. J., Speed, R. M., Thomson, E. J., Jobling, M., Taylor, K., et al. (1993) A Y chromosome gene family with RNA-binding protein homology: Candidates for the azoospermia factor AZF controlling human spermatogenesis, *Cell 75*, 1287–1295.
171 Lovestone, S., Reynolds, C. H., Latimer, D., Davis, D. R., Anderton, B. H., Gallo, J. M., Hanger, D., Mulot, S., Marquardt, B., Stabel, S., et al. (1994) Alzheimer's disease-like phosphorylation of the microtubule-associated protein tau by glycogen synthase kinase-3 in transfected mammalian cells, *Curr Biol 4*, 1077–1086.
172 Beyer, K., Domingo-Sabat, M., Humbert, J., Carrato, C., Ferrer, I., and Ariza, A. (2008) Differential expression of alpha-synuclein, parkin, and synphilin-1 isoforms in Lewy body disease, *Neurogenetics 9*, 163–172.
173 Beyer, K., Domingo-Sabat, M., Lao, J. I., Carrato, C., Ferrer, I., and Ariza, A. (2008) Identification and characterization of a new alpha-synuclein isoform and its role in Lewy body diseases, *Neurogenetics 9*, 15–23.
174 Venables, J. P. (2004) Aberrant and alternative splicing in cancer, *Cancer Res 64*, 7647–7654.
175 Wang, P., Yan, B., Guo, J. T., Hicks, C., and Xu, Y. (2005) Structural genomics analysis of alternative splicing and application to isoform structure modeling, *Proc Natl Acad Sci U S A 102*, 18920–18925.
176 Furnham, N., Ruffle, S., and Southan, C. (2004) Splice variants: A homology modeling approach, *Proteins 54*, 596–608.
177 Romero, P. R., Zaidi, S., Fang, Y. Y., Uversky, V. N., Radivojac, P., Oldfield, C. J., Cortese, M. S., Sickmeier, M., LeGall, T., Obradovic, Z., and Dunker, A. K. (2006) Alternative splicing in concert with protein intrinsic disorder enables increased functional diversity in multicellular organisms, *Proc Natl Acad Sci U S A 103*, 8390–8395.
178 Dunker, A. K. and Uversky, V. N. (2008) Signal transduction via unstructured protein conduits, *Nat Chem Biol 4*, 229–230.
179 Dunker, A. K., Silman, I., Uversky, V. N., and Sussman, J. L. (2008) Function and structure of inherently disordered proteins, *Curr Opin Struct Biol 18*, 756–764.

REFERENCES

180 Dosztanyi, Z., Csizmok, V., Tompa, P., and Simon, I. (2005) IUPred: Web server for the prediction of intrinsically unstructured regions of proteins based on estimated energy content, *Bioinformatics 21*, 3433–3434.

181 Uversky, V. N. and Dunker, A. K. (2008) Biochemistry. Controlled chaos, *Science 322*, 1340–1341.

182 Gsponer, J., Futschik, M. E., Teichmann, S. A., and Babu, M. M. (2008) Tight regulation of unstructured proteins: From transcript synthesis to protein degradation, *Science 322*, 1365–1368.

183 Ward, J. J., Sodhi, J. S., McGuffin, L. J., Buxton, B. F., and Jones, D. T. (2004) Prediction and functional analysis of native disorder in proteins from the three kingdoms of life, *J Mol Biol 337*, 635–645.

184 Grimmler, M., Wang, Y., Mund, T., Cilensek, Z., Keidel, E. M., Waddell, M. B., Jakel, H., Kullmann, M., Kriwacki, R. W., and Hengst, L. (2007) Cdk-inhibitory activity and stability of p27Kip1 are directly regulated by oncogenic tyrosine kinases, *Cell 128*, 269–280.

185 Tompa, P. (2005) The interplay between structure and function in intrinsically unstructured proteins, *FEBS Lett 579*, 3346–3354.

186 Paliy, O., Gargac, S. M., Cheng, Y., Uversky, V. N., and Dunker, A. K. (2008) Protein disorder is positively correlated with gene expression in *Escherichia coli*, *J Proteome Res 7*, 2234–2245.

187 Fernandez, A. and Scheraga, H. A. (2003) Insufficiently dehydrated hydrogen bonds as determinants of protein interactions, *Proc Natl Acad Sci U S A 100*, 113–118.

188 Fernandez, A. (2004) Keeping dry and crossing membranes, *Nat Biotechnol 22*, 1081–1084.

189 Kelly, J. W. (1998) The alternative conformations of amyloidogenic proteins and their multi-step assembly pathways, *Curr Opin Struct Biol 8*, 101–106.

190 Dobson, C. M. (1999) Protein misfolding, evolution and disease, *Trends Biochem Sci 24*, 329–332.

191 Bellotti, V., Mangione, P., and Stoppini, M. (1999) Biological activity and pathological implications of misfolded proteins, *Cell Mol Life Sci 55*, 977–991.

192 Uversky, V. N., Talapatra, A., Gillespie, J. R., and Fink, A. L. (1999) Protein deposits as the molecular basis of amyloidosis. I. Systemic amyloidoses, *Med Sci Monit 5*, 1001–1012.

193 Uversky, V. N., Talapatra, A., Gillespie, J. R., and Fink, A. L. (1999) Protein deposits as the molecular basis of amyloidosis. II. Localized amyloidosis and neurodegenerative disordres, *Med Sci Monit 5*, 1238–1254.

194 Rochet, J. C. and Lansbury, P. T., Jr (2000) Amyloid fibrillogenesis: Themes and variations, *Curr Opin Struct Biol 10*, 60–68.

195 Uversky, V. N. and Fink, A. L. (2004) Conformational constraints for amyloid fibrillation: The importance of being unfolded, *Biochim Biophys Acta 1698*, 131–153.

196 Uversky, V. N. and Fink, A. L. (2005) Pathways to amyloid fibril formation: Partially folded intermediates in fibrillation of unfolded proteins. In: Sipe, J. D. (ed.), *Amyloid proteins: The beta pleated sheet conformation and disease*. Wiley-VCH, Verlag GmbH & Co. KGaA, Weinheim, Germany, pp. 247–265.

197 Glenner, G. G. and Wong, C. W. (1984) Alzheimer's disease and Down's syndrome: Sharing of a unique cerebrovascular amyloid fibril protein, *Biochem Biophys Res Commun 122*, 1131–1135.

198 Masters, C. L., Multhaup, G., Simms, G., Pottgiesser, J., Martins, R. N., and Beyreuther, K. (1985) Neuronal origin of a cerebral amyloid: Neurofibrillary tangles of Alzheimer's disease contain the same protein as the amyloid of plaque cores and blood vessels, *Embo J 4*, 2757–2763.

199 Lee, V. M., Balin, B. J., Otvos, L., Jr, and Trojanowski, J. Q. (1991) A68: A major subunit of paired helical filaments and derivatized forms of normal Tau, *Science 251*, 675–678.

200 Ueda, K., Fukushima, H., Masliah, E., Xia, Y., Iwai, A., Yoshimoto, M., Otero, D. A., Kondo, J., Ihara, Y., and Saitoh, T. (1993) Molecular cloning of cDNA encoding an unrecognized component of amyloid in Alzheimer disease, *Proc Natl Acad Sci U S A 90*, 11282–11286.

201 Wisniewski, K. E., Dalton, A. J., McLachlan, C., Wen, G. Y., and Wisniewski, H. M. (1985) Alzheimer's disease in Down's syndrome: Clinicopathologic studies, *Neurology 35*, 957–961.

202 Dev, K. K., Hofele, K., Barbieri, S., Buchman, V. L., and van der Putten, H. (2003) Part II: Alpha-synuclein and its molecular pathophysiological role in neurodegenerative disease, *Neuropharmacology 45*, 14–44.

203 Prusiner, S. B. (2001) Shattuck lecture—Neurodegenerative diseases and prions, *N Engl J Med 344*, 1516–1526.

204 Zoghbi, H. Y. and Orr, H. T. (1999) Polyglutamine diseases: Protein cleavage and aggregation, *Curr Opin Neurobiol 9*, 566–570.

205 Okazawa, H. (2003) Polyglutamine diseases: A transcription disorder? *Cell Mol Life Sci 60*, 1427–1439.

206 Cummings, C. J. and Zoghbi, H. Y. (2000) Fourteen and counting: Unraveling trinucleotide repeat diseases, *Hum Mol Genet 9*, 909–916.

207 Gusella, J. F. and MacDonald, M. E. (2000) Molecular genetics: Unmasking polyglutamine triggers in neurodegenerative disease, *Nat Rev Neurosci 1*, 109–115.

208 Orr, H. T. (2001) Beyond the Qs in the polyglutamine diseases, *Genes Dev 15*, 925–932.

209 Fischbeck, K. H. (2001) Polyglutamine expansion neurodegenerative disease, *Brain Res Bull 56*, 161–163.

210 Lee, H., Mok, K. H., Muhandiram, R., Park, K. H., Suk, J. E., Kim, D. H., Chang, J., Sung, Y. C., Choi, K. Y., and Han, K. H. (2000) Local structural elements in the mostly unstructured transcriptional activation domain of human p53, *J Biol Chem 275*, 29426–29432.

211 Adkins, J. N. and Lumb, K. J. (2002) Intrinsic structural disorder and sequence features of the cell cycle inhibitor p57Kip2, *Proteins 46*, 1–7.

212 Chang, B. S., Minn, A. J., Muchmore, S. W., Fesik, S. W., and Thompson, C. B. (1997) Identification of a novel regulatory domain in Bcl-X(L) and Bcl-2, *Embo J 16*, 968–977.

213 Campbell, K. M., Terrell, A. R., Laybourn, P. J., and Lumb, K. J. (2000) Intrinsic structural disorder of the C-terminal activation domain from the bZIP transcription factor Fos, *Biochemistry 39*, 2708–2713.

REFERENCES

214 Sunde, M., McGrath, K. C., Young, L., Matthews, J. M., Chua, E. L., Mackay, J. P., and Death, A. K. (2004) TC-1 is a novel tumorigenic and natively disordered protein associated with thyroid cancer, *Cancer Res 64*, 2766–2773.

215 Uversky, V. N., Oldfield, C. J., and Dunker, A. K. (2008) Intrinsically disordered proteins in human diseases: Introducing the D2 concept, *Annu Rev Biophys 37*, 215–246.

216 Uversky, V. N., Oldfield, C. J., Midic, U., Xie, H., Xue, B., Vucetic, S., Iakoucheva, L. M., Obradovic, Z., and Dunker, A. K. (2009) Unfoldomics of human diseases: Linking protein intrinsic disorder with diseases, *BMC Genomics 10*, Suppl 1, S7.

217 Garner, E., Romero, P., Dunker, A. K., Brown, C., and Obradovic, Z. (1999) Predicting binding regions within disordered proteins, *Genome Inform Ser Workshop Genome Inform 10*, 41–50.

218 Cheng, Y., Oldfield, C. J., Meng, J., Romero, P., Uversky, V. N., and Dunker, A. K. (2007) Mining alpha-helix-forming molecular recognition features with cross species sequence alignments, *Biochemistry 46*, 13468–13477.

219 Chen, C. T., Wagner, H., and Still, W. C. (1998) Fluorescent, sequence-selective peptide detection by synthetic small molecules, *Science 279*, 851–853.

220 Chene, P. (2004) Inhibition of the p53-hdm2 interaction with low molecular weight compounds, *Cell Cycle 3*, 460–461.

221 Chene, P. (2004) Inhibition of the p53-MDM2 interaction: Targeting a protein–protein interface, *Mol Cancer Res 2*, 20–28.

222 Anderson, C. W. and Appella, E. (2004) Signaling to the p53 tumor suppressor through pathways activated by genotoxic and nongenotoxic stress. In: Bradshaw, R. A. and Dennis, E. A. (eds.), *Handbook of cell signaling*. Academic Press, New York, pp. 237–247.

223 Hollstein, M., Sidransky, D., Vogelstein, B., and Harris, C. C. (1991) p53 mutations in human cancers, *Science 253*, 49–53.

224 Balint, E. E. and Vousden, K. H. (2001) Activation and activities of the p53 tumour suppressor protein, *Br J Cancer 85*, 1813–1823.

225 Kussie, P. H., Gorina, S., Marechal, V., Elenbaas, B., Moreau, J., Levine, A. J., and Pavletich, N. P. (1996) Structure of the MDM2 oncoprotein bound to the p53 tumor suppressor transactivation domain, *Science 274*, 948–953.

226 Bottger, A., Bottger, V., Sparks, A., Liu, W. L., Howard, S. F., and Lane, D. P. (1997) Design of a synthetic Mdm2-binding mini protein that activates the p53 response in vivo, *Curr Biol 7*, 860–869.

227 Wasylyk, C., Salvi, R., Argentini, M., Dureuil, C., Delumeau, I., Abecassis, J., Debussche, L., and Wasylyk, B. (1999) p53 mediated death of cells overexpressing MDM2 by an inhibitor of MDM2 interaction with p53, *Oncogene 18*, 1921–1934.

228 Chene, P., Fuchs, J., Bohn, J., Garcia-Echeverria, C., Furet, P., and Fabbro, D. (2000) A small synthetic peptide, which inhibits the p53-hdm2 interaction, stimulates the p53 pathway in tumour cell lines, *J Mol Biol 299*, 245–253.

229 Garcia-Echeverria, C., Chene, P., Blommers, M. J., and Furet, P. (2000) Discovery of potent antagonists of the interaction between human double minute 2 and tumor suppressor p53, *J Med Chem 43*, 3205–3208.

230 Klein, C. and Vassilev, L. T. (2004) Targeting the p53-MDM2 interaction to treat cancer, *Br J Cancer 91*, 1415–1419.

231 Vassilev, L. T. (2004) Small-molecule antagonists of p53-MDM2 binding: Research tools and potential therapeutics, *Cell Cycle 3*, 419–421.

232 Vassilev, L. T., Vu, B. T., Graves, B., Carvajal, D., Podlaski, F., Filipovic, Z., Kong, N., Kammlott, U., Lukacs, C., Klein, C., Fotouhi, N., and Liu, E. A. (2004) In vivo activation of the p53 pathway by small-molecule antagonists of MDM2, *Science 303*, 844–848.

233 Cheng, Y., LeGall, T., Oldfield, C. J., Mueller, J. P., Van, Y. Y., Romero, P., Cortese, M. S., Uversky, V. N., and Dunker, A. K. (2006) Rational drug design via intrinsically disordered protein, *Trends Biotechnol 24*, 435–442.

234 Arkin, M. R. and Wells, J. A. (2004) Small-molecule inhibitors of protein–protein interactions: Progressing towards the dream, *Nat Rev Drug Discov 3*, 301–317.

235 Arkin, M. (2005) Protein–protein interactions and cancer: Small molecules going in for the kill, *Curr Opin Chem Biol 9*, 317–324.

236 Cochran, A. G. (2000) Antagonists of protein–protein interactions, *Chem Biol 7*, R85–R94.

237 Hammoudeh, D. I., Follis, A. V., Prochownik, E. V., and Metallo, S. J. (2009) Multiple independent binding sites for small-molecule inhibitors on the oncoprotein c-Myc, *J Am Chem Soc 131*, 7390–7401.

I

CONFORMATIONAL ANALYSIS OF UNFOLDED STATES

2

EXPLORING THE ENERGY LANDSCAPE OF SMALL PEPTIDES AND PROTEINS BY MOLECULAR DYNAMICS SIMULATIONS

Gerhard Stock, Abhinav Jain, Laura Riccardi, and Phuong H. Nguyen

2.1. INTRODUCTION: FREE ENERGY LANDSCAPES AND HOW TO CONSTRUCT THEM

During the last decades we have witnessed tremendous progress in the experimental and theoretical characterization of biomolecular processes such as molecular recognition, folding, and aggregation. On the experimental side, techniques such as magnetic resonance and optical and infrared spectroscopy have provided a wealth of information on biomolecular structure, dynamics, and functions with ever-improving resolution in space and time (see the chapters in this book). On the theoretical side, classical molecular dynamics (MD) simulations using all-atom force fields and explicit solvent are able to capture the structure and motion of biomolecules in microscopic detail [1]. Thanks to increasing computational power and significant effort in the development of algorithms, we are nowadays able to directly simulate biomolecular processes on a microsecond time scale and to account even for millisecond processes when enhanced sampling strategies are employed. Generating a huge amount of data, however, these simulations pose the problem of

---

Protein and Peptide Folding, Misfolding, and Non-Folding, First Edition. Edited by Reinhard Schweitzer-Stenner.
© 2012 John Wiley & Sons, Inc. Published 2012 by John Wiley & Sons, Inc.

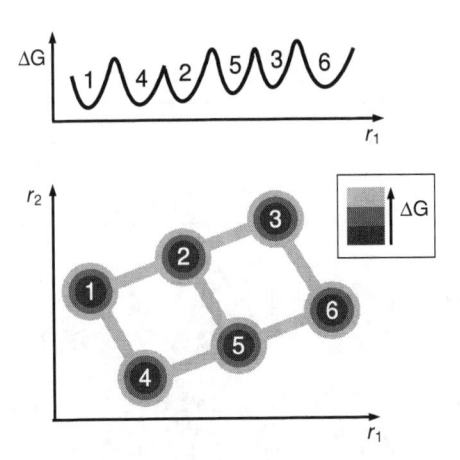

Figure 2.1. Schematic one- and two-dimensional representations of a model free energy landscape. Although the reduced-dimensionality representation reproduces the correct number of minima and their energies, the connectivity of these states and their barriers are obscured in a single dimension. Reprinted with permission from Reference 5.

analyzing the data set obtained. A key concept to interpret and illustrate biomolecular processes is provide by the molecule's free energy landscape

$$\Delta G(r) = -k_B T \ln P(r), \qquad (2.1)$$

where P is the probability distribution of the molecular system along some (in general multidimensional) coordinate r. Popular choices for the coordinate r include the fraction of native contacts, the radius of gyration, and the root mean square deviation of the molecule with respect to the native state. Characterized by its minima (which represent the metastable conformational states of the systems) and its barriers (which connect these states), the energy landscape allows us to account for the pathways and their kinetics occurring in a biomolecular process [2–4].

Clearly, a suitable representation of the energy landscape should (at least) reproduce the correct number, energy, and location of the metastable states and barriers. Unfortunately, these crucial quantities often get lost when the energy landscape is projected on a low-dimensional subspace. To illustrate the problem, Figure 2.1 shows schematic one- and two-dimensional representations of a model free energy landscape. The two-dimensional energy landscape $\Delta G(r_1, r_2)$ exhibits six minima of energy ΔG_i corresponding to metastable conformational states of the system. The minima are connected by barriers of height ΔG_{ij}. The projection of the two-dimensional surface on its first coordinate is given by

$$\Delta G(r_1) \propto -k_B T \ln \int dr_2 P(r_1, r_2). \qquad (2.2)$$

The one-dimensional representation is found to reproduce the correct number of minima and their energies. The former is clearly a consequence of the fact that all minima are located at different values of r_1. In general, however, we may obtain less minima in lower dimensions because several minima may overlap along the reduced coordinate r_1. More importantly, though, Figure 2.1 reveals that the true nature of the barriers may be obscured in reduced dimensionality. As a typical example, consider minima **2** and **4**. In two dimensions, there exist two pathways of minimal energy between these two states, $2 \to 1 \to 4$ and $2 \to 5 \to 4$. Projecting on a single dimension, however, this connectivity gets lost. Now states **2** and **4** are direct neighbors connected by a single barrier, and states **1** and **2** are only connected via state **4**. The energies ΔG_{24} and ΔG_{12} of these spurious barriers and the corresponding transition rates k_{24} and k_{12} may be smaller or larger than in full dimensionality [5].

The simple example explains why commonly used one- and two-dimensional representations may lead to serious artifacts and oversimplifications of the free energy landscape of biomolecules [6]. Numerous groups with various scientific background have therefore be concerned with the development of systematic dimensionality reduction methods [7–22]. The most common approach is principal component analysis [23] (PCA), also called quasiharmonic analysis or essential dynamics method [7–10]. Considering the dynamics of M atoms, the basic idea is that the correlated internal motions are represented by the covariance matrix

$$\sigma_{ij} = \langle (q_i - \langle q_i \rangle)(q_j - \langle q_j \rangle) \rangle, \qquad (2.3)$$

where $q_1, \ldots, q_{3M}$ are the mass-weighted Cartesian coordinates of the molecule and $\langle \ldots \rangle$ denotes the average over all sampled conformations [7–10]. The PCA represents a linear transformation that diagonalizes the covariance matrix and thus removes the instantaneous linear correlations among the variables. Ordering the eigenvalues of the transformation decreasingly, it has been shown that a large part of the system's fluctuations can be described in terms of only a few PCA eigenvectors or principal components. Indeed, recent MD simulations combined with a suitable PCA have revealed that the free energy landscape of small peptides, proteins, and nucleic acids can be quite complex, showing numerous numerous multidimensional minima and saddle points [13, 17, 18]. In particular, it has been found that the unfolded region of these systems is not random but can be well characterized by a number of metastable conformational states.

In this chapter, we review our recent advances to construct biomolecular free energy landscape from an MD simulation. First we show that it may be important to use internal coordinates such as the backbone dihedral angles ϕ, ψ in a PCA rather than the usual Cartesian coordinates. Hence, the basic idea of the dihedral angle principal component analysis (dPCA) is to perform a PCA on sin- and cos-transformed dihedral angles [13]

$$\begin{aligned} q_{2n-1} &= \cos\psi_n, \\ q_{2n} &= \sin\psi_n, \end{aligned} \qquad (2.4)$$

where $n = 1, \ldots, N$, and N is the total number of peptide backbone and side-chain dihedral angles used in the analysis. Equation 2.4 represents a transformation from the space of dihedral angles $\{\varphi_n\}$ to a linear *metric* coordinate space (i.e., a vector space with the usual Euclidean distance), thus avoiding problems arising from the circularity of angular variables [24, 25]. Employing the dPCA to study the energy landscape of various polyalanines, we find that there are numerous free energy minima, which correspond to metastable conformational states with well-defined structures and characteristic hydrogen bonding patterns. This is in contrast to the simple and smooth energy landscape obtained from the Cartesian PCA, which is an artifact of the mixing of internal and overall motion. The dPCA is furthermore appealing because other internal coordinates such as bond lengths and bond angles usually do not undergo changes of large amplitudes. Hence the analysis already starts with the relevant part of the dynamics, thus avoiding unnecessary noise.

On the basis of the dPCA, we next develop a systematic and general strategy to characterize the thus obtained free energy landscape. We briefly discuss various practical clustering methods to determine and characterize the system's metastable states and barriers. It is shown that the conformational states obtained from the dPCA correspond to well-defined intermediate structures of the peptide's folding and unfolding process. Moreover, we discuss various ways to visualize the main features (states, barriers, connectivities, energy basins, etc.) of the landscape. In a second step, the dPCA energy landscape is employed to facilitate a low-dimensional description of the *dynamics* of the system. This can be done by either invoking a Markov state model [11, 17–21, 26] (using the metastable states), by employing a Langevin equation [27–30] (using directly the dPCA coordinates), or by constructing a nonlinear deterministic model of the dynamics [31, 32]. To this end, we derive criteria to determine the dimensionality of the landscape such that it contains all slow large-amplitude motions of the molecule, while the remaining "bath" coordinates only account for its high-frequency fluctuations [30, 32]. It is shown that a sufficiently large dimension of the model is essential to ensuring a clear time scale separation of system and bath variables.

Moving on to larger systems such as small proteins, it is found that residual nonlinear correlations between various principal components can seriously hamper a low-dimensional representation of the system [33]. We show that this problem can be overcome by partitioning the protein into parts (e.g., its secondary-structure elements), for each of which we perform a separate PCA. In a second step, we then construct the full free energy landscapes as a direct product of the energy surfaces of each part. Adopting extensive MD simulations of the villin headpiece by Pande and coworkers, it is shown that this "PCA by parts" allows us to characterize the free energy landscape of the protein with unprecedented detail [34]. Finally, we briefly demonstrate that these concepts can also be applied to the study of the free energy landscape of several molecules, for example, to describe the aggregation of peptides. Recent advances that for reasons of space limitation will hardly be mentioned in this review include the improvement of MD force fields and sampling [1], the development of methods to identify Markov states [18–20], various nonlinear approaches [16, 32, 35, 36], as well as the use of network theory to characterize energy landscapes [12, 22].

2.2. DIHEDRAL ANGLE PCA ALLOWS US TO SEPARATE INTERNAL AND GLOBAL MOTION

To demonstrate the need to use internal rather than Cartesian coordinates, we consider penta-alanine Ac-Ala$_5$-NHMe (Ala$_5$) in explicit water as a simple example. As detailed in Ref. 13, we performed an MD simulation of Ala$_5$ at 300 K, using the GROMACS program suite [37] and the GROMOS force field 45A3 [38]. Due to the short length of the peptide, the MD trajectory of Ala$_5$ is found to undergo numerous folding and unfolding events within the simulation time of 100 ns. Employing a standard PCA using the Cartesian coordinates of all atoms, Figure 2.2A shows the free energy surface of Ala$_5$ as a function of the first two eigenvectors. The energy landscape exhibits a single prominent minimum which is found to correspond to the α-helical structure. In agreement with previous works [39, 40], this finding suggests a simple, relatively smooth free energy landscape.

Employing the dPCA, which uses the backbone dihedral angles of the peptide, on the other hand, Figure 2.2B,C show that the free energy landscape of Ala$_5$ actually exhibits numerous free energy minima. That is, the true free energy landscape is actually quite rugged and its smooth appearance in the Cartesian PCA represents an

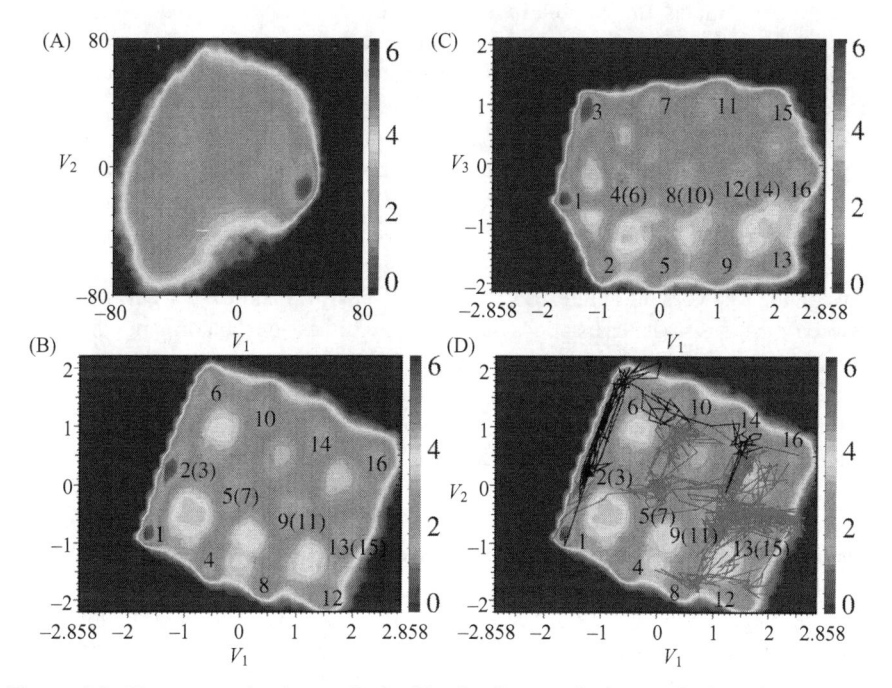

Figure 2.2. Free energy landscape (in kcal/mol) of penta-alanine as obtained from various PCAs of the 100 ns MD simulation. Compared are the results of (A) the standard PCA using Cartesian coordinates and (B–D) the dPCA using dihedral angles. In (A), (B), and (D) the energy is plotted in the plane of the first two principal components V_1 and V_2; in (C) the first and third principal components are used. The red and black lines in (D) reflect two representative folding pathways of the peptide. Reprinted with permission from Reference 13. See color insert.

artifact of the mixing of internal and overall motion. As internal coordinates naturally provide a correct separation of internal and overall dynamics, the dPCA is an important improvement over a Cartesian PCA, when one considers large-amplitude motions of biomolecules. Recently, the theoretical foundations of the dPCA was established [25], and the method was implemented in the GROMACS MD program [37].

To characterize the energy landscape, we employed a distance criterion in the space of the first three dPCA eigenvectors, which was used to identify the most important free energy minima [13]. This way, 16 conformational states of Ala$_5$ were obtained, which cover nearly 90% of the configurations sampled in the simulation. The corresponding molecular structures cover a broad range of the conformational space of the peptide, from the all-extended conformer, structure 16, to the all-α-helical conformer, structure 1. Table 2.1 contains a detailed description of these conformational states including average dihedral angles, relative populations, free energies, average lifetimes, and hydrogen bonds. It reveals that the so-called unfolded conformational space of the peptide contains numerous metastable conformations with well-defined hydrogen bonding, for example, structures 1, 3, 6, 4, and 2 contain three, two, and one $(i,i + 4)$ H-bonds, respectively. Compared with recent experiments on short polyalanines [41–43] the GROMOS force field (and most other force fields) overestimates the populations of α_R conformations [44–46]. We note in passing that the dPCA amounts to a one-to-one representation of the original angle distribution [25], which means that two distinct minima on the dPCA free energy surface must correspond to conformational states of different structure [47, 48].

Having identified its minima, the next step to characterize the free energy landscape is to study the possible transitions between these minima. To this end, we have constructed a connectivity map that reports on all conformational transitions of Ala$_5$ within the simulation time of 100 ns [13]. It reveals that most conformational states are connected to either three or four next neighbors. By calculating the distances between the various states in the dPCA space, we obtain an almost perfect correlation between the state-to-state distances and the probabilities of the corresponding conformational transitions; that is, virtually all connected minima lie close together. Most interestingly, the connectivity map can be employed to identify possible pathways of folding and unfolding of Ala$_5$. In brief, we have found that the main folding pathways for helix nucleation are path A, which leads to an α-helical H-bond at the N-terminal; path B, which leads to an α-helical H-bond at the center of the peptide; path C, which has an α-helical H-bond at the C-terminal; and path D, which exhibits a global hairpin structure. As an illustration, Figure 2.2D shows two representative examples of the complete folding pathway from state 16 to state 1. Although the process essentially takes the above described folding pathways A and B, the figure clearly elucidates the diffusive character of the folding, involving numerous excursions to side states.

2.3. DIMENSIONALITY OF THE FREE ENERGY LANDSCAPE

In the case of penta-alanine, the free energy landscape is mainly described by only three principal components. For this simple example, the conformational states and

TABLE 2.1. Characterization of the Most Prominent Conformational States of Ala$_5$ in Water

State	P (%)	τ (ps)	ΔG	ϕ_1, ψ_1	ϕ_2, ψ_2	ϕ_3, ψ_3	ϕ_4, ψ_4	ϕ_5, ψ_5	1$_\alpha$	2$_\alpha$	3$_\alpha$	1$_{3_{10}}$	2$_{3_{10}}$	1$_t$	2$_t$
1	14	35	0.05	−57, −48	−64, −43	−63, −43	−65, −39	−72, −34	11	27	60	28	7	2	0
2	3	18	0.98	−58, −44	−72, −36	−68, −36	−88, 102	−67, −35	70	2	0	29	10	33	3
3	15	25	0	−58, −47	−65, −40	−66, −39	−69, −39	−90, 115	25	67	1	26	6	28	0
4	4	20	0.84	−61, 59	−58, −18	−68, −10	−63, −39	−76, −36	29	33	0	30	9	24	1
5	3	17	0.90	−65, 7	−70, −4	−80, 31	−71, 110	−68, −37	5	0	0	27	4	41	11
6	7	23	0.44	−60, −43	−70, −37	−70, −38	−93, 113	−75, 117	68	1	0	24	8	39	7
7	3	14	0.89	−63, 75	−56, −28	−68, −14	−69, −42	−91, 103	46	2	0	30	12	40	7
8	1	12	1.58	−71, 110	−59, 32	−68, 55	−60, −43	−85, −35	22	0	0	27	7	36	6
9	2	16	1.13	−72, 91	−70, 53	−65, 54	−62, 116	−72, −42	4	0	0	9	1	40	11
10	6	20	0.51	−67, 10	−71, −5	−81, 23	−78, 111	−78, 118	5	0	0	29	4	43	14
11	8	18	0.41	−69, 79	−73, 24	−79, 95	−60, −36	−89, 113	2	0	0	15	1	40	12
12	1	13	1.52	−74, 121	−73, 121	−69, 119	−56, −46	−87, −36	3	0	0	26	1	37	10
13	1	19	1.41	−74, 120	−74, 117	−60, 115	−67, 119	−74, −43	1	0	0	5	0	42	13
14	8	23	0.35	−73, 78	−66, 45	−67, 70	−66, 110	−77, 117	2	0	0	11	2	42	16
15	6	26	0.58	−72, 118	−72, 117	−67, 118	−50, −36	−90, 116	0	0	0	15	3	41	17
16	5	24	0.69	−70, 116	−74, 117	−64, 116	−60, 111	−81, 117	1	0	0	9	1	40	18

Shown are the population probability P (%), the average lifetime τ (in ps), the free energy ΔG (in kcal/mol) relative to the minimum-energy state 3, the average dihedral angles (ϕ_i, ψ_i), as well as the occurrences (in %) of various hydrogen bonding types: n_α denotes the occurrence of n α-helical ($i, i + 4$) bonds, while $n_{3_{10}}$ and n_t refer to 3$_{10}$-helical ($i, i + 3$) bonds and to ($i, i + 2$) turns, respectively. Reprinted with permission from Reference 13.

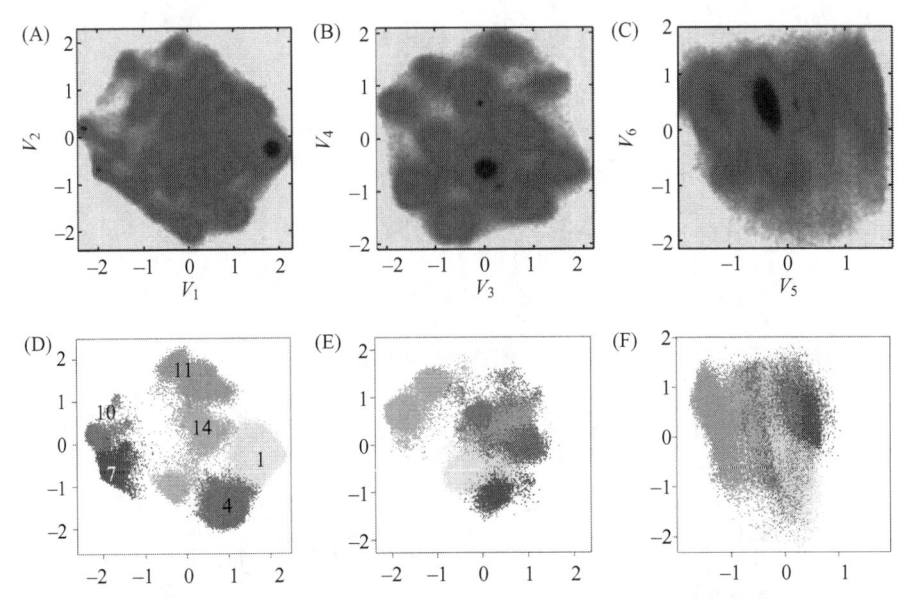

Figure 2.3. Two-dimensional representations of the free energy landscape of Ala$_7$ as obtained by dPCA: (A) $\Delta G(V_1, V_2)$, (B) $\Delta G(V_3, V_4)$, and (C) $\Delta G(V_5, V_6)$. The color coding in panels (D–F) illustrates some prominent conformational states of the system. Reprinted with permission from Reference 5.

transitions are readily identified by visual inspection. The characterization of the free energy landscape of larger systems is less straightforward and therefore needs to be carried out in a systematic manner. With this end in mind, in the following we consider the free energy landscape of hepta-alanine (Ala$_7$), which was obtained from an 800-ns MD simulation in aqueous solution at 300 K [5]. To get a first impression, Figure 2.3 shows two-dimensional representations of the free energy landscape of Ala$_7$, as obtained from an dPCA of the backbone dihedral angles. Similarly to penta-alanine, the energy landscape of Ala$_7$ is found to be quite rugged and exhibits numerous minima.

Generally speaking, the goal of any reduced-dimensionality representation is to appropriately describe a given problem by using a minimum number of dimensions. For the dPCA representation of the free energy landscape, this amounts to the question of how many principal components are needed in order to (at least) reproduce the correct number, energy, and location of the metastable states and barriers. From Figure 2.3 we find that the free energy exhibits several minima along the first five principal components, while there is only a single minimum found along V_6. As a further indication of the number of "essential" principal components, we may consider the percentage of overall fluctuations covered by the first n principal components (i.e., the sum of the first n eigenvalues of the PCA). Interestingly, Figure 2.4A reveals three kinds of principal components: The first one covers 22% of all fluctuations, each of the next four contribute about 10%, while the remaining principal components contribute less than 4% each.

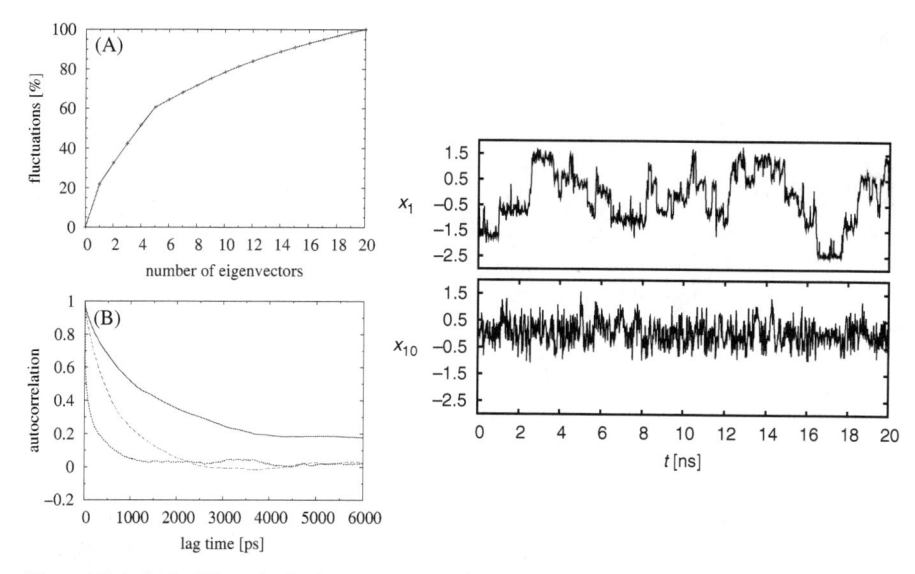

Figure 2.4. Left: The principal components of Ala₇ as obtained by the dPCA, characterized by (A) their cumulative fluctuations and (B) their normalized fluctuation autocorrelation functions. The latter is shown for the principal components V_1 (full red line), V_2 (dashed green line), and V_6 (dotted blue line). Right: Short time trace of the first (top) and the tenth (bottom) principal component of Ala₇. Reprinted with permission from References 5 and 30.

A similar behavior is found for the time scales of the fluctuations, revealed by the normalized fluctuation autocorrelation function $\left(\langle V_n(t)V_n\rangle - \langle V_n\rangle^2\right)/\left(\langle V_n^2\rangle - \langle V_n\rangle^2\right)$ shown in Figure 2.4B. Judged by their initial time evolution, the first principal component decays within several nanoseconds, the next four decay on a time scale of 1 ns, and the decay time of the higher principal components is clearly shorter. This time scale separation is also nicely illustrated by considering a short time trace of the first and the tenth principal component of Ala₇. Figure 2.4 (right panels) clearly contrasts the slow large-amplitude motion of the "reaction coordinate" $x_1(t)$ with the high-frequency fluctuations of the "bath coordinate" $x_{10}(t)$. The former describes transitions between metastable conformational states of the peptide, while the latter accounts for fluctuations within such a metastable state. Taken together, we expect that a five-dimensional dPCA representation of the free energy surface of Ala₇ suffices to correctly describe its main features.

2.4. CHARACTERIZATION OF THE FREE ENERGY LANDSCAPE: STATES, BARRIERS, AND TRANSITIONS

In a next step, we wish to identify and characterize the metastable states of the reduced free energy landscape of Ala₇ shown in Figure 2.3. To this end, we employ the *k*-means algorithm [49] as a well-established simple and fast geometric clustering method. *k*-means aims at finding a partition of a given data set into *k* subsets that minimizes the sum of squares of distances between the objects and their corresponding cluster centroids. As the number of clusters must be known beforehand in *k*-means, we first need to decide how many clusters should be considered in the analysis. From a visual inspection of Figure 2.3A it is already clear that we should include at least ~20 clusters to distinguish all states shown by the $\Delta G(V_1, V_2)$ surface. However, since the two-dimensional representations in Figure 2.3 do not reveal possible correlations between each other, we cannot tell if the ~20 states in $\Delta G(V_1, V_2)$ split up further in $\Delta G(V_3, V_4)$ or not. To test if a clustering in *k* states is suitable, we request that such a clustering should give a large fraction (say, larger that 90%) of good clusters. Here we call a cluster "good" when the average circular variance of all dihedral angles is less than a certain threshold, thus discriminating fluctuations *within* a conformational state from transitions *between* different conformational states. Doing so, we found that $k = 23$ are a suitable number of clusters (see Ref. [5] for details). This findings are reconfirmed by the results of a *kinetic* clustering of the Ala$_7$ trajectory, that is, a clustering that defines its states through their metastability rather than through geometric similarity. The lower panels of Figure 2.3 show some of the prominent clusters of Ala$_7$ obtained this way. Remarkably, the clusters correlate with the minima of the free energy and therefore indeed represent the metastable conformational states of the system.

The above results clearly demonstrated that the five-dimensional dPCA free energy landscape is a suitable and accurate representation of the full-dimensional landscape of Ala$_7$. That is, by using only five dimensions, we correctly account for all populations and metastabilities of the conformational states as well as for all slow motions of the system. In a second step, this reduced-dimensionality representation may be employed to schematically illustrate the main features (states, barriers, connectivities, energy basins, etc.) of the biomolecular system. As an example, Figure 2.5 shows several popular ways to visualize the energy landscape. Panel (A) displays the most important conformations and transitions as a function of the first two principal components. We see that the arrows mostly connect neighboring states, that is, kinetically well separated clusters are also geometrically distinct in the first two principal components of the dPCA. Note, however, that the distances of the cluster centers in (V_1, V_2) subspace do not reflect the true distances in full-dimensional space. Alternatively, one may therefore define a plane on which the distances obtained through the projection of the cluster centers deviates minimally from the original distances [50, 51]. Figure 2.5 reveals that the distances in the resulting representation (panel B) may differ from the distances in the (V_1, V_2) subspace (panel A). As a popular alternative, one may construct a free energy disconnectivity graph [4, 52, 53], which directly displays the connectivity and the barriers between all states, see Figure 2.5C. Dividing up the energy landscape in six "basins," this representation readily reveals the hierarchy of the states.

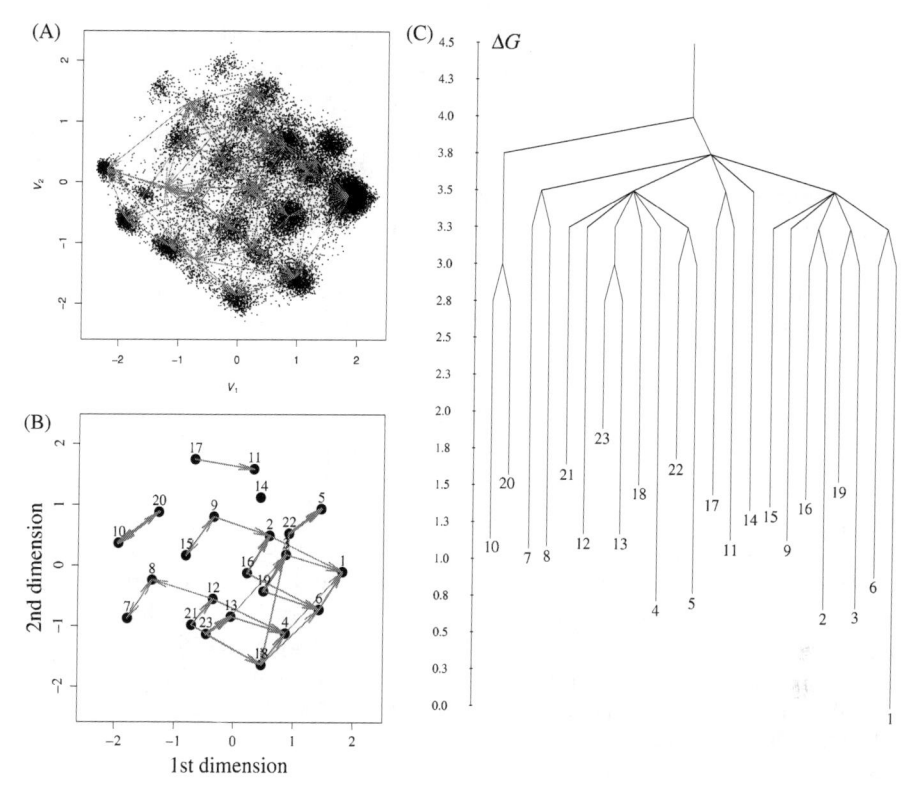

Figure 2.5. Visualization of the free energy landscape of Ala$_7$. Shown are (A) a two-dimensional cut $\Delta G(V_1, V_2)$ along the first two components including possible transitions between cluster centers, (B) a two-dimensional principal coordinate representation where only transitions with probability $T_{ij} > 1.5\%$ are indicated (using a linewidth that is proportional to T_{ij}), and (C) a disconnectivity graph of the system. Reprinted with permission from Reference 5.

2.5. LOW-DIMENSIONAL SIMULATION OF BIOMOLECULAR DYNAMICS TO CATCH SLOW AND RARE PROCESSES

Apart from the "static" characterization of the equilibrium free energy surface, moreover, the above analysis can be employed to facilitate a low-dimensional description of the *dynamics* of the system. This can be done by either invoking a Markov state model [11, 17–21, 26] (using the metastable states), by employing a Langevin equation [27–30] (using directly the dPCA coordinates), or by constructing a nonlinear deterministic model of the dynamics [31, 32]. Considering only the system's essential degrees of freedom rather than the multitudinous atomic coordinates of an MD simulation, these approaches allow us to extend the simulation time to the millisecond range and therefore can also account for slow and rare biomolecular processes.

If the process under consideration can be described by a Markov chain of metastable states, a suitable clustering combined with a simple master equation provides the complete information of the time evolution of the system. In biomolecular systems, however, the underlying assumption of a time scale separation between fast intrastate and slow interstate transitions may break down. This was also found to be the case for our dPCA states of hepta-alanine [5], where the minimal time scale to assure Markovian dynamics is already in the order of the conformational state's life time. The problem can be circumvented by invoking a large number of conformations in order to construct the Markov model. In particular, the groups of Pande and Noe have recently presented impressive examples of Markov modeling of protein folding [19, 20].

Working in continuous phase space, the Langevin approach does not require to define suitable metastable states. The applicability of a memory-free Langevin equation requires, however, that the "system" (described by the included principal components) contains all slow large-amplitude motions of the molecule, while the remaining "bath" coordinates only account for its high-frequency fluctuations. In the case of Ala$_7$, this separation of time scales is nicely illustrated in Figure 2.4 (right panels), which clearly contrasts the slow large-amplitude motion of the "reaction coordinate" $x_1(t)$ with the high-frequency fluctuations of the "bath coordinate" $x_{10}(t)$. Applying methods from nonlinear time series analysis, Hegger et al. have developed a practical Langevin algorithm that performs a *local* estimation of the multidimensional Langevin vector fields describing deterministic drift and stochastic driving [30]. Applied to hepta-alanine, it was shown that a five-dimensional Langevin model correctly reproduces the structure and conformational dynamics of the system. To demonstrate the need to include all five degrees of freedom in the calculation (in contrast to common one-dimensional calculations), we consider the life time distributions of specific conformational states of Ala$_7$. Choosing the most stable state 1 and the least stable state 23 as representative examples, Figure 2.6 compares the life time distributions as obtained from MD and Langevin simulations, respectively. We find that a one-dimensional Langevin model fails to account for the conformational

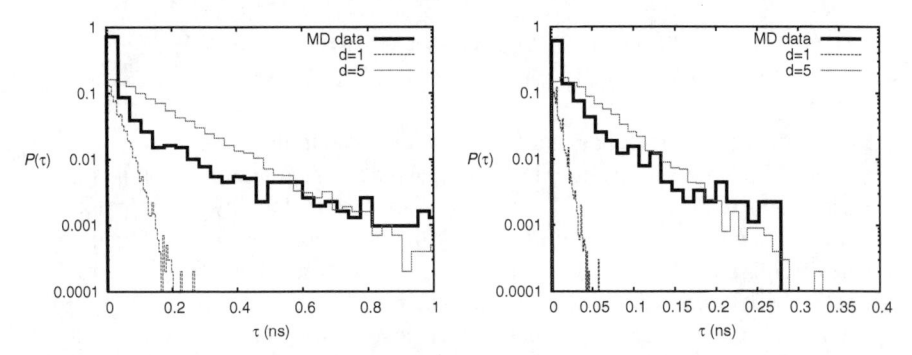

Figure 2.6. Lifetime distribution of the most stable (left) and least stable (right) conformational state of Ala$_7$. Compared are original data from MD simulation (thick black line) and Langevin data using one dimension (dotted line) and full dimensionality (thin line). Reprinted with permission from Reference 30.

dynamics of the system, as it considerably underestimates its life times. The full-dimensional Langevin model, on the other hand, nicely reproduces the life time distributions of the MD reference data and easily provides a significantly improved sampling of the processes under consideration.

Apart from a stochastic description, it may be also desirable to construct a *deterministic* model of biomolecular dynamics [31]. To this end, first the influence of the bath variables are eliminated by using a noise reduction scheme. The resulting approximately deterministic time evolution of the reaction coordinate can then be modeled by using methods from nonlinear time series theory. Applied to the conformational dynamics of small peptides [32], this strategy has been found to yield similar agreement with the original MD data as found for the Langevin modeling. As a virtue, a deterministic model allows us to use techniques from nonlinear dynamics theory to analyze the system. For example, a Lyapunov analysis of a deterministic model of peptide dynamics showed that the effective dimension of the dynamic system is rather small and may even decrease with chain length [32].

2.6. PCA BY PARTS: THE FOLDING PATHWAYS OF VILLIN HEADPIECE

We have demonstrated above that the free energy landscape of small peptides can be quite complex, showing numerous metastable conformational states [13, 17, 18]. On the other hand, various computational studies of proteins have given a comparatively simple picture of their energy landscape [14–16]. Although this may be expected at low energies where the protein fluctuates around its native state, it appears surprising at higher energies where the protein can reversibly fold and unfold. A protein consists of several secondary-structure segments such as α-helices, β-sheets, and turns, which need to be formed in the folding process. Assuming that the energy landscape of each part is already complex, one would naively expect that the complete protein landscape is even more involved. As discussed, possible reasons for this "hidden complexity" include a too small dimensionality of landscape [5, 6, 15], and the use of Cartesian coordinates. In addition, it turns out that for larger systems such as proteins the residual nonlinear correlation between various principal components may contribute to the problem.

As an illustrative example, we adopt a fast folding variant of the villin headpiece subdomain (HP-35 NleNle) [21, 54–58] and use extensive (about 350 μs in total) simulations of HP-35 in explicit solvent carried out by Pande and coworkers [57]. Figure 2.7 (left panels) shows the structure of the native state of HP-35 as well as the starting structure used in the folding MD simulations of HP-35. To get an overall impression of the free energy landscape of HP-35, we first performed a standard PCA of the complete protein, using the Cartesian coordinates of the backbone atoms (cPCA). The first four principal components clearly exhibit non-Gaussian distributions and contain ~60% of the total fluctuations of the system. The resulting cPCA energy landscape along the first two PCs is shown in Figure 2.7. It reveals the dominant native state of the system as well as a few metastable conformational

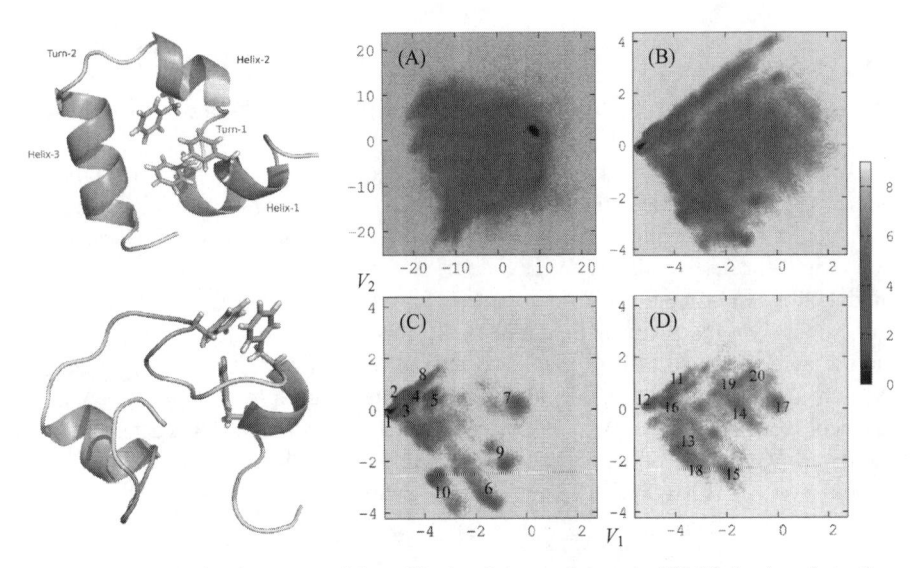

Figure 2.7. Left: Native state of the villin headpiece subdomain HP-35 (top) and starting structure of the folding MD simulations of HP-35 (bottom). Right: Two-dimensional representations of the free energy landscape $\Delta G(V_1, V_2)$ of HP-35 as obtained by the cPCA (A), dPCA (B), and pPCA (C and D). The latter shows the twenty most metastable conformational states (1–10 in A and 11–20 in B), using the eigenvectors of the overall dPCA in (B). Reprinted with permission from Reference 34. See color insert.

states, which show up as purple-colored minima in the figure. However, since the energy landscape is rather diffuse, only a few conformational states can be identified from an analysis of the cPCA landscape. As already discussed, this lack of structure may be caused by the mixing of internal and overall motion in Cartesian coordinates. Employing a dihedral angle PCA (dPCA), the energy landscape shown in Figure 2.7 is similarly diffuse and structureless as in the case of the cPCA. Even worse, it is found that the dPCA converges quite slowly with the number of included PCs; one needs 15 components to account for 60% of the system's fluctuations. A closer examination reveals that this effect is caused by significant nonlinear correlations between the individual PCs, which reflects the fact that the motion of adjacent backbone dihedral angles is necessarily correlated in a protein with relatively rigid secondary structures [33]. Hence many PCs are needed to account for the system's conformational dynamics, which hampers a conclusive low-dimensional representation of the free energy landscape.

To overcome these problems, we have recently suggested a "divide and conquer" strategy called PCA by parts (pPCA) [34]. The main idea of the pPCA is based on the assumption that each individual part of the molecule shows a structured free energy landscape. In principle, this can be always enforced by making the parts smaller. Here we partition HP-35 into five parts (Fig. 2.7): helix-1 (residues 4–10), turn-1 (residues 11–14), helix-2 (residues 15–19), turn-2 (residues 20–22), and helix-

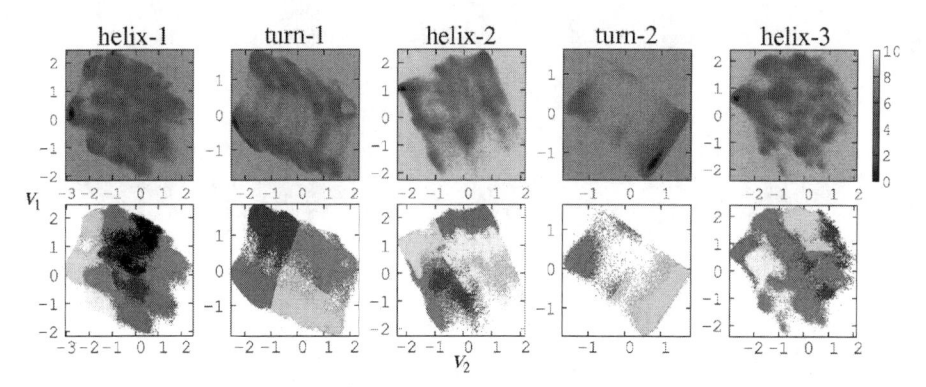

Figure 2.8. Top: Free energy landscapes $\Delta G(V_1, V_2)$ obtained from a dPCA of the various secondary-structure parts of HP-35. Bottom: K-means clustering of the conformational states in these landscapes. Reprinted with permission from Reference 34.

3 (residues 23–32). The residues 1–3 and 33–35 at the termini of the protein are not considered in the analysis. Next, we perform a dPCA individually for each part. Interestingly, only a few (2–4) PCs are needed to cover more that 60% of the fluctuations of each individual part. That is, the dPCA works well for the secondary-structure parts, but not so for the complete protein, which requires many PCs. As shown in Figure 2.8, the resulting free energy landscapes of the secondary-structure parts are well resolved and exhibit numerous well-defined minima. Employing k-means clustering, we obtain 9, 4, 6, 2, and 7 clusters for helix-1, turn-1, helix-2, turn-2, and helix-3, respectively. Their position in the energy landscape (Fig. 2.8) reveals that for the most part the clusters coincide with the free energy minima and therefore represent metastable conformational states. For later reference, we label the states of each part by a letter code: U for unfolded, I_j ($j = 1, \ldots, 5$) for intermediates, N_j ($j = 1, 2$) for nearly native, and F for folded.

Based on the clustering of the five parts, we now construct a time series $C(t)$ from the MD trajectory of HP-35, where $C = \{C_i\}$ is a five-dimensional vector with C_i denoting the conformational state of part i. For example, at time $t = 0$ we start in the unfolded state, $C(0) = (UUUFI_4)$, and at long times a typical trajectory will end up in the folded state $C(\infty) = (FFFFF)$. In this way, each different combination of the states of the secondary-structure parts corresponds to a specific and unique conformation of the complete protein. We note that this second step of the pPCA is crucial because it restores all structural correlations between the various parts. Although $9 \times 4 \times 6 \times 2 \times 7 = 3024$ combinations are possible in the chosen clustering, only 1197 combinations are sampled by the MD trajectories due to conformational and energetic constraints of HP-35. Even much fewer states are sampled notably, for example, the 10 and 20 most sampled states already cover 50% and 60% of the overall population, respectively. As a further illustration, Figure 2.7C,D show the free energy landscape constructed from the twenty most metastable clusters. Hence, the pPCA provides a means to identify with

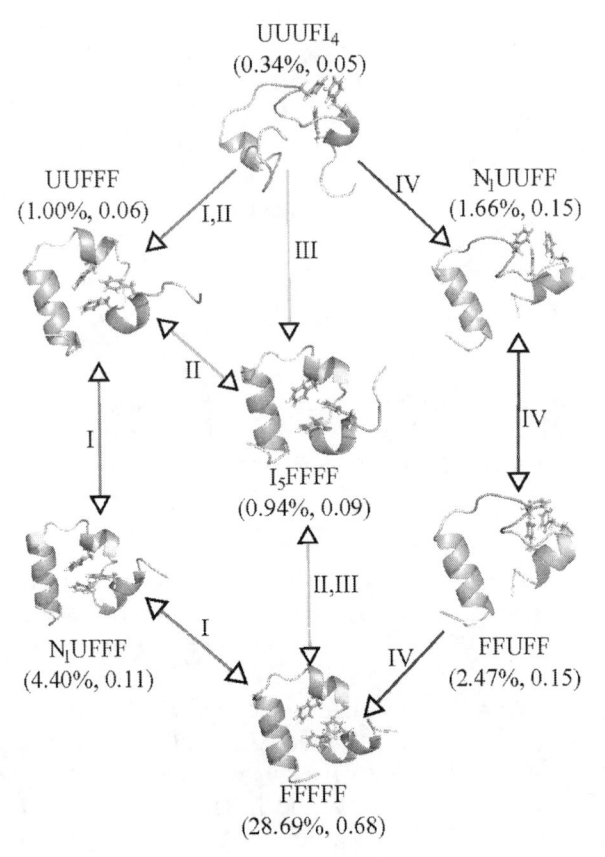

Figure 2.9. Major folding pathways of HP-35. All conformational states are characterized by a five-letter structure code, their population (%), and their metastability.

unprecedented resolution the metastable conformational states, which may be hidden in a conventional PCA.

As an instructive application of the pPCA, we finally construct the main folding pathways of HP-35 found in the MD simulations. For facile graphic representation, we restrict the discussion to the 25 mostly populated pPCA states (covering more than 60%) and show only connecting paths between metastable states with transition counts ≥10. Figure 2.9 displays the structures of all pPCA states along the four major folding pathways of HP-35. Paths I, II, and IV undergo folding in three steps, while path III represents a two-step folding process. Both paths I and II fold helix-2 and helix-3 in the first step, followed by a partial and then complete folding of helix-1 in the intermediate and the final step, respectively. The two paths differ in the rearrangement of turn-1, which occurs in the final step in path I and the intermediate step in path II. Path III is similar to path II, with the first two steps combined together in a single step, followed by a complete folding of helix-1 to reach the native state. Alternatively, path IV folds helix-3 and partially folds helix-1 in the first step. This is followed by a complete folding of helix-1 and rearrangement of turn-1 in the intermediate step, and the folding of helix-2 in the final step. Paths II and III require an average folding time of about 100 ns and are thus faster than paths I and IV, which take about 300 ns. While several simulations have indicated that helix-2 and helix-3 form before helix-1 [21, 56], the alternative path IV has not yet been reported.

2.7. THE ENERGY LANDSCAPE OF AGGREGATING Aβ-PEPTIDES

There has been considerable interest to investigate the structure and dynamics of amyloid fibrils, because they are thought to be a key factor in various diseases including Alzheimer's and Parkinson's disease [59]. To study the mechanism of the formation of fibrils from peptide monomers, Nguyen et al. recently performed extensive MD simulations (in total 6.9 μs) of the growth and assembly of Aβ_{16-22} oligomers [60]. Interestingly, the study revealed that a monomer adds to a preformed structured oligomer by a two-stage dock–lock mechanism. To illustrate the various aspects occurring in the aggregation of peptides, one would again like to construct a free energy landscape that accounts for the intermediate states and the barriers of the process. The task is complicated because intramolecular as well as intermolecular degrees of freedom of the various monomers need to be considered. To this end, several order parameters such as the radius of gyration, the root mean square distance, and the orientational order parameter P_2 have been adopted [61, 62].

Here we present our first attempts to apply a PCA to the study of the aggregation of peptides. Due to the lack of a physically meaningful reference structure of the full system, it is clear that it is not possible to perform a PCA using the Cartesian atomic coordinates of the system. The dPCA using the backbone dihedral angles of the monomers, on the other hand, is straightforward. As an example, Figure 2.10 shows the dPCA free energy landscape of the trimer (Aβ_{16-22})$_3$. Although the antiparallel β-sheet structure is found to be the most stable conformational state, there are numerous other metastable states reflecting quite different structures.

A closer examination of the dPCA energy landscape of the trimer reveals, however, that a single energy minimum may contain MD snapshots with quite different structures. This is because, so far we have only considered the intramolecular coordinates of the monomers but neglected the intermolecular degrees of freedom such as their relative orientations and distances. As a first step, we have represented the orientation of each monomer by a vector connecting the two end points and calculated all angles between these vectors. Converting to the cosine of these angles for consistency, we then performed a PCA of these orientational variables (oPCA). Figure 2.11 shows the resulting energy landscape of the oPCA obtained for a 2 μs simulation of the pentamer (Aβ_{16-22})$_5$. It reveals two minima along the first principal component V_1, refelcting a coplanar pentamer aggregate for large values of V_1 and pentamer aggregates with considerable orientational disorder for small values of V_1. Combining both intramolecular and intermolecular degrees of freedom in a single PCA, we find a significantly higher resolution of the energy landscape.

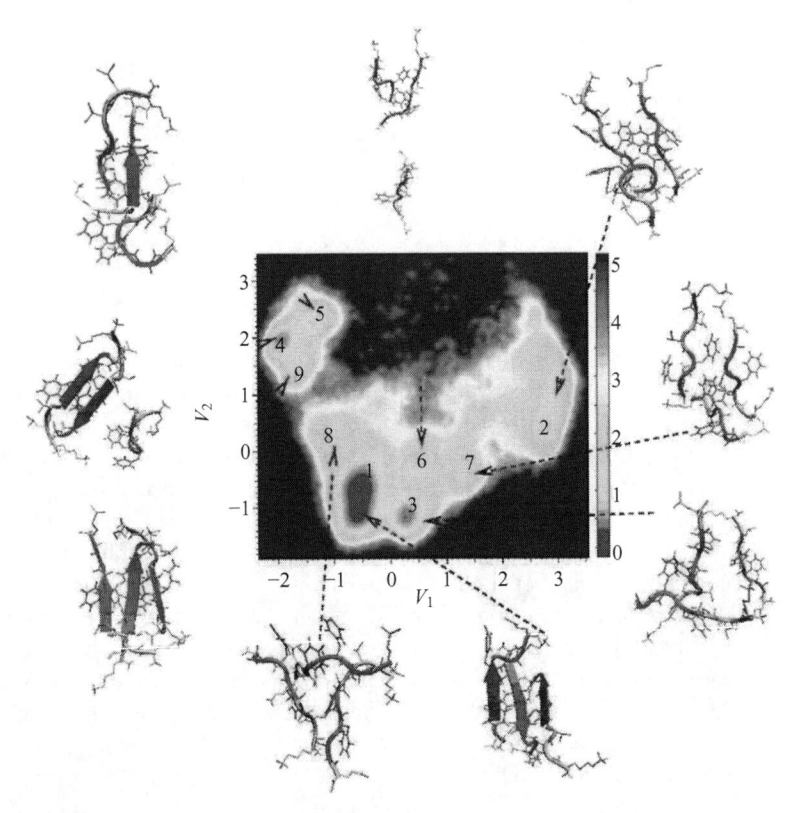

Figure 2.10. Free energy landscape $\Delta G(V_1, V_2)$ obtained from a dPCA of the aggregation process of the trimer $(A\beta_{16-22})_3$. Reprinted with permission from Reference 60.

2.8. CONCLUDING REMARKS

We have reviewed our recent efforts to employ a systematic PCA approach in order to reduce the vast number of atomic coordinates of a MD trajectory to a few collective degrees of freedom. Three important issues have been discussed: (1) the use of internal coordinates such as the backbone dihedral angles in the dPCA in order to avoid artifact due to the mixing of internal and overall motion, (2) the choice of an appropriate dimensionality of the energy landscape such that it contains all slow large-amplitude motions of the molecule while it neglects coordinates describing high-frequency fluctuations, and (3) the divide and conquer strategy of the PCA by parts (pPCA), which significantly reduces residual nonlinear correlation between various principal components and hence allows us to study the free energy landscape with unprecedented resolution. Applying this methodology, it has been shown that the energy landscape of folding and aggregating peptides and proteins can be quite complex, showing numerous multidimensional minima and saddle points. In particular, it has been found that the unfolded region of these systems is not random but

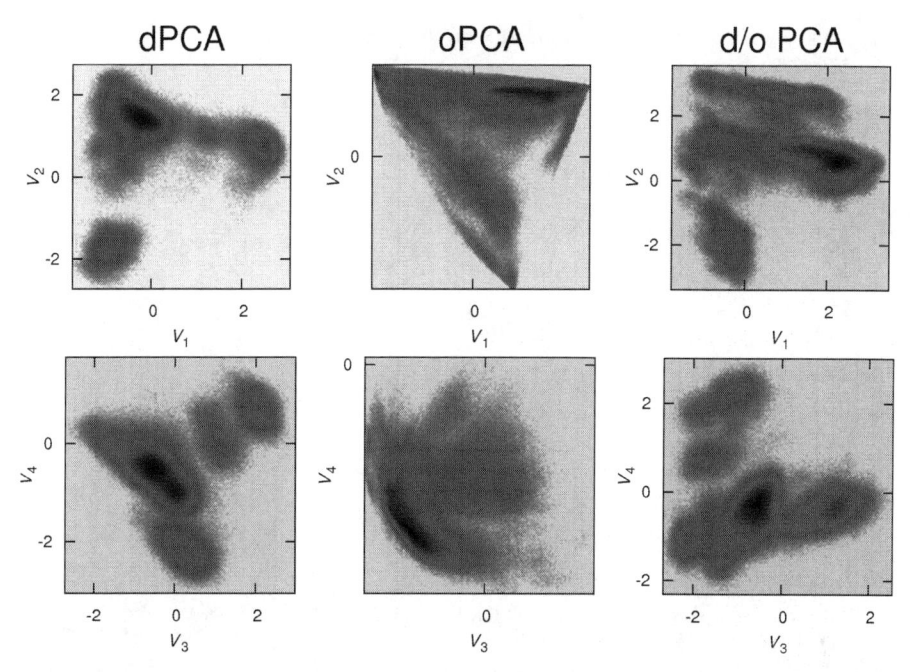

Figure 2.11. Free energy landscapes $\Delta G(V_1, V_2)$ of the pentamer $(A\beta_{16-22})_5$, obtained from a PCA using the monomers' backbone dihedral angles (dPCA), the orientational angles (oPCA), and both kinds of angles (d/o PCA).

can be well characterized by a number of metastable conformational states. Moreover, small proteins such as villin headpiece have been found to exhibit a multitude of folding pathways. In conclusion, it seems fair to say that the generation and characterization of biomolecular energy landscapes is an active field of research that is rapidly expanding.

ACKNOWLEDGMENTS

We thank Alessandros Altis, Rainer Hegger, and Yuguang Mu for numerous inspiring and helpful discussions, and Vijay Pande for providing the MD trajectories of HP-35. This work has been supported by the Frankfurt Center for Scientific Computing, the Fonds der Chemischen Industrie, and the Deutsche Forschungsgemeinschaft.

REFERENCES

1. van Gunsteren, W. F., Bakowies, D., Baron, R., Chandrasekhar, I., Christen, M., Daura, X., Gee, P., Geerke, D. P., Glättli, A., Hünenberger, P. H., Kastenholz, M. A., Oostenbrink, C., Schenk, M., Trzesniak, D., van der Vegt, N. F. A., and Yu, H. B. (2007) *Angew Chem Int Ed 45*, 4064–4092.

2. Onuchic, J. N., Schulten, Z. L., and Wolynes, P. G. (1997) *Annu Rev Phys Chem 48*, 545–600.
3. Dill, K. A. and Chan, H. S. (1997) *Nat Struct Biol 4*, 10–19.
4. Wales, D. J. (2003) *Energy landscapes*, Cambridge University Press, Cambridge.
5. Altis, A., Otten, M., Nguyen, P. H., Hegger, R., and Stock, G. (2008) *J Chem Phys 128*, 245102.
6. Krivov, S. V. and Karplus, M. (2004) *Proc Natl Acad Sci U S A 101*, 14766–14770.
7. Ichiye, T. and Karplus, M. (1991) *Proteins 11*, 205–217.
8. Garcia, A. E. (1992) *Phys Rev Lett 68*, 2696–2699.
9. Amadei, A., Linssen, A. B. M., and Berendsen, H. J. C. (1993) *Proteins 17*, 412–425.
10. Kitao, A. and Gō, N. (1999) *Curr Opin Struct Biol 9*, 164–169.
11. de Groot, B. L., Daura, X., Mark, A. E., and Grubmüller, H. (2001) *J Mol Biol 309*, 299–313.
12. Rao, F. and Caflisch, A. (2004) *J Mol Biol 342*, 299.
13. Mu, Y., Nguyen, P. H., and Stock, G. (2005) *Proteins 58*, 45.
14. Lange, O. F. and Grubmüller, H. (2006) *J Phys Chem B 110*, 22842–22852.
15. Maisuradze, G. G., Liwo, A., and Scheraga, H. A. (2009) *Phys Rev Lett 102* (23), 238102.
16. Das, P., Moll, M., Stamati, H., Kavraki, L. E., and Clementi, C. (2006) *Proc Natl Acad Sci U S A 103*, 9885–9890.
17. Noe, F., Horenko, I., Schütte, C., and Smith, J. C. (2007) *J Chem Phys 126*, 155102.
18. Chodera, J. D., Singhal, N., Pande, V. S., Dill, K. A., and Swope, W. C. (2007) *J Chem Phys 126*, 155101.
19. Noé, F., Schütte, C., Vanden-Eijnden, E., Reich, L., and Weikl, T. (2009) *Proc Natl Acad Sci U S A 106*, 19011–19016.
20. Bowman, G. R., Beauchamp, K. A., Boxer, G., and Pande, V. S. (2009) *J Chem Phys 131* (12), 124101.
21. Bowman, G. R. and Pande, V. S. (2010) *Proc Natl Acad Sci U S A 107*, 10890–10895.
22. Rao, F. (2010) *J Phys Chem Lett 1*, 1580–1583.
23. Jolliffe, I. T. (2002) *Principal component analysis*, Springer, New York.
24. Fisher, N. I. (1996) *Statistical analysis of circular data*, Cambridge University Press, Cambridge.
25. Altis, A., Nguyen, P. H., Hegger, R., and Stock, G. (2007) *J Chem Phys 126*, 244111.
26. Swope, W., Pitera, J., and Suits, F. (2004) *J Phys Chem B 108* (21), 6571–6581.
27. Lange, O. F. and Grubmüller, H. (2006) *J Chem Phys 124*, 214903.
28. Horenko, I., Hartmann, C., Schütte, C., and Noe, F. (2007) *Phys Rev E 76*, 016706.
29. Yang, S., Onuchic, J. N., and Levine, H. (2006) *J Chem Phys 125*, 054910.
30. Hegger, R. and Stock, G. (2009) *J Chem Phys 130*, 034106.
31. Kantz, H. and Schreiber, T. (1997) *Nonlinear time series analysis*, Cambridge University Press, Cambridge, UK.
32. Hegger, R., Altis, A., Nguyen, P. H., and Stock, G. (2007) *Phys Rev Lett 98*, 028102.
33. Omori, S., Fuchigami, S., Ikeguchi, M., and Kidera, A. (2010) *J Chem Phys 132*, 115103.
34. Jain, A., Hegger, R., and Stock, G. (2010) *J Phys Chem Lett 1*, 2769–2773.
35. Lange, O. F. and Grubmüller, H. (2006) *Proteins 62*, 1053–1061.

REFERENCES

36. Nguyen, P. H. (2006) *Proteins 65*, 898.
37. van der Spoel, D., Lindahl, E., Hess, B., Groenhof, G., Mark, A. E., and Berendsen, H. J. C. (2005) *J Comput Chem 26*, 1701–1718.
38. Schuler, L. D., Daura, X., and van Gunsteren, W. F. (2001) *J Comput Chem 22*, 1205–1218.
39. Hummer, G., Garcia, A. E., and Garde, S. (2001) *Proteins 42*, 77–84.
40. Margulis, C. J., Stern, H. A., and Berne, B. J. (2002) *J Phys Chem B 106*, 10748–10752.
41. Woutersen, S. and Hamm, P. (2000) *J Phys Chem B 104*, 11316.
42. Graf, J., Nguyen, P. H., Stock, G., and Schwalbe, H. (2007) *J Am Chem Soc 129*, 1179–1189.
43. Schweitzer-Stenner, R. and Measey, T. J. (2007) *Proc Natl Acad Sci U S A 104*, 6649–6654.
44. Mu, Y., Kosov, D. S., and Stock, G. (2003) *J Phys Chem B 107*, 5064.
45. Gnanakaran, S. and Garcia, A. E. (2003) *J Phys Chem B 107*, 12555–12557.
46. Best, R. B. and Hummer, G. *J Phys Chem B* 2009, *113* (26), 9004–9015.
47. Hinsen, K. (2006) *Proteins 64*, 795–797.
48. Mu, Y., Nguyen, P. H., and Stock, G. (2006) *Proteins 64*, 798–799.
49. Hartigan, J. A. and Wong, M. A. (1979) *Appl Stat 28*, 100–108.
50. Gower, J. C. (1967) *Statistician 17*, 13–28.
51. Becker, O. M. (1997) *Proteins 27*, 213–226.
52. Becker, O. M. and Karplus, M. (1997) *J Chem Phys 106* (4), 1495–1517.
53. Krivov, S. V. and Karplus, M. (2002) *J Chem Phys 117*, 10894–10903.
54. Duan, Y. and Kollman, P. A. (1998) *Science 282*, 740–744.
55. Snow, C. D., Nguyen, H., Pande, V. S., and Gruebele, M. (2002) *Nature (London) 420*, 102.
56. Lei, H., Wu, C., Liu, H., and Duan, Y. (2007) *Proc Natl Acad Sci U S A 104*, 4925–4930.
57. Ensign, D. L., Kasson, P. M., and Pande, V. S. (2007) *J Mol Biol 374*, 806–816.
58. Rajan, A., Freddolino, P. L., and Schulten, K. (2010) *PLoS ONE 5*, e9890.
59. Uversky, V. N. and Fink, A. L. (2006) *Protein misfolding, aggregation and conformational disease*, Springer, New York.
60. Nguyen, P. H., Li, M. S., Stock, G., Straub, J. E., and Thirumalai, D. (2007) *Proc Natl Acad Sci U S A 104*, 111–116.
61. Cecchini, M., Rao, F., Seeber, M., and Caflisch, A. (2004) *J Chem Phys 121*, 10748.
62. Strodel, B., Whittleston, C. S., and Wales, D. J. (2007) *J Am Chem Soc 129*, 16005–16014.

3

LOCAL BACKBONE PREFERENCES AND NEAREST-NEIGHBOR EFFECTS IN THE UNFOLDED AND NATIVE STATES

Joe DeBartolo, Abhishek Jha, Karl F. Freed, and Tobin R. Sosnick

3.1. INTRODUCTION

After being synthesized in cells, nascent proteins perform a range of functions that are typically associated with conformational changes. The largest conformational change executed by a protein during its lifetime is the folding to its native state, the reversible transition from a disordered to a uniquely ordered conformation. This essential folding transition has been the focus of experimental and theoretical studies for more than half a century.

Levinthal posed one of the earliest and fundamental questions regarding the folding transition: if proteins in the disordered state have no preference for a particular conformation, then folding to the native conformation would occur only on astronomical time scales. For example, assume that each residue can adopt any one of three equally probable rotameric conformations in the Ramachandran distribution of backbone ϕ, ψ dihedral angles (Fig. 3.1) [1, 2]. Then a 100-residue protein would have $3^{100} \approx 10^{47}$ possible conformers. Because backbone dihedral angles rotate on subpicosecond time scales, a 100-residue protein would take longer than the age of the universe to find its unique native conformation. In reality, however, the folding time scales for proteins range from microseconds to minutes.

Protein and Peptide Folding, Misfolding, and Non-Folding, First Edition. Edited by Reinhard Schweitzer-Stenner.
© 2012 John Wiley & Sons, Inc. Published 2012 by John Wiley & Sons, Inc.

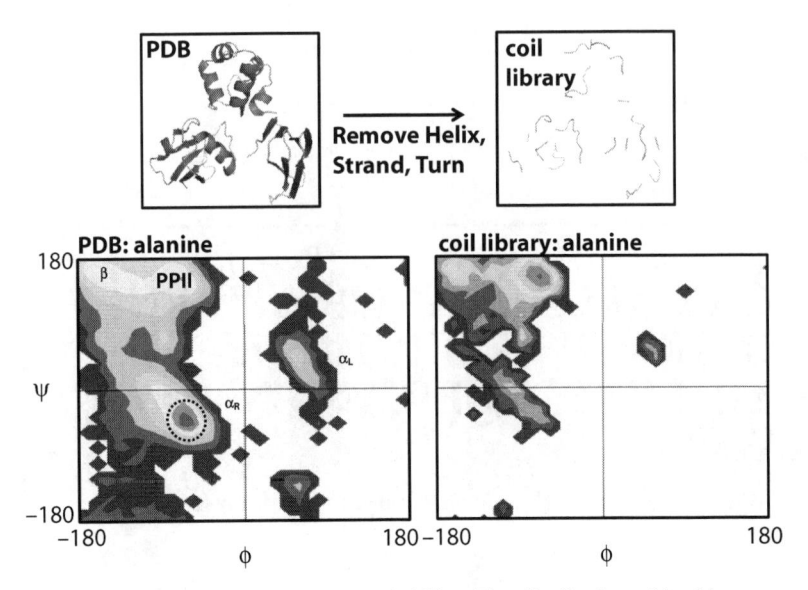

Figure 3.1. Coil library Ramachandran probability. The distribution of backbone φ, ψ torsional angles varies for the entire PDB and for a subset of the PDB that is restricted to coil regions, as depicted for alanine in each respective library. Structure images generated using the program PyMOL. See color insert.

Because "Levinthal's paradox" is predicated on the assumption of a random unbiased search but disagrees with experiment, the logical conclusion is that the conformational search during protein folding must be biased. Indeed, even a small energy bias can rationalize the time scales observed for protein folding [3]. One contributing bias is the preference of the backbone of each amino acid to adopt certain sets of dihedral angles for rotation about the backbone single bonds.

Since the seminal work of Pauling and Wu [4, 5], our views concerning the conformational preferences of amino acids have evolved. We begin by briefly tracing the path of that evolution and then continue with a discussion of our recent applications that explain previous enigmatic experiments concerning the unfolded state by incorporating the influence of local conformational preferences in the unfolded state and along the folding pathway to the native structure. Specific focus is placed on the conformational biases for each of the amino acids and how these biases are influenced by the chemical identity and conformation of the neighboring residues.

3.2. EARLY DAYS: RANDOM COIL—THEORY AND EXPERIMENT

The earliest work considers a self-avoiding random coil model for the unfolded state. However, some recent theoretical and experimental results have been interpreted as suggesting the existence of significant conformational biases in the unfolded state [6–8]. The classical random coil model fails to explain the recent experimental observations including nuclear magnetic resonance (NMR) measurements that indicate that the denatured state has non-random native and non-native structure [7–15]. Nevertheless, a statistical-coil model is shown here to provide a more realistic framework for describing the chemically denatured state of proteins.

Protein denaturation was established as an important phenomenon worthy of research following the experiments of Kauzmann and Simpson in the early 1950s [16]. Their studies of the kinetics of protein denaturation confirm the hypothesis of Wu [5] and Mirsky and Pauling [4] that physical and chemical properties of proteins are a consequence of the structure of protein and are lost upon denaturation. Kauzmann and Simpson characterize the reversible thermal transition of a protein between its native and denatured states and observe that many physical and chemical properties are lost upon melting [16]. The experimental findings of Kauzmann and Simpson clearly establish the need to characterize this melted, denatured state of proteins.

Soon after, the consensus emerged that the denatured state of proteins can effectively be described as self-avoiding random coils. Flory, among others, developed this viewpoint [17]. By definition, a random coil polymer has no preferred backbone conformations because of the assumption that all sterically accessible backbone conformations are separated by barriers with heights of the order of thermal energy.

Kuhn and Guth, as well as Mark, independently provide the earliest theoretical investigations of the properties of polymers in the 1930s by using a statistical treatment for the polymer's configurations [18]. Among other things, the dimension of a polymer chain generally may depend on the nature of solvent. In simplest terms, solvents are classified into three classes vis-à-vis their effect on the dimensions of the polymer. If the solvent promotes chain expansion by favoring chain–solvent interactions over intrachain interactions, it is known as good solvent. On the other hand, a poor solvent promotes chain contraction by favoring intrachain interactions over chain–solvent interactions. Finally, in the early 1950s Flory [19] introduced the idea of a theta solvent in which, on average, these interactions exactly counterbalance each other, leading to unperturbed chain behavior. Flory showed that the dimensions (e.g., the radius of gyration, R_g) of a polypeptide in good solvents often follow the simple scaling law $R_g = aN^b$ for a long polymer with N monomers, where the prefactor a depends on local structure, and the exponent b depends on solvent quality. The exponent b is ~1/3 for collapsed polymers (poor solvent conditions), ~0.6 in good solvent conditions, and 1/2 in ideal θ-solvents.

Although the atomistic modeling of the interaction of proteins with solvent is very complex, a simplified treatment considers denaturing conditions, which contain agents such as urea or guanidinium chloride and are known to produce good solvent conditions. Thus, following Flory's work, the experimental strategy for studying unfolded proteins begins with the measurement of the dimensions of proteins of different sizes under denaturing conditions, where the data are fit to $R_g = aN^b$ in order to determine the exponent b. For example, the observed b of close to 0.6 would imply that the proteins can be treated under denaturing conditions as self-avoiding random coil polymers. The classic studies by Tanford et al. demonstrate by using hydrodynamic methods that the global dimensions of denatured proteins exhibit the size dependence expected for self-avoiding random coil polymers [20, 21]. More recent measurements of the radius of gyration, R_g, using small-angle scattering methods exhibit the same self-avoiding random coil scaling behavior with length, $R_g \propto N^{0.60}$ [22]. These observations are consistent with denatured proteins being self-avoiding random coils in good solvent conditions.

3.3. DENATURED PROTEINS AS SELF-AVOIDING RANDOM COILS

An assumption at the core of random coil models—that all the accessible backbone conformations of the polymer are separated by barriers of the order of thermal energy—is supported by the measurements by Tanford [20, 21] and later by others [22, 23]. These measurements suggest that denatured proteins behave like self-avoiding random coils in good solvent conditions and thus lead to the inference that proteins under denaturing conditions, on an average, are devoid of any structure. However, early NMR studies provide some evidence for residual structure in certain chain regions for both proteins and peptides in the denatured state [24–26]. The limited success of these early NMR experiments failed to alter the perception of the field that denaturing conditions abolish all structure in proteins and that denatured proteins are truly random coils. However, the development of novel NMR methods that are capable of providing site-resolved structural information [7, 9–14] has stimulated a resurgence in the search for evidence of residual structure in denatured state of proteins. One prominent result by Shortle and Ackerman argues for the persistence of some native-like structure in staphylococcal nuclease in strongly denaturing conditions (8 M urea), possibly even encoding the native topology [7]. Subsequently, many more studies [8, 15] have arrived at the same conclusion: proteins in the denatured state contain far richer structural diversity than earlier believed.

The set of two experimental observations, the self-avoiding random coil scaling behavior and the presence of significant amounts of local structure in the unfolded state, lead to what Millett et al. [23] call the "reconciliation problem." However, in his classic review on protein denaturation, Tanford notes that the R_g scaling behavior alone is an insufficient criterion for identifying random coil behavior. More recently, Rose and coworkers explained that even a "deliberately extreme" model of chains composed of native-like segments connected by flexible residues can reproduce self-avoiding random coil scaling behavior [27]. Hence, the recapitulation of the scaling behavior for the self-avoiding chain dimensions provides only a weak test for any unfolded state model. Nevertheless, spectroscopic measurements, such as circular dichroism, indicate that most unfolded states, particularly chemical denatured proteins [22, 28–30], contain little if any hydrogen-bonded secondary structure. Accordingly, the unrealistic native-like segment model is ruled out. More exacting tests are needed, particularly those that involve site-resolved information.

3.4. MODELING THE UNFOLDED STATE

The classical view that denatured proteins can be described by a random coil model may be understood in terms of the simple rotational isomeric state model that applies to chains under theta solvent conditions where it provides the basic insight into their conformational characteristics. There are two primary underlying assumptions of the model. First, all backbone conformations of a random coil are accessible with no bias because the barriers separating them have heights of the order of the thermal energy. The random coil model further assumes the absence of correlations between the local interactions of two monomers of the polymer chain on any length scale. This assumption is also known as Flory's isolated-pair hypothesis. However, strong evidence exists to suggest that both underlying assumptions are flawed.

A fundamental descriptor of a polypeptide's conformation is the set of its allowed backbone dihedral or torsional angles ϕ and ψ for rotation about the single bonds in the backbone of a single amino acid. These angles specify for each residue a location in the Ramachandran plot of ϕ and ψ angles. Ramachandran plots of the frequency of observation of ϕ and ψ differ for each amino acid, first of all because the ϕ,ψ distribution is affected by interactions between the backbone and the side chains. For example, the Ramachandran plots for proline and glycine are quite distinct due to their unusual side chains. The other 18 amino acids exhibit a smaller variation and mainly populate the same three regions (helical, β, and polyproline II [PPII]) but with varying propensity (Fig. 3.2). Hence, amino acids do not adopt all accessible backbone conformations with equal bias. Second, the conformational

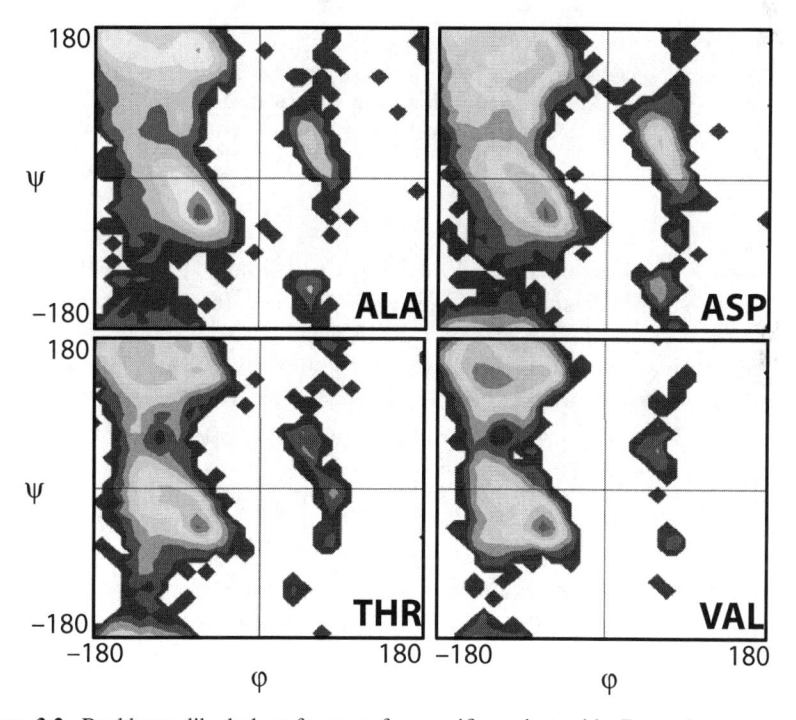

Figure 3.2. Backbone dihedral preferences for specific amino acids. Ramachandran preferences evaluated from experimentally determined protein structures from the PDB display subtle differences between different amino acid types.

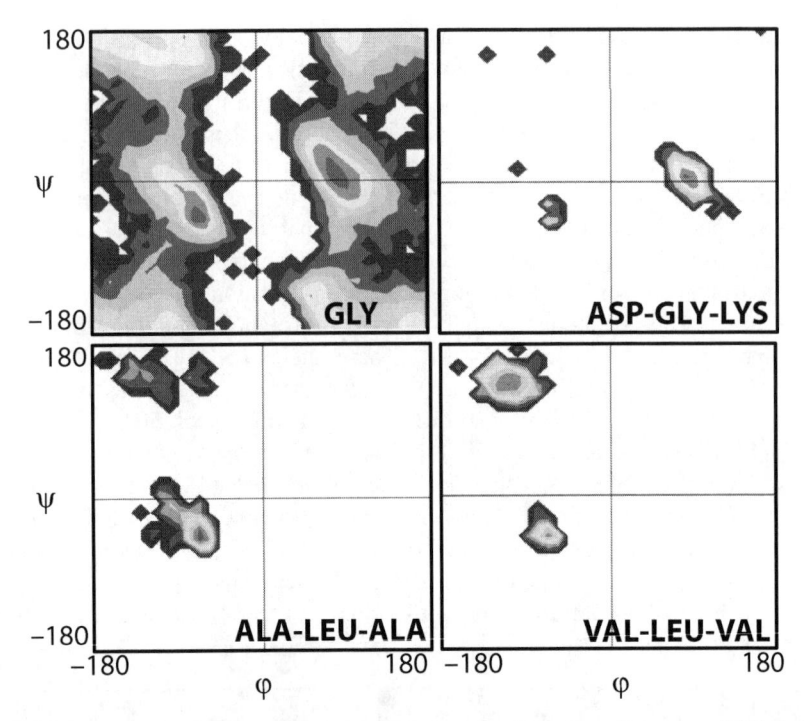

Figure 3.3. Nearest-neighbor effects on native state backbone configuration. PDB-derived Ramachandran distribution of glycine with all neighbors and with N-terminal aspartic acid and C-terminal lysine and leucine when neighbored by alanines (left) and valines (right). Images generated at http://godzilla.uchicago.edu/cgi-bin/rama_all.cgi.

preferences of a residue are also influenced by the chemical identity and conformation of the neighboring residues (Fig. 3.3) [31–40]. These observations of a neighbor dependence contradict the zeroth order Flory isolated-residue hypothesis [19], which states that the conformations adopted by any residue are independent of the chemical identity and conformation of its neighbors.

The rotational isomeric state model provides a useful recipe for computer simulations of the denatured state of proteins, provided excluded-volume (self-avoiding) interactions are retained in addition to the intrinsic conformational preferences of the amino acids and the correlations due to nearest neighbors (NN). Molecular dynamics simulations of peptides have the potential to identify backbone preferences and delineate NN effects. However, such procedures are computationally challenging because of the large number of distinct combinations of amino acids with their immediate NNs (8000 for triplets). In addition, the results of these simulations are very sensitive to the choice of force fields [31, 41–47].

A more fruitful approach is to utilize a subset of residues located outside of regular secondary structures in crystal structures [33, 48–50]. Early studies [33, 36, 48, 49, 51–53] assume that the use of statistical averages from the entire database of folded protein structures would average over many environments, thereby largely eliminating the influence of context. Although such an approximation cannot completely account for contextual influences, the database contains sterically allowed conformations that directly reflect the chemical character of the individual amino acids. Subsequently, only a subset of the database, termed the "coil library" and containing data only for residues not in hydrogen bonded secondary structure (i.e., sheets, helices and turns), has been retained to provide a better reproduction of the intrinsic conformational preferences and to quantify the NN effects [54, 55].

Ramachandran preferences for individual amino acid types vary in subtle ways (Fig. 3.2), but even more dramatic is the difference observed between libraries constructed from data in the Protein Data Bank (PDB) based on the retention of information from defined secondary structures and on the (non-hydrogen bonded) coil regions (Fig. 3.1). Indeed, the Ramachandran distribution of the coil library displays a strong preference for the extended PPII region for all amino acid types [54]. The PPII preference is also experimentally observed in tripeptides [56–64] and in denatured proteins, naturally inviting modeling of the unfolded state based on the coil library distributions.

We have generated ensembles of unfolded structures for which the Ramachandran basin frequencies for each amino acid are assigned according to the identity of the residue, using the explicit ϕ, ψ angles obtained from a PDB-based coil library and a non-overlapping excluded-volume interaction [55]. The Ramachandran basin assignment for each amino acid either uses or neglects information on the chemical and conformational identity of neighboring residues. A comparison of these ensembles for several protein sequences with experimental data provides insight both into the ability of these ensembles to recapitulate the experimental results and into the importance of NN effects.

The influence of NNs on the unfolded state is evident when comparing unfolded state models generated with and without NN effects with experimental residual dipolar couplings (RDC) of denatured proteins (Fig. 3.4). RDCs provide a powerful tool for probing the structure of denatured proteins [6, 7, 65–69]. The RDC is a measure of the orientation of bond vectors (generally backbone N–H bonds) relative to an alignment tensor fixed in the molecular frame. Although RDCs generally average to zero when the molecules freely tumble, RDCs no longer vanish when proteins are confined in a weakly aligning anisotropic media, such as compressed acrylamide gels [7, 9–14, 70, 71].

The RDCs calculated from models built using NN effects [55] exhibit a significantly stronger correlation with experimental RDCs for chemically denatured apomyoglobin [69] (Fig. 3.4) and other proteins [6, 67] compared with models built without NN effects. This finding emphasizes the extent to which the local amino acid sequence imparts a conformational preference to the local backbone of the chain, a preference that can be modeled using knowledge-based statistical libraries and energies.

It is important to note that the conformations found in the denatured state often are not strongly biased toward the native structure [55, 72], which only exacerbates the Levinthal search problem. Since the unfolded state contains a high abundance

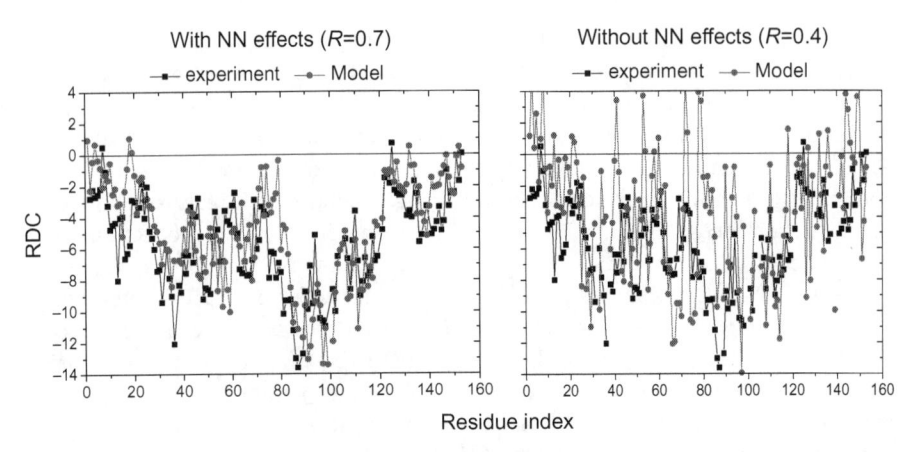

Figure 3.4. Nearest-neighbor effects on the unfolded state. Calculated (gray) and experimental (black) RDC values for apoMb in 10% acrylamide (experimental data from [69]). The calculated RDCs are obtained by averaging 5000 structures generated by using a PDB-based statistical potential derived from the coil library and by including excluded-volume constraints. RDCs are calculated without (left) and with (right) nearest-neighbor effects on backbone conformations. RDCs are presented for all of the residues except prolines and those for which there are no amide proton (NH) resonances.

of PPII and β conformers, this search problem is most pronounced for helical proteins because the unfolded chain spends considerable time searching unproductively through non-native states. Possible solutions to the search problem include sequential stabilization wherein tertiary interactions support weak local biases that enable the buildup of structure [73–76], a possibility that is discussed in the following section.

3.5. NN EFFECTS IN PROTEIN STRUCTURE PREDICTION

An understanding of the influence of NN amino acid and secondary-structure identity on the properties of proteins in the native state has important implications for protein structure prediction and the analysis of protein folding pathways. Specifically, the large conformational search space accessible to an amino acid, as observed in experimentally determined protein structures, is greatly reduced by considering the identity of neighboring amino acids. For example, data for glycine gathered without specifying NN information exhibit an equal preference for left- and right-handed helices and turns (Fig. 3.3). However, after specifying that the glycine residue has an N-terminal aspartic acid and a C-terminal lysine, left-handed conformations are strongly preferred over right-handed ones. Thus, certain scenarios have the NN information reducing the accessible areas of Ramachandran search space available to a position by over 50%. In other cases, the different neighbors can cause an amino acid preference to switch from one preferred Ramachandran region to another, as found for leucine, which prefers a helical backbone when neighbored by alanines, but an extended backbone when neighbored by valines (Fig. 3.3). Thus, a statistical analysis of real protein structures guides our knowledge of local conformational preference.

The combination of first principles methods with statistical information has enabled us to construct a computational model to predict protein pathways and structure that accommodates both a knowledge-based approach and a more fundamental methodology. Our focus is on whether protein folding can be accurately depicted using only the heavy atoms in the backbone and the C_β side chain atom. This simplified representation greatly diminishes the conformational search as it only involves the backbone dihedral angles, ϕ and ψ, with no need for the very costly consideration of side-chain rotamers. To recapture the lost information concerning the side chains, local interactions are incorporated by sampling the backbone dihedral angles using a backbone rotamer library, constructed from the PDB, that tabulates dihedral angles for residues according to their amino acid identities and Ramachandran basin assignments (e.g., helical, β, or PPII). Secondly, the tertiary interactions are scored using a statistical potential that is constructed from residue-dependent pair interactions for all the atoms in the backbone and for all the C_β atoms [77].

An initial test of this methodology uses a simplified description in which structures are obtained by minimizing the scoring function with a computationally rapid Monte Carlo simulated annealing (MCSA) algorithm using the PDB-based backbone rotamer sampling where each residue is constrained to their native Ramachandran basins. Although advanced knowledge of the native Ramachandran basin precludes this study from being an *ab initio* structure prediction, the protein may still adopt a huge number of non-native, highly extended conformations with individual residues that are constrained to a single broad Ramachandran basin. Hence, this simplified treatment provides an excellent first test of the general approach and assumptions [78]. Only when NN effects are included do the lowest energy structures generally fall within 4 Å of the native backbone RMSD, despite the initial configuration being highly expanded with an average root mean square deviation (RMSD) $\geq$ 10 Å.

3.6. UTILIZING FOLDING PATHWAYS FOR STRUCTURE PREDICTION

The enormous conformational search space available to the polypeptide chain implied by the Levinthal paradox [79] suggests that proteins fold along a defined pathway. Additionally, native state hydrogen exchange experiments demonstrate that subunits of secondary structures, called foldons, form cooperatively [80] and sequentially assemble along a folding pathway [73, 74, 76, 81]. Considering this evidence, structure prediction methods may take advantage of the sequential process by incorporating information concerning folding pathways. Conversely, it may be possible that pathway information can be extracted from folding simulations in order to obtain a better understanding of the physical mechanism through which proteins fold.

Keeping folding pathways in mind, we have developed a significantly improved version of our structure prediction algorithm which can fold single domain proteins without foreknowledge of the native Ramachandran basins and which uses an approach intended to mimic an authentic folding pathway. Our "ItFix" method [72] incorporates information about neighboring amino acids in the sampling of ϕ,ψ angles and by iteratively specifying NN secondary structure to restrict the Ramachandran sampling distribution at each position. Successive rounds of folding are used to couple the formation of secondary structure to the formation of tertiary structure elements, resulting in a refined definition of secondary structure for the final round of folding. This method is significantly enhanced in ItFix–SPEED (Fig. 3.5), which averages the Ramachandran distribution and NN identities across a multiple sequence alignment, producing a ϕ,ψ distribution that is more likely to contain the native angles [82].

ItFix permits the extraction of a folding pathway by not requiring input of any exogenous 2° structure prediction or homology information, and by allowing the whole chain to interact throughout the entire folding process. Furthermore, the sampling moves involve changes only in a single pair of dihedral angles (ϕ,ψ), with the angles taken from the PDB and depending on the neighboring residues' identity and secondary structure. Indeed, the order of fixing the secondary structure in ItFix mimics the sequence of foldons that add to existing structures in a process of sequential stabilization that may resemble the pathway taken by authentic proteins. In contrast to methods that use preformed fragments or exogenous 2° structure predictions where the connection to the authentic pathway is murky at best, the Itfix protocol begins with an initial unstructured chain, and the buildup of structure evolves out of the folding process. Hence, the order of fixing of structural elements may recapitulate major features of the authentic pathway that is followed as the real chain progresses along the free energy surface.

The order of fixing of structural subunits in the α/β protein ubiquitin accords with the experimental folding pathway (Fig. 3.6). A notable feature in the order of events is the early formation of the parallel β-strand interaction between the amino and the carboxy termini. This long-range contact occurs prior to the 2° structure assignment of 30 intervening residues and is a possible occurrence with our method because the simulation includes the entire chain at all times.

The ability to model the influence of NN effects on protein folding pathways has important implications for the resolution of Levinthal's paradox. Local conformational biases influence the structure of unfolded and native proteins and therefore are involved in the pathway-directed search through an enormous conformational search space.

3.7. NATIVE STATE MODELING

The accuracy achieved in modeling proteins in their native state is enhanced through the inclusion of NN information in a manner similar to that discussed earlier. The native state preferences for regions of the Ramachandran map for individual amino

NATIVE STATE MODELING 89

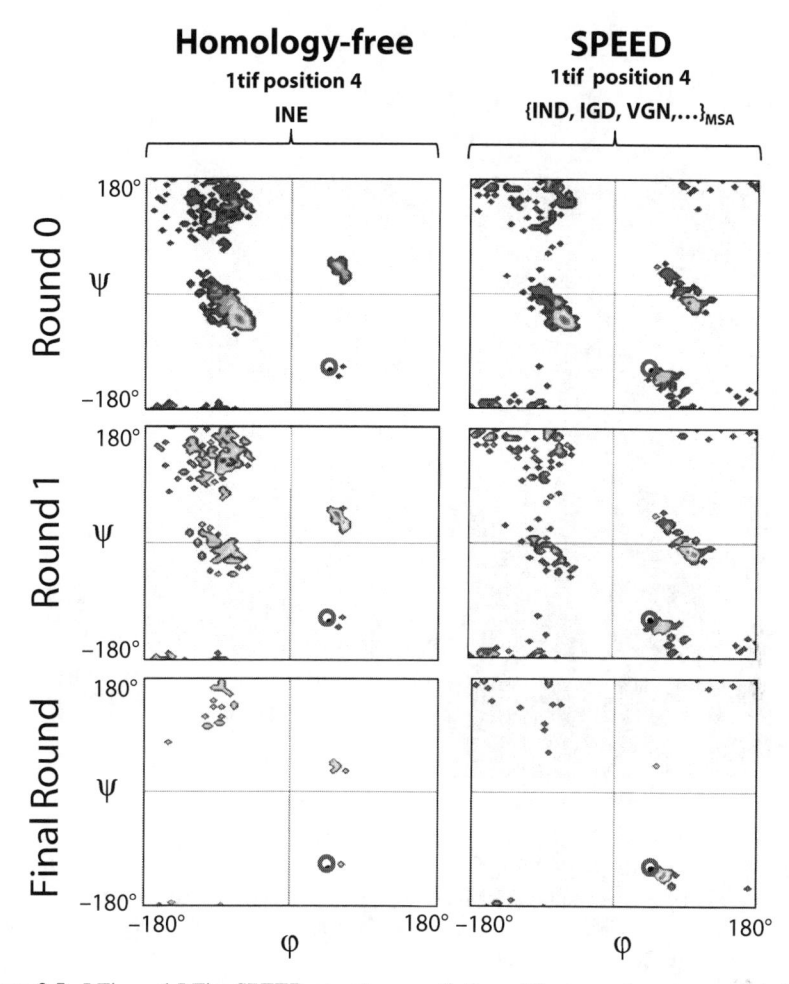

Figure 3.5. ItFix and ItFix–SPEED structure predictions. The secondary structure is iteratively determined through successive rounds of folding in the ItFix method. The NN-dependent Ramachandran distribution at a position is also conditional on the secondary-structure definition of the current round. Homology-free Ramachandran distributions consider the target sequence amino acid identities of the position and its neighbors. The addition of evolutionary information by averaging the Ramachandran distribution across a sequence alignment (ItFix–SPEED) produces a more native distribution at each position. The native angle is highlighted with a circle. In the example shown, the native angles are located in the left-handed PPII region of the Ramachandran map, which has a low probability in the homology-free distribution. In the ItFix–SPEED distribution, the native region has a higher probability in the first round (no secondary structure) and increases in probability as more secondary structure is defined at that position.

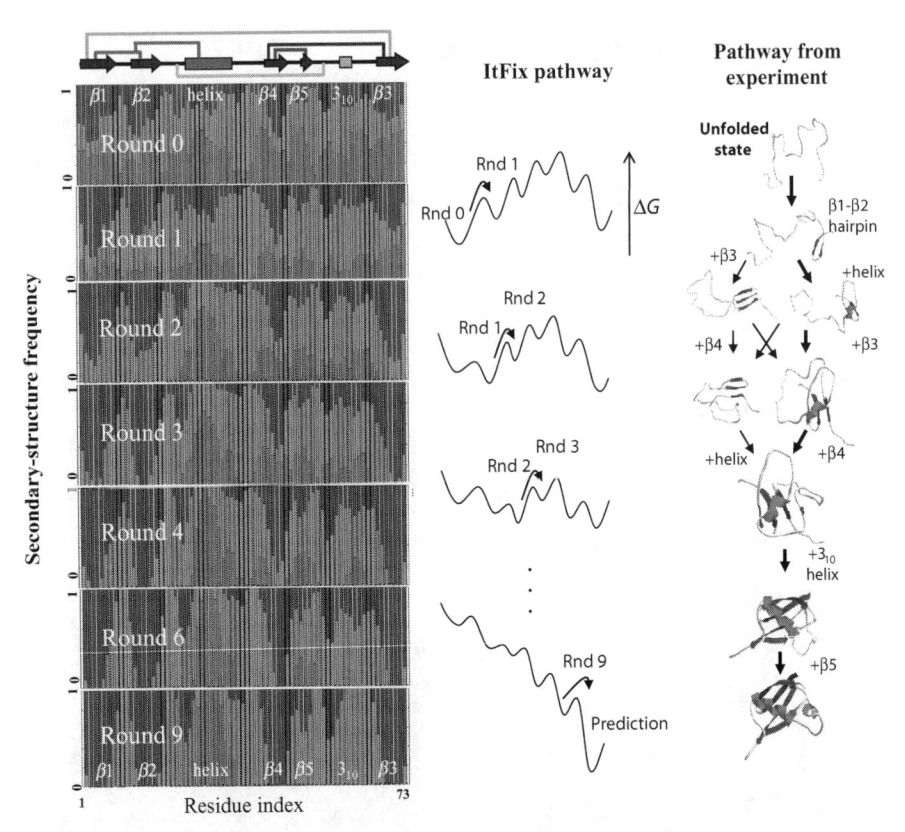

Figure 3.6. Computational and experimental folding pathways. The folding and prediction of protein structure using the ItFix algorithm begins with no secondary-structure assignments, but through subsequent rounds of folding incorporates tertiary-structure information in the formation of secondary-structure specifications for subsequent rounds. The position dependence of the secondary-structure frequencies at the end of each round, extended (blue), helix (red), and coil (green), are shown on the left. A single color bar represents a residue assigned to a single secondary-structure type (native secondary structure shown at top, along with long-range contacts). As the rounds progress, uncertainties in 2° structure diminish. This introduces a bias in the free energy surface which progressively tilts toward the native state through the folding rounds (middle diagram). The major steps in the proposed folding pathway [73, 81] are similar to the order of structure fixing over the multiple rounds: The hairpin forms, followed by the helix and $\beta 3$ strand, and then $\beta 4$. The final two events are the folding of the 3–10 helix and $\beta 5$. Their formation appears in some trajectories but not at a high enough frequency to be fixed. See color insert.

acids are, as for the unfolded state, dependent on NN identity. In some examples, this dependence induces a switch of the preferences from one region of Ramachandran space to another, such as the situation where a leucine prefers a helical backbone when neighbored by alanines, but prefers an extended backbone when neighbored by valines (Fig. 3.2). A suite of web-based applications (available

NATIVE STATE MODELING

at http://godzilla.uchicago.edu/pages/projects.html) demonstrates this and other aspects of the NN effects by displaying the Ramachandran distribution given the amino acid identity and the secondary structures and identities of its neighbors. The amino acids can be chosen as any of the 20 types or as averages over amino acids, and the secondary structures are defined as helix, strand, or the five different coil types defined by the algorithm Define Secondary Structure of Proteins (DSSP) [83]. Alternatively, the secondary structure can be left unspecified in determining the Ramachandran distribution at a position or its neighbors.

A related application inverts the above process by calculating the probability that each of the 20 amino acids is found for a specific region of the Ramachandran plot (Fig. 3.7). For example, by selecting the left-handed helix region, the application identifies an asparagine with an N-terminal leucine and a C-terminal glycine as the sequence most likely to be found with this geometry in the PDB. It is important to note that this sequence may not have a strong bias toward this conformation (e.g., it may prefer other conformations in the unfolded state). Rather, this tool identifies the sequence most likely to be found with this conformation in native proteins.

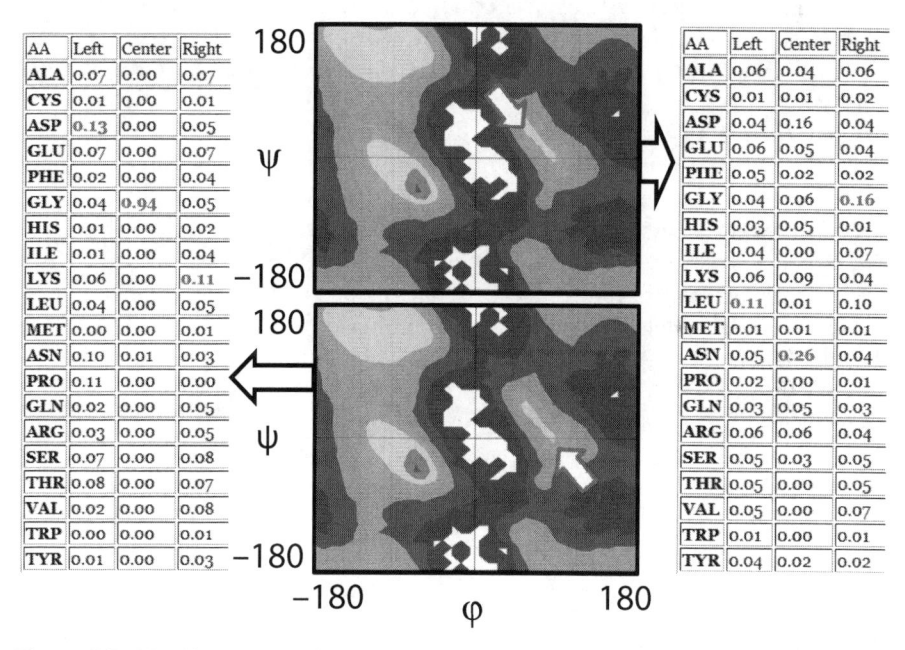

AA	Left	Center	Right
ALA	0.07	0.00	0.07
CYS	0.01	0.00	0.01
ASP	0.13	0.00	0.05
GLU	0.07	0.00	0.07
PHE	0.02	0.00	0.04
GLY	0.04	0.94	0.05
HIS	0.01	0.00	0.02
ILE	0.01	0.00	0.04
LYS	0.06	0.00	0.11
LEU	0.04	0.00	0.05
MET	0.00	0.00	0.01
ASN	0.10	0.01	0.03
PRO	0.11	0.00	0.00
GLN	0.02	0.00	0.05
ARG	0.03	0.00	0.05
SER	0.07	0.00	0.08
THR	0.08	0.00	0.07
VAL	0.02	0.00	0.08
TRP	0.00	0.00	0.01
TYR	0.01	0.00	0.03

AA	Left	Center	Right
ALA	0.06	0.04	0.06
CYS	0.01	0.01	0.02
ASP	0.04	0.16	0.04
GLU	0.06	0.05	0.04
PHE	0.05	0.02	0.02
GLY	0.04	0.06	0.16
HIS	0.03	0.05	0.01
ILE	0.04	0.00	0.07
LYS	0.06	0.09	0.04
LEU	0.11	0.01	0.10
MET	0.01	0.01	0.01
ASN	0.05	0.26	0.04
PRO	0.02	0.00	0.01
GLN	0.03	0.05	0.03
ARG	0.06	0.06	0.04
SER	0.05	0.03	0.05
THR	0.05	0.00	0.05
VAL	0.05	0.00	0.07
TRP	0.01	0.00	0.01
TYR	0.04	0.02	0.02

Figure 3.7. Identifying the preferred amino acid sequence for a particular native geometry. A web server presents the probability of finding each of the 20 amino acids for a given $10° \times 10°$ position in the Ramachandran map. A left-handed helix prefers asparagines, whereas left-handed turns prefer glycine. The most probable neighbor amino acid identities also vary given the position in the map. Maps and tables are generated at http://godzilla.uchicago.edu/rama_AA.html. See color insert.

3.8. SECONDARY-STRUCTURE PROPENSITIES: NATIVE BACKBONES IN UNFOLDED PROTEINS

The preference for an amino acid to reside in a helix or β-sheet can be predicted solely from its intrinsic tendency to adopt this conformation in a stringent coil library composed only of residues that lie outside of helices, strands, and hydrogen bonded turns (Fig. 3.1). The helical and β-basin preferences for the amino acids in this coil library are compared with their relative frequencies in authentic helical and sheet structures, as originally described by Chou and Fasman, where the preference is defined as the frequency with which a residue appears in a given secondary structure [51].

The correlation between the stringent coil library and Chou and Fasman frequencies for helices is poor. This perplexing deviation between the coil frequencies and the Chou and Fasman helical frequencies is removed when the former are derived using the much narrower ϕ,ψ region appropriate for regular helical structures, rather than the larger region employed for the full helical basin, an extended region that also contains turn geometries. Given this more restricted helix region (as identified in Fig. 3.1), the correlation becomes strong and accounts for the Chou and Fasman helical propensity scale as well as for the experimental propensity (ΔG) scale observed in guest–host systems of all non-Pro residues [84] (Fig. 3.8). Therefore, the helical propensity in Figure 3.8 equates to the preference for the chain to adopt a true helical geometry, rather than any angle in the helical basin, a set that includes other geometries.

Compared with helical propensities, β-sheet propensities are believed to be less well defined [85, 86], and even the existence of intrinsic β propensities has been questioned [87, 88]. The correlation between our coil library and Chou and Fasman frequencies is even stronger for β-sheets than for α-helical frequencies (Fig. 3.8). The correlation between β-basin populations in the coil library and the frequency of true β-conformers in β-sheets excludes residues found in β-sheets but which actually occupy the PPII region. This strong correlation found for β-sheets does not contradict the importance of context in influencing β-propensities. The coil library presumably averages out such contextual effects, thereby unmasking the intrinsic backbone preferences.

In summary, we are able to accurately recapitulate both helix and sheet frequencies for the amino acids in structured regions using their conformation biases in a stringent coil library. Therefore, structural propensities for α-helices and β-sheets can be rationalized solely by local effects.

3.9. CONCLUSIONS

Locally determined backbone conformational preferences exert a strong influence on protein structure, folding, and energetics. In combination with a tertiary context, these local conformational preferences help determine the structure and stability of the native state. As such, a complete understanding of the mechanism through which

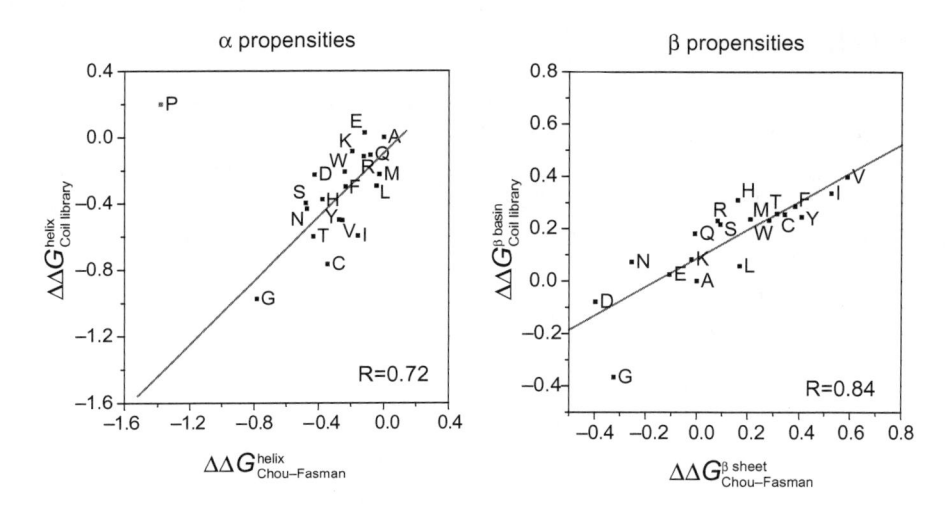

Figure 3.8. Correlation between coil library and structural propensities. Left: Correlation between the observed frequencies in the helical basin $\Delta\Delta G_{i,coil\ library}^{helix} = -RT\ln(\text{Prob}_{i,coil\ library}^{helix}/\text{Prob}_{alanine,coil\ library}^{helix})$, as derived from the coil library, and the Chou and Fasman (C–F) helical propensities, normalized to the alanine frequencies, and defined by $\Delta\Delta G_{i,C-F}^{helix} = -RT\ln(\text{Prob}_{i,helix}^{authentic\ helix}/\text{Prob}_{alanine,helix}^{authentic\ helix})$. The index i refers to the ith amino acid. Right: Corresponding plot for the β-basin and β-sheet propensities. Units are kcal/mol.

amino acid sequence encodes local conformational preferences is a vital component of the resolution of the protein folding problem. This fact is underscored by Levinthal's paradox, which suggests that proteins fold along a pathway due to the theoretically enormous conformational search space to be sampled in an empirically short amount of time. Thus, any consideration of local conformational preference in the unfolded state must appreciate its contribution toward the pathway-based nature through which proteins reach their native fold.

For this reason, we have traced the history of the influence of local conformation on the unfolded state from groundbreaking theoretical work to present-day modeling of the unfolded state and pathway-based protein folding. What is most clear is that the unfolded state is tremendously influenced by the effect of neighboring residues on local backbone conformations, and likewise the pathway through which proteins fold is in part dictated by NNs as well. Consequently, local conformational preferences of the protein backbone clearly narrow and bias the chain's search on the folding energy landscape.

ACKNOWLEDGMENTS

We thank members of our group and colleagues for enlightening discussions. This work was supported by National Institutes of Health research grants.

REFERENCES

1. Ramachandran, G. N. and Sasisekharan, V. (1968) Conformation of polypeptides and proteins, *Adv Protein Chem 23*, 283–438.
2. Ramachandran, G. N., Ramakrishnan, C., and Sasisekharan, V. (1963) Stereochemistry of polypeptide chain configurations, *J Mol Biol 7*, 95–99.
3. Zwanzig, R., Szabo, A., and Bagchi, B. (1992) Levinthal's paradox, *Proc Natl Acad Sci U S A 89*, 20–22.
4. Mirsky, A. E. and Pauling, L. (1931) On the structure of native, denatured, and cogulated proteins, *Proc Natl Acad Sci U S A 22*, 439.
5. Wu, H. (1931) Studies on denaturation of proteins. XIII. A theory of denaturation, *Chin J Physiol 5*, 321–344.
6. Ohnishi, S., Lee, A. L., Edgell, M. H., and Shortle, D. (2004) Direct demonstration of structural similarity between native and denatured eglin C, *Biochemistry 43*, 4064–4070.
7. Shortle, D. and Ackerman, M. S. (2001) Persistence of native-like topology in a denatured protein in 8 M urea, *Science 293*, 487–489.
8. Cho, J. H. and Raleigh, D. P. (2009) Experimental characterization of the denatured state ensemble of proteins, *Methods Mol Biol 490*, 339–351.
9. Prestegard, J. H., Al-Hashimi, H. M., and Tolman, J. R. (2000) NMR structures of biomolecules using field oriented media and residual dipolar couplings, *Q Rev Biophys 33*, 371–424.
10. Bax, A. (2003) Weak alignment offers new NMR opportunities to study protein structure and dynamics, *Protein Sci 12*, 1–16.
11. Tjandra, N. (1999) Establishing a degree of order: Obtaining high-resolution NMR structures from molecular alignment, *Struct Fold Des 7*, R205–R211.
12. Meiler, J., Blomberg, N., Nilges, M., and Griesinger, C. (2000) A new approach for applying residual dipolar couplings as restraints in structure elucidation, *J Biomol NMR 16*, 245–252.
13. Skrynnikov, N. R. and Kay, L. E. (2000) Assessment of molecular structure using frame-independent orientational restraints derived from residual dipolar couplings, *J Biomol NMR 18*, 239–252.
14. Delaglio, F., Kontaxis, G., and Bax, A. (2000) Protein structure determination using molecular fragment replacement and NMR dipolar couplings, *J Am Chem Soc 122*, 2142–2143.
15. McCarney, E. R., Kohn, J. E., and Plaxco, K. W. (2005) Is there or isn't there? The case for (and against) residual structure in chemically denatured proteins, *Crit Rev Biochem Mol Biol 40*, 181–189.
16. Simpson, R. B. and Kauzmann, W. (1953) The kinetics of protein denaturation, *J Am Chem Soc 75*, 5139–5192.
17. Flory, P. J. (1969) *Statistical mechanics of chain molecules*, Wiley, New York.
18. Yamakawa, H. (1971) *Modern theory of polymer solutions*, Harper & Row, San Francisco.
19. Flory, P. J. (1953) *Principles of polymer chemistry*, Cornell University Press, Ithaca, NY.
20. Tanford, C. (1970) Protein denaturation. C. Theoretical models for the mechanism of denaturation, *Adv Protein Chem 24*, 1–95.

REFERENCES

21. Tanford, C., Aune, K. C., and Ikai, A. (1973) Kinetics of unfolding and refolding of proteins. III. Results for lysozyme, *J Mol Biol 73*, 185–197.
22. Kohn, J. E., Millett, I. S., Jacob, J., Zagrovic, B., Dillon, T. M., Cingel, N., Dothager, R. S., Seifert, S., Thiyagarajan, P., Sosnick, T. R., Hasan, M. Z., Pande, V. S., Ruczinski, I., Doniach, S., and Plaxco, K. W. (2004) Random-coil behavior and the dimensions of chemically unfolded proteins, *Proc Natl Acad Sci U S A 101*, 12491–12496.
23. Millett, I. S., Doniach, S., and Plaxco, K. W. (2002) Toward a taxonomy of the denatured state: Small angle scattering studies of unfolded proteins, *Adv Protein Chem 62*, 241–262.
24. Neri, D., Billeter, M., Wider, G., and Wuthrich, K. (1992) NMR determination of residual structure in a urea denatured protein, the 434 repressor, *Science 257*, 1559–1563.
25. Osterhout, J. J., Baldwin, R. L., York, E. J., Stewart, J. M., Dyson, H. J., and Wright, P. E. (1989) 1H NMR studies of the solution conformations of an analogue of the C-peptide of ribonuclease A, *Biochemistry 28*, 7059–7064.
26. Dyson, H. J. and Wright, P. E. (1991) Defining solution conformations of small linear peptides, *Annu Rev Biophys Biophys Chem 20*, 519–538.
27. Fitzkee, N. C. and Rose, G. D. (2004) Reassessing random-coil statistics in unfolded proteins, *Proc Natl Acad Sci U S A 101*, 12497–12502.
28. Tanford, C. (1968) Protein denaturation, *Adv Protein Chem 23*, 121–282.
29. Jacob, J., Krantz, B., Dothager, R. S., Thiyagarajan, P., and Sosnick, T. R. (2004) Early collapse is not an obligate step in protein folding, *J Mol Biol 338*, 369–382.
30. Millet, I. S., Townsley, L. E., Chiti, F., Doniach, S., and Plaxco, K. W. (2002) Equilibrium collapse and the kinetic "foldability" of proteins, *Biochemistry 41*, 321–325.
31. Zaman, M. H., Shen, M. Y., Berry, R. S., Freed, K. F., and Sosnick, T. R. (2003) Investigations into sequence and conformational dependence of backbone entropy, inter-basin dynamics and the Flory isolated-pair hypothesis for peptides, *J Mol Biol 331*, 693–711.
32. Pappu, R. V., Srinivasan, R., and Rose, G. D. (2000) The Flory isolated-pair hypothesis is not valid for polypeptide chains: Implications for protein folding, *Proc Natl Acad Sci U S A 97*, 12565–12570.
33. Smith, L. J., Bolin, K. A., Schwalbe, H., MacArthur, M. W., Thornton, J. M., and Dobson, C. M. (1996) Analysis of main chain torsion angles in proteins: Prediction of NMR coupling constants for native and random coil conformations, *J Mol Biol 255*, 494–506.
34. Keskin, O., Yuret, D., Gursoy, A., Turkay, M., and Erman, B. (2004) Relationships between amino acid sequence and backbone torsion angle preferences, *Proteins 55*, 992–998.
35. Penkett, C. J., Redfield, C., Dodd, I., Hubbard, J., McBay, D. L., Mossakowska, D. E., Smith, R. A., Dobson, C. M., and Smith, L. J. (1997) NMR analysis of main-chain conformational preferences in an unfolded fibronectin-binding protein, *J Mol Biol 274*, 152–159.
36. Gibrat, J. F., Garnier, J., and Robson, B. (1987) Further developments of protein secondary structure prediction using information theory: New parameters and consideration of residue pairs, *J Mol Biol 198*, 425–443.
37. Kang, H. S., Kurochkina, N. A., and Lee, B. (1993) Estimation and use of protein backbone angle probabilities, *J Mol Biol 229*, 448–460.

38 Hagarman, A., Measey, T. J., Mathieu, D., Schwalbe, H., and Schweitzer-Stenner, R. (2010) Intrinsic propensities of amino acid residues in GxG peptides inferred from amide I' band profiles and NMR scalar coupling constants, *J Am Chem Soc 132*, 540–551.

39 Pizzanelli, S., Forte, C., Monti, S., Zandomeneghi, G., Hagarman, A., Measey, T. J., and Schweitzer-Stenner, R. (2010) Conformations of phenylalanine in the tripeptides AFA and GFG probed by combining MD simulations with NMR, FTIR, polarized Raman, and VCD spectroscopy, *J Phys Chem B 114*, 3965–3978.

40 Chen, K., Liu, Z., Zhou, C., Shi, Z., and Kallenbach, N. R. (2005) Neighbor effect on PPII conformation in alanine peptides, *J Am Chem Soc 127*, 10146–10147.

41 Zaman, M. H., Shen, M. Y., Berry, R. S., and Freed, K. F. (2003) Computer simulation of met-enkephalin using explicit atom and united atom potentials: Similarities, differences, and suggestions for improvement, *J Phys Chem B 107*, 1685–1691.

42 Mu, Y. G., Kosov, D. S., and Stock, G. (2003) Conformational dynamics of trialanine in water. 2. Comparison of AMBER, CHARMM, GROMOS, and OPLS force fields to NMR and infrared experiments, *J Phys Chem B 107*, 5064–5073.

43 Hu, H., Elstner, M., and Hermans, J. (2003) Comparison of a QM/MM force field and molecular mechanics force fields in simulations of alanine and glycine "dipeptides" (Ace-Ala-Nme and Ace-Gly-Nme) in water in relation to the problem of modeling the unfolded peptide backbone in solution, *Proteins 50*, 451–463.

44 Garcia, A. E. and Sanbonmatsu, K. Y. (2002) Alpha-helical stabilization by side chain shielding of backbone hydrogen bonds, *Proc Natl Acad Sci U S A 99*, 2782–2787.

45 Duan, Y., Wu, C., Chowdhury, S., Lee, M. C., Xiong, G., Zhang, W., Yang, R., Cieplak, P., Luo, R., Lee, T., Caldwell, J., Wang, J., and Kollman, P. (2003) A point-charge force field for molecular mechanics simulations of proteins based on condensed-phase quantum mechanical calculations, *J Comput Chem 24*, 1999–2012.

46 Zagrovic, B., Lipfert, J., Sorin, E. J., Millett, I. S., van Gunsteren, W. F., Doniach, S., and Pande, V. S. (2005) Unusual compactness of a polyproline type II structure, *Proc Natl Acad Sci U S A 102*, 11698–11703.

47 Best, R. B. and Hummer, G. (2009) Optimized molecular dynamics force fields applied to the helix-coil transition of polypeptides, *J Phys Chem B 113*, 9004–9015.

48 Munoz, V. and Serrano, L. (1994) Intrinsic secondary structure propensities of the amino acids, using statistical phi-psi matrices: Comparison with experimental scales, *Proteins 20*, 301–311.

49 Swindells, M. B., MacArthur, M. W., and Thornton, J. M. (1995) Intrinsic phi, psi propensities of amino acids, derived from the coil regions of known structures, *Nat Struct Biol 2*, 596–603.

50 Avbelj, F. and Baldwin, R. L. (2003) Role of backbone solvation and electrostatics in generating preferred peptide backbone conformations: Distributions of phi, *Proc Natl Acad Sci U S A 100*, 5742–5747.

51 Chou, P. Y. and Fasman, G. D. (1974) Conformational parameters for amino acids in helical, beta-sheet, and random coil regions calculated from proteins, *Biochemistry 13*, 211–222.

52 Smith, L. J., Fiebig, K. M., Schwalbe, H., and Dobson, C. M. (1996) The concept of a random coil. Residual structure in peptides and denatured proteins, *Fold Des 1*, R95–106.

REFERENCES

53 Stites, W. E. and Pranata, J. (1995) Empirical evaluation of the influence of side chains on the conformational entropy of the polypeptide backbone, *Proteins 22*, 132–140.

54 Jha, A. K., Colubri, A., Zaman, M. H., Koide, S., Sosnick, T. R., and Freed, K. F. (2005) Helix, sheet, and polyproline II frequencies and strong nearest neighbor effects in a restricted coil library, *Biochemistry 44*, 9691–9702.

55 Jha, A. K., Colubri, A., Freed, K. F., and Sosnick, T. R. (2005) Statistical coil model of the unfolded state: Resolving the reconciliation problem, *Proc Natl Acad Sci U S A 102*, 13099–13104.

56 Shi, Z., Olson, C. A., Rose, G. D., Baldwin, R. L., and Kallenbach, N. R. (2002) Polyproline II structure in a sequence of seven alanine residues, *Proc Natl Acad Sci U S A 99*, 9190–9195.

57 Shi, Z., Woody, R. W., and Kallenbach, N. R. (2002) Is polyproline II a major backbone conformation in unfolded proteins? *Adv Protein Chem 62*, 163–240.

58 Eker, F., Griebenow, K., and Schweitzer-Stenner, R. (2003) Stable conformations of tripeptides in aqueous solution studied by UV circular dichroism spectroscopy, *J Am Chem Soc 125*, 8178–8185.

59 Pappu, R. V. and Rose, G. D. (2002) A simple model for polyproline II structure in unfolded states of alanine-based peptides, *Protein Sci 11*, 2437–2455.

60 Creamer, T. P. and Campbell, M. N. (2002) Determinants of the polyproline II helix from modeling studies, *Adv Protein Chem 62*, 263–282.

61 Dukor, R. K. and Keiderling, T. A. (1991) Reassessment of the random coil conformation: Vibrational CD study of proline oligopeptides and related polypeptides, *Biopolymers 31*, 1747–1761.

62 Krimm, S. and Tiffany, M. L. (1974) The circular dichroism spectrum and structure of unordered polypeptides and proteins, *Isr J Chem 12*, 189–200.

63 Wilson, G., Hecht, L., and Barron, L. D. (1996) Residual structure in unfolded proteins revealed by Raman optical activity, *Biochemistry 35*, 12518–12525.

64 Keiderling, T. A. and Xu, Q. (2002) Unfolded peptides and proteins studied with infrared absorption and vibrational circular dichroism spectra, *Adv Protein Chem 62*, 111–161.

65 Ohnishi, S. and Shortle, D. (2003) Observation of residual dipolar couplings in short peptides, *Proteins 50*, 546–551.

66 Ackerman, M. S. and Shortle, D. (2002) Molecular alignment of denatured states of staphylococcal nuclease with strained polyacrylamide gels and surfactant liquid crystalline phases, *Biochemistry 41*, 3089–3095.

67 Ackerman, M. S. and Shortle, D. (2002) Robustness of the long-range structure in denatured staphylococcal nuclease to changes in amino acid sequence, *Biochemistry 41*, 13791–13797.

68 Fieber, W., Kristjansdottir, S., and Poulsen, F. M. (2004) Short-range, long-range and transition state interactions in the denatured state of ACBP from residual dipolar couplings, *J Mol Biol 339*, 1191–1199.

69 Mohana-Borges, R., Goto, N. K., Kroon, G. J., Dyson, H. J., and Wright, P. E. (2004) Structural characterization of unfolded states of apomyoglobin using residual dipolar couplings, *J Mol Biol 340*, 1131–1142.

70 Bernado, P., Blanchard, L., Timmins, P., Marion, D., Ruigrok, R. W., and Blackledge, M. (2005) A structural model for unfolded proteins from residual dipolar couplings and small-angle x-ray scattering, *Proc Natl Acad Sci U S A 102*, 17002–17007.

71. Mukrasch, M. D., Markwick, P., Biernat, J., Bergen, M., Bernado, P., Griesinger, C., Mandelkow, E., Zweckstetter, M., and Blackledge, M. (2007) Highly populated turn conformations in natively unfolded tau protein identified from residual dipolar couplings and molecular simulation, *J Am Chem Soc 129*, 5235–5243.

72. DeBartolo, J., Colubri, A., Jha, A. K., Fitzgerald, J. E., Freed, K. F., and Sosnick, T. R. (2009) Mimicking the folding pathway to improve homology-free protein structure prediction, *Proc Natl Acad Sci U S A 106*, 3734–3739.

73. Krantz, B. A., Dothager, R. S., and Sosnick, T. R. (2004) Discerning the structure and energy of multiple transition states in protein folding using psi-analysis, *J Mol Biol 337*, 463–475.

74. Sosnick, T. R. (2008) Kinetic barriers and the role of topology in protein and RNA folding, *Protein Sci 17*, 1308–1318.

75. Bai, Y. (2006) Energy barriers, cooperativity, and hidden intermediates in the folding of small proteins, *Biochem Biophys Res Commun 340*, 976–983.

76. Maity, H., Maity, M., Krishna, M. M., Mayne, L., and Englander, S. W. (2005) Protein folding: The stepwise assembly of foldon units, *Proc Natl Acad Sci U S A 102*, 4741–4746.

77. Shen, M. Y. and Sali, A. (2006) Statistical potential for assessment and prediction of protein structures, *Protein Sci 15*, 2507–2524.

78. Colubri, A., Jha, A. K., Shen, M. Y., Sali, A., Berry, R. S., Sosnick, T. R., and Freed, K. F. (2006) Minimalist representations and the importance of nearest neighbor effects in protein folding simulations, *J Mol Biol 363*, 835–857.

79. Levinthal, C. (1968) Are there pathways for protein folding, *J Chim Phys 65*, 44–45.

80. Bai, Y. and Englander, S. W. (1996) Future directions in folding: The multi-state nature of protein structure, *Proteins 24*, 145–151.

81. Sosnick, T. R., Krantz, B. A., Dothager, R. S., and Baxa, M. (2006) Characterizing the protein folding transition state using psi analysis, *Chem Rev 106*, 1862–1876.

82. DeBartolo, J., Hocky, G., Wilde, M., Xu, J., Freed, K. F., and Sosnick, T. R. (2010) Protein structure prediction enhanced with evolutionary diversity: SPEED, *Protein Sci 19*, 520–534.

83. Kabsch, W. and Sander, C. (1983) Dictionary of protein secondary structure: Pattern recognition of hydrogen-bonded and geometrical features, *Biopolymers 22*, 2577–2637.

84. Pace, C. N. and Scholtz, J. M. (1998) A helix propensity scale based on experimental studies of peptides and proteins, *Biophys J 75*, 422–427.

85. Smith, C. K., Withka, J. M., and Regan, L. (1994) A thermodynamic scale for the beta-sheet forming tendencies of the amino acids, *Biochemistry 33*, 5510–5517.

86. Smith, C. K. and Regan, L. (1995) Guidelines for protein design: The energetics of beta sheet side chain interactions, *Science 270*, 980–982.

87. Minor, D. L., Jr and Kim, P. S. (1994) Context is a major determinant of beta-sheet propensity, *Nature 371*, 264–267.

88. Dill, K. A. (1990) Dominant forces in protein folding, *Biochemistry 29*, 7133–7155.

4

SHORT-DISTANCE FRET APPLIED TO THE POLYPEPTIDE CHAIN

Maik H. Jacob and Werner M. Nau

4.1. A SHORT TIMELINE OF RESONANCE ENERGY TRANSFER APPLIED TO THE POLYPEPTIDE CHAIN

Förster resonance energy transfer (FRET) [1–3] is a phenomenon of an optically excited fluorophore transferring its excitation energy through space to an acceptor chromophore—with a rate that depends on the distance between donor and acceptor. The transfer is caused by the Coulombic interaction of the donor and acceptor transition dipoles, and the rate of transfer was therefore thought to decrease with the third power of the distance. In 1948, Theodor Förster used both a classical and a quantum mechanical approach and predicted the rate to level off with the sixth power of separation [4]. He derived how the optical properties of a FRET pair—the donor emission spectrum and radiative lifetime, in addition to the acceptor absorption spectrum—determine a critical distance and the range of distances around it, which can be detected by using that pair. Because of the steep distance dependence, this range is rather narrow.

In 1967, Förster's predictions were experimentally confirmed by Stryer and Haugland, who appended a FRET donor–acceptor pair to the chain ends of synthetic

---

Protein and Peptide Folding, Misfolding, and Non-Folding, First Edition. Edited by Reinhard Schweitzer-Stenner.
© 2012 John Wiley & Sons, Inc. Published 2012 by John Wiley & Sons, Inc.

polyproline peptides that were known to adopt helical structures of predictable lengths. FRET was recommended as a molecular ruler [5]. In fact, the polyproline helix is close to the ideal case of intramolecular distances that vary little among molecules of an ensemble and that fluctuate little with time. Such a distance can then be measured by comparing the fluorescence intensity of a donor-only labeled peptide to the quenched fluorescence of the doubly labeled, donor–acceptor variant. When the same procedure is applied to a flexible chain with varying distances, the result is significantly shorter than the real average donor–acceptor spacing, simply because energy transfer is more effective at shorter distances. Between 1972 and 1975, Katchalski-Kazir, Steinberg, and Haas demonstrated—first in a simulation then in an experiment—how probability distributions of distances can be accessed by recording the kinetic trace of a donor emission after a short excitation pulse [6, 7]. They utilized highly viscous solvents, assumed intramolecular distances to be time-invariant under such conditions, and analyzed the donor-fluorescence decay traces in terms of a probability distribution of transfer rates.

In 1978, Steinberg and Haas included mutual donor–acceptor diffusion in the analysis [8]. Brownian motion enhances the probability of energy transfer events, since it drives donor and acceptor close to each other during the lifetime of the donor. The opposite process—diffusion separating the pair further—does not compensate the first: The probability is high that, at short distance, the transfer event has already taken place as a consequence of the strong distance bias toward very effective energy transfer at short donor–acceptor separations. Diffusion contributes to energy transfer to an extent that depends on the coefficient of mutual donor–acceptor diffusion as much as on the donor's radiative lifetime—both determine the time available for diffusion to contribute to energy transfer. But the large number of degrees of freedom of the diffusion model requires a series of emission traces to be analyzed simultaneously; several methods and algorithms have been proposed [9, 10], but they lack general applicability and have not yet been cross-tested among laboratories and models.

Laboratories in Israel remained a major place of FRET being applied to the polypeptide chain. Haas and coworkers measured distance distributions in the chain of stably unfolded proteins [11, 12] as well as in the chain of a protein at different stages of folding into the native state [13, 14]. To enable such a "double kinetics" experiment, as dubbed by its inventors, Ratner, Haas' coworker, coupled time-resolved FRET to stopped-flow mixing instrumentation [15]. This technique, as well, has still to be made accessible to a broader community of users. Interest in FRET spread widely, when research on the elementary protein folding mechanism shifted to increasingly smaller and simpler proteins [16, 17]—when it became clear that Levinthal's paradox [18] does not necessarily imply an early reduction of chain entropy caused by a random hydrophobic collapse [19] but can be resolved, alternatively, by considering local or even long-range conformational propensities of the "unfolded" chain under folding conditions [20, 21]. Evidently, the conformational entropy of an unfolded polypeptide chain is much smaller than the entropy of a random coil [22].

To distinguish, via FRET, a low-entropy, partially ordered state of a chain from a random state requires detecting a distance between two chain positions that differs for both states. The distance resolution of FRET detection becomes critical already for chains with less than 100 amino acids. The FRET donor–acceptor pair should be selected for a Förster radius that is significantly shorter than the gyration radius of the unfolded chain [23, 24].

Three research groups used steady-state FRET detection coupled to stopped-flow mixing to test the unfolded chain for structure [24–26]. In all cases, tryptophan was chosen as donor as it is a natural, fluorescent and infrequent amino acid that, via mutagenesis, can be conveniently placed at virtually any selected position. When tryptophan is not a suitable choice, two external FRET probes have to be appended to the chain, an art that has only started to be turned into science [27, 28]. Regardless of which donor is chosen, the shortest Förster radii previously achievable were limited to about 20 Å. The research group with the lowest ratio of Förster to gyration radius could report that the chain forms a long loop instantly upon denaturant removal but could only vaguely point to the node of the loop—to the specific interactions composing it [24]. All three studies would have gained in substance from a FRET pair providing a significantly shorter Förster distance. Short-distance FRET pairs shall gain general importance in tackling the protein folding problem of translating a one-dimensional amino acid sequence into a three-dimensional structure, because they also support the "bottom-up" approach that is aimed toward understanding how conformational proclivities of short chain segments depend on their amino acid composition. Short peptides are more than proteins accessible to molecular dynamics (MD) simulations; the discrepancies between theoretical and experimental results provide a continuous source of motivation to improve their quality [29].

In 2006, we introduced 2,3-diazabicyclo(2.2.2)oct-2-ene (DBO; Fig. 4.1) as a FRET acceptor that in combination with tryptophan exhibits a Förster radius of ~10 Å [30]. We applied DBO to (1) polyproline peptides of variable length [31]—to study the rigid chain and to compare the distance resolution provided by the Trp/DBO pair with other pairs used in the past, to (2) peptide segments contained in signaling proteins [32]—to demonstrate that DBO can be used to monitor minute but biologically consequential chain conformational transitions induced by enzymatic phosphorylation, and to (3) glycine–serine repeat peptides [33] (polyGS)—to obtain reference distance distributions of extremely flexible chains with no or little structural bias. We summarize our findings with a focus on the suitability of DBO for short-distance determinations: this question has been addressed by other researchers in an independent investigation that arrived at a positive conclusion [34]. Before, we comment on the mathematical structure of FRET analysis by using models of growing complexity, and derive the central equations that we can employ afterwards.

4.2. A SHORT THEORY OF FRET APPLIED TO THE POLYPEPTIDE CHAIN

In our presentation of FRET theory, we focus on the rate constant of energy transfer—the FRET rate [1], from which all relations used in FRET analysis can be derived with ease.

4.2.1. The Rate of Energy Transfer

The interaction energy, W, between the transition dipole moment, μ_D, of the excited donor and the transition dipole, μ_A, of the acceptor in the ground state depends on the dipole moments and their distance (R_{DA}) according to $W \propto \mu_D\,\mu_A/R_{DA}^3$. The rate of energy transfer, k_{FRET}, depends on the square of the interaction energy: $k_{FRET} \propto W^2 \propto (\mu_D\mu_A)^2/R_{DA}^6$. Förster expressed the rate in terms of the spectroscopic properties of donor and acceptor:

$$k_{FRET} = c \cdot \kappa^2 / n^4 \cdot J \cdot k_{D,rad} / R_{DA}^6. \tag{4.1}$$

The numerical constant c equals $9000 \ln 10/(128\pi^5 N)$ with N being Avogadro's number. The orientation factor, κ^2, takes into account that the dipole moments can adopt various orientations toward each other and can vary between 0 (perpendicular vectors) and 4 (collinear vectors). It adopts a value of 2/3, when donor and acceptor dipoles sample all possible orientations randomly and rapidly in comparison to the time scale of the donor emission decay [35]. The refractive index, n, is a measure of the polarizability of the donor environment and can most often be taken as the refractive index of the measurement solution. The constant J is determined by the extent to which the donor emission spectrum, normalized to an area of one, that is, ($F_D(\lambda)$), overlaps with the spectrum of the acceptor absorption coefficient: $J = \int d\lambda \cdot F_D(\lambda) \cdot \varepsilon(\lambda) \cdot \lambda^4$. The rate $k_{D,rad}$ is the rate of donor emission, when the quantum yield reaches unity, that is, when each photon absorbed by the donor leads to one being emitted, and when no deactivation by FRET or any other nonradiative pathway occurs. Experimentally, the rate of energy transfer can be obtained from the rate of donor deactivation, k_D, in the donor-only labeled peptide and from the rate of donor deactivation, k_{DA}, in the donor–acceptor labeled peptide by assuming additivity according to $k_{DA} = k_D + k_{FRET}$.

4.2.2. The Förster Critical Distance

At a sufficiently short donor–acceptor distance, R_F, the donor is as likely to be deactivated by photon emission as by FRET. Equation 4.1 then becomes: $k_{FRET} = k_{D,rad} = c \cdot \kappa^2/n^4 \cdot J \cdot k_{D,rad}/R_F^6$. When this equation is solved for R_F^6 ($R_F^6 = c \cdot \kappa^2/n^4 \cdot J$) and substituted into Equation 4.1, the FRET rate adopts the simple form $k_{FRET} = k_{D,rad} \cdot (R_F/R_{DA})^6$. Förster introduced R_F, but more commonly reported is the Förster radius R_0, the distance, at which k_{FRET} equals the measured donor decay rate k_D that in contrast to $k_{D,rad}$ includes deactivation by nonradiative processes such as contact quenching: $k_D = k_{D,rad} + k_{D,nonrad}$. The same arguments as above lead to Equation 4.2:

$$k_{FRET} = k_D \cdot (R_0/R_{DA})^6. \tag{4.2}$$

The rate constant k_D and the Förster distance R_0 vary with the donor quantum yield, Φ_D, defined by $\Phi_D = k_{D,rad}/(k_{D,rad} + k_{D,nonrad})$. It follows that $k_D = k_{D,rad}/\Phi_D$ and

$R_0^6 = \Phi_D R_F^6$, leading to Equation 4.3, which is commonly used for the determination of Förster radii.

$$R_0^6 = c \cdot \kappa^2 / n^4 \cdot J \cdot \Phi_D. \tag{4.3}$$

Thus, R_0^6 depends sensitively on experimental conditions such as oxygen content, co-additive concentration, and temperature. It is commonly accepted that the value of the distance to be determined, R_{DA}, should not exceed the limits of the range 0.5 R_0 – 1.5 R_0 as otherwise the rate constants k_{FRET} and k_D would differ by more than an order of magnitude (see Eq. 4.2). When R_{DA} exceeds $2 \cdot R_0$, the decay traces of the donor in the donor-only and the donor–acceptor peptide become indistinguishable even within the smallest experimental error (±1.5%).

4.2.3. The Efficiency of Energy Transfer

The energy transfer efficiency, E, is dimensionless and defined as $E = k_{FRET}/(k_D + k_{FRET})$. It is 50%, when the Förster and the actual donor–acceptor distance coincide. Using Equation 4.2, one obtains

$$E = R_0^6 / (R_0^6 + R_{DA}^6). \tag{4.4}$$

The transfer efficiency is a central quantity in FRET analysis as it can be accessed by time-resolved (*tr*) as well as by steady-state (*ss*) fluorescence measurements carried out on the donor-only labeled peptide and the double-labeled peptide: using its definition together with the assumption $k_{DA} = k_D + k_{FRET}$, one obtains $E_{tr} = (k_{DA} - k_D)/k_{DA}$. The analysis software of standard time-resolved fluorometers reports time constants or fluorescence lifetimes (τ) not rate constants (k), but both are simply reciprocally related ($\tau = 1/k$). The efficiency $E = E_{tr}$ is then determined via Equation 4.5. The efficiency $E = E_{ss}$ can also be obtained from steady-state measurements of the fluorescence intensity, I, of peptides (Eq. 4.6):

$$E_{tr} = (\tau_D - \tau_{DA}) / \tau_D. \tag{4.5}$$

$$E_{ss} = (I_D - I_{DA}) / I_D. \tag{4.6}$$

The equivalence of Equations 4.5 and 4.6 becomes apparent by considering that the area under a fluorescence decay curve is proportional to the total emission. In case of a mono-exponential decay, it is also proportional to the lifetime (constant). Decay curves in donor-only and donor–acceptor peptides, and particularly for peptides containing tryptophan as donor, frequently have to be fitted to at least bi-exponential functions that yield at least two lifetime components and two pre-exponential factors, α_I, that sum to one $\sum \alpha_I = 1$. In such cases, the amplitude-weighted lifetime [3], τ_{avg}, calculated from Equation 4.7, corresponds to the total emission and has to be used in Equation 4.5:

$$\tau_{avg} = \sum_i \alpha_i \tau_i. \tag{4.7}$$

4.2.4. Distributions of Time-Invariant Distances

Any intrachain distance accessible to FRET detection adopts a variety of values in the molecules of an ensemble. If measurements are carried out by using a donor with short-lived fluorescence and if additionally a sufficiently viscous solvent is used, distance separations in any given chain molecule can be considered time invariant [7]. The measured FRET efficiency depends then on the distance probability distribution $P(R)$ according to $E = \int dR \cdot E(R) \cdot P(R)$. If, despite this fact, the donor–acceptor distance is still estimated via Equation 4.4 in the form $R_{DA} = R_0 (1/E - 1)^{1/6}$, the result is an apparent distance that is shorter than the average distance of the real distribution—short distances contribute more to energy transfer and enter the average with a higher weight.

Information on the distance distribution in chains is accessible via time-resolved FRET spectroscopy by analyzing the donor emission decay $I_{DA}(t)$ in the double-labeled peptide [7]. The decay $I_{DA}(t)$ is described by a distribution of rate constants. With $k_{DA}(R) = k_D + k_{FRET}(R)$ and Equation 4.2, one obtains: $I_{DA}(t) = I_0 \cdot \int dR P(R) \cdot \exp[-k_D \cdot t - k_D \cdot (R_0/R)^6 \cdot t]$. If the donor fluorescence decay in the donor-only peptide cannot be modeled by a single exponential, one uses: $I_D(t) = I_0 \cdot \sum_i \alpha_i \exp(-k_{D,i} t)$, where the pre-exponential factors, again, sum up to unity, $\sum_I \alpha_I = 1$ [3]. The donor decay, $I_{DA}(t)$, is then fitted to Equation 4.8:

$$I_{DA}(t) = I_0 \cdot \int_0^\infty dR \cdot P(R) \cdot \sum_i \alpha_i \exp\left(-k_{D,i} \cdot t - k_{D,i} \cdot (R_0/R)^6 \cdot t\right). \quad (4.8)$$

A single time course of donor emission may not carry sufficient information to pinpoint a specific distribution function. But even a simple Gaussian function, when used for fitting, yields values of two informative parameters, the average distance, R_{mean}, and the standard deviation, σ, measuring the spread of distances around the mean (Eq. 4.9). The distance spread is a specifically important measure of chain flexibility and conformational propensity. Alternatively, the polypeptide can be modeled as a semi-stiff chain as done in the worm-like chain model. The distribution function of a worm-like chain model variant [36] yields values for the contour length, l_c, related to the length of the stretched peptide, and for the persistence length, l_p, related to the distance between two chain-segmental vectors that ensures the vectors can adopt all orientations toward each other with equal probability (Eq. 4.10). Thus, distance spread and persistence length carry the same kind of information on the chain's flexibility. Under systematically varied experimental conditions, the parameter values obtained with both functions reveal consistent chain-structural trends:

$$P(R) = \frac{1}{\sigma\sqrt{2\pi}} \exp\left(-\frac{(R - R_{mean})^2}{2\sigma^2}\right). \quad (4.9)$$

$$P(R) = \frac{4\pi N R^2}{l_c^2 (1 - (R/l_c)^2)^{9/2}} \exp\left(\frac{-3l_c}{4l_p (1 - (R/l_c)^2)}\right). \quad (4.10)$$

4.2.5. Distributions of Fluctuating Distances

To account for Brownian motion, Steinberg derived a linear partial differential equation (Eq. 4.11) that has to be solved by numerical methods [8]. The number of excited-donor chains, N^*, varies with time, t, and with the donor–acceptor distance, $R_{DA} = r$. When solved, this equation yields the coefficient of mutual donor–acceptor diffusion, D, as well as the initial distance distribution of excited chains, $N^*(t = 0)$, which perfectly mirrors the distance distribution in the entire ensemble:

$$\frac{\partial N^*(r,t)}{\partial t} = -\left(k_D + k_D \frac{R_0^6}{r^6}\right) N^*(r,t) + \frac{\partial}{\partial r}\left(N_0^*(r) D \frac{\partial (N^*/N_0^*)}{\partial r}\right). \quad (4.11)$$

If, in practice, the impact of diffusion on energy transfer is ignored, the effective distances obtained from efficiency analysis and the average distances obtained from distribution analysis are shorter than the real average distances. Furthermore, the width of the obtained distribution is smaller than the real distance distribution. To minimize diffusion effects, it is advisable to employ donors with short fluorescence lifetime and work in a highly viscous solvent, for instance, in propylene glycol, which is 45 times more viscous than water [37].

4.3. DBO AND Dbo

DBO (Fig. 4.1) is a remarkably stable chromophore that resists degradation by temperature, light, and hydrolysis. It is soluble in organic solvents but best soluble in water [38]. It is possible to covalently link the DBO chromophore to an asparagine residue and thereby construct building blocks for standard solid-phase peptide synthesis. We have referred to the resulting labeled amino acid as "Dbo" (Fig. 4.1), which can be conveniently incorporated at any position of the chain [39], and which, incidentally, enhances the hydrophilicity of the peptides of which it has been made part [40]. The synthesis of Dbo itself requires more than 10 steps as well as either a costly precursor or phosgene at an early stage [41], but the compound is now commercially available in a protected form directly suitable for peptide synthesis. We have started to use the pair of Dbo and tryptophan in 2002 [39], at first, to assess peptide-chain flexibility in experiments complementary to FRET, based on almost identical samples and measurement instrumentation as used in FRET, but yielding the mutual collision rate of the labels—a measure of how fast the chain can form a loop. This experiment is made possible by the long fluorescence lifetime of DBO—half a microsecond in degassed water [42, 43]— which we expect to prove useful in cases in which FRET analysis has to be based on acceptor fluorescence: When the measurement sample is less transparent to ultraviolet (UV) radiation, the donor fluorescence signal is often obscured by scattered light. But the fluorescence of DBO is detectable long after donor fluorescence and noise have decayed [44].

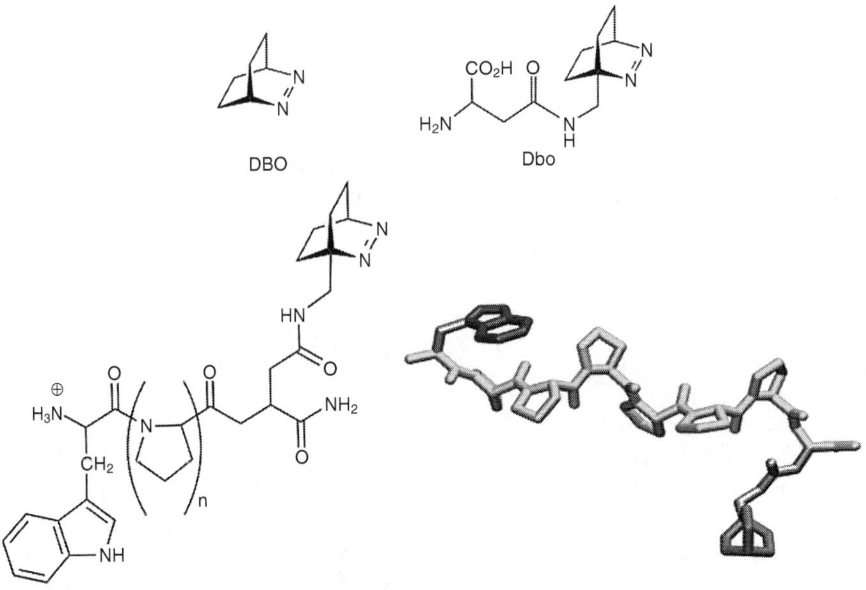

Figure 4.1. The parent DBO and the synthetic amino acid Dbo (top row), and Dbo and Trp appended to the ends of the hexaproline helix (bottom row, indole group in dark gray, DBO in medium gray).

4.3.1. The Point-Dipole Approximation in Förster Theory

A FRET dye (either donor or acceptor) to be used for short-distance detection must primarily meet a single requirement; it must be small. Only if the dipole vectors are much shorter than the distance between donor and acceptor, the point-dipole approximation holds and Förster's theory can be applied [4]. The transition dipole of DBO is located on two atoms only, on the N=N bond, which defines the azo chromophore. The quantum mechanics of the Trp/DBO FRET pair has been extensively studied [34], with the conclusion that DBO as acceptor is a superior choice. Tryptophan has a higher transition dipole spread than, for instance, tyrosine, but remains the intrinsic donor of choice because of its scarcity in natural sequences.

4.3.2. A FRET Pair with a Short and Robust Förster Distance

The absorption spectrum of DBO (Fig. 4.2, left panel, red line) is nicely centered inside the emission spectrum of tryptophan (blue line). For most other acceptors such a "good overlap" would result in large absolute values of the overlap integral and a large Förster distance. But DBO's absorption coefficient is extremely small, never exceeding 70/M/cm in protic solvents. This results in very small values for the overlap integral despite excellent spectral overlap, and the Förster distance calculated according to Equation 4.3 is therefore only about 10 Å (Table 4.1) [33]. Arguably, short Förster distances could also be achieved by screening donor and acceptor dyes for very small overlap areas that could result from a side-by-side

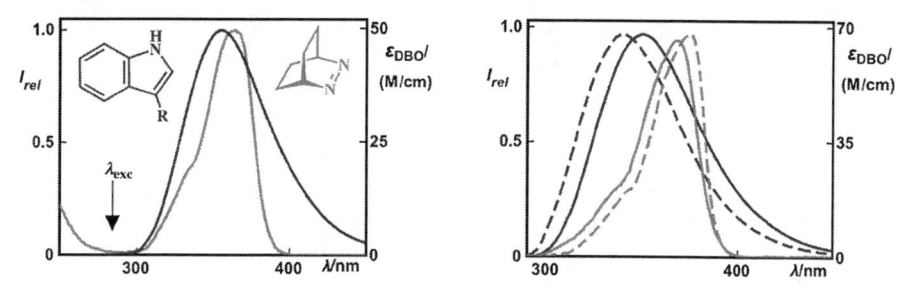

Figure 4.2. Spectral overlap for the Trp/DBO FRET pair (left) and for the pair of peptides, W–P$_6$–NH2/P$_4$–Dbo–NH$_2$, in water (solid lines) and in propylene glycol (dashed lines).

TABLE 4.1. Photophysical Parameters and Förster[a] Radii of Free and Peptide-Incorporated Labels in Water as Well as in Propylene Glycol (Values in Parentheses)

Donor–acceptor	$J/(10^{11}/\text{M/cm nm}^4)$	Φ_D	n	$R_0/\text{Å}$
Trp/DBO	4.2 (4.5)	0.13 (0.22)	1.333 (1.423)	10.0 (10.2)
W–P$_6$–NH2/P$_4$–Dbo–NH$_2$	5.1 (4.0)	0.06 (0.11)	1.340 (1.432)	9.0 (8.7)
W–(GS)$_4$/(GS)$_6$–Dbo–NH$_2$	5.1 (4.0)	0.07 (0.16)	1.340 (1.423)	9.2 (9.5)

[a] The Förster radius, R_0, was computed via Equation 4.3 from the overlap integral J, the quantum yield of the donor Φ_D, and the refractive index n, with the orientation factor κ^2 set to 2/3.

arrangement of the spectra. This approach, while frequently attempted, has numerous disadvantages, since either energy transfer becomes endergonic and thermally sensitive, energy transfer to higher excited states of the acceptor becomes competitive, or the excitation of the donor becomes unselective. Another severe drawback of FRET pairs with poor spectral overlap is that the values of the integral and the Förster radius would strongly fluctuate even with slight spectral shifts that cannot be avoided when experimental conditions are varied. In other words, many applications require a robust Förster distance. The "top-on-top" arrangement of the spectra displayed by Trp and DBO leads to almost identical Förster radii in water and in propylene glycol and to a variation of only 1 Å when the probes become part of the peptide chain (right panel of Fig. 4.2; Table 4.1) [31, 33]. Such a situation is ideal, for instance, in protein folding experiments with immeasurable spectra of short-lived intermediate states [23].

4.4. SHORT-DISTANCE FRET APPLIED TO THE STRUCTURED POLYPEPTIDE CHAIN

4.4.1. Poly-L-Proline Peptides and the Polyproline II Helical Motif

In 1955 Cowan and McGavin obtained the structure of crystallized poly-L-proline, specifically, of a poly-L-proline sample synthesized in the laboratory of

Katchalski-Kazir, with an average of 20 prolines per chain [45]. X-ray diffraction revealed a threefold screw axis along an extended backbone, a left-handed helix with three residues per turn, and backbone dihedral angles of $\phi = -75°$ and $\psi = 145°$. So far, the polyproline II helix (PPII) is the only example of a well-defined polypeptide structure in the absence of any intrachain pattern of hydrogen bonds. For that reason, it is inconveniently accessible to solution nuclear magnetic resonance (NMR) spectroscopy and is best studied by methods based on optical activity and distance-dependent fluorescence [46, 47]. The PPII structural motif was found to often describe the preferred conformation of additional chain segments that contain no proline at all [22, 48], and will thus contribute in the future to explain the low entropy of unfolded proteins [49]. Real polyproline segments in proteins play central roles in protein recognition. For example, the formation of F-actin structure is regulated by Profilin, an actin-binding protein that can simultaneously bind to PPII-structured proline-repeat sequences and thus can become localized at sites of F-actin demand [50].

4.4.2. Energy Transfer in Trp/Dbo-Labeled Polyproline Peptides in Water and in Propylene Glycol

We analyzed donor–acceptor labeled chains composed of n proline units, W–P_n–Dbo, with $n = 1, 2, 4,$ and 6 in water, and, as a control, in propylene glycol, to assess the extent of FRET caused by Brownian motion. In chains with more than six proline residues, FRET is insignificant. We will later see that this limitation to the number of amino acids applies only to the rigidly extended and not to the flexible chain, where FRET can be observed for up to 20 residues. The donor lifetimes, τ_D, and steady-state donor fluorescence intensities, I_D, in the absence of FRET were measured by using the peptide W–P_6–NH$_2$; these were used as reference values to obtain the energy transfer efficiency in double-labeled peptides according to Equations 4.5 ($E_{tr} = 1 - \tau_{DA}/\tau_D$) and 4.6 ($E_{ss} = 1 - I_{DA}/I_D$). The fluorescence spectra of the double-labeled peptides were normalized to the spectrum of the reference peptide, whose maximum was set to one (Fig. 4.3, black line). Going from the longest double-

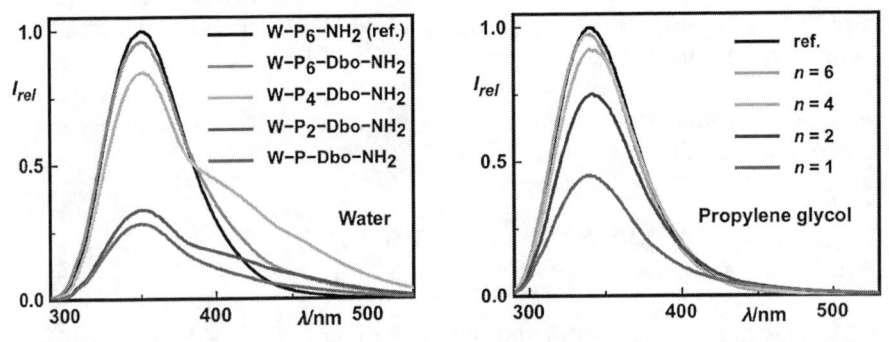

Figure 4.3. Fluorescence spectra after excitation at 280 nm, of donor-only and donor–acceptor-labeled polyproline peptides in water and in propylene glycol. Modified from Ref. [30, 31].

TABLE 4.2. Steady-State and Time-Resolved FRET Analysis of Polyproline Peptides

Peptide	Steady-State Analysis[a]			Time-Resolved Analysis[b]		
	I_{rel}	$E_{ss}/\%$	$R_{ss}/Å$	τ/ns	$E_{tr}/\%$	$R_{tr}/Å$
W–P$_6$–NH$_2$	1.0 (1.0)[c]			1.94 (2.15)		
W–P–Dbo–NH$_2$	0.28 (0.45)	72 (55)	7.8 (8.4)	0.70 (1.30)	64 (39)	8.2 (9.7)
W–P$_2$–Dbo–NH$_2$	0.36 (0.74)	64 (26)	8.2 (10.3)	0.76 (1.62)	61 (25)	8.4 (10.8)
W–P$_4$–Dbo–NH$_2$	0.87 (0.92)	13 (8.3)	12.3 (13.1)	1.74 (1.98)	10 (7.9)	13.0 (13.6)
W–P$_6$–Dbo–NH$_2$	0.97 (0.97)	2.6 (2.4)	16.6 (16.5)	1.89 (2.10)	2.6 (2.3)	16.5 (16.8)

[a] Relative fluorescence intensities, I, of donor-only and double-labeled peptides were used to calculate the energy transfer efficiencies, $E = E_{ss}$, and effective donor–acceptor distances, $R_{DA} = R_{ss}$, from Equations 4.6 and 4.4.
[b] Average donor lifetimes were used to calculate the energy transfer efficiencies, $E = E_{tr}$, and donor–acceptor distances, $R_{DA} = R_{tr}$, from Equations 4.5 and 4.4.
[c] Values measured in propylene glycol are given in parentheses.

labeled peptide to the shorter chains, FRET becomes increasingly apparent: the tryptophan emission with its maximum at 355 nm decreases, while Dbo fluorescence with its maximum at 420 nm increases, giving rise to a somewhat scurrile shape of the spectra that is best seen for W–P$_4$–Dbo (left panel of Fig. 4.3). Spectra recorded in propylene glycol are smoothly shaped (right panel of Fig. 4.3), partly because this solvent quenches Dbo fluorescence to a certain extent [38].

In these experiments, the donor was excited with an LED laser at 280 nm, at a wavelength at which the acceptor is fully transparent. As an added benefit to this selective excitation, the Trp/Dbo FRET pair adds as another advantage that the acceptor does not fluoresce below 370 nm and contributes no fluorescence at the wavelength of the donor emission peak. As is not often the case, the emission value at 355 nm can be taken without correction as the donor intensity in the presence of acceptor, I_{DA}. Steady-state efficiencies were determined in water as well as in propylene glycol (Table 4.2), with the values in propylene glycol given in parentheses in Table 4.2 and all successive tables. Equation 4.4, in the form $R_{DA} = R_0 (1/E-1)^{1/6}$, yielded the donor–acceptor distances, $R_{DA} = R_{ss}$, varying from 8 Å for the shortest chain, W–P–Dbo, to 17 Å for the longest one, W–P$_6$–Dbo (Table 4.2).

Since FRET causes a decrease of donor fluorescence intensity in double-labeled peptides, it shortens also the donor fluorescence lifetime (Fig. 4.4). Using time-correlated single-photon counting, we recorded donor decay traces at 350 nm upon excitation at 280 nm. In all cases, decays could be fitted to a bi-exponential model, and the obtained time constants and amplitudes were used to calculate the average (amplitude-weighted) lifetime from Equation 4.7, $\tau_{avg} = \Sigma_i \alpha_i \tau_i$. The efficiencies E_{tr} and E_{ss} obtained from time-resolved and steady-state experiments, respectively, would ideally be identical. In practice, however, the efficiencies E_{tr} were slightly smaller for all peptides than the corresponding E_{ss} values. Accordingly, the distances R_{tr} were slightly larger than R_{ss}. The reason for this discrepancy (an apparently static

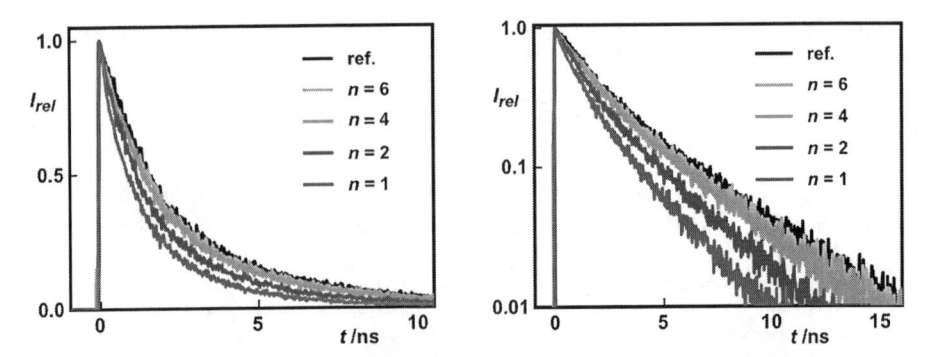

Figure 4.4. Time-resolved fluorescence decays of W–P_n–Dbo–NH_2 peptides in propylene glycol (normalized, linear intensity scale on the left and logarithmic one on the right) relative to the fluorescence decay of the W–P_6–NH_2 reference peptide lacking the acceptor (black trace). Modified from Ref. [31]. See color insert.

fluorescence quenching) lies in the limited time-resolution of the lifetime spectrometer. When donor and acceptor are close to each other in the very moment donor excitation occurs, the probability of energy transfer is high, and this very fast donor deactivation by FRET escapes detection, that is, is not separable from the instrument response function. However, the differences in distance were small and, for the longest peptide, negligible. Specifically, in the longer and presumably most rigid peptides, the stably extended PPII structure kept donor and acceptor apart from each other.

Compared with the FRET efficiencies determined in water, the efficiencies, E_{ss} and E_{tr}, measured in propylene glycol (Table 4.2) were smaller—considerably smaller in the case of the two shortest peptides, negligibly smaller in the case of the two longest peptides. The measured distances in the short peptides are close to the Förster radius, where diffusional distance variations have the strongest impact on transfer probability and efficiency. When expressed in terms of distances, R_{ss} and R_{tr} (Eq. 4.4), the solvent effect appears much less dramatic: The R_{tr} values of the two shorter peptides are larger by about 2 Å, the values of the two longer peptides by only about 0.5 Å. In the case of a highly flexible chain and a long lifetime of donor fluorescence, diffusion can become the dominant factor for the FRET efficiency. But in the case of the investigated polyproline peptides labeled with tryptophan, the solvent effect on the measured distances was moderate, which suggests that these peptides are indeed rather rigid and that the fluorescence lifetime of tryptophan is sufficiently short to keep diffusion effects small. Such conditions enable a meaningful analysis of the distance distributions. But first we conclude the efficiency analysis by placing our results in the context of the literature.

Since the seminal work of Stryer and Haugland [5], who used tryptophan and dansyl (Dans) as donor and acceptor, polyproline chains have been studied with a variety of FRET labels, including pairs of large Alexa probes, to enable single-molecule spectroscopy [51]. Here, we just compare the results obtained for hexaproline and plot the effective distances obtained with different FRET pairs against their Förster radii (Fig. 4.5). Ideally, the recovered distance should depend little on

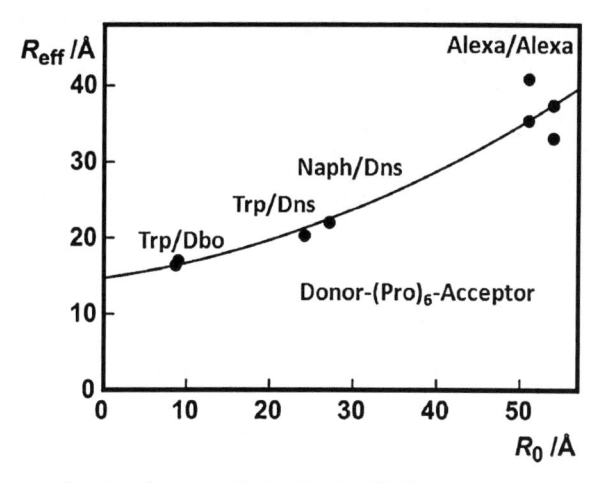

Figure 4.5. Hexaproline has been studied with the FRET pairs Trp/Dbo [31], Trp/Dansyl (Dns) [52], napthylalanine (Naph)/Dansyl [5], and Alexa/Alexa [51, 53] pairs. Here we plot the distances that resulted from efficiency analysis against the Förster radius of each pair. The solid line presents an arbitrary fitting to expose the trend. Modified from Ref. [31].

the labels and not at all on the Förster radius. In practice, we observe that larger probes lead to larger effective distances. This might be due to a breakdown of the point-dipole approximation when the size of the labels becomes comparable to the distance between them, but could also be traced back to excluded-volume effects exerted by the labels, leading to larger distance separations as they would occur in non-perturbed systems. It clearly transpires that any short-distance detection by FRET requires small probes with a short Förster radius.

4.4.2. Polyproline Distance Distributions in Water and in Propylene Glycol

We analyzed the donor decays of double-labeled peptides by assuming a Gaussian distribution (Eq. 4.9) as well as a worm-like chain distribution (Eq. 4.10) of time-invariant donor–acceptor distances. The Gaussian distributions of the peptides in water as well as in propylene glycol are shown in the left column of Figure 4.6, and worm-like chain distributions in the right column. Both distribution functions afforded very similar results. As expected for short polyproline helices, the distance distributions were narrow. Consequently, the effective distances obtained from efficiency analysis and the mean distances obtained from distribution analysis were very similar as well. The spread of distances, that is, the width of the distributions, was correlated with the peptide length.

Surprisingly, propylene glycol led to such narrow distributions that the distinction between a single effective distance and a distance distribution became irrelevant. The largest effect was seen for the longest peptide, whose persistence length (rigidity) increased from 30 to 180 Å (Table 4.3). This phenomenon cannot be explained by suppressed diffusion that would lead to the opposite effect: consider the case of

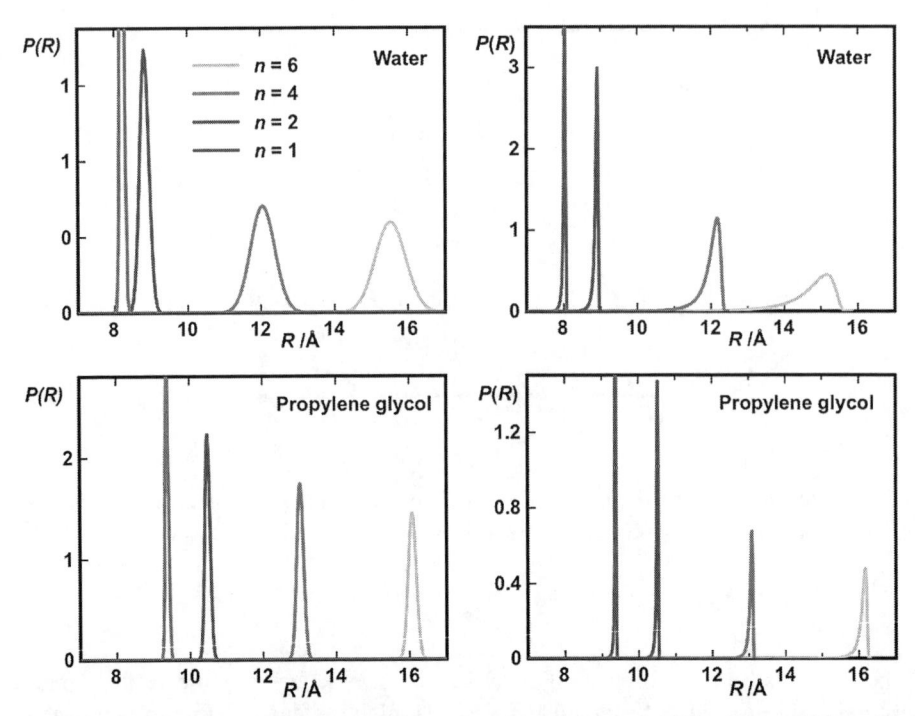

Figure 4.6. Probability distributions of donor–acceptor distances in W–P$_n$–Dbo peptides with $n = 1$ (red), 2 (blue), 4 (green), and 6 (yellow), in water (top row) and propylene glycol (bottom row) obtained from a Gaussian distribution function (left column) and a worm-like chain distribution function (right column) given by Equations 4.9 and 4.10. The area under each profile is one; the sharpest spike for $n = 1$ has been removed for clarity. Modified from Ref. [31]. See color insert.

TABLE 4.3. Distribution Parameters and Average Distances Obtained from MD Simulations

	Gaussian		Worm-Like Chain		MD	
Peptide	R_{mean}/Å	FWHM^a/Å	l_c/Å	l_p/Å	R_{DA}/Å	R_{CC}/Å
W–P–Dbo–NH$_2$	8.2 (9.4)	0.52 (0.05)	8.3 (9.4)	130 (290)	7.3	6.3
W–P$_2$–Dbo–NH$_2$	8.3 (10.5)	0.82 (0.07)	8.5 (10.6)	66 (250)	8.9	8.9
W–P$_4$–Dbo–NH$_2$	12.7 (13.0)	1.1 (0.09)	13.2 (13.2)	45 (210)	11.3	15.0
W–P$_6$–Dbo–NH$_2$	15.3 (16.1)	1.25 (0.11)	15.7 (16.3)	30 (180)	18.0	20.7

Average distances, R_{mean}, and values of the full width at half maximum, FWHM = 2.35·σ, obtained with a Gaussian distribution function (Eq. 4.9). Contour lengths, l_c, and persistence lengths, l_p, obtained from a worm-like chain distribution function (Eq. 4.10). Donor–acceptor distances, R_{DA}, and distances, R_{CC}, between the C$_\alpha$ atoms of Trp and Dbo, obtained from MD simulations. See Ref. [31].

very rapid donor–acceptor diffusion in flexible chains with an average donor–acceptor distance larger than R_0. In almost all chains, the energy transfer event would take place in a narrow range of distances close to the Förster radius, where the probability of transfer, the rate of FRET, is large. Thus, when an analysis ignores diffusion, it yields an apparent distance distribution that is narrower than the real distribution. With decreasing diffusion, the apparent distribution should become broader and approach the real one [3]. Therefore, the observed narrow distribution is not due to the suppression of diffusion in this highly viscous solvent.

In fact, the narrow distributions in propylene glycol can have another, more fundamental, reason. Water molecules can stabilize, via hydrogen bridges, twisted or bent conformations of the helix that are not supported by the less polar, less dynamic, and more bulky propylene glycol molecules. In the course of their biological activity, polyproline segments in proteins experience aqueous and non-aqueous environments and their adaptable flexibility might play a biological role. Before we settle on this hypothesis, we have to consider an alternative explanation for the narrowing effect of propylene glycol: conformational heterogeneity in water but homogeneity in propylene glycol. If several stable conformational populations are present in the solution that conform to similar and narrow distance distribution profiles, a single-distribution analysis would be successful but would result in an artificially broadened profile. Possibilities for conformational heterogeneity in polyproline peptides are multifold. In a polyproline peptide, the internal Pro–Pro bonds as well as the Trp–Pro bond can adopt a *trans* ($\omega = 180°$) or a *cis* ($\omega = 0°$) configuration [54, 55]. Thus, although our focus is on short-distance FRET, we included a study on the chiral properties of the peptides here, in which we found that all Pro–Pro bonds are expected to be *trans* but that some of the Trp–Pro bonds are also to be in the *cis* conformation. These results will be described in the next section.

4.4.3. Circular Dichroism Spectra of Donor–Acceptor-Labeled Polyproline Peptides

In electronic circular dichroism spectra, the PPII helix is recognizable by a minimum at 203 nm with an ellipticity that is proportional to the number of prolines that compose the helix [56, 57]. Spectra measured in water (left column of Fig. 4.7) and in propylene glycol (right column) show the expected trough and a positive band near 225 nm for all double-labeled peptides. Whether Dbo contributes to the spectral shape was tested by comparing hexaproline peptides with and without Dbo; the spectra coincide: Dbo has no influence. The integrity of the PPII structure in peptides becomes most apparent when the spectra are normalized with respect to the number of prolines (bottom row of Fig. 4.7). The value of the minimum obtained in this way, $-4.6\ 10^5$ degrees·cm^2/dmol , is in the upper range of values reported for polyprolines [57, 58]. The similarity of the normalized spectra of peptides with two or more prolines, measured in water and in propylene glycol, is remarkable and establishes spectroscopically that these peptides form a PPII helix composed of proline residues in their *trans* configuration. For hexaproline, but not for the shorter ones, this has already been concluded from earlier studies [5, 57, 59].

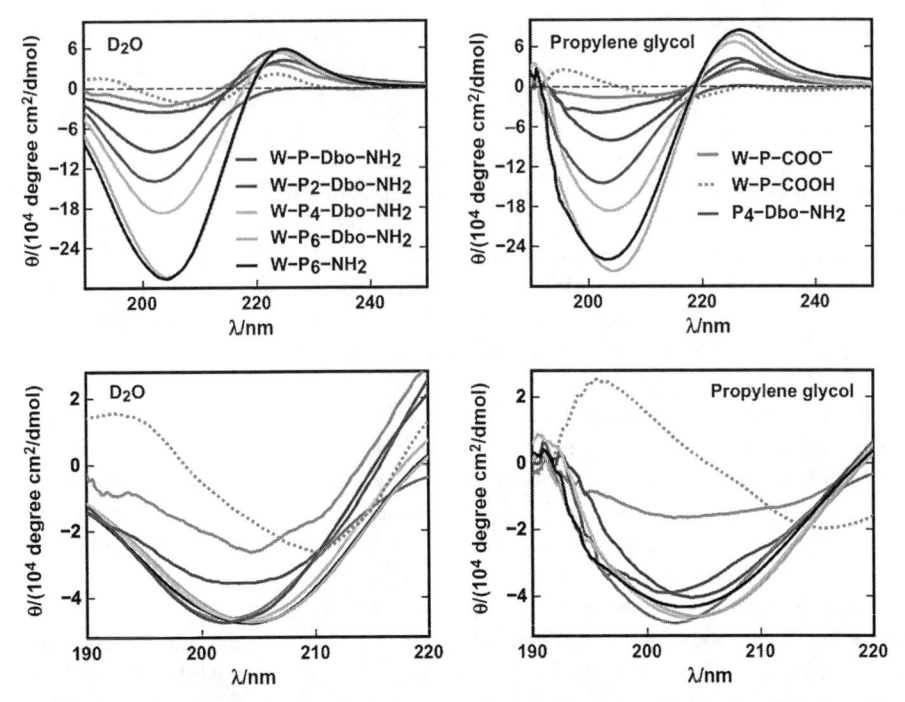

Figure 4.7. CD spectra of proline peptides in D_2O (left column) and propylene glycol (right column). The upper spectra show the molar ellipticity (per mole of peptide), and the lower spectra are normalized with respect to the number of proline residues in the peptide chain (per mole of proline residue). See color insert.

The aminoterminal W–P bond, in contrast, does ocurr in both *cis* and *trans* forms. This is exemplified by the peptide W–P–COOH, which is known to be predominantly in its *cis* form at pH 2 (Fig. 4.7, dotted red line) and predominantly in its *trans* form at pH 7 (solid red line) and which affords drastically different spectra at different pH [60, 61]. We studied the shortest double-labeled peptide, W–P–Dbo, by NMR in water (D_2O), and found that about 30% of the molecules adopt a *cis* conformation. As similar values are to be expected for the longer peptides, the polyproline peptides in solution are indeed conformationally heterogenous; both isomerization states of the amino-terminal peptide bond are populated, and isomerization is slow compared with the time scale of the experiment [62]. To estimate the distance variation caused by *cis–trans* isomerization, we studied the peptides in MD simulations in collaboration with Roccatano [31].

4.4.4. MD Simulations on Donor–Acceptor-Labeled Polyproline Peptides

Using the GROMOS96 force field [63], 50-ns simulations of the double-labeled peptides were set up with all proline–proline bonds in a *trans* conformation and the tryptophan–proline bond in either a *trans* or a *cis* conformation. The peptides were solvated with water molecules in periodic cubic boxes sufficiently large to contain the peptide and 0.9 nm of solvent on all sides.

The simulation time was too short to allow isomerization to occur but sufficiently long to allow efficient sampling of the accessible conformational space. The average donor–acceptor distances in the *cis* peptides were always longer—by 0.7 to 1.7 Å—than the distances in the corresponding *trans* peptides (Fig. 4.8). The most populated structures revealed a tendency of tryptophan to fold back along the proline backbone, presumably, as a consequence of hydrophobic interactions.

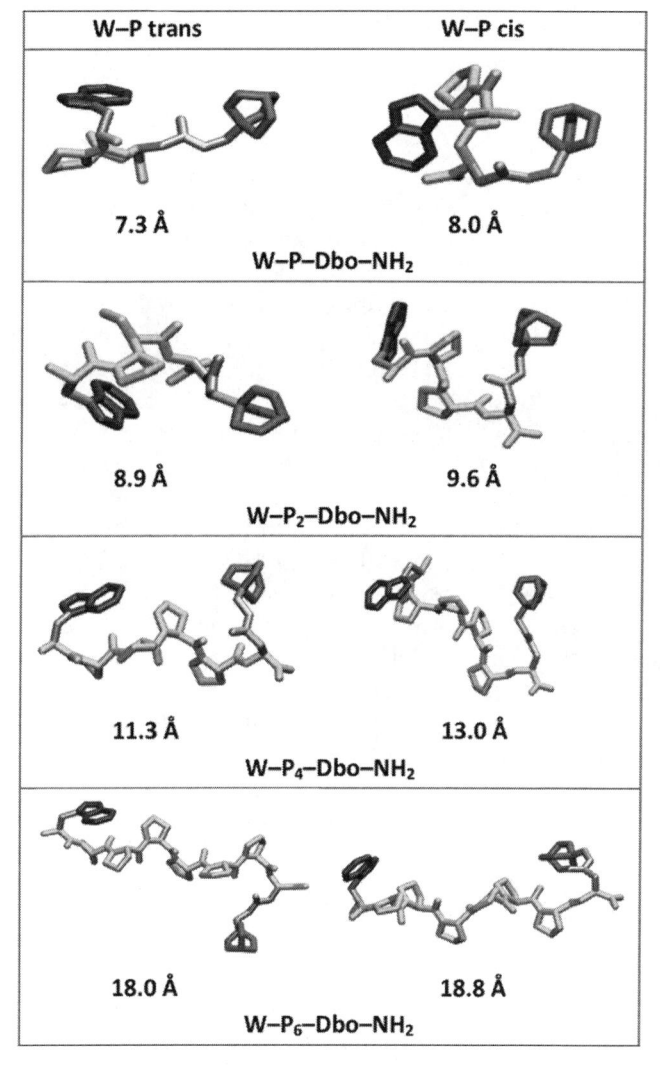

Figure 4.8. Average donor–acceptor distances and representative structures for the most populated clusters obtained from MD simulation, modified from Ref. [31]. Proline residues are shown in green, the donor chromophore in blue, and the acceptor chromophore in red.

We now return to the question why propylene glycol leads to such sharpened distance distributions. One could speculate that the less polar solvent weakens a possible hydrophobic interaction of Trp with backbone and induces a preferred conformation of the Trp–Pro bond such that only one narrow distance distribution remains in solution. However, *cis* and *trans* populations are so similar in distance that their simultanous presence in solution should at best perturb the distribution analysis of the shortest peptides, much less the analysis of the long peptides. What was observed was the opposite: the longer the peptide, the more striking was the effect of propylene glycol on its persistence length. Therefore, the most likely explanation is that the flexibility of the terminal residues, Trp and Dbo, depends weakly on the solvent, and the flexibility of the PPII helix depends strongly on the solvent. When peptides become longer, the relative contribution of the terminal residues to flexibility and persistence length decreases, and the rigidity of the PPII helix in solvents other than water becomes increasingly apparent.

The distance difference between the *cis* and *trans* forms is small but sufficiently large to allow future steady-state and time-resolved, double-kinetics experiments that are based on the Trp/Dbo FRET pair and that follow *cis–trans* isomerization reactions to be followed by fluorescence. A short calculation shall corroborate this idea: starting from Equation 4.3, $E = R_0^6 /(R_0^6 + R_{DA}^6)$ and evaluating the derivative of the efficiency with respect to the distance at the point $R_{DA} = R_0$, the result is $dE/dR_{DA}(R_0 = R_{DA}) = 2/3/R_{DA}$. At a Förster distance of 9.0 Å (Table 4.1), a distance variation in R_{DA} of 1 Å, as expected from *cis–trans* isomerization, would lead to an efficiency variation of 7.4%. This value is sufficiently large to follow the fluorescence of even fast transformations completed within submillisecond times. The simulations answered a further question important for future applications. In FRET distance determinations, it is often neglected that the distance between donor and acceptor is usually not equal to the intrachain distance adressed by the experiment. The larger the fluorescence labels, the larger is the difference between the donor–acceptor distance and the distance between, for instance, the C_α carbons that carry the chromophoric side chains. Values of donor–acceptor and $C_\alpha C_\alpha$ distances are compared for peptides in their all-*trans* form in Table 4.3. For all peptides but the shortest one, the $C_\alpha C_\alpha$ distance is larger by minimally 0 Å and maximally 3.7 Å. Thus, the size of the labels itself puts a limit to the accuracy to which an absolute distance can be determined. But, as shown by the above calculations and by experiments described below, short-distance FRET opens a path to monitor distance variations that are minute but consequential in a many intra- and intercellular processes. Specifically, we were able to analyze how short peptide segments structurally reorganize when a single internal residue is phosphorylated.

4.5. SHORT-DISTANCE FRET TO MONITOR CHAIN-STRUCTURAL TRANSITIONS UPON PHOSPHORYLATION

The phopshorylation of a protein by a protein kinase has been recognized as a key step in a wealth of processes that enable and sustain live, for example, cell differentiation, signal transduction, muscle contraction [64]). Any amino acid residue that carries a hydroxyl group—serine, threonine, tyrosine—can be phosphorylated, becomes negatively charge, and is then able to interact with basic amino acids such as arginine. We studied the structural consequences of serine and threonine phosphorylation on two model peptides that represent the recognition motifs of protein kinase A and C. Their sequences are, respectively, LRRW<u>S</u>LG (sequence **S**) and WKR<u>T</u>LRR (sequence **T**). Both peptides contain a single residue, serine or threonine, as target of enzymatic phosphorylation. Both contain a natural tryptophan that can serve as donor in FRET experiments. To probe peptide structure before and after phosphorylation, we appended Dbo to the C-terminus and measured effective distances and distance distributions in peptides containing serine (**S**) versus phosphoserine (**pS**) and threonine (**T**) versus phosphothreonine (**pT**). Fluorescence lifetimes and intensities of the donor in the absence of acceptor were obtained from the donor-only peptides, LRRWSLG–NH$_2$, and WK$_6$–NH$_2$. The individual peptides used in the analysis are denoted as in the following chart:

Peptide	Sequence S	Sequence T
Unphosphorylated	LRR–W–<u>S</u>–LG–Dbo **S**$_{DA}$	W–KR–<u>T</u>–LRR–Dbo **T**$_{DA}$
Phosphorylated	LRR–W–<u>pS</u>–LG–Dbo **pS**$_{DA}$	W–KR–<u>pT</u>–LRR–Dbo **pT**$_{DA}$
Reference	LRR–<u>W</u>–SLG-NH$_2$ **S**$_D$	<u>W</u>–K$_6$-NH$_2$ **T**$_D$

To simplify our experimental design, we did not adhere to the puristic protocol that would have demanded the use of two different donor-only peptides, one with unphosphorylated, one with phosphorylated residue. We did not select the peptide WKRTLRR as donor-only peptide **T**$_D$ in the analysis of sequence T but used WK$_6$–NH$_2$ instead: the fluorescence intensity and lifetime of tryptophan depends on its immediate microenvironment, which can be assumed to be virtually identical in the actually used and the ideal donor-only peptides. Further, the donor-only values enter the analysis as constants and, as such, might slightly affect the outcome of absolute values of efficiencies and distribution parameters but never their relative variation observed upon phosphorylation.

4.5.1. Efficiency Analysis

The steady-state and time-resolved FRET analysis was again performed in water and in propylene glycol. For segment **S**, the intensity of donor fluorescence decreased steeply upon going from the donor-only peptide **S**$_D$ to the donor–acceptor peptide, in which energy transfer was apparently efficient (Fig. 4.9, Table 4.3). It was, however, less efficient in **pS**$_{DA}$ than in **S**$_{DA}$, which indicates a larger donor–acceptor

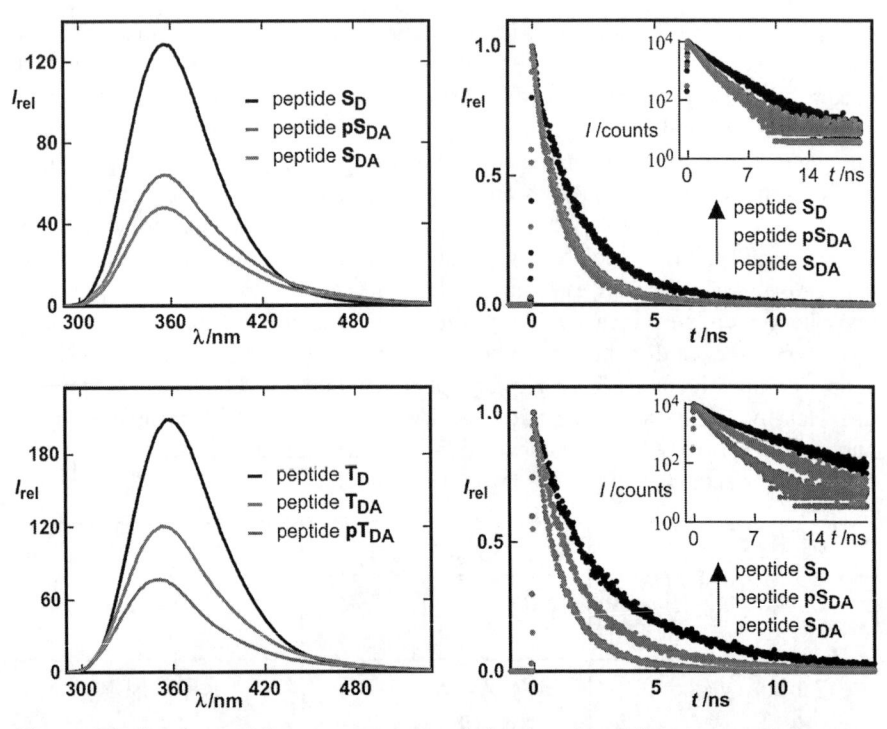

Figure 4.9. Steady-state fluorescence spectra (left column) and fluorescence decay traces (right column) of sequence **S** peptides (top row) and sequence **T** peptides (bottom row, inset: logarithmic intensity scale). Modified from Ref. [32].

TABLE 4.4. Transfer Efficiencies, Effective Distances, and Gaussian Distribution Parameters in Water (Propylene Glycol)[a]

Peptide	E_{ss}/%	E_{tr}/%	R_{ss}/Å	R_{tr}/Å	R_{mean}/Å	FWHM/Å
S_{DA}	62 (46)	42 (33)	8.6 (9.7)	9.9 (10.5)	9.7 (11.2)	3.0 (1.6)
pS_{DA}	50 (36)	36 (26)	9.4 (10.4)	10.3 (11.1)	10.5 (12.2)	3.1 (2.0)
T_{DA}	42 (33)	40 (22)	10.7 (11.6)	10.0 (11.5)	10.9 (12.4)	3.1 (1.2)
pT_{DA}	63 (50)	61 (30)	9.3 (10.3)	8.7 (10.8)	9.3 (11.5)	0.7 (0.2)

[a] Values in propylene glycol are given in parentheses.

spacing when serine is phosphorylated. For segment **T**, the opposite was observed, transfer was much less pronounced in T_{DA} than in pT_{DA}: Phosphorylation of threonine contracts the donor–acceptor spacing. The same conclusions were reached by time-resolved FRET analysis. The donor fluorescence decay of each peptide (Fig. 4.9) was modeled by a bi-exponential function yielding two lifetime and amplitude values, from which average lifetimes were calculated (Table 4.4).

The difference between the FRET efficiency values obtained from the fluorescence intensities and lifetimes, $\Delta E = E_{ss} - E_{tr}$, was more pronounced than in the case of the polyproline peptides, whose relative rigidity hindered the labels to come within contact distance. We shortly recall that if instantly after excitation the donor and the acceptor are near van der Waals contact, the donor is deactivated by FRET with a rate beyond the temporal resolution of the time-resolved instrument; the corresponding chain population remains invisible and shows up only as an apparently static quenching component in the steady-state measurements. The size of this "invisible" population correlates with ΔE and is characterized by an average van der Waals distance of $R_{vdw} = 4$ Å, as taken from MD simulations. The fraction of the visible donor–acceptor population correlates with $1 - \Delta E$ and is characterized by the distance R_{tr}. Thus, one can obtain a characteristic distance for the whole ensemble of molecules by weighing and summing the individual characteristic distances according to Equation 4.12. We found the result to be in all cases very close or identical to the measured, steady-state distance R_{ss}:

$$R_{ss} = \Delta E \cdot R_{vdW} + (1 - \Delta E) R_{tr}. \qquad (4.12)$$

The efficiency variation upon phosphorylation is highly significant, about 10% for segment **S** and 20% for segment **T**, but it corresponds to a variation of the effective donor–acceptor distance, which is rather small in absolute terms, that is, 0.8 Å for segment **S** and 1.4 Å for segment **T** (see Table 4.3). These experiments clearly prove the high spatial resolution (sub-Å) of short-distance FRET, but they do not yet enable an unambiguous statement on what exactly happens when the internal serine or threonine residue becomes phosphorylated. The next level of FRET analysis, the analysis of distance distributions, allowed such a statement.

4.5.2. Distribution Analysis

Distance distributions were obtained for all donor–acceptor peptides by fitting their time courses of donor fluorescence with a simple Gaussian distribution function (Eq. 4.9). The "invisible" chain population with donor–acceptor distances in contact distance was modeled by a further Gaussian distribution. This treatment was arbitrarily chosen to illustrate the information obtained from steady-state measurements in the distribution profiles (Fig. 4.10). But several MD simulations on short peptides indeed point to an additional pronounced peak of distance probability at contact distance [65, 66]. Propylene glycol led to sharper distributions, shifted to somewhat larger distances compared with the profiles in water. While the distance shift is most easily interpreted as a result of suppressed diffusion, the narrowing of distributions can, once again, not be explained in that way.

Compared with the profile of peptide **S**$_{DA}$ (left panel of Fig. 4.10, solid lines), the profile of the phosphorylated peptide **pS**$_{DA}$ (dotted lines) is slightly shifted to larger distances, in agreement with the result from efficiency analysis: phosphorylation induces donor–acceptor separation. This is obviously not due to a specific structure formation: the distributions of **S**$_{DA}$ and **pS**$_{DA}$ are relatively broad and of very similar widths. This is in stark contrast to the results obtained for segment **T** (right panel). Phosphorylation shifts the distribution to shorter distances, again in agreement with

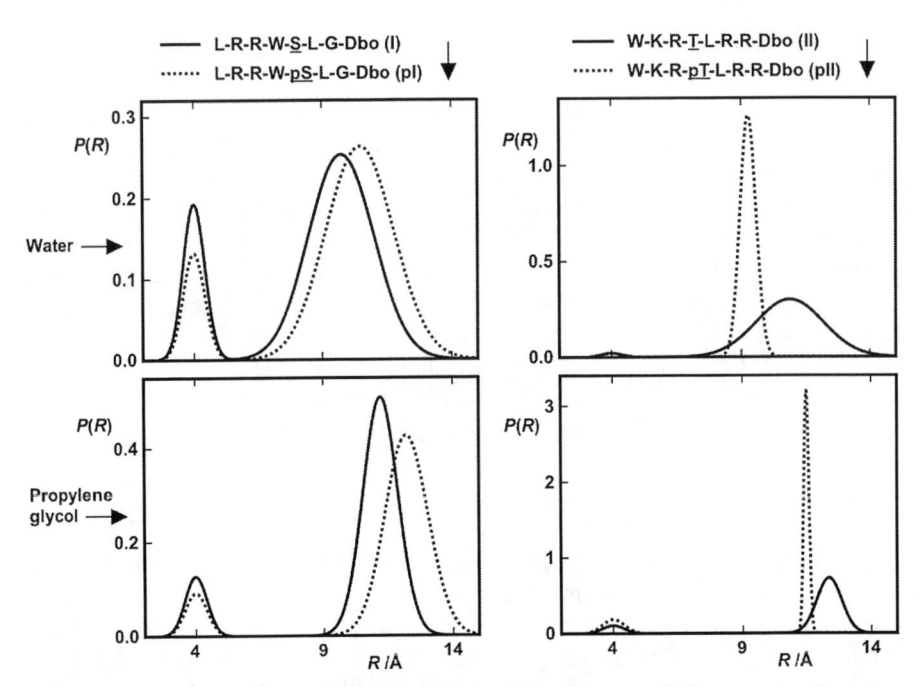

Figure 4.10. Gaussian distribution profiles of the donor–acceptor distances in peptides S_{DA} (solid line) and pS_{DA} (dotted line) on the left and in peptides T_{DA} and pT_{DA} on the right in water (top row) and propylene glycol (bottom row). Modified from Ref. [32].

the efficiency analysis, but it also leads to dramatically narrowed distributions (Table 4.3). It seems that upon phosphorylation the acceptor is pulled into a specific distance to the acceptor, and we assumed an intramolecular salt bridge to be responsible for this fixation. With the presumption of a specific salt bridge between the phosphothreonine residue and one of the arginine side chains in peptide pT_{DA}, we repeated the experiments under conditions that destabilize charge–charge interactions. Indeed, in the presence of 0.3 M sodium phosphate (pH 7.8) or 1–2 M sodium chloride (pH 6.8), no comparable effects induced by phosphorylation could be observed. The magnitudes of distance and distribution shifts can thus be modeled by assuming a salt bridge between phosphothreonine and the arginine residue nearest to Dbo (Fig. 4.11), as only this salt bridge could sufficiently displace the acceptor toward the donor as well as markedly rigidify the chain, as experimentally observed.

4.6. SHORT-DISTANCE FRET APPLIED TO THE STRUCTURELESS CHAIN

4.6.1. The Glycine–Serine Repeat Chain

So far it appeared that the lifetime of the donor, tryptophan, is too short to allow more than moderate contributions of donor–acceptor diffusion to FRET efficiency.

peptide T$_{DA}$

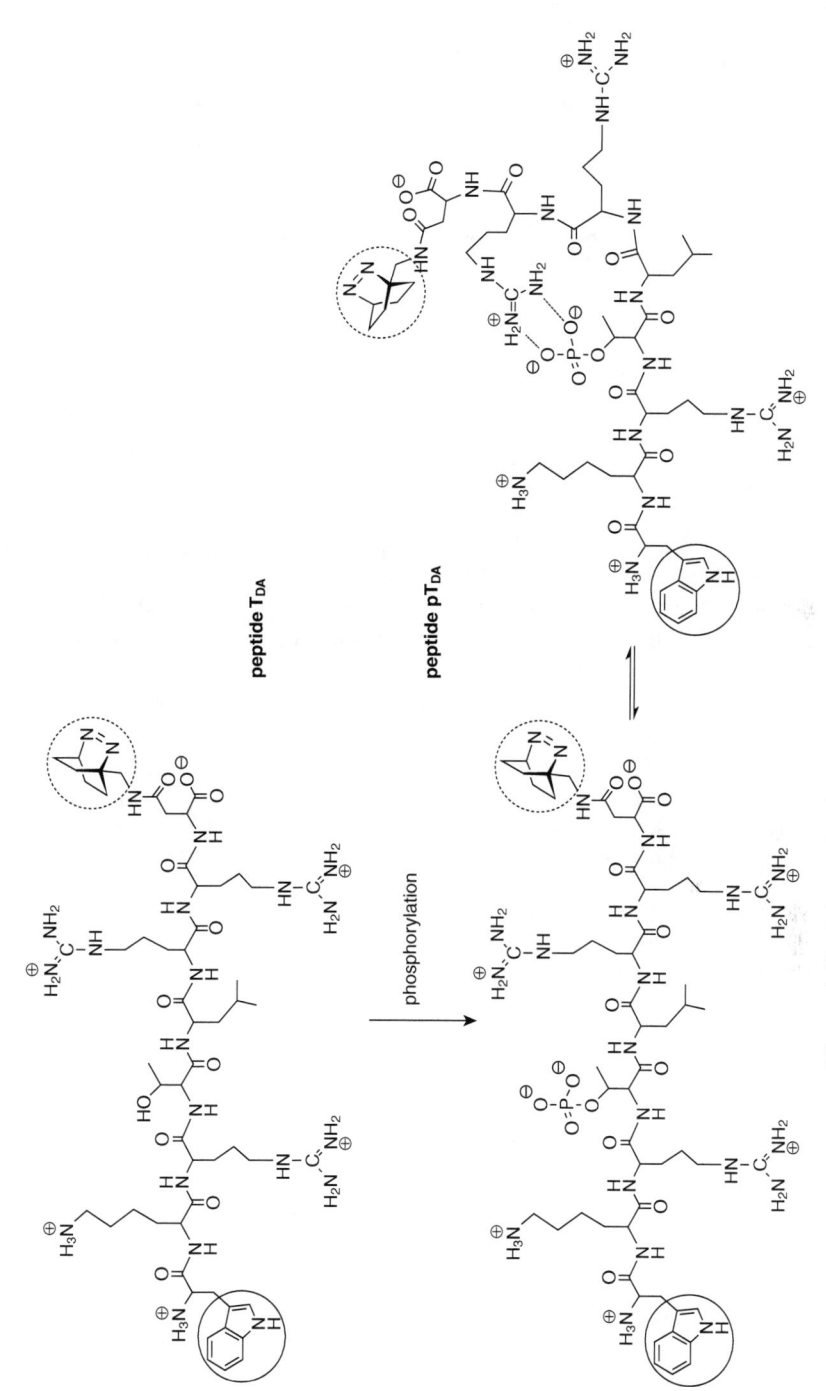

peptide pT$_{DA}$

Figure 4.11. Salt-bridge formation upon phosphorylation in peptide T$_{DA}$.

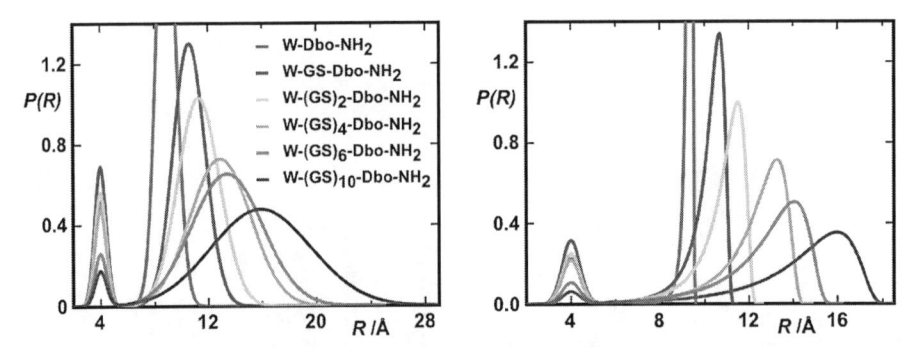

Figure 4.12. Gaussian (left) and worm-like chain (right) probability density distributions in propylene glycol, of donor–acceptor distances in double-labeled glycine–serine repeat peptides obtained from a combined steady-state and time-resolved FRET analysis. The short-distance distribution accounts for the chain population with instant donor deactivation, the long-distance distribution was obtained from fitting the time course of donor deactivation (Eq. 4.8). Modified from Ref. [30].

In the cases discussed so far, effective distances measured in propylene glycol were longer than in water as expected from suppressed diffusion, but, unexpectedly, the distribution profiles became narrower and not broader. Specific effects of propylene glycol—or water absence—are best studied on maximally flexible chains without structural bias. The polyproline peptide and the polyGS mark the extremes of minimal and maximal backbone flexibility. The GS building block guarantees that chains are highly hydrophilic but have a backbone whose conformational freedom is minimally impeded by side-chain bulk. The polyGS chain was introduced in 1999 by Kiefhaber and coworkers to study how fast a chain can form a loop—the minimal event of protein folding [67].

4.6.2. The W–Dbo and W–(GS)$_n$–Dbo Peptide

We investigated energy transfer in the polyGS peptides W–(GS)$_n$–Dbo–NH$_2$ with the number of GS repeats varying from $n = 0$ to 10. The W–Dbo peptide ($n = 0$) was prepared to establish FRET as the dominant mechanism of donor deactivation even when fluorophores are forced into distances close to the contact distance. Other mechanisms than the Förster dipole–dipole interaction could easily become effective, for example, the Dexter mechanism, deactivating the donor almost instantly and leading to apparent FRET efficiencies close to 100%: we found the energy transfer efficiency in this peptide to be accurately measurable and well below 100% (Table 4.5). The structural and dynamic dissimilarity of the polyproline and polyGS peptide becomes most apparent when one compares the longest peptides that still allow short-distance FRET: The longest polyproline sequence consists of six residues, the longest polyGS sequence of more than 22 residues.

TABLE 4.5. FRET Efficiency Analysis of W–(GS)$_n$–Dbo–NH$_2$ Peptides[a]

Peptide	E_{ss}/%	E_{tr}/%	R_{ss}/Å	R_{tr}/Å
W–Dbo–NH$_2$	84 (72)	69 (60)	7.0 (8.1)	8.1 (8.9)
W–GS–Dbo–NH$_2$	65 (56)	56 (40)	8.3 (9.1)	8.9 (10.2)
W–(GS)$_2$–Dbo–NH$_2$	46 (45)	42 (32)	9.5 (9.9)	9.7 (10.8)
W–(GS)$_4$–Dbo–NH$_2$	39 (32)	28 (21)	9.9 (10.8)	10.9 (11.9)
W–(GS)$_6$–Dbo–NH$_2$	27 (24)	22 (18)	10.9 (11.5)	11.4 (12.3)
W–(GS)$_{10}$–Dbo–NH$_2$	13 (12)	10 (8)	12.7 (13.3)	13.3 (14.4)

[a] Steady-state and time-resolved analysis for peptides in water and in propylene gylcol (values in parentheses).

TABLE 4.6. Distribution Parameters in Gly-Ser Repeat Peptides

Peptide	Gaussian[a] R_{mean}/Å	FWHM/Å	Worm-Like Chain[b] l_c/Å	l_p/Å
W–Dbo–NH$_2$	8.9	1.7	9.6	46.5
W–GS–Dbo–NH$_2$	10.6	2.8	11.5	13.8
W–(GS)$_2$–Dbo–NH$_2$	11.3	3.7	12.6	12.1
W–(GS)$_4$–Dbo–NH$_2$	12.9	5.3	14.9	11.4
W–(GS)$_6$–Dbo–NH$_2$	13.4	6.2	16.2	10.0
W–(GS)$_{10}$–Dbo–NH$_2$	15.9	15.9	18.8	10.4

[a] Donor fluorescence decays analyzed with a Gaussian distribution function (Eq. 4.9).
[b] Decays analyzed with a distribution function derived from a variant of the worm-like chain model (Eq. 4.10).

4.6.3. Solvent and Diffusion Effects

The transfer efficiencies measured in water and propylene glycol seem to be very similar and for some peptides virtually identical. One is tempted to conclude that diffusion contributes little to energy transfer as its suppression leads to little change. We have to recall, however, that the Förster radius increases from 9.2 Å in water to 9.5 Å in propylene glycol (Table 4.1). These radii seem to differ marginally, but raised to a power of 6 in the efficiency equation (Eq. 4.4) they would lead to efficiency values that are considerably larger in the viscogene. Instead, the values are up to 16% decreased, which is a first indication that suppressing diffusion decreases energy transfer. The considerable contribution of diffusion to transfer, in water, became evident, when we tried to analyze the distance distributions. Even when the analysis succeeded, the obtained distance profile was sharply peaked, mimicking the case of a fixed distance in a structured peptide. Reasonably broad distributions were obtained in propylene glycol (Fig. 4.12). Here, the Gaussian and the worm-like chain analysis conveyed similar information as both led to profiles with an almost identical distance of maximal probability. The average mean distance and its counterpart, the contour length, increased gradually with peptide length (Table 4.6). The persistence length, in contrast, reached plateau values around 10 Å, which is characteristic for semi-flexible biopolymers [36]. The shortest peptide, W–Dbo, is the stiffest (l_p = 47 Å) as it lacks the flexible GS linker. This rigidity is independently confirmed by a very low rate of loop formation, which has been determined in earlier experiments by exploiting collision-induced fluorescence quenching of Dbo by tryptophan [39]. The Gaussian distribution of the longest polyGS peptide displays almost identical values of the width and the mean distance—the rule-of-thumb benchmark for a random coil [68].

It is a question of fundamental importance to protein folding whether the bare amide backbone is inherently inclined to form hydrogen bonds in water, which, although appearing in patterns of only momentary stability, would nonetheless keep the chain in compact conformations. Kiefhaber and coworkers, who selected the polyGS sequence for its flexibility [67], tested it later for such a structural bias by using two regular FRET pairs (naphtalene/dansyl with R_0 = 23.3 Å, and pyrene/dansyl with 20.5 Å in water) [69]. They measured FRET in water in the absence and presence of guanidinium chloride, up to a very high concentration (8 M) of this highly water-soluble denaturant. In an inventive approach, they used two donors with different lifetimes (napthalene with 37 ns and pyrene with 226 ns) and based the simultaneous analysis of the donor fluorescence decays in the donor–acceptor peptides on Steinberg's equation (Eq. 4.11). In a correction of their initial results [70], they reported average donor–acceptor distances of 38 Å (initially 18.9 Å) in the absence and 50 Å (initially 39.2 Å) in the presence of 8 M guanidinium chloride as well as mutual diffusion coefficients of 50 Å²/ns (initially 4 Å²/ns) and 60 Å²/ns (initially 15 Å²/ns). The error margins in these measurements are presumably large because the Förster distances, determined in water, of the two employed FRET pairs were only 23.3 Å and 20.5 Å (recall that FRET analysis becomes inaccurate when the distances approach either $0.5R_0$ or $2R_0$). Although not reported, it is likely that the Förster radii were larger in the presence of the denaturant.

The negligible difference of the donor–acceptor diffusion coefficients, normalized with respect to solution viscosity, indicates that the denaturant had little effect on the internal diffusion and the internal friction in polyGS. Such results do certainly not support the initial conclusion drawn from the uncorrected analysis that the chain has a propensity to form internal hydrogen bonds [69]. In an earlier detailed analysis, Kiefhaber and coworkers evidenced that the guanidinium cation binds weakly to the chain backbone [71]. Its bulkiness might stiffen the chain, its charge might lead to internal electrostatic repulsion, and both of these effects can easily explain the observed moderate increase in end-to-end distance. To summarize, the image of polyGS as an almost paradigmatic example for the structurally unbiased polypeptide remains intact; the near future promises more debates and detailed investigations of this important issue. In the course of this process, the FRET analysis of diffusional distance distributions has to be extended and refined.

4.7. THE FUTURE OF SHORT-DISTANCE FRET

The bicyclic DBO molecule is a tiny fluorophore with a minimal spread of transition dipole density; these are two decisive factors for the detection and determination of short distances by FRET. DBO is hydrophilic and not prone to stick to hydrophobic regions of peptides or proteins. The risk of label-induced structural perturbations is minimal as well as the risk of constraining the orientations of the transition dipole vector: the value of the orientation factor, one of the determinants of the Förster distance, remains constant. Conjugated to asparagine, DBO can be easily inserted during peptide synthesis at every location of a chain. For post-column labeling, we have already prepared DBO derivatives that react selectively with thiol and amino groups. So far, the synthesis of DBO has been costly and complicated; an optimized, economical procedure will be published soon.

In peptides labeled with Trp and Dbo, FRET remains the main mechanism of donor fluorescence deactivation even at donor–acceptor distances close to the van der Waals distance. The FRET pair has a Förster radius of about 9 Å, with minimal dependence on the environmental conditions. With such a short radius, it is possible to monitor even minute distance variations with maximal resolution, and this paves the way for the study of the imminent structural consequences of, for example, amide *cis–trans* isomerization and enzymatic phosphorylation. Exciting "double-kinetics" experiments are also within reach. Perhaps most important, the Trp/Dbo pair offers the possibility to simultaneously study the same amino acid sequence by short-distance FRET and collision-induced fluorescence quenching (CIFQ). FRET quenching of Trp fluorescence by DBO and diffusion-controlled, collision-induced quenching of DBO fluorescence by Trp are processes that do not interfere with each other but provide independent and complementary, structural and dynamic, insights that can be accessed under almost identical experimental conditions. The additional information provided by CIFQ could be decisive in devising an analysis that affords more accurate distance distributions as well as more accurate diffusion coefficients. When folding reactions of peptides and proteins are studied by time-resolved FRET, the long fluorescence lifetime of DBO allows one to easily separate the acceptor signal from noise such that FRET analysis can alternatively be based on acceptor fluorescence [24]. The unresolved discrepancies between theory (simulation) and experiment will continue to attract attention [29].

ACKNOWLEDGMENTS

We would like to thank the Deutsche Forschungsgemeinschaft (DFG, NA 686/6, "Polypeptide dynamics and structure studied simultaneously by collision-induced fluorescence quenching and resonance energy transfer in the 10-Å domain" and the Fonds der Chemischen Industrie for financial support. We thank Prof. Dr. Danilo Roccatano for his continued support on MD simulations of Dbo-labeled peptides published in the original studies [30, 31, 72].

DEDICATION

M.H.J dedicates this chapter to the memory of Vladimir Ratner († 23.11.2004), a scientist with a formidable human soul.

REFERENCES

1. May, V. and Kühn, O. (2000) *Charge and energy transfer dynamics in molecular systems*, Wiley-VCH, Berlin.
2. van Der Meer, B. W., Coker, G., and Chen, S.-Y. S. (1994) *Resonance energy transfer: Theory and data*, John Wiley and Sons Ltd., New York.
3. Lakowicz, J. R. (2006) *Principles of fluorescence spectroscopy*, 3rd ed., Kluwer Academic/Plenum, New York.
4. Förster, T. (1948) Zwischenmolekulare Energiewanderung und Fluoreszenz, *Ann Phys* 437, 55–75.
5. Stryer, L. and Haugland, R. P. (1967) Energy transfer: A spectroscopic ruler, *Proc Natl Acad Sci U S A* 58, 719–726.
6. Grinvald, A., Haas, E., and Steinberg, I. Z. (1972) Evaluation of the distribution of distances between energy donors and acceptors by fluorescence decay, *Proc Natl Acad Sci U S A* 69, 2273–2277.
7. Haas, E., Wilchek, M., Katchalski-Katzir, E., and Steinberg, I. Z. (1975) Distribution of end-to-end distances of oligopeptides in solution as estimated by energy transfer, *Proc Natl Acad Sci U S A* 72, 1807–1811.
8. Haas, E., Katchalski-Katzir, E., and Steinberg, I. Z. (1978) Brownian motion of the ends of oligopeptide chains in solution as estimated by energy transfer between the chain ends, *Biopolymers* 17, 11–31.
9. Beechem, J. M. and Haas, E. (1989) Simultaneous determination of intramolecular distance distributions and conformational dynamics by global analysis of energy transfer measurements, *Biophys J* 55, 1225–1236.
10. Gryczynski, I., Wiczk, W., Johnson, M. L., Cheung, H. C., Wang, C.-K., and Lakowicz, J. R. (1988) Resolution of end-to-end distance distributions of flexible molecules using quenching-induced variations of the Förster distance for fluorescence energy transfer, *Biophys J* 54, 577–586.
11. Amir, D. and Haas, E. (1987) Estimation of intramolecular distance distributions in bovine pancreatic trypsin inhibitor by site-specific labeling and nonradiative excitation energy-transfer measurements, *Biochemistry* 26, 2162–2175.
12. Amir, D. and Haas, E. (1988) Reduced bovine pancreatic trypsin inhibitor has a compact structure, *Biochemistry* 27, 8889–8893.
13. Ratner, V., Sinev, M., and Haas, E. (2000) Determination of intramolecular distance distribution during protein folding on the millisecond timescale, *J Mol Biol* 299, 1363–1371.
14. Ratner, V., Kahana, E., and Haas, E. (2002) The natively helical chain segment 169–188 of *Escherichia coli* adenylate kinase is formed in the latest phase of the refolding transition, *J Mol Biol* 320, 1135–1145.

15 Ratner, V. and Haas, E. (1998) An instrument for time resolved monitoring of fast chemical transitions: Application to the kinetics of refolding of a globular protein, *Rev Sci Instrum 69*, 2147–2154.

16 Plaxco, K. W., Simons, K. T., Ruczinski, I., and Baker, D. (2000) Topology, stability, sequence, and length: Defining the determinants of two-state protein folding kinetics, *Biochemistry 39*, 11177–11183.

17 Jacob, M., Schindler, T., Balbach, J., and Schmid, F. X. (1997) Diffusion control in an elementary protein folding reaction, *Proc Natl Acad Sci U S A 94*, 5622–5627.

18 Dill, K. A. and Chan, H. S. (1997) From Levinthal to pathways to funnels, *Nat Struct Biol 4*, 10–19.

19 Jacob, J., Krantz, B., Dothager, R. S., Thiyagarajan, P., and Sosnick, T. R. (2004) Early collapse is not an obligate step in protein folding, *J Mol Biol 338*, 369–382.

20 Kohn, J. E., Millett, I. S., Jacob, J., Zagrovic, B., Dillon, T. M., Cingel, N., Dothager, R. S., Seifert, S., Thiyagarajan, P., Sosnick, T. R., Hasan, M. Z., Pande, V. S., Ruczinski, I., Doniach, S., and Plaxco, K. W. (2004) Random-coil behavior and the dimensions of chemically unfolded proteins, *Proc Natl Acad Sci U S A 101*, 12491–12496.

21 Chen, Y., Wedemeyer, W. J., and Lapidus, L. J. (2010) A general polymer model of unfolded proteins under folding conditions, *J Phys Chem B 114*, 15969–15975.

22 Cortajarena, A. L., Lois, G., Sherman, E., O'Hern, C. S., Regan, L., and Haran, G. (2008) Non-random-coil behavior as a consequence of extensive PPII structure in the denatured state, *J Mol Biol 382*, 203–212.

23 Magg, C. and Schmid, F. X. (2004) Rapid collapse precedes the fast two-state folding of the cold shock protein, *J Mol Biol 335*, 1309–1323.

24 Orevi, T., Ben Ishay, E., Pirchi, M., Jacob, M. H., Amir, D., and Haas, E. (2009) Early closure of a long loop in the refolding of adenylate kinase: A possible key role of non-local interactions in the initial folding steps, *J Mol Biol 385*, 1230–1242.

25 Magg, C., Kubelka, J., Holtermann, G., Haas, E., and Schmid, F. X. (2006) Specificity of the initial collapse in the folding of the cold shock protein, *J Mol Biol 360*, 1067–1080.

26 Sinha, K. K. and Udgaonkar, J. B. (2007) Dissecting the non-specific and specific components of the initial folding reaction of barstar by multi-site FRET measurements, *J Mol Biol 370*, 385–405.

27 Ratner, V., Kahana, E., Eichler, M., and Haas, E. (2002) A general strategy for site-specific double labeling of globular proteins for kinetic FRET studies, *Bioconjug Chem 13*, 1163–1170.

28 Jacob, M. H., Amir, D., Ratner, V., Gussakowsky, E., and Haas, E. (2005) Predicting reactivities of protein surface cysteines as part of a strategy for selective multiple labeling, *Biochemistry 44*, 13664–13672.

29 Verbaro, D., Gosh, I., Nau, W. M., and Schweitzer-Stenner, R. (2010) Discrepancies between conformational distributions of a polyalanine peptide in solution obtained from molecular dynamics force fields and amide I' band profiles, *J Phys Chem B 114*, 17201–17208.

30 Sahoo, H., Roccatano, D., Zacharias, M., and Nau, W. M. (2006) Distance distributions of short polypeptides recovered by fluorescence resonance energy transfer in the 10 Å domain, *J Am Chem Soc 128*, 8118–8119.

31. Sahoo, H., Roccatano, D., Hennig, A., and Nau, W. M. (2007) A 10-Å spectroscopic ruler applied to short polyprolines, *J Am Chem Soc 129*, 9762–9772.
32. Sahoo, H. and Nau, W. M. (2007) Phosphorylation-induced conformational changes in short peptides probed by fluorescence resonance energy transfer in the 10-Å domain, *ChemBioChem 8*, 567–573.
33. Sahoo, H., Hennig, A., and Nau, W. M. (2006) Temperature-dependent loop formation kinetics in flexible peptides studied by time-resolved fluorescence spectroscopy, *Int J Photoenergy 8*, 1–9.
34. Khan, Y. R., Dykstra, T. E., and Scholes, G. D. (2008) Exploring the Förster limit in a small FRET pair, *Chem Phys Lett 461*, 305–309.
35. Haas, E., Katchalski-Katzir, E., and Steinberg, I. Z. (1978) Effect of the orientation of donor and acceptor on the probability of energy transfer involving electronic transitions of mixed polarization, *Biochemistry 17*, 5064–5070.
36. Thirumalai, D. and Ha, B.-Y. (1998) Statistical mechanics of semiflexible chains: A meanfield variational approach. In: Grossberg, A. (ed.), *Theoretical and mathematical models in polymer research*. Academia, New York, pp. 1–35.
37. Lide, D. R. (2003) *Handbook of chemistry and physics*, 84th ed., CRC Press, Boca Raton.
38. Nau, W. M., Greiner, G., Rau, H., Wall, J., Olivucci, M., and Scaiano, J. C. (1999) Fluorescence of 2,3-diazabicyclo[2.2.2]oct-2-ene revisited: Solvent-induced quenching of the n,pi*-excited state by an aborted hydrogen atom transfer, *J Phys Chem A 103*, 1579–1584.
39. Hudgins, R. R., Huang, F., Gramlich, G., and Nau, W. M. (2002) A fluorescence-based method for direct measurement of submicrosecond intramolecular contact formation in biopplymers: An exploratory study with polypeptides, *J Am Chem Soc 124*, 556–564.
40. Huang, F. and Nau, W. M. (2003) A conformational flexibility scale for amino acids in peptides, *Angew Chem Int Ed 42*, 2269–2272.
41. Engel, P. S., Horsey, D. W., Scholz, J. N., Karatsu, T., and Kitamura, A. (1992) Intramolecular triplet energy-transfer in ester-linked bichromophoric azoalkanes and naphthalenes, *J Phys Chem B 96*, 7524–7535.
42. Nau, W. M. (1998) A fluorescent probe for antioxidants, *J Am Chem Soc 120*, 12614–12618.
43. Nau, W. M. and Zhang, X. Y. (1999) An exceedingly long-lived fluorescent state as a distinct structural and dynamic probe for supramolecular association: An exploratory study of host-guest complexation by cyclodextrins, *J Am Chem Soc 121*, 8022–8032.
44. Nau, W. M., Huang, F., Wang, X., Bakirci, H., Gramlich, G., and Marquez, C. (2003) Exploiting long-lived molecular fluorescence, *Chimia 57*, 161–167.
45. Cowan, P. M. and McGavin, S. (1955) Structure of poly-L-proline, *Nature 176*, 501–503.
46. Bochicchio, B. and Tamburro, A. M. (2002) Polyproline II structure in proteins: Identification by chiroptical spectroscopies, stability, and functions, *Chirality 14*, 782–792.
47. Eker, F., Griebenow, K., Cao, X., Nafie, L. A., and Schweitzer-Stenner, R. (2004) Preferred peptide backbone conformations in the unfolded state revealed by the structure analysis of alanine-based (AXA) tripeptides in aqueous solution, *Proc Natl Acad Sci U S A 101*, 10054–10059.

REFERENCES

48 Schweitzer-Stenner, R., Eker, F., Griebenow, K., Cao, X., and Nafie, L. A. (2004) The conformation of tetraalanine in water determined by polarized Raman, FT-IR, and VCD spectroscopy, *J Am Chem Soc 126*, 2768–2776.

49 Rath, A., Davidson, A. R., and Deber, C. M. (2005) The structure of "unstructured" regions in peptides and proteins: Role of the polyproline II helix in protein folding and recognition, *Biopolymers 80*, 179–185.

50 Mahoney, N. M., Janmey, P. A., and Almo, S. C. (1997) Structure of the profilin-poly-L-proline complex involved in morphogenesis and cytoskeletal regulation, *Nat Struct Biol 4*, 953–960.

51 Schuler, B., Lipman, E. A., Steinbach, P. J., Kumke, M., and Eaton, W. A. (2005) Polyproline and the "spectroscopic ruler" revisited with single-molecule fluorescence, *Proc Natl Acad Sci U S A 102*, 2754–2759.

52 Lakowicz, J. R., Wiczk, W., Gryczynski, I., and Johnson, M. L. (1990) Influence of oligopeptide flexibility on donor-acceptor distance distribution by frequency-domain fluorescence spectroscopy, *Proc SPIE 1204*, 192–205.

53 Ruttinger, S., Macdonald, R., Kramer, B., Koberling, F., Roos, M., and Hildt, E. (2006) Accurate single-pair Forster resonant energy transfer through combination of pulsed interleaved excitation, time correlated single-photon counting, and fluorescence correlation spectroscopy, *J Biomed Opt 11*, 024012.

54 Reimer, U., Scherer, G., Drewello, M., Kruber, S., Schutkowski, M., and Fischer, G. (1998) Side-chain effects on peptidyl-prolyl cis/trans isomerisation, *J Mol Biol 279*, 449–460.

55 Best, R. B., Merchant, K. A., Gopich, I. V., Schuler, B., Bax, A., and Eaton, W. A. (2007) Effect of flexibility and cis residues in single-molecule FRET studies of polyproline, *Proc Natl Acad Sci U S A 104*, 18964–18969.

56 Helbecque, N. and Loucheux-Lefebvre, M. H. (1982) Critical chain length for polyproline-II structure formation in H-Gly-(Pro)$_n$-OH, *Int J Pept Protein Res 19*, 94–101.

57 Wierzchowski, K. L., Majcher, K., and Poznanski, J. (1995) CD investigations on conformation of H-X-(Pro)$_n$-Y-OH peptides (X = Trp, Tyr; Y = Tyr, Met); models for intramolecular long range electron transfer, *Acta Biochim Pol 42*, 259–268.

58 Kakinoki, S., Hirano, Y., and Oka, M. (2005) On the stability of polyproline-I and II structures of proline oligopeptides, *Polym Bull 53*, 109–115.

59 Poznański, J., Ejchart, A., Wierzchowski, K. L., and Ciurak, M. (1993) ^{1}H- and ^{13}C-NMR investigations on cis-trans isomerization of proline peptide bonds and confromation of aromatic side chains in H-Trp-(Pro)$_n$-Tyr-OH peptides, *Biopolymers 33*, 781–795.

60 Grathwohl, C. and Wüthrich, K. (1976) NMR studies of the molecular conformations in the linear oligopeptides H-(L-Ala)$_n$-L-Pro-OH, *Biopolymers 15*, 2043–2057.

61 Grathwohl, C. and Wüthrich, K. (1976) The X-Pro peptide bond as an NMR probe for conformational studies of flexible linear peptides, *Biopolymers 15*, 2025–2041.

62 Grathwohl, C. and Wüthrich, K. (1981) NMR studies of the rates of proline cis-trans isomerization in oligopeptides, *Biopolymers 20*, 2623–2633.

63 van Gunsteren, W. F., Daura, X., and Mark, A. E. (1998) GROMOS96, *Encycl Comput Chem 2*, 1211–1216.

64 Adams, J. A. (2001) Kinetic and catalytic mechanisms of protein kinases, *Chem Rev 101*, 2271–2290.

65 Yeh, I.-C. and Hummer, G. (2002) Peptide loop-closure kinetics from microsecond molecular dynamics simulations in explicit solvent, *J Am Chem Soc 124*, 6563–6568.

66 Roccatano, D., Nau, W. M., and Zacharias, M. (2004) Structural and dynamic properties of the CAGQW peptide in water: A molecular dynamics simulation study using different force fields, *J Phys Chem B 108*, 18734–18742.

67 Bieri, O., Wirz, J., Hellrung, B., Schutkowski, M., Drewello, M., and Kiefhaber, T. (1999) The speed limit for protein folding measured by triplet-triplet energy transfer, *Proc Natl Acad Sci U S A 96*, 9597–9601.

68 Rubinstein, M. and Colby, R. (2003) *Polymer physics*, Oxford University Press, Oxford.

69 Möglich, A., Joder, K., and Kiefhaber, T. (2006) End-to-end distance distributions and intrachain diffusion constants in unfolded polypeptide chains indicate intramolecular hydrogen bond formation, *Proc Natl Acad Sci U S A 103*, 12394–12399.

70 Möglich, A., Joder, K., and Kiefhaber, T. (2008) Correction for Möglich et al., End-to-end distance distributions and intrachain diffusion constants in unfolded polypeptide chains indicate intramolecular hydrogen bond formation, *Proc Natl Acad Sci U S A 105*, 6787.

71 Möglich, A., Krieger, F., and Kiefhaber, T. (2005) Molecular basis for the effect of urea and guanidinium chloride on the dynamics of unfolded polypeptide chains, *J Mol Biol 345*, 153–162.

72 Roccatano, D., Sahoo, H., Zacharias, M., and Nau, W. M. (2007) Temperature dependence of looping rates in a short peptide, *J Phys Chem B 111*, 2639–2646.

5

SOLVATION AND ELECTROSTATICS AS DETERMINANTS OF LOCAL STRUCTURAL ORDER IN UNFOLDED PEPTIDES AND PROTEINS

Franc Avbelj

5.1. LOCAL STRUCTURAL ORDER IN UNFOLDED PEPTIDES AND PROTEINS

One of the most difficult problems in chemistry is how a protein molecule folds from an unfolded state to its native conformation. It has been suggested that the local structural order (i.e., residual structure) may guide a polypeptide chain from the denatured to the native state [1, 2]. To understand the process of protein folding and misfolding, it is important to understand the nature of the local structural order in unfolded peptides and proteins.

Chemically unfolded proteins (8 M urea or guanidinium hydrochloride) are generally highly open and solvent-exposed molecules with no detectable α_R-helix or β-sheet secondary structures. Proteins under such strong denaturing conditions are considered to be completely unfolded [3]. The nuclear magnetic resonance (NMR) coupling constants $^3J(H^\alpha, H^N)$ and chemical shifts of residues in completely unfolded proteins, in general, deviate little from the respective values in small peptides in aqueous solution [4–13]. Small peptides in water were presumed to be completely unstructured (random coil); therefore, it has been assumed that chemically denatured

Protein and Peptide Folding, Misfolding, and Non-Folding, First Edition. Edited by Reinhard Schweitzer-Stenner.
© 2012 John Wiley & Sons, Inc. Published 2012 by John Wiley & Sons, Inc.

proteins are true random coils that can be modeled by Flory's random coil model (RCM)[14–16].

But experimental studies have shown that small peptides, which are too short to form any structure that is stabilized by long-range peptide hydrogen bonds, have a considerable local structural order [17–45]. Moreover, a growing number of experimental studies have shown that unfolded proteins display a certain degree of local structural order even under the most severe denaturing conditions. The local structural order in unfolded proteins is demonstrated by the following four indicators: the backbone conformational preferences [4–13, 46], the nearest-neighbor effect [47–53], the cooperative formation of larger local structures [5, 7–13, 46, 54–57], and the hydrophobic clusters [58–60]. These indicators differ in the level of cooperativity, that is, the number of adjacent residues involved.

The backbone conformational preference of a residue is an intrinsic property of a residue, determined by its side chain. For example, alanine residues in unfolded polypeptides prefer to adopt the P_{II} conformation (Fig. 5.1). The backbone conformational preferences of blocked peptides with a single amino acid residue (dipeptides) are similar to those found in short peptides and the "coil" residues in the database of native proteins [37]. The "coil" residues are those amino acid residues in the database of native proteins that are not found in α-helix and β-sheet secondary structures. This similarity indicates that the backbone conformational preferences of residues in unfolded and folded polypeptides are determined already at the dipeptide level. The conformational preferences of dipeptides thus represent true intrinsic backbone propensities of residues [37].

The nearest-neighbor effect is a consequence of interactions between two nearest-neighbor residues [47–53]. Its structural role was discovered by Penkett et al. [49] by measuring the NMR $^3J_{HN\alpha}$ coupling constants of unfolded polypeptides. These constants are directly related to the backbone angle φ by the Karplus relation. When a neighboring residue ($i - 1$ or $i + 1$) belongs to class L (aromatic and β-branched amino acid residues, FHITVWY) rather than class S (all others, G and P excluded), then the backbone angle φ of residue i is more negative for all amino acid residues.

Some completely unfolded proteins show distinct patterns of backbone conformational preferences and flexibility along sequences. The reason for these patterns is cooperative formation of larger local structures involving more than two adjacent residues [5, 7–13, 46, 54–57]. These patterns sometimes coincide with secondary structures in the native states. For example, the residues in completely unfolded immunoglobulin superfamily domain IgSF that correspond to the β-strands in native protein have been found to be stiffer than the other residues [8].

The hydrophobic clusters among residues distant in sequence have been observed in some completely unfolded proteins [58–60]. It has been shown that urea and guanidinium hydrochloride at large concentrations considerably decrease the strengths of hydrophobic interactions and backbone–backbone hydrogen bonding in polypeptides [61–63]. Nevertheless, it appears that the strength of hydrophobic interactions is substantial even at the very large concentrations of chemical denaturants. We will not discuss this subject here.

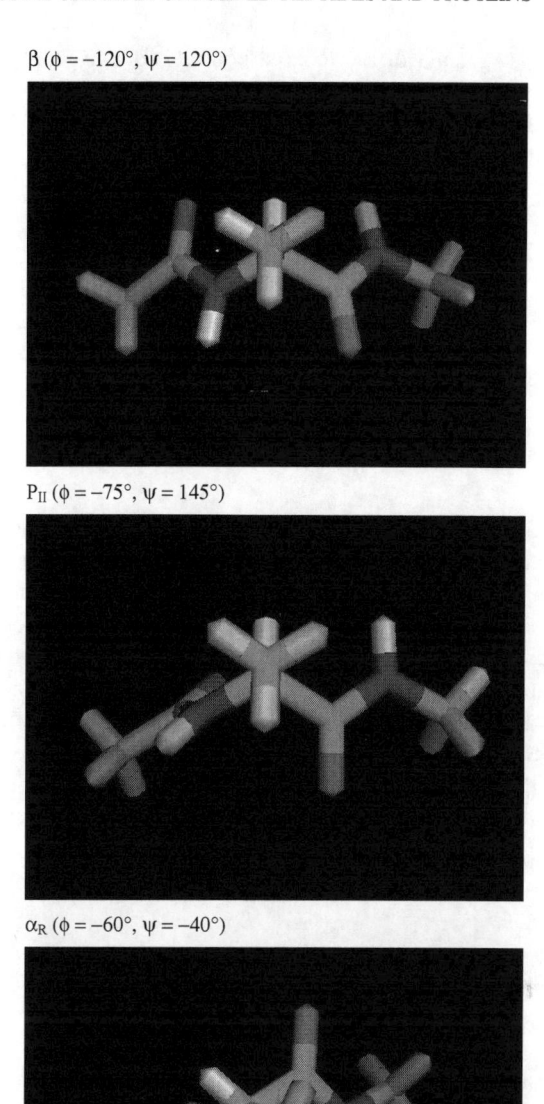

Figure 5.1. Alanine dipeptide in the β ($\varphi = -120°$, $\psi = 120°$), P_{II} ($\varphi = -75°$, $\psi = 145°$), and α_R ($\varphi = -60°$, $\psi = -40°$) conformations. See color insert.

The physical background of the local structural order in unfolded polypeptides is a highly controversial issue. Two theoretical models of unfolded proteins, Flory's RCM [14–16] and the database propensity model (DPM) [10, 49, 64–66], are unable to model the local structural order observed in unfolded proteins [53, 67]. The *ab initio* quantum mechanical calculations and simulations using empirical force fields generally fail to predict the backbone conformational preferences of even small systems such as alanine dipeptide in aqueous solution [68–74]. However, it has been shown that the electrostatic screening model (ESM) of unfolded peptides and proteins can explain the backbone conformational preferences [37, 67, 75–77], the nearest-neighbor effect [53], and the cooperative formation of local structures in unfolded ubiquitin [46].

5.2. ESM

The ESM [37, 43, 46, 53, 67, 75–82] utilizes an extraordinary ability of water to screen electrostatic interactions. Bringing two cations from infinity to contact distance in vapor phase is a highly unfavorable process. In the solvent, however, water dipoles almost completely compensate (screen) unfavorable electrostatic interaction by stabilizing cations in close contact much more than separated ions [83]. Analogously, water dipoles also screen an increase of the local main-chain electrostatic energy E_{local} in the unfavorable transition from the β to α_R conformation. The transition from the β to α_R conformation is unfavorable because the alignment of NH and CO dipole moments within a residue is anti-parallel in the β-strand conformation but parallel in the α_R conformation (Fig. 5.2). The difference in E_{local} for

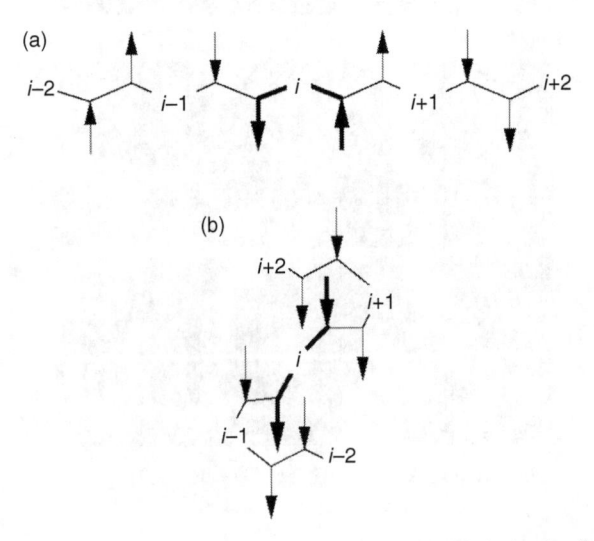

Figure 5.2. Orientation of dipole moments of a polypeptide chain in the β (A) and α_R (B) conformations.

the transition from the β conformation to the α_R conformation is large, that is, ~4.8 kcal/mol.

E_{local} is the electrostatic energy of a residue arising from interactions of main-chain CO and NH groups of the same residue with each other and with the first neighbor peptide bonds [14–16, 75]. Flory and coworkers have shown that the electrostatic interactions of second neighbor peptide bonds in unfolded proteins are smaller than those of the first neighbors by at least an order of magnitude and can be ignored [14–16]. Splitting of backbone–backbone electrostatic energy into local E_{local} and less significant non-local $E_{nonlocal}$ contributions considerably simplifies the analysis of backbone electrostatic interactions in unfolded polypeptides. E_{local} is equal to V_d used by Flory and coworkers in the RCM of unfolded polypeptides, provided that the same point atomic charges are used. The local main-chain electrostatic energy of a residue E_{local} is calculated using point atomic charges and Coulomb's law with a dielectric constant of 1. The value of E_{local} of a residue i depends on the backbone conformations of a triplet, that is, the six backbone dihedral angles of residues i, $i - 1$, and $i + 1$.

The *ab initio* quantum mechanical calculations of alanine dipeptide in the gas phase confirm that the β conformation is by ~3.3 kcal/mol more stable than the α_R conformation [73, 84–89]. Moreover, there is a significant correlation between the relative *ab initio* energy $E_{ab\,initio}$ with the electrostatic energy E_{electr} (Fig. 5.3) of the most important conformations of alanine dipeptide in the gas phase [37]. The E_{electr} is a sum of electrostatic energies E_{local} of alanine residue and blocking acetyl and N-methyl groups. The correlation coefficient is 0.991 for 14 points. This correlation shows that the simple Coulomb's point atomic electrostatic energy dominates the conformational equilibrium of alanine dipeptide in the gas phase.

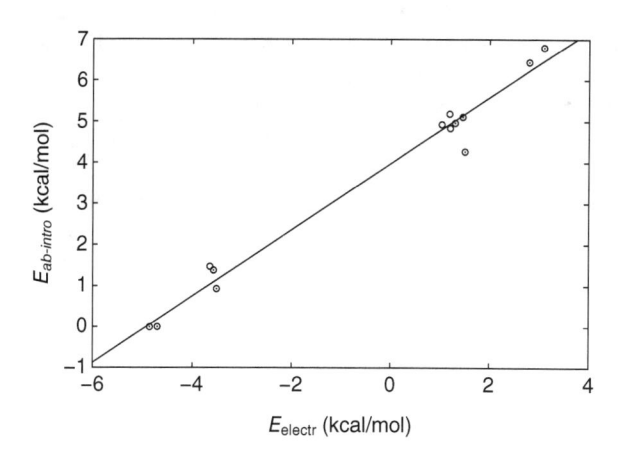

Figure 5.3. Correlation between the relative *ab initio* energy $E_{ab\,initio}$ and the electrostatic energy E_{electr} of the following conformations of alanine dipeptide in the gas phase: $C7_{eq}$, β, α_R, α_L, 3–10, and δ_L. The correlation coefficient is 0.991. The number of data points is 14. The relative values of $E_{ab\,initio}$ have been calculated using the following levels of theory: MP2/CBS limit//MP2/aug-cc-pVDZ [86], MP2/aug-cc-pVTZ* [89], and LMP2/cc-VQZ(-g) [88].

The large energy difference between the β and α_R conformations is considerably reduced in an aqueous environment. The electric fields produced by the parallel dipoles of peptide bonds flanking a residue in the α_R conformation reinforce each other, resulting in strong interactions with the dipoles of water molecules. The resulting energy offsets the unfavorable intra-backbone contribution. Conversely, for a residue in the β conformation, the peptide dipoles are anti-parallel and result in a weak electric field and thus weaker interactions with water in the vicinity. The change in the values of E_{local} for a particular conformational transition of a residue is partly or completely compensated (screened) by the backbone electrostatic solvation free energy (ESF). Screening of E_{local} by water depends on the sizes and shapes of nearby side chains, which cause the backbone conformational preferences [37, 43, 67, 75–77, 80, 81], the nearest-neighbor effect [53], and the cooperative formation of local structure in proteins [46].

The solvation free energy of an amino acid residue ΔG^*_{solv} is the free energy change of transferring an amino acid residue from a fixed position in an ideal gaseous phase to a fixed position in a peptide or a protein in the liquid phase [90]. The solvation free energies of residues in proteins and peptides are not accessible to the direct experimental measurements because their concentrations in the gas phase are extremely small. Theoretical methods are therefore an extremely important resource of obtaining the solvation free energies of residues in peptides and proteins. The solvation free energy of a residue i $\Delta G^{*,i}_{solv}$ is a sum of electrostatic, cavity formation, van der Waals, rotational, and vibrational terms [90–93]. The solvation free energy of a residue in some fixed conformation can be decomposed into contributions from backbone atoms and the remaining atoms [94–96]. The largest contribution to the solvation free energy of backbone atoms of a residue is from the electrostatic term ESF (the electrostatic solvation free energy of backbone atoms). The ESF value of a residue in a polypeptide chain depends strongly on the conformation of backbone and side-chain atoms of all residues that are close in space. The DelPhi algorithm has been commonly used to calculate ESF [92]. The PARSE parameters (partial charges and atomic radii) have been calibrated against a database of experimental solvation free energies for model compounds. Calculated solvation free energies $\Delta G^{*,i}_{solv}$ and ESF values of amino acid residues are not accurate [90]. The error of calculating ESF is estimated to be between 1 and 2 kcal/mol per residue [77].

There is a linear anti-correlation between the electrostatic backbone solvation free energy ESF and the local backbone electrostatic energy E_{local} of residues in peptides and unfolded polypeptides [77]. Figure 5.4 shows the anti-correlation between ESF and E_{electr} of alanine dipeptide in aqueous solution of the most important backbone conformations. The E_{electr} is a sum of electrostatic energies E_{local} of the alanine residue and both blocking groups: acetyl and N-methyl. The correlation coefficient is −0.996 for 14 points. Similar anti-correlation was found for residues in tripeptides and other polypeptides [77]. The slope of the approximately linear relation between ESF and E_{local} (denoted here by $\Delta[ESF]/\Delta[E_{local}]$) differs between residue types and reflects the screening coefficients γ_{local} of amino acid residues [77]. The screening coefficient γ_{local} is related to the slope of the approximately linear relation between ESF and E_{local} of a particular residue type by the following equation: $\gamma_{local} = \Delta(ESF)/\Delta(E_{local}) + 1$

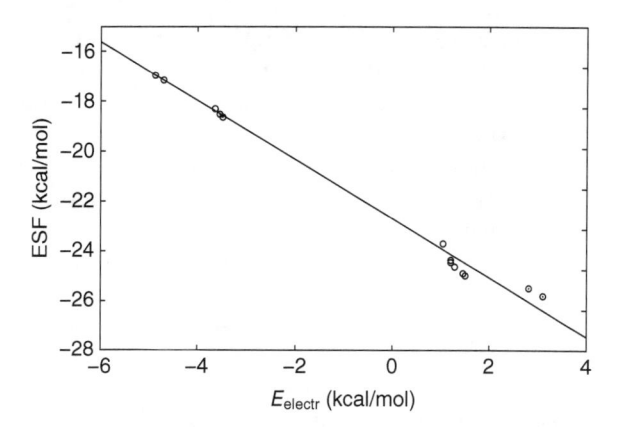

Figure 5.4. Correlation between the electrostatic energy E_{electr} and the *ESF* energy of alanine dipeptide in the following conformations: $C7_{eq}$, β, α_R, α_L, 3–10, δ_L. The correlation coefficient is 0.996. The number of data points is 14. The conformations are obtained by the *ab initio* calculations using the following levels of theory: MP2/CBS limit//MP2/aug-cc-pVDZ [86], MP2/aug-cc-pVTZ* [89], and LMP2/cc-VQZ(-g) [88].

[77]. Larger side chains cause less negative slopes and thus larger screening coefficients and consequently stronger electrostatics.

There are two methods of estimating electrostatic screening in polypeptides. In the first method the solvation free energy $\Delta G^{*,i}_{solv}$ of a residue used in the energy function is approximated by the *ESF* [67, 77]. In the second method the solvation free energy $\Delta G^{*,i}_{solv}$ of a residue is approximated using the screening coefficients γ_{local} [46, 75–77]. For each residue type there is a different screening coefficient. The screening coefficients are derived from the dependence of *ESF* on E_{local} of tripeptides [77] or from the potentials of mean force based on the high-resolution protein structures [75–77, 97].

The ESM is similar to the RCM of Flory and coworkers [14–16]. The most important difference between these two models is in the treatment of screening of backbone electrostatic interactions. In the RCM, the backbone electrostatic interactions are screened uniformly for all residue types by dividing the local electrostatic energy E_{local} with the dielectric constant of 3.5. In the ESM the screening of backbone electrostatic interactions depends strongly on residue type.

5.3. THE ESM AND STRAND-COIL TRANSITION MODEL

The strand-coil transition model has been developed to model very strong electrostatic coupling between neighbor residues in a polypeptide chain [76]. This coupling causes the nearest-neighbor effect and cooperative formation of β-strands in unfolded proteins. Strong coupling between neighboring residues is caused by local backbone electrostatic interactions and backbone solvation, which both depend on the conformations of at least two neighbor residues.

The energetics of such complex systems, in which the free energy of a residue depends on the conformations of at least three residues, is cooperative [98]. This cooperativity in polypeptides in the strand-coil transition model is treated using the mathematics of the Lifson–Roig helix-coil theory [76, 99]. Each residue in a polypeptide sequence is considered to be in equilibrium between two states: the β-strand conformation and the coil. In the former, three consecutive residues are in the β conformation. The coil state is composed predominantly of residues in the P_{II} conformation. The statistical weights o and t, which are analogous to the statistical weights v and w used in the Lifson–Roig helix-coil transition theory [76, 99], depend only on E_{local} and its screening by backbone solvation determined by the screening coefficients [76]. The statistical weights of a residue in water are then obtained by multiplying the differences in E_{local} by the corresponding screening coefficients γ_{local}.

The partition function and the β-strand free energy profile G_{strand} of a polypeptide chain are obtained using the matrix method [76]. G_{strand} is the free energy difference between the β-strand and the coil conformations of each residue in a sequence. A negative value of G_{strand} indicates that the β-strand conformation of that residue is more favorable than the coil. A strong minimum in the free energy profile G_{strand} indicates that residues in this region prefer backbone conformations with low E_{local} (triplets $\beta\beta\beta$; β-strand conformation). A strong maximum in the free energy profile G_{strand} indicates that residues prefer backbone conformations with high E_{local} values (turns). The E_{local} values of the central residue in a triplet increase in the following order: $\beta\beta\beta$ (E_{local} = –3.4 kcal/mol; β-strand conformation) < $\pi\pi\pi$ (E_{local} = –2.1 kcal/mol; polyproline-helix conformation) < $\alpha\alpha\alpha$ (E_{local} = 1.4 kcal/mol; α_R-helix turn) [82]. The symbols β, π, and α represent the β, P_{II}, and α_R conformations, respectively. The β-strand is the lowest energy conformation of a polypeptide chain ("ground state"), because the alignment of neighboring peptide dipole moments is anti-parallel. The difference in E_{local} between the higher energy α_R-helix and β-strand is very large (~4.8 kcal/mol). The difference in E_{local} between the polyproline-helix and the β-strand is smaller but still significant ~1.3 kcal/mol.

The strand-coil transition model has been implemented in the algorithm for predicting secondary structures in native proteins [76]. The three-state accuracy of the algorithm, which contains only the free energy terms due to the main-chain electrostatics modeled by 40 coefficients, is 68.7%. This accuracy is close to that of the best secondary structure prediction algorithm based on neural networks; however, many thousands of parameters have to be optimized during the training of the neural networks to reach this level of accuracy.

5.4. THE ESM AND BACKBONE CONFORMATIONAL PREFERENCES

Amino acid residues in folded and unfolded polypeptides display significant variations of preferences (propensities) for various backbone conformations (α_R-helices and β-sheets, α_R, β, and P_{II} conformations, etc.). The backbone conformational preferences have been characterized by various propensity scales based on data obtained from statistical surveys of experimental X-ray structures of proteins

[64, 65, 75, 100–102], host–guest systems [103, 104], site-directed mutagenesis [105–108], and model peptides [45, 109–117]. The propensity scales are not uniform because they reflect intrinsic propensities of the respective residues and also the influence of neighbor residues, particularly due to the nearest-neighbor effect [47–53]. Because the nearest-neighbor effect is absent in dipeptides, the conformational preferences of dipeptides represent true intrinsic backbone propensities of residues [37].

The population distributions of the three major backbone conformations (P_{II}, β, and α_R) of 13 dipeptides in aqueous solution have been determined by infrared (IR), Raman, and NMR spectroscopy [37, 43]. These data suggest that dipeptides display clear structural preferences in aqueous solutions. The conformations of dipeptides are generally in equilibrium between only the two conformations: β and P_{II}. The population of the α_R conformation is relatively small in all dipeptides. Alanine dipeptide adopts predominantly the P_{II} conformation. The population of the β conformation increases in other dipeptides and reaches the highest level in valine dipeptides. A large population of the P_{II} conformation of alanine residue in peptides in aqueous solution is in accord with the results of many other experimental studies [17–21, 23–26, 28, 29, 31–35, 40–42, 118, 119].

The physical background of the backbone conformational preferences is controversial. The following physical factors have been suggested to be responsible for the backbone conformational preferences of residues in folded and unfolded polypeptides: screening of backbone electrostatic interactions with backbone solvation [37, 43, 46, 53, 67, 75–77, 79–81, 120–122], van der Waals interactions [123–125], side-chain conformational entropy [126–129], hydrophobic interactions [130–132], and side-chain hydrogen bonding [133]. The *ab initio* quantum mechanical calculations and simulations using empirical and semi-empirical force fields generally fail to predict prevalence of the P_{II} conformation of alanine dipeptide in aqueous solution [68–74]. These methods generally overestimate the population of α_R conformations. Ad hoc adjustments of the force fields have been performed to obtain agreement with experimental data for alanine dipeptide in aqueous solutions [134, 135]

The ESM explains the experimental backbone conformational preferences of dipeptides [37, 43], oligopeptides [77], unfolded [46, 67], and folded proteins [75, 76, 78, 79, 81]. In the ESM the change in the values of E_{local} for a particular conformational transition of a residue is partly or completely compensated (screened) by the backbone solvation *ESF*. The *ESF* of a peptide group is amino acid specific because it depends on the access of the peptide group to water, which causes different preferences of residues for the most important backbone conformations. For example, the transition of alanine and valine dipeptides from the β to the P_{II} or α_R conformations is a highly unfavorable process *in vacuo* (see above). The level of electrostatic screening by water dipoles depends on the access of water to the backbone. A small residue like alanine provides better access to water and thus larger screening than a β-branched residue like valine. The screening of ΔE_{local} in alanine dipeptide is thus very effective. Therefore only a small fraction of alanine residues are in the β conformation. The screening of ΔE_{local} in valine dipeptide is less effective

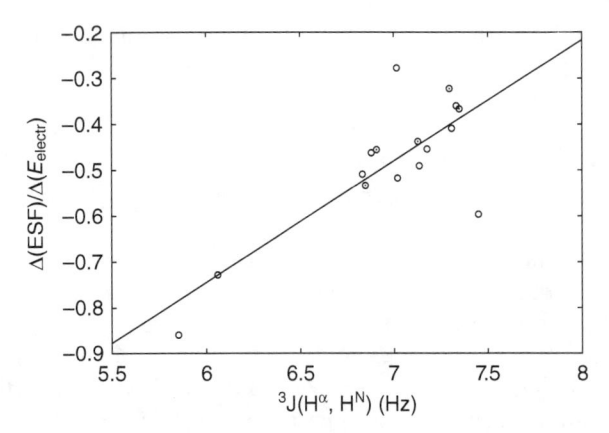

Figure 5.5. Relationship between the NMR coupling constant $^3J(H^\alpha,H^N)$ of dipeptides and the slope of ESF versus E_{local} (denoted by $\Delta[ESF]/\Delta[E_{electr}]$) for central residues of randomly generated tripeptides. Slope $\Delta(ESF)/\Delta(E_{electr})$ is amino acid-specific and linearly related to the screening coefficient γ_{local} of the amino acid residue. The correlation coefficient is 0.80. The $^3J(H^\alpha,H^N)$ values for Asp, Glu, and His are omitted because the $^3J(H^\alpha,H^N)$ values were measured for the partly or fully ionized forms, whereas the ESF values were calculated for the neutral forms. There is a strong dependence of $^3J(H^\alpha,H^N)$ on the extent of ionization for Asp and Glu, whose charged groups are near the backbone, and a similar effect is expected for His; the charged groups of Lys and Arg are more remote.

and the effect of local backbone electrostatics on backbone conformation is stronger. Therefore, a large fraction of valine residues are in the β conformation.

Figure 5.5 shows correlation between the NMR coupling constant $^3J(H^\alpha,H^N)$ of dipeptides and the slope of ESF versus E_{local} denoted by $\Delta(ESF)/\Delta(E_{electr})$. Coupling constant $^3J(H^\alpha,H^N)$ is related to dihedral angle φ by the Karplus relation and thus measures the conformational preference for the dihedral angle φ. A large coupling constant indicates a large population of the β conformation. Slope $\Delta(ESF)/\Delta(E_{electr})$ is amino acid-specific and linearly related to the screening coefficient γ_{local} of the amino acid residue (see above). There is a clear linear relation with the coefficient of 0.80. The correlation indicates that the backbone preferences expressed in the $^3J(H^\alpha,H^N)$ values are related to those predicted by the ESM. The values of $\Delta(ESF)/\Delta(E_{electr})$ are taken from table 2 of Ref. [77] for the central residues of a large number of randomly generated tripeptides.

The electrostatic screening effect can be utilized to rationalize the observed increase in the value of $^3J(H^\alpha,H^N)$ of dipeptides in non-polar solvents [17, 136]. The values of $^3J(H^\alpha,H^N)$ increase with the bulkiness of side chains and with non-polarity of solvents. Weaker screening of backbone electrostatics in non-polar solvents causes larger populations of residues for the β conformation with a larger value of $^3J(H^\alpha,H^N)$. Clarke and coworkers have found that the β propensities are enthalpic in origin [137, 138] and arise from changes in backbone solvation as proposed by the ESM. Scheraga and coworkers have studied the helix-coil transition of alanine-based peptides and found that its behavior is in accord with the ESM [120]. Makhatadze and coworkers have shown that hydration of the peptide backbone defines the thermodynamic propensity scale of the C-capping box of α-helices [121] and that the changes of the relative helix propensities with temperature have the sign expected if side chains interfere with backbone solvation [122].

5.5. THE NEAREST-NEIGHBOR EFFECT

The nearest-neighbor effect has been observed in unfolded polypeptides and in the "coil" library of Protein Data Bank structures of residues not in α_R-helices and not in β-sheets [47–53]. When a first neighboring residue ($i - 1$ or $i + 1$) belongs to large aromatic and β-branched amino acid of class L (FHITVWY) rather than class S (all others, G and P excluded), then a residue i is more inclined to be in the β conformation. This effect applies to essentially all amino acid residues.

The nearest-neighbor effect is in accord with the ESM [53]. The effect is a consequence of strong electrostatic coupling between residues. Nearest-neighbor residues are coupled together through backbone solvation and local backbone electrostatic interactions. A bulky side chain of neighboring residue ($i - 1$, $i + 1$) causes reduced access of water to backbone atoms. This effect stabilizes the β-strand conformation of a residue i because it leads to less effective screening of the local backbone electrostatics.

The nearest-neighbor effect has been investigated with two different systems. The first system is a model peptide system, acetyl-A_4XA_4-amide, where X is any amino acid and the peptide has the β conformation or the P_{II} conformation. The second system is the coil library of residue structures from the Protein Data Bank, which are neither α-helical nor β-sheet. It has been shown that the size of the neighboring residue effect seen by *ESF* is large. For the substitution of Ala by Val in an extended (Ala)$_9$ peptide, *ESF* change occurs at the substitution site i, the first neighbors: $i + 1$, $i - 1$, and also the second neighbors: $i + 2$, $i - 2$. The energetic cost of the substitution is nearly 2 kcal/mol, if the substituted residue is in the P_{II} conformation, or ~1 kcal/mol if it is in the β conformation. One group of amino acid residues (FHITVWY) shows values of ΔESF that are almost twice as large as the others. These two side chain groups correspond to the class L and class S amino acid residues of Penkett et al. [49]. On average, residues with L neighbors have more negative φ values than residues with S neighbors by 2.2° in the coil library. On average, the changes in *ESF* values are less negative by 0.21 kcal/mol with L than with S neighbors. The dipole–dipole interactions in the peptide backbone determine that *ESF* depends directly on φ, and the dependence has the correct sign to account for the neighboring residue effect.

5.6. THE ESM AND COOPERATIVE LOCAL STRUCTURES— FLUCTUATING β-STRANDS

Distinct patterns of backbone conformational preferences and flexibility, which sometimes coincide with secondary structures in the native states, have been observed in chemically denatured proteins by NMR spectroscopy [5, 7–12, 54] [13, 46, 55–57]. Such patterns are observed in the absence of ordinary secondary structures (α_R-helices, β-sheets) and disulfide bridges and indicate that a certain level of cooperativity exists between neighbor residues in unfolded proteins. Such patterns are inconsistent with the RCM [14–16] or by the DPM of unfolded proteins [10, 49, 64–66].

Physical reasons for the patterns of flexibility in unfolded proteins are unclear. It has been suggested that the regions of restricted backbone flexibility in unfolded proteins are associated with formation of hydrophobic clusters [7, 9, 12, 55, 60] because NMR parameters associated with flexibility sometimes correlate with increased buried surface area and hydrophobicity [9, 54, 139]. Reduced backbone mobility has been also associated with regions of extreme polarity in a protein sequence [9].

The ESM predicts the existence of such patterns in unfolded proteins [46]. The patterns are caused by cooperative formation of fluctuating β-strands in completely unfolded proteins driven by electrostatic interactions. These β-strands should form in the absence of stabilizing long-range tertiary interactions of β-sheets and in those regions of an unfolded polypeptide chain that contain mainly non-polar residues because large non-polar side chains shield the backbone and prevent screening by water dipoles. The local backbone electrostatic interactions are thus stronger, stabilizing the β-strand conformation. The ESM also predicts that the backbone of those regions in a sequence that forms fluctuating β-strands in unfolded proteins is stiffer than the backbone of other regions. Those residues that can easily exchange between conformations should be more flexible than others. The electrostatic screening effect should persevere even at very large concentrations of denaturant.

Compelling experimental evidence for the fluctuating β-strands has been recently found in unfolded ubiquitin [46]. The distinct patterns of backbone conformational preferences, measured by the coupling constant $^3J(H^\alpha,H^N)$ [10], the ratio of Nuclear Overhauser Effect (NOE) connectivities Q^{NOE} [46], and the cross-correlated relaxation rates $\Gamma^c_{HN,C\alpha H\alpha}$ [10], have been observed in unfolded ubiquitin. These NMR parameters correlate with the β-strand free energy profile G_{strand}. The Pearson correlation coefficients between the G_{strand}, calculated by the ESM, and the experimental NMR data of urea-denatured ubiquitin: $^3J(H^\alpha,H^N)$, Q^{NOE}, and $\Gamma^c_{HN,C\alpha H\alpha}$ [10] are –0.67 for 70 residues, 0.70 for 40 residues (Fig. 5.6), and 0.61 for 50 residues, respectively. There is also a strong correlation between the minima of the β-strand free energy profile G_{strand} and the occurrence of the native β-strands in folded ubiquitin (Fig. 5.7).

These correlations strongly support the hypothesis that screening of backbone electrostatic interactions by water is responsible for determining the backbone structure in urea-denatured ubiquitin. Urea-denatured ubiquitin contains no detectable β-sheet secondary structure; nevertheless, the fluctuating β-strands in urea-denatured ubiquitin coincide with the β-strands in the native state (Fig. 5.7). The observed β-strands in urea-denatured ubiquitin are not fully populated. The central sections of the β-strands are better ordered and disorder increases with distance along the sequence from the center, which is in accord with the strand-coil transition model. The residues that are located at the centers of disordered β-strands almost fully

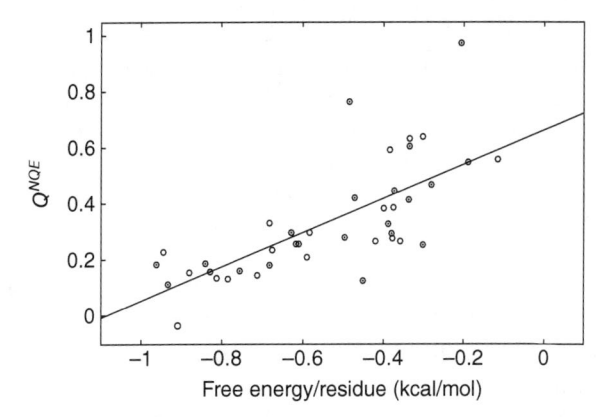

Figure 5.6. The correlation between the ratio of NOE connectivities Q^{NOE} and the β-strand free energy profile G_{strand} (kcal/mol) of urea-denatured ubiquitin. The value of Q^{NOE} depends predominantly on the value of ψ for residue i. The G_{strand} was calculated using the ESM (Model I, [76]). The correlation coefficient is 0.70 for 40 residues.

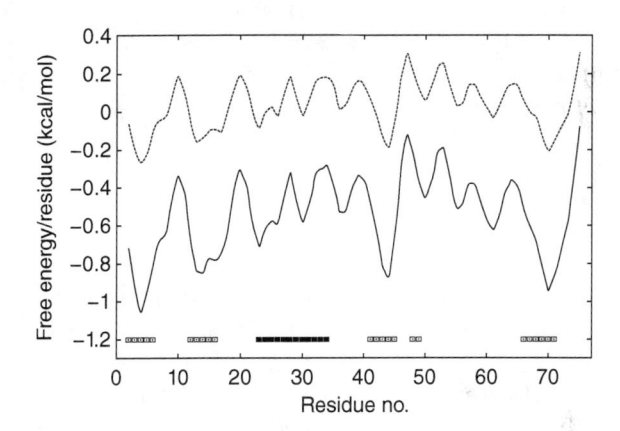

Figure 5.7. The β-strand free energy profile G_{strand} (kcal/mol) along the polypeptide chain of urea-denatured ubiquitin. The profile G_{strand} has been calculated using the ESM (Model I, [76]). Two values of the average local electrostatic energy of a residue in the interior of the β-strand are used: –3.0 kcal/mol (dashed line) and –3.5 kcal/mol (solid line). The native α_R-helix and β-strands [160] are marked by filled and open squares, respectively.

(~90%) populate the β-region of φ,ψ space (as distinct from the extended region, which also includes the P_{II} conformation). The residues in disordered β-strands show extensive conformational averaging, mainly between the β and P_{II} conformations. The transition between these two backbone conformations is fast because there is no significant energetic barrier between these two conformations. There is no correlation between the NMR data of urea-denatured ubiquitin and the average buried surface area [140] or between the NMR data and the Kyte–Doolittle hydrophobicity index [141].

NMR chemical shifts of $^1H_\alpha$, $^{13}C_\alpha$, and $^{13}C_O$ atoms of some regions in chemically denatured proteins exhibit small but systematic deviations from their values in small peptides. For example, the residues of urea-denatured plastocyanin that correspond to the β-sheet in the folded state exhibit small upfield shifts of $^{13}C_\alpha$ and $^{13}C_O$ resonances relative to their values in small peptides, indicating larger populations of the β conformation in the former [11]. It has been suggested that reduced sensitivity of the chemical shifts to variations in backbone conformation of solvent-exposed residues reveal actually much larger variations of conformational preferences in chemically denatured proteins than estimated earlier [82]. The patterns of upfield $^{13}C_\alpha$ secondary shifts exhibited by several adjacent residues of urea-denatured plastocyanin [11] and other chemically denatured proteins may thus indicate formation of fluctuating β-strands.

It has been proposed that the aggregation into fibrils is caused by fluctuating β-strands in denatured proteins. These local structures present a nucleus in the denatured state that seeds nonspecific assembly of other parts of a polypeptide chain into a large β-sheet structure, presumably by a zipper mechanism. The hypothesis is based on the observation that strong minima exist in the β-strand free energy profiles G_{strand} in the regions corresponding to the amyloid-forming nuclei in three proteins: β_2-microglobulin, hen lysozyme, and α-synuclein [46]. The hypothesis has been successfully tested in predicting the fibrillization propensity of the parent stefins and their chimeric forms [142].

5.7. THE ESM AND β-SHEET PREFERENCES IN NATIVE PROTEINS—SIGNIFICANCE OF UNFOLDED STATE

There are large differences among the preferences of amino acid residues for the β-sheet. Thermodynamic experiments in which protein stability has been measured for mutants produced by substituting 20 different amino acid residues at a single site give conflicting results for the β-sheet preferences in four systems studied [105–108]. Kim and Berg used the metal-dependent folding of a zinc finger protein to measure spectrophotometrically the free energy change for its unfolding reaction by varying the free Co(II) concentration, for which there is a competition between the protein and a chromophoric indicator [105].

It has been found that peptide backbone solvation is a major factor determining thermodynamic β propensities measured by Kim and Berg [81]. There is a strong correlation (correlation coefficient is 0.94) between the relative stability $\Delta\Delta G$ and the change in backbone solvation ΔESF for the zinc finger mutants studied by Kim and Berg (Fig. 5.8). However, there is a paradox concerning the β-sheet preferences of the amino acid residues. The amino acid residues with the highest β-sheet preferences (valine and isoleucine) have the highest tendency to desolvate the peptide backbone, which should result in a loss of stability. This inverse correlation between stability and *ESF* can be explained in terms of the mutant *ESF* differences being larger in the unfolded than in the native protein. Consequently, mutations such as Ala to Val destabilize the unfolded form more than the native protein.

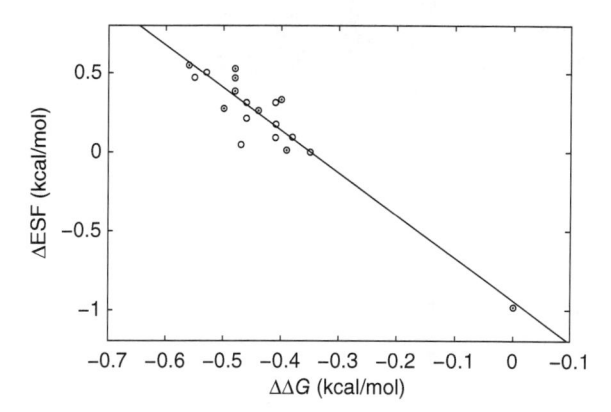

Figure 5.8. Linear relation between the mutant ΔESF value (kcal/mol) in the native state and mutant $\Delta\Delta G$ value (kcal/mol) in the zinc finger system studied by Kim and Berg [105]. The correlation coefficient is −0.94. ΔESF is the difference between the mutant ESF and that of the alanine variant, both in the native form. $\Delta\Delta G$ is the difference in free energy of unfolding between the mutant and the glycine variant.

By comparing mutant ΔESF values in isolated β-strands versus β-sheets, it has been concluded that amino acid residues with high β-propensities should exert their stabilizing effects at early stages in folding. This deduction agrees with the studies by Clarke and coworkers [137, 138] of the thermodynamics of folding of the β-sheet protein CD2.d1. They found that most mutations with favorable β propensities stabilize the molten globule intermediate even though they destabilize the folded protein. Thus, β-propensity mutations are most effective at an early stage in folding. The authors conclude that β-propensities are probably enthalpic in origin [137, 138] and arise from changes in backbone solvation as proposed by the ESM.

5.8. THE ESM AND SECONDARY CHEMICAL SHIFTS OF POLYPEPTIDES

The NMR chemical shifts of $^1H_\alpha$, $^{13}C_\alpha$, and $^{13}C_\beta$, and other atoms are commonly used to determine secondary structures of folded proteins. This method is based on the secondary structure shift, which is the difference between the observed chemical shift and the random coil value assigned to this amino acid type in the unfolded conformation (chemical shift index method) [143]. The physical origin of the different chemical shifts of nuclei in α_R-helices versus β-sheets is an unsolved problem of considerable interest.

The chemical shifts of $^1H_\alpha$, $^{13}C_\alpha$, and $^{13}C_\beta$, and other atoms in unfolded proteins generally deviate little from the so-called random coil values; therefore, it has been suggested that the residues in unfolded proteins do not form any ordered backbone structure, particularly the β-strands [11]. A characteristic downfield shift is expected for the $^{13}C_\beta$ resonances of such β-strands [144]. To resolve this issue, a large number

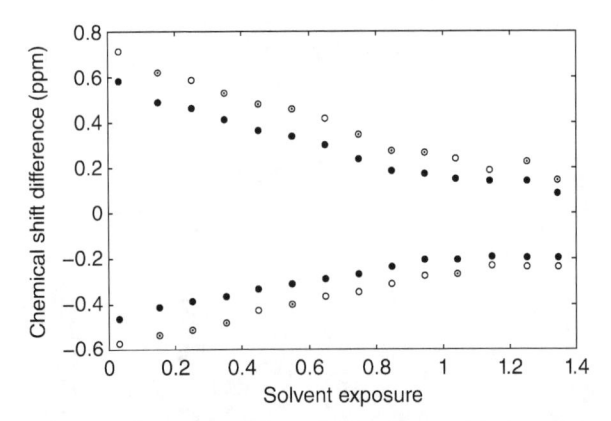

Figure 5.9. Chemical shift difference (observed value minus random coil value) is plotted against solvent exposure for $^1H_\alpha$ nuclei. Residues are included both within and outside the secondary structure, and all amino acid residues, except proline, are included. The data are binned with a bin size of 0.1 solvent exposure and mean values (•) are shown, as well as the root mean square deviations (°). The root mean square (RMS) deviation values are multiplied by −1 when the chemical shift difference is negative. Data for 1010 proteins and 103,084 residues are averaged.

of experimental chemical shifts in the 1010 native proteins of known three-dimensional structures deposited in the BioMagResBank database have been analyzed [82].

It has been shown that the chemical shift contributions arising from secondary structure (secondary-structure shifts) depend strongly on the extent of exposure to solvent. The chemical shifts of $^1H_\alpha$, $^{13}C_\alpha$, and $^{13}C_\beta$ atoms are reliable indicators of backbone structure only if a residue is buried in the protein interior. Most of the residues in unfolded proteins are completely exposed to solvent; therefore, the chemical shifts of atoms in unfolded proteins are unreliable indicators of polypeptide structure. When random coil values are subtracted from the chemical shifts of all $^1H_\alpha$ nuclei (Pro residues excluded) and the residual chemical shifts are summed to plot the mean values against solvent exposure, the results give a funnel-shaped curve that approaches a small value at full-solvent exposure (Fig. 5.9).

When chemical shifts are plotted instead against E_{local}, the electrostatic contribution to conformational energy produced by local dipole–dipole interactions, a well-characterized dependence of chemical shifts of $^1H_\alpha$, $^{13}C_\alpha$, and $^{13}C_\beta$ atoms in all residue types except proline on E_{local} is found (Fig. 5.10). The slope of this plot varies with both the type of amino acid and the extent of solvent exposure. E_{local} is very useful in analyzing the chemical shifts of atoms because it simplifies the relationships between the chemical shifts and the backbone structure. Remarkably, the slopes of the plot of $^1H_\alpha$ chemical shift versus E_{local} are correlated with thermodynamic β-structure propensity, which is thought to be completely unrelated (Fig. 5.11). This extraordinary behavior of the chemical shifts cannot be explained using the current theoretical models of chemical shifts. Many empirical and quantum

THE ESM AND SECONDARY CHEMICAL SHIFTS OF POLYPEPTIDES 147

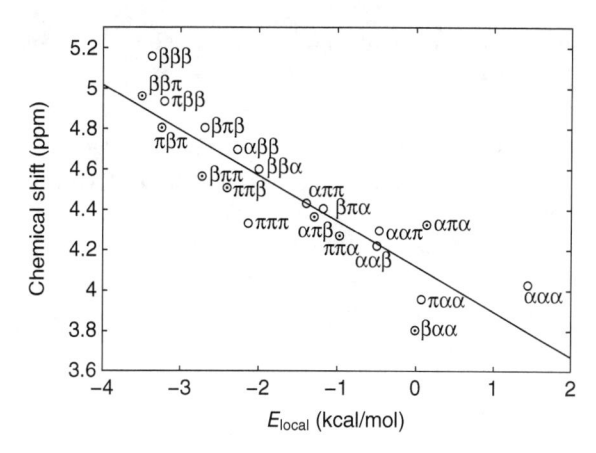

Figure 5.10. Relationship between mean chemical shift values and E_{local} for the central alanine residues of selected triplet types composed of only three main core backbone conformations: α_R, β, and P_{II} (π) [82].

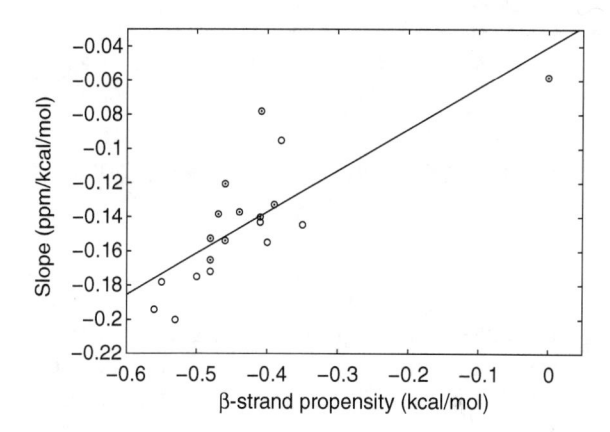

Figure 5.11. The slope of chemical shift values versus E_{local} [82] is displayed against and the thermodynamic β-propensity [105] of the amino acid types.

chemical methods have been used to study the relationship between chemical shifts and secondary structure of proteins [145–153]. The correlation between the slope of the function of chemical shifts with E_{local} and the thermodynamic β-sheet-forming propensity (Fig. 5.11) indicates that the same physical factor determines these two apparently completely unrelated variables. It has been found that the screening of the backbone electrostatic interactions by the water dipoles determines the thermodynamic β-sheet-forming propensity of residues [75, 76, 81]. The behavior of the chemical shifts of $^1H_\alpha$, $^{13}C_\alpha$, and $^{13}C_\beta$ atoms is consistent with the hypothesis in which the conformation-dependent part of the chemical shifts is determined mainly by the electric field of the protein, which is screened by water dipoles at residues in contact with solvent. This result contradicts the current theoretical models of chemical shifts in which the conformation-dependent part of chemical shifts is assigned predominantly to the peptide magnetic anisotropy contributions [145, 147, 150].

5.9. ROLE OF BACKBONE SOLVATION IN DETERMINING HYDROGEN EXCHANGE RATES OF UNFOLDED POLYPEPTIDES

Englander and coworkers [154, 155] showed that the reaction rate of the OH$^-$ ion with the peptide NH proton (i.e., the hydrogen exchange rate) drops fourfold from Ala to Ile and Val, and Leu has an intermediate rate. They also found the correlation between the hydrogen exchange rates of model peptides and the β-sheet propensities of 13 blocked amino acid residues. They proposed that the side chains can modulate the strength of the main-chain hydrogen bonds by the side-chain steric blocking effect.

The dependence of the hydrogen exchange rates log k_{HX} on neighboring CO-NH groups has been explained by a through-bonds inductive effect [156, 157]. However, recent calculations show that the effect can be calculated by using the electrostatic model with fixed partial charges and a continuum solvent [53, 158, 159]. The bulkiness of a non-polar side chain affects the rate of hydrogen exchange of the adjacent peptide NH proton because the side chain reduces the access of solvent to the peptide group, which is measured by the *ESF*.

5.10. OTHER THEORETICAL MODELS OF UNFOLDED POLYPEPTIDES

The RCM was introduced by Flory and coworkers [14–16], who argued that the backbone conformations and chain dimensions of denatured proteins can be predicted by specifying a small set of energy terms. They found that agreement between the measured dimensions of unfolded polypeptides and theoretical predictions is achieved only by taking into account dipole–dipole interactions of the nearest-neighbor peptide bonds (CO-NH groups) [14, 15]. The energy function of the RCM contains only the following terms: non-bond energy, intrinsic torsion potentials for φ and ψ torsion angles, and local electrostatic energy (E_{local}). Solvation of the peptide backbone is included in the energy function by uniform screening by solvent of the peptide dipole–dipole interactions using the dielectric constant of 3.5. The RCM has not been formulated in a manner that allows quantitative predictions of experimentally observed parameters of unfolded proteins. The main problem is the lack of van der Waals parameters for the side chains other than Ala. The random coil method predicts that alanine residues in unfolded proteins adopt predominantly the two extended conformations β and P_{II} with about equal probability. The RCM uses the Flory's isolated-pair hypothesis in which conformation of a residue is independent of conformations of other residues in a polypeptide chain; therefore, this model cannot predict the nearest-neighbor effect and formation of other local structures.

The DPM [10, 49, 64–66] is an empirical model of the highly denatured proteins. This model proposes that the distributions of φ and ψ angles of residues in the highly denatured proteins are equivalent to those of the "coil" residues in the database of native proteins. The DPM of unfolded proteins is based on the NMR studies, which have shown that native-like conformational propensities of residues observed in the "coil" library are retained in unfolded proteins. The coil library model suggests that residues in unfolded proteins sample the α_R conformation with relatively high probability (32% for Ala, 20% for Val) [65]. This model is not in accord with the experimental data on small peptides [22–24, 26, 29, 31, 32, 40, 43–45] and some unfolded proteins [9, 11], which show that populations of the α_R conformation are relatively small. The residues in the "coil" library are biased toward compact structures because they are frequently located in turns that link secondary structures together. This bias probably causes the exaggerated populations of residues in the α_R conformation. The nearest-neighbor effect and other local ordered structures in unfolded polypeptides cannot be predicted by the DPM because Flory's isolated-pair hypothesis is used.

ACKNOWLEDGMENTS

I would like to thank Robert L. Baldwin and Jože Grdadolnik for reading the manuscript and providing helpful suggestions.

REFERENCES

1. Baldwin, R. L. (1986) Seeding protein folding, *Trends Biochem Sci 11*, 6–9.
2. McGee, W. A., et al. (1996) Thermodynamic cycles as probes of structure in unfolded proteins, *Biochemistry 35*, 1995–2007.
3. Dill, K. A. and Shortle, D. (1991) Denatured states of proteins, *Annu Rev Biochem 60*, 795–825.
4. Logan, T. M., Theriault, Y., and Fesik, S. W. (1994) Structural characterization of the FK506 binding protein unfolded in urea and guanidine hydrochloride, *J Mol Biol 236*, 637–648.
5. Frank, M. K., Clore, G. M., and Gronenborn, A. M. (1995) Structural and dynamic characterization of the urea denatured state of the immunoglobulin binding domain of streptococcal protein G by multidimensional NMR spectroscopy, *Protein Sci 4*, 2605–2615.
6. Arcus, V. L., et al. (1995) A comparison of the pH, urea, and temperature-denatured states of barnase by heteronuclear NMR: Implications for the initiation of protein folding, *J Mol Biol 254*, 305–321.
7. Schwalbe, H., et al. (1997) Structural and dynamical properties of a denatured state. Heteronuclear 3D NMR experiments and theoretical simulations of lysozyme in 8 M urea, *Biochemistry 36*, 8977–8991.
8. Fong, S., et al. (1998) Characterization of urea-denatured states of an immunoglobulin superfamily domain by heteronuclear NMR, *J Mol Biol 278*, 417–429.
9. Meekhof, A. E. and Freund, S. M. V. (1999) Probing residual structure and backbone dynamics on the milli- to picosecond timescale in a urea-denatured fibronectin type III domain, *J Mol Biol 286*, 579–592.

10. Peti, W., et al. (2000) NMR spectroscopic investigation of psi torsion angle distribution in unfolded ubiquitin from analysis of 3J(Calpha,Calpha) coupling constants and cross-correlated relaxation rates, *J Am Chem Soc 122*, 12017–12018.

11. Bai, Y., et al. (2001) Structural and dynamic characterization of an unfolded state of poplar Apo-plastocyanin formed under non-denaturing conditions, *Protein Sci 10*, 1056–1066.

12. Schwarzinger, S., Wright, P. E., and Dyson, H. J. (2002) Molecular hinges in proteins folding: The urea-denatured state of apomyoglobin, *Biochemistry 41*, 12681–12686.

13. Kumar, A., et al. (2006) Local structural preferences and dynamics restrictions in the urea-denatured state of SUMO-1: NMR characterization, *Biophys J 90*, 2498–2509.

14. Brant, D. A. and Flory, P. J. (1965) The configuration of random polypeptide chains. I. Experimental results, *J Am Chem Soc 87*, 2788–2791.

15. Brant, D. A. and Flory, P. J. (1965) The role of dipole interactions in determining polypeptide conformation, *J Am Chem Soc 87*, 663–664.

16. Flory, P. J. (1969) *Statistical mechanics of chain molecules*, Oxford University Press, New York, p. 30.

17. Madison, V. and Kopple, K. D. (1980) Solvent-dependent conformational distributions of some dipeptides, *J Am Chem Soc 102*, 4855–4863.

18. Deng, Z., et al. (1996) Solution-phase conformations of N-Acetyl-L-alanine N'-Methylamide from vibrational Raman optical activity, *J Phys Chem B 100*, 2025–2034.

19. Poon, C.-D. and Samulski, E. T. (2000) Do bridging water molecules dictate the structure of a model dipeptide in aqueous solution, *J Am Chem Soc 122*, 5642–5643.

20. Woutersen, S. and Hamm, P. (2000) Structure determination of trialanine in water using polarization sensitive two-dimensional vibrational spectroscopy, *J Phys Chem B 104*, 11316–11320.

21. Schweitzer-Stenner, R., et al. (2001) Dihedral angles of trialanine in D2O determined by combining FTIR and polarized visible Raman spectroscopy, *J Am Chem Soc 123*, 9628–9633.

22. Eker, F., et al. (2002) Tripeptides adopt stable structures in water. A combined polarized visible Raman, FTIR, and VCD spectroscopy study, *J Am Chem Soc 124*, 14330–14341.

23. Shi, Z., et al. (2002) Polyproline II structure in a sequence of seven alanine residues, *Proc Natl Acad Sci U S A 99*, 9190–9195.

24. Ding, L., et al. (2003) The pentapeptide GGAGG has PII conformation, *J Am Chem Soc 125*, 8092–8093.

25. Weise, C. F. and Weisshaar, J. C. (2003) Conformational analysis of alanine dipeptide from dipolar couplings in a water based liquid crystal, *J Phys Chem B 107*, 3265–3277.

26. Eker, F., Griebenow, K., and Schweitzer-Stenner, R. (2003) Stable conformations of tripeptides in aqueous solution studied by UV circular dichroism spectroscopy, *J Am Chem Soc 125*, 8178–8175.

27. McColl, I. H., et al. (2004) Vibrational Raman optical activity characterization of Poly(L-proline) II helix in alanine oligopeptides, *J Am Chem Soc 126*, 5076–5077.

28. Liu, Z., et al. (2004) Solvent dependence of PII conformation in model alanine peptides, *J Am Chem Soc 126*, 15141–15150.

REFERENCES

29. Eker, F., et al. (2004) Preferred peptide backbone conformations in the unfolded state revealed by the structure analysis of alanine-based (AXA) tripeptides in aqueous solution, *Proc Natl Acad Sci U S A 101*, 10054–10059.
30. Eker, F., et al. (2004) Tripeptides with ionizable side chains adopt a perturbed polyproline II structure in water, *Biochemistry 43*, 613–621.
31. Schweitzer-Stenner, R., et al. (2004) The conformation of tetraalanine in water determined by polarized Raman, FT-IR, and VCD spectroscopy, *J Am Chem Soc 126*, 2768–2776.
32. Chen, K., Liu, Z., and Kallenbach, N. R. (2004) The polyproline II conformation in short alanine peptides is noncooperative, *Proc Natl Acad Sci U S A 101*, 15352–15357.
33. Takekiyo, T., et al. (2004) Temperature and pressure effects on conformational equilibria of alanine dipeptide in aqueous solution, *Biopolymers 73*, 283–290.
34. Mehta, M. A., et al. (2004) Structure of the alanine dipeptide in condensed phases determined by 13C NMR, *J Phys Chem B 108*, 2777–2780.
35. Kim, Y. S., Wang, J., and Hochstrasser, R. H. (2005) Two-dimensional infrared spectroscopy of the alanine dipeptide in aqueous solution, *J Phys Chem B 109*, 7511–7521.
36. Chen, K., et al. (2005) Neighbor effect on PPII conformation in alanine peptides, *J Am Chem Soc 127*, 10146–10147.
37. Avbelj, F., et al. (2006) Intrinsic backbone preferences are fully present in blocked amino acids, *Proc Natl Acad Sci U S A 103* (5), 1277–1277.
38. Hagarman, A., et al. (2006) Conformational analysis of XA and AX dipeptides in water by electronic circular dichroism and 1H NMR spectroscopy, *J Phys Chem B 110*, 6979–6986.
39. Schweitzer-Stenner, R. and Measey, T. J. (2007) The alanine-rich XAO peptide adopts a heterogeneous population, including turn-like and polyproline II conformations, *Proc Natl Acad Sci U S A 104*, 6649–6654.
40. Graf, J., et al. (2007) Structure and dynamics of the homologous series of alanine peptides: A joint molecular dynamics/NMR study, *J Am Chem Soc 129*, 1179–1189.
41. Lee, K.-K., et al. (2007) Dipeptide structure determination by vibrational circular dichroism combined with the quantum chemistry calculations, *Chemphyschem 8*, 2218–2226.
42. Mukhopadhyay, P., Zuber, G., and Beratan, D. N. (2008) Characterizing aqueous solution conformations of a peptide backbone using Raman optical activity computations, *Biophys J 95*, 5574–5586.
43. Grdadolnik, J., Grdadolnik, S. G., and Avbelj, F. (2008) Determination of conformational preferences of dipeptides using vibrational spectroscopy, *J Phys Chem B 112*, 2712–2718.
44. Schweitzer-Stenner, R. (2009) Distribution of conformations sampled by the central amino acid residue in tripeptides inferred from amide I band profiles and NMR scalar coupling constants, *J Phys Chem B 113*, 2922–2932.
45. Hagarman, A., et al. (2010) Intrinsic propensities of amino acid residues in GXG peptides inferred from amide I' band profiles and NMR scalar coupling constants, *J Am Chem Soc 132*, 540–551.
46. Avbelj, F. and Grdadolnik, S. G. (2007) Electrostatic screening and backbone preferences of amino acid residues in urea-denatured ubiquitin, *Protein Sci 16*, 273–284.

47. Braun, D., Wider, G., and Wuthrich, K. (1994) Sequence-corrected 15-N random coil chemical shifts, *J Am Chem Soc 116*, 8466–8469.
48. Wishart, D. S., et al. (1995) 1H 13C and 15N random coil NMR chemical shifts of the common amino acids. I. Investigation of nearest-neighbor effects, *J Biomol NMR 5*, 67–81.
49. Penkett, C. J., et al. (1997) NMR analysis of main-chain conformational preferences in an unfolded fibronectin-binding Protein, *J Mol Biol 274*, 152–159.
50. Griffiths-Jones, S. R., et al. (1998) Modulation of intrinsic phi,psi propensities in the coil regions of protein structures: NMR analysis and dissection of beta-hairpin peptide, *J Mol Biol 284*, 1597–1609.
51. Peti, W., et al. (2001) Chemical shifts in denatured proteins: Resonance assignment for denatured ubiquitin and comparisons with other denatured proteins, *J Biomol NMR 19*, 153–165.
52. Schwarzinger, S., et al. (2001) Sequence-dependent correction of random coil NMR chemical shifts, *J Am Chem Soc 123*, 2970–2978.
53. Avbelj, F. and Baldwin, R. L. (2004) Origin of the neighboring residue effect on peptide backbone conformation, *Proc Natl Acad Sci U S A 101*, 10967–10972.
54. Farrow, N. A., et al. (1997) Characterization of the backbone dynamics of folded and denatured states of an SH3 domain, *Biochemistry 36*, 2390–2402.
55. Ohnishi, S. and Shortle, D. (2003) Observation of residual dipolar couplings in short peptides, *Proteins 50*, 546–551.
56. Mohana-Borges, R., et al. (2004) Structural characterization of unfolded states of apomyoglobin using residual dipolar couplings, *J Mol Biol 340*, 1131–1142.
57. Wirmer, J., Peti, W., and Schwalbe, H. (2006) Motional properties of unfolded ubiquitin: A model for a random coil protein, *J Biomol NMR 35*, 175–186.
58. Neri, D., et al. (1992) NMR determination of residual structure in a urea-denatured protein, the 434-repressor, *Science 257*, 1559–1563.
59. Pan, H., et al. (1995) Extensive nonrandom structure in reduced and unfolded bovine pancreatic trypsin inhibitor, *Biochemistry 34*, 13974–13981.
60. Klein-Seetharaman, J. K., et al. (2002) Long-range interactions within a non-native protein, *Science 295*, 1719–1722.
61. Scholtz, J. M., et al. (1995) Urea unfolding of peptide helices as a model for interpreting protein unfolding, *Proc Natl Acad Sci U S A 92*, 185–189.
62. Myers, J. K., Pace, C. N., and Scholtz, J. M. (1995) Denaturant m values and heat capacity changes: Relation to changes in accessible surface areas of protein folding, *Protein Sci 4*, 2138–2148.
63. Zou, Q., Habermann-Rottinghous, S. M., and Murphy, K. P. (1998) Urea effects on protein stability: Hydrogen bonding and the hydrophobic effect, *Proteins 31*, 107–115.
64. Serrano, L. (1995) Comparison between the phi distribution of the amino acids in the protein database and NMR data indicates that amino acids have various phi propensities in the random coil conformation, *J Mol Biol 254*, 322–333.
65. Smith, L. J., et al. (1996) Analysis of main chain torsion angles in proteins: Prediction of NMR coupling constants for native and random coil conformations, *J Mol Biol 255*, 494–506.
66. Fiebig, K. M., et al. (1996) Toward a description of the conformations of denatured states of proteins. Comparison of a random coil model with NMR measurements, *J Phys Chem 100*, 2661–2666.

REFERENCES

67 Avbelj, F. and Baldwin, R. L. (2003) Role of backbone solvation and electrostatics in generating preferred peptide backbone conformations: Distributions of phi, *Proc Natl Acad Sci U S A 100*, 5742–5747.

68 Roterman, I. K., et al. (1989) A comparison of the CHARMM, AMBER and ECCEP potentials for peptides. II. phi-psi maps for N-Acetyl alanine N'-methyl amide: Comparisons, contrasts and simple experimental tests, *J Biomol Struct Dyn 7*, 421–453.

69 Gould, I. R., Cornell, W. D., and Hillier, I. H. (1994) A quantum mechanical investigation of the conformational energetics of the alanine and glycine dipeptides in the gas phase and in aqueous solution, *J Am Chem Soc 116*, 9250–9256.

70 Shang, H. S. and Head-Gordon, T. (1994) Stabilization of helices in glycine and alanine dipeptides in a reaction field model of solvent, *J Am Chem Soc 116*, 1528–1532.

71 Hu, H., Elstner, M., and Hermans, J. (2003) Comparison of a QM/MM force filed and molecular mechanics force fields in simulations of alanine and glycine dipeptides (Ace-Ala-Nme and Ace-Gly-Nme) in water in relation to the problem of modeling the unfolded peptide backbone in solution, *Proteins 50*, 451–463.

72 Hudaky, I., Hudaky, P., and Perczel, A. (2004) Solvation model induced structural changes in peptides. A quantum chemical study on Ramachandran surfaces and conformers of alanine diamide using the polarizable continuum model, *J Comput Chem 25*, 1522–1531.

73 Wang, Z.-X. and Duan, Y. (2004) Solvations effects on alanine dipeptide: A MP2/cc-pVTZ//MP2/6-31G** study of (Phi,Psi) energy maps and conformers in the gas phase, ether, and water, *J Comput Chem 25*, 1699–1716.

74 Seabra, G. D. M., Walker, R. C., and Roitberg, A. E. (2009) Are current semiempirical methods better than force fields? A study from the thermodynamics perspective, *J Phys Chem A 113*, 11938–11948.

75 Avbelj, F. and Moult, J. (1995) Role of electrostatic screening in determining protein main chain conformational preferences, *Biochemistry 34*, 755–764.

76 Avbelj, F. and Fele, L. (1998) Role of main-chain electrostatics, hydrophobic effect, and side-chain conformational entropy in determining the secondary structure of proteins, *J Mol Biol 279*, 665–684.

77 Avbelj, F. (2000) Amino acid conformational preferences and solvation of polar backbone atoms in peptides and proteins, *J Mol Biol 300*, 1337–1361.

78 Avbelj, F. and Moult, J. (1995) The conformation of folding initiation sites in proteins determined by computer simulation, *Proteins Struct Funct Genet 23*, 129–141.

79 Avbelj, F. and Fele, L. (1998) Prediction of the three dimensional structure of proteins using the electrostatic screening model and hierarchic condensation, *Proteins Struct Funct Genet 31*, 74–96.

80 Avbelj, F., Luo, P., and Baldwin, R. L. (2000) Energetics of the interaction between water and the helical peptide group and its role in determining helix propensities, *Proc Natl Acad Sci U S A 97*, 10786–10791.

81 Avbelj, F. and Baldwin, R. L. (2002) Role of backbone solvation in determining thermodynamic beta-propensities of the amino acids, *Proc Natl Acad Sci U S A 99*, 1309–1313.

82 Avbelj, F., Kocjan, D., and Baldwin, R. L. (2004) Protein chemical shifts arising from alpha-helices and beta-sheets depend on solvent exposure, *Proc Natl Acad Sci U S A 101*, 17394.

83 Warshel, A. and Russell, S. T. (1984) Calculation of electrostatic interactions in biological systems and in solutions, *Q Rev Biophys 17*, 283–422.

84 Schafer, L., et al. (1993) Evaluation of the dipeptide approximation in peptide modeling by ab initio geometry optimization of oligopeptides, *J Am Chem Soc 115*, 272–280.

85 Beachy, M. D., et al. (1997) Accurate ab initio quantum chemical determination of the relative energetics of peptide conformations and assessment of empirical force fields, *J Am Chem Soc 119*, 5908–5920.

86 Vargas, R., et al. (2002) Conformational study of the alanine dipeptide at the MP2 and DFT levels, *J Phys Chem A 106*, 3213–3218.

87 Perczel, A., et al. (2003) Peptide models. XXXIII. Extrapolation of low-level Hartree-Fock data of peptide conformation to large basis set SCF, MP2, DFT, and CCSD(T) results. The Ramachandran surface of alanine dipeptide computed at various levels of theory, *J Comput Chem 24*, 1026–1042.

88 Mackerell, A. D., Feig, M., and Brooks, C. L. III (2004) Extending the treatment of backbone energetics in protein force fields: Limitations of gas-phase quantum mechanics in reproducing protein conformational distributions in molecular dynamics simulations, *J Comput Chem 25*, 1400–1415.

89 Improta, R. and Barone, V. (2004) Assessing the reliability of density functional methods in the conformational study of polypeptides: The treatment of intraresidue nonbonding interactions, *J Comput Chem 25*, 1333–1341.

90 Ben-Naim, A. (1987) *Solvation thermodynamics*, Plenum, New York and London.

91 Tomasi, J. and Persico, M. (1994) Molecular interactions in solution: An overview of methods based on continuous distributions of the solvent, *Chem Rev 94*, 2027–2094.

92 Sitkoff, D., Sharp, K. A., and Honig, B. (1994) Accurate calculations of hydration free energies using macroscopic solvent models, *J Phys Chem 98*, 1978–1988.

93 Florian, J. and Warshel, A. (1997) Langevin dipoles model for ab initio calculations of chemical processes in solution: Parameterization and application to hydration free energies of neutral and ionic solutes and conformational analysis in aqueous solution, *J Phys Chem B 101*, 5583–5595.

94 Wolfenden, R., et al. (1981) Affinities of amino acid side chain for solvent water, *Biochemistry 20*, 849–855.

95 Fauchere, J.-L. and Pliska, V. (1983) Hydrophobic parameters of amino acid side chains from partitioning of N-acetyl-amino-acid amides, *Eur J Med Chem - Chim Ther 18*, 369–375.

96 Eisenberg, D. and McLachlan, A. D. (1986) Solvation energy in protein folding and binding, *Nature 319*, 199–203.

97 Avbelj, F. (1992) Use of a potential of mean force to analyze free energy contributions in protein folding, *Biochemistry 31*, 6290–6297.

98 Poland, D. and Scheraga, H. A. (1967) *Poly-alpha-amino acids protein models for conformational studies—Theory of noncovalent structure in polyamino acids*, Marcel Dekker, New York, pp. 391–497.

99 Lifson, S. and Roig, A. (1961) On the theory of helix-coil transition in polypeptides, *J Chem Phys 34*, 1963–1974.

100 Chou, P. Y. and Fasman, G. D. (1974) Conformational parameters for amino acid in helical, beta-sheet, and random coil regions calculated from proteins, *Biochemistry 13*, 211–222.

REFERENCES

101 Swindellls, M. B., MacArthur, M. W., and Thornton, J. M. (1995) Intrinsic phi,psi propensities if amino acids, derived from the coil regions of known structures, *Nat Struct Biol 2*, 596–603.

102 Pace, C. N. and Scholtz, J. M. (1998) A helix propensity scale based on experimental studies of peptides and proteins, *Biophys J 75*, 422–427.

103 Wojcik, J., Altmann, K.-H., and Scheraga, H. A. (1990) Helix-coil stability constants for the naturally occurring amino acids in water. XXIV. Half-cysteine parameters from random Poly(Hydroxybutylglutamine-co-S-Methylthio-l-Cysteine), *Biopolymers 30*, 121–134.

104 Padmanabham, S., et al. (1994) Helix-forming tendencies of amino acids in short (Hydroxybutyl)-L-glutamine peptides: An evaluation of the contradictory results from host-guest studies and short alanine-based peptides, *Biochemistry 33*, 8604–8609.

105 Kim, C. A. and Berg, J. M. (1993) Thermodynamic beta-sheet propensities measured using a zinc finger host peptide, *Nature 362*, 267–270.

106 Minor, D. L. and Kim, P. S. (1994) Context is a major determinant of beta-sheet propensity, *Nature 371*, 264–267.

107 Smith, C. K., Withka, J. M., and Regan, L. (1994) A thermodynamic scale for the beta-sheet forming tendencies of the amino acids, *Biochemistry 33*, 5510–5517.

108 Minor, D. L. and Kim, P. S. (1994) Measurement of the beta-sheet-forming propensities of amino acids, *Nature 367*, 660–663.

109 O'Neal, K. T. and DeGrado, W. F. (1990) A thermodynamics scale for the helix-forming tendencies of the commonly occurring amino acids, *Science 250*, 646–651.

110 Chakrabartty, A., Shellman, J. A., and Baldwin, R. L. (1991) Large differences in the helix propensities of alanine and glycine, *Nature 351*, 586–588.

111 Lyu, P. C., et al. (1990) Side chain contribution to the stability of alpha-helical structure in proteins, *Science 250*, 669–673.

112 Merutka, G., et al. (1990) Effect of central-residue replacement on the helical stability of a monomeric peptide, *Biochemistry 29*, 7511–7515.

113 Scholtz, J. M., et al. (1991) Calorimetric determination of the enthalpy change for the alpha-helix to coil transition of an alanine peptide in water, *Proc Natl Acad Sci U S A 88*, 2854–2858.

114 Kemp, D. S., Boyd, J. G., and Muendel, C. C. (1991) The helical s constant for alanine in water derived from template-nucleated helices, *Nature 352*, 451–454.

115 Chakrabartty, A., Kortemme, T., and Baldwin, R. L. (1994) Helix propensities of the amino acids measured in alanine-based peptides without helix-stabilizing side-chain interactions, *Protein Sci 3*, 843–852.

116 Shi, Z., et al. (2005) Polyproline II propensities from GGXGG peptides reveal anticorrelation with beta-sheet scales, *Proc Natl Acad Sci U S A 102*, 17964–17968.

117 Moreau, R. J., et al. (2009) Context-independent, temperature-dependent helical propensities for amino acid residues, *J Am Chem Soc 131*, 13107–13116.

118 Han, W.-G., et al. (1998) Theoretical study of aqueous N-Acetyl-L-alanine N'-Methylamide: Structure and Raman VCD, and ROA spectra, *J Phys Chem B 102*, 2587–2602.

119 Woutersen, S., et al. (2002) Peptide conformational heterogeneity revealed from nonlinear vibrational spectroscopy and molecular-dynamics simulations, *J Chem Phys 117*, 6833–6840.

120. Vila, J. A., Ripoll, D. R., and Scheraga, H. A. (2000) Physical reasons for the unusual alpha-helix stabilization afforded by charged or neutral polar residues in alanine-rich peptides, *Proc Natl Acad Sci U S A 97*, 13075–13079.
121. Thomas, S. T., Loladze, V. V., and Makhatadze, G. I. (2001) Hydration of the peptide backbone largely defines the thermodynamic propensity scale of residues at the C' position of the C-capping box of alpha-helices, *Proc Natl Acad Sci U S A 98*, 10670–10675.
122. Lopez, M. M., et al. (2002) The enthalpy of the alanine peptide helix measured by isothermal titration calorimetry using metal-binding to induce helix formation, *Proc Natl Acad Sci U S A 99*, 1298–1302.
123. Yun, R. H. and Hermans, J. (1991) Conformational equilibria of valine studied by dynamics simulation, *Protein Eng 4*, 761–766.
124. Hermans, J., Anderson, A. G., and Yun, R. H. (1992) Differential helix propensity of small apolar side chains studied by molecular dynamics simulations, *Biochemistry 31*, 5646–5653.
125. Drozdov, A. N., Grossfield, A., and Pappu, R. V. (2004) Role of solvent in determining conformational preferences of alanine dipeptide in water, *J Am Chem Soc 126*, 2574–2581.
126. Padmanabhan, S. and Baldwin, R. L. (1991) Straight-chain non-polar amino acids are good helix-formers in water, *J Mol Biol 219*, 135–137.
127. Creamer, T. P. and Rose, G. D. (1992) Side-chain entropy opposes alpha-helix formation but rationalizes experimentally determined helix-forming propensities, *Proc Natl Acad Sci U S A 89*, 5937–5941.
128. Creamer, T. C. and Rose, G. D. (1994) Alpha-helix-forming propensities in peptides and proteins, *Proteins Struct Funct Genet 19*, 85–97.
129. Street, A. G. and Mayo, S. L. (1999) Intrinsic beta-sheet propensities result from van der Waals interactions between side chains and the local backbone, *Proc Natl Acad Sci U S A 96*, 9074–9076.
130. Blaber, M., Zhang, X., and Matthews, B. W. (1993) Structural basis of amino acid alpha-helix propensity, *Science 260*, 1637–1640.
131. Blaber, M., et al. (1994) Determination of alpha-helix propensity within the context of a folded protein, *J Mol Biol 235*, 600–624.
132. Creamer, T. P. and Rose, G. D. (1995) Interactions between hydrophobic side chains within alpha-helices, *Protein Sci 4*, 1305–1314.
133. Huyghues-Despointes, B. M. P., Klinger, T. M., and Baldwin, R. L. (1995) Measuring the strength of side-chain hydrogen bonds in peptide helices: The Gln Asp (i,i + 4) interaction, *Biochemistry 34*, 13267–13271.
134. Gnanakaran, S. and Garcia, A. E. (2003) Validation of an all-atom protein force field: From dipeptides to larger proteins, *J Phys Chem B 107*, 12555–12557.
135. Kwac, K., et al. (2008) Classical and quantum mechanical/molecular mechanical molecular dynamics simulations of alanine dipeptide in water: Comparisons with IR and vibrational circular dichroism spectra, *J Chem Phys 128*, 105106-1–105106-13.
136. Fermandjian, S., et al. (1990) Local interactions in peptides: H-H and 13C-H coupling constants for the conformational analysis of N-acetyl-N'-methylamides of aliphatic amino acids, *Int J Pept Protein Res 35*, 473–480.
137. Lorch, M., et al. (1999) Effects of core mutations on the folding of a beta-sheet protein: Implications for backbone organization in the I-state, *Biochemistry 38*, 1377–1385.

REFERENCES

138 Lorch, M., et al. (2000) Effects of mutants on the thermodynamics of a protein folding reactions: Implications for the mechanism of formation of the intermediate and transition states, *Biochemistry 39*, 3480–3485.

139 Yao, J., et al. (2001) NMR structural and dynamic characterization of the acid-unfolded state of apomyoglobin provides insight into the early events in protein folding, *Biochemistry 40*, 3561–3571.

140 Rose, G. D., et al. (1985) Hydrophobicity of amino acid residues in globular proteins, *Science 229*, 834–838.

141 Kyte, J. and Doolittle, R. F. (1992) A simple method for displaying the hydropathic character of a protein, *J Mol Biol 12*, 345–364.

142 Kenig, M., et al. (2006) Folding and amyloid-fibril formation for a series of human stefins' chimeras: Any correlation? *Proteins 62*, 918–927.

143 Wishart, D. S., Sykes, B. D., and Richards, F. M. (1992) The chemical shift index: A fast and simple method for the assignment of protein secondary structure through NMR spectroscopy, *Biochemistry 31*, 1647–1651.

144 Spera, S. and Bax, A. (1991) Empirical correlation between protein backbone conformation and C-alpha and C-beta 13C nuclear magnetic resonance chemical shifts, *J Am Chem Soc 113*, 5490–5492.

145 Osapay, K. and Case, D. A. (1991) A new analysis of proton chemical shifts in proteins, *J Am Chem Soc 113*, 9436–9444.

146 Dios, A. C. D., Pearson, J. G., and Oldfield, E. (1993) Secondary and tertiary structural effects on proteins NMR chemical shifts: An ab initio approach, *Science 260*, 1491–1496.

147 Williamson, M. P. and Asakura, T. (1993) Empirical comparison of models for chemical-shift calculation in proteins, *J Magn Reson B 101*, 63–71.

148 Osapay, K. and Case, D. A. (1994) Analysis of proton chemical shifts in regular secondary structure of proteins, *J Biomol NMR 4*, 215–230.

149 Asakura, T., et al. (1995) The relationship between amide proton chemical shifts and secondary structure in proteins, *J Biomol NMR 6*, 227–236.

150 Sitkoff, D. and Case, D. A. (1997) Density functional calculations of proton chemical shifts in model peptides, *J Am Chem Soc 119*, 12262–12273.

151 Sitkoff, D. and Case, D. A. (1998) Theories of chemical shift anisotropies in proteins and nucleic acids, *Prog Nucl Magn Reson Spectrosc 32*, 165–190.

152 Wishart, D. S. and Nip, A. M. (1998) Protein chemical shift analysis: A practical guide, *Biochem Cell Biol 76*, 153–163.

153 Wishart, D. S. and Case, D. A. (2001) Use of chemical shifts in macromolecular structure determination, *Methods Enzymol 338*, 3–34.

154 Bai, Y., et al. (1993) Primary structure effects on peptide group hydrogen exchange, *Proteins 17*, 75–86.

155 Bai, Y. and Englander, S. W. (1994) Hydrogen bond strength and beta-sheet propensities: The role of a side chain blocking effect, *Proteins 18*, 262–266.

156 Sheinblatt, M. (1970) Determination of an acidity scale for peptide hydrogens from nuclear magnetic resonance kinetic studies, *J Am Chem Soc 92*, 2505–2509.

157 Molday, R. S. and Kallen, R. G. (1972) Substituent effects on amide hydrogen exchange rates in aqueous solution, *J Am Chem Soc 94*, 6739–6745.

158 Fogolari, F., et al. (1998) pKa shift effects on backbone amide base-catalyzed hydrogen exchange rates in peptides, *J Am Chem Soc 120*, 3735–3738.

159 Avbelj, F. and Baldwin, R. L. (2009) Origin of the change in solvation enthalpy of the peptide group when neighboring peptide groups are added, *Proc Natl Acad Sci U S A 106*, 3137–3141.

160 Kabsch, W. and Sander, C. (1983) Dictionary of protein structure: Pattern recognition of hydrogen-bonded and geometrical features, *Biopolymers 22*, 2577–2637.

6

EXPERIMENTAL AND COMPUTATIONAL STUDIES OF POLYPROLINE II PROPENSITY

W. Austin Elam, Travis P. Schrank, and Vincent J. Hilser

6.1. INTRODUCTION

Classic experiments performed by Anfinsen and colleagues showed that, in the proper solvent, certain proteins spontaneously and reproducibly fold into their active and functional form. This discovery led to the conclusion that all of the information required for protein folding and function is encoded in the primary sequence [1]. Additionally, Levinthal posited that proteins cannot fold in a biologically reasonable amount of time through exhaustive, random search of conformational space [2], suggesting that conformational bias must exist in the denatured state of polypeptides. Indeed, early circular dichroism (CD) spectroscopy experiments by Tiffany and Krimm using homopolymers of proline, glutamine, and lysine suggested that conformational bias exists in these polymers, as their CD signal differed from that of a random coil, and they proposed that this difference was due to the presence of left-handed polyproline II (PII) helix [3], which can be observed even in polymers that do not contain proline [4].

The canonical PII helix formed by a homopolymer of proline is an extended left-handed helix with exactly three residues per turn, having characteristic phi (φ), psi

---

Protein and Peptide Folding, Misfolding, and Non-Folding, First Edition. Edited by Reinhard Schweitzer-Stenner.
© 2012 John Wiley & Sons, Inc. Published 2012 by John Wiley & Sons, Inc.

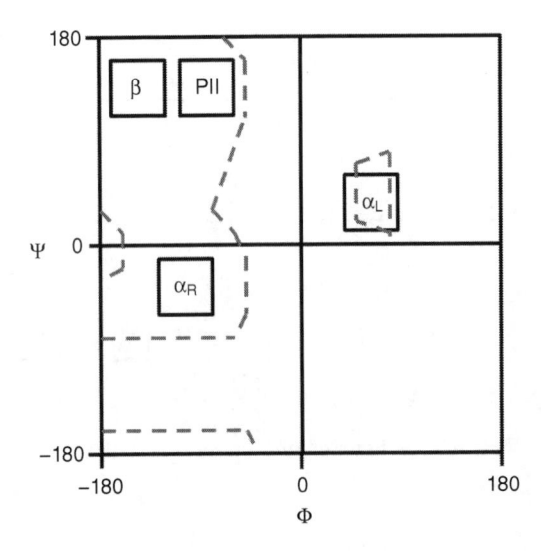

Figure 6.1. Schematic of Ramachandran space showing the region of the PII conformation relative to other secondary-structure elements: α-helix, β-sheet, and left-handed helix. The labeled regions of φ,ψ space correspond to α-helix (αR), β-sheet (β), left-handed helix (αL), and polyproline II helix (PII). The gray dashed line approximates the sterically accessible space in an alanine dipeptide modeled with full van der Waals radii [91].

(ψ) dihedral angles (−75, +145) and containing no intramolecular hydrogen bonds [5]. Conformational bias at a single position toward PII, corresponding to a region of Ramachandran space shown in Figure 6.1, has been demonstrated independently using a variety of methods. Numerous CD studies have investigated the presence of PII within the conformational ensemble of different peptide models [3, 4, 6–15]. Computational [16–19], and spectroscopic [20–27] methods have been used to investigate conformational bias for PII in unfolded peptide systems. Surveys and other studies of PII in the native states of globular proteins have also been conducted [28–30], demonstrating that PII helices can be found within native protein folds both in proline-rich regions and in non-proline sites.

Importantly, the PII conformation can be vital for protein function. Structural proteins such as collagen [31] and plant cell wall proteins [32], and other proteins that play roles in cellular processes such as cell motility [33], immune response [34], and cell signaling [35], have been shown to contain a propensity toward the PII conformation. One example of PII helix formation is in the cell signaling activation of the Ras pathway, which is involved in cellular processes such as cell growth [36, 37]. In this pathway, modular Src-homology 3 (SH3) domains located in an adaptor protein bind a proline-rich loop region of a nucleotide-exchanger protein (mSos1 in mammals) that triggers the Ras signaling cascade. Importantly, the crystal structure (2.0 Å resolution) of the C-terminal Sem-5 SH3 domain from *Caenorhabditis elegans* bound to an mSos-derived proline-rich peptide showed that the peptide is in the PII conformation in the bound complex [38].

The C-terminal Sem-5 SH3 domain and Sos peptide system have been studied previously in a variety of contexts [38–43]. Nuclear magnetic resonance (NMR) spectroscopy and isothermal titration calorimetry (ITC) have been used to characterize the solution structure, dynamics, and thermodynamics of the Sem-5 SH3 domain, both alone and binding Sos ligand [39, 40]. The impact of ligand PII formation on the thermodynamics of binding has been calorimetrically assessed using mutants of both the SH3 domain and Sos ligand [41, 42]. Recent computational work has sought to elucidate the extent to which mutations in the SH3 domain modulate the native ensemble and influence Sos ligand binding [43], as well as the impact of PII conformational bias on the binding of Sos peptides to SH3 [44].

Here, we present a critical review of previous work on the PII conformation, including PII propensity scales and computational models. We describe how the well-characterized SH3:Sos system can be implemented for host–guest study of PII formation. Correlations of PII scales to each other and to a database of physicochemical property scales and secondary-structure propensity scales are reported.

6.2. EXPERIMENTAL MEASUREMENT OF PII PROPENSITIES

A number of experimental approaches have been used to investigate conformational bias in the denatured state. However, limited consensus regarding the biological PII propensities of each amino acid has resulted from previously published work. In this section, we briefly summarize key experimental contributions to quantifying amino acid PII propensities. The strengths of these studies as well as their conceptual and technical limitations are addressed.

Creamer and colleagues employed CD and other complimentary techniques to experimentally interrogate the PII conformation in model peptide systems. Using a proline-rich model peptide (Ac-PPPXPPPGY-NH$_2$) as a host–guest system, the first, nearly complete PII propensity scale was developed [10, 12]. The numerical PII propensities listed in Table 6.1 [12] are obtained using Equation 6.1 taken from Ref. [10]:

$$\text{PII\%} = ((\theta_{\text{max}} - 6100)/13{,}700) \times 100. \tag{6.1}$$

The value θ_{max} is the highest molar ellipticity measured in the wavelength range of 220–230 nm, where a characteristic peak is observed and ascribed to the PII conformation [9]. The θ_{max} for each host–guest peptide may occur at different wavelengths, and the collective peptide spectra lack an isodichroic point. Additionally, some assumptions were necessary to quantify PII with their approach and host–guest system. Numerical values in Equation 6.1 correspond to a global, empirical θ_{max} (13,700) observed for polyproline in 8.4 M guanidine hydrochloride and a θ_{max} (6100) observed for a completely disordered model peptide [10], which they assume as values for 100% and 0% PII propensity, respectively. Further, they assume that no other secondary structure contributes to the absorbance in this wavelength range. Although these studies represent the first important steps toward measurement of biological PII propensities, technical limitations imposed by CD prohibited

TABLE 6.1. PII Propensity Scales

PII Propensity Scales (% PII)

Amino Acid	PPXPP[a]	GGXGG[b]	GXG[c]	PPXPP[d]	GGXGG[d]	GGXGG[e]
A	61	81.8	79	32.1	42.3	14.6
C	55	55.7	–	34.8	34.2	10.6
D	63	55.2	–	48.3	52.3	4.8
E	62	68.4	54	44.9	49.7	14.75
F	58	63.9	42	29.9	43.6	12.4
G	58	–	–	10.3	9.7	4.7
H	55	42.8	–	33.9	47.3	10.6
I	50	51.9	–	5.7	4.4	24.3
K	59	58.1	50	44.0	49.1	10.2
L	58	57.4	56	0.4	0.1	11.4
M	55	49.8	64	44.5	48.9	10.5
N	55	66.7	–	49.0	46.8	6.6
P	67	–	–	92.2	74.6	58.5
Q	66	65.4	–	43.4	50.8	10.9
R	61	63.8	–	45.1	49.6	11.0
S	58	77.4	45	35.8	48.5	23.2
T	53	55.3	–	29.6	29.2	24.5
V	49	74.3	40	5.0	3.0	20.3
W	–	76.4	–	30.5	45.6	8.1
Y	–	63.0	–	29.8	43.5	10.6

[a] Rucker et al. 2003, Table 1. Errors in PII% are ±1–2% estimated from a measured error in molar ellipticities of ±3% [12].
[b] Shi et al. 2005, Table 1 [25].
[c] Hagarman et al. 2010, Table 1. Errors reported are ±2–5% [27].
[d] Tran et al. 2005, Tables 4 and 7, Supplemental Materials [19].
[e] Beck et al. 2010, Table 1 [67]. Some values represent averages of multiple simulations.

interpretation of tryptophan and tyrosine propensities, leaving the PII scale incomplete. A proline-rich peptide host is necessary for reproducible and measurable signal within the wavelength range of 220–230 nm. However, the context of the host (flanking prolines) and resolution of the CD measurements result in a PII propensity scale that has a high and narrow range as shown in Table 6.1. Regardless of the numerical values reported (see Table 6.1; Ref. [12]), examination of the rank order of amino acid PII propensities obtained from the host–guest study provides some insights. Generally, proline and long, charged side chains possess the highest PII bias, while β-branched amino acids isoleucine and valine are at the bottom of the scale [12]. The high propensity for proline, glutamine, and lysine to adopt the PII conformation is consistent with other CD studies [3, 4, 11, 14], bolstering the credibility of CD for assessing the presence of PII, although the exact propensities of intermediate residues in the scale are within experimental error of each other and difficult to distinguish.

Attention has also been given to studies of alanine-based peptides as model systems for investigations of PII conformational bias. A seminal but controversial study of the peptide $AcX_2A_7O_2NH_2$ (where X is diaminobutyric acid and O is ornithine), hereafter referred to as the XAO peptide, found that sequences of alanine contained significant PII content on the basis of CD spectra and J-coupling analysis [22]. Spectroscopic studies by Schweitzer-Stenner and colleagues using alanine peptides and XAO support the concept that alanine frequently adopts the PII conformation [45, 46], although they acknowledge that the conformations adopted by the XAO peptide are heterogeneous [47]. Kallenbach and colleagues also performed a Raman optical activity spectroscopy study on short alanine peptides and claimed that PII can be a dominant conformation of these peptides [48].

Other groups dispute claims that alanine has such a high PII propensity. Scheraga and coworkers have presented evidence to challenge the previous claims of Kallenbach and others, reporting that PII is one of many conformations available to alanine in XAO and that PII is not a dominant global conformation of the XAO peptide [49, 50]. Supporting this result, small angle X-ray scattering (SAXS) also demonstrated that the XAO peptide adopts a much smaller radius of gyration (~7.4 Å) in solution compared with that expected of an ideal, fully extended PII helix (~13 Å), indicating that PII bias is local and that PII helix is not a global conformation of the peptide [51]. In contrast to SAXS results, spin relaxation enhancement experiments suggested that extended PII conformations dominate in short alanine peptides; however, a model peptide containing polyproline stretches was used in this particular study rather than the XAO peptide [52]. Further evidence in support of high PII bias in the XAO peptide is also presented in a review [26], much of which is focused on the dispute regarding the prevalence of the PII helix in alanine-based models. In an attempt to reconcile the arguments surrounding the XAO peptide, it is important to point out that Kallenbach and colleagues have never claimed that XAO is a single PII helix in solution, explicitly stating in their original (and subsequent) publications: "significant fluctuations from the idealized structure shown here (depicting an extended PII helix [22]) probably occur."

In addition to work in the XAO peptide system, Kallenbach and colleagues conducted a host–guest study using the peptide Ac-GGXGG-NH_2 and reported a PII scale of ~40–80% (Table 6.1), with alanine having a PII propensity of 82.8% in this context [25], maintaining the debated, high propensity in the XAO system. The approach used by Shi et al. [25] in the GGXGG system leverages $^3J_{\alpha N}$-coupling and CD spectra to quantify position-specific PII at the guest site in these model peptides. To briefly summarize their method, experimental $^3J_{\alpha N}$-coupling constants ($J_{measured}$) for each amino acid in the guest position are compared with ideal $^3J_{\alpha N}$-coupling constants (J_{PII} and J_β) for the matching amino acid calculated from a coil library [53], where the $^3J_{\alpha N}$-coupling constants are obtained using the Karplus equation parameterized by Vuister and Bax [54]. Equation 6.2, below, frames their scheme [25]:

$$\text{PII}\% \times J_{PII} + (1 - \text{PII}\%) \times J_\beta = J_{measured}. \tag{6.2}$$

The experimental $^3J_{\alpha N}$-coupling constants ($J_{measured}$) are interpreted in terms of only the PII and β conformations, as most CD spectra obtained in this system display isodichroic points and Nuclear Overhauser Effects (NOEs) are not observed for α-helix formation [23, 25]. As only PII and β conformations are considered in their scheme, the PII propensities for many of the amino acids are high, even compared with propensities in a proline-rich host (Table 6.1, [12]). The body of work generated by Kallenbach and colleagues [22, 23, 25, 26] introduced the use of NMR and an alternative, glycine-rich host–guest system to the study of PII propensity, while spurring other groups to investigate conformational bias in the denatured state using NMR [55, 56].

Raman optical activity (ROA) spectroscopy is yet another technique that can be used to characterize the conformational behavior of peptides in solution [21]. Schweitzer-Stenner and colleagues have brought ROA and a host of other spectroscopic techniques to bear on the PII propensities of amino acids in solution using a number of model systems [24, 27, 45–47, 57, 58]. Of particular note are two host–guest studies performed in different host–guest systems, AXA [24] and, most recently, GXG [27]. Although the GXG scale is incomplete (Table 6.1), its strength lies in the number of techniques employed to quantify PII bias, which yield internally consistent results. Furthermore, the use of tripeptides eases the interpretation of guest-site conformational bias, as these host systems lack long-range interactions.

Work in our laboratory has taken a different approach than previously described. The strategy to calorimetrically measure PII propensities using the binding of the Sem-5 SH3 domain to the Sos peptide (Ac-VPPXVPPRRRY-NH$_2$) warrants detailed explanation. The association of SH3 and Sos peptides is an equilibrium process, and the unbound peptide is assumed to adopt an ensemble of conformations. A fraction of these conformers will be in PII, the binding competent state. The apparent binding free energy (ΔG_{app}), obtained through best-fit parameterization of ITC experiments, contains free energy contributions as shown in Equation 6.3. This includes contributions from the binding interaction (ΔG_{int}), and conformational free energies of SH3 ($\Delta G_{con,SH3}$), and Sos peptide ($\Delta G_{con,Sos}$) folding into binding competent conformations, the latter of which we seek to analyze:

$$\Delta G_{app} = \Delta G_{int} - \Delta G_{con,SH3} - \Delta G_{con,Sos}. \qquad (6.3)$$

To access solely the conformational free energy of the peptides, the difference in the apparent binding free energy of each Sos mutant is measured relative to the wild-type Sos (X=PRO). The resulting expression for the difference in apparent free energy of binding is shown below:

$$\Delta\Delta G_{app,Sos-X} = (\Delta G_{int,Sos} - \Delta G_{int,X}) - (\Delta G_{con,SH3} - \Delta G_{con,SH3}) - (\Delta G_{con,Sos} - \Delta G_{con,X}).$$

$$(6.4)$$

where the index X denotes the respective free energies involved in mutant peptide binding. To demonstrate that the calculated difference in the apparent binding free energies are directly reporting conformational differences in the Sos peptide, the first and second terms of Equation 6.4, corresponding to the free energy differences of the binding interface and conformation of SH3, respectively, must be the same for

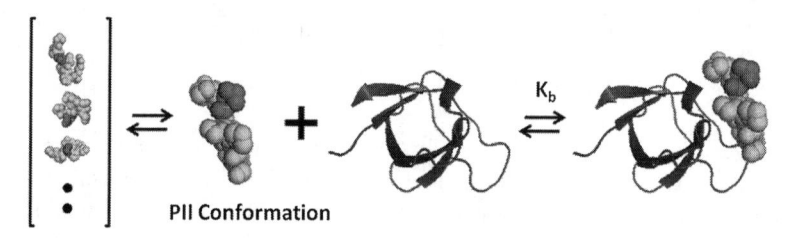

Figure 6.2. The SH3:Sos model system employs a mutation strategy that allows calorimetric determination of a residue-specific PII propensity. Schematic of SH3 domain and SosY peptide (VPPPVPPRRRY) binding using the crystal structure (1SEM) of Lim et al. [38]. The mutation site in the SosY peptide, which is surface-exposed in the bound complex, is highlighted in red. The mutation is expected to perturb only the conformational equilibrium of the peptide between PII and other (binding incompetent) conformations. Images generated using the PyMOL Molecular Graphics System, Version 1.2r3pre, Schrödinger, LLC. See color insert.

the wild-type and mutant peptides. Evidence suggests that the free energy of the binding interface is not perturbed by mutations to the Sos peptide. The site of the mutation in the third position (counting from zero) in the peptide sequence is a position along the peptide PII helix that is completely solvent-exposed in the crystal structure of the bound complex as shown in Figure 6.2 [38]. NMR experiments with isotopically labeled SH3 domain in saturated ligand solutions of either SosY or Ala or Gly peptides showed superimposable ^{1}H-^{15}N HSQC spectra between SH3 bound to Ala and Gly mutant ligands. The lack of perturbation to the chemical environment of the residues in SH3, as seen in the ^{1}H-^{15}N HSQC overlays, suggests that the binding interface does not change between Sos mutants [41]. Since the evidence demonstrates that the SH3 protein remains unchanged throughout the ITC experiments, we assume the conformational free energy for SH3 to adopt a binding competent conformation is not changed, that is, $\Delta G_{con,SH3} - \Delta G_{con,SH3} = 0$. Accepting that the differences in the apparent free energy of binding are attributable only to differences in the conformational free energy of the peptides, the apparent binding free energy may be simplified:

$$\Delta\Delta G_{app,Sos-X} \sim \Delta G_{con,Sos} - \Delta G_{con,X}. \qquad (6.5)$$

As discussed previously in the work of Whitten et al. [44], if no difference is observed in the free energy of binding between wild-type and mutant Sos peptides, then the unbound peptide exists in a "prefolded," binding competent PII state. In contrast, non-zero differences in binding free energy suggest that the unbound peptide exists in equilibrium between binding competent (PII) conformations and other binding incompetent (random coil) conformations. Decrease in the binding free energy difference observed for Ala and Gly mutations indicates a shift in the peptide ensemble toward binding incompetent (random coil) conformations [41]. Conformational partition functions for the wild-type (Q_{WT}) and mutant (Q_{P3X}) peptides can be expressed as follows:

$$Q_{WT} = \ldots(1+K_1)(1+K_2)(1+K_4)(1+K_5)\ldots \qquad (6.6)$$

$$Q_{P3X} = \ldots(1+K_1)(1+K_2)(\mathbf{1+K_{3X}})(1+K_4)(1+K_5)\ldots \qquad (6.7)$$

where K_i are the conformational equilibrium constants between PII and other disordered conformations. As it is formulated, this partition function contains several assumptions that have been validated both computationally and experimentally. Implicit in the conformational partition function for each residue is the assumption that there are no conformational states other than PII for which the peptide has significant bias. This assumption has been shown to be valid using an algorithm based on hard sphere [59] collisions, which generates random conformers of any specified peptide sequence through unbiased search of Ramachandran space, accepting only conformations that do not contain steric violations [44]. Also implicit in this scheme are the assumptions that PII formation is locally driven [60] and non-cooperative [61] and that a proline following another proline is significantly restricted in available backbone dihedral angle space [5, 8, 62–64]. Therefore, the only difference between Equations 6.6 and 6.7 is conformational equilibrium for the site of mutation. Given these reasonable assumptions, the difference in the calculated apparent free energies of binding can be expressed as shown in Equation 6.8:

$$\Delta\Delta G_{app,Sos-X} = -RT \ln(Q_{P3X}/Q_{WT}) = -RT \ln(1+K_{3X}). \qquad (6.8)$$

from which the position-specific PII propensity at the guest site can directly be calculated. PII propensities obtained for Ala (~37%) and Gly (~13%) using the SH3:Sos system, correspond to $\Delta\Delta G$ values previously published [41]. Importantly, this scheme allows for the development of a complete, calorimetrically determined PII scale.

Comparison of three previously published experimental PII propensity scales listed in Table 6.1 [12, 25, 27] shows a striking lack of correlation between any of the scales. Differences in the host–guest systems and the techniques employed to measure the PII bias of each amino acid make comparison of the numerical values in each scale difficult to interpret. The numerical PII propensities for each amino acid are depicted in Figure 6.3A, where close inspection reveals that certain amino acids appear to have similar PII propensities between the three scales (Leu), while others may differ by more than 10% (Ser). Comparison of the rank order of amino acid PII propensities from two of the scales of equivalent completion, as shown in Figure 6.3B, provides clear evidence of the lack of correlation between scales. Results of linear correlations in the data are shown in Table 6.2, demonstrating no statistically significant correlation between scales. Disparate PII propensities might be expected in different host–guest systems, as proline and glycine flanking residues provide different contexts for the guest. The discrepancy in the conformational propensities between GGXGG [25] and GXG [27] shown in Figure 6.3A may be attributable to assumptions made that exclude conformations and the effect of those assumptions on interpretation of $^3J_{\alpha N}$-coupling values [25].

EXPERIMENTAL MEASUREMENT OF PII PROPENSITIES 167

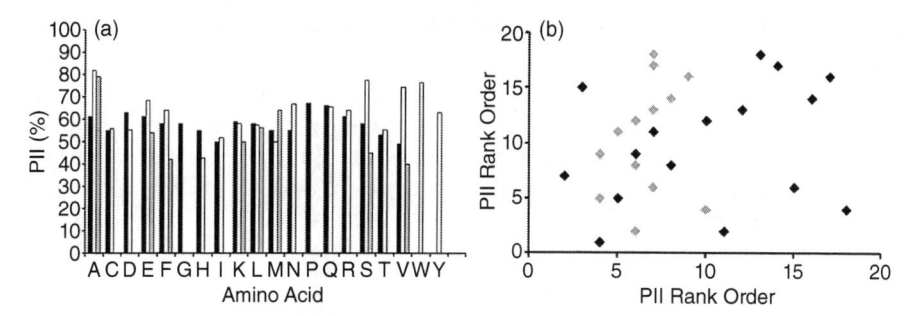

Figure 6.3. Comparison of experimentally determined PII propensity scales demonstrates little consensus. (A) Numerical PII propensities are reported for proline-rich (black) [12], and for glycine-rich GGXGG (white) [25] and GXG (gray) [27] contexts. (B) Comparison of the rank order of PII propensities of Rucker et al. [12] (abscissa) and Shi et al. [25] (ordinate), two scales of equivalent completeness. Rank order is from highest PII bias (#1) to lowest (#18) (black), as two amino acids are missing from each scale. Points shown in gray are for the rank order comparison allowing for equivalent PII propensities creating "ties" in the rank order.

TABLE 6.2. Statistical Correlation of PII Scales

Statistical Correlation of PII Propensity Scales

	Scale Comparison	Peptides	P-value[e]
Rucker et al. 2003 PPXPP[a]			
	Shi et al. 2005	GGXGG	0.758
	Tran et al. 2005	PPXPP	0.014
	Tran et al. 2005	GGXGG	<0.01
	Beck et al. 2010[d]	GGXGG	0.713
Shi et al. 2005 GGXGG[b]			
	Tran et al. 2005	PPXPP	0.919
	Tran et al. 2005	GGXGG	0.951
	Beck et al. 2010	GGXGG	0.446
Tran et al. 2005 PPXPP[c]			
	Tran et al. 2005	GGXGG	<0.01
	Beck et al. 2010	GGXGG	0.347
Tran et al. 2005 GGXGG[c]			
	Beck et al. 2010	GGXGG	0.609

[a] [12].
[b] [25].
[c] [19].
[d] [67].
[e] Spearman linear correlation P-value.

6.3. COMPUTATIONAL STUDIES OF DENATURED STATE CONFORMATIONAL PROPENSITIES

Perhaps the single most complete and ambitious study of conformational bias in the denatured state is the work of Pappu and colleagues [19], where they computationally developed conformational propensity scales (including PII) in five host–guest systems (AAXAA, FFXFF, GGXGG, PPXPP, and VVXVV). Here, we will focus specifically on their results pertaining to bias for the PII conformation (scales shown in Table 6.1). Essentially, the approach of Tran et al. [19] is to computationally generate conformers using excluded-volume interactions and Metropolis Monte Carlo method to select torsional angles for the peptide backbone [19]. The puckering of the proline pyrrolidine ring is explicitly included in the model, but Tran et al. simplify their systems by assuming that the peptide unit is always *trans* with $\omega = 179.5°$ [19]. Clustering of Tran et al.'s conformational ensemble allows for the calculation of the conformational propensities of their guest residues [19]. Summarizing their results, Tran et al. find that (1) conformational propensities show a general bias toward extended conformational space in the upper left regions of the Ramachandran map corresponding to β and PII; (2) the host context affects the conformational propensity of the guest, although as expected glycine has the smallest effect; and (3) surprisingly, the proline host diminishes the bias for some amino acids, such as alanine, to form PII [19]. Importantly, Tran et al.'s study also reveals the steric persistence length, or "steric unit," of a polypeptide to be five residues long, a finding that reconciles the observations of high position-specific conformational bias with the overall random coil behavior of a polypeptide chain [19].

Tran et al. [19] claim agreement with trends in experimental measurements discussed above [12, 24, 41], with few outliers that are specifically addressed in their publication. Figure 6.4A,C show the numerical values of two experimental scales [12, 25] compared with those obtained by Pappu and colleagues [19] in the same host–guest system, PPXPP and GGXGG, respectively. Similar trends can be observed between experimental scales in Figure 6.3A, where some amino acids appear to have similar PII propensities between the scales. However, other amino acid PII propensities are in clear disagreement. Comparison of the simple rank order of experimentally measured PII propensities compared with those determined using excluded-volume computational models shown in Figure 6.4B,D demonstrates that limited consensus exists regarding which amino acids have high and low PII propensities. However, there does appear to be some agreement in Figure 6.4B, with some amino acids having similar, but not identical, rank order. Linear correlations of the PPXPP scale of Tran et al. [19] to other PII scales show that the excluded-volume PII scales statistically correlate ($P < 0.05$) with the PII propensity scale reported by Creamer and colleagues [12] using CD (Table 6.2). Interestingly, when the rank order of PII propensities from Tran et al. [19] are compared between two different host systems (GGXGG and PPXPP), Figure 6.5 shows that the general rank order is preserved, suggesting that the host context provides subtle bias but does not drastically alter the position-specific bias in a peptide sequence, even for hosts as different as proline and glycine. Linear correlation of the PPXPP and

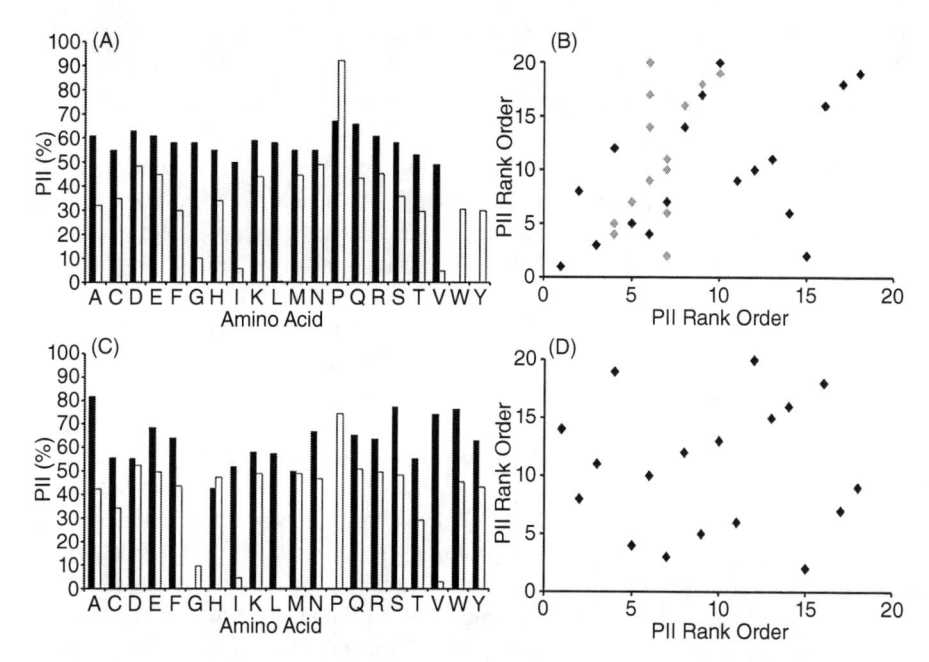

Figure 6.4. Comparison of PII propensities obtained from an excluded-volume model to experimental propensities reported in the same host–guest systems. (A) Numerical PII propensities reported for a proline-rich peptide determined using CD (black) [12] and an excluded-volume model (white) [19]. (B) Comparison of the rank order of PII propensities of Rucker et al. [12] (abscissa) and Tran et al. [19] (ordinate). Rank order is from highest PII bias (#1) to lowest (#20) (black). Points shown in gray are for the rank order comparison allowing for equivalent PII propensities creating "ties" in the rank order. (C) Numerical PII propensities reported for a glycine-rich peptide determined using NMR (black) [25] and an excluded-volume model (white) [19]. (D) Comparison of the rank order of PII propensities of Shi et al. [25] (abscissa) and Tran et al. [19] (ordinate). Rank order is from highest PII bias (#1) to lowest (#20) (black).

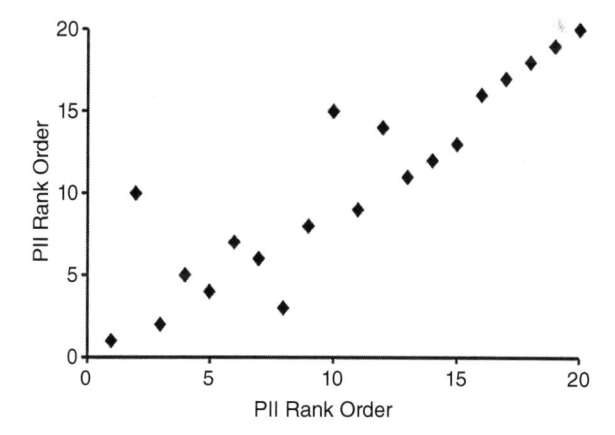

Figure 6.5. Rank order comparison of PII propensities in different peptide contexts reveals similar trends in PII bias. Comparison of the rank order of PII propensities of Tran et al. [19] observed for the PPXPP host (abscissa) and GGXGG host (ordinate) using an excluded-volume model. Rank order is from highest PII bias (#1) to lowest (#20) (black).

GGXGG excluded-volume scales shown in Table 6.2 is statistically significant ($P < 0.05$).

Other groups have also endeavored to computationally challenge experimentally measured conformational propensities. Two studies by Scheraga and colleagues used computationally generated conformational ensembles to recreate expected experimental parameters (^{13}C chemical shifts). From these chemical shifts, $^3J_{\alpha N}$-coupling constants and $^1J_{C\alpha H\alpha}$ vicinal coupling constants were calculated and interpreted in terms of PII propensity in the XAO peptide [65] and in a proline-rich host [18]. In the XAO peptide system, Vila et al. [65] concluded that the PII conformational bias of alanine should be ~30%, which is in agreement with other computational [19] and calorimetric [41] studies, but contrary to other spectroscopic data [24, 48]. Results obtained following a conceptually similar approach in a proline-rich host revealed two trends that has also been observed experimentally via CD [10]. Inspection of the reported PII propensities of Vila et al. [18] showed little numerical correlation to those obtained by CD with the exception of proline (~67%), although they did observe a general drop in guest PII bias relative to proline, consistent with earlier results [10, 66].

Simulations by Daggett and colleagues [67] on GGXGG peptides showed no correlation to experimental results [25] and claim that the PII bias at the guest position is much lower (~5–25%, see Table 6.1) than previous experiments indicated (~40–80%) [67]. Molecular dynamics simulations starting from fully extended (φ and $\psi = 180°$) peptides in explicit solvent at 298 K were run for 100 ns and information on the torsional angles of the peptide backbone was collected [67]. From these data, population frequencies for various conformations were computed. In Figure 6.6A, we compare the numerical PII bias experimentally measured by Shi et al. [25], with the computationally determined values of Daggett and colleagues [19, 67]. No numerical correlation exists. Inspection of the rank order of these scales reveals no correlation, as shown in Figure 6.6B,C. Linear correlation of the molecular dynamics simulation-derived PII propensity scale [67] showed no statistically significant correlation to any other PII scale as shown in Table 6.2. Both Table 6.1 and Figure 6.2 of Beck et al. [67] may provide some explanation of the observed low PII biases. The reported population frequencies and Ramachandran maps [67] indicate a large propensity ($\geq 40\%$) for α-helix or "near α-helix" regions not observed in the experimental system [25] and only modestly populated ($\leq 20\%$) in the excluded-volume model [19].

Figure 6.7A summarizes the experimental and computationally derived PII propensity scales, from which we draw several conclusions. First, the numerical PII propensities of amino acids cover a wide range of bias from nearly 0% to 80% and the ranges of computational and experimental PII scales scarcely overlap in some cases. Next, different techniques and host–guest systems yield different estimates for PII bias, despite the apparent lack of context dependence shown in Figure 6.5. Lastly, computational PII scales have limited to no correlation with experimental scales using identical model host–guest systems, with the notable exception of the correlation between the PPXPP excluded-volume model [19] and the CD-derived scale [12]. Generally, the ranges of experimental scales are compressed relative to computational models.

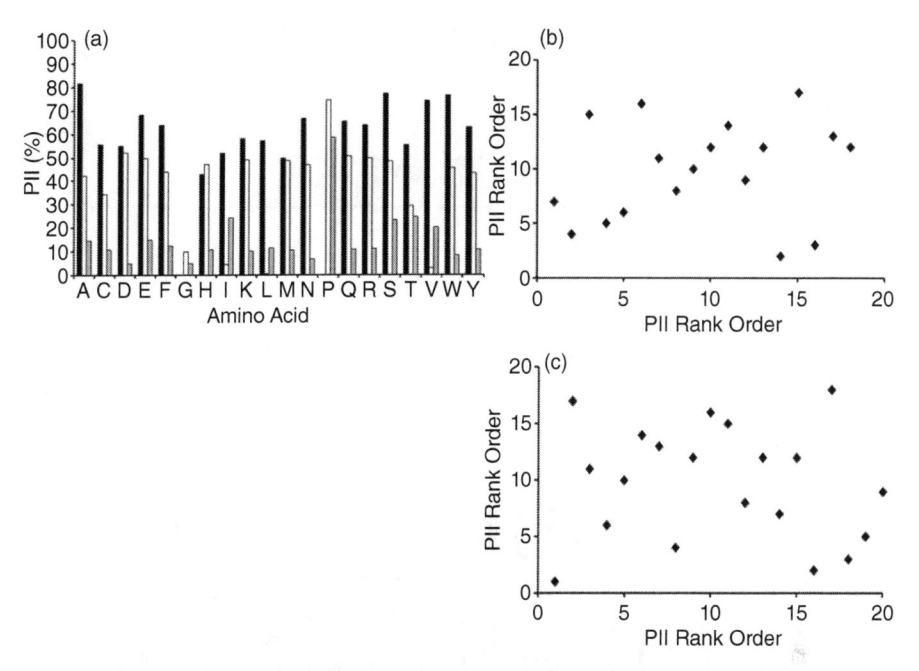

DENATURED STATE CONFORMATIONAL PROPENSITIES   171

Figure 6.6. Comparison of PII propensities obtained from a molecular dynamics simulation to those obtained experimentally and with an excluded-volume model all in the glycine-rich host–guest system (GGXGG). (A) Numerical PII propensities obtained using NMR (black) [25], an excluded-volume model (white) [19], and molecular dynamics simulation (gray) [67]. (B) Comparison of the rank order of PII propensities of Shi et al. [25] (abscissa) and Beck et al. [67] (ordinate). Rank order is from highest PII bias (#1) to lowest (#20) (black). (C) Comparison of the rank order of PII propensities of Tran et al. [19] (abscissa) and Beck et al. [67] (ordinate). Rank order is from highest PII bias (#1) to lowest (#20) (black).

Importantly, some consensus may be reached regarding which amino acids possess high or low propensity for the PII conformation. Figure 6.7B shows the average rank order of amino acids taken from several PII propensity scales [12, 19, 25, 67], from which we can obtain general trends in PII propensities for some amino acids. The top five amino acids in terms of average PII propensity rank order are Pro, Glu, Ser, Gln, and Arg from highest to lowest. Furthermore, the standard deviation of the average ranking of these amino acids is small, suggesting a general agreement between PII propensity scales. The amino acids Cys, Tyr, Ile, Leu, and Gly have the lowest average rank orders (high to low); however, only Tyr and Gly have low standard deviations that place them firmly at the bottom of the PII propensity scale. Other remaining amino acids have average rank orders ranging from ~7 to 11 and have large standard deviations (3.2–7.3), making it difficult to establish a true rank order for these amino acids. With the exception of Pro, which is known to have high PII propensity because of steric hinderance [5, 8, 62–64], nonpolar and aromatic amino acids tend to have intermediate or low PII propensities, while polar and charged amino acids tend to be intermediate or high. However, this does not

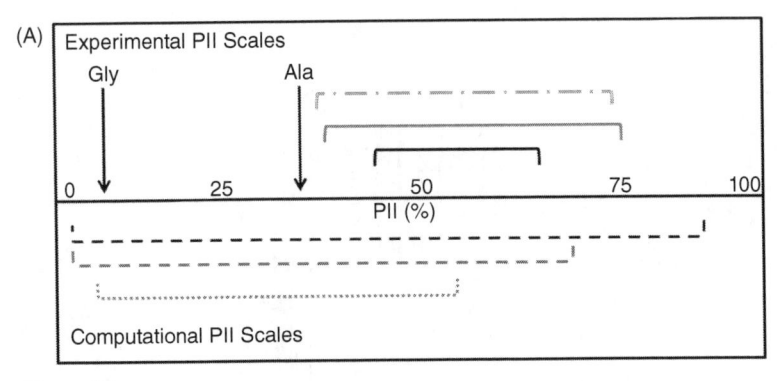

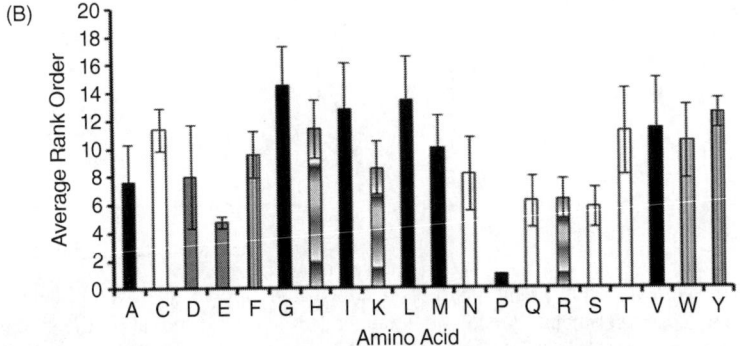

Figure 6.7. Schematic of the ranges of both computational and experimental PII propensity scales. (A) Ranges are shown by brackets for PPXPP obtained using CD (black solid) [12], an excluded-volume model (black dashed) [19] and in GGXGG/GXG from NMR (gray solid) [25], an excluded-volume model (gray dashed) [19], molecular dynamics simulation (gray dotted) [67], and by a combination of spectroscopic techniques (gray dash-dotted) [27]. Positions marked by black arrows for Ala and Gly PII propensities have been experimentally measured [41], and computationally validated [18, 19]. (B) Average rank order of each amino acid taken from experimental [12, 25] and computational [19, 67] PII propensity scales containing at least 8 of 20 amino acids. Bars are color-coded by amino acid type: nonpolar (black), polar (white), aromatic (vertical stripes), positively charged (horizontal stripes), and negatively charged (gray). Error bars represent the standard deviation of the average ranking of each amino acid.

imply that PII propensity is driven by a single amino acid feature, as will be discussed later.

6.4. A STERIC MODEL REVEALS COMMON PII PROPENSITY OF THE PEPTIDE BACKBONE

Employing an excluded-volume model from previous studies, Mini-Protein Modeler (MPMOD) [43, 44], the PII propensities of Ala and Gly are determined in the context

A STERIC MODEL REVEALS COMMON PII PROPENSITY

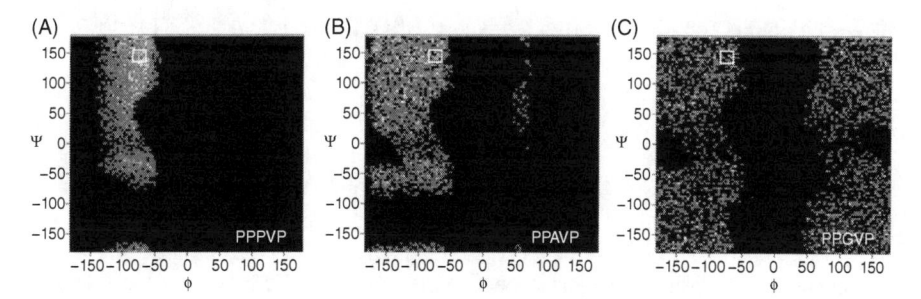

Figure 6.8. Ramachandran plot of dihedral angles adopted by the guest residue of PPXVP pentapeptides generated by MPMOD. (A) PPPVP, (B) PPAVP, and (C) PPGVP. Counts of the number of conformers in the PII region ($-75 \pm 10°$, $145 \pm 10°$) boxed in white are given in Table 6.3. Conformers are not energy weighed. See color insert.

TABLE 6.3. MPMOD Quantitatively Reproduces PII Propensities for Amino Acids Where Conformational Entropy of the Backbone Is the Sole Contributor

MPMOD Predicted PII Propensities

Model	Amino Acid (X)	Counts PII[a]	Count Ratio[b]	Measured PII[c]
PPXVP				
	Proline	244	1	
	Alanine	90	0.37 ± 0.06	0.37 ± 0.03
	Glycine	30	0.12 ± 0.04	0.13 ± 0.01
GGXGG				
	Proline	60	1	
	Alanine	22	0.36 ± 0.06	0.81[d]

[a] Counts of guest residue (X) having φ and ψ dihedral angles within the range of ($-75 \pm 10°$, $145 \pm 10°$).
[b] Errors estimated from 95% confidence interval of the measured proportion of Ala or Gly in PII compared with Pro.
[c] Errors in PII% propagated from experimental error in measuring difference in the free energy of binding [41].
[d] Experimental PII propensity measured for Ala in the GGXGG host, although the reported value (81.8%) is not relative to proline [25].

of the Sos peptide (Ac-VPPXVPPRRRY-NH$_2$). MPMOD builds conformers by randomly selecting torsional angles (φ and ψ) for the peptide backbone and randomly inserting rotamers for side chains. An MPMOD-generated conformer is accepted or rejected based on the presence of steric collisions using standard van der Waals radii [68]. Three thousand conformers were generated for PPXVP pentapeptides where X was Pro, Ala, and Gly. As shown in Figure 6.8A–C, the accessible conformations for Pro, Ala, and Gly are significantly limited by sterics, but sampling of all allowed space is accomplished. The number of peptides with torsional angles in the PII region ($-75 \pm 10°$, $145 \pm 10°$) was counted for each guest residue. The counts of Pro, Ala, and Gly in the PII region are presented in Table 6.3. Importantly, when the counts for Ala and Gly are taken relative to Pro (as in the SH3:Sos calorimetric scheme, discussed earlier), the steric model reproduces experimentally measured PII propensities within error [41]. The importance of this agreement cannot be understated, as it implies that the peptide backbone has some sterically driven baseline propensity for the PII conformation that is common for all polypeptides. In the cases of Pro, Ala, and Gly, the conformational entropy of the peptide backbone is the sole contributor to the conformational bias, enabling accurate modeling by sterics alone. Other side chains contain charged and polar moieties that have numerous interactions and cannot be captured by MPMOD. Furthermore, these propensities (Table 6.3) are in agreement with Ala and Gly PII propensities calculated by Pappu and colleagues [19]. It is also consistent with the interpretation of CD spectra initially published by Shi et al. [22] but interpreted by Scheraga and colleagues [18]. However, the latter correlation should be taken cautiously, as it is difficult to precisely quantify PII content by CD alone.

As MPMOD is a simple and easily modified tool, it can be used to model any peptide system or probe for conformational bias over any region of Ramachandran space. Using MPMOD, we modeled Pro and Ala in the context of the GGXGG host system and sought to answer (1) how does the host context affect the conformational bias of the guest site; (2) what is the PII bias of Ala relative to Pro; and (3) what region of Ramachandran space must be evaluated to yield a PII bias of ~80% for Ala relative to Pro?

Figure 6.9A,B shows the conformational space accessed by Ala and Pro, respectively, in the GGXGG system. In the glycine-rich host, the guest site visits the central region of the Ramachandran map (0 ± 20°, 0 ± 20°) more frequently than in the Sos peptide context (Fig. 6.8). Remarkably, although the counts for Pro and Ala in the PII region (defined as $\varphi = -75 \pm 10°$, $\psi = 145 \pm 10°$) are fewer in the GGXGG model, the ratio of counts for Ala relative to Pro is actually the same (36%) as that observed in Sos (37%, Table 6.3), as calculated from this crude, steric model. Importantly, that nearly identical conformational bias toward PII is observed for Ala relative to Pro in these models supports the apparent lack of context dependence for a simple steric model (Fig. 6.5) and provides further evidence of an intrinsic bias of

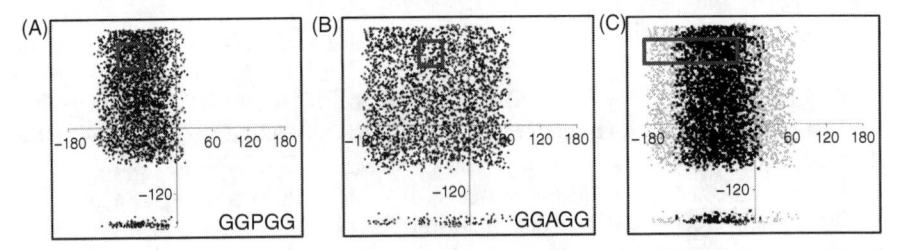

Figure 6.9. Ramachandran plot of dihedral angles adopted by the guest residue of GGXGG pentapeptides generated by MPMOD. (A) GGPGG, (B) GGAGG, and (C) overlay of GGPGG (black) with GGAGG (gray). Counts of the number of conformers in the PII region ($-75 \pm 10°$, $145 \pm 10°$) boxed in gray are given in Table 6.3. Conformers are not energy weighed. The PII region (gray box) in C is expanded to ($-180°$ to $-55°$, $145 \pm 10°$).

the peptide backbone for the PII conformation. Additionally, these results for Ala PII bias (36%) relative to Pro quantitatively agrees with Pappu and colleagues, within error [19]. Figure 6.9C shows the smallest region of Ramachandran space necessary to achieve a PII conformational bias of 82% for Ala relative to Pro. Briefly, our approach was simply to expand the PII region systematically and count the number of conformers that reside in the space with each step. We performed this counting of Ala and Pro while expanding both φ and ψ dimensions of the PII region and also while holding either the φ or ψ dimension constant. Maintaining $\varphi = -75 \pm 10°$ and varying the size of the PII region in terms of ψ had no effect on the ratio of Ala conformations present relative to Pro. Only when φ was varied and ψ held constant (145 ± 10°), did the populations of conformers of Ala begin to increase relative to Pro. This steric model demonstrates that there is no way to obtain a high PII bias for Ala relative to Pro without sampling some region outside of PII. Furthermore, it underscores the significance of the ability of MPMOD to reproduce calorimetrically measured PII bias, as it strongly suggests our results report specifically to PII propensity and not to conformational bias corresponding to another region of Ramachandran space.

6.5. CORRELATION OF PII PROPENSITY TO AMINO ACID PROPERTIES

Although many experimental and computational studies have been performed to quantify conformational bias toward PII in the denatured state of polypeptides, little is known as to whether amino acid PII propensity correlates to any general physicochemical property or to any propensity for secondary structure (α-helix, β-sheet, coil, or turn). In the host–guest study using the GGXGG system [25], Shi et al. claim that their PII scale has an anti-correlation with a β-sheet scale [69]. To understand how this anti-correlation is obtained, one must examine the equation used to interpret their observed $^3J_{\alpha N}$-coupling constants in terms of PII propensities (Eq. 6.2).

Because they assume that the guest position only populates PII or β conformations, PII should be anti-correlated with β by definition. With this assumption, to obtain a statistically significant anti-correlation with a β-sheet scale, it is necessary to treat over 25% (5 out of 18 amino acids: Ile, Phe, Trp, Tyr, and Val) of their data separately [25].

Access to the AAIndex Database, GenomeNet Japan [70], allows for a complete and unbiased comparison of experimentally and computationally determined PII propensities with a large data set (544 scales) of previously published amino acid property scales encompassing secondary-structure propensities (α-helix, β-sheet, coil, or turn) and physicochemical properties. The linear correlation of the most complete experimental [12, 25] and computational [19, 67] PII scales with all secondary-structure propensity scales in the AAIndex (~190 total) are shown in Figure 6.10A–D. All five scales exhibit a wide range of Pearson correlation coefficients for all secondary-structure types, which were manually grouped from the 544 total scales in the AAIndex. For example, for the α-helix structure type, over 70

176 EXPERIMENTAL AND COMPUTATIONAL STUDIES OF POLYPROLINE II PROPENSITY

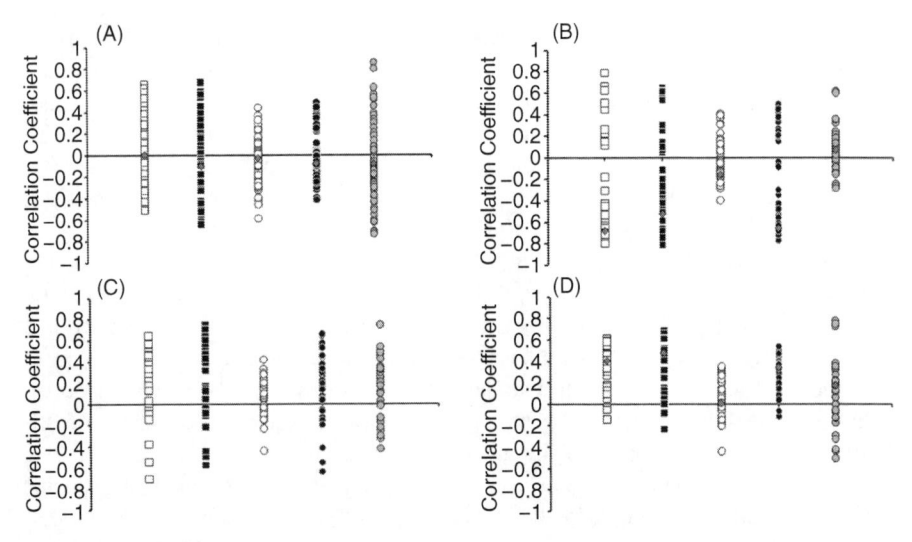

Figure 6.10. Range of Pearson linear correlation coefficients of PII propensity scales to secondary-structure propensity scales. Linear correlation coefficients were calculated by an in-house code written in C and Perl. Secondary-structure propensity scales were manually grouped by secondary-structure element (A) α-helix, (B) β-sheet, (C) coil, and (D) turn. The correlated PII scales are for the PPXPP (white squares) [12], (black squares) [19] and GGXGG (white circles) [25], (black circles) [19], and (gray circles) [67]. Data points highlighted in gray (diamond) are correlation coefficients to frequencies obtained by Chou and Fasman [71]. The database of scales, AAIndex, is available from GenomeNet [70]. The extent to which a given secondary-structure propensity scale correlated with others in the same group varied.

different scales were selected, with each of the chosen scales reporting on a variety of parameters that characterize α-helix propensity including normalized frequency to be in an α-helix, position-specific propensities along the length of an α-helix, and frequency to be in an α-helix termini. Dissecting these large structure types into more specific sets of scales of high internal correlation did not produce additional significant correlations (data not shown). Out of all secondary-structure propensity scale correlations for each PII scale, shown in Figure 6.10, few are statistically significant (Table 6.4). Clearly, there is no discernible pattern for any of the PII scales to be correlated or anti-correlated with any type of secondary structure. However, as a caveat, many of the secondary-structure scales contained in the AAIndex do not correlate with each other.

Analogous correlations can be performed with different physicochemical property scales in the AAIndex. Figure 6.11 shows the range of Pearson coefficients for charge, polarity, hydrophobicity, and size (A–D) for the most complete PII scales [12, 25, 67]. Other physicochemical properties including pK, isoelectric point, melting point, and optical rotation were examined, and they showed no significant correlation (data not shown). Similar to correlations with secondary-structure propensities, there seems to be no appreciable pattern for correlation or anti-correlation

TABLE 6.4. PII Scale Correlations to Secondary Structure Propensity and Physicochemical Property Scales from AAIndex (GenomeNet)

PII Scale Correlations to Database of Amino Acid Scales from GenomeNet

	Num	Rucker[a]	Tran[b]	Shi[c]	Tran[b]	Beck[d]
	Scale	PPXPP	PPXPP	GGXGG	GGXGG	GGXGG
Secondary structure						
α-helix	78					
Low/High[e]		−0.51/0.66	−0.64/0.68	−0.59/0.44	−0.42/0.49	−0.73/0.86
Average[f]		0.05	−0.03	−0.05	0.00	−0.09
β-sheet	46					
Low/high		−0.80/0.79	−0.81/0.65	−0.40/0.41	−0.77/0.49	−0.28/0.62
Average		−0.34	−0.26	0.00	−0.31	0.08
Coil	34					
Low/high		−0.70/0.65	−0.57/0.75	−0.44/0.42	−0.64/0.66	−0.42/0.75
Average		0.20	0.24	0.06	0.17	0.15
Turn	38					
Low/High		−0.14/0.62	−0.24/0.69	−0.44/0.35	−0.12/0.53	−0.50/0.78
Average		0.35	0.36	0.06	0.29	0.12
Physicochemical						
Charge	10					
Low/high		−0.25/0.31	−0.21/0.24	−0.28/0.25	−0.35/0.37	−0.35/0.42
Average		0.08	0.07	0.07	0.13	−0.12
Polarity	5					
Low/High		−0.57/0.47	−0.45/0.38	−0.22/0.35	−0.47/0.37	−0.21/0.16
Average		0.06	0.02	0.03	0.01	−0.01
Hydrophobicity	60					
Low/high		−0.67/0.55	−0.69/0.50	−0.54/0.31	−0.71/0.56	−0.35/0.47
Average		−0.33	−0.24	−0.14	−0.29	0.06
Size	23					
Low/high		−0.48/0.37	−0.18/0.35	−0.51/0.07	−0.43/0.40	−0.30/0.34
Average		−0.08	0.07	−0.36	−0.02	0.07
Significance[g]		Y (1)	Y (1)	N	Y (4)	Y (3)

[a] [12] Significantly anti-correlated with "Weights for beta-sheet at the window position of 1" [84].

[b] [19] PPXPP is significantly anti-correlated with "Weights for beta-sheet at the window position of 2" [84] and GGXGG is significantly anti-correlated with "AA composition of MEM of single-spanning proteins" [85], "AA composition of MEM of multispanning proteins" [85], "Transmembrane regions of non-mt-proteins" [86], and "Average relative probability of inner beta-sheet" [87].

[c] [25].

[d] [67] Significantly correlated with "Helix initiation parameter at position i,i,i" [88], "Helix formation parameters ΔΔG" [89], and "Normalized frequency of chain reversal R" [90].

[e] Lowest and highest calculated Pearson linear correlation coefficients.

[f] Average correlations were calculated by manually grouping scales from GenomeNet according to a given property and then averaging the coefficients for each group.

[g] A stringent Pearson correlation threshold of $P < 9 \times 10^5$ was used to access significance. $(Y/N)(X)$ denotes presence and number (X) of statistically significant correlations or anti-correlations among the 544 AAIndex scales available on GenomeNet (Japan) [70].

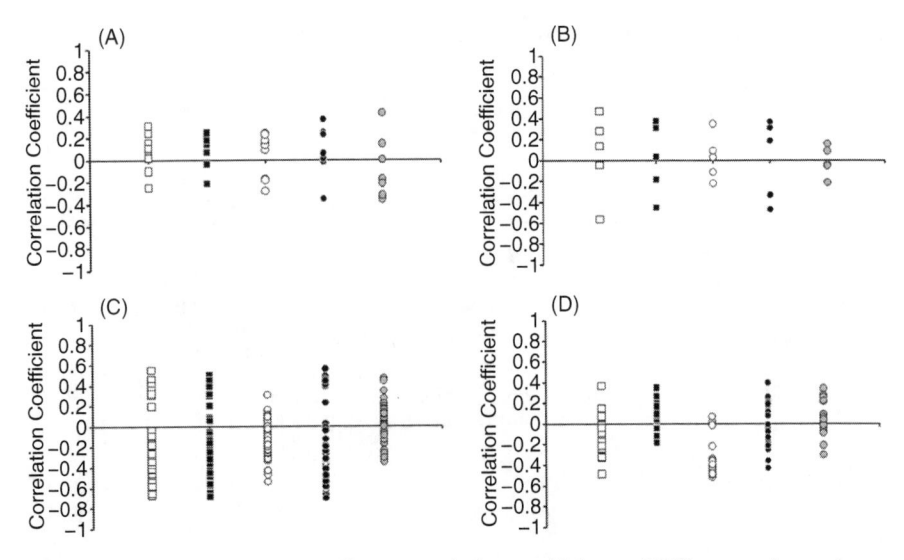

Figure 6.11. Range of Pearson linear correlation coefficients of PII propensity scales to physicochemical properties. Linear correlation coefficients were calculated by an in-house code written in C and Perl. Physicochemical property scales were grouped by (A) charge, (B) polarity, (C) hydrophobicity, and (D) size. The correlated PII scales are for the PPXPP (white squares) [12], (black squares) [19] and GGXGG (white circles) [25], (black circles) [19], and (gray circles) [67]. The database of scales, AAIndex, is available from GenomeNet [70]. The extent to which a given physicochemical property scale correlated with others in the same group varied.

of any of the considered PII scales with any other property scale type. Out of all linear comparisons, none were statistically significant (Table 6.4).

Taken together, the data presented in Figures 6.10 and 6.11 suggest several important conclusions. First, as different PII propensity scales [12, 25, 67] do not correlate with each other (Figs. 6.3, 6.4, and 6.6), they also correlate differently with other scales from the AAIndex. The highlighted data points in Figure 6.10, for example, show that the Pearson linear correlation coefficients for each scale correlated with normalized amino acid secondary-structure frequencies taken from the classic work of Chou and Fasman [71]. However, although we have highlighted this one AAIndex scale to demonstrate inter-scale discrepancies, we caution those who would draw insights from the correlation of PII conformational bias with any other single scale. The range of correlations presented in Figures 6.10 and 6.11 suggests that claims for the correlation of PII to a single scale of any type might not be meaningful, as choice of another scale that claims to report on the same property may result in correlation, anti-correlation, or even no correlation. Lastly, the significance of the apparent lack of correlation of PII conformational bias with any other amino acid property cannot be understated. The lack of correlation strongly suggests that PII propensity in the denatured state is a novel, independent property of amino acids and not a surrogate for any other known amino acid feature.

If PII is indeed an independent property of amino acids, what can we learn from these unbiased, large-scale comparisons? What does the lack of correlation with a single amino acid property tell us about the nature of the molecular driving forces behind PII conformational bias? Although amino acids with long, charged side chains such as glutamine and lysine have been observed to have a high PII character [4, 10–12, 14] by CD, other techniques suggest that lysine tripeptides have a higher tendency to form turns [72]. Other charged side chains such as asparagine typically have lower PII propensities. Comparison of the reported PII propensities of glutamine and asparagine, which possess the same chemical moiety but differ by one carbon in the side-chain length, reveals conflicting results. Two published results, one experimental [25] and another computational [19], report similar PII propensities for glutamine and asparagine, while two other studies report that asparagine is ~10% lower than glutamine [12, 67]. Another curious discrepancy lies in the correlations of amino acid size to PII propensity. Close inspection of amino acid PII propensity scales presents conflicting results regarding size and the ability of β-branched amino acids to favor the PII conformation. Work by Creamer and colleagues showed by CD that the β-branched amino acids, isoleucine and valine, have low PII propensities [12], which were confirmed by Pappu and colleagues in several host contexts [19] (Table 6.1). In contrast, simulations of Daggett and colleagues obtain PII propensities for isoleucine and valine that rank these amino acids among the highest on their scale [67] (Table 6.1), while the interpretation of experimental data on the same system (GGXGG) claims that isoleucine and valine have very different PII propensities, ~52% and 74%, respectively. The above discrepancies in observed PII propensities for similar amino acids represent two cases that suggest that not one property, but an interplay of many molecular driving forces, are responsible for local conformational bias to PII and that the influence of these driving forces may differentially impact the observable end points of different approaches used to interrogate PII conformational bias.

Computational models have attempted to explain the effects of solvent on the stability of the PII conformation [17, 73–78]. One proposal stemming from these computational models has been that transient water bridges stabilize PII [73], although others disagree [74, 76, 78]. Experimentally, propensity to adopt the PII conformation has been shown to increase in the presence of the denaturant, guanidine hydrochloride [10, 79, 80], and in D_2O [13], but decrease in trifluoroethanol and other alcohols [81].

Pappu and Rose have proposed a model where PII formation is driven primarily by minimization of side-chain packing density [17], though they agree that solvent interactions with the peptide backbone also promote PII bias. Indeed, were sterics the only driving force for PII formation, one might expect a correlation between PII propensity and side chain size, but this is not observed in any published PII scale (Fig. 6.12D). The extent to which β-branched or large, cyclic side chains may shield the peptide backbone from solvent interactions that would stabilize the PII conformation is unknown. The interplay between avoidance of steric collisions and establishing solvent interactions proposed by Pappu and Rose [17] is consistent with excluded-volume models (published later) that showed that sterics bias the backbone to extended conformations [19]. They are also consistent with experimental observations that solvents which stabilize the denatured state, such as guanidine hydrochloride and urea, promote the PII conformation [10, 79, 80]. Furthermore, this simple interplay model [17] is consistent with the results of our AAIndex database correlations, as no single property appears to favor or oppose bias to the PII conformation. Importantly, this same interplay effectively reduces the size of the denatured state ensemble [82].

6.6. SUMMARY

As discussed previously by Hamburger et al. in the context of a funnel-shaped folding landscape [83], an apparently small bias of an unfolded peptide toward a conformation (PII) can have a significant impact on the size of the denatured state ensemble. Indeed, the difference in the number of states adopted by a 100 amino acid protein with a smooth (unbiased) folding landscape (~10^{90} possible conformations) compared with a peptide with a rugged (~30% PII biased) landscape (10^{20} possible conformations) is profound [83]. However, this reduction in conformational search space caused by PII bias does not necessarily mean that proteins fold *through* PII mechanistically. Instead PII may function as a "place holder" of the local conformation, preventing short segments of proteins from misfolding. Regions of high PII bias in a protein sequence may serve to bias the backbone to extended conformations, while areas of low PII bias collapse to form loops or the protein interior. Investigations into the possible functional roles that PII may serve in the native and denatured state are currently underway in our lab.

ACKNOWLEDGMENTS

We would like to thank James O. Wrabl for assistance in generating figures, statistical analysis, and PII scale correlations, as well as Jiin-Yu Chen and Andrew Campagnolo for reviewing the manuscript. We also acknowledge D. Wayne Bolen for helpful discussions. Support for this research was provided by National Science Foundation Grant MCB-0446050.

REFERENCES

1. Anfinsen, C. (1973) Principles that govern the folding of protein chains, *Science 181*, 223–230.
2. Levinthal, C. (1968) Are there pathways for protein folding? *Extrait du Journal de Chimie Physique 65*, 44–45.
3. Tiffany, M. and Krimm, S. (1968) Circular dichroism of poly-L-proline in an unordered conformation, *Biopolymers 6*, 1767–1770.
4. Tiffany, M. and Krimm, S. (1968) New chain conformations of poly(glutamic acid) and polylysine, *Biopolymers 6*, 1379–1382.
5. Cowan, P. and McGavin, S. (1955) Structure of poly-L-proline, *Nature 176*, 501–503.

REFERENCES

6 Krimm, S. and Tiffany, M. L. (1974) The circular dichroism spectra and structure of unordered polypeptides and proteins, *Israeli Journal of Chemistry 12*, 68–84.

7 Drake, A., Siligardi, G., and Gibbons, W. (1988) Reassessment of the electronic circular dichroism criteria for random coil conformations of poly(L-lysine) and implications for protein folding and denaturation studies, *Biophysical Chemistry 31*, 143–146.

8 Dukor, R. and Keiderling, T. (1991) Reassessment of the random coil conformation: Vibrational CD study of proline oligopeptides and related polypeptides, *Biopolymers 31*, 1747–1761.

9 Woody, R. (1992) Circular dichroism and conformation of unordered peptides, *Advances in Biophysical Chemistry 2*, 37–79.

10 Kelly, M., Chellgren, B., Rucker, A., Troutman, J., Fried, M., Miller, A., et al. (2001) Host-guest study of left-handed polyproline II helix formation, *Biochemistry 40*, 14376–14383.

11 Rucker, A. and Creamer, T. (2002) Polyproline II helical structure in protein unfolded states: Lysine peptides revisited, *Protein Science 11*, 980–985.

12 Rucker, A., Pager, C., Campbell, M., Qualls, J., and Creamer, T. (2003) Host-guest scale of left-handed polyproline II helix formation, *Proteins 53*, 68–75.

13 Chellgren, B. and Creamer, T. (2004) Effects of HO and DO on polyproline II helical structure, *Journal of the American Chemical Society 126* (45), 14734–14735.

14 Chellgren, B., Miller, A., and Creamer, T. (2006) Evidence of polyproline II helical structure in short polyglutamine tracts, *Journal of Molecular Biology 361*, 362–371.

15 Woody, R. (2009) Circular dichroism spectrum of peptides in the poly(pro) II conformation, *Journal of the American Chemical Society 131*, 8234–8245.

16 Sreerama, N. and Woody, R. (2009) Molecular dynamics simulations of polypeptide conformations in water: A comparison of α, β, and poly(pro) II conformations, *Proteins 36*, 400–406.

17 Pappu, R. and Rose, G. (2002) A simple model for polyproline II structure in unfolded states of alanine-based peptides, *Protein Science 11*, 2437–2455.

18 Vila, J., Baldoni, H., Ripoll, D., Ghosh, A., and Scheraga, H. (2004) Polyproline II helix conformation in a proline-rich environment: A theoretical study, *Biophysical Journal 6*, 731–742.

19 Tran, H., Wang, X., and Pappu, R. V. (2005) Reconciling observations of sequence-specific conformational propensities with the generic polymeric behavior of denatured proteins, *Biochemistry 44*, 11369–11380.

20 Wilson, G., Hecht, L., and Barron, L. (1996) Residual structure in unfolded proteins revealed by Raman optical activity, *Biochemistry 35*, 12518–12525.

21 Barron, L., Hecht, L., Blanch, E., and Bell, A. (2000) Solution structure and dynamics of biomolecules from Raman optical activity, *Progress in Biophysics and Molecular Biology 73*, 1–49.

22 Shi, Z., Olson, C., Rose, G., Baldwin, R., and Kallenbach, N. (2002) Polyproline II structure in a sequence of seven alanine residues, *Proceedings of the National Academy of Sciences of the United States of America 99* (14), 9190–9195.

23 Ding, L., Chen, K., Santini, P., Shi, Z., and Kallenbach, N. (2003) The pentapeptide GGAGG has PII conformation, *Journal of the American Chemical Society 125*, 8092–8093.

24 Eker, F., Griebenow, K., Cao, X., Nafie, L., and Schweitzer-Stenner, R. (2004) Preferred peptide backbone conformations in the unfolded state revealed by the structure analysis

of alanine-based (AXA) tripeptides in aqueous solution, *Proceedings of the National Academy of Sciences of the United States of America 101* (27), 10054–10059.

25 Shi, Z., Chen, K., Liu, Z., Ng, A., Bracken, W., and Kallenbach, N. (2005) Polyproline II propensities from GGXGG peptides reveal an anticorrelation with β-sheet scales, *Proceedings of the National Academy of Sciences of the United States of America 102* (50), 17964–17968.

26 Shi, Z., Chen, K., Liu, Z., and Kallenbach, N. (2006) Conformation of the backbone in unfolded proteins, *Chemical Reviews 106*, 1877–1897.

27 Hagarman, A., Measey, T., Mathieu, D., Schwalbe, H., and Schweitzer-Stenner, R. (2010) Intrinsic propensities of amino acids in GXG peptides inferred from amide I' band profiles and NMR scalar coupling constants, *Journal of the American Chemical Society 132*, 540–551.

28 Adzhubei, A. and Sternberg, M. (1993) Left-handed polyproline II helices commonly occur in globular proteins, *Journal of Molecular Biology 229*, 472–493.

29 Sreerama, N. and Woody, R. (1994) Poly(pro)II helixes in globular proteins: Identification and circular dichroic analysis, *Biochemistry 33*, 10022–10025.

30 Stapley, B. and Creamer, T. (1999) A survey of left-handed polyproline II helices, *Protein Science 8*, 587–595.

31 Pauling, L. and Corey, R. (1951) The structure of fibrous proteins of the collagen-gelatin group, *Proceedings of the National Academy of Sciences of the United States of America 37*, 272–281.

32 Owens, N., Stetefeld, J., Lattova, E., and Schweizer, F. (2010) Contiguous O-galactosylation of 4(R)-hydroxy-L-proline residues forms very stable polyproline II helices, *Journal of the American Chemical Society 132*, 5036–5042.

33 Mahoney, N., Janmey, P., and Almo, S. (1997) Structure of the profilin-poly-L-proline complex involved in morphogenesis and cytoskeletal regulation, *Nature 4* (11), 953–960.

34 Jardetzky, T., Brown, J., Gorga, J., Stern, L., Urban, R., Strominger, J., et al. (1996) Crystallographic analysis of endogenous peptides associated with HLA-DR1 suggest a common, polyproline II-like conformation for bound peptides, *Proceedings of the National Academy of Sciences of the United States of America 93*, 734–738.

35 Kay, B., Williamson, M., and Sudol, M. (2000) The importance of being proline: The interaction of proline-rich motifs in signaling proteins with their cognate domains, *The FASEB Journal 14*, 231–241.

36 Schlessinger, J. (1993) How receptor tyrosine kinases activate Ras, *Trends in Biochemical Sciences 18*, 273–275.

37 Williamson, M. (1994) The structure and function of proline-rich regions in proteins, *The Biochemical Journal 297*, 249–260.

38 Lim, W., Richards, F., and Fox, R. (1994) Structural determinants of peptide-binding orientation and of sequence specificity in SH3 domains, *Nature 372*, 375–379.

39 Ferreon, J., Volk, D., Luxon, B., Gorenstein, D., and Hilser, V. (2003) Solution structure, dynamics, and thermodynamics of the native state ensemble of the Sem-5 c-terminal SH3 domain, *Biochemistry 42*, 5582–5591.

40 Ferreon, J. and Hilser, V. (2003) Ligand-induced changes in dynamics in the RT loop of the c-terminal SH3 domain of the Sem-5 indicate cooperative conformational coupling, *Protein Science 12*, 982–996.

41 Ferreon, J. and Hilser, V. (2003) The effect of the polyproline II (PPII) conformation on the denatured state entropy, *Protein Science 12*, 447–457.

REFERENCES

42 Ferreon, J. and Hilser, V. (2004) Thermodynamics of binding to SH3 domains: The energetic impact of polyproline II (PII) helix formation, *Biochemistry 43*, 7787–7797.

43 Manson, A., Whitten, S., Ferreon, J., Fox, R., and Hilser, V. (2009) Characterizing the role of ensemble modulation in mutation-induced changes in binding affinity, *Journal of the American Chemical Society 131*, 6785–6793.

44 Whitten, S., Yang, H., Fox, R., and Hilser, V. (2008) Exploring the impact of polyproline II (PII) conformational bias on the binding of peptides to the Sem-5 SH3 domain, *Protein Science 17*, 1200–1211.

45 Schweitzer-Stenner, R., Eker, F., Griebenow, K., Cao, X., and Nafie, L. (2004) The conformation of tetraalanine in water determined by polarized Raman, FT-IR, and VCD spectroscopy, *Journal of the American Chemical Society 126*, 2768–2776.

46 Schweitzer-Stenner, R. and Measey, T. (2007) The alanine-rich XAO peptide adopts a heterogeneous population, including turn-like polyproline II conformations, *Proceedings of the National Academy of Sciences of the United States of America 104* (16), 6649–6654.

47 Schweitzer-Stenner, R., Measey, T., Kakalis, L., Jordan, F., Pizzanelli, S., Forte, C., et al. (2007) Conformations of alanine-based peptides in water probed by FTIR, Raman, vibrational circular dichroism, electronic dichroism, and NMR spectroscopy, *Biochemistry 46*, 1587–1596.

48 McColl, I., Blanch, E., Hecht, L., Kallenbach, N., and Barron, L. (2004) Vibrational Raman optical activity characterization of poly(L-proline) II helix in alanine oligopeptides, *Journal of the American Chemical Society 126*, 5076–5077.

49 Makowska, J., Rodziewicz-Motowidlo, S., Baginska, K., Vila, J., Liwo, A., Chmurzynski, L., et al. (2006) Polyproline II conformation is one of many local conformational states and is not an overall conformation of unfolded peptides and proteins, *Proceedings of the National Academy of Sciences of the United States of America 103* (6), 1744–1749.

50 Makowska, J., Rodziewicz-Motowidlo, S., Baginska, K., Makowska, M., Vila, J., Liwo, A., et al. (2007) Further evidence for the absence of polyproline II stretch in the XAO peptide, *Biophysical Journal 92*, 2904–2917.

51 Zagrovic, B., Lipfert, J., Sorin, E., Millett, I., van Gunsteren, W., Doniach, S., et al. (2005) Unusual compactness of a polyproline type II structure, *Proceedings of the National Academy of Sciences of the United States of America 102* (33), 11698–11703.

52 Chen, K., Liu, Z., Zhou, C., Bracken, W., and Kallenbach, N. R. (2007) Spin relaxation enhancement confirms dominance of extended conformations in short alanine peptides, *Angewandte Chemie (International ed. in English) 46*, 9036–9039.

53 Avbelji, F. and Baldwin, R. (2004) Origin of the neighboring residue effect on peptide conformation, *Proceedings of the National Academy of Sciences of the United States of America 101* (30), 10967–10972.

54 Vuister, G., Wang, A., and Bax, A. (1993) Measurement of three-bond nitrogen-carbon J couplings in proteins uniformly enriched and 15N and 13C, *Journal of the American Chemical Society 115*, 5334–5335.

55 Lam, S. and Hsu, V. (2003) NMR identification of left-handed polyproline type II helices, *Biopolymers 69*, 270–281.

56 Nodet, G., Salmon, L., Ozenne, V., Meier, S., Jensen, M., and Blackledge, M. (2009) Quantitative description of backbone conformational sampling of unfolded proteins at amino acid resolution from NMR residual dipolar couplings, *Journal of the American Chemical Society 131*, 17908–17918.

57. Schweitzer-Stenner, R., Eker, F., Perez, A., Griebenow, K., Cao, X., and Nafie, L. (2003) The structure of tri-proline in water probed by polarized Raman, Fourier transform infrared, vibrational circular dichroism, and electric ultraviolet circular dichroism spectroscopy, *Biopolymers 71*, 558–568.
58. Eker, F., Griebenow, K., Cao, X., Nafie, L., and Schweitzer-Stenner, R. (2004) Tripeptides with ionizable side chains adopt a perturbed polyproline II structure in water, *Biochemistry 43*, 613–621.
59. Richards, F. (1977) Areas, volumes, packing, and protein structure, *Annual Review of Biophysics and Bioengineering 6*, 151–176.
60. Creamer, T. (1998) Left-handed polyproline II helix formation is (very) locally driven, *Proteins 33*, 218–226.
61. Chen, K., Liu, Z., and Kallenbach, N. (2004) The polyproline II conformation in short alanine peptides is noncooperative, *Proceedings of the National Academy of Sciences of the United States of America 101* (43), 15352–15357.
62. Okabayashi, H., Isemura, T., and Sakakibara, S. (1968) Steric structure of L-proline oligopeptides. II. Far-ultraviolet absorption spectra and optical rotations of L-proline oligopeptides, *Biopolymers 6*, 323–330.
63. Deber, C., Bovey, F., Carver, J., and Blout, E. (1970) Nuclear magnetic resonance evidence for cis-peptide bonds in proline oligomers, *Journal of the American Chemical Society 92*, 6191–6198.
64. Helbecque, N. and Loucheux-Lefebvre, M. (1982) Critical chain length for polyproline-II structure formation in H-Gly-(Pro)n-OH, *International Journal of Peptide and Protein Research 19*, 94–101.
65. Vila, J., Baldoni, H., Ripoll, D., and Scheraga, H. (2004) Fast and accurate computation of the 13C chemical shifts for an alanine-rich peptide, *Proteins 57*, 87–98.
66. Petrella, E., Machesky, L., Kaiser, D., and Pollard, T. (1996) Structural requirements and thermodynamics of the interaction of proline peptides with profilin, *Biochemistry 35*, 16535–16543.
67. Beck, D., Alonso, D., Inoyama, D., and Daggett, V. (2008) The intrinsic conformational propensities of the 20 naturally occurring amino acids and reflection of these propensities in proteins, *Proceedings of the National Academy of Sciences of the United States of America 105* (34), 12259–12264.
68. Ramakrishnan, C. and Ramachandran, G. (1965) Stereochemical criteria for polypeptide and protein chain conformations, *Biophysical Journal 5* (6), 909–933.
69. Kim, C. and Berg, J. (1993) Thermodynamic beta-sheet propensities measured using a zinc-finger host peptide, *Nature 362* (6417), 267–270.
70. Kawashima, S., Ogata, H., and Kanehisa, M. (1999) AAIndex: Amino acid index database, *Nucleic Acids Research 27*, 368–369.
71. Chou, P. and Fasman, G. (1978) Prediction of the secondary structure of proteins from their amino acid sequence, *Advances in Enzymology and Related Areas of Molecular Biology 47*, 45–148.
72. Eker, F., Cao, X., Nafie, L., and Schweitzer-Stenner, R. (2002) Tripeptides adopt stable structures in water: A combined polarized visible Raman, FTIR, and VCD spectroscopy study, *Journal of the American Chemical Society 124*, 14330–14341.
73. Sreerama, N. and Woody, R. (1999) Molecular dynamics simulations of polypeptide conformations in water: A comparison of a, B, and poly(pro) II conformations, *Proteins 35*, 400–406.

REFERENCES

74 Drozdov, A., Grossfield, A., and Pappu, R. (2004) Role of solvent in determining conformational preferences of alanine dipeptide in water, *Journal of the American Chemical Society 126*, 2574–2581.

75 Garcia, A. (2004) Characterization of non-alpha helical conformations in Ala peptides, *Polymer 45*, 669–676.

76 Mezei, M., Fleming, P., Srinivasan, R., and Rose, G. (2004) Polyproline II helix is the preferred conformation for unfolded polyalanine in water, *Proteins 55*, 502–507.

77 Fleming, P., Fitzkee, N., Mezei, M., Srinivasan, R., and Rose, G. (2005) A novel method reveals that solvent water favors polyproline II over β-strand conformation in peptides and unfolded proteins: Conditional hydrophobic accessible surface area (CHASA), *Protein Science 14*, 111–118.

78 Law, P. and Daggett, V. (2010) The relationship between water bridges and the polyproline II conformation: A large-scale analysis of molecular dynamics simulations and crystal structures, *Protein Engineering, Design, and Selection 23* (1), 27–33.

79 Tiffany, M. and Krimm, S. (1973) Extended conformations of polypeptides and proteins in urea and guanidine hydrochloride, *Biopolymers 12*, 575–587.

80 Whittington, S., Chellgren, B., Hermann, V., and Creamer, T. (2005) Urea promotes polyproline II helix formation: Implications for protein denatured states, *Biochemistry 44* (16), 6269–6275.

81 Liu, Z., Chen, K., Ng, A., Shi, Z., Woody, R., and Kallenbach, N. (2004) Solvent dependence of PII conformation in model alanine peptides, *Journal of the American Chemical Society 126*, 15141–15150.

82 Fitzkee, N. and Rose, G. (2005) Sterics and solvation winnow accessible conformational space for unfolded proteins, *Journal of Molecular Biology 353*, 873–887.

83 Hamburger, J., Ferreon, J., Whitten, S., and Hilser, V. (2004) Thermodynamic mechanism and consequences of the polyproline II (PII) structural bias in the denatured states of proteins, *Biochemistry 43*, 9790–9799.

84 Qian, N. and Sejnowski, T. (1988) Predicting the secondary structure of globular proteins using neural networks, *Journal of Molecular Biology 202* (4), 865–884.

85 Nakashima, H. and Nishikawa, K. (1992) The amino acid composition is different between the cytoplasmic and extracellular sides in membrane proteins, *FEBS Letters 303* (2–3), 141–146.

86 Nakashima, H., Nishikawa, K., and Ooi, T. (1990) Distinct character in hydrophobicity of amino acid composition of mitochondrial proteins, *Proteins 8* (2), 173–178.

87 Kanehisa, M. and Tsong, T. (1980) Local hydrophobicity stabilizes secondary structures in proteins, *Biopolymers 19* (9), 1617–1628.

88 Finkelstein, A., Badretdinov, A., and Ptitsyn, O. (1991) Physical reasons for secondary structure stability: Alpha-helices in short peptides, *Proteins 10* (4), 287–299.

89 O'Neil, K. and DeGrado, W. (1990) A thermodynamic scale for the helix-forming tendencies of the commonly occurring amino acids, *Science 250* (4981), 646–651.

90 Tanaka, S. and Scheraga, H. (1977) Statistical mechanical treatment of protein conformation. 5. A multi-state model for sequence-specific copolymers of amino acids, *Macromolecules 10* (2), 291–304.

91 Ramachandran, G. and Sasisekharan, V. (1968) Conformation of polypeptides and proteins, *Advances in Protein Chemistry 23*, 283–438.

7

MAPPING CONFORMATIONAL DYNAMICS IN UNFOLDED POLYPEPTIDE CHAINS USING SHORT MODEL PEPTIDES BY NMR SPECTROSCOPY

Daniel Mathieu, Karin Rybka, Jürgen Graf, and Harald Schwalbe

7.1. INTRODUCTION

The study of conformational preferences of amino acids in small peptides, in unfolded polypeptides, and in unfolded states of proteins is a topic of immense interest. Different spectroscopic (Fourier transform infrared [FTIR] [1–3], ultraviolet-circular dichroism [UV-CD] [4, 5]) and computational (molecular dynamics [MD] simulations [6, 7], *ab initio* calculations [8]) methods are being applied to determine the conformational dynamics present in small peptides. Comparing the results of the different methods provides interesting, sometimes controversial, insights and addresses a number of long-standing key questions in protein sciences including the following:

1. Biophysical studies have shown that sites of non-random structure in unfolded states coincide with nucleation sites of productive protein; folding occurs at sites of preferential, non-random structure. In order to detect non-random structure of amino acids in a polypeptide chain, information about the intrinsic conformation and dynamics of a given amino acid needs to be delineated.

Protein and Peptide Folding, Misfolding, and Non-Folding, First Edition. Edited by Reinhard Schweitzer-Stenner.
© 2012 John Wiley & Sons, Inc. Published 2012 by John Wiley & Sons, Inc.

Small peptides, so-called random coil peptides, have long been used in spectroscopic studies for the characterization of such intrinsic amino acid properties [9].

2. Besides the native folded state, it is now evident that a considerable fraction of the human proteome is not folded in its functional form but possesses little persistent structure [10, 11]. Properties including local structure and overall compaction of these intrinsically unstructured states of proteins (IUPs) can be predicted to very good approximation by models that utilize the conformational preferences of amino acids to define the ensemble characteristics of IUPs. Here, the prediction of "random coil" behavior always forms the baseline for the detection of non-random clusters in these IUPs.

3. Misfolding and aggregation are dictated by the chemical and physical properties of individual amino acids; particularly striking in this context is the effect of single-point mutations on the onset of neurodegenerative diseases associated with IUPs [12–15]. Single-point mutations change the global properties of polypeptide chains, for example, their compactness, which in turn changes their aggregation propensities *in vitro* and *in vivo*. To understand the influence of such single-point mutations, the intrinsic conformational properties of amino acids need to be delineated.

4. Understanding the conformational preferences of amino acids and stretches of small polypeptides also forms an important aspect of the conceptual framework for the *de novo* design of proteins with novel folds [14–19].

The unfolded state of a protein or a polypeptide chain is characterized by a large number of conformers that change torsion angles rapidly, the individuals members interconvert and have to be described by time-averaged properties. The interconversion occurs at least on a submillisecond to a submicrosecond time scale, and the spectroscopic signals are time-averaged.

A striking aspect of the points discussed above is the notion that the ϕ, ψ, and χ_1 preferences of amino acids are considerably more restricted than initially predicted by statistical-coil models [20, 21]. The ϕ, ψ, and χ_1 preferences of each amino acid are different and, in addition, the conformational preferences of the torsion angles are correlated. In addition, the ϕ, ψ, and χ_1 preferences show a dependence on the context; in other words, the nearest neighbor is important when testing the torsion angle preferences; the amino acids are especially influenced by the N-terminal preceding residues in the polypeptide chain [22–24]. Therefore, the exact definition of the "random coil" model of peptides is a matter of considerable debate for experimental studies. In addition, the choice of solvent has pronounced effects on the conformational sampling, often bringing into question the validity of calculations in a vacuum.

In general, the prediction of a spectroscopic signature for a given set of conformations is straightforward, the inverse problem, however, the unambigous derivation of the conformational ensemble leading to an averaged signal represents a typical so-called ill-defined problem. Common to all characterizations of unfolded states of proteins therefore is the stimulating discussion of how to derive models of the conformational averaging that occurs in the unfolded state of proteins.

The determination of the conformational preferences in model peptides has a long tradition for the definition of the intrinsic conformational preferences of an amino acid in the absence of any long-range bias. In this contribution, we therefore focus mainly on model peptides while also discussing aspects of longer polypeptide chains. For a more detailed account of nuclear magnetic resonance (NMR) spectroscopic studies of longer polypeptide chains and of IUPs, we refer to previously published review articles [9, 25].

7.2. GENERAL ASPECTS OF NMR SPECTROSCOPY

Among the many techniques for characterizing the conformation and dynamics of small peptides and unfolded polypeptides chains, NMR parameters yield important information about the local and global structures of proteins and peptides. In contrast to other experimental techniques, NMR spectroscopy can assess local structural information in atomic detail. NMR spectroscopic measurements are performed on samples at concentrations ranging from 10 μM to 1 mM, at ambient temperature and in solution. A large number of chemical denaturants can be added to the solution without impeding the experimental measurement, and allow delineation of the effect of, for example, pH, temperature, chemical denaturants including urea and guanidinium hydrochloride, as well as pressure or additives that simulate molecular crowding, to name a few. Due to the high concentration required to perform NMR experiments (between 10^{13} and 10^{17} molecules), NMR data are ensemble-averaged. In addition, the NMR parameters are time-averaged. However, the averaging time regime varies by NMR parameters.

Biophysical NMR studies most often utilize the magnetic properties of spin ½ nuclei; isotope enrichment of low natural abundance nuclei including carbon (^{13}C) and nitrogen (^{15}N) is possible and also often required to measure as many NMR parameters as possible. The need to collect numerous NMR parameters is particularly important for conformational averaged systems including unfolded protein states and small random coil peptides to investigate multistate conformational equilibria. In line with approaches in other biophysical techniques, small peptides are often studied to provide information about the intrinsic conformational properties of single amino acids.

In this contribution, we will discuss the large number of different NMR parameters and how they report on key conformational and dynamic properties of small peptides or unfolded polypeptide chains. We will also discuss various models, stimulated by the NMR community, to describe the unfolded states of proteins.

7.2.1. Overview of NMR Parameters

NMR determination of protein secondary and tertiary structure is based on a variety of parameters (see Table 7.1) of spin-1/2-nuclei (^{1}H, ^{13}C, ^{15}N).

TABLE 7.1. NMR Parameters and Their Information Content about Structure

NMR Parameter	Structural Information	Averaging Time Regime
Chemical shifts	Secondary structure	μs–ms
J-coupling constants	Torsion angles	ms–s
NOEs	Inter-atom distances	ps–ns
		Longer for exchange
Heteronuclear relaxation rates R_1, R_2, and hetNOE	Global rotational tumbling, local flexibility, global shape	ps–ns Longer for exchange (R_2)
Residual dipolar couplings	Bond vector orientation relative to an external alignment frame	ms–s
Pseudocontact shifts	Relative orientation and distance to a paramagnetic probe	μs–ms
Paramagnetic relaxation enhancement	Distance to a paramagnetic probe	ps–ns
Cross-correlated relaxation rates	Projection angles related to torsion angles	ms–s
Hydrogen/deuterium exchange	Hydrogen bond stability	s–days
Diffusion constants from diffusion-ordered spectroscopy spectra	Global shape, oligomeric state	Order of diffusion time

1. Using empirical database correlations, we can use chemical shift data of these nuclei (^{1}H, ^{13}C, ^{15}N) to derive mainly secondary structure [26–28]. Three-dimensional (3-D) structure determinations of small proteins solely based on chemical shifts have been performed [29, 30].
2. J-coupling constant data can be converted into dihedral angle information utilizing empirical or theoretically predicted Karplus equations [31–36]. By using these coupling constants, it is, for example, possible to determine the backbone angles ϕ and ψ as well as the side-chain angle χ_1.
3. Nuclear Overhauser Effects (NOEs) yield atomic distances of up to 5 Å. They are the parameters most often used in structure calculations of biomolecules.

In addition, more recently developed NMR methods derive projection-angle restraints from cross-correlated relaxation rates Γ^{cc} [37, 38] and long-range distances and global orientation from pseudo-contact shifts (PCS) [39, 40], paramagnetic relaxation enhancement (PRE) [41–43], and residual dipolar couplings (RDCs) [44–46].

To derive information about the global tumbling or the oligomeric state of a protein by NMR spectroscopy, the rotational correlation time (τ_c) can be extracted from heteronuclear relaxation rates [47]. Via Stokes–Einstein relations, τ_c is correlated with the size of the polymeric chain. These heteronuclear relaxation rates also provide information about local dynamics on the sub-τ_c time scales, that is, time scales faster than the overall rotational correlation time τ_c [48, 49]. Diffusion-ordered spectroscopy (DOSY) provides the radius of gyration or the diffusion coefficient, information linked to the aggregation state or the compactness of a protein [50, 51].

For structure determination of the folded state of a protein, numerous experimental restraints are collected to serve as input utilizing a simulated annealing protocol to provide a converging family of structures that represents the fold of a protein [52–54]. This fold had been represented as the single low-energy structure of the protein, where only in some areas such as loops or termini were more substantial dynamics observed. Only recently has a more dynamic representation of the native state of a protein integrating dynamic information on a time scale longer than nanoseconds been discussed [55–57].

However, short linear peptides in solution most often do not adopt a single, well-defined conformation, mostly because they are too short to form the necessary interactions to adopt stable secondary structures like α-helices or β-sheets. These linear peptides can therefore be considered the best test systems to determine the intrinsic, context-independent conformational preferences of an amino acid.

As stated above, the time regime over which averaging occurs differs for each individual NMR parameter. It is longest for chemical shifts and shortest for relaxation-induced NMR parameters including NOEs, heteronuclear relaxation rates, and paramagnetic relaxation enhancements. These differences in the time regime of conformational averaging imply that chemical shift values, averaged over multiple conformations, will be closest to the values expected for random coiled conformations in most cases.

In the following, we will introduce the main NMR parameters that yield important information about the conformational dynamics of small model peptides and unfolded polypeptide chains. We will also discuss the difficulties and potential systematic problems for NMR spectroscopic investigations of conformational dynamics in small peptides and unfolded polypeptide chains.

7.3. NMR PARAMETERS AND THEIR MEASUREMENT

7.3.1. Scalar Coupling Constants, J Couplings

The scalar ($^3J[I,S]$) coupling constant between two NMR-active spin-1/2-nuclei I and S three covalent bonds apart is transmitted through bonds (not only covalent but also hydrogen bonds) and depends on the torsion angle between the two nuclei. Scalar coupling splits the NMR lines into submultiplets (doublet, triplet, quartet. etc.) depending on the number of chemically equivalent S-nuclei coupled to a spin I. In the case of an isolated I,S spin system, the NMR is split into a doublet where the J-coupling constant equals the difference between the two multiple lines (Fig. 7.1).

In the following, we introduce different methods of measuring J-coupling constants. Coupling constant experiments can be divided into one-dimensional (1-D) and multidimensional (two-dimensional [2-D] and 3-D) experiments, and they have different requirements in terms of isotope-labeling pattern for the investigated

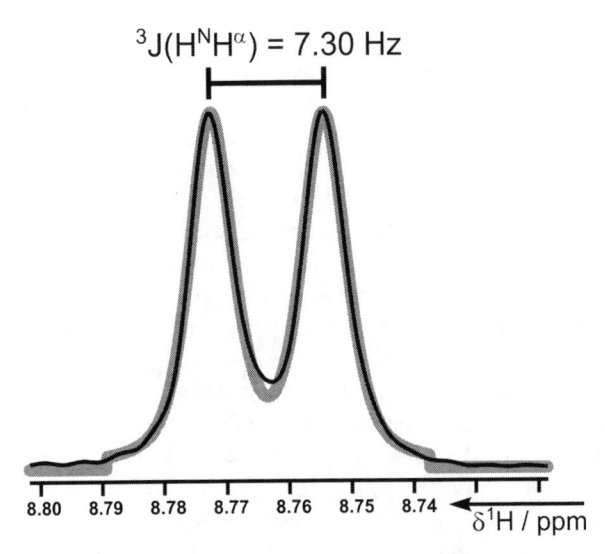

Figure 7.1. 1-D amide peak measured for the short peptide GCG. The two doublet components shown for the amide resonance of the central cysteine are split by the $^3J(H^N,H^\alpha)$ coupling. Deconvolution is shown as the underlying gray curve. The spectrum was recorded on 1 mM sample of NH_3^+-GCG-COOH in 90% H_2O/10% D_2O pH = 2 using a 400-MHz Avance spectrometer. The spectrum was recorded using 16 transient scans and digitized using 8k points for a spectral width of 4800 Hz. The spectrum was zero filled to 32k points prior to processing. The linewidth of the two components is 2.7 Hz in this case.

peptides. For carbon and nitrogen, whose major isotopes (^{12}C and ^{14}N) are either non-NMR active or have a spin quantum number >|½|, respectively, enrichment with spin-1/2 isotopes (^{13}C and ^{15}N) has to be carried out by chemical or biochemical means. In the case of small peptides, isotope enrichment is most conveniently achieved by either Boc- or Fmoc-based peptide synthesis [58, 59] using isotope-labeled protected precursor amino acids, which are commercially available, although often quite expensive. With such isotope-labeled peptides, heteronuclear multidimensional experiments can be performed. Such heteronuclear experiments need to be performed given the complexity of the conformational sampling to derive additional information from heteronuclear coupling constants to be able to differentiate between different models for conformational sampling, in particular for the backbone angle ψ.

In principle, experiments to measure J-coupling constants can be separated into two different classes, namely frequency-based methods and intensity-based methods. Aspects that depend on the exact NMR method to determine the coupling constant of interest as well as the size of the protein influence both the precision and the accuracy of J-coupling constant determination.

Coupling constant determination in frequency-based methods is limited by the spectral resolution. The most straightforward approach involves reading off the size of a coupling from the submultiplet splittings in a 1-D experiment. The experimental setup results in a certain digital resolution, which can easily be increased to the desired resolution by digitizing a larger of number of points without a detrimental effect on measurement time. In the case of a 1-D experiment, recording a larger number of points for a given spectral width does not significantly increase the experimental time as the acquisition time is usually short compared with the total experimental time. In the case of multidimensional NMR experiments, the experimental time scales linearly with the number of increments in every indirect dimension, which implies that increasing the digital resolution significantly increases the experimental time. New methods that utilize nonlinear sampling in Fourier transform NMR spectroscopy also mostly alleviate restrictions due to insufficient sampling of indirect dimension in multidimensional NMR spectroscopy [60–62]. Typically, the digital resolution can be further increased by zero filling and yields accurate precision if the line shape is Lorentzian. The final resolution is typically increased by one order of magnitude compared with digitized resolution. However, the maximum achievable resolution is limited by the natural linewidth for an NMR in a given molecule, which is dependent on the relaxation properties and their global rotational tumbling rate.

In the case of small peptides, the determination is usually straightforward because the small number of resonances in the amide region results in well-resolved spectra. Also, the linewidths for short peptides are narrow so it is possible to determine coupling constants with high precision. The coupling constants are usually measured by deconvolution of the resulting spectra. In this deconvolution method, Lorentzian lines are fitted to the experimental data. Deconvolution takes into account aspects of insufficient digital resolution; for example, the point with the highest intensity is not necessarily sampled, which would lead to a minor error if the peak position is determined just by choosing the highest intensity. Deconvolution also takes into account that the apparent coupling constant may be smaller in case of overlapping doublet components (see Fig. 7.2).

The most widely used 2-D frequency-based method to determine J-coupling constants utilize the E.COSY principle [63, 64]. This experiment is particularly useful if the coupling constant to be measured is of the order of the linewidth, because the coupling constant can then be read off from two submultiplets that are displaced in two dimensions by a so-called passive, typically large coupling constant. One example is the HNCO[C^α]-E.COSY experiment to determine the coupling constant $^3J(H_i^N, C_{i-1}^\alpha)$ (Fig. 7.3). The expected coupling constant is small and would not be detectable from a regular doublet split by the J coupling. In an E.COSY-type experiment, an additional spin is used as a coupling partner, and the state of that spin is conserved throughout the experiment to yield the E.COSY pattern. In case of the HNCO[C^α]-E.COSY experiment, the C^α spin is used as a passive spin. The magnetization is transferred from the H^N proton to nitrogen and finally to the carbonyl. During the chemical shift evolution of the carbonyl, the $^1J(C_{i-1}^\alpha, C_{i-1}')$ is allowed to evolve. The size of this coupling is ~55 Hz, so it can be fully resolved in the evolution period t_1 (along the y-axis of the 2-D plot). Magnetization is transferred back to the amide protons for detection. During the acquisition, carbonyl nuclei are selectively decoupled, allowing the $^3J(H^N, C^\alpha)$ coupling to evolve. The

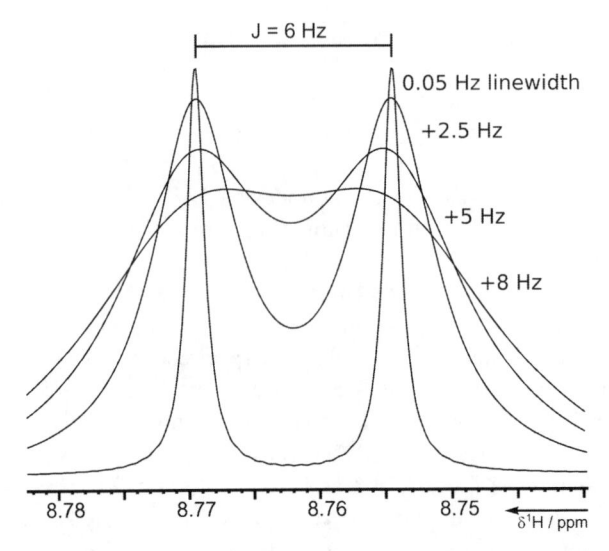

Figure 7.2. Simulated peak pattern for a doublet split by a J coupling of 6 Hz. Peaks were simulated using a linewidth of 0.05 Hz. In the case of the well-resolved resulting spectrum, the determination of the coupling constant by determining the peak positions at the highest intensity yields a coupling constant of 6 Hz. In the case of additionally broadened lines, the apparent coupling constant becomes smaller, which in the case of 8 Hz line broadening is as small as 4.2 Hz. However, deconvolution of the peak patterns assuming Lorentzian line shapes yields a coupling constant of 6 ± 0.01 Hz for all of the four spectra. Broadened was scaled for easier comparison. As long as individual peaks are well resolved, the spectral resolution does not have an effect on the measured coupling constant.

described experiment will result in the desired E.COSY pattern where the peak is split in the indirect dimension by the passive coupling (in this case, $^1J[C^\alpha_{i-1}, C'_{i-1}]$). Each of the two components that are separated in the indirect dimension will show only one component of the doublet caused by the $^3J(H^N_i, C^\alpha_{i-1})$ coupling in the direct dimension. This separation of the two submultiplet components allows the determination of a coupling that is smaller than the linewidth. The coupling constant can be measured precisely by taking the 1-D slices with maximum intensity of the two doublet components and shifting them toward each other to minimize the difference integral (see Fig. 7.4). Another advantage of this type of experiment is that the signal-to-noise ratio in the experiments is independent of the size of the coupling to be determined. Therefore, large and small couplings, as in the example shown, can be determined with the same precision.

The DQ/ZQ experiment is a variation of the E.COSY method to measure coupling constants; in the DQ/ZQ experiment, the sum and the difference of a large coupling and a small coupling (DQ: $^1J + {^3J}$; ZQ: $^1J - {^3J}$; $DQ - ZQ = 2 \cdot {^3J}$) are measured in a single dimension. For the application to larger molecules, this DQ/ZQ experiment [65–67] has significant advantages since it suppresses a systematic relaxation-induced

NMR PARAMETERS AND THEIR MEASUREMENT 195

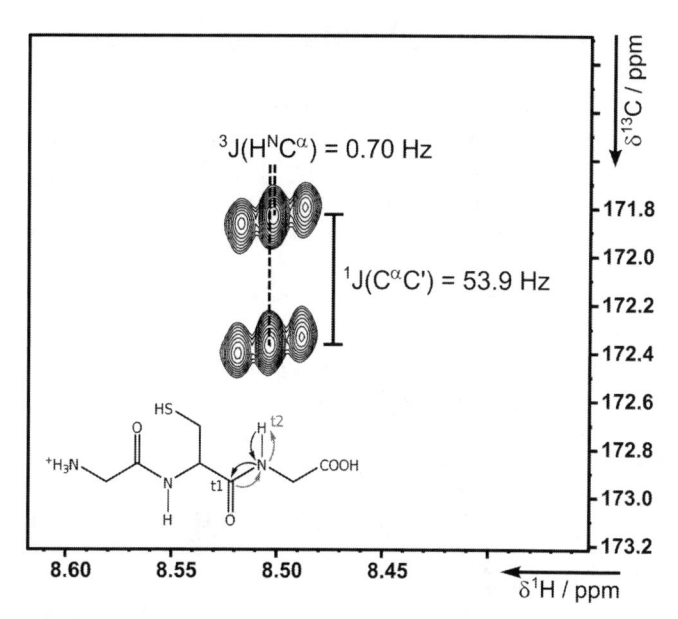

Figure 7.3. 2-D HNCO[C^α]-E.COSY H^NC' cross-peak shown for 1 mM sample of ($^{13}C'$) G($^{13}C^{15}$N)C(^{15}N)G in 90% H_2O/10% D_2O pH = 2. The peaks are split by the passive $^1J(C^\alpha,C')$ coupling in the indirect dimension (y-axis). Upfield and downfield components show only one component of the expected doublet for the $^3J(H^N,C^\alpha)$ coupling, resulting in the desired E.COSY peak pattern. The spectrum was recorded on a Bruker 400-MHz Avance spectrometer. The spectrum was recorded for a spectral width of 4800 × 1460 Hz, digitized using 4k × 256 complex points, and zero filled to 8k × 512 points prior to Fourier transformation. The magnetization transfer used for the 2-D HNCO[C^α]-E.COSY experiment is shown at the bottom left.

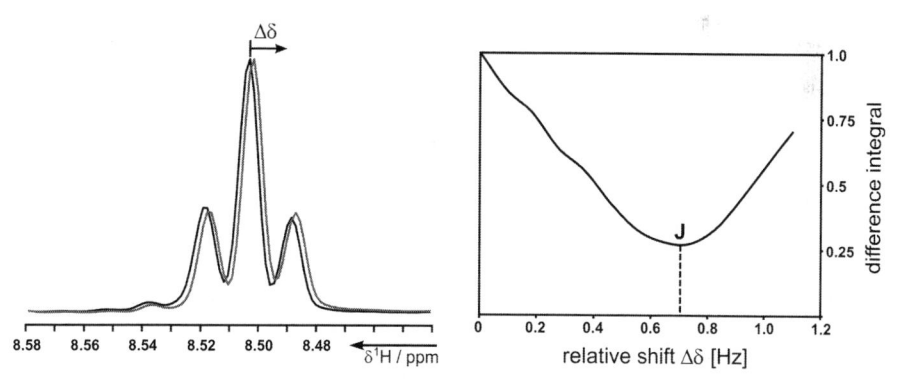

Figure 7.4. E.COSY cross-sections extracted from a 2-D spectrum are shown on the left. The two cross-sections taken for the maximum intensities for the downfield and upfield submultiplets are shown in black and gray. The graph on the right shows the integral of the absolute value taken for the difference of both cross-sections depending on the relative shift. The actual J coupling can be determined from the minimum.

error to coupling constants, especially for proton–proton ^3J couplings as discussed in more detail later.

Intensity-based methods extract the coupling constant by comparing the intensities of a reference and a cross-peak in 2-D spectra [68]. The experiments utilize coherence transfer modules that depend on the size of the J coupling of interest. The transfer amplitude leading to diagonal peaks and cross-peaks are differently modulated with respect to the coupling of interest. The spectra correlate two coupled spins by allowing the coupling of interest evolve for a certain time τ. Since the diagonal peak intensity is then modulated by $\cos^2(\pi J \tau)$ and the cross-peak is modulated by $\sin^2(\pi J \tau)$, I_{cross} divided by $I_{diagonal}$ is modulated by $\tan^2(\pi J \tau)$, which makes it easy to extract the coupling constant from the two peak intensities. The accuracy of the extracted coupling constant depends mainly on the signal-to-noise ratio. Intensity-based measurements of J couplings can be achieved using either quantitative J methods [68] or J-modulated experiments [69]. One example of quantitative J methods is the HNHA experiments (see Fig. 7.5). The coupling constant is measured by comparing the H^N diagonal-peak and the $H^N H^\alpha$ cross peak intensities.

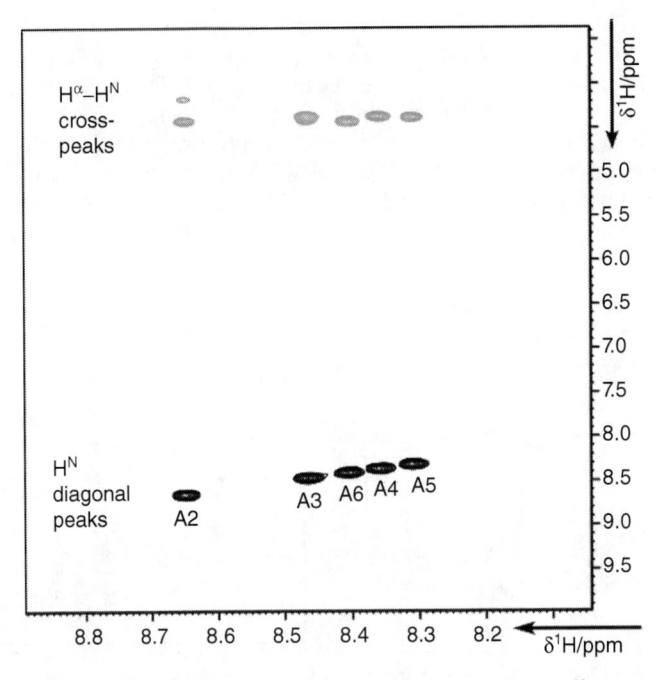

Figure 7.5. 2-D HNHA experiment shown for a short peptide. The H^N diagonal peaks are shown in black, the $H^N H^\alpha$ cross-peaks of the opposite sign are shown in gray. The spectrum was recorded on 2 mM uniformly $^{13}C^{15}N$-labeled sample of Ala$_6$ in 90% H_2O/10% D_2O at pH = 2. The spectrum was recorded using 2k × 256 complex points and 32 transient scans on a Bruker 600 MHz DRX spectrometer. Zero filling was applied to 2k × 1k points prior to processing.

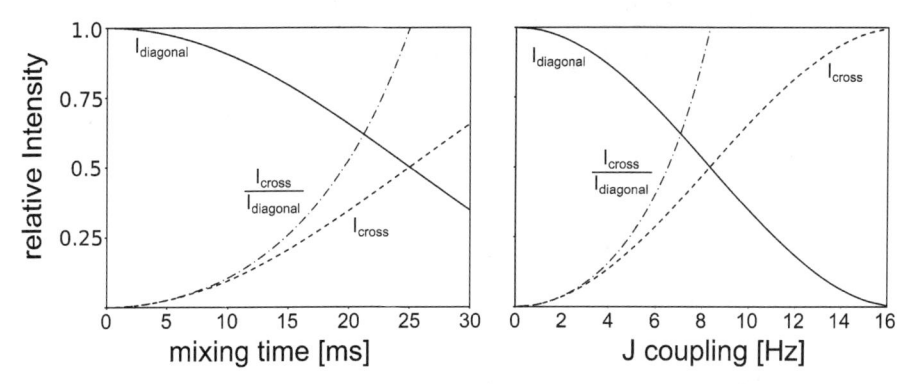

Figure 7.6. Relative intensities of diagonal (~$\cos^2 \pi J$) and cross-peaks (~$\sin^2 \pi J$) in quantitative J experiments. Shown on the left is the intensity dependent on the mixing time for a J coupling of 8 Hz. Shown on the right is the intensity dependent on the size of the J coupling for a mixing time of 30 ms.

In general, quantitative J measurement of a coupling is carried out by allowing the coupling evolve for a certain mixing time. The diagonal peak is then modulated with the cosine of the coupling constant, whereas the cross-peak is modulated with the sine of the coupling constant. After determination of the mixing time, the coupling constant can be extracted from the intensity ratio of diagonal and cross-peak, which is proportional to the tan of the coupling constant. Since both peaks, diagonal and cross-peak, need a sufficient signal-to-noise ratio, mixing time is crucial for an accurate measurement and depends on the size of the expected coupling constant (see Fig. 7.6).

J-modulated experiment uses a series of experiments in which one or more couplings are allowed to evolve for a certain time that is incremented. The evolution of J coupling leads to a partially dephased signal which lowers the intensity. The plot of intensities versus coupling evolution times can be fitted by a cosine function to yield the coupling constants (see Fig. 7.7).

All the above-mentioned methods can lead to precise measurements of J-coupling constants, particularly for short peptides, which usually have narrow lines and no significant spectral overlap.

7.3.1.1. Statistic and Systematic Errors on J-Coupling Determination

The precision and accuracy of measured J couplings largely depend on the method used. With increasing molecular weight, the 1-D spectra will suffer from increasing overlap as well as from systematic effects due to increasing linewidths. The linewidth $\Gamma^{1/2} = (1/\pi T_2^*)$ of an NMR signal is determined by the intrinsic magnetic homogeneity obtainable for a specific NMR instrument and probe ($\Gamma^{1/2,\text{intrinsic}} \sim 0.3$–$0.5$ Hz), and the linewidths determined by rotational correlation time of the molecule. $\Gamma^{1/2,\text{intrinsic}}$ can independently be determined from the linewidth of a singlet signal, for example, from reference molecules including tetramethylsilane. For increasing linewidths in molecules with larger molecular weights, the two multiplet lines may not

$$I(\tau)=I_0\cos(\pi J\tau)e^{-\tau/T_2^*}$$

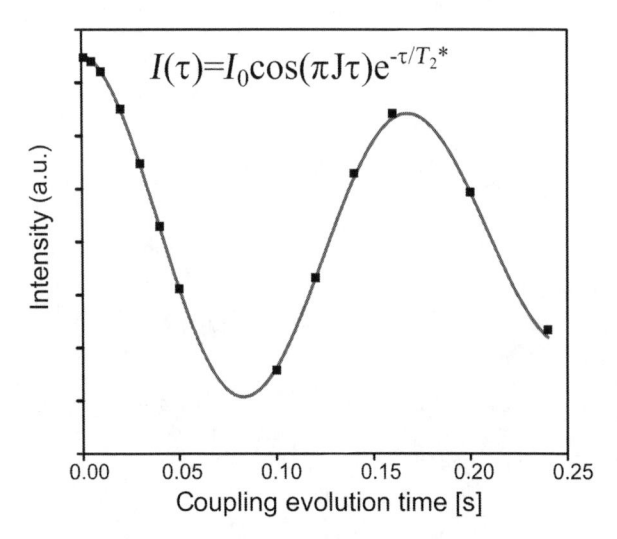

Figure 7.7. Plot of intensities versus coupling evolution time measured with a J-modulated HSQC-type experiment to measure $^1J(NC^\alpha)$ and $^2J(NC^\alpha_{i-1})$ [94] on 1 mM sample of ($^{13}C'$) G($^{13}C^{15}N$)C(^{15}N)G in 90% H_2O/10% D_2O pH = 2. Because of the labeling scheme, only the $^1J(NC^\alpha)$ coupling is observed for the central amide peak.

be baseline separated. In this case, the apparent coupling constant measured as peak-to-peak distance between the maxima of the doublet NMR signal is smaller than the true coupling constant. Multiplet splitting is also observed as antiphase splitting in 2-D correlation spectra such as DQF-COSY experiments. In this situation, the apparent linewidth is larger than the true linewidth [70]. Methods to correct for this deviation in 2-D COSY have been published [71]. The $^3J(H^N,H_\alpha)$ coupling constant is most routinely used to extract information about the angle ϕ (see Fig. 7.1), since it can be measured in small peptides without the need for isotope labeling. The smallest values for this coupling constant are ~ 3 Hz. For such small coupling constants, the linewidth can already be of the order of the coupling constant, especially if measuring at low temperature. Therefore, corrections need to be performed.

Additional effects that can lead to systematic errors include the effects of auto- and cross-correlated relaxation on the position of a submultiplet component. Let us assume a two-spin system I-S with a J coupling J(I,S) and observe the resonance of spin I. If spin I is weakly coupled to spin S, the active relaxation rates for the detection of I are the ones for I^+ and I^+S_z. If the relaxation rates for I^+ and I^+S_z are equal or at least close, the two components of a doublet will appear at positions $\pm\pi J$, which means that the determined coupling constant is exact. However, if the difference in relaxation rates of the two operators become large, this changes not only the linewidths but also the peak positions. In this case, the apparent coupling constant is also smaller than the true one [72]. The effect is important for proteins with long rotational correlation times (>5 ns), but can safely be ignored for small peptides with extremely long T_1 times.

The measurement of coupling constants from intensity-based methods are also affected by T_1 time effects of the coupled S spin, since the relaxation properties of the two operators leading to cross- and diagonal peak may be different. Ad hoc methods for the correction of this error have been proposed [73].

A third effect that may prevent the accurate determination of J couplings is the fact that cross- correlated relaxation also affects the intensities of the two components of a doublet. This differential attenuation of multiplet components in fact forms the basis of the so-called TROSY effect. This cross-correlated relaxation leads to a weakening of one of the two multiplet compounds but not to a frequency shift (as opposed to the above discussed effect). For small peptides, although visible, the effect does not systematically affect coupling constant determination.

7.3.2. NOE Distance Information

The so-called Nuclear Overhauser Effect (NOE) is among the most important NMR parameters for structure determination. In protein structure determination, usually the proton–proton NOEs are measured. Cross-peak intensities are typically referenced to a cross-peak with known reference distance and can then be translated into proton–proton distances; typically cross-peaks between two protons with distances of 5 Å or less can be observed. In case of short peptides, one has to keep in mind that the NOE is a dipolar cross-relaxation effect and is therefore dependent on the overall motion of the molecule, which is described by the rotational correlation time τ_c. The expected NOE transfer efficiency is also dependent on the Larmor frequency ω_0 and has a zero crossing at $\omega_0 \tau_c = 1.12$. Such a situation is often encountered for smaller peptides (6–10mers) at an NMR frequency of 300–400 MHz at room temperature.

The rotational correlation time is dependent on molecular size (hydrodynamic radius), viscosity of the solvent, and temperature. Since temperature and Larmor frequency are the only variables that can be tuned, and those two only in a small range, this leaves the problem that molecules of a certain size have nearly no NOE transfer efficiency. In this case one can still collect distance information by exploiting the Rotating Frame Overhauser Effect (ROE).

The cross-correlated relaxation rate that gives rise to a cross-peak in a NOESY or ROESY spectrum is given by

$$\sigma_{H,H}^{NOE} = \frac{d^2 \tau_c}{5}\left[-1 + \frac{6}{1+4\omega_0^2 \tau_c^2}\right]$$
$$\sigma_{H,H}^{ROE} = \frac{d^2 \tau_c}{5}\left[2 + \frac{3}{1+\omega_0^2 \tau_c^2}\right], \qquad (7.1)$$

where d is the dipolar interaction, which can be described as

$$d = \mu_0 h \gamma_H^2 / (\sqrt{8} r_{H,H}^3 \pi^2), \qquad (7.2)$$

where γ_H is the gyromagnetic ratio, μ_0 the vacuum permeability, h Planck's constant, and r the interspin distance. This formalism is only strictly valid when a few assumptions can actually be made for a given system. The overall tumbling is assumed to be isotropic, and the formula does not take into account local motions. For a folded protein these assumptions are usually valid; however, in the case of a rather flexible peptide they may not. In case of an elongated structure, the overall tumbling will most likely not be isotropic, which leads to different contributions for rotation along the short and long axis to τ_c. In addition to that, any local flexibility τ_e will also contribute to the effective cross-correlated relaxation leading to an NOE. This means that the condition where no NOE can be observed ($\omega_0 \tau_c = 1.12$) is not necessarily fulfilled for the entire molecule at once but largely depends on the relative orientation with respect to the molecular tumbling and also on the local dynamics. Therefore the absence of a cross-peak in a NOESY can only with great care be interpreted as indicating an averaged distance larger than 5 Å.

7.3.3. RDCs

Residual dipolar couplings have proven to be a valuable tool when gathering protein structural information [45, 74]. The measurement of RDCs, in contrast to most of the other NMR observables, requires special sample preparation. RDCs are orientation-dependent couplings that arise whenever the tumbling in solution does not average all possible orientations and is not completely isotropic. To measure residual dipolar couplings, partial alignment of the molecule of interest is required. In case of folded proteins, many different alignment media are well established. Bicelles or stretched gels align biomolecules through sterical interactions; macrophages such as Pf1 are used mainly to align biomolecules that carry negative charges on the surface through electrostatic interaction; and covalently attached paramagnetic ions such as lanthanide ions that have a large anisotropy in magnetic susceptibility also lead to an alignment of the attached molecule.

In general, the Hamiltonian for the dipolar coupling for two spins i and j is given by the formula

$$H_{ij}^D(t) = -\frac{\gamma_i \gamma_j \mu_0 h}{8\pi^3 r_{ij}^3(t)} I_{iz} I_{jz} \frac{(3\cos^2 \theta_{ij}(t) - 1)}{2}, \quad (7.3)$$

where $\gamma_{i/j}$ are the gyromagnetic ratios, μ_0 the vacuum permeability, h the Planck constant, r_{ij} the interspin distance, $I_{iz/jz}$ the angular momentum operators, and θ_{ij} the angle defined by the interspin vector and the external magnetic field B_0. Since the magnitude of the dipolar coupling depends on $3\cos^2\theta - 1$, it averages to zero for a molecule tumbling fast in solution. Only second-order perturbation theory and incomplete averaging over time scales shorter than the overall correlation time τ_c lead to relaxation effects, observable for example in NOE effects. However, if there is a preferential orientation, it is not completely averaged out.

The orientation of an individual spin pair (two spins connected via a bond) is usually described not by the angles with respect to the laboratory frame (where B_0 represents the z axis) but by angles with respect to a molecular frame that represents

NMR PARAMETERS AND THEIR MEASUREMENT

the averaged orientation of the whole molecule with respect to the magnetic field. In this case the orientation can be described by

$$\cos\theta_{ij} = \begin{pmatrix} \cos\zeta_x \\ \cos\zeta_y \\ \cos\zeta_z \end{pmatrix} \begin{pmatrix} \cos\xi_x \\ \cos\xi_x \\ \cos\xi_x \end{pmatrix} = \cos\zeta_x \cos\xi_x + \cos\zeta_y \cos\xi_y + \cos\zeta_z \cos\xi_z, \quad (7.4)$$

where ζ_{xyz} are the angles defined by the interspin vector and the axis of the molecular frame and ξ_{xyz} are the angles defined by the axis of the molecular frame and the B_0 vector. The average orientation can then be described by a symmetric 3 × 3 matrix A, the so-called Saupe matrix [75]:

$$A = \begin{pmatrix} \cos\xi_x \cos\xi_x - \dfrac{1}{2} & \cos\xi_x \cos\xi_y & \cos\xi_x \cos\xi_z \\ \cos\xi_y \cos\xi_x & \cos\xi_y \cos\xi_y - \dfrac{1}{2} & \cos\xi_y \cos\xi_z \\ \cos\xi_z \cos\xi_x & \cos\xi_z \cos\xi_y & \cos\xi_z \cos\xi_z - \dfrac{1}{2} \end{pmatrix}. \quad (7.5)$$

This matrix in Equation 7.5 can be diagonalized yielding the principal axis system with the values A_{xx}, A_{yy}, and A_{zz}. Any individual residual dipolar coupling for a bond vector in this system can be described by

$$D_{ij}(\theta,\varphi) = -\dfrac{\gamma_i \gamma_j \mu_0 h}{8\pi^3 r_{ij,eq}^3}(A_{zz}\cos^2\theta + A_{xx}\sin^2\theta\cos^2\varphi + A_{yy}\sin^2\theta\sin^2\varphi), \quad (7.6)$$

where θ is the angle defined by the bond vector and the axis A_{zz}, and φ is the angle defined by the bond angle projection in the A_{xx}/A_{yy} plane and the axis A_{xx}. The alignment tensor is usually expressed as the axial component $A_a = A_{zz}/2$ and the rhombicity $A_r = (A_{xx} - A_{yy})/3$, which results in a simplified equation for the individual dipolar coupling:

$$D_{ij}(\theta,\varphi) = -\dfrac{\gamma_i \gamma_j \mu_0 h}{16\pi^3 r_{ij,eq}^3}\left(A_a(3\cos^2\theta - 1) + \dfrac{3}{2}A_r \sin^2\theta \cos 2\varphi\right). \quad (7.7)$$

The alignment tensor can then be described by five independent variables, A_a and A_r as well as the Euler rotation $R(\alpha,\beta,\gamma)$, where the three angles α,β, and γ describe the 3-D orientation of the alignment tensor.

For short peptides, the number of alignment media that can be applied is rather limited due to the following considerations. First, the alignment media must be compatible with the buffer conditions; in case of unfolded peptides, this often involves low pH, where most bicelles fail. Second, the alignment media must not interact with the molecule changing, for example, its conformation. Because short peptides will be more or less unfolded, they usually present a combination of charged and hydrophobic side chains to the solution, which makes them more likely to interact with alignment media such as charged phages or the hydrophobic part of bicelles.

One alignment medium that can be applied in general, independent of the buffer conditions or the molecule of interest, is a stretched polyacrylamide (PAA) gel. These gels are stable over a wide range of temperature and pH values, and do not seem to interact even with unfolded peptides or proteins. Usually, these gels are polymerized outside the NMR tube with a diameter slightly larger then the inner diameter of an NMR tube. The gel is then inserted through a funnel into the NMR tube and is thereby stretched. The cavities inside the tube, which are assumed to be symmetrical, are deformed during the stretching, leading to a preferential orientation of included molecules. In the case of shorter peptides, the degree of alignment required to yield reasonably large RDCs requires a high percentage of PAA gels, which can make it difficult to stretch them.

The measurement of residual dipolar couplings is carried out the same way as for the corresponding J couplings. The coupling is measured for the aligned and for the non-aligned sample; the difference between those two measured couplings yields the RDC. To extract angular information, usually the residual dipolar couplings are measured along one bond of known bond length. For structure determination, proton–proton RDCs can also be used, but as the proton–proton distances and orientations may vary, these RDCs are an additional parameter to refine against in structure determination protocols and do not provide direct orientation information.

Measurement of ^1J (or $^1J + {}^1D$) couplings is usually carried out by recording non-decoupled hetero single quantum correlation (HSQC)-type experiments. The spectra can show splitting in either the direct or the indirect dimension, where both of these methods have their disadvantages. Measurement in the indirect dimension might suffer from the fact that the indirect dimension has a lower resolution most of the time. The measurement in the direct dimension might be difficult due to proton–proton dipolar coupling being active during the acquisition, which might lead to significant line broadening in case of dense proton networks. In case of larger systems where the splitting of signals would lead to additional overlap, usually the above-mentioned IPAP, S^3E, or J-modulated techniques are applied.

7.4. TRANSLATING NMR PARAMETERS TO STRUCTURAL INFORMATION

7.4.1. From J-Coupling Constants to Torsion Angles

There are multiple ways of interpreting NMR measurements. The method of choice depends mainly on the nature of the peptide being observed as it does on the available data obtained from the NMR measurements and complementary methods.

The structural information provided by J couplings has been known for a long time. Most often the information is used to determine dihedral angles from ^3J-coupling constants utilizing the Karplus equation. In its general form the Karplus equation is given by

$$J(\theta) = A\cos^2(\theta) + B\cos(\theta) + C, \tag{7.8}$$

where J is the measured scalar coupling constant, θ is the involved dihedral angle, and A, B, and C are empirically derived parameters depending on the respective coupling constant. As one can already see from Equation 7.8, depending on the actual size of the measured coupling constant, there are up to four possible solutions. In principle, every ^3J-coupling constant can be used to determine dihedral angles. In the case of proteins or peptides, Karplus parameters exist for all backbone-related J-coupling constants. For the backbone angle ϕ, six different coupling constants can be measured between five different spin-1/2-nuclei to define the dihedral angle. The most generally used one is $^3J(H^N,H^\alpha)$. Because it involves two protons, which have the highest gyromagnetic ratios, it has the largest absolute value and can be measured using non-labeled peptides. It also has the largest spread in possible values, ranging from 2 to 10 Hz. For the angle ϕ, there are five other ^3J-coupling constants, $^3J(H^\alpha,C')$, $^3J(H^N,C^\beta)$, $^3J(H^N,C^\beta)$, $^3J(H^N,C')$, $^3J(C',C')$, and $^3J(C',C^\beta)$, which can be used to resolve the degeneracy of the Karplus equation (see Fig. 7.8). They differ not only in their absolute values and the spread, but also in the phase offset of the respective Karplus equations, which further helps to resolve the degeneracy. Depending on the involved nuclei and the chosen experiment to measure these couplings, they all do require ^{13}C and/or ^{15}N stable isotope labeling.

For the backbone angle ψ, it is much more difficult to obtain a large number of dependent J-coupling constants and therefore resolve the degeneracy; however, one can exploit the fact that not only ^3J couplings but also ^1J and ^2J couplings depend on dihedral angles. It is possible to determine the angle ψ by measuring the $^1J(N_i, C_i^\alpha)$ and the $^2J(N_i, C_{i+1}^\alpha)$ coupling as those two coupling constants both depend on the same angle ψ (see Fig. 7.9).

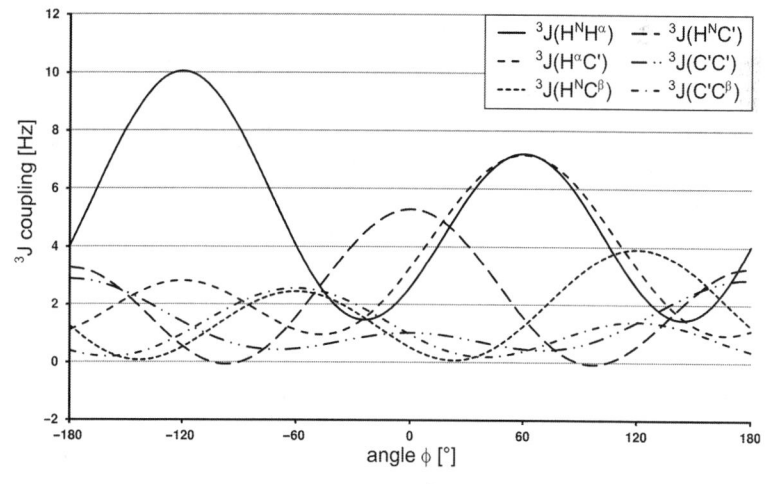

Figure 7.8. ϕ angle dependency of all ϕ-related ^3J-coupling constants. As parameters, the refined solution Karplus parameters published by Bax et al. were used [95].

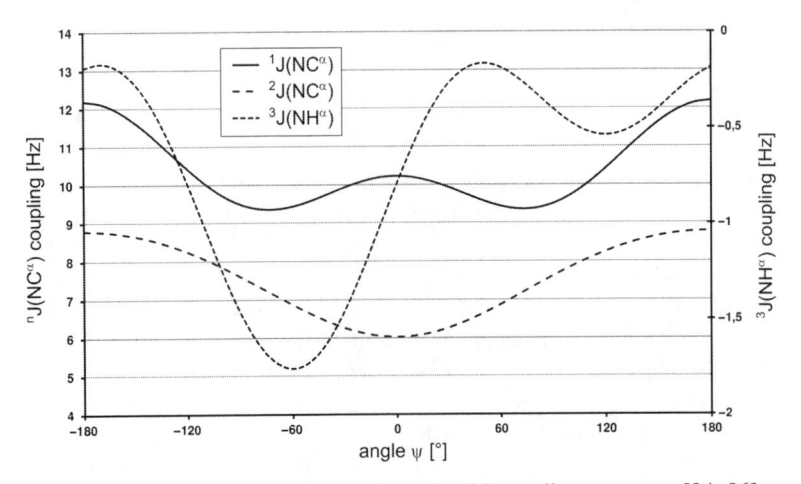

Figure 7.9. ψ angle dependency of ψ-related J-coupling constants [94, 96].

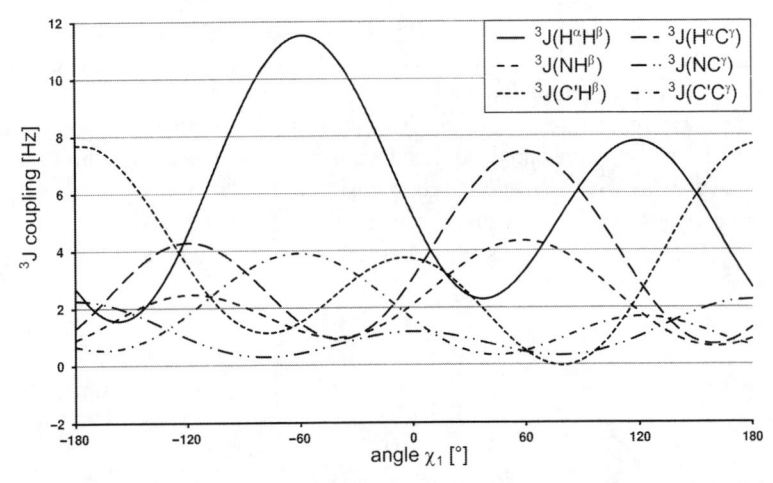

Figure 7.10. χ_1 angle dependency of all χ_1-related ^{3}J-coupling constants. Shown are data for the consensus Karplus parameters published by Schmidt [97]. All H^β-related coupling constants refer to $H^{\beta 1}$. Curves for $H^{\beta 2}$ for non-beta-branched amino acids are shifted by +120°.

In principle, every dihedral angle can be determined using a set of J-coupling constants in combination with an appropriate Karplus equation and parameterization. For most amino acids this holds true for the first side-chain angle χ_1 (see Fig. 7.10). In principle, there are again six different coupling constants that can be measured, but this number may be reduced depending on the amino acid type, or by overlapping H_β resonances. Although a general Karplus equation was originally proposed for the side-chain angle χ_1, it has been shown that individual parameterization for every amino acid needs to be used.

If the measured coupling constants are translated to structural information, the accuracy depends not only on the error of the J couplings themselves but also on the Karplus parameters. As already mentioned, first Karplus parameters are derived empirically, then subjected to continuous refinement (see Table 7.2), by applying, for example, density functional theory (DFT) calculations, or by fitting NMR data to existing high-resolution crystal structures.

In addition to potential imprecision in the Karplus parameters used to derive information on the conformation of peptides, conformational averaging between the reference systems used to derive the Karplus parameters and the system being studied can differ. Typically, Karplus parameterization is derived from experimental data of proteins and high-resolution X-ray structures with limited local librational motions of, for example, the N–H and C_α–H_α bonds. However, the simultaneous determination of conformational averaging and averaged Karplus parameters has been proposed, for example, in Ref. [76, 77].

In contrast to the general idea that a ^{3}J-coupling constant is dependent mainly on the rotation with respect to the central bond, it has been shown that it may also depend on the adjacent torsion angles, which lead to Karplus-like equations that involve more than a single torsion angle (see Fig. 7.11), and to a higher degeneracy. Those equations can be useful, particularly when trying to determine the backbone angle ψ, where all related ^{3}J-coupling constants, including ψ as the central torsion angle, involve at least one nitrogen spin and are therefore small.

For short peptides, there are now several ways of applying the measured J-coupling constants to determine the conformational distribution for a given peptide. The simplest way is to translate the J-coupling measurements directly to dihedral angle.

TABLE 7.2. Overview of Published Karplus Parameters

Coupling Constant	Karplus Parameterization	Ref.
$^3J(H^N,H^\alpha)$	$^3J = 6.51\cos^2(\phi-60°) - 1.76\cos(\phi-60°) + 1.60$	[99]
	$^3J = 6.40\cos^2(\phi-60°) - 1.40\cos(\phi-60°) + 1.60$	[100]
	$^3J = 6.60\cos^2(\phi-60°) - 1.30\cos(\phi-60°) + 1.50$	[101]
	$^3J = 7.90\cos^2(\phi-60°) - 1.05\cos(\phi-60°) + 0.65$	[102]
	$^3J = 6.64\cos^2(\phi-60°) - 1.43\cos(\phi-60°) + 1.86$	[103]
	$^3J = 7.09\cos^2(\phi-60°) - 1.42\cos(\phi-60°) + 1.55$	[95]
$^3J(H^N,C')$	$^3J = 4.01\cos^2(\phi) - 1.09\cos(\phi) + 0.07$	[104]
	$^3J = 4.02\cos^2(\phi) - 1.12\cos(\phi) + 0.07$	[103]
	$^3J = 4.29\cos^2(\phi+180°) - 1.01\cos(\phi+180°)$	[95]
$^3J(H^N,C^\beta)$	$^3J = 4.70\cos^2(\phi+60°) - 1.76\cos(\phi+60°) - 0.20$	[105]
	$^3J = 2.78\cos^2(\phi+60°) - 0.37\cos(\phi+60°) - 0.03$	[103]
	$^3J = 3.06\cos^2(\phi+60°) - 0.74\cos(\phi+60°) + 0.13$	[95]

(*continued overleaf*)

TABLE 7.2. (*Continued*)

Coupling Constant	Karplus Parameterization	Ref.
$^3J(H^\alpha,C')$	$^3J = 4.50\cos^2(\phi+120°) - 1.30\cos(\phi+120°) - 1.20$	[105]
	$^3J = 3.72\cos^2(\phi+120°) - 1.71\cos(\phi+120°) + 1.07$	[106]
	$^3J = 3.62\cos^2(\phi-60°) - 2.11\cos(\phi-60°) + 1.29$	[103]
	$^3J = 3.72\cos^2(\phi+120°) - 2.18\cos(\phi+120°) + 1.28$	[95]
$^3J(C',C^\beta)$	$^3J = 1.66\cos^2(\phi-120°) - 0.66\cos(\phi-120°) + 0.26$	[107]
	$^3J = 1.28\cos^2(\phi-120°) - 1.02\cos(\phi-120°) + 0.30$	[108]
	$^3J = 2.54\cos^2(\phi-120°) - 0.55\cos(\phi-120°) + 0.37$	[106]
	$^3J = 1.74\cos^2(\phi-120°) - 0.57\cos(\phi-120°) + 0.25$	[95]
$^3J(C',C')$	$^3J = 1.33\cos^2(\phi) - 0.88\cos(\phi) + 0.62$	[109]
	$^3J = 1.57\cos^2(\phi) - 1.07\cos(\phi) + 0.49$	[109]
	$^3J = 1.36\cos^2(\phi) - 0.93\cos(\phi) + 0.60$	[95]
$^1J(N_i,C^\alpha_i)$	$^1J = 1.70\cos^2(\psi_i) - 0.98\cos(\psi_i) + 9.51$	[28]
	$^1J = 2.85\cos^2(\psi_i) - 1.21\cos(\psi_i) + 8.65$	[110]
$^2J(N_i, C^\alpha_{(i-1)})$	$^2J = -0.37\cos^2(\psi_{i-1}) - 0.64\cos^2(\phi_{i-1}) - 1.39\cos(\psi_{i-1}) - 0.17\cos(\phi_{i-1}) + 7.82$	[28]
	$^2J = -0.66\cos^2(\psi_{i-1}) - 1.52\cos(\psi_{i-1}) + 7.85$	[110]
$^3J(H^\alpha,H^\beta)$	$^3J = 9.5\cos^2(\chi_1-120°) - 1.6\cos(\chi_1-120°) + 1.8$	[111]
	$^3J = 5.83 - 1.37\cos(\chi_1-120°) + 3.61\cos(2(\chi_1-120°))$	[112]
	$^3J = 5.86 - 1.86\cos(\chi_1-120°) + 3.81\cos(2(\chi_1-120°)) - 0.37\sin(\chi_1-120°)$	[97]
$^3J(N_i,H^\beta)$	$^3J = 2.22 - 0.75\cos(\chi_1+120°) + 1.15\cos(2(\chi_1+120°))$	[112]
	$^3J = 2.15 - 0.93\cos(\chi_1+120°) + 1.26\cos(2(\chi_1+120°)) + 0.17\sin(\chi_1+120°)$	[97]
$^3J(C',H^\beta)$	$^3J = 3.32 - 1.58\cos(\chi_1) + 2.01\cos(2\chi_1)$	[112]
	$^3J = 3.24 - 1.99\cos(\chi_1) + 2.48\cos(2\chi_1) - 0.59\sin(\chi_1)$	[97]
$^3J(H^\alpha,C^\gamma)$	$^3J = 3.46 - 0.96\cos(\chi_1+120°) + 2.67\cos(2(\chi_1+120°))$	[112]
	$^3J = 3.41 - 1.58\cos(\chi_1+120°) + 2.46\cos(2(\chi_1+120°)) - 0.10\sin(\chi_1+120°)$	[97]
$^3J(N_i,C^\gamma)$	$^3J = 1.02 - 0.49\cos(\chi_1) + 0.65\cos(2\chi_1)$	[112]
	$^3J = 1.05 - 0.55\cos(\chi_1) + 0.68\cos(2\chi_1) - 0.02\sin(\chi_1)$	[97]
$^3J(C',C^\gamma)$	$^3J = 1.70 - 0.87\cos(\chi_1-120°) + 1.15\cos(2(\chi_1-120°))$	[112]
	$^3J = 1.69 - 1.11\cos(\chi_1-120°) + 1.11\cos(2(\chi_1-120°)) - 0.10\sin(\chi_1-120°)$	[97]

The list is not complete.

There are a number of approaches to interpret the experimental J coupling into the description of an (averaged) conformation. One approach for analyzing the main-chain conformations of unfolded peptides and proteins was proposed by Luis Serrano in 1995 [22]. The work focused on the comparison of experimental data, namely J couplings and chemical shifts, with statistical or empirical data derived from the

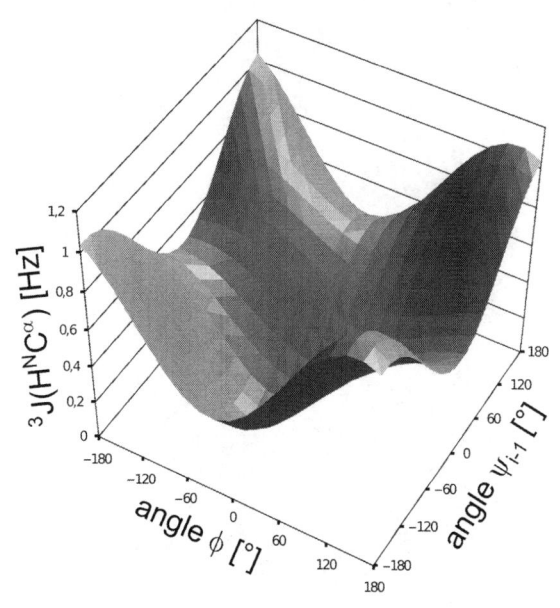

Figure 7.11. 2-D Karplus-like dependency of the $^3J(H^NC^\alpha)$ coupling constant on the adjacent angles ϕ and ψ_{i-1} [98].

protein database. The results clearly showed a correlation of experimental and theoretical data. However, it was also shown that amino acids have different intrinsic propensities regarding the ϕ angle, which is mainly influenced by the side chain. If the protein database used for this approach is large enough, there is no significant difference if the unstructured parts of proteins or all regions are used since the neighbor effects will be averaged out. This is true especially when looking only at the backbone angle ϕ, where the effect of neighboring residues is assumed to be much smaller than that in the ψ angle. A similar approach was used by Dobson and coworkers [23, 78–80] to describe peptides, chemical denatured states of proteins, and also an IUP, a 130 amino acid protein, using different overlapping fragments. The measured NMR parameters were compared with random coil values generated by averaging over predictions for a set of selected proteins from the protein database. Again, the experimental values were in good agreement with the statistical model. Only on such a definition of random coil behavior, can deviations from the random coil definition be interpreted as a structure propensity and non-random structure.

J couplings were also used by Graf et al. [81] to determine the conformational propensities of homopolymeric polyalanine peptides. The work used an extensive set of eight ϕ and φ dependent J couplings for a series of polyalanine peptides (A_3–A_7) in conjunction with an all atom molecular dynamics (MD) simulation in explicit solvent (see Table 7.3).

The work showed that the MD simulation can reproduce accurately the conformations present in solution but with a considerable uncertainty regarding the actual

TABLE 7.3. Experimental and MD-Derived Coupling Constants for Ala$_7$. Table Modified from [113]

#	Coupling Constant	Angle	\alpha	\beta	PP$_{II}$	MD	Fit	Exp	P$_\alpha$	P$_\beta$	P$_{PP_{II}}$	
A2	$^3J(H_N,H_\alpha)$	(ϕ_2)	4.7 ± 2.3	9.3 ± 1.0	4.8 ± 1.7	5.5 ± 2.6	5.6	5.61 ± 0.04	40	17	35	(MD)
	$^3J(H_N,C')$	(ϕ_2)	1.7 ± 1.2	0.8 ± 0.8	1.3 ± 0.9	1.5 ± 1.2	1.2	1.15 ± 0.02	0	17	83	(Fit)
	$^3J(H_{\alpha},C')$	(ϕ_2)	1.5 ± 0.5	2.6 ± 0.3	1.4 ± 0.4	1.8 ± 1.2	1.6	1.89 ± 0.32				
	$^3J(H_N,C_\beta)$	(ϕ_2)	2.1 ± 0.6	0.6 ± 0.4	2.4 ± 0.2	1.8 ± 0.8	2.0	2.31 ± 0.05				
	$^1J(N,C_\alpha)$	(ψ_2)	9.7 ± 0.2	10.7 ± 0.9	11.0 ± 0.8	10.4 ± 0.9	10.9	11.37 ± 0.01				
	$^2J(N,C_\alpha)$	(ψ_1)	7.6 ± 1.0	7.6 ± 1.0	7.7 ± 1.0	8.6 ± 0.2		9.17 ± 0.02				
	$^3J(H_N,C_\alpha)$	(ϕ_2,ψ_1)	0.6 ± 0.1	0.8 ± 0.1	0.6 ± 0.1	0.6 ± 0.1	0.6	0.71 ± 0.02				
A3	$^3J(H_N,H_\alpha)$	(ϕ_3)	4.5 ± 2.1	9.4 ± 1.0	4.9 ± 1.7	5.4 ± 2.5	5.6	5.66 ± 0.01	45	15	29	(MD)
	$^3J(H_N,C')$	(ϕ_3)	1.7 ± 1.1	0.8 ± 0.8	1.3 ± 0.8	1.5 ± 1.2	1.2	1.20 ± 0.02	0	16	84	(Fit)
	$^3J(H_{\alpha},C')$	(ϕ_3)	1.4 ± 0.5	2.6 ± 0.3	1.4 ± 0.4	1.9 ± 1.4	1.6	1.85 ± 0.20				
	$^3J(H_N,C_\beta)$	(ϕ_3)	2.2 ± 0.5	0.7 ± 0.4	2.3 ± 0.2	1.9 ± 0.7	2.0	2.20 ± 0.10				
	$^1J(N,C_\alpha)$	(ψ_3)	9.7 ± 0.2	10.6 ± 0.9	10.8 ± 0.8	10.2 ± 0.8	10.8	11.27 ± 0.02				
	$^2J(N,C_\alpha)$	(ψ_2)	7.3 ± 1.0	7.4 ± 1.1	7.6 ± 1.0	7.7 ± 1.0	7.6	8.52 ± 0.03				
	$^3J(H_N,C_\alpha)$	(ϕ_3,ψ_2)	0.4 ± 0.2	0.6 ± 0.2	0.4 ± 0.2	0.5 ± 0.2	0.5	0.66 ± 0.01				
A4	$^3J(H_N,H_\alpha)$	(ϕ_4)	4.7 ± 2.0	9.4 ± 1.0	5.0 ± 1.7	5.5 ± 2.4	5.7	5.77 ± 0.02	57	12	21	(MD)
	$^3J(H_N,C')$	(ϕ_4)	1.5 ± 1.0	0.8 ± 0.8	1.2 ± 0.9	1.5 ± 1.1	1.1	1.20 ± 0.05	0	15	85	(Fit)
	$^3J(H_{\alpha},C')$	(ϕ_4)	1.4 ± 0.5	2.6 ± 0.3	1.4 ± 0.4	2.0 ± 1.5	1.6	1.80 ± 0.14				
	$^3J(H_N,C_\beta)$	(ϕ_4)	2.2 ± 0.5	0.7 ± 0.4	2.3 ± 0.2	1.9 ± 0.7	2.0	2.23 ± 0.02				
	$^1J(N,C_\alpha)$	(ψ_4)	9.7 ± 0.2	10.6 ± 0.9	10.7 ± 0.8	10.1 ± 0.7	10.7	11.22 ± 0.02				
	$^2J(N,C_\alpha)$	(ψ_3)	6.9 ± 0.9	7.0 ± 1.1	7.4 ± 1.1	7.4 ± 1.0	7.6	8.29 ± 0.03				
	$^3J(H_N,C_\alpha)$	(ϕ_4,ψ_3)	0.3 ± 0.2	0.6 ± 0.2	0.4 ± 0.2	0.4 ± 0.3	0.5	0.56 ± 0.04				

A5	$^3J(H_N,H_\alpha)$	(ϕ_5)	5.2 ± 2.1	9.4 ± 1.0	5.1 ± 1.7	5.9 ± 2.4	5.7	5.92 ± 0.02	52	16	22	(MD)
	$^3J(HN,C')$	(ϕ_5)	1.3 ± 0.9	0.8 ± 0.8	1.2 ± 0.8	1.3 ± 1.0	1.1	1.19 ± 0.06	0 14	86	(Fit)	
	$^3J(H_\alpha,C')$	(ϕ_5)	1.5 ± 0.5	2.6 ± 0.3	1.4 ± 0.4	2.0 ± 1.4	1.6	1.56 ± 0.25				
	$^3J(N,C_\beta)$	(ϕ_5)	2.2 ± 0.5	0.7 ± 0.4	2.3 ± 0.2	1.9 ± 0.8	2.0	2.23 ± 0.08				
	$^1J(N,C_\alpha)$	(ψ_5)	9.8 ± 0.2	10.5 ± 0.9	10.7 ± 0.8	10.1 ± 0.7	10.6	11.29 ± 0.01				
	$^2J(N,C_\alpha)$	(ψ_4)	7.0 ± 1.0	7.0 ± 1.0	7.4 ± 1.1	7.1 ± 1.0	7.3	8.22 ± 0.04				
A6	$^3J(H_N,H_\alpha)$	(ϕ_6)	5.5 ± 2.2	9.3 ± 1.2	5.3 ± 1.8	6.2 ± 2.4	5.9	6.04 ± 0.03	43	21	25	(MD)
	$^3J(H_N,C')$	(ϕ_6)	1.2 ± 0.9	0.9 ± 0.9	1.1 ± 0.9	1.3 ± 1.1	1.1	1.10 ± 0.04	0	17	83	(MD$_{fit}$)
	$^3J(H_\alpha,C')$	(ϕ_6)	1.5 ± 0.5	2.6 ± 0.4	1.5 ± 0.4	2.2 ± 1.6	1.6	1.67 ± 0.20				
	$^3J(N,C_\beta)$	(ϕ_6)	2.1 ± 0.6	0.7 ± 0.4	2.3 ± 0.2	1.7 ± 0.8	2.0	2.21 ± 0.04				
	$^1J(N,C_\alpha)$	(ψ_6)	9.8 ± 0.2	10.4 ± 0.9	10.6 ± 0.8	10.1 ± 0.7	10.6	11.29 ± 0.01				
	$^2J(N,C_\alpha)$	(ψ_5)	6.4 ± 0.4	8.0 ± 0.9	8.2 ± 0.7	7.2 ± 1.1	7.4	8.24 ± 0.01				
A7	$^3J(H_N,H_\alpha)$	(ϕ_7)	6.5 ± 2.6	6.3 ± 2.7	6.4 ± 2.7	6.4 ± 2.7		6.60 ± 0.03				
	$^3J(H_N,C')$	(ϕ_7)	1.3 ± 1.1	1.4 ± 1.3	1.4 ± 1.2	1.3 ± 1.2		1.25 ± 0.05				
	$^3J(H_\alpha,C')$	(ϕ_7)	2.1 ± 1.3	2.3 ± 1.6	2.2 ± 1.4	2.2 ± 1.4		2.03 ± 0.14				
	$^3J(N,C_\beta)$	(ϕ_7)	1.6 ± 0.9	1.6 ± 0.8	1.6 ± 0.9	1.6 ± 0.9		1.99 ± 0.12				
	$^1J(N,C_\alpha)$	(ψ_7)	10.5 ± 1.0	10.4 ± 0.9	10.4 ± 0.9	10.5 ± 0.9		11.51 ± 0.01				
	$^2J(N,C_\alpha)$	(ψ_6)	6.4 ± 0.4	7.9 ± 0.8	8.2 ± 0.7	7.3 ± 1.1	8.1	8.18 ± 0.02				
	$^3J(H_N,C_\alpha)$	(ϕ_7,ψ_6)	0.3 ± 0.1	0.6 ± 0.1	0.6 ± 0.1	0.5 ± 0.2		0.59 ± 0.00				

populations. It was therefore proposed, based on the experimental data, to reweight the exact population. Polyalanine peptides populate extended conformations with a high polyproline type II (PP$_{II}$) helical content of about 90%. The estimated error of the determined populations is less the 5%. These findings were discussed in the context of new DFT-derived Karplus parameterizations by Best et al. [82]. MD simulations using these refined parameters yield a significantly populated α-helical region for the same systems. However, additional spectroscopic techniques by the Schweitzer-Stenner group reinforced the proposed conformational distributions published by Graf et al. [83]. In addition, work by Kallenbach et al. [84] using paramagnetic relaxation effects in related peptides also reinforced Graf et al.'s findings.

The work by Graf et al. provides a model for determining the temperature dependence of the conformational averaging process. Seven different coupling constants have been measured for four different temperatures ranging from 275 to 350 K. It could be shown that the ϕ-dependent coupling constants change linearly with increasing temperature, which indicates rising β-sheet content for the used model Ala$_3$. On the other hand, the ψ-dependent coupling constants show only minor changes, which indicates that the peptide still occupies almost exclusively the stretched conformations. This result is in agreement with other observations carried out on similar peptides [85–87]. To further investigate the temperature effect, the coupling constants have been used to determine the exact conformational distributions by using the main conformations from MD simulations carried out for 300 K and minimizing the difference between experimental and back-calculated coupling constants. This approach is valid under the assumption that the backbone angles for the main conformations (in this case α-helix, β-sheet, and P$_{II}$ helix) do not change with rising temperature. In case of Ala$_3$, only the two elongated conformations are present in solution, which makes it possible to describe the system with a two-state model. Using the determined populations for different temperatures, it was possible to determine the difference in Gibbs free energy (ΔG) for the transition from β-sheet to P$_{II}$ helix. Furthermore, this value was used to determine the temperature coefficients for the J-coupling constants, which were in very good agreement with the experimental values. The strict linear dependency of the coupling constants on temperature changes challenges previous reports proposing a transition point with increasing temperature [85, 86]. As suggested by Graf, it seems more likely that short peptides adopting the P$_{II}$ helical structure in solution show an exponential rise in β-sheet content with increasing temperatures, which is observed as a linear increase, for example, in the $^3J(H^NH^\alpha)$ coupling constants.

7.4.2. Interpretation of NOE Distance Information

Despite the above-mentioned difficulties in interpreting the absolute NOE intensities for flexible peptides, observable NOE contacts provide unique information. Unlike for example J couplings, NOE contacts will not simply average for different conformations that are present in solution, but different conformations will have different contributions, which will add to the observable NOE. This makes it possible to obtain unique insights into the cooperativity of conformational transitions.

Figure 7.12. H^N-H^N distances in an α-helix (left) and a β-sheet (right). The H^N protons are shown as black spheres.

If we assume a two-state model in which a stretch of four amino acids within a peptide can adopt only an α-helical or β-sheet conformation, there are eight combinations for the conformations of the first three residues: $\alpha\alpha\alpha$, $\alpha\alpha\beta$, $\alpha\beta\alpha$, $\beta\alpha\alpha$, $\beta\beta\alpha$, $\beta\alpha\beta$, $\alpha\beta\beta$, and $\beta\beta\beta$. If each residue populates α and β conformations equally (50%), the conformational averaging observed for three residues can be cooperative or non-cooperative. In the cooperative case, the conformations $\alpha\alpha\alpha$ and $\beta\beta\beta$ will be both populated by 50%. In the non-cooperative case, all eight combinations will be populated by 12.5%. NMR parameters like J couplings, which depend only on one torsion angle, yield average coupling constants of 50% for α-helix and 50% for β-sheet in both cooperative and non-cooperative cases for all the amino acids. If one interprets NOE contacts of the amino acids i and $i + 3$, it is possible to distinguish between both cases, since the distance depends on the configuration of three amino acids (i, $i + 1$, and $i + 2$). The H^N-H^N distance for an α-helix is about 4.6 Å, while for a β-sheet it is about 10.4 Å (see Fig. 7.12).

For the cooperative case, where $\alpha\alpha\alpha$ and $\beta\beta\beta$ are being populated, only for the first combination is the proton distance close enough to contribute to the NOE cross-peak intensity, which leads to a peak that has 50% of the expected intensity, predicted for an α-helix.

In the non-cooperative case, $\alpha\alpha\alpha$ is populated only by 12.5%, and the only other conformation that contributes to the cross-peak intensity is $\beta\alpha\alpha$, which has a higher proton distance, leading to a lower contribution. In sum, the observed cross-peak intensity is only 17% of the intensity predicted for an α-helix [9].

7.4.3. Interpretation of Residual Dipolar Couplings

Residual dipolar couplings have proven to be a valuable tool for determining structure and dynamics in folded and more or less rigid molecules. However, describing unfolded peptides or proteins is more difficult.

If an unfolded peptide was totally flexible, the averaged shape would always be spherical, which would make it impossible to align the molecule through steric interactions. The initial assumption that an unfolded polymer would not show any RDCs was proven wrong [88]. This phenomenon was later on explained by the fact that the shape of a peptide chain of a certain length will not average to a sphere, since not every angle between two chain segments is possible. Mathematical formulae to predict RDCs were introduced by Annila and coworkers using a random flight-chain to describe the unfolded peptide chain [89]. Their approach used a 1-D random flight-chain, placed in between two barriers of arbitrary distance, to predict the RDCs for an unfolded peptide, which is dependent on the chain length, the position within the chain, and the barrier distance. The result showed that RDCs became smaller for segments close to the chain ends and were generally lower for longer chains, which is consistent with the idea that infinitely long chains will adopt a spherical shape.

The initial model has been further expanded to a more realistic 3-D random flight-chain by Obolensky et al. [90]. The RDCs are calculated from chain length, position along the chain, and the barrier distance. A closed-form analytical result for the calculation of RDCs in unfolded peptide chains is presented:

$$D_{PQ} = \frac{\mu_0 \gamma_P \gamma_Q}{4\pi R_{PQ}^3} \frac{3\cos^2 \alpha_{PQ} - 1}{2} \frac{-\frac{8}{15\sqrt{\pi}}\sqrt{\frac{3}{2(N_1+N_2)}} + \frac{1}{4\pi}\sqrt{\frac{1}{N_1 N_2}} + \frac{4}{35\sqrt{\pi}}\sqrt{\frac{3}{2(N_1+N_2)^3}} - \frac{N_1+N_2}{32\pi}\sqrt{\frac{1}{N_1^3 N_2^3}}}{2L - \frac{4}{\sqrt{\pi}}\sqrt{\frac{2(N_1+N_2)}{3}} - \frac{4}{3\sqrt{\pi}}\sqrt{\frac{3}{2(N_1+N_2)}} + \frac{1}{2\pi}\sqrt{\frac{1}{N_1 N_2}} + \frac{1}{5\sqrt{\pi}}\sqrt{\frac{3}{2(N_1+N_2)^3}} - \frac{N_1+N_2}{20\pi}\sqrt{\frac{1}{N_1^3 N_2^3}}},$$

(7.9)

where D_{PQ} is the RDC for nuclei P and Q, μ_0 is the vacuum permeability, $\gamma_{P/Q}$ are the gyromagnetic ratios, R_{PQ} is the distance between spins P and Q, α_{PQ} is the angle defined by the interspin vector and the segment main axis, L is the distance between barriers, and N_1 and N_2 are the numbers of segments preceding and following the segment of interest. The overall chain length then equals $N = N_1 + N_2 + 1$. Assuming that $L \gg \sqrt{N} \gg 1$, the equation was simplified, yielding

$$D_{PQ} = \frac{\mu_0 \gamma_P \gamma_Q}{4\pi R_{PQ}^3} \frac{3\cos^2 \alpha_{PQ} - 1}{2} \frac{1}{L}\left[\frac{4}{15\sqrt{\pi}}\sqrt{\frac{3}{2(N_1+N_2)}} - \frac{1}{8\pi}\sqrt{\frac{1}{N_1 N_2}}\right], \quad (7.10)$$

which directly indicates that shorter chains yield larger RDCs (because of $(N_1 + N_2)^{-1}$ being dependent only on the overall chain length and becoming smaller for larger values of N) and that segments closer to chain ends yield smaller RDCs (because $(N_1 N_2)^{-1}$ is largest for either N_1 or N_2 approaching zero). The predicted RDCs again show a bell-shaped distribution along the chain. The two approaches utilizing the random flight-chain model represent the completely unfolded state of a protein and are neglecting any amino acid-specific preferences or restrictions. However, Obolensky et al. demonstrated that their approach very well characterizes the fully denatured state of a protein in the presence of a denaturing agent such as urea.

An approach that took into account the properties of individual amino acids was published by a number of important contributions. Blackledge et al., for example, used a two-domain protein model, which contains one natively unfolded domain [46]. It was shown that the agreement between experimental data and predicted RDCs could be increased substantially when going from a random sampling of the conformational space to a database-derived conformational ensemble, using the loop regions of folded proteins extracted from 500 high resolution x-ray structures. In addition, small angle X-ray scattering (SAXS) data was used to confirm that this model also predicts the overall shape of the unfolded ensemble.

A similar study was recently carried out by Dames et al. for a series of short peptides [91]. Their work shows that the previously suggested coil model resembles the investigated peptides very well. However, these studies also observe a significant influence of the close-by termini in the case of short peptides, depending on the residue of interest.

In case of short peptides, we also propose a much more simplified approach which was used to validate the results for polyalanine peptides, published by Graf et al. From the homologues series of poly alanine peptides Ala_6 was chosen as a test system. The peptide was aligned using a 0.5 mM sample of uniformly $^{13}C,^{15}N$-labeled Ala_6 dissolved in 90% H_2O/10% D_2O pH = 2 containing 10% DIODPC/CHAPSO bicelles [92]. Using HSQC based experiments, 5 RDCs per residue could be obtained (see Fig. 7.13).

To compare the measured RDCs with the previously determined populations for α-helix, β-sheet, and PP_{II} helix, three structures were generated, representing an ideal α-helix, β-sheet, and PP_{II} helix, respectively. The RDCs for the three secondary-structure elements were predicted using the program PALES [93]. The RDCs were weighted using different populations minimizing the RMSD between calculated weighted averages and experimental values. The couplings fit with an overall RMSD of 0.2 Hz for populations of 82% PP_{II} helix and 18% β-sheet, which is in very good agreement with the values obtained by Graf et al. using a joint MD/J-coupling approach. This result, using only three different major conformations, may also be an indication of the cooperative sampling of the conformational space. In addition, the RDCs were used to confirm the MD simulations carried out previously. RDCs for 1200 structures, uniformly sampled from the MD simulation, were predicted using PALES, averaged, and compared with the experimental values (see Fig. 7.14).

Since RDCs are, in contrast to J couplings, free of empirical parameters, which can be an additional source of errors, this simplified approach may be an easy way to validate results obtained by complementary methods, such as MD simulations or J-coupling analysis.

7.5. CONCLUSIONS

In this contribution, we introduced the use of NMR spectroscopy to derive conformational dynamics of small random coil peptides. We put a special emphasis on possible systematic errors for the determination of the primary NMR parameters. In our mind, NMR spectroscopy has an unmatched capacity to derive precise information on the conformational averaging of every torsion in every amino acid in small

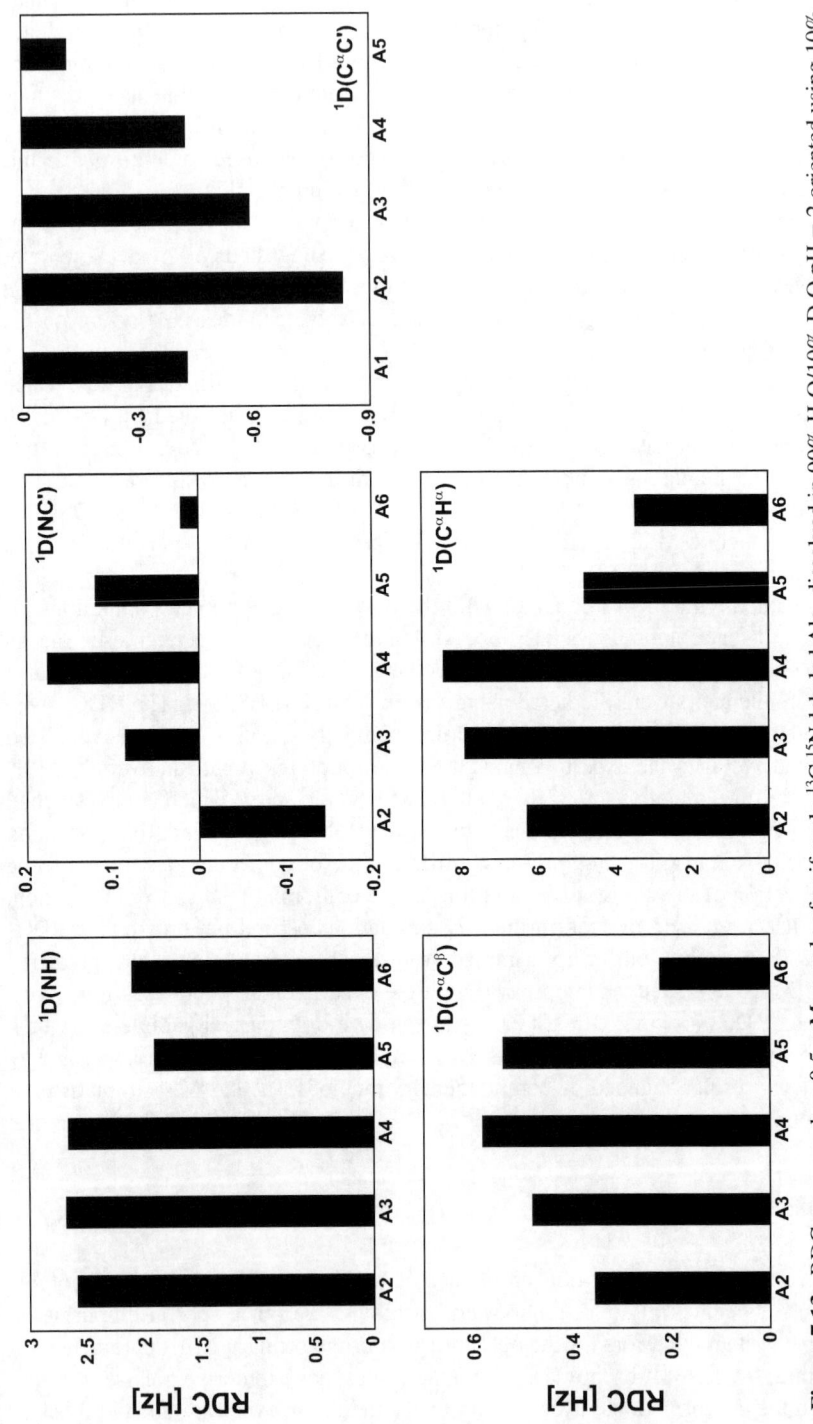

Figure 7.13. RDCs measured on 0.5 mM sample of uniformly ^{13}C, ^{15}N-labeled Ala$_6$ dissolved in 90% H$_2$O/10% D$_2$O pH = 2 oriented using 10% DIODPC/CHAPSO bicelles.

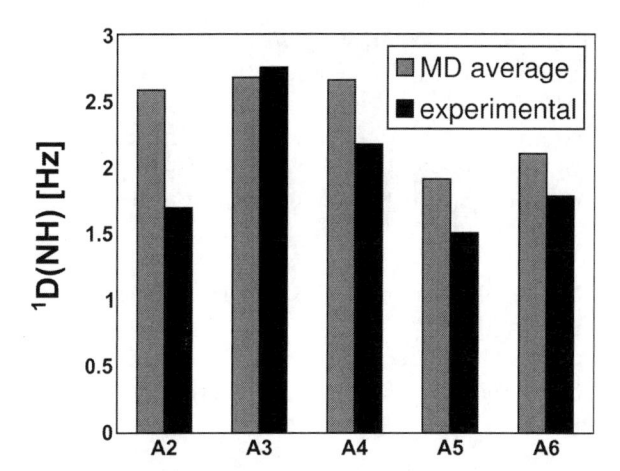

Figure 7.14. Comparison between experimental RDCs and averaged RDCs for 1200 structures uniformly sampled from the MD simulation carried out for Ala$_6$ using J couplings.

peptides as well as unfolded polypeptide chains, for example, in intrinsically unstructured proteins in a noninvasive, label-free approach under a large variety of different experimental conditions. The comparison of different experimental and theoretical results the topic of unfolded state characteristics remains a stimulating field of research with tremendous implications for our understanding of protein folding and misfolding.

ACKNOWLEDGMENTS

The authors wish to thank Reinhard Schweitzer-Stenner, Gerhard Stock, and Martin Blackledge for insightful discussions. The Center for Biomolecular Magnetic Resonance (BMRZ) is funded by the state of Hesse. H. S. is member of the Deutsche Forschungsgemeinschaft (DFG)-funded cluster of excellence: macromolecular complexes. D. M. was funded by a stipend from Degussa. We apologize for not being able to cite all important work in this research.

REFERENCES

1. Schweitzer-Stenner, R. (2009) *J Phys Chem B 113*, 2922.
2. Hagarman, A., Measey, T. J., Mathieu, D., Schwalbe, H., and Schweitzer-Stenner, R. (2010) *J Am Chem Soc 132*, 540.
3. Eker, F., Cao, X., Nafie, L., and Schweitzer-Stenner, R. (2002) *J Am Chem Soc 124*, 14330.
4. Tiffany, M. L. and Krimm, S. (1968) *Biopolymers 6*, 1379.
5. Woutersen, S. and Hamm, P. (2000) *J Phys Chem B 104*, 11316.
6. Mu, Y. G., Kosov, D. S., and Stock, G. (2003) *J Phys Chem B 107*, 5064.

7   Garcia, A. E. and Sanbonmatsu, K. Y. (2002) *Proc Natl Acad Sci U S A 99*, 2782.
8   Beachy, M. D., Chasman, D., Murphy, R. B., Halgren, T. A., and Friesner, R. A. (1997) *J Am Chem Soc 119*, 5908.
9   Wirmer, J., Schlörb, C., and Schwalbe, H. (2008) In: Buchner, J. and Kiefhaber, T. (eds.), *Protein folding handbook*, vol. 2. WILEY-VCH, Weinheim, p. 737.
10 Sickmeier, M., Hamilton, J. A., LeGall, T., Vacic, V., Cortese, M. S., Tantos, A., Szabo, B., Tompa, P., Chen, J., Uversky, V. N., Obradovic, Z., and Dunker, A. K. (2007) *Nucleic Acids Res 35*, D786.
11 Dosztanyi, Z., Chen, J., Dunker, A. K., Simon, I., and Tompa, P. (2006) *J Proteome Res 5*, 2985.
12 Gerum, C., Schlepckow, K., and Schwalbe, H. (2010) *J Mol Biol 401*, 7.
13 Kumar, J., Sreeramulu, S., Schmidt, T. L., Richter, C., Vonck, J., Heckel, A., Glaubitz, C., and Schwalbe, H. (2010) *Chembiochem 11*, 1208.
14 Chiti, F. and Dobson, C. M. (2009) *Nat Chem Biol 5*, 15.
15 Tartaglia, G. G., Pawar, A. P., Campioni, S., Dobson, C. M., Chiti, F., and Vendruscolo, M. (2008) *J Mol Biol 380*, 425.
16 Dobson, N., Dantas, G., Baker, D., and Varani, G. (2006) *Structure 14*, 847.
17 Dahiyat, B. I. and Mayo, S. L. (1997) *Proc Natl Acad Sci U S A 94*, 10172.
18 Go, A., Kim, S., Baum, J., and Hecht, M. H. (2008) *Protein Sci 17*, 821.
19 Wei, Y., Kim, S., Fela, D., Baum, J., and Hecht, M. H. (2003) *Proc Natl Acad Sci U S A 100*, 13270.
20 Brant, D. A. and Flory, P. J. (1965) *J Am Chem Soc 87*, 2788.
21 Brant, D. A. and Flory, P. J. (1965) *J Am Chem Soc 87*, 2791.
22 Serrano, L. (1995) *J Mol Biol 254*, 322.
23 Smith, L. J., Bolin, K. A., Schwalbe, H., MacArthur, M. W., Thornton, J. M., and Dobson, C. M. (1996) *J Mol Biol 255*, 494.
24 Peti, W., Smith, L. J., Redfield, C., and Schwalbe, H. (2001) *J Biomol NMR 19*, 153.
25 Meier, S., Blackledge, M., and Grzesiek, S. (2008) *J Chem Phys 128*, 052204.
26 Shen, Y., Delaglio, F., Cornilescu, G., and Bax, A. (2009) *J Biomol NMR 44*, 213.
27 Wishart, D. S. and Nip, A. M. (1998) *Biochem Cell Biol 76*, 153.
28 Wishart, D. S., Sykes, B. D., and Richards, F. M. (1991) *J Mol Biol 222*, 311.
29 Shen, Y., Vernon, R., Baker, D., and Bax, A. (2009) *J Biomol NMR 43*, 63.
30 Robustelli, P., Kohlhoff, K., Cavalli, A., and Vendruscolo, M. (2010) *Structure 18*, 923.
31 Karplus, M. (1959) *J Chem Phys 30*, 11.
32 Karplus, M. (1963) *J Am Chem Soc 85*, 2870.
33 Bystov, V.F. (1976) *Prog. NMR Spectros 10*, 41.
34 Schmidt, J. M., Blümel, M., Löhr, F., and Rüterjans, H. (1999) *J Biomol NMR 14*, 1.
35 Bystrov, V. F., Gavrilov, Y. D., Ivanov, V. T., and Ovchinnikov, Y. A. (1977) *Eur J Biochem 78*, 63.
36 Pardi, A., Billeter, M., and Wüthrich, K. (1984) *J Mol Biol 180*, 741.
37 Schwalbe, H., Carlomagno, T., Hennig, M., Junker, J., Reif, B., Richter, C., and Griesinger, C. (2001) *Methods Enzymol 338*, 35.
38 Reif, B., Diener, A., Hennig, M., Maurer, M., and Griesinger, C. (2000) *J Magn Reson 143*, 45.

REFERENCES

39 Gochin, M. and Roder, H. (1995) *Protein Sci 4*, 296.
40 Otting, G. (2010) *Annu Rev Biophys 39*, 387.
41 Gillespie, J. R. and Shortle, D. (1997) *J Mol Biol 268*, 170.
42 Gillespie, J. R. and Shortle, D. (1997) *J Mol Biol 268*, 158.
43 Boisbouvier, J., Gans, P., Blackledge, M., Brutscher, B., and Marion, D. (1999) *J Am Chem Soc 121*, 7700.
44 Tolman, J. R., Flanagan, J. M., Kennedy, M. A., and Prestegard, J. H. (1995) *Proc Natl Acad Sci U S A 92*, 9279.
45 Tjandra, N. and Bax, A. (1997) *Science 278*, 1111.
46 Bernado, P., Blanchard, L., Timmins, P., Marion, D., Ruigrok, R. W., and Blackledge, M. (2005) *Proc Natl Acad Sci U S A 102*, 17002.
47 Lee, L. K., Rance, M., Chazin, W. J., and Palmer, A. G., 3rd (1997) *J Biomol NMR 9*, 287.
48 Lipari, G. and Szabo, A. (1982) *J Am Chem Soc 104*, 4546.
49 Lipari, G. and Szabo, A. (1982) *J Am Chem Soc 104*, 4559.
50 Wu, D. H. and Johnson, C. S. (1995) *J Magn Reson A 116*, 270.
51 Chen, A. D., Wu, D. H., and Johnson, C. S. (1995) *J Phys Chem 99*, 828.
52 Linge, J. P., Habeck, M., Rieping, W., and Nilges, M. (2003) *Bioinformatics 19*, 315.
53 Schwieters, C. D., Kuszewski, J. J., Tjandra, N., and Clore, G. M. (2003) *J Magn Reson 160*, 65.
54 Guntert, P. (2004) *Methods Mol Biol 278*, 353.
55 Vendruscolo, M. and Dobson, C. M. (2006) *Science 313*, 1586.
56 Lakomek, N. A., Carlomagno, T., Becker, S., Griesinger, C., and Meiler, J. (2006) *J Biomol NMR 34*, 101.
57 Markwick, P. R. L., Bouvignies, G., Salmon, L., McCammon, J. A., Nilges, M., and Blackledge, M. (2009) *J Am Chem Soc 131*, 16968.
58 Merrifield, R. B. (1963) *J Am Chem Soc 85*, 2149.
59 Carpino, L. A. and Han, G. Y. (1972) *J Org Chem 37*, 3404.
60 Motáčkova, V., Nováček, J., Zawadzka-Kazimierczuk, A., Kazimierczuk, K., Žídek, L., Šanderová, H., Krásný, L., Kozmiński, W., and Sklenář, V. (2010) *J Biomol NMR 48*, 169.
61 Kazimierczuk, K., Zawadzka, A., Kozminski, W., and Zhukov, I. (2006) *J Biomol NMR 36*, 157.
62 Jaravine, V. A., Zhuravleva, A. V., Permi, P., Ibraghimov, I., and Orekhov, V. Y. (2008) *J Am Chem Soc 130*, 3927.
63 Griesinger, C., Sørensen, O. W., and Ernst, R. R. (1986) *J Chem Phys 85*, 6837.
64 Bax, A. and Freeman, R. (1981) *J Magn Reson 45*, 177.
65 Rexroth, A., Schmidt, P., Szalma, S., Sørensen, O. W., Schwalbe, H., and Griesinger, C. (1995) *J Cell Biochem 75*, 75.
66 Otting, G. (1997) *J Magn Reson 124*, 503.
67 Vögeli, B., Ying, J., Grishaev, A., and Bax, A. (2007) *J Am Chem Soc 129*, 9377.
68 Bax, A., Vuister, G. W., Grzesiek, S., Delaglio, F., Wang, A. C., Tschudin, R., and Zhu, G. (1994) *Methods Enzymol 239*, 79.

69 Billeter, M., Neri, D., Otting, G., Qian, Y. Q., and Wüthrich, K. (1992) *J Biomol NMR* 2, 257.
70 Szyperski, T., Güntert, P., Otting, G., and Wüthrich, K. (1992) *J Magn Reson 99*, 552.
71 Delaglio, F., Wu, Z., and Bax, A. (2001) *J Magn Reson 149*, 276.
72 Harbison, G. S. (1993) *J Am Chem Soc 115*, 3026.
73 Vuister, G. W. and Bax, A. (1993) *J Am Chem Soc 115*, 7772.
74 Sanders, C. R., Hare, B. J., Howard, K. P., and Prestegard, J. H. (1994) *Prog NMR Spectrosc 26*, 421.
75 Saupe, A. (1968) *Angew Chem Int Ed Engl 7*, 97.
76 Case, D. A., Scheurer, C., and Brüschweiler, R. (2000) *J Am Chem Soc 122*, 10390.
77 Markwick, P. R. L., Showalter, S. A., Bouvignies, G., Brüschweiler, R., and Blackledge, M. (2009) *J Biomol NMR 45*, 17.
78 Penkett, C. J., Redfield, C., Jones, J. A., Dodd, I., Hubbard, J., Smith, R. A., Smith, L. J., and Dobson, C. M. (1998) *Biochemistry 37*, 17054.
79 Schwalbe, H., Fiebig, K. M., Buck, M., Jones, J. A., Grimshaw, S. B., Spencer, A., Glaser, S. J., Smith, L. J., and Dobson, C. M. (1997) *Biochemistry 36*, 8977.
80 Smith, L. J., Fiebig, K. M., Schwalbe, H., and Dobson, C. M. (1996) *Fold Des 1*, R95.
81 Graf, J., Nguyen, P. H., Stock, G., and Schwalbe, H. (2007) *J Am Chem Soc 129*, 1179.
82 Best, R. B., Buchete, N. V., and Hummer, G. (2008) *Biophys J 95*, L07.
83 Verbaro, D., Ghosh, I., Nau, W. M., and Schweitzer-Stenner, R. (2010) *J Phys Chem B 114*, 17201.
84 Chen, K., Liu, Z., Zhou, C., Bracken, W. C., and Kallenbach, N. R. (2007) *Angew Chem Int Ed Engl 46*, 9036.
85 Chen, K., Liu, Z. G., Zhou, C. H., Shi, Z. S., and Kallenbach, N. R. (2005) *J Am Chem Soc 127*, 10146.
86 Chen, K., Liu, Z. G., and Kallenbach, N. R. (2004) *Proc Natl Acad Sci U S A 101*, 15352.
87 Shi, Z. S., Olson, C. A., Rose, G. D., Baldwin, R. L., and Kallenbach, N. R. (2002) *Proc Natl Acad Sci U S A 99*, 9190.
88 Shortle, D. and Ackerman, M. S. (2001) *Science 293*, 487.
89 Louhivuori, M., Pääkkönen, K., Fredriksson, K., Permi, P., Lounila, J., and Annila, A. (2003) *J Am Chem Soc 125*, 15647.
90 Obolensky, O. I., Schlepckow, K., Schwalbe, H., and Solov'yov, A. V. (2007) *J Biomol NMR 39*, 1.
91 Dames, S. A., Aregger, R., Vajpai, N., Bernado, P., Blackledge, M., and Grzesiek, S. (2006) *J Am Chem Soc 128*, 13508.
92 Cavagnero, S., Dyson, H. J., and Wright, P. E. (1999) *J Biomol NMR 13*, 387.
93 Zweckstetter, M. and Bax, A. (2000) *J Am Chem Soc 122*, 3791.
94 Wirmer, J. and Schwalbe, H. (2002) *J Biomol NMR 23*, 47.
95 Hu, J. S. and Bax, A. (1997) *J Am Chem Soc 119*, 6360.
96 Wang, A. C. and Bax, A. (1995) *J Am Chem Soc 117*, 1810.
97 Schmidt, J. M. (2007) *J Biomol NMR 37*, 287.
98 Hennig, M., Bermel, W., Schwalbe, H., and Griesinger, C. (2000) *J Am Chem Soc 122*, 6268.

REFERENCES

99 Zhang, O. and Forman-Kay, J. D. (1995) *Biochemistry 34*, 6784.
100 Farrow, N. A., Zhang, O., Forman-Kay, J. D., and Kay, L. E. (1995) *Biochemistry 34*, 868.
101 Farrow, N. A., Zhang, O. W., Szabo, A., Torchia, D. A., and Kay, L. E. (1995) *J Biomol NMR 6*, 153.
102 Farrow, N. A., Zhang, O., Forman-Kay, J. D., and Kay, L. E. (1997) *Biochemistry 36*, 2390.
103 Yang, D. W., Mok, Y. K., FormanKay, J. D., Farrow, N. A., and Kay, L. E. (1997) *J Mol Biol 272*, 790.
104 Zhang, O. and Forman-Kay, J. D. (1997) *Biochemistry 36*, 3959.
105 Zhang, O., Forman-Kay, J. D., Shortle, D., and Kay, L. E. (1997) *J Biomol NMR 9*, 181.
106 Mok, Y. K., Kay, C. M., Kay, L. E., and Forman-Kay, J. (1999) *J Mol Biol 289*, 619.
107 Kortemme, T., Kelly, M. J., Kay, L. E., Forman-Kay, J., and Serrano, L. (2000) *J Mol Biol 297*, 1217.
108 Choy, W. Y. and Forman-Kay, J. D. (2001) *J Mol Biol 308*, 1011.
109 Tollinger, M., Skrynnikov, N. R., Mulder, F. A., Forman-Kay, J. D., and Kay, L. E. (2001) *J Am Chem Soc 123*, 11341.
110 Mok, Y. K., Elisseeva, E. L., Davidson, A. R., and Forman-Kay, J. D. (2001) *J Mol Biol 307*, 913.
111 Demarco, A., Llinas, M., and Wüthrich, K. (1978) *Biopolymers 17*, 2727.
112 Perez, C., Lohr, F., Ruterjans, H., and Schmidt, J. M. (2001) *J Am Chem Soc 123*, 7081.
113 Graf, J. (2006) PhD thesis, Goethe University.

8

SECONDARY STRUCTURE AND DYNAMICS OF A FAMILY OF DISORDERED PROTEINS

Pranesh Narayanaswami and Gary W. Daughdrill

8.1. INTRODUCTION

The amino acid sequence of a protein contains the necessary information to specify its three-dimensional structure, and protein function is determined by this 3-D structure [1–4]. This close link between protein structure and function has played a crucial role in the various efforts to resolve the three-dimensional structures of numerous proteins in the past five decades. However, there are many proteins and protein domains that are entirely or partially disordered under physiological conditions and yet are still able to perform important biological functions [5–19]. Recent studies have indicated that intrinsically disordered proteins (IDPs) with greater than 50 amino acid residues are common and that IDPs form a diverse set of protein families that can perform their molecular functions using a variety of mechanisms. For instance, analysis of the protein sequence databases has revealed a prevalence of IDPs in human signaling and cancer-associated proteins [11]. Additionally, the prion protein responsible for the pathogenesis of mad cow disease contains an unstructured "evolutionarily active" N-terminal domain [13]. Characterizing the structure and dynamics of IDPs will enhance our understanding of protein function and evolution.

Protein and Peptide Folding, Misfolding, and Non-Folding, First Edition. Edited by Reinhard Schweitzer-Stenner.
© 2012 John Wiley & Sons, Inc. Published 2012 by John Wiley & Sons, Inc.

In accordance with Uversky [6], IDPs can be classified into two structurally distinct subgroups—intrinsic coils and pre-molten globules. The set of proteins that comprise the intrinsic coils group possess very little to no ordered secondary structure and have hydrodynamic dimensions typical of random coils in poor solvent. The second set of proteins that comprise the pre-molten globules group exhibit some amount of residual secondary structure, are more compact than the intrinsic coil group, but are still less compact than native or molten globule proteins [6].

IDPs lack compact globular structure and form a rapidly interconverting ensemble of structures that are resistant to crystallization as well as structure determination using traditional Nuclear Overhauser Effect (NOE)-based nuclear magnetic resonance (NMR) approaches [7, 20]. The rapidly interconverting ensemble of structures of IDPs is characterized by differing backbone torsion angles and in some cases resembles the denatured state of ordered proteins [7, 14, 21]. The current lack of reliable atomic models for IDP structures complicates the development of relationships between IDP structure and function. In order to expand the current model of protein structure and function to include structure–function relationships for IDPs, it is important to first determine if IDPs within the same functional families adopt similar ensembles of structures whose properties are collectively consistent with experimental measurements. The absence of long-range interactions between amino acid residues in IDPs allows for greater sequence variation compared with ordered proteins, which means that sequence alignments may not correctly identify evolutionary relationships for IDPs [10, 22–24]. We propose that identifying the presence of conserved ensemble-average structures for IDPs in the same functional families could provide the necessary metrics to define evolutionary relationships for functionally similar yet sequentially dissimilar proteins. NMR relaxation and chemical shift measurements serve as excellent tools for defining these metrics and studying the dynamic structures of IDPs [6, 7, 25, 26]. NMR relaxation measurements can be used to specify intramolecular distances and the time scales of molecular motion, while chemical shift measurements can be utilized to estimate the average backbone structure of the IDP's dynamic ensemble.

We have developed an IDP model from a specific functional family. This IDP model forms an intrinsically unstructured linker domain (IULD) in the 70-kDa subunit of Replication Protein A (RPA70). Replication Protein A is a heterotrimeric single-stranded DNA binding protein. It is composed of three subunits: 70 kDa, 32 kDa, and 14 kDa. RPA70 consists of five domains (Fig. 8.1). They are, in order from N- to C-terminus, as follows: a protein interaction and weak ssDNA binding domain (DBD F), the IULD, two high-affinity ssDNA binding domains (DBD A and DBD B), and a damaged DNA recognition and binding domain (DBD C) [27–33].

The IULD separates DBD F from DBD A and DBD B, which are structured domains that form five stranded anti-parallel beta-barrels. The IULD is typically in the range of 65 to 85 amino acids in length and may function to enhance the weak ssDNA binding affinity of DBD F, facilitating the competitive inhibition of DBD F–protein interactions [34–36]. The IULD accomplishes this by tethering DBD F to DBD A and DBD B. This tethering provides mobility for DBD F to interact with

RPA 70

protein/protein interactions DBD F	flexible linker IULD	ssDNA-binding domain1 DBD A	ssDNA-binding domain2 DBD B	Damaged DNA binding and dimerization domain DBD C
aa# 100	200	300	400	500 600

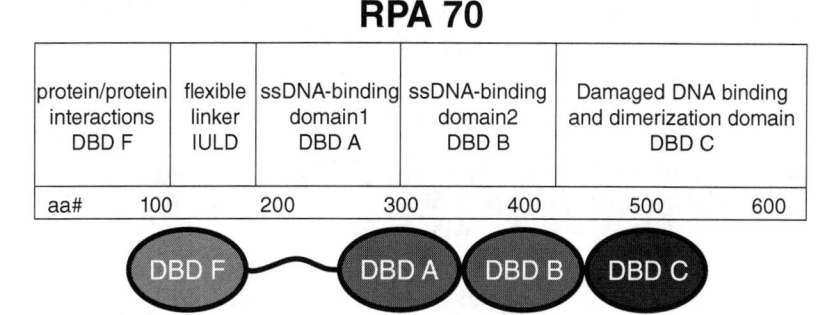

Figure 8.1. Schematic and cartoon showing the domain structure of RPA70.

TABLE 8.1. RPA70 Linker Residue Region, Linker Length, and Nomenclature

RPA70 Homologue	IULD Residue Region	Linker Length (Number of Amino Acids)	Abbreviation Used throughout This Chapter
Homo sapiens	105–180	76	hsl
Rattus norvegicus	105–180	76	rnl
Oryza sativa	105–180	76	osl
Arabidopsis thaliana	106–193	88	atl
Neurospora crassa	105–167	63	ncl

other proteins, the other subunits of the Replication Protein A heterotrimer, or DNA, as RPA70 functions during replication, recombination, and repair [10, 25, 34, 35].

In order to better understand IDP structure and function, we have investigated the secondary structure and dynamics of the RPA70 IULD from six species spanning three kingdoms: two animals, two plants, and two fungi. For this study we chose IULDs from *Homo sapiens*, *Rattus norvegicus*, *Oryza sativa*, *Arabidopsis thaliana*, *Neurospora crassa*, and *Saccharomyces cerevisiae*. The *S. cerevisiae* IULD formed a gel at concentrations of ~0.3 mM and was not suitable for analysis by NMR spectroscopy. The other five IULDs were suitable for analysis by NMR spectroscopy. Our findings indicate that the RPA70 IULDs form random coil-like structures and conserve their dynamic behavior in the absence of significant sequence similarity.

8.2. MATERIALS AND METHODS

8.2.1. Selection of the RPA70 IULDs

Multiple sequence alignments were used to identify the IULDs from different species. Because the IULD sequences are highly divergent, it was necessary to initially perform alignments on the flanking folded domains (see Fig. 8.1; DBD F and DBD A). Based on this analysis it was determined that the IULDs for the different RPA70 homologues were of varying length. Table 8.1 shows the RPA70 residues that correspond to the IULD, the length of the IULD, and the abbreviation used for the different IULDs throughout this chapter.

8.2.2. Expression and Purification of the IULDs

Stable transformants of *Escherichia coli* BL21:DE3 harboring the pET15b plasmid containing the cDNA for the various IULDs were inoculated into two flasks containing 50 mL of M9 minimal media with 50 mg/L ampicillin to maintain plasmid selection and 1 g/L $^{15}NH_4Cl$ as a nitrogen source. These cultures were grown overnight and reinoculated into 1 L of the same M9 minimal media to an OD_{600} of 0.02–0.04. The 1 L cultures were then grown to an OD_{600} of 0.6–0.8. Next, 250 mg of isopropyl β-D-1-thiogalactopyranoside (IPTG) was added per liter of media to induce protein expression, and the cultures were grown for 3 more hours.

Following growth and protein expression the cells were harvested by centrifugation at 6164 × g for 15 minutes. The cell pellets were resuspended in 30 mL of a buffer containing 50 mM NaH_2PO_4 at pH 8.0, 300 mM NaCl, and 10 mM imidazole (B1). Since IDPs are extremely sensitive to protease degradation [5], 1 mL of a protease inhibitor cocktail containing 2 mM 4-(2-aminoethyl)benzenesulfonyl fluoride (AEBSF), 14 μM E-64, 130 μM bestatin, 1 μM leupeptin, 0.3 μM aprotinin, and 1 mM sodium EDTA was added to the cell suspension. The cell suspension was then lysed by two passes through a french press. Next, the lysate was cleared by centrifugation for 1 hour at 25,175 × g. The supernatant from the lysate was decanted and loaded on a 1.6 × 20-cm column containing Ni-NTA agarose equilibrated with 5 column volumes of B1.

After the lysate was loaded, the column was washed with B1 to remove unbound proteins. The column was then washed with 3.5 column volumes of B1 plus 40 mM imidazole to remove weakly bound proteins. The linker was eluted with 5 column volumes of B1 plus 240 mM imidazole. The fractions containing protein were detected by ultraviolet (UV) absorbance at 280 nm, and the fractions containing the linkers were identified by polyacrylamide gel electrophoresis. Fractions containing the linkers were dialyzed into buffer B2 containing 50 mM NaH_2PO_4 at pH 7.0, 300 mM NaCl, 1 mM EDTA, and 0.02% azide. The linkers were concentrated to a volume of 2.0 mL using a centrifugal concentrator, and loaded onto a 1.6 × 60-cm size-exclusion column (bead size 24–44 μm) equilibrated with B2. The sample was eluted with 1 column volume of B2, and the fractions were analyzed by gel electrophoresis to identify those containing the linkers. Fractions containing the linker were pooled and dialyzed two times against 1 L of a buffer containing 50 mM NaH_2PO_4 at pH 6.5, 50 mM NaCl, 1 mM EDTA, and 0.02% azide (B3). This protein sample was concentrated to 0.5–1 mM using a YM-3 centriprep, after which the sample was ready for NMR data collection.

8.2.3. NMR Data Collection and Analysis

All NMR experiments were performed at a ^1H resonance frequency of 600 MHz and a sample temperature of 298 K. The amide ^1H, ^{15}N, C_α, and C_β resonance assignments of hsl were described in a previous study and utilized in the current experiments [34]. Resonance assignments were made for rnl, osl, atl, and ncl using sensitivity-enhanced HNCACB experiments performed on uniformly ^{15}N- and ^{13}C-labeled samples of rnl, osl, atl, and ncl in 90% H$_2$O/10% D$_2$O using B3 for rnl and atl, and B3 plus 1 mM dithiothreitol (DTT) for osl and ncl [37]. For the HNCACB experiment, data in the ^{1}H dimension was acquired using a sweep width of 8000 Hz and 512 complex t3 points; data in the ^{13}C dimension was acquired using a sweep width of 12,065.5 Hz and 128 complex t1 points; and data in the ^{15}N dimension was acquired using a sweep width of 2200 Hz and 32 complex t2 points. Following transformation, analysis of the HNCACB data resulted in 65 ^{1}H$_N$, ^{15}N, ^{13}C$_\alpha$, and ^{13}C$_\beta$ resonance assignments for rnl; 64 ^{1}H$_N$, ^{15}N, ^{13}C$_\alpha$, and ^{13}C$_\beta$ resonance assignments for osl; 53 ^{1}H$_N$, ^{15}N, ^{13}C$_\alpha$, and C$_\beta$ resonance assignments for ncl; and 69 ^{1}H$_N$, ^{15}N, ^{13}C$_\alpha$, and ^{13}C$_\beta$ resonance assignments for atl.

After obtaining the chemical shifts for the protein of interest, the next step is to obtain the chemical shift difference. The chemical shift difference, also known as the secondary chemical shift ($\Delta\delta$), is the difference of the observed chemical shift for an amino acid and a random coil standard value for the amino acid. When the alpha-carbon and beta-carbon chemical shifts are obtained, they can be plotted on the y-axis with the amino acids on the x-axis. Deviations from zero yield clues to the structure in that region. The random coil chemical shifts employed in this study were determined for short peptides in 8 M urea at pH 2.3 and were adjusted to account for sequence-specific effects [38, 39]. The adjusted random coil chemical shifts were then subtracted from the chemical shifts for hsl, rnl, osl, atl, and ncl to yield the $\Delta\delta$. All NMR data were processed and analyzed using the Felix software from the Accelerys Corporation (Cambridge, MA, USA). The Felix software is a program for offline data processing, spectral visualization, and analysis of high-resolution, one- to four-dimensional, homonuclear and heteronuclear NMR data. Apodization was achieved in ^{1}H, ^{13}C, and ^{15}N dimensions using a squared sine bell function shifted by 90°. Apodization was followed by zero filling to twice the number of real data points, and mirror image linear prediction was used in processing the ^{15}N dimension of the HNCACB. The referencing method used in this report is based on International Union of Pure and Applied Chemists (IUPAC) recommendations for using 2,2-Dimethyl-2-silapentane-5-sulfonic acid (DSS) in a highly polar solvent such as water [40].

8.2.4. Relaxation Data Collection and Analysis

The spin-lattice relaxation rates (R_1), spin-spin relaxation rates (R_2), and rotating frame relaxation rates ($R_{1\rho}$) were measured by inverse-detected two-dimensional NMR methods [41–43]. Spin–lattice relaxation rates were determined by collecting 10 two-dimensional spectra using relaxation delays of 10, 50, 110, 190, 310, 500, 650, 1000, 1500, and 1900 ms. The spin–spin relaxation rates and rotating-frame relaxation rates were each determined by collecting 10 two-dimensional spectra using relaxation delays of 10, 30, 50, 90, 110, 150, 190, 210, 230, and 250 ms. A 65° off resonance spin-lock pulse was used for the $R_{1\rho}$ experiments.

Spectra for the five IULD homologues were acquired on 0.5–1.0 mM uniformly ^{15}N-labeled samples with the osl and ncl B3 also containing DTT. Peak heights from each series of relaxation experiments were fitted to a single decaying exponential function. Peak heights uncertainties were estimated from baseline noise level and were typically less than 1% of the peak heights from the first R_1, R_2, and $R_{1\rho}$ delay points. In general, errors were within 5% of the measured relaxation rates.

8.3. RESULTS AND DISCUSSION

8.3.1. Sequence Conservation of RPA70 IULDs

Understanding the evolutionary relationships among proteins helps with understanding the relationship between their structure and function. Multiple sequence alignment and compositional analysis can be used to investigate these evolutionary relationships. Figure 8.2a shows the sequence alignment for the five RPA70 IULDs used in this study and illustrates that the linkers are highly divergent between major taxonomic groups. However, there is a reasonable level of similarity within the groups. For instance, there is limited sequence conservation across the five RPA70 IULD homologues we studied. However, there is considerable sequence conservation between the hsl and rnl homologues (43% identity). We previously showed that the RPA70 IULD homologues contain an abundance of highly divergent segments [10]. In general, it appears that IDPs have a higher rate of evolution in comparison with compact globular proteins, and genetic distance measurements of the RPA70

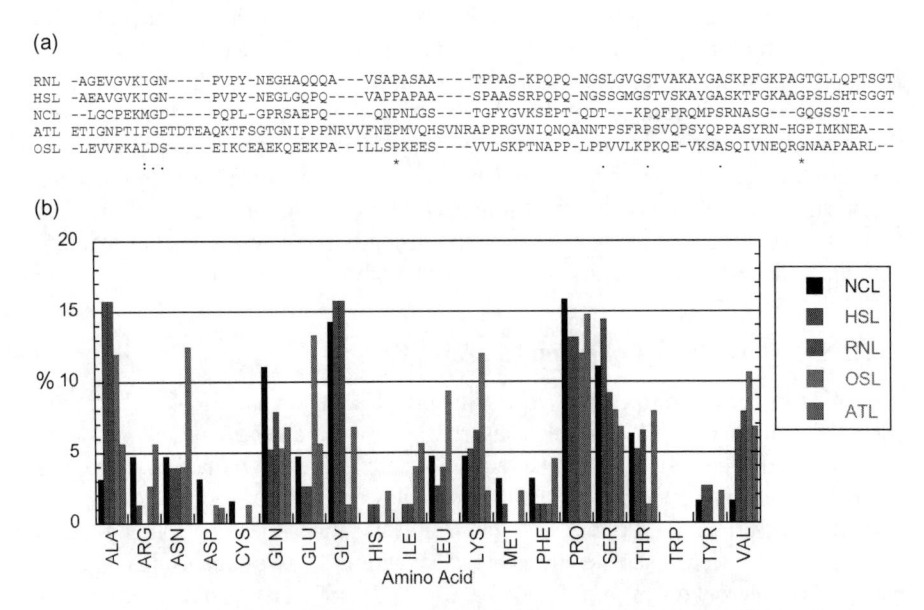

(a)
```
RNL -AGEVGVKIGN-----PVPY-NEGHAQQQA---VSAPASAA----TPPAS-KPQPQ-NGSLGVGSTVAKAYGASKPFGKPAGTGLLQPTSGT
HSL -AEAVGVKIGN-----PVPY-NEGLGQPQ----VAPPAPAA----SPAASSRPQPQ-NGSSGMGSTVSKAYGASKTFGKAAGPSLSHTSGGT
NCL --LGCPEKMGD-----PQPL-GPRSAEPQ-----QNPNLGS----TGFYGVKSEPT-QDT---KPQFPRQMPSRNASG---GQGSST-----
ATL ETIGNPTIFGETDTEAQKTFSGTGNIPPPNRVVFNEPMVQHSVNRAPPRGVNIQNQANNTPSFRPSVQPSYQPPASYRN-HGPIMKNEA---
OSL -LEVVFKALDS-----EIKCEAEKQEEKPA--ILLSPKEES----VVLSKPTNAPP-LPPVVLKPKQE-VKSASQIVNEQRGNAAPAARL--
        :..                                *                                .   .    .       *
```

(b)

Figure 8.2. Amino acid sequence alignment of the five IULD homologues. See color insert.

homologues suggest that the IULD has evolved at a rate that is 1.5 to 5 times faster than the compact globular domains of RPA70 [22, 24].

Figure 8.2b shows the percent composition for amino acid types in the RPA70 IULD homologues. In general, the values are consistent with those expected for an IDP [44–46]. One notable exception is the relatively low frequency of charged residues for all five IULDs. Mean net charge was initially recognized as an important feature of IDPs and is expected to prevent the chain from collapsing [6, 47, 48]. Many IDPs have a high mean net charge with values greater than 0.1 being common, but the IULDs have a low mean net charge with values ranging from 0.004 for osl to 0.0375 for hsl and rnl. One interesting difference in sequence composition between the IULD homologues is the increased frequency of glycine residues in ncl, hsl, and rnl compared with osl and atl. We previously showed that this high frequency of glycine is maintained in other mammalian IULDs and helps to explain the dynamic differences observed between the mammalian and plant IULDs [10]. To further investigate the differences between the IULD homologues suggested by an analysis of sequence similarity and amino acid composition, NMR spectroscopy was used to determine their secondary structure and backbone dynamics.

8.3.2. Secondary Chemical Shifts of the RPA70 Linkers Suggest an Absence of Compact Globular Structure

The procedure for determining the secondary chemical shifts ($\Delta\delta$) is described in Section 8.2. Analysis of $\Delta\delta$ provides a rapid and reliable assessment of protein secondary structure [49–53]. The magnitude of $\Delta\delta$ in ordered proteins is higher than that of IDPs and the rapid conformational averaging observed for IDPs only permits the estimation of the relative populations of α-helix or β-strand and does not represent the presence of stable secondary structure [7]. Analysis of $\Delta\delta$ for IDPs also appears to be complicated by the effect of dynamics on the value of the chemical shift [54]. This statement is speculative and the putative phenomenon has not been rigorously investigated. For ordered proteins, $\Delta\delta$ values for $^{13}C^\alpha$ that are less than –0.5 ppm indicate the presence of a β-strand structure, and a $\Delta\delta$ value for $^{13}C^\alpha$ nuclei that is greater than 0.8 ppm indicates an α-helical structure [49, 51, 53–55]. For folded proteins, $\Delta\delta$ values for $^{13}C^\beta$ that are positive indicate the presence of a β-strand structure; $\Delta\delta$ value for $^{13}C^\beta$ nuclei that are near zero indicates the presence of an α-helical structure, and negative $\Delta\delta$ values for $^{13}C^\beta$ indicate the presence of structures with positive φ and ψ angles. Similar trends are expected for IDPs, but as mentioned above the magnitude of $\Delta\delta$ is expected to be lower for IDPs when compared with ordered proteins.

Figure 8.3a,b respectively shows the $^{13}C^\alpha$ $\Delta\delta$ values for atl and osl. Figure 8.3c,d respectively shows the $^{13}C^\beta$ $\Delta\delta$ values for atl and osl. Based on the $^{13}C^\alpha$ $\Delta\delta$ values, a transient helical segment is observed near the N-terminus of atl and the C-terminus of osl. Most of the other residues have $\Delta\delta$ values consistent with the presence of extended structures. Overall the deviations from zero ppm are greater for osl than for atl. The meaning of simultaneous negative $\Delta\delta$ values for both $^{13}C^\alpha$ and $^{13}C^\beta$ is unclear but may be related to some uncharacterized contribution of dynamics on the

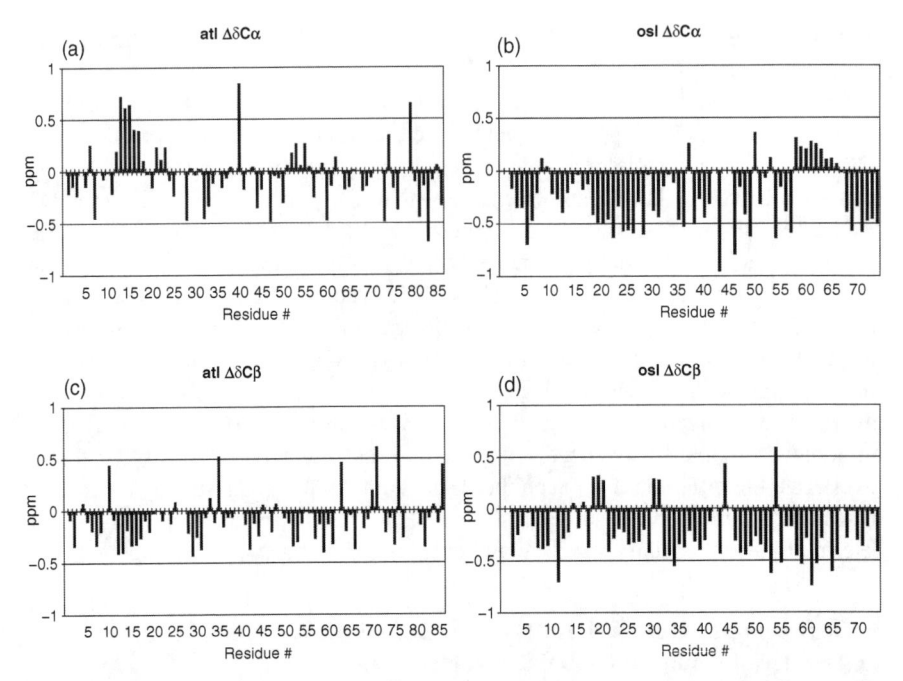

Figure 8.3. Plots showing secondary chemical shifts (Δδ) for atl and osl. Chemical shift differences in parts per million (ppm) are plotted on the vertical axis and residue number is plotted on the horizontal axis. (a) $^{13}C_\alpha$ secondary chemical shifts for atl; (b) $^{13}C_\alpha$ secondary chemical shifts for osl; (c) $^{13}C_\beta$ secondary chemical shifts for atl; and (d) $^{13}C_\beta$ secondary chemical shifts for osl.

value of the chemical shifts. Figure 8.4a,b respectively shows the $^{13}C^\alpha$ Δδ values for hsl and rnl. Figure 8.4c,d respectively shows the $^{13}C^\beta$ Δδ values for hsl and rnl. Similar patterns in the Δδ values are observed for both the $^{13}C^\alpha$ and $^{13}C^\beta$ nuclei of hsl and rnl. This is expected because they share 43% sequence identity. No transient helical segments are observed for either hsl or rnl, and the pattern of Δδ values observed for both mammalian linkers is consistent with the presence of mostly extended structures. Figure 8.5a,b respectively shows the $^{13}C^\alpha$ and $^{13}C^\beta$ Δδ values for ncl. The pattern observed for the ncl Δδ values also suggests the presence of mostly extended structures. In general, the small magnitude of the Δδ values observed for all the IULDs demonstrate a lack of stable secondary structure.

8.3.3. Backbone Dynamics of the RPA70 Linkers Suggest the Presence of Segmental Motion

NMR studies have confirmed that IDPs possess nonuniform structural properties that are not consistent with the existence of a narrow conformational ensemble of low-energy structures. Instead, IDPs contain varied amounts of residual structure ranging from no secondary and tertiary structure to the presence of dynamic structure

RESULTS AND DISCUSSION

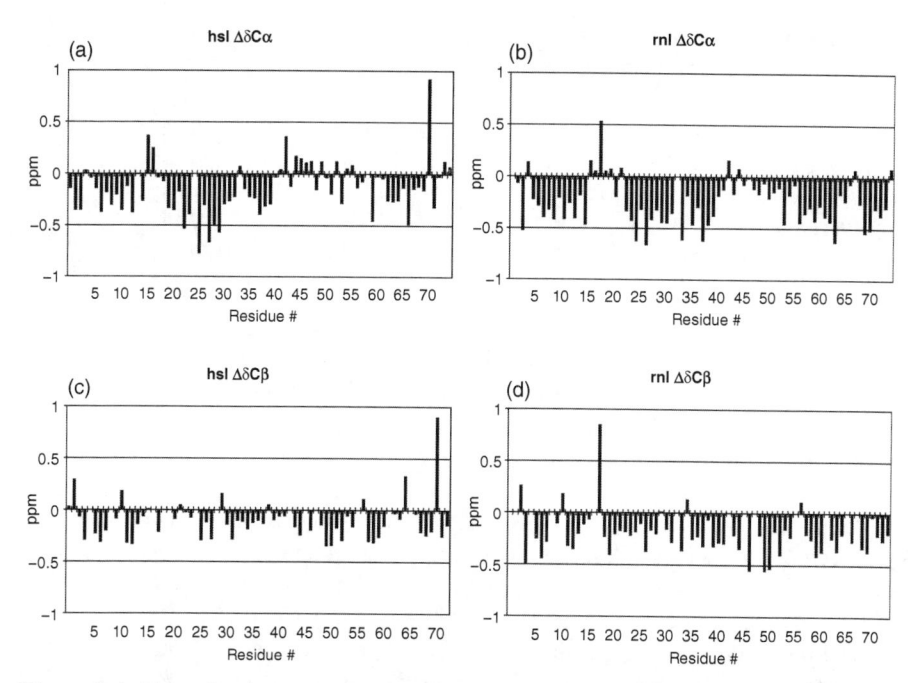

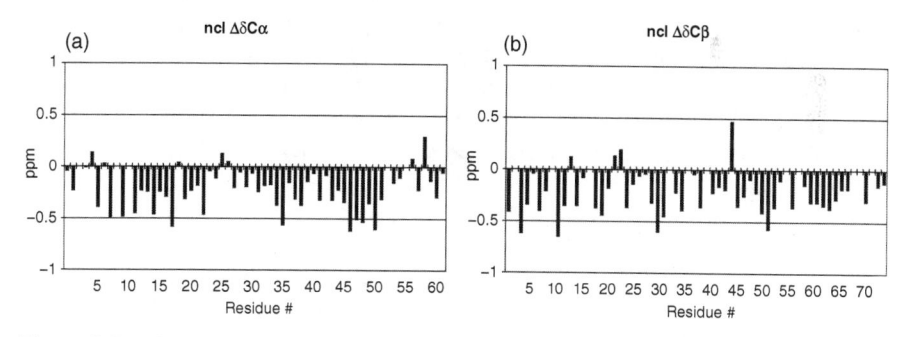

Figure 8.4. Plots showing secondary chemical shifts ($\Delta\delta$) for hsl and rnl. Chemical shift differences in parts per million (ppm) are plotted on the vertical axis and residue number is plotted on the horizontal axis. (a) $^{13}C_\alpha$ secondary chemical shifts for hsl. (b) $^{13}C_\alpha$ secondary chemical shifts for rnl. (c) $^{13}C_\beta$ secondary chemical shifts for hsl. (d) $^{13}C_\beta$ secondary chemical shifts for rnl.

Figure 8.5. Plots showing secondary chemical shifts ($\Delta\delta$) for ncl. Chemical shift differences in parts per million (ppm) are plotted on the vertical axis and residue number is plotted on the horizontal axis. (a) $^{13}C_\alpha$ secondary chemical shifts for ncl. (b) $^{13}C_\beta$ secondary chemical shifts for ncl.

TABLE 8.2. Mean Values for R_1, R_2, and $R_{1\rho}$

IULD Homologue	R_1 (s^{-1})	R_2 (s^{-1})	$R_{1\rho}$ (s^{-1})	R_1/R_2
O. sativa	2.41 (±0.17)	4.29 (±0.14)	4.58 (±0.16)	0.562
A. thaliana	2.42 (±0.11)	3.68 (±0.11)	4.09 (±0.08)	0.657
R. norvegicus	1.69 (±0.09)	2.58 (±0.18)	3.26 (±0.1)	0.655
H. sapiens	1.67 (±0.08)	2.84 (±0.12)	3.25 (±0.14)	0.588
N. crassa	1.91 (±0.11)	3.18 (±0.11)	3.61 (±0.1)	0.601

favoring α-helical or β conformations [7, 21, 56]. NMR relaxation measurements provide a powerful tool for analyzing this variable structure because they report on the time scales of rotational diffusion and provide a very sensitive way to discriminate between folded, unfolded, and partially folded regions in a protein [7, 34, 36, 56, 57]. In particular, the spin-lattice (R_1) and spin-spin (R_2) relaxation rates vary systematically with the rotational diffusion time, which is on the nanosecond to picosecond time scale [43]. In addition, R_2 is sensitive to conformational exchange on the microsecond–millisecond timescale, and this contribution can be assessed directly by measuring the rotating-frame relaxation rate ($R_{1\rho}$) [43, 58].

The procedure for determining R_1, R_2, and $R_{1\rho}$ for the five IULD homologues is presented in the Section 8.2. Table 8.2 shows the average values R_1, R_2, and $R_{1\rho}$ for the five IULD homologues. Average values for R_1 range from 1.67 for hsl to 2.42 for atl; average values for R_2 range from 2.58 for rnl to 4.29 for osl; and average values for $R_{1\rho}$ range from 3.25 for hsl to 4.58 for osl. Values for $R_{1\rho}$ should be systematically higher than R_2 due to the 70° off-resonance spin-lock pulse that is used to suppress chemical exchange. Values of R_2 greater than $R_{1\rho}$ indicate the presence of conformational exchange process on the microsecond–millisecond time scale. Specific residues that exhibit this behavior are discussed below.

Figure 8.6 compares the R_1 and R_2 values for the two plant linkers. The R_1 values for osl and atl are plotted in Figure 8.6a and the R_2 values are plotted in Figure 8.6b. The dynamics of both IULDs is complex. Oscillations in the R_1 and R_2 values suggest the presence of segmental motion. Osl has a region near the N-terminus that has higher than average R_2 values. In fact, the R_2 values for several of these residues are larger than the corresponding $R_{1\rho}$ values, indicating the presence of conformation exchange on the microsecond–millisecond timescale (this includes residues 2–9, 12, 15, and 16. Atl does not exhibit this behavior in this region. Osl also shows two sharp dips in the R_1 values around residue 25 and 50. Inspection of the sequence near residue 25 shows a hydrophobic cluster flanked by two proline residues (PAILLSP) and a PP motif near residue 50. The dip in R_1 near residue 25 corresponds to a peak in the R_2 values, which suggest that the rotational correlation time for this segment has passed the inflection point for R_1. If one assumes isotropic tumbling and an order parameter (S^2) of 1, this would correspond to a rotational correlation time of ~5 ns. Of course these assumptions are not valid and it is necessary to recognize that any estimates of the correlation time are qualitative and mostly useful for comparison with other IULDs.

RESULTS AND DISCUSSION

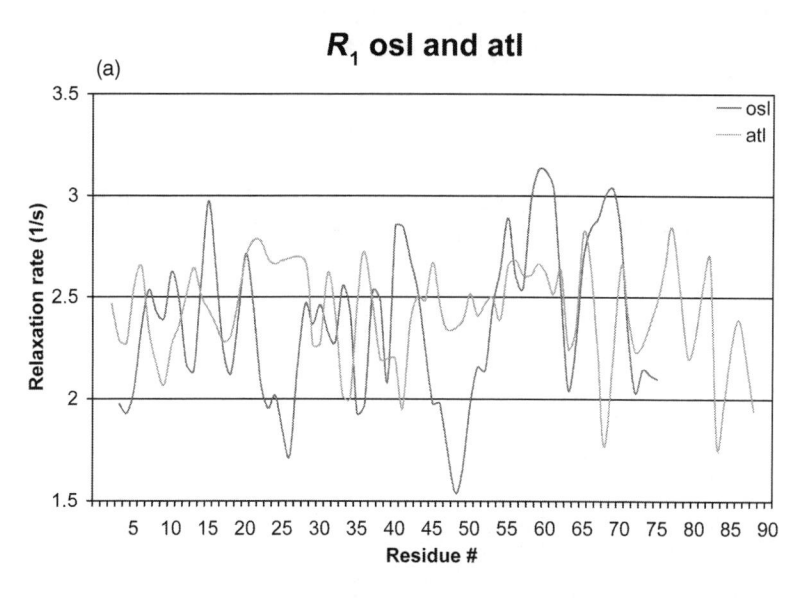

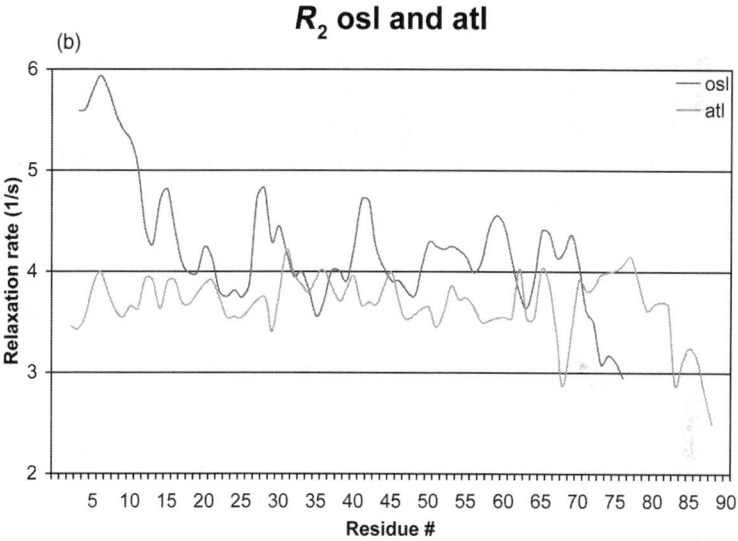

Figure 8.6. Plots showing (a) spin-lattice (R_1) and (b) spin-spin (R_2) relaxation rates for rnl (blue) and hsl (red). The relaxation rates are plotted on the vertical axis and residue number is plotted on the horizontal axis. Relaxation rates were averaged over a 2 residue window. See color insert.

Figure 8.7 compares the R_1 and R_2 values for the two mammalian linkers. The R_1 values for rnl and hsl are plotted in Figure 8.7a and the R_2 values are plotted in Figure 8.7b. The dynamics of the mammalian IULDs is also complex, with oscillations in the R_1 and R_2 values suggesting the presence of segmental motion. For both hsl and rnl, no evidence of conformational exchange on the microsecond–millisecond

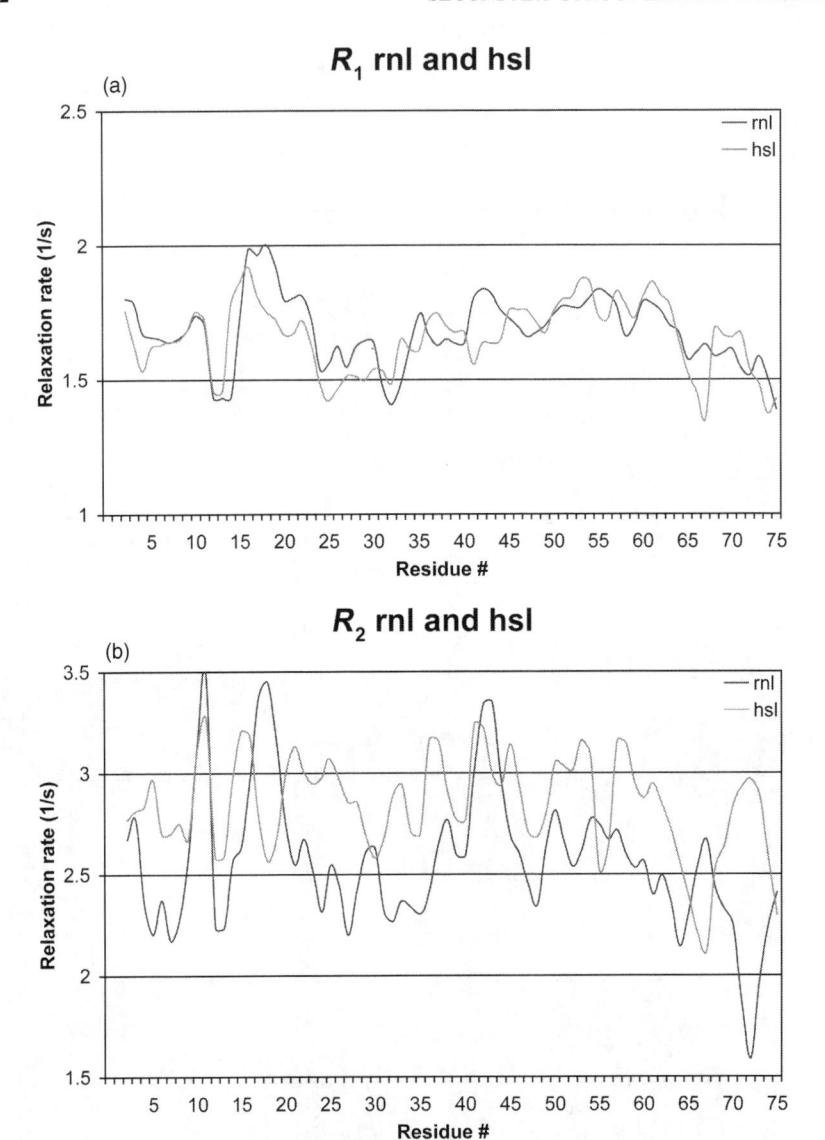

Figure 8.7. Plots showing (a) spin-lattice (R_1) and (b) spin-spin (R_2) relaxation rates for rnl (violet) and hsl (orange). The relaxation rates are plotted on the vertical axis and residue number is plotted on the horizontal axis. Relaxation rates were averaged over a two-residue window. See color insert.

timescale was observed. As expected, the R_1 values for the two mammalian homologues are similar. This is consistent with their relatively high sequence identity (43%). In contrast, there are significant differences in the R_2 values. These differences are potentially interesting but difficult to explain given the sequence identity between the two IULDs. For instance, the differences between the R_2 values for

RESULTS AND DISCUSSION

residues 5–10 do not correspond to any sequence differences. However, some of the differences can be interpreted based on sequence differences. For instance, there is a peak in R_2 at position 18 of rnl and a corresponding dip in R_2 for hsl. This residue is a histidine in rnl and a leucine in hsl. NMR data for all of the IULDs was collected at pH 6.5. At this pH, the titratable group on the histidine side chain would be partially ionized. This partial ionization could result in conformational exchange on the microsecond–millisecond timescale, which would explain the higher than expected R_2 values in this region. However, comparison of R_2 and $R_{1\rho}$ values for this region showed no evidence of conformational exchange on the microsecond–millisecond time scale. The peak in the R_2 values for hsl near the C-terminus also corresponds to a histidine residue.

Figure 8.8 shows the R_1 and R_2 values for the one fungal linker. The R_1 values for ncl are plotted in Figure 8.8a and the R_2 values are plotted in Figure 8.8b. Unfortunately, the solubility problems with the *S. cerevisiae* linker did not allow a comparison of the backbone dynamics within this taxonomic group. The larger than average R_1 and R_2 values at the N-terminus of ncl are near the only cysteine residue in this linker, and it is located adjacent to a proline. A comparison of the R_2 and $R_{1\rho}$ values in this region indicates the presence of conformational exchange on the microsecond–millisecond time scale. DTT was added to this sample to reduce the cysteine residue so the conformational exchange should not be due to disulfide bond formation.

8.3.4. Conclusions

There is a reasonable correlation between the relaxation rates R_1, R_2, $R_{1\rho}$, and the segregated evolutionary history of the IULD homologues. The two mammalian homologues have very similar relaxation rates and the two plant homologues have relatively similar relaxation rates. Similar relaxation rates for the two mammalian IULDs were expected based on their level of sequence identity. However, the two plant IULDs do not share significant sequence identity and diverged from one another during a whole genome duplication event that occurred in plants ~300 million years ago [59, 60]. The relaxation rates for the fungus IULD are different from those observed for the plant and mammalian homologues. The relaxation data obtained from the five homologues that were characterized in our study indicate that the mammalian homologues are the most flexible, followed by the fungus homologue, and that the plant homologues are the least flexible. This conclusion is supported by reduced spectral density mapping of the relaxation data presented in this report [10]. In the latter study, we tested the evolutionary conservation of dynamic behavior for the RPA70 IULDs. Evolutionary rate measurements were performed on a set of nine mammalian RPA70 IULDs. The results of this analysis identified the molecular basis for the dynamic differences observed between the IULDs of vertebrates and plants and demonstrated the IULDs from vertebrates have accumulated and conserved glycine residues relative to the plant IULDs. The frequency of glycine in the vertebrate IULDs was $13.6 \pm 2.0\%$ compared with $5.0 \pm 2.0\%$ in plants.

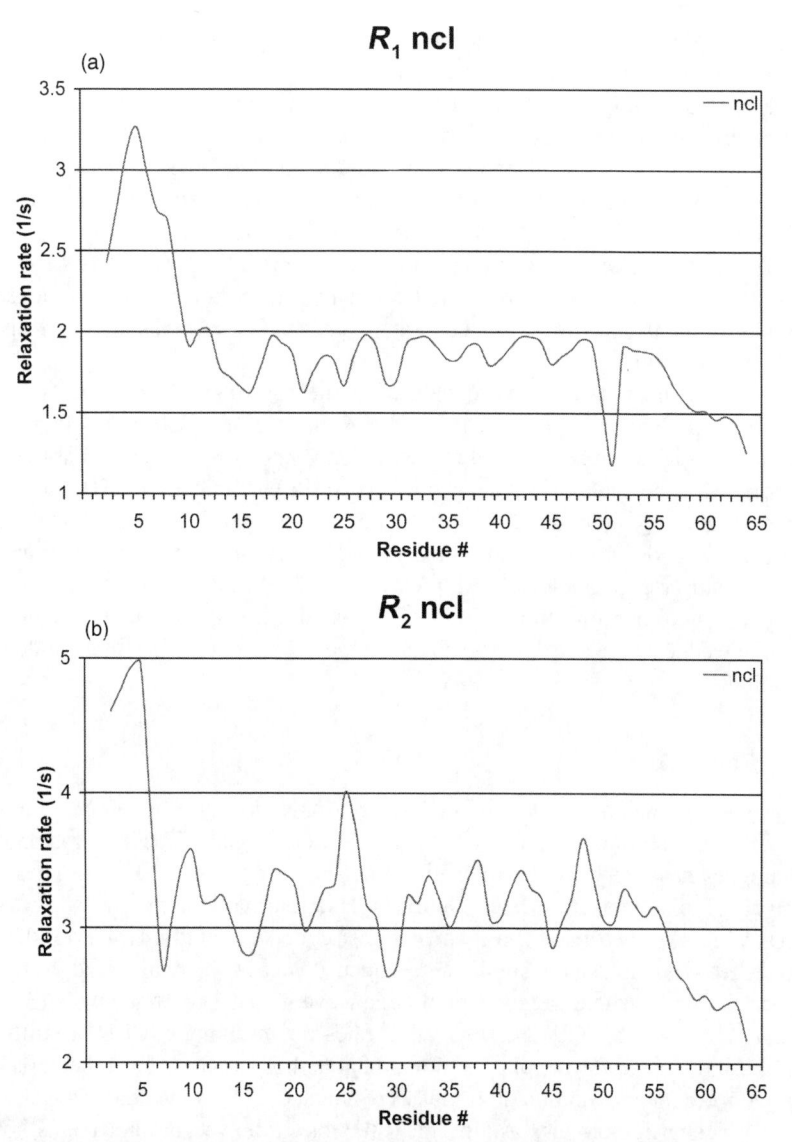

Figure 8.8. Plots showing (a) spin-lattice (R_1) and (b) spin-spin (R_2) relaxation rates for ncl (green). The relaxation rates are plotted on the vertical axis and residue number is plotted on the horizontal axis. Relaxation rates were averaged over a two-residue window.

ACKNOWLEDGMENTS

G. W. D is supported by the National Science Foundation (Award #0939014) and the American Cancer Society (RSG-07-289-01-GMC).

REFERENCES

1. Mirsky, A. E. and Pauling, L. (1936) On the structure of native, denatured, and coagulated proteins, *Proc Natl Acad Sci U S A 22*, 439–447.
2. Kendrew, J. C., Dickerson, R. E., and Strandberg, B. E. (1960) Structure of myoglobin: A three-dimensional Fourier synthesis at 2 angstrom resolution, *Nature 206*, 757–763.
3. Perutz, M. F., Rossmann, M. P., Cullis, A. F., Muirhead, H., Will, G., and North, A. C. (1960) Structure of haemoglobin: A three dimensional Fourier synthesis at 5.5 angstrom resolution, obtained by X-ray analysis, *Nature 185*, 416–422.
4. Anfinsen, C. B. (1973) Principles that govern the folding of protein chains, *Science 181*, 223–230.
5. Wright, P. E. and Dyson, H. J. (1999) Intrinsically unstructured proteins: Re-assessing the protein structure-function paradigm, *J Mol Biol 293*, 321–331.
6. Uversky, V. N. (2002) Natively unfolded proteins: A point where biology waits for physics, *Protein Sci 11*, 739–756.
7. Dyson, H. J. and Wright, P. E. (2002) Insights into the structure and dynamics of unfolded proteins from nuclear magnetic resonance, *Adv Protein Chem 62*, 311–340.
8. Dyson, H. J. and Wright, P. E. (2002) Coupling of folding and binding for unstructured proteins, *Curr Opin Struct Biol 12*, 54–60.
9. Dunker, A. K., Brown, C. J., Lawson, J. D., Iakoucheva, L. M., and Obradovic, Z. (2002) Intrinsic disorder and protein function, *Biochemistry 41*, 6573–6582.
10. Daughdrill, G. W., Narayanaswami, P., Gilmore, S. H., Belczyk, A., and Brown, C. J. (2007) Dynamic behavior of an intrinsically unstructured linker domain is conserved in the face of negligible amino acid sequence conservation, *J Mol Evol 65*, 277–288.
11. Iakoucheva, L. M., Brown, C. J., Lawson, J. D., Obradovic, Z., and Dunker, A. K. (2002) Intrinsic disorder in cell-signaling and cancer-associated proteins, *J Mol Biol 323*, 573–584.
12. Tompa, P. (2002) Intrinsically unstructured proteins, *Trends Biochem Sci 27*, 527–533.
13. Tompa, P. (2003) Intrinsically unstructured proteins evolve by repeat expansion, *Bioessays 25*, 847–855.
14. Tompa, P. (2005) The interplay between structure and function in intrinsically unstructured proteins, *FEBS Lett 579*, 3346–3354.
15. Dyson, H. J. and Wright, P. E. (2005) Intrinsically unstructured proteins and their functions, *Nat Rev Mol Cell Biol 6*, 197–208.
16. Tompa, P. (2010) *Structure and function of intrinsically disordered proteins*, 1st ed., Taylor and Francis Group, Boca Raton.
17. Romero, P., Obradovic, Z., Kissinger, C. R., Villafranca, J. E., Garner, E., Guilliot, S., and Dunker, A. K. (1998) Thousands of proteins likely to have long disordered regions, *Pac Symp Biocomput*, 437–448.

18. Dunker, A. K., Obradovic, Z., Romero, P., Garner, E. C., and Brown, C. J. (2000) Intrinsic protein disorder in complete genomes, *Genome Inform Ser Workshop Genome Inform 11*, 161–171.

19. Dunker, A. K., Brown, C. J., Lawson, J. D., Iakoucheva-Sebat, L. M., Vucetic, S., and Obradovic, Z. (2002) The protein trinity: Structure/function relationships that include intrinsic disorder, *ScientificWorldJournal 2*, 49–50.

20. Daughdrill, G. W., Pielak, G. J., Uversky, V. N., Cortese, M. S., and Dunker, A. K. (2005) Natively disordered proteins. In: Buchner, J. and Kiefhaber, T. (eds.), *Protein folding handbook*. WILEY-VCH, Darmstadt, pp. 275–357.

21. Dyson, H. J. and Wright, P. E. (1998) Equilibrium NMR studies of unfolded and partially folded proteins, *Nat Struct Biol 5 Suppl*, 499–503.

22. Brown, C. J., Takayama, S., Campen, A. M., Vise, P., Marshall, T. W., Oldfield, C. J., Williams, C. J., and Dunker, A. K. (2002) Evolutionary rate heterogeneity in proteins with long disordered regions, *J Mol Evol 55*, 104–110.

23. Radivojac, P., Obradovic, Z., Brown, C. J., and Dunker, A. K. (2002) Improving sequence alignments for intrinsically disordered proteins, *Pac Symp Biocomput*, 589–600.

24. Brown, C. J., Johnson, A. K., and Daughdrill, G. W. (2009) Comparing models of evolution for ordered and disordered proteins, *Mol Biol Evol 27*, 609–621.

25. Vise, P. D., Baral, B., Latos, A. J., and Daughdrill, G. W. (2005) NMR chemical shift and relaxation measurements provide evidence for the coupled folding and binding of the p53 transactivation domain, *Nucleic Acids Res 33*, 2061–2077.

26. Kay, L. E. (1998) Protein dynamics from NMR, *Nat Struct Biol 5 Suppl*, 513–517.

27. Wold, M. S. (1997) Replication protein A: A heterotrimeric, single-stranded DNA-binding protein required for eukaryotic DNA metabolism, *Annu Rev Biochem 66*, 61–92.

28. Gomes, X. V. and Wold, M. S. (1996) Functional domains of the 70-kilodalton subunit of human replication protein A, *Biochemistry 35*, 10558–10568.

29. Henricksen, L. A., Umbricht, C. B., and Wold, M. S. (1994) Recombinant replication protein A: Expression, complex formation, and functional characterization, *J Biol Chem 269*, 11121–11132.

30. Lao, Y., Gomes, X. V., Ren, Y., Taylor, J. S., and Wold, M. S. (2000) Replication protein A interactions with DNA. III. Molecular basis of recognition of damaged DNA, *Biochemistry 39*, 850–859.

31. Lao, Y., Lee, C. G., and Wold, M. S. (1999) Replication protein A interactions with DNA. 2. Characterization of double-stranded DNA-binding/helix-destabilization activities and the role of the zinc-finger domain in DNA interactions, *Biochemistry 38*, 3974–3984.

32. Walther, A. P., Gomes, X. V., Lao, Y., Lee, C. G., and Wold, M. S. (1999) Replication protein A interactions with DNA. 1. Functions of the DNA-binding and zinc-finger domains of the 70-kDa subunit, *Biochemistry 38*, 3963–3973.

33. Wold, M. S. and Kelly, T. (1988) Purification and characterization of replication protein A, a cellular protein required for in vitro replication of simian virus 40 DNA, *Proc Natl Acad Sci U S A 85*, 2523–2527.

34. Olson, K. E., Narayanaswami, P., Vise, P. D., Lowry, D. F., Wold, M. S., and Daughdrill, G. W. (2005) Secondary structure and dynamics of an intrinsically unstructured linker domain, *J Biomol Struct Dyn 23*, 113–124.

REFERENCES

35 Zhou, H. X. (2001) The affinity-enhancing roles of flexible linkers in two-domain DNA-binding proteins, *Biochemistry 40*, 15069–15073.

36 Daughdrill, G. W., Ackerman, J., Isern, N. G., Botuyan, M. V., Arrowsmith, C., Wold, M. S., and Lowry, D. F. (2001) The weak interdomain coupling observed in the 70 kDa subunit of human replication protein A is unaffected by ssDNA binding, *Nucleic Acids Res 29*, 3270–3276.

37 Wittekind, M. and Mueller, L. (1993) HNCACB, a high-sensitivity 3D NMR experiment to correlate amide-proton and nitrogen resonances with the alpha- and beta-carbon resonances in proteins, *J Magn Reson B 101* (2), 201–205.

38 Schwarzinger, S., Kroon, G. J., Foss, T. R., Chung, J., Wright, P. E., and Dyson, H. J. (2001) Sequence-dependent correction of random coil NMR chemical shifts, *J Am Chem Soc 123*, 2970–2978.

39 Schwarzinger, S., Kroon, G. J., Foss, T. R., Wright, P. E., and Dyson, H. J. (2000) Random coil chemical shifts in acidic 8 M urea: Implementation of random coil shift data in NMRView, *J Biomol NMR 18*, 43–48.

40 Wishart, D. S., Bigam, C. G., Yao, J., Abildgaard, F., Dyson, H. J., Oldfield, E., Markley, J. L., and Sykes, B. D. (1995) 1H, 13C and 15N chemical shift referencing in biomolecular NMR, *J Biomol NMR 6*, 135–140.

41 Dedmon, M. M., Lindorff-Larsen, K., Christodoulou, J., Vendruscolo, M., and Dobson, C. M. (2005) Mapping long-range interactions in alpha-synuclein using spin-label NMR and ensemble molecular dynamics simulations, *J Am Chem Soc 127*, 476–477.

42 Kay, L. E., Keifer, P., and Saarinen, T. (1992) Pure absorption gradient enhanced heteronuclear single quantum correlation spectroscopy with improved sensitivity, *J Am Chem Soc 114*, 10663.

43 Kay, L. E., Torchia, D. A., and Bax, A. (1989) Backbone dynamics of proteins as studied by 15N inverse detected heteronuclear NMR spectroscopy: Application to staphylococcal nuclease, *Biochemistry 28*, 8972–8979.

44 Romero, P., Obradovic, P., and Dunker, K. (1997) Sequence data analysis for long disordered regions prediction in the calcineurin family, *Genome Inform Ser Workshop Genome Inform 8*, 110–124.

45 Romero, P., Obradovic, Z., Li, X., Garner, E. C., Brown, C. J., and Dunker, A. K. (2001) Sequence complexity of disordered protein, *Proteins 42*, 38–48.

46 Obradovic, Z., Peng, K., Vucetic, S., Radivojac, P., Brown, C. J., and Dunker, A. K. (2003) Predicting intrinsic disorder from amino acid sequence, *Proteins 53*, Suppl 6, 566–572.

47 Uversky, V. N., Gillespie, J. R., and Fink, A. L. (2000) Why are "natively unfolded" proteins unstructured under physiologic conditions? *Proteins 41*, 415–427.

48 Uversky, V. N. (2002) What does it mean to be natively unfolded? *Eur J Biochem 269*, 2–12.

49 Wishart, D. S., Sykes, B. D., and Richards, F. M. (1991) Relationship between nuclear magnetic resonance chemical shift and protein secondary structure, *J Mol Biol 222*, 311–333.

50 Wishart, D. S., Sykes, B. D., and Richards, F. M. (1991) Simple techniques for the quantification of protein secondary structure by 1H NMR spectroscopy, *FEBS Lett 293*, 72–80.

51 Wishart, D. S., Sykes, B. D., and Richards, F. M. (1992) The chemical shift index: A fast and simple method for the assignment of protein secondary structure through NMR spectroscopy, *Biochemistry 31*, 1647–1651.

52 Wishart, D. S. and Sykes, B. D. (1994) The 13C chemical-shift index: A simple method for the identification of protein secondary structure using 13C chemical-shift data, *J Biomol NMR 4*, 171–180.

53 Wishart, D. S. and Sykes, B. D. (1994) Chemical shifts as a tool for structure determination, *Methods Enzymol 239*, 363–392.

54 Wishart, D. S. and Case, D. A. (2001) Use of chemical shifts in macromolecular structure determination, *Methods Enzymol 338*, 3–34.

55 Wishart, D. S. and Nip, A. M. (1998) Protein chemical shift analysis: A practical guide, *Biochem Cell Biol 76*, 153–163.

56 Dyson, H. J. and Wright, P. E. (2001) Nuclear magnetic resonance methods for elucidation of structure and dynamics in disordered states, *Methods Enzymol 339*, 258–270.

57 Eliezer, D. (2007) Characterizing residual structure in disordered protein states using nuclear magnetic resonance, *Methods Mol Biol 350*, 49–67.

58 Lefevre, J. F., Dayie, K. T., Peng, J. W., and Wagner, G. (1996) Internal mobility in the partially folded DNA binding and dimerization domains of GAL4: NMR analysis of the N-H spectral density functions, *Biochemistry 35*, 2674–2686.

59 Simillion, C., Vandepoele, K., Van Montagu, M. C., Zabeau, M., and Van de Peer, Y. (2002) The hidden duplication past of *Arabidopsis thaliana*, *Proc Natl Acad Sci U S A 99*, 13627–13632.

60 Bowers, J. E., Chapman, B. A., Rong, J., and Paterson, A. H. (2003) Unravelling angiosperm genome evolution by phylogenetic analysis of chromosomal duplication events, *Nature 422*, 433–438.

II

DISORDERED PEPTIDES AND MOLECULAR RECOGNITION

9

BINDING PROMISCUITY OF UNFOLDED PEPTIDES

Christopher J. Oldfield, Bin Xue, A. Keith Dunker, and Vladimir N. Uversky

9.1. PROTEIN–PROTEIN INTERACTION NETWORKS

Interactions between proteins play an essential role in living cells. Proteins are involved in the multitude of interactions creating sophisticated networks that define and control cellular-biological processes. Protein–protein interaction (PPI) networks provide a means to integrate various biological signals including those used for, for example, growth, energy generation, and cell division. The architecture of these PPI networks suggests that they are approximately scale-free [1–8]. For scale-free networks, a log-log plot of the number of nodes versus the number of links (or interactions) at each node gives a straight line with a negative slope. Thus, such networks contain a few proteins with many links, that is, hub proteins, and many proteins with only a few links, that is, non-hub proteins. The designation of hub protein is given to proteins with a large number of interactions relative to the other proteins in a given PPI network; there is no clear threshold for the number of links separating hubs and non-hubs.

Scale-free architecture has been observed for several networks such as the Internet, cellular phone systems, social interactions, and author citations. With regard to PPIs, scale-free network architecture is suggested to provide biological

---

Protein and Peptide Folding, Misfolding, and Non-Folding, First Edition. Edited by Reinhard Schweitzer-Stenner.
© 2012 John Wiley & Sons, Inc. Published 2012 by John Wiley & Sons, Inc.

advantages. Since the overall fraction of hub proteins is small, random deleterious mutations will more likely occur in non-hub proteins. Elimination of the functions of such non-hub proteins typically would have little overall effect on the network and so, generally, would not be serious. On the other hand, a deleterious mutation of a hub protein is more likely to be lethal [4–9]. Another advantage of scale-free architecture is that signals can traverse these networks in a small number of steps, so the efficiency of signal transduction is improved compared with that expected for random networks [7].

Several studies have focused on understanding PPI network evolution across different species [10–13]. From this body of work, hub proteins have been shown to evolve more slowly than non-hub proteins, an observation that is consistent with Fisher's classic proposal that pleiotropy constrains evolution [14, 15]. With respect to temporal limitations on PPIs, some proteins have multiple, simultaneous interactions ("party hubs") [16] while others have multiple sequential interactions ("date hubs") [16]. From a functional perspective, date hubs may connect biological modules to each other [17] while party hubs may form scaffolds that enable the assembly of functional modules [16].

While the idea of a scale-free network topology for PPI networks is receiving considerable attention, some words of caution are in order. Current PPI networks have a great deal of noise, due to the presence of a large proportion of both false-positive and false-negative interactions [8, 18–20]. Also, the current level of coverage of PPI network for various model organisms [21–24] is not sufficient to prove that these PPI networks possess a scale-free architecture. That is, a recent simulation study suggests that current data are insufficient to unambiguously distinguish scale-free architecture from random networks [21–24]. Indeed, improving the accuracy of network models is a very active area of research. In summary, whether PPI networks are truly scale-free or only approximately so, it is still clear that a relatively small number of proteins interact with many partners, either as date hubs or as party hubs, while many proteins interact with just a few partners.

9.2. ROLE OF INTRINSIC DISORDER IN PPI NETWORKS

Obviously, the ability to bind to multiple partners involves a mechanism for PPI not contained within the classical molecular recognition mechanisms [25]. Indeed, neither the lock-and-key [26] nor the original induced-fit [27] mechanism can readily explain how one protein can bind to multiple partners. Note that the original induced fit mechanism was defined as changes of a structured protein, such as domain shifts, upon binding to the partner [27] or as changes analogous to a glove changing shape to fit a hand. Thus, as originally proposed, the starting point for induced fit was a structured protein.

On the other hand, several previous studies, both theoretical and experimental, suggested that natively unstructured or intrinsically disordered proteins (IDPs) are plastic and can adopt different structures upon binding to different partners [28–34]. IDPs or intrinsically disordered regions (IDRs) play a number of crucial roles in mediating protein interactions [28–49]. Given this prior body of work, we suggested that molecular recognition via disorder-to-order transitions upon binding would be a reasonable mechanism for binding by hub proteins [35]. We have previously noted examples suggesting that intrinsic disorder could enable one protein to bind with multiple partners (one-to-many signaling) or to enable multiple partners to bind to one protein (many-to-one signaling) [34].

Several additional recent bioinformatics publications support the importance of protein disorder for hubs [36–40]. Disorder appears to be more clearly associated with date hubs [38, 40] than with party hubs. However, since some protein complexes clearly use long regions of disorder as a scaffold for assembling an interacting group of proteins [41–49], the potential importance of disorder for party hubs needs to be examined further. Additional evidence for the importance of disorder for highly connected hub proteins comes from a structure-based study of the yeast–protein interaction network [50]. The authors considered only interactions that could be mediated by domains with known structures and found that the degree distribution of the resulting network contained no proteins with more than 14 interactions, which is more than an order of magnitude less than one unfiltered, high-confidence data set. This result indicates that a structure-based view of hub proteins is insufficient to explain the multitude of partners that interact with hub proteins.

9.3. TRANSIENT STRUCTURAL ELEMENTS IN PROTEIN-BASED RECOGNITION

Different types of PPIs have been defined [51]. These definitions include the distinction between obligate and non-obligate complexes, which are protomers that only exist in complex and protomers that exist both independently and in complex, respectively. Another distinction can be made between permanent and transient complexes. Permanent PPIs are strong and irreversible, whereas a complex qualifies as transient if it readily undergoes changes in oligomeric state [52]. Biophysical characterization of IDPs and IDRs made it evident that although these proteins/segments do not possess a single, well-folded state, they often exhibit transient, residual organization either at secondary-structure or at tertiary-structure level. Some of these transient elements are briefly in the following sections.

9.3.1. Molecular Recognition Elements/Features

Intrinsic conformational preferences of IDPs/IDRs can be utilized to predict the sites of binding regions within IDPs/IDRs [53]. We have previously reported that, in some cases, PONDR® VL-XT predicts short regions of predicted order bounded by regions of predicted disorder, where the predicted ordered region corresponds to a region known to interact with a partner [54]. An example of this behavior is shown in Figure 9.1, which shows the complex between the 4E binding protein 1 (4EBP1) and the eukaryotic translation initiation factor 4E [55] (Fig. 9.1A) and the PONDR VL-XT prediction for 4EBP1 (Fig. 9.1B). In unbound form, 4EBP1 is completely disordered

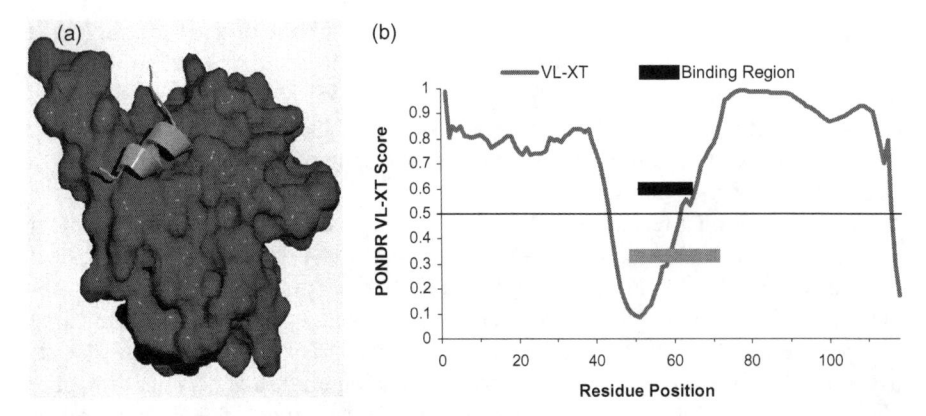

Figure 9.1. Example of binding regions and their positions relative to the regions of predicted order (PONDR® VL-XT score) and α-MoRF. (A) Eukaryotic initiation factor (blue) and the binding region of 4EBP1 (red). (B) The PONDR VL-XT prediction for 4EBP1 with the binding region (blue bar) and the predicted α-MoRF region (pink bar) shown. Reproduced from Ref. [170].

by NMR [56]. However, a short stretch undergoes a disorder-to-order transition upon interaction with its binding partner [55]. Figure 9.1B shows that there is a sharp dip in the PONDR VL-XT plot in the area of the binding region. This drop is flanked by long regions of predicted disorder.

Additional work has validated the use of these distinctive downward spikes in PONDR VL-XT curves to locate functional binding regions within IDRs [53]. These elements consist of a short region that undergoes coupled binding and folding within a longer region of disorder. Initially, we referred to these as "molecular recognition elements," (MoREs). Later, these regions were renamed molecular recognition features (MoRFs) to emphasize their unique polymorphic nature. An algorithm has been designed [53] for identification of regions that have a propensity for α-helix-forming molecular recognition features (α-MoRFs) based on a discriminant function that indicates such regions while giving a low false-positive error rate on a collection of structured proteins. Application of this predictor to proteomes and functionally annotated proteins indicates that α-MoRFs are likely to play important roles in PPIs involved in signaling events [53].

This first α-MoRF identifier was developed using a training data set of a limited size (a set of 12 proteins containing 14 potential α-MoRFs). Recently, the prediction algorithm was improved by [1] including additional α-MoRF examples and their homologues in the positive training set [2], carefully extracting monomer structure chains from the Protein Data Bank (PDB) as the negative training set [3], including attributes from recently developed disorder predictors, secondary-structure predictions, and amino acid indices, and [4] constructing neural network-based predictors and performing validation [57]. The sensitivity, specificity, and accuracy of the resulting predictor, α-MoRF-PredII, were 0.87 ± 0.10, 0.87 ± 0.11, and 0.87 ± 0.08, respectively, over 10 cross-validations [57].

Systematic studies of PDB entries revealed that many protein complexes comprise a short protein segment bound to a larger globular protein [58]. Analysis of literature data showed that some of these short peptides, being specifically folded into α-helix, β-hairpin, β-strand, polyproline II helix (PPII), or irregular structure, and so on within their complexes with globular partners, are intrinsically disordered prior the corresponding complex formation [58]. Thus, all these regions can be considered as illustrative members of the subset of PPIs involving disorder-to-order transitions during the complex formation and hence can be considered as MoRFs. In a recent study, short polypeptide chains with lengths between 10 and 70 residues and were bound to a globular partner (with chains ≥ 100 residues) were extracted from PDB [58]. This process resulted in a data set comprising 372 non-redundant protein chains (9093 residues). The secondary-structure assignment showed that 27% of this data set consisted of α-helical residues, 12% were β-sheet residues and approximately 48% of the residues had an irregular conformation. The remaining 13% of the residues were found to be disordered, characterized by missing coordinate information in their respective PDB files [58]. This data set can be used for further studies to find sequence attributes for discriminating these different MoRFs from each other and from ordered proteins.

9.3.2. Retro-MoRFs

Recently, we proposed the combination of sequence alignment and disorder prediction as an approach to improve the confidence of identifying MoRFs [59]. In addition to direct alignment, reverse alignments were additionally investigated to improve MoRF prediction performance. The rationale behind reverse alignment is that while reversing the sequence of a globular protein often disrupts its structure, the flexibility of IDRs is sufficient to accommodate the altered geometry of a reverse sequence. This idea leads us to propose the retro-MoRFs concept, which have the reversed sequence of an identified MoRF. It was proposed that the partner of a MoRF would also like to bind to the corresponding retro-MoRF [59]. This hypothesis provided new grounds for exploring the complexities of PPI networks [59].

Theoretically, it was expected that a retro-protein, that is, a protein obtained as a result of reading the sequence backwards, might adapt a topological equivalent of the mirror image of the three-dimensional structure of its parent protein [60–62]. However, the lattice model simulations of the retro-sequence of the B domain of Staphylococcal protein A revealed that the secondary-structure elements in the retro-protein did not exactly match their counterparts in the original protein structure [62], and later the full-atom simulation analysis showed that this retro-protein was essentially unfolded [63]. Based on the analyses of inverse sequence similarity in proteins, it has been concluded that the tertiary structures of retro-proteins did not imply folds comparable with their parent protein [64]. Furthermore, it was shown that the sequence inversion affected the foldability of some model peptides and proteins in such a way that retro-proteins were generally no more similar to their parent sequences than any random sequence, despite their common hydrophobic/hydrophilic pattern, global amino acid composition, and possible tertiary contacts [65]. Therefore, it has been concluded that the direction of protein sequence is a critical factor in the formation of a unique structure. This directionality explains why the sequences of ordered proteins are generally not palindromic [65]. The differences between the parent and retro-proteins likely originate from the differences in the local, detailed structures of amino acids, their dihedral angles, side-chain orientations, and packing inside a protein structure.

In contrast, careful analysis of a biologically active retro-protein, retro-human metallothionein-2 α domain, revealed that despite the significant alterations in the protein structure induced by the reversal direction of the domain sequence backbone, this retro-domain retained its metal binding ability and foldability, which is mostly due to the fact that reversion of a sequence was not critical to the interaction between Cys side chains and metal ions [65].

Other potential exceptions from the spatial restrictions on reverse sequences include PPII helices, which tend to occur on the surfaces of proteins, and PPII-based binding motifs. These structural motifs are left-handed, all-*trans* extended helices with average backbone dihedral angles of $(\Phi, \Psi) = (-75°, +145°)$. Each PPII helix has precisely three residues per turn, compared with 3.6 residues per turn in an α-helix. This results in a considerably extended helical structure, with PPII helices translating 3.12 Å per residue compared with 1.50 Å per residue in the α-helix. Each turn of a PPII helix spans approximately 9 Å, resulting in a perfect threefold rotational symmetry [66, 67]. Furthermore, residues in PPII helices are significantly more solvent-exposed than the average for all residues in ordered proteins, with polar residues in PPII helices 60% more solvent-exposed and hydrophobic residues 50% more exposed than the average for all residues [66]. This high surface exposure of both the hydrophobic and polar side chains of residues in PPII conformation provides for an easily accessible hydrophobic or polar interaction surface. As a result, proline-rich sequences are very common recognition sites for PPI modules such as the SH3 domain, the WW domain, and the EVH1 domain [68]. For example, the consensus ligand peptides interacting with various SH3-domain-containing proteins in yeast were assigned to class I (RXXPXXP) or class II (PXXPXR) motifs [69], which are both regarded as a Pro-rich core LPPLP motif, with the position of the R residue (N- or C-terminal to the Pro core) dictating whether the ligand falls in class I or class II [70]. Furthermore, PPII-based binding motifs can be inverted, due to the high symmetry, and due to the fact that the PPII helix has three residues per turn, where residues at positions i and $i + 3$ lie on the same edge of the ligand structure. In fact, class I and class II ligands bind to the SH3 domain in reverse orientations relative to each other [71], where a class I ligand binds with its N-terminus at the RT loop and a class II ligand with its C-terminus at this site.

We propose that the restrictions in fine structure, dihedral angles, and side-chain packing details imposed by backbone directionality can be avoided if intrinsic disorder is taken into account. The intrinsic flexibility of IDPs and IDRs might allow them to gain specific structures needed for successful and specific binding to their partners. Therefore, the spatial hindrance of binding between a reversed fragment and a partner can be overcome by the flexibility of the flanking regions or the binding region itself. Hence, the combination of reverse sequence alignment with disorder analysis might provide very useful information for identifying the possible interaction regions.

Given that region **F** in protein **A** can bind to partner **P** and that protein **B** contains a region **rF**, which has a reversed sequence of region **F**, then the question is: can protein **B** interact with partner **P**? To answer this question, the software package, PONDR-RIBS, was developed [59]. This new tool provides a synthetic analysis of the binding capability of a reversed fragment and the partners. PONDR-RIBS aligns sequences by CLUSTALW [72] and predicts intrinsic disorder by PONDR-FIT [73]. Here, we restricted the criteria as follows: [1] the sequence identity between reversed fragment **rF** and the original fragment **F** is higher than 60% [2]; the aligned fragment **rF** has similar hydrophobic/charge pattern to that of fragment **F** [3]; and the aligned fragment **rF** is disordered or is flanked by disordered regions. If all these conditions are satisfied, then reversed fragment **rF** of protein **B** might interact with partner **P** with a high probability.

The principles of PONDR-RIBS may be used not only for the reversed alignment, but also for the normal-order alignment. Suppose a segment **F** binds to partner **P** and **F*** is a segment sequentially similar to **F**, the probability of **F*** binding to **P** should be much higher if **F*** is flanked by disordered regions or locates in a disordered tail. Hence, by combining the normal-order sequence alignment and disorder prediction, the certainty of identifying binding segments can be significantly improved.

9.3.3. Preformed Structural Elements

IDPs often undergo disorder-to-order transition upon binding to their partners, which raises the question of whether the structure adopted in the bound form is enforced by the partner molecule or reflects inherent conformational preferences of IDPs. In other words, the binding-coupled folding of IDPs may be induced by the template or, alternatively, selected from the ensemble of conformations. To answer this question, the structures of 26 IDPs in complex with their globular partners were analyzed for the predictability of their secondary structural elements [74]. The results of the analysis, using three algorithms, GOR, ALB, and PROF, suggested that the accuracy of predicting secondary structural elements in IDPs is higher than that either in their partner proteins or randomized sequences. This observation suggests that IDPs may have strong conformational preferences for their bound conformations; that is, they probably use elements for recognition that are partially or transiently preformed in the solution state. This preference is strongest for helices and is weakest for extended structures.

Although the insight from studying complexes under steady-state conditions is not fully conclusive on the role of these preformed elements in binding, in certain cases a similar structure in the free and bound states was observed when IDPs have been characterized in the solution state by NMR. Such correlation is apparent in the case of the KID domain of CREB [75, 76], p21Cip1/p27^{Kip2} [32, 77], p53 [78], FlgM [79, 80], PKI alpha [81], thymosin β4 [82], Bad (PDB: 2bzw), and measles virus nucleoprotein [83]. Such preformed elements might also serve as initial contact points of interaction, but this issue requires further study. A very similar concept (intrinsically folded structural units [IFSUs]) has also been suggested, based on a study of the molecular function and binding of p27^{Kip1} [84].

9.3.4. Primary Contact Sites

Primary contact sites (PCSs) are short recognition motifs that are distinguished by rapid binding kinetics. The concept of PCSs was derived from the observation that the large-scale structural reorganization concomitant to binding of IDPs is usually realized very rapidly. In fact, structural disorder is thought to confer the advantage of rapid binding [85]. Hence, it was reasoned that certain regions within the disordered ensemble are more exposed than others and thereby may serve as the first sites of contact with the partner.

The idea that limited proteolysis at extremely low concentrations of proteases may preferentially affect regions of an IDP that are exposed relative to other regions was experimentally checked in the case of two IDPs, calpastatin and MAP2 [86]. At very low concentrations, narrow (trypsin, chymotrypsin, and plasmin) or broad (subtilisin and proteinase K) substrate specificity proteasespreferentially cleave both proteins in regions thought to make the first contact with their respective partners: subdomains A, B, and C in calpastatin, and the central Pro-rich region (PRR) in MAP2c. This non-random structural behavior was further probed by CD spectroscopy and NMR relaxation spectroscopy. In the case of calpastatin, the CD spectra and hydration of the two halves are not additive, which suggest long-range tertiary interactions within the protein [86]. In MAP2c, no such tertiary interactions could be identified, but exposure of the PCSs could be accounted for by local structural constraints. Urea and temperature dependence of the CD spectrum of its central PRR pointed to the presence of PPII helix conformation in this region, which is rather stretched out and keeps the interaction site exposed [86]. Figure 9.2 illustrates these observations by presenting far-ultraviolet (UV) circular dichroism (CD) spectra of calpastatin, MAPC2, and their halves.

Some additional observations in the literature are also in line with the concept of PCSs. For example, rapid assembly of large membrane-bound complexes of highly repetitive and disordered membrane-associated proteins, such as AP180, epsin1, and auxilin in exocytosis [87, 88], is essential for proper execution of the membrane fusion. It was suggested that a large capture radius of specific, exposed recognition elements enabled by the disordered nature of these proteins provides the key ingredient of this mechanism [89]. A structural study on p53, the tumor-suppressor transcription factor, also provides relevant information on the concept of PCSs. p53 has a long, disordered N-terminal transactivator domain (TAD), which has two binding sites, for the E3 ubiquitin ligase MDM2, and the 70-kDa subunit of replication protein A, RPA70. Paramagnetic relaxation enhancement (PRE) experiments identified distance constraints in TAD [90] and suggested that TAD is rather compact and dynamic, with the two binding motifs separated by an average distance of 10–15 Å. Prior to binding, a more extended conformation of the ensemble must be populated to expose binding for sites for either MDM2 or RPA70, in agreement with the PCS concept [90].

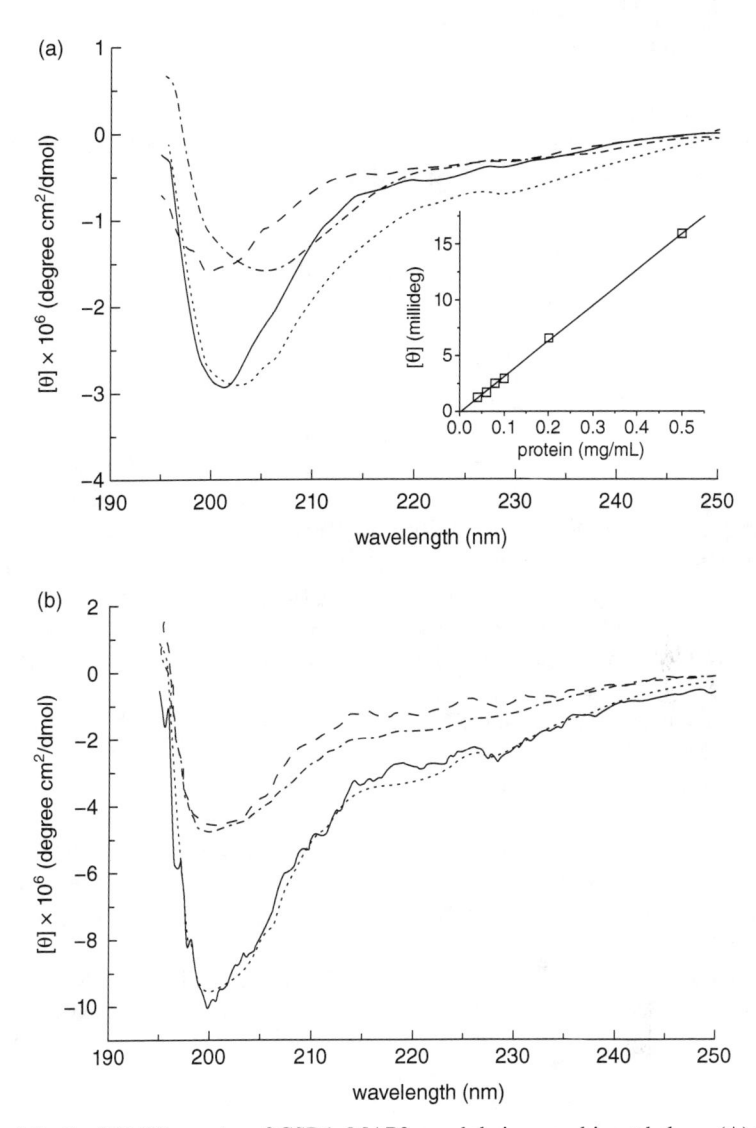

TRANSIENT STRUCTURAL ELEMENTS IN PROTEIN-BASED RECOGNITION 249

Figure 9.2. Far-UV CD spectra of CSD1, MAP2c, and their recombinant halves. (A) Molar ellipticity of CSD1 (–), its N-terminal half (---), C-terminal half (-··-), and the sum of the spectra of the N- and C-terminal halves (···). The CD spectrum of the C-terminal half was also recorded at several concentrations and the absolute value of ellipticity values at 208 nm is plotted as a function of the protein concentration (inset). (B) Molar ellipticity of MAP2c (–), its N-terminal half (---), C-terminal half (-··-), and the sum of the two halves (···). Reproduced with permission from Ref. [86].

9.3.5. Short Linear Motifs

Analyses of sequences that were observed to mediate specific PPIs, whether they result in a stable complex or enzymatic modification, suggested that the element of recognition is often a short motif of discernible conservation, also denoted as a "consensus" sequence. The generality of this relation has led to the concept of linear motifs (LMs, also denoted as eukaryotic linear motifs [ELMs], and short linear motifs [SLiMs]). Such elements were first implicated in recognition by kinases or binding sites of SH3 domains [91]. LMs are usually defined by a short consensus pattern, with conserved residues that are interspersed with rather freely exchangeable, variable positions. The first set of residues serve as specificity determinants, whereas the second likely act as spacers. Due to their evolutionary variability, LMs may constitute dynamic switch-like elements, frequently generated and erased in evolution. The typical length of LMs is between 5 and 25 residues, and their specificity is determined by a few conserved residues while the embedding sequence environment is hardly constrained. Due to the resulting limited information content, LMs are much more difficult to identify by sequence comparisons than domains. Traditional BLAST searches cannot positively identify LMs, but special algorithms that focus on non-globular regions and that combine large-scale interaction data can tackle this problem, such as DILIMOT [92] and SLiMDisc [93].

LMs have been collected in the ELM database available through the ELM server [94], which contains about 800 ELM examples that belong to more than 100 ELM classes. LMs are thought to fall into locally disordered regions [94, 95], which was corroborated in a systematic analysis by bioinformatics predictors [96]. It was found that disordered 20-residue long flanking segments contribute to the plasticity of LMs (Fig. 9.3A). LMs also have a peculiar amino acid composition, in that they resemble the characteristic composition of IDPs (Fig. 9.3B), but they are enriched in certain hydrophobic (Trp, Leu, Phe, and Tyr) and charged (Arg and Asp) residues. Furthermore, LMs are depleted in Gly and Ala, perhaps to limit the tendencies for both flexibility and secondary structure. In addition, Pro dominates in both LMs and their flanking segments, probably due to its direct involvement in interactions and also due to providing the extended secondary structural motif, PPII helix.

The amino acid composition of the specificity determinant (restricted) and variable (non-restricted) sites markedly differ, which explains the observed peculiarity of amino acid propensities [96]. Conserved positions are occupied by either hydrophobic and rigid, or charged and flexible residues, whereas non-restricted positions abound in flexible residues, similarly to IDPs in general. The only exception is Pro, which prefers both restricted positions and flanking regions, indicating its dual role as a contact residue and promoter of an open structure. In all, the unique amino acid composition suggests a mixed nature of LMs, with a few specificity-determinant residues strongly favoring order, grafted on a completely disordered carrier sequence flanking and intervening in the region critical for interaction [96]. Table 9.1 lists some of the illustrative examples of linear motifs with corresponding levels of predicted disorder [96].

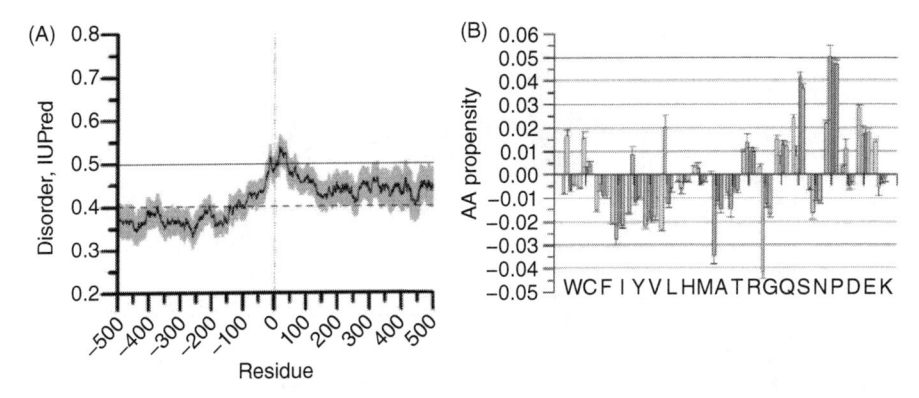

Figure 9.3. Disorder profiles and amino acid compositions of linear motifs (LMs, also denoted in the literature as ELMs, and SLiMs). (A) Disorder profiles by the IUPred algorithm were computed and averaged. A thin horizontal line at 0.5 shows the threshold of disorder, whereas a dashed line at 0.4 shows the average score for experimentally verified disordered proteins in DisProt. Values of the standard error of the mean (SEM) are displayed in light gray. (B) Amino acid propensities of IDPs of the DisProt database (yellow), LMs (cyan), 20-residue-long LM flanking segments (magenta), and LMs plus 20-residue-long flanking segments (light gray) are shown in reference to the composition of globular proteins. Reproduced from Ref. [170].

TABLE 9.1. Some Linear Motifs and Their Predicted Disorder (Reproduced from Ref. [96])

LM	Number of Examples	Average LM Length	Average LM Disorder	SEM of LM Disorder	Pattern
LIG_GYF	1	9.0	0.989	±0.000	[QHR].{0,1}P[PL]PP[GS]H[RH]
MOD_NMyristoyl	1	6.0	0.920	±0.000	^MG|^G[^EDRKHPFYW]..[STAGCN][^P]
LIG_SH3_3	6	7.0	0.876	±0.089	...[PV]..P
LIG_SH3_2	11	6.0	0.784	±0.242	P..P.[KR]
CLV_TASPASE1	2	7.0	0.740	±0.071	Q[MLVI]DG..[DE]
LIG_Clathr_ClatBox_2	2	5.0	0.734	±0.169	PWDLW
LIG_TPR	3	4.0	0.723	±0.120	EEVD$
LIG_Dynein_DLC8_1	4	5.0	0.720	±0.024	[KR].TQT
LIG_SH3_5	2	5.0	0.710	±0.288	P..DY
LIG_AP2alpha_2	48	3.0	0.706	±0.157	DP[FW]
TRG_ER_diArg_1	3	4.7	0.699	±0.192	^M[DAL][VNI]R[RK]|^M[HL]RR
TRG_ER_KDEL_1	12	4.0	0.690	±0.200	[KRHQSAP][DENQT]EL$
LIG_COP1	4	8.0	0.684	±0.182	[DE][DE]...VP[DE]

(continued overleaf)

TABLE 9.1. (*Continued*)

LM	Number of Examples	Average LM Length	Average LM Disorder	SEM of LM Disorder	Pattern
LIG_CtBP	33	5.0	0.659	±0.178	[PG][LVIPME][DENS]L[VASTRGE]
LIG_14-3-3_2	3	7.0	0.652	±0.051	R.[SYFWTQAD].[ST].[PLM]
LIG_SH3_1	5	7.0	0.642	±0.302	[RKY]..P..P
LIG_TRAF6	16	9.0	0.638	±0.250	..P.E..[FYWHDE].
MOD_CDK	1	7.0	0.636	±0.000	...([ST])P.[KR]
LIG_AP_GAE_1	12	7.0	0.631	±0.200	[DE][DES][DEGAS]F[SGAD][DEAP][LVIMFD]
LIG_14-3-3_3	12	6.0	0.614	±0.201	[RHK][STALV].[ST].[PESRDIF]
LIG_SH2_GRB2	13	4.0	0.586	±0.303	Y.N.
LIG_14-3-3_1	5	6.0	0.572	±0.170	R[SFYW].S.P
LIG_CYCLIN_1	23	4.6	0.565	±0.177	[RK].L.{0,1}[FYLIVMP]
LIG_HOMEOBOX	16	4.0	0.557	±0.140	[FY][DEP]WM
LIG_Sin3_3	2	8.0	0.540	±0.149	[FA].[LA][LV][LVI]..[AM]
MOD_SUMO	39	4.0	0.539	±0.283	[VILMAFP]K.E
LIG_TRAF2_1	13	4.0	0.535	±0.307	[PSAT].[QE]E
LIG_WW_1	1	4.0	0.527	±0.000	PP.Y
LIG_CORNRBOX	4	9.0	0.526	±0.071	L[^P]{2}[HI]I[^P]{2}[IAV][IL]
MOD_PKA_1	1	7.0	0.524	±0.000	[RK][RK].[ST]...
LIG_AP2alpha_1	4	5.0	0.520	±0.047	F.D.F
LIG_Clathr_ClatBox_1	14	5.0	0.511	±0.210	L[IVLMF].[IVLMF][DE]
MOD_PKB_1	3	9.0	0.510	±0.359	R.R..([ST])...
LIG_NRBOX	23	7.0	0.508	±0.174	[^P](L)[^P][^P](L)(L)[^P]
LIG_SH2_STAT6	1	5.0	0.488	±0.000	GY[KQ].F
LIG_PIP2_ENTH_1	7	10.0	0.478	±0.036	[DE]AT.{2}[DE]PWG[PA]
MOD_TYR_CSK	7	8.8	0.471	±0.070	[TAD][EA].Q(Y)[QE].[GQA][PEDLS]
LIG_SH2_STAT3	8	4.0	0.469	±0.176	Y..Q
LIG_HP1_1	7	5.0	0.466	±0.215	P[MVLIRWY]V[MVLIAS][LM]
LIG_SH2_STAT5	11	4.0	0.460	±0.171	Y[VLTFIC]..
MOD_TYR_ITSM	11	8.0	0.451	±0.101	..T.(Y)..[IV]
LIG_WRPW_1	60	4.6	0.451	±0.165	[WFY]RP[WFY].{0,7}$
LIG_SH3_4	2	8.0	0.449	±0.039	KP.[QK]...
TRG_ENDOCYTIC_2	6	4.0	0.448	±0.161	Y..[LMVIF]
LIG_RGD	16	3.0	0.445	±0.288	RGD

Linear motifs from the ELM database are enlisted, with their level of average disorder as predicted by the IUPred predictor and with their defining amino acid patterns.

9.3.6. Binding Regions Found by the ANCHOR Algorithm

A general method, ANCHOR, was designed for recognizing disordered binding regions in proteins [97]. ANCHOR utilizes the basic biophysical properties of disordered binding regions using estimated energy calculations. Here, estimated energies were assigned to each residue in a sequence and were shown to well approximate the corresponding energies calculated from known structures of globular proteins [98]. Generally, disordered regions can be discriminated from ordered proteins by unfavorable estimated energies, a concept utilized in the IUPred server for the prediction of protein disorder [99]. IUPred is based on the assumption that IDPs/IDRs have specific amino acid compositions that do not allow the formation of a stable well-defined structure. The method utilizes statistical potentials to calculate the pairwise interaction energy from known coordinates. IUPred was developed using a data set of globular proteins only. It estimates the pairwise interaction energy of proteins directly from the amino acid sequence. Furthermore, IUPred can take into account that the disorder tendency of residues can be modulated by their environment through the amino acid composition of the sequential environment [99].

According to IUPred concepts, disordered residues are predicted by having unfavorable estimated pairwise energies [99]. The estimated energies can also detect regions that are likely to gain energetically by interacting with globular proteins; predictions in ANCHOR combine the general disorder tendency with the sensitivity to the structural environment [100]. ANCHOR was able to recognize disordered binding regions with almost 70% accuracy at the segment level on various data sets. It was also able to discriminate the disordered binding regions from generally disordered regions. False-positive rates were further reduced by eliminating segments with IUPred scores that were too low to be compatible with disordered binding regions [97]. The performance of ANCHOR was largely independent from the amino acid composition and adopted secondary structure. Longer binding sites generally were predicted to be segmented, in agreement with available experimentally characterized examples. Applying ANCHOR to several hundred proteomes revealed that the occurrence of disordered binding sites increased with the complexity of the organisms even compared with disordered regions in general [100]. Furthermore, the length distribution of binding sites was different from disordered protein regions in general and was dominated by shorter segments. These results clearly showed the importance of disordered proteins and protein segments in molecular recognition [100].

In order to predict disordered binding regions, ANCHOR identifies segments that are in disordered regions, cannot form enough favorable intrachain interactions to fold on their own, and are likely to gain stabilizing energy by interacting with a globular protein partner [100]. To illustrate the principles of ANCHOR, Figure 9.4 represents the prediction output for human p27 [100]. Four interacting regions are identified with the first one [27–37] clearly corresponding to D1. The gap between the first two regions [38–58] coincides with the weakly interacting LH domain. The last three regions [32, 59–67, 74–76, 78–89] cover the strongly interacting D2. Figure 9.4 also shows the number of atomic contacts/residue for p27 (averaged in a window

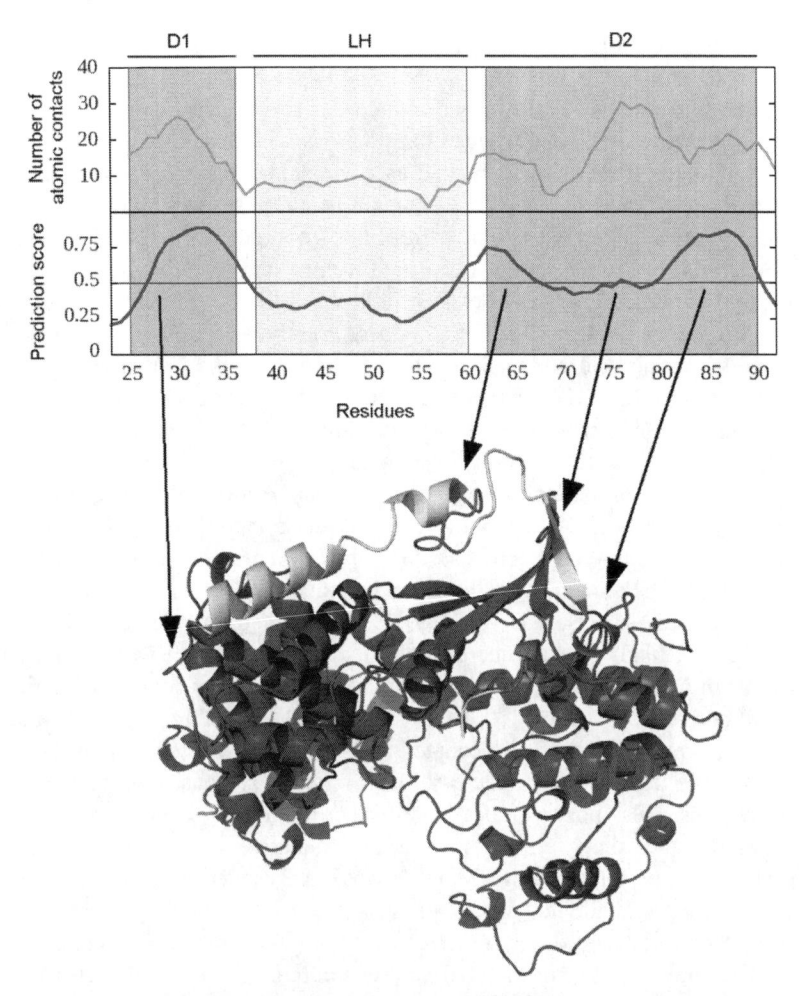

Figure 9.4. ANCHORing human p27. Number of atomic contacts (green) and prediction output (blue) and for the N-terminal binding region of human p27. "D1"and "D2" denote the two strongly interacting domains (red boxes) and "LH" denotes the weakly interacting linker domain between them (yellow box). Bottom: Crystal structure of human p27 (red and yellow) complexed with CDK2 (magenta) and cyclin A (blue) (PDB ID: 1jsu). Red parts denote regions that are predicted to bind by the predictor. These regions correspond to the experimentally verified strongly binding regions of p27. The figure was generated by PyMOL. Reproduced from Ref. [100]. See color insert.

of size 3). This contact number profile exhibits well-pronounced peaks that line up with the regions that were predicted by the ANCHOR algorithm. The figure also shows the four predicted regions mapped to the crystal structure of the complex.

Similar to MoRF predictors, ANCHOR predicts disordered binding regions without any information about the binding partners. A complementary approach identifies protein binding regions using motif searches. It was suggested that interaction with certain proteins or protein families are mediated through specific linear motifs that capture key residues responsible for binding. The information about such linear motifs is present in the ELM server [94]. It has been pointed out that the presence of sequence motifs reduces the complex task of finding putative protein binding sites to a simple pattern matching problem. However, such matches can contain many false positives, suggesting that the definition of the binding motif should include information about the specific structural context. Since linear motifs frequently occur within disordered regions, disordered binding regions could help to filter out false-positive matches [97]. Therefore, complementing the prediction of disordered binding regions with specific motif searches can prove useful in many cases and help to explore other motifs [97].

9.3.7. Anchors at the Binding Interface

Anchor residues at the interaction interface are the residues that are deeply buried upon binding [101]. The interface of the interacting ordered proteins is not flat; some residues from one protein are deeply inserted in the complementary grooves of the binding partner at the interface. This deep burial of anchor residues produces the most significant changes in the solvent-accessible surface areas among all the residues at the interface [101]. One of the most remarkable features of anchor residues is their strong preference for the bound state conformation in the unbound state [101, 102]. Since anchor residues promote binding kinetics and contribute to the stability of the complexes, they are important for molecular recognition [101–107]. Recently, anchor residues of IDPs were investigated [108]. To this end, anchor residues in the p53 N-terminal domain (p53N, which is known to interact specifically with at least two unrelated partners, MDM2 and Taz2) were identified and their conformational behavior in the unbound state was investigated. Study revealed that p53N utilizes different anchor residues for interaction with binding partners (Phe19, Trp 23, and Leu 26 in the p53N–MDM2 complex and Leu22 in the p53N–Taz2 complex). The p53 helix involved in binding rotates about 90° in two complexes, and, therefore, p53N uses different surfaces of the helix to bind to these two targets. Furthermore, anchor residues in the unbound state were shown to frequently sample conformations similar to those observed in the bound state where anchor residues acted as anchors and rarely sampled conformations observed in the alternative complexes [108]. NMR experiments revealed that although the p53N region was flexible in solution, its residues 18–26 formed a transiently stable helix, which underwent further stabilization at the complex formation [78].This same region of p53 was identified as a prototypical α-MoRF [53]. Based on these observations it has been proposed that the bound-like conformations of IDP anchor residues in unbound state determine specific interactions of IDPs with binding partners stabilizing transient encounter complexes [108]. A new mechanism combining MoRF and anchor residue concepts was also proposed to explain the binding promiscuity of IDPs. In this mechanism, anchor residues in MoRF regions frequently sample bound conformations, and a particular binding partner selects a specific group of anchor residues packing of which enables them to adopt conformations compatible with the partner's interface [108].

9.4. CHAMELEONS AND ADAPTORS: BINDING PROMISCUITY OF UNFOLDED PEPTIDES

To improve our understanding of how intrinsic disorder is used to facilitate binding diversity, we investigated two prototypical examples: p53 and 14-3-3. Both are well-studied hubs that are clearly implicated in important biological functions. For example, p53 is at the center of a large signaling network, regulating expression of genes involved in cellular processes such as cell cycle progression, apoptosis induction, DNA repair, and response to cellular stress [109]. When p53 function is lost, either directly through mutation or indirectly through several other mechanisms, the cell often undergoes cancerous transformation [110]. Cancers showing mutations in p53 are found in colon, lung, esophagus, breast, liver, brain, reticuloendothelial tissues, and hemopoietic tissues [110]. It has been shown that p53 induces or inhibits over 150 genes, including *p21*, *GADD45*, *MDM2*, *IGFBP3*, and *BAX* [111]. There are three structural domains in p53: N-terminal translational activation domain, central DNA binding domain (DBD), and C-terminal tetramerization and regulatory domain. At the transactivation region, it interacts with TFIID, TFIIH, MDM2, RPA, CBP/p300, and CSN5/Jab1 [109]. At the C-terminal domain, it interacts with GSK3β, PARP-1, TAF1, TRRAP, hGcn5, TAF, 14-3-3, S100B($\beta\beta$), and many other proteins [109].

As for 14-3-3 proteins, they were identified to contribute to a wide range of crucial regulatory processes, including signal transduction, apoptosis, cell cycle progression, DNA replication, and cell malignant transformation [112]. These numerous activities are accomplished via 14-3-3 interactions with various proteins in a phosphorylation-dependent manner. The number of proteins shown to interact with members of 14-3-3 family has now surpassed 200 [113–115], and it has been speculated that 14-3-3-interacting proteins potentially amount to approximately 0.6% of the human proteome [115]. One proposed functional model is that 14-3-3 binds to the specific target as a "molecular anvil" causing conformational changes in the partner. In their turn, these changes can affect enzymatic (biological) activity of a target protein, or mask or reveal specific motifs that regulate its localization, activity, phosphorylation state, and/or stability [116].

The 14-3-3 protein has several sequence isomers, including α, β, γ, δ, ε, η, σ, τ, and ζ [117]. These 14-3-3 isomers are structured dimer proteins with grooves that bind to more than 200 different partners having different sequences for their binding regions. Laboratory screening experiments have identified peptides that bind to all the different isomers, suggesting that the binding grooves in the different isomers have some common features [118]. A recent bioinformatics study suggests that the partners of 14-3-3 utilize intrinsic disorder for binding [119].

Below we represent some details of the interactions of p53 and 14-3-3 with their partners from an order-disorder viewpoint. Major observations are compared with the previously published data reporting on the peculiarities of interactions of these two proteins with their binding partners [117–135]. In the case of p53, different regions in the disordered tails enable this protein to bind to multiple partners at the same time. In addition, one single region of disorder adopts clearly different secondary structures and uses the same amino acids to different extents in different binding interactions. For this case the plasticity of the disordered region clearly enables the binding to multiple partners. In the case of 14-3-3, the two different partners have distinct sequences. Their interactions with 14-3-3 show characteristics, such as hydrogen bonds between side chains of 14-3-3 and the backbone of the partners and such as hydrogen bonds between the backbone of the partners and water, indicating that the two partners were very likely unfolded in water just prior to association with 14-3-3. The distinct sequences of the partners do not adopt identical backbone structures and the various side chain interactions between 14-3-3 and the two different partners involve interesting adjustments of the 14-3-3 structure. Overall, these detailed examinations provide insight into how the plasticity of disordered proteins can be used to enable the binding diversity of hub proteins, both for the case in which a single disordered region binds to multiple partners and for the case in which multiple disordered regions bind to the same partner.

9.4.1. Molecular Basis for Binding Promiscuity of p53

The p53 molecule interacts with a large number of other proteins in order to carry out its signal transduction function. Many of these proteins are downstream targets, that is, transcription factors, and many more are activators or inhibitors of p53's transactivation function. These interactions have been mapped to regions of the p53 sequence (see Fig. 9.5): the N-terminal domain (i.e., the transactivation domain), the C-terminal domain (i.e., the regulatory domain), and the DBD. These domains have also been characterized in terms of their intrinsic order–disorder state [136, 137], where the DBD is intrinsically structured and the terminal domains are intrinsically disordered [78, 138]. Additionally, many distinct sites of multiple types of posttranslational modification have been identified in p53 (see Fig. 9.5).

Comparing experimentally characterized regions of order and disorder reveals a strong bias toward mediation of interactions by IDRs. Overall, 60/84 = 71% of the interactions are mediated by IDRs in p53. A bias toward IDRs is even more pronounced in the sites of posttranslational modifications, with 86%, 90%, and 100% of observed acetylation, phosphorylation, and protein conjugation sites, respectively, found in IDRs. This is consistent with a previous observation of a strong bias for posttranslational modifications toward IDRs. This concentration of functional elements within IDRs compares with just 29% of the residues being disordered [35]. Clearly, p53 extensively uses disordered regions to mediate and modulate interactions with other proteins.

In addition to experimentally characterized disorder, predictions of intrinsic disorder for p53 by PONDR VL-XT [139] and VSL2 predictions [140] were examined. The latter is one of the highest accuracy prediction algorithms available [141], whereas the former has been observed to be useful in identifying binding regions within longer regions of disorder. Both predictors have been found to be in good agreement with

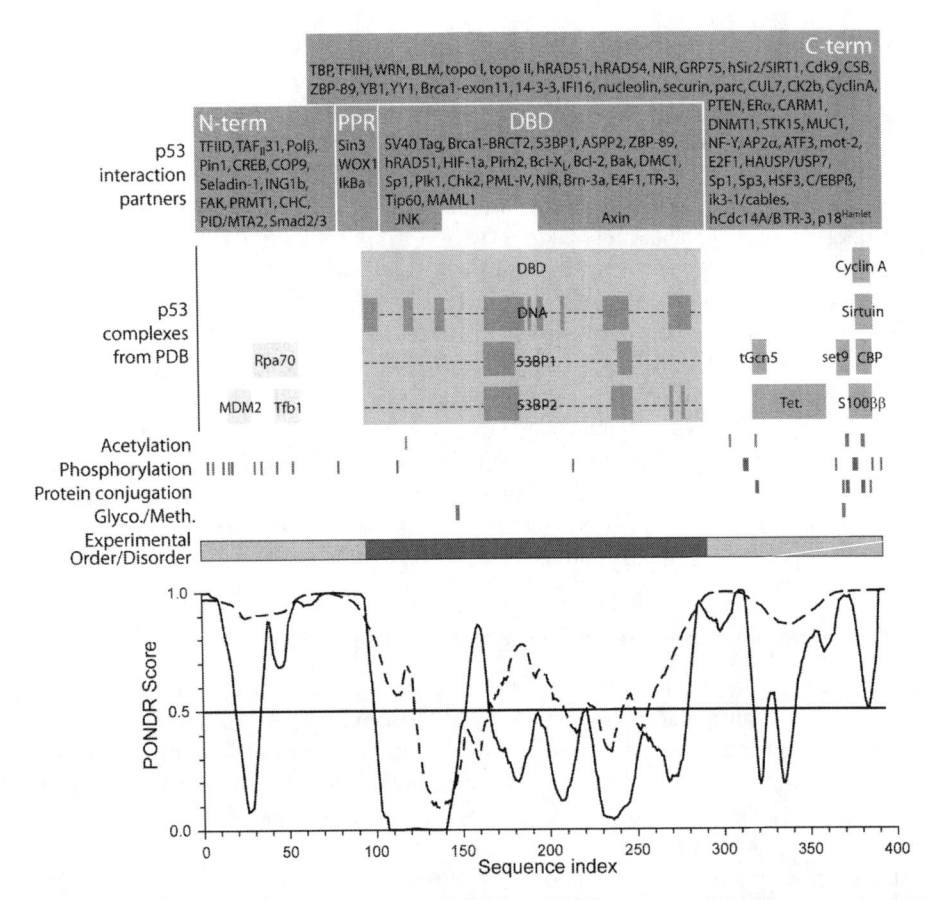

Figure 9.5. Summary of p53 interactions and structure. Gray boxes indicate the approximate binding regions of p53's known binding partners. The regions of p53 represented in structure complexes in PDB are represented by horizontal bars, labeled with the name of the binding partner. For the DBD, the extent of the globular domain is indicated by the tan box, where the internal horizontal bars indicate regions involved in binding to a particular partner. Posttranslational modifications sites are represented by vertical ticks. Experimentally characterized regions of disorder (red) and order (blue) are indicated by the horizontal bar. Finally, predictions of disorder (scores > 0.5) and order (scores < 0.5) are shown for two PONDR predictors: VLXT (solid line) and VSL2P (dashed line). All features are presented to scale, as indicated by the horizontal axis. The p53 interaction partners and posttranslational modification sites were adapted from Anderson and Appella [171]. Reproduced from Ref. [137]. See color insert.

experimental determination of intrinsic disorder [83, 142–160], and in the case of p53 they also agree well with experimental characterization (see Fig. 9.5).

In PDB, there are at least 14 complexes between various p53 regions and unique binding partners (Fig. 9.6). The interactions with 10 of these partners are mediated by region experimentally characterized as intrinsically disordered, where PONDR VL-XT detects the majority of these binding regions as short predictions of order

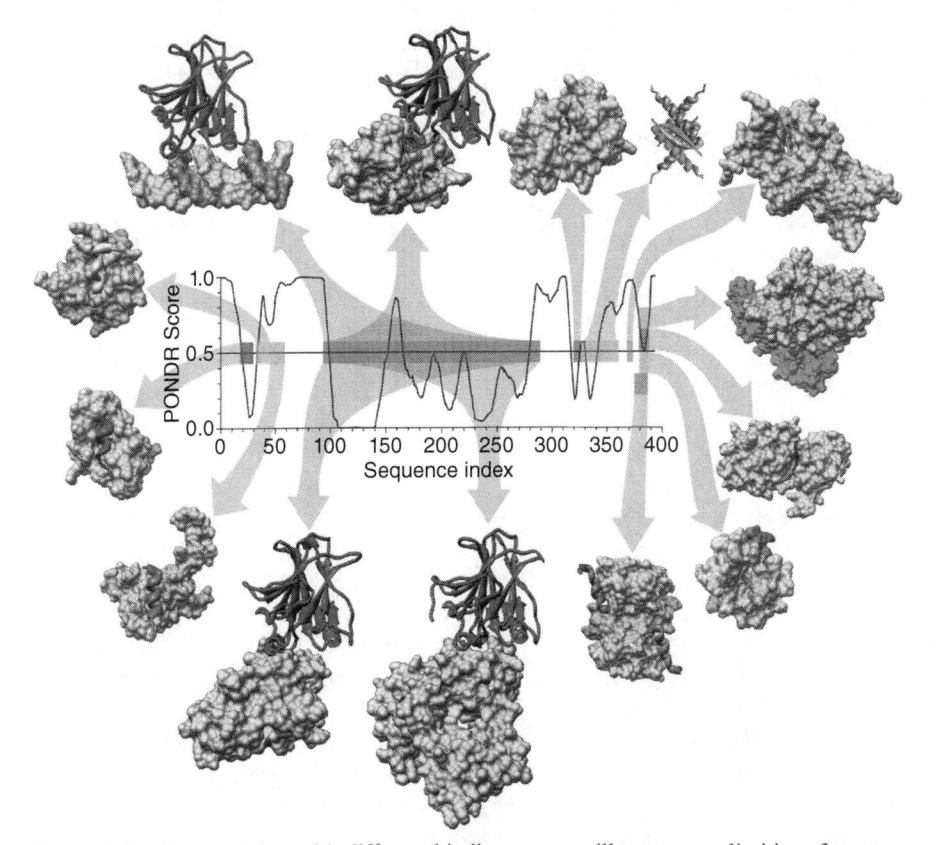

Figure 9.6. p53 interaction with different binding partners illustrate peculiarities of one-to-many signaling. A structure versus disorder prediction on the p53 amino acid sequence is shown in the center of the figure (up = disorder, down = order) along with the structures of various regions of p53 bound to 14 different partners. The predicted central region of structure with the predicted amino and carbonyl termini as being disordered was confirmed experimentally for p53. The various regions of p53 are color-coded to show their structures in the complex and to map the binding segments to the amino acid sequence. The large, centrally located DNA-binding domain is predicted and observed to be structured, both when it is bound to DNA (upper left, note the DNA molecule as a partner) and when it binds in a similar fashion to three different protein partners (one above the prediction curve and two below, all are indicated by the color magenta or the similarly folded p53 central domain). Many partnerships that involve the disordered regions of p53are formed. An interesting aspect of these many partnerships is that, for each interaction, typically only a short region of p53 becomes structured upon binding. For one particular complex (upper center, light blue), one region of disorder self-associates to form a dimer (with nearly all of the buried residues in the dimer interface), and this dimer further aggregates into a tetramer. Thus, this association involves the coupled binding and folding of a disordered region. For another set of four complexes (right side, light yellow, red, and light and dark green), the same short segment near the C-terminus binds to four different partners. Because this segment is unstructured to begin with, it can use different conformations when binding to the different partners. For this particular example, the disordered segment adopts a helix, a sheet, and two different coils upon binding with its four different partners. Starting with the p53–DNA complex (top, left, magenta protein, blue DNA), and moving in a clockwise direction, the Protein Data Bank IDs and partner names are given as follows for the 14 complexes: (1tsr, DNA), (1gzh, 53BP1), (1q2d, gcn5), (3sak, p53 [tetdom]), (1xqh, set9), (1h26, cyclinA), (1ma3, sirtuin), (1jsp, CBP bromo domain), (1dt7, s100$\beta\beta$), (2h1l, sv40 Large T antigen), (1ycs, 53BP2), (2gs0, PH), (1ycr, MDM2), and (2b3g, rpa70). See color insert.

within a longer prediction of disorder. These structures are complexes between p53 and cyclin A [120], sirtuin [121], CBP [122], S100$\beta\beta$ [123], set9 [124], tGcn5 [125], Rpa70 [126], MDM2 [127], Tfb1 [128], and itself [129]. The remaining three interactions are mediated by the structured DBD, between p53 and DNA [130], 53BP1 [131], and 53BP2 [132].

Importantly, structures of the same region of the p53 sequence bound to four different partners have been determined and deposited in the PDB so far (cyclin A [120], sirtuin [121], CBP [122], and S100$\beta\beta$ [123]). This region is from residue 374 to 388 in the p53 sequence. The regions that mediate these interactions and their respective secondary structures were mapped precisely to the p53 sequence (Fig. 9.7A). While slightly different regions of the p53 sequence are used in each interaction, there is a very high degree of overlap, with a span of seven core residues being involved in all four interactions. Interestingly, the four complexes display all three major secondary-structure types. The core span becomes a helix when binding to S100$\beta\beta$, a sheet when binding to sirtuin, and a coil with two distinct backbone trajectories when binding to CBP and cyclin A2 (see Fig. 9.7A).

Because the secondary-structure elements of the bound p53 are distinct, it seems likely that this protein utilizes different residues for the interactions with these four different partners. To examine this, the buried surface area for each residue in each interaction was quantified by calculating the changes in the accessible surface area (ΔASA) (Fig. 9.7B). Different amino acid interaction profiles are seen for each of the interactions, showing that the same residues are used to different extents in the four interfaces. The particularly large ΔASA peaks for K382 in complexes with CBP and sirtiun are due to extra buried areas arising from acetylation of this residue. This highlights that posttranslational modification within disordered regions can be used to modify PPI networks.

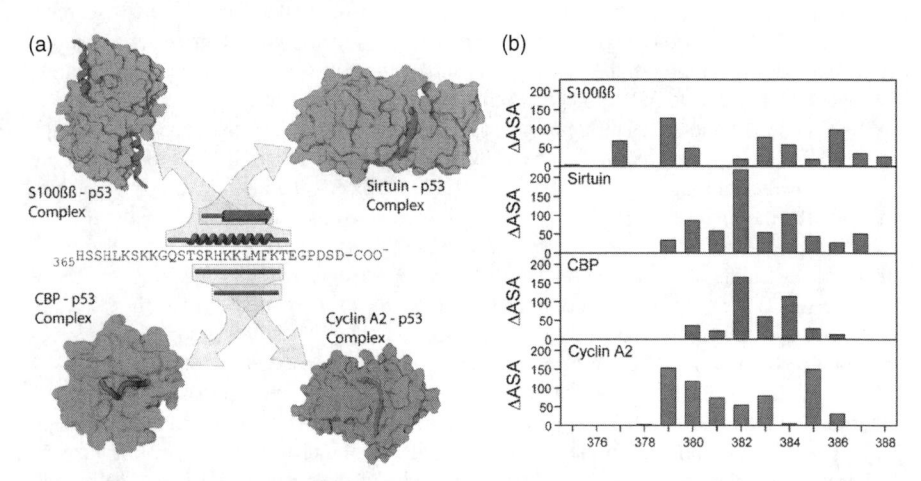

Figure 9.7. Sequence and structure comparison for the four overlapping complexes in the C-terminus of p53. (A) Primary, secondary, and quaternary structure of p53 complexes. (B) The ΔASA for rigid association between the components of complexes for each residue in the relevant sequence region of p53. Reproduced from Ref. [137]. See color insert.

9.4.2. Mechanisms of the 14-3-3 Multiple Specificity

In PDB, five structures of sequence identical to 14-3-3ζ bound to five distinct partners: a peptide from the tail of histone H3 [133], serotonin N-acetyltransferase (AANAT) [134], a phage display-derived peptide (R18) [135], and motif 1 and 2 peptides (m1 and m2, respectively) [117]. For AANAT, only the region bound to the canonical 14-3-3 binding site is considered and the globular region is neglected. Two additional structures are available but are either unsuitable for structural analysis or are highly redundant with another structure. All peptides are phosphorylated in their respective structures, except R18.

The sequences of these five peptides were aligned structurally. The 14-3-3 domain structures were multiply aligned, without considering the bound peptides. Then the alignment was anchored manually by observed correspondence of peptide C_α atoms at the 0 and −1 positions and extending the alignment without gaps from the anchor positions (Fig. 9.8A). In terms of sequence, the R18 sequence has no identical positions to any other peptide. The number of identities between the other peptides ranges from 1 to 4.

Figure 9.7 shows that there is little divergence in the backbone trajectories of the five peptides from positions −3 to 1 but large divergences at either end of the alignment (Fig. 9.8A), which is apparent qualitatively in the superimposed structures of the five peptides (Fig. 9.8B). This divergence is loosely correlated with sequence similarity, where positions with three identical residues have a lower divergence than those with no identical residues. This suggests that 14-3-3 may use different binding pocket residues to bind to different peptide residues.

To gain further insight into 14-3-3 multiple specificities, we compared a pair of 14-3-3 binding peptides, m1 and m2, in detail. These peptides were derived from two motifs, identified through the screening of peptide libraries for sequences that

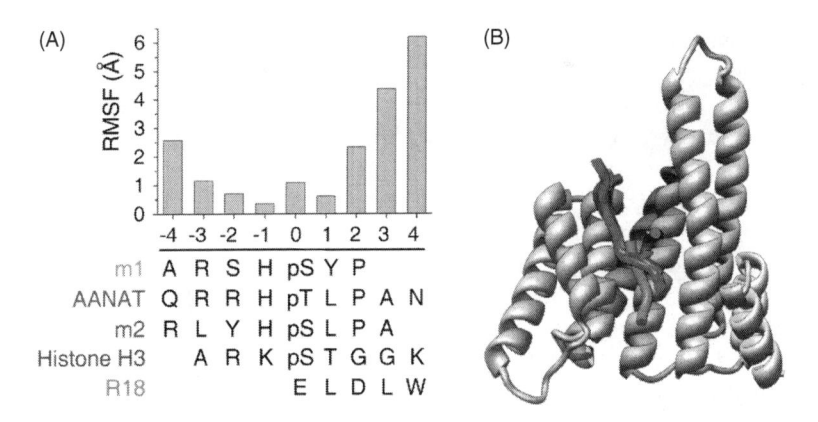

Figure 9.8. Sequence and structure for five peptides bound to 14-3-3ζ. (A) Sequence alignment of the peptides and the root mean square fluctuation (RMSF) of their conformations. (B) Aligned ribbon representations of the structures of the five peptides, which were aligned through multiple alignments of their respectively bound 14-3-3 domains, show along with a representative ribbon representation of a 14-3-3 domain. Reproduced from Ref. [137].

bound to all 14-3-3 isoforms [118]. These two peptide structures have been compared previously [117], but we reanalyze these structural data from the order–disorder point of view [137]. As noted previously [117], the backbone traces of the two peptides are noticeably different, even though the m1 and m2 peptides bind to essentially the same region of 14-3-3ζ (Fig. 9.9A,B, respectively). Examining the side-chain interactions of these peptides with specific 14-3-3 residues shows that there is a difference in the location and identity of residues involved. Similarly, distinctive hydrogen bonding patterns are exhibited between the two peptides and 14-3-3ζ and between the two peptides and bound water (Fig. 9.9C,D). Since a cardinal feature of a structured protein is internal satisfaction of hydrogen bond donors and acceptors, these data are both consistent with the peptides being from unstructured regions of protein before binding.

Therefore, 14-3-3ζ has distinct conformations when bound to the two different peptides. Overlaying the backbone structures of the four binding helices from both complexes—based on a pairwise alignment of the complete domains—shows only minor variability in conformation, with the most occurring at the N-terminus of the helix spanning residues 216 to 228 (Fig. 9.9E). Finally, comparison of side-chain conformations in the two complexes shows significant differences in several of 14-3-3ζ side chains (Fig. 9.9F); residues outlined in red show significant movement) and several other minor differences. Overall, these data suggest that a difference in the conformations of some side chains with rather less difference in backbone conformations is sufficient to accommodate the binding of two different phosphopeptides by the 14-3-3ζ molecule.

9.5. PRINCIPLES OF USING THE UNFOLDED PROTEIN REGIONS FOR BINDING

In general, unfolded peptides are commonly utilized in one-to-many and in many-to-one binding scenarios, both of which are specific cases of the date hubs, which can bind different proteins, but not at the same time. In the first mechanism, one unfolded segment is used by a protein to interact with multiple unrelated binding partners (e.g., N-terminal region of p53). In the second mechanism, many unrelated unfolded fragments are used by unrelated proteins to interact with the same partner (e.g., 14-3-3).

The p53 data discussed above as well as results of recent bioinformatics studies [36–40]clearly show that intrinsic disorder is heavily utilized for protein partner binding. Many of the binding sites on p53 are indicated on the order–disorder predictions as dips, meaning short segments with order tendency flanked by regions of disorder tendency on either side (see Figs. 9.5 and 9.6). Although the observed binding sites in the disordered regions of p53 are associated with a localized tendency for ordered structure, not all disorder-associated binding sites have such features in their disorder prediction plots. We have found many binding sites that are associated with high disorder prediction values across the entire spans of the binding sites. Many of these form irregular structures upon binding with their

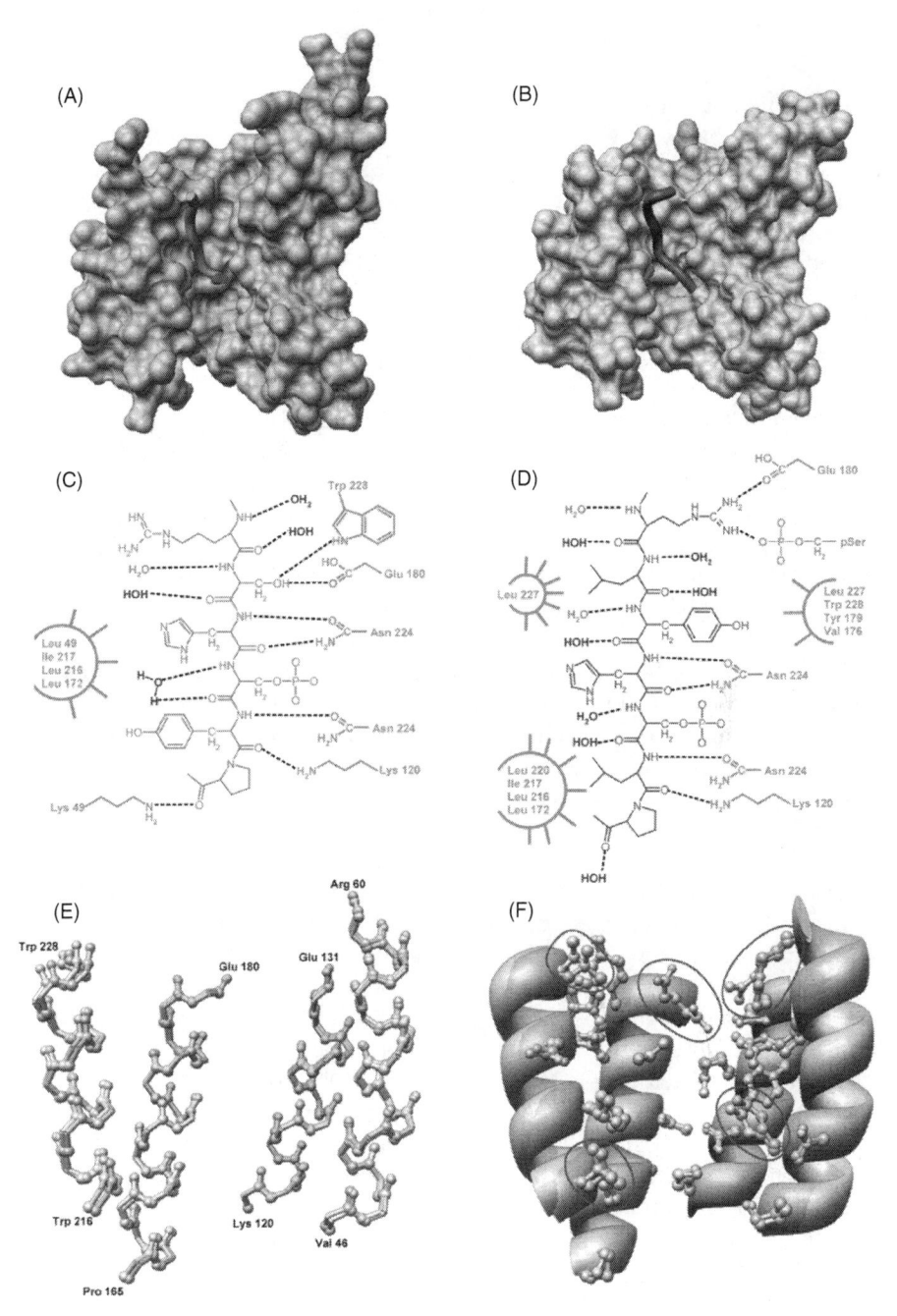

Figure 9.9. Detailed analysis of 14-3-3ξ peptide binding. The m1 peptide (A, orange ribbon) and m2 peptide (B, red ribbon) bound to 14-3-3 (A and B, show by the green and blue surface, respectively). Details of 14-3-3 peptide binding are shown by a chemical schematic for the m1 peptide (C) and the m2 peptide (D), where both crystallographic waters (blue) and implicit waters (red) are shown. (E) Superposition of the backbone atoms from the four helices with the primary peptide binding residues for m1 (green) and m2 (blue) bound 14-3-3. (F) Superposition of ribbons of the same four helices showing the side chains of the residues that participate in m1 (green) and/or m2 (blue) binding. Reproduced from Ref. [137].

partners, and often such binding sites are also rich in proline. Our recent study of the complexes that form when various disordered segments bind to ordered partners indicates that the disorder-associated binding regions have distinct sequence features, even when the bound structure is irregular or sheet instead of helix, and so it should be feasible to develop predictors for the different types of MoRFs [161].

9.5.1. One-to-Many Signaling

Figure 9.6 shows how a single region of p53 is used to interact with four different partners. The amino acids involved in each interaction show a significant overlap and no two of these interactions could exist simultaneously. Furthermore, the same residues adopt helix, sheet, and two different irregular structures when associated with the different partners. Finally, the same amino acids are buried to different extents in each of the molecular associations. These results show very clearly how one segment of disordered protein can bind to multiple partners via the ability to adopt distinct conformations. In essence, these pliable unfolded regions allow IDPs to be "poly-linguistic" and communicate to different partners using different "languages."

The idea that one segment of protein can adopt different secondary structures depending on the context is not new. Many unrelated proteins have identical subsequences of length six, and sometimes even up to length eight, with the same sequences often adopting different secondary structures in different contexts [162–164]. Such sequences have been called chameleons for their ability to adopt different structures in different environments [163–169]. Chameleon behavior could be an important feature that enables one disordered region to bind to multiple partners. With different secondary structures and with different side-chain participation in the different complexes, it is as if one sequence can be "read" in multiple ways by the various binding partners.

The chameleon behavior is typical not only for short peptides (octamers), but also for longer protein fragments and even for entire proteins. For example, it has been found that the 17 residues-long arginine-rich RNA binding domain (residues 65–81) of the Jembrana disease virus (JDV) Tat protein is able to recognize two different transactivating response element (TAR) RNA sites, from human and bovine immunodeficiency viruses (HIV and BIV, respectively), adopting different conformations in the two RNA contexts and using different amino acids for recognition [166]. In addition to the conformational differences, the JDV domain requires the cyclin T1 protein for high-affinity binding to HIV TAR, but not to BIV TAR [166]. Another example is human α-synuclein, a protein implicated in Parkinson's disease and in a number of other neurodegenerative disorders known as synucleinopathies. This protein may either stay substantially unfolded, or adopt an amyloidogenic partially folded conformation, or fold into α-helical or β-structural species, both monomeric and oligomeric. Furthermore, it might form several morphologically different types of aggregates, including oligomers (spheres or doughnuts), amorphous aggregates, or amyloid-like fibrils [33].

Furthermore, it is likely that these chameleon sequences, or at least multiple specificity binding sites, are common in p53. For example, the disordered C-terminus (~100 residues) is known to interact with 44 distinct partners. The average length of a binding site in this region is ~14 residues, which means that on average only 100/14 = ~7 partners could bind distinct regions of the C-terminus. Therefore, at minimum, each 14 residue segment of the N-terminus must recognize 6–7 distinct proteins. This simple back-of-the-envelope calculation suggests that multiple specificity sequences may be the rule for p53 interactions, rather than a curiosity of a single region [136, 137].

9.5.2. Many-to-One Signaling

In 14-3-3, which illustrates the many-to-one binding scenario, a common binding groove in a structured dimeric protein is used to bind to multiple, distinct sequences provided by different binding partners. A recent bioinformatics study revealed that the only common feature connecting about 200 various 14-3-3-binding proteins is the fact that their14-3-3-recognitionregions were predicted to be highly disordered by multiple disorder prediction methods [119]. This clearly suggested that the 14-3-3 recognition generally involved coupled binding and folding of the recognition region. The flexibility of disordered segments was previously suggested as a mechanism enabling slightly different sequences to spatially rearrange so as to be able to bind to a common binding site, thereby facilitating many-to-one signaling [34]. This mechanism is partially responsible for multiple recognition of 14-3-3, since some peptides take different paths through the binding cleft and interaction with binding site residues to different extents.

9.5.3. Unfolded Peptides and One-to-Many versus Many-to-One Signaling

The examples of the p53 C-terminus and 14-3-3 considered above demonstrated different roles of intrinsic disorder in enabling multiple specificities. In p53, drastic conformational changes enable distinct surfaces of an IDR to be exposed to binding partners and, as a consequence, to form structurally different elements in the bound state. In 14-3-3, subtle differences in 14-3-3 conformation and peptide binding locations enable multiple specificities. Why would nature use one mechanism rather than the other for a particular biological role? The interactions of p53 serve to activate or inhibit its primary role as a transcription regulator, while 14-3-3 alters the functions or subcellular localization of many proteins. From this, some highly speculative proposals were made [136, 137]: (1) disordered binding regions play a passive role in regulation by simply providing a specific binding site (e.g., such disordered regions can serve as the identification sites of the protein to be regulated); and (2) ordered proteins play the active role in regulation by altering the activity of the proteins to which they bind (e.g., recognition of disordered regions by an ordered partner allows for a generalized specificity since a single ordered protein can alter the activity of many others).

9.6. CONCLUSIONS

Disordered or unfolded regions may be an extremely common mechanism by which hub proteins bind to their multitude of partners [36–40]. Several algorithms have been developed for predicting disordered regions involved in molecular recognition, including the MoRF, ANCHOR, and ELM predictors. Several common features of disordered binding regions have been identified, such as preformed structural elements, PCSs, and anchor residues, which furthers our understanding of disorder-mediated binding. The specific examples of p53 and 14-3-3 contrast the mechanisms by which disorder facilitates multiple recognition, where the former involves drastic conformational differences in a single disordered region and the latter involves a variety of subtler changes in order to recognize multiple disordered regions. Furthermore, the differences between one-to-many interactions (e.g., the binding of the disordered region of p53 to unrelated, structurally different ordered partners) and many-to-one signaling scenario (e.g., the binding of different disordered regions of unrelated partners to the ordered 14-3-3 dimer) may have crucial biological implications.

ACKNOWLEDGMENTS

This work was supported in part by grant EF 0849803 from the National Science Foundation to A. K. D and V. N. U., and the Program of the Russian Academy of Sciences for the Molecular and Cellular Biology to V. N. U. This chapter is partially based on previous publications [59, 136, 137, 170].

REFERENCES

1. Goh, K. I., Oh, E., Jeong, H., Kahng, B., and Kim, D. (2002) Classification of scale-free networks, *Proc Natl Acad Sci U S A 99*, 12583–12588.
2. Watts, D. J. and Strogatz, S. H. (1998) Collective dynamics of "small-world" networks, *Nature 393*, 440–442.
3. Erdös, P. and Rényi, A. (1960) On the evolution of random graphs, *Publ Math Inst Hung Acad Sci 5*, 17–61.
4. Barabasi, A. L. and Bonabeau, E. (2003) Scale-free networks, *Sci Am 288*, 60–69.
5. Albert, R., Jeong, H., and Barabasi, A. L. (2000) Error and attack tolerance of complex networks, *Nature 406*, 378–382.
6. Jeong, H., Mason, S. P., Barabasi, A. L., and Oltvai, Z. N. (2001) Lethality and centrality in protein networks, *Nature 411*, 41–42.
7. Milgram, S. (1967) The small world problem, *Psycol Today 2*, 60–67.
8. Bork, P., Jensen, L. J., von Mering, C., Ramani, A. K., Lee, I., and Marcotte, E. M. (2004) Protein interaction networks from yeast to human, *Curr Opin Struct Biol 14*, 292–299.
9. Barabasi, A. L. and Oltvai, Z. N. (2004) Network biology: Understanding the cell's functional organization, *Nat Rev Genet 5*, 101–113.

REFERENCES

10 Wu, C. H., Huang, H., Nikolskaya, A., Hu, Z., and Barker, W. C. (2004) The iProClass integrated database for protein functional analysis, *Comput Biol Chem 28*, 87–96.

11 Huang, T. W., Tien, A. C., Huang, W. S., Lee, Y. C., Peng, C. L., Tseng, H. H., Kao, C. Y., and Huang, C. Y. (2004) POINT: A database for the prediction of protein-protein interactions based on the orthologous interactome, *Bioinformatics 20*, 3273–3276.

12 Kelley, B. P., Yuan, B., Lewitter, F., Sharan, R., Stockwell, B. R., and Ideker, T. (2004) PathBLAST: A tool for alignment of protein interaction networks, *Nucleic Acids Res 32*, W83–W88.

13 von Mering, C., Jensen, L. J., Snel, B., Hooper, S. D., Krupp, M., Foglierini, M., Jouffre, N., Huynen, M. A., and Bork, P. (2005) STRING: Known and predicted protein-protein associations, integrated and transferred across organisms, *Nucleic Acids Res 33*, D433–D437.

14 Hahn, M. W. and Kern, A. D. (2005) Comparative genomics of centrality and essentiality in three eukaryotic protein-interaction networks, *Mol Biol Evol 22*, 803–806.

15 Huang, S. (2004) Back to the biology in systems biology: What can we learn from biomolecular networks? *Brief Funct Genomic Proteomic 2*, 279–297.

16 Han, J. D., Bertin, N., Hao, T., Goldberg, D. S., Berriz, G. F., Zhang, L. V., Dupuy, D., Walhout, A. J., Cusick, M. E., Roth, F. P., and Vidal, M. (2004) Evidence for dynamically organized modularity in the yeast protein-protein interaction network, *Nature 430*, 88–93.

17 Hartwell, L. H., Hopfield, J. J., Leibler, S., and Murray, A. W. (1999) From molecular to modular cell biology, *Nature 402*, C47–C52.

18 Cesareni, G., Ceol, A., Gavrila, C., Palazzi, L. M., Persico, M., and Schneider, M. V. (2005) Comparative interactomics, *FEBS Lett 579*, 1828–1833.

19 von Mering, C., Krause, R., Snel, B., Cornell, M., Oliver, S. G., Fields, S., and Bork, P. (2002) Comparative assessment of large-scale data sets of protein-protein interactions, *Nature 417*, 399–403.

20 Bader, G. D. and Hogue, C. W. (2002) Analyzing yeast protein-protein interaction data obtained from different sources, *Nat Biotechnol 20*, 991–997.

21 Ito, T., Chiba, T., Ozawa, R., Yoshida, M., Hattori, M., and Sakaki, Y. (2001) A comprehensive two-hybrid analysis to explore the yeast protein interactome, *Proc Natl Acad Sci U S A 98*, 4569–4574.

22 Uetz, P., Giot, L., Cagney, G., Mansfield, T. A., Judson, R. S., Knight, J. R., Lockshon, D., Narayan, V., Srinivasan, M., Pochart, P., Qureshi-Emili, A., Li, Y., Godwin, B., Conover, D., Kalbfleisch, T., Vijayadamodar, G., Yang, M., Johnston, M., Fields, S., and Rothberg, J. M. (2000) A comprehensive analysis of protein-protein interactions *in Saccharomyces cerevisiae*, *Nature 403*, 623–627.

23 Li, S., Armstrong, C. M., Bertin, N., Ge, H., Milstein, S., Boxem, M., Vidalain, P. O., Han, J. D., Chesneau, A., Hao, T., Goldberg, D. S., Li, N., Martinez, M., Rual, J. F., Lamesch, P., Xu, L., Tewari, M., Wong, S. L., Zhang, L. V., Berriz, G. F., Jacotot, L., Vaglio, P., Reboul, J., Hirozane-Kishikawa, T., Li, Q., Gabel, H. W., Elewa, A., Baumgartner, B., Rose, D. J., Yu, H., Bosak, S., Sequerra, R., Fraser, A., Mango, S. E., Saxton, W. M., Strome, S., Van Den Heuvel, S., Piano, F., Vandenhaute, J., Sardet, C., Gerstein, M., Doucette-Stamm, L., Gunsalus, K. C., Harper, J. W., Cusick, M. E., Roth, F. P., Hill, D. E., and Vidal, M. (2004) A map of the interactome network of the metazoan *C. elegans*, *Science 303*, 540–543.

24. Giot, L., Bader, J. S., Brouwer, C., Chaudhuri, A., Kuang, B., Li, Y., Hao, Y. L., Ooi, C. E., Godwin, B., Vitols, E., Vijayadamodar, G., Pochart, P., Machineni, H., Welsh, M., Kong, Y., Zerhusen, B., Malcolm, R., Varrone, Z., Collis, A., Minto, M., Burgess, S., McDaniel, L., Stimpson, E., Spriggs, F., Williams, J., Neurath, K., Ioime, N., Agee, M., Voss, E., Furtak, K., Renzulli, R., Aanensen, N., Carrolla, S., Bickelhaupt, E., Lazovatsky, Y., DaSilva, A., Zhong, J., Stanyon, C. A., Finley, R. L., Jr, White, K. P., Braverman, M., Jarvie, T., Gold, S., Leach, M., Knight, J., Shimkets, R. A., McKenna, M. P., Chant, J., and Rothberg, J. M. (2003) A protein interaction map of *Drosophila melanogaster*, *Science 302*, 1727–1736.

25. Hasty, J. and Collins, J. J. (2001) Protein interactions. Unspinning the web, *Nature 411*, 30–31.

26. Fischer, E. (1894) Einfluss der configuration auf die wirkung derenzyme, *Ber Dt Chem Ges 27*, 2985–2993.

27. Koshland, D. E., Jr., Ray, W. J., Jr., and Erwin, M. J. (1958) Protein structure and enzyme action, *Fed Proc 17*, 1145–1150.

28. Landsteiner, K. (1936) *The specificity of serological reactions*, Courier Dover Publications, Mineola.

29. Pauling, L. (1940) A theory of the structure and process of formation of antibodies, *J Am Chem Soc 62*, 2643–2657.

30. Karush, F. (1950) Heterogeneity of the binding sites of bovine serum albumin, *J Am Chem Soc 72*, 2705–2713.

31. Meador, W. E., Means, A. R., and Quiocho, F. A. (1993) Modulation of calmodulin plasticity in molecular recognition on the basis of x-ray structures, *Science 262*, 1718–1721.

32. Kriwacki, R. W., Hengst, L., Tennant, L., Reed, S. I., and Wright, P. E. (1996) Structural studies of p21Waf1/Cip1/Sdi1 in the free and Cdk2-bound state: Conformational disorder mediates binding diversity, *Proc Natl Acad Sci U S A 93*, 11504–11509.

33. Uversky, V. N. (2003) A protein-chameleon: Conformational plasticity of alpha-synuclein, a disordered protein involved in neurodegenerative disorders, *J Biomol Struct Dyn 21*, 211–234.

34. Dunker, A. K., Garner, E., Guilliot, S., Romero, P., Albrecht, K., Hart, J., Obradovic, Z., Kissinger, C., and Villafranca, J. E. (1998) Protein disorder and the evolution of molecular recognition: Theory, predictions and observations, *Pac Symp Biocomput*, 473–484.

35. Dunker, A. K., Cortese, M. S., Romero, P., Iakoucheva, L. M., and Uversky, V. N. (2005) Flexible nets. The roles of intrinsic disorder in protein interaction networks, *FEBS J 272*, 5129–5148.

36. Patil, A. and Nakamura, H. (2006) Disordered domains and high surface charge confer hubs with the ability to interact with multiple proteins in interaction networks, *FEBS Lett 580*, 2041–2045.

37. Haynes, C., Oldfield, C. J., Ji, F., Klitgord, N., Cusick, M. E., Radivojac, P., Uversky, V. N., Vidal, M., and Iakoucheva, L. M. (2006) Intrinsic disorder is a common feature of hub proteins from four eukaryotic interactomes, *PLoS Comput Biol 2*, e100.

38. Ekman, D., Light, S., Bjorklund, A. K., and Elofsson, A. (2006) What properties characterize the hub proteins of the protein-protein interaction network of *Saccharomyces cerevisiae*? *Genome Biol 7*, R45.

REFERENCES

39. Dosztanyi, Z., Chen, J., Dunker, A. K., Simon, I., and Tompa, P. (2006) Disorder and sequence repeats in hub proteins and their implications for network evolution, *J Proteome Res 5*, 2985–2995.
40. Singh, G. P., Ganapathi, M., Sandhu, K. S., and Dash, D. (2006) Intrinsic unstructuredness and abundance of PEST motifs in eukaryotic proteomes, *Proteins 62*, 309–315.
41. Marinissen, M. J. and Gutkind, J. S. (2005) Scaffold proteins dictate Rho GTPase-signaling specificity, *Trends Biochem Sci 30*, 423–426.
42. Jaffe, A. B., Aspenstrom, P., and Hall, A. (2004) Human CNK1 acts as a scaffold protein, linking Rho and Ras signal transduction pathways, *Mol Cell Biol 24*, 1736–1746.
43. Jaffe, A. B. and Hall, A. (2005) Rho GTPases: Biochemistry and biology, *Annu Rev Cell Dev Biol 21*, 247–269.
44. Hohenstein, P. and Giles, R. H. (2003) BRCA1: A scaffold for p53 response? *Trends Genet 19*, 489–494.
45. Luo, W. and Lin, S. C. (2004) Axin: A master scaffold for multiple signaling pathways, *Neurosignals 13*, 99–113.
46. Rui, Y., Xu, Z., Lin, S., Li, Q., Rui, H., Luo, W., Zhou, H. M., Cheung, P. Y., Wu, Z., Ye, Z., Li, P., Han, J., and Lin, S. C. (2004) Axin stimulates p53 functions by activation of HIPK2 kinase through multimeric complex formation, *Embo J 23*, 4583–4594.
47. Salahshor, S. and Woodgett, J. R. (2005) The links between axin and carcinogenesis, *J Clin Pathol 58*, 225–236.
48. Wong, W. and Scott, J. D. (2004) AKAP signalling complexes: Focal points in space and time, *Nat Rev Mol Cell Biol 5*, 959–970.
49. Carpousis, A. J. (2007) The RNA Degradosome of *Escherichia coli*: A multiprotein mRNA-degrading machine assembled on RNase E, *Annu Rev Microbiol 61*, 71–87.
50. Kim, P. M., Lu, L. J., Xia, Y., and Gerstein, M. B. (2006) Relating three-dimensional structures to protein networks provides evolutionary insights, *Science 314*, 1938–1941.
51. Nooren, I. M. and Thornton, J. M. (2003) Diversity of protein-protein interactions, *EMBO J 22*, 3486–3492.
52. Perkins, J. R., Diboun, I., Dessailly, B. H., Lees, J. G., and Orengo, C. (2010) Transient protein-protein interactions: Structural, functional, and network properties, *Structure 18*, 1233–1243.
53. Oldfield, C. J., Cheng, Y., Cortese, M. S., Romero, P., Uversky, V. N., and Dunker, A. K. (2005) Coupled folding and binding with alpha-helix-forming molecular recognition elements, *Biochemistry 44*, 12454–12470.
54. Garner, E., Romero, P., Dunker, A. K., Brown, C., and Obradovic, Z. (1999) Predicting binding regions within disordered proteins, *Genome Inform Ser Workshop Genome Inform 10*, 41–50.
55. Mader, S., Lee, H., Pause, A., and Sonenberg, N. (1995) The translation initiation factor eIF-4E binds to a common motif shared by the translation factor eIF-4 gamma and the translational repressors 4E-binding proteins, *Mol Cell Biol 15*, 4990–4997.
56. Fletcher, C. M. and Wagner, G. (1998) The interaction of eIF4E with 4E-BP1 is an induced fit to a completely disordered protein, *Protein Sci 7*, 1639–1642.
57. Cheng, Y., Oldfield, C. J., Meng, J., Romero, P., Uversky, V. N., and Dunker, A. K. (2007) Mining alpha-helix-forming molecular recognition features with cross species sequence alignments, *Biochemistry 46*, 13468–13477.

58. Mohan, A., Oldfield, C. J., Radivojac, P., Vacic, V., Cortese, M. S., Dunker, A. K., and Uversky, V. N. (2006) Analysis of molecular recognition features (MoRFs), *J Mol Biol 362*, 1043–1059.

59. Xue, B., Dunker, A. K., and Uversky, V. N. (2010) Retro-MoRFs: Identifying protein binding sites by normal and reverse alignment and intrinsic disorder prediction, *Int J Mol Sci 11*, 3725–3747.

60. Guptasarma, P. (1992) Reversal of peptide backbone direction may result in the mirroring of protein structure, *FEBS Lett 310*, 205–210.

61. Schoniger, M. and Waterman, M. S. (1992) A local algorithm for DNA sequence alignment with inversions, *Bull Math Biol 54*, 521–536.

62. Olszewski, K. A., Kolinski, A., and Skolnick, J. (1996) Does a backwardly read protein sequence have a unique native state? *Protein Eng 9*, 5–14.

63. Lacroix, E., Viguera, A. R., and Serrano, L. (1998) Reading protein sequences backwards, *Fold Des 3*, 79–85.

64. Preissner, R., Goede, A., Michalski, E., and Frommel, C. (1997) Inverse sequence similarity in proteins and its relation to the three-dimensional fold, *FEBS Lett 414*, 425–429.

65. Pan, P. K., Zheng, Z. F., Lyu, P. C., and Huang, P. C. (1999) Why reversing the sequence of the alpha domain of human metallothionein-2 does not change its metal-binding and folding characteristics, *Eur J Biochem 266*, 33–39.

66. Rath, A., Davidson, A. R., and Deber, C. M. (2005) The structure of "unstructured" regions in peptides and proteins: Role of the polyproline II helix in protein folding and recognition, *Biopolymers 80*, 179–185.

67. Creamer, T. P. and Campbell, M. N. (2002) Determinants of the polyproline II helix from modeling studies, *Adv Protein Chem 62*, 263–282.

68. Kay, B. K., Williamson, M. P., and Sudol, M. (2000) The importance of being proline: The interaction of proline-rich motifs in signaling proteins with their cognate domains, *FASEB J 14*, 231–241.

69. Cesareni, G., Panni, S., Nardelli, G., and Castagnoli, L. (2002) Can we infer peptide recognition specificity mediated by SH3 domains? *FEBS Lett 513*, 38–44.

70. Dalgarno, D. C., Botfield, M. C., and Rickles, R. J. (1997) SH3 domains and drug design: Ligands, structure, and biological function, *Biopolymers 43*, 383–400.

71. Feng, S., Chen, J. K., Yu, H., Simon, J. A., and Schreiber, S. L. (1994) Two binding orientations for peptides to the Src SH3 domain: Development of a general model for SH3-ligand interactions, *Science 266*, 1241–1247.

72. Thompson, J. D., Higgins, D. G., and Gibson, T. J. (1994) CLUSTAL W: Improving the sensitivity of progressive multiple sequence alignment through sequence weighting, position-specific gap penalties and weight matrix choice, *Nucleic Acids Res 22*, 4673–4680.

73. Xue, B., Dunbrack, R. L., Williams, R. W., Dunker, A. K., and Uversky, V. N. (2010) PONDR-FIT: A meta-predictor of intrinsically disordered amino acids, *Biochim Biophys Acta 1804*, 996–1010.

74. Fuxreiter, M., Simon, I., Friedrich, P., and Tompa, P. (2004) Preformed structural elements feature in partner recognition by intrinsically unstructured proteins, *J Mol Biol 338*, 1015–1026.

REFERENCES

75 Parker, D., Rivera, M., Zor, T., Henrion-Caude, A., Radhakrishnan, I., Kumar, A., Shapiro, L. H., Wright, P. E., Montminy, M., and Brindle, P. K. (1999) Role of secondary structure in discrimination between constitutive and inducible activators, *Mol Cell Biol 19*, 5601–5607.

76 Radhakrishnan, I., Perez-Alvarado, G. C., Dyson, H. J., and Wright, P. E. (1998) Conformational preferences in the Ser133-phosphorylated and non-phosphorylated forms of the kinase inducible transactivation domain of CREB, *FEBS Lett 430*, 317–322.

77 Lacy, E. R., Filippov, I., Lewis, W. S., Otieno, S., Xiao, L., Weiss, S., Hengst, L., and Kriwacki, R. W. (2004) p27 binds cyclin-CDK complexes through a sequential mechanism involving binding-induced protein folding, *Nat Struct Mol Biol 11*, 358–364.

78 Lee, H., Mok, K. H., Muhandiram, R., Park, K. H., Suk, J. E., Kim, D. H., Chang, J., Sung, Y. C., Choi, K. Y., and Han, K. H. (2000) Local structural elements in the mostly unstructured transcriptional activation domain of human p53, *J Biol Chem 275*, 29426–29432.

79 Daughdrill, G. W., Hanely, L. J., and Dahlquist, F. W. (1998) The C-terminal half of the anti-sigma factor FlgM contains a dynamic equilibrium solution structure favoring helical conformations, *Biochemistry 37*, 1076–1082.

80 Dedmon, M. M., Patel, C. N., Young, G. B., and Pielak, G. J. (2002) FlgM gains structure in living cells, *Proc Natl Acad Sci U S A 99*, 12681–12684.

81 Hauer, J. A., Barthe, P., Taylor, S. S., Parello, J., and Padilla, A. (1999) Two well-defined motifs in the cAMP-dependent protein kinase inhibitor (PKIalpha) correlate with inhibitory and nuclear export function, *Protein Sci 8*, 545–553.

82 Domanski, M., Hertzog, M., Coutant, J., Gutsche-Perelroizen, I., Bontems, F., Carlier, M. F., Guittet, E., and van Heijenoort, C. (2004) Coupling of folding and binding of thymosin beta4 upon interaction with monomeric actin monitored by nuclear magnetic resonance, *J Biol Chem 279*, 23637–23645.

83 Longhi, S., Receveur-Brechot, V., Karlin, D., Johansson, K., Darbon, H., Bhella, D., Yeo, R., Finet, S., and Canard, B. (2003) The C-terminal domain of the measles virus nucleoprotein is intrinsically disordered and folds upon binding to the C-terminal moiety of the phosphoprotein, *J Biol Chem 278*, 18638–18648.

84 Sivakolundu, S. G., Bashford, D., and Kriwacki, R. W. (2005) Disordered p27Kip1 exhibits intrinsic structure resembling the Cdk2/cyclin A-bound conformation, *J Mol Biol 353*, 1118–1128.

85 Pontius, B. W. (1993) Close encounters: Why unstructured, polymeric domains can increase rates of specific macromolecular association, *Trends Biochem Sci 18*, 181–186.

86 Csizmok, V., Bokor, M., Banki, P., Klement, É., Medzihradszky, K. F., Friedrich, P., Tompa, K., and Tompa, P. (2005) Primary contact sites in intrinsically unstructured proteins: The case of calpastatin and microtubule-associated protein 2, *Biochemistry 44*, 3955–3964.

87 Kalthoff, C., Alves, J., Urbanke, C., Knorr, R., and Ungewickell, E. J. (2002) Unusual structural organization of the endocytic proteins AP180 and epsin 1, *J Biol Chem 277*, 8209–8216.

88 Scheele, U., Alves, J., Frank, R., Duwel, M., Kalthoff, C., and Ungewickell, E. (2003) Molecular and functional characterization of clathrin- and AP-2-binding determinants within a disordered domain of auxilin, *J Biol Chem 278*, 25357–25368.

89 Dafforn, T. R. and Smith, C. J. (2004) Natively unfolded domains in endocytosis: Hooks, lines and linkers, *EMBO Rep 5*, 1046–1052.

90 Vise, P., Baral, B., Stancik, A., Lowry, D. F., and Daughdrill, G. W. (2007) Identifying long-range structure in the intrinsically unstructured transactivation domain of p53, *Proteins 67*, 526–530.

91 Neduva, V. and Russell, R. B. (2005) Linear motifs: Evolutionary interaction switches, *FEBS Lett 579*, 3342–3345.

92 Neduva, V. and Russell, R. B. (2006) DILIMOT: Discovery of linear motifs in proteins, *Nucleic Acids Res 34*, W350–W355.

93 Davey, N. E., Shields, D. C., and Edwards, R. J. (2006) SLiMDisc: Short, linear motif discovery, correcting for common evolutionary descent, *Nucleic Acids Res 34*, 3546–3554.

94 Puntervoll, P., Linding, R., Gemund, C., Chabanis-Davidson, S., Mattingsdal, M., Cameron, S., Martin, D. M., Ausiello, G., Brannetti, B., Costantini, A., Ferre, F., Maselli, V., Via, A., Cesareni, G., Diella, F., Superti-Furga, G., Wyrwicz, L., Ramu, C., McGuigan, C., Gudavalli, R., Letunic, I., Bork, P., Rychlewski, L., Kuster, B., Helmer-Citterich, M., Hunter, W. N., Aasland, R., and Gibson, T. J. (2003) ELM server: A new resource for investigating short functional sites in modular eukaryotic proteins, *Nucleic Acids Res 31*, 3625–3630.

95 Linding, R., Russell, R. B., Neduva, V., and Gibson, T. J. (2003) GlobPlot: Exploring protein sequences for globularity and disorder, *Nucleic Acids Res 31*, 3701–3708.

96 Fuxreiter, M., Tompa, P., and Simon, I. (2007) Structural disorder imparts plasticity on linear motifs, *Bioinformatics 23*, 950–956.

97 Dosztanyi, Z., Meszaros, B., and Simon, I. (2009) ANCHOR: Web server for predicting protein binding regions in disordered proteins, *Bioinformatics 25*, 2745–2746.

98 Dosztanyi, Z., Csizmok, V., Tompa, P., and Simon, I. (2005) The pairwise energy content estimated from amino acid composition discriminates between folded and intrinsically unstructured proteins, *J Mol Biol 347*, 827–839.

99 Dosztanyi, Z., Csizmok, V., Tompa, P., and Simon, I. (2005) IUPred: Web server for the prediction of intrinsically unstructured regions of proteins based on estimated energy content, *Bioinformatics 21*, 3433–3434.

100 Meszaros, B., Simon, I., and Dosztanyi, Z. (2009) Prediction of protein binding regions in disordered proteins, *PLoS Comput Biol 5*, e1000376.

101 Rajamani, D., Thiel, S., Vajda, S., and Camacho, C. J. (2004) Anchor residues in protein-protein interactions, *Proc Natl Acad Sci U S A 101*, 11287–11292.

102 Kimura, S. R., Brower, R. C., Vajda, S., and Camacho, C. J. (2001) Dynamical view of the positions of key side chains in protein-protein recognition, *Biophys J 80*, 635–642.

103 Li, X., Keskin, O., Ma, B., Nussinov, R., and Liang, J. (2004) Protein-protein interactions: Hot spots and structurally conserved residues often locate in complemented pockets that pre-organized in the unbound states: Implications for docking, *J Mol Biol 344*, 781–795.

104 Camacho, C. J. (2005) Modeling side-chains using molecular dynamics improve recognition of binding region in CAPRI targets, *Proteins 60*, 245–251.

105 Smith, G. R., Sternberg, M. J., and Bates, P. A. (2005) The relationship between the flexibility of proteins and their conformational states on forming protein-protein complexes with an application to protein-protein docking, *J Mol Biol 347*, 1077–1101.

REFERENCES

106 Yogurtcu, O. N., Erdemli, S. B., Nussinov, R., Turkay, M., and Keskin, O. (2008) Restricted mobility of conserved residues in protein-protein interfaces in molecular simulations, *Biophys J 94*, 3475–3485.

107 Ben-Shimon, A. and Eisenstein, M. (2010) Computational mapping of anchoring spots on protein surfaces, *J Mol Biol 402*, 259–277.

108 Huang, Y. and Liu, Z. (2011) Anchoring intrinsically disordered proteins to multiple targets: Lessons from p53 N-terminal domain, *Int J Mol Sci 12*, 1410–1430.

109 Anderson, C. W. and Appella, E. (2004) Signaling to the p53 tumor suppressor through pathways activated by genotoxic and nongenotoxic stress. In: Bradshaw, R. A. and Dennis, E. A. (eds.), *Handbook of cell signaling*. Academic Press, New York, pp. 237–247.

110 Hollstein, M., Sidransky, D., Vogelstein, B., and Harris, C. C. (1991) p53 mutations in human cancers, *Science 253*, 49–53.

111 Zhao, R., Gish, K., Murphy, M., Yin, Y., Notterman, D., Hoffman, W. H., Tom, E., Mack, D. H., and Levine, A. J. (2000) Analysis of p53-regulated gene expression patterns using oligonucleotide arrays, *Genes Dev 14*, 981–993.

112 Dougherty, M. K. and Morrison, D. K. (2004) Unlocking the code of 14-3-3, *J Cell Sci 117*, 1875–1884.

113 Pozuelo Rubio, M., Geraghty, K. M., Wong, B. H., Wood, N. T., Campbell, D. G., Morrice, N., and Mackintosh, C. (2004) 14-3-3-affinity purification of over 200 human phosphoproteins reveals new links to regulation of cellular metabolism, proliferation and trafficking, *Biochem J 379*, 395–408.

114 Meek, S. E., Lane, W. S., and Piwnica-Worms, H. (2004) Comprehensive proteomic analysis of interphase and mitotic 14-3-3-binding proteins, *J Biol Chem 279*, 32046–32054.

115 Jin, J., Smith, F. D., Stark, C., Wells, C. D., Fawcett, J. P., Kulkarni, S., Metalnikov, P., O'Donnell, P., Taylor, P., Taylor, L., Zougman, A., Woodgett, J. R., Langeberg, L. K., Scott, J. D., and Pawson, T. (2004) Proteomic, functional, and domain-based analysis of in vivo 14-3-3 binding proteins involved in cytoskeletal regulation and cellular organization, *Curr Biol 14*, 1436–1450.

116 Yaffe, M. B. (2002) How do 14-3-3 proteins work?—Gatekeeper phosphorylation and the molecular anvil hypothesis, *FEBS Lett 513*, 53–57.

117 Rittinger, K., Budman, J., Xu, J., Volinia, S., Cantley, L. C., Smerdon, S. J., Gamblin, S. J., and Yaffe, M. B. (1999) Structural analysis of 14-3-3 phosphopeptide complexes identifies a dual role for the nuclear export signal of 14-3-3 in ligand binding, *Mol Cell 4*, 153–166.

118 Yaffe, M. B., Rittinger, K., Volinia, S., Caron, P. R., Aitken, A., Leffers, H., Gamblin, S. J., Smerdon, S. J., and Cantley, L. C. (1997) The structural basis for 14-3-3:phosphopeptide binding specificity, *Cell 91*, 961–971.

119 Bustos, D. M. and Iglesias, A. A. (2006) Intrinsic disorder is a key characteristic in partners that bind 14-3-3 proteins, *Proteins 63*, 35–42.

120 Lowe, E. D., Tews, I., Cheng, K. Y., Brown, N. R., Gul, S., Noble, M. E., Gamblin, S. J., and Johnson, L. N. (2002) Specificity determinants of recruitment peptides bound to phospho-CDK2/cyclin A, *Biochemistry 41*, 15625–15634.

121 Avalos, J. L., Celic, I., Muhammad, S., Cosgrove, M. S., Boeke, J. D., and Wolberger, C. (2002) Structure of a Sir2 enzyme bound to an acetylated p53 peptide, *Mol Cell 10*, 523–535.

122 Mujtaba, S., He, Y., Zeng, L., Yan, S., Plotnikova, O., Sachchidanand, Sanchez, R., Zeleznik-Le, N. J., Ronai, Z., and Zhou, M. M. (2004) Structural mechanism of the bromodomain of the coactivator CBP in p53 transcriptional activation, *Mol Cell 13*, 251–263.

123 Wu, H., Maciejewski, M. W., Marintchev, A., Benashski, S. E., Mullen, G. P., and King, S. M. (2000) Solution structure of a dynein motor domain associated light chain, *Nat Struct Biol 7*, 575–579.

124 Chuikov, S., Kurash, J. K., Wilson, J. R., Xiao, B., Justin, N., Ivanov, G. S., McKinney, K., Tempst, P., Prives, C., Gamblin, S. J., Barlev, N. A., and Reinberg, D. (2004) Regulation of p53 activity through lysine methylation, *Nature 432*, 353–360.

125 Poux, A. N. and Marmorstein, R. (2003) Molecular basis for Gcn5/PCAF histone acetyltransferase selectivity for histone and nonhistone substrates, *Biochemistry 42*, 14366–14374.

126 Bochkareva, E., Kaustov, L., Ayed, A., Yi, G. S., Lu, Y., Pineda-Lucena, A., Liao, J. C., Okorokov, A. L., Milner, J., Arrowsmith, C. H., and Bochkarev, A. (2005) Single-stranded DNA mimicry in the p53 transactivation domain interaction with replication protein A, *Proc Natl Acad Sci U S A 102*, 15412–15417.

127 Kussie, P. H., Gorina, S., Marechal, V., Elenbaas, B., Moreau, J., Levine, A. J., and Pavletich, N. P. (1996) Structure of the MDM2 oncoprotein bound to the p53 tumor suppressor transactivation domain, *Science 274*, 948–953.

128 Di Lello, P., Jenkins, L. M., Jones, T. N., Nguyen, B. D., Hara, T., Yamaguchi, H., Dikeakos, J. D., Appella, E., Legault, P., and Omichinski, J. G. (2006) Structure of the Tfb1/p53 complex: Insights into the interaction between the p62/Tfb1 subunit of TFIIH and the activation domain of p53, *Mol Cell 22*, 731–740.

129 Kuszewski, J., Gronenborn, A. M., and Clore, G. M. (1999) Improving the packing and accuracy of NMR structures with a pseudopotential for the radius of gyration, *J Am Chem Soc 121*, 2337–2338.

130 Cho, Y., Gorina, S., Jeffrey, P. D., and Pavletich, N. P. (1994) Crystal structure of a p53 tumor suppressor-DNA complex: Understanding tumorigenic mutations, *Science 265*, 346–355.

131 Joo, W. S., Jeffrey, P. D., Cantor, S. B., Finnin, M. S., Livingston, D. M., and Pavletich, N. P. (2002) Structure of the 53BP1 BRCT region bound to p53 and its comparison to the Brca1 BRCT structure, *Genes Dev 16*, 583–593.

132 Gorina, S. and Pavletich, N. P. (1996) Structure of the p53 tumor suppressor bound to the ankyrin and SH3 domains of 53BP2, *Science 274*, 1001–1005.

133 Macdonald, N., Welburn, J. P., Noble, M. E., Nguyen, A., Yaffe, M. B., Clynes, D., Moggs, J. G., Orphanides, G., Thomson, S., Edmunds, J. W., Clayton, A. L., Endicott, J. A., and Mahadevan, L. C. (2005) Molecular basis for the recognition of phosphorylated and phosphoacetylated histone h3 by 14-3-3, *Mol Cell 20*, 199–211.

134 Obsil, T., Ghirlando, R., Klein, D. C., Ganguly, S., and Dyda, F. (2001) Crystal structure of the 14-3-3zeta: Serotonin N-acetyltransferase complex. A role for scaffolding in enzyme regulation, *Cell 105*, 257–267.

135 Petosa, C., Masters, S. C., Bankston, L. A., Pohl, J., Wang, B., Fu, H., and Liddington, R. C. (1998) 14-3-3zeta binds a phosphorylated Raf peptide and an unphosphorylated peptide via its conserved amphipathic groove, *J Biol Chem 273*, 16305–16310.

136 Oldfield, C. J., Meng, J., Yang, J. Y., Uversky, V. N., and Dunker, A. K. (2007) Intrinsic disorder in protein-protein interaction networks: Case studies of complexes involving

p53 and 14-3-3. In: Arabnia, H. R., Yang, M. Q., and Yang, J. Y. (eds.), *The 2007 international conference on bioinformatics and computational biology*. CSREA Press, Las Vegas, pp. 553–564.

137 Oldfield, C. J., Meng, J., Yang, J. Y., Yang, M. Q., Uversky, V. N., and Dunker, A. K. (2008) Flexible nets: Disorder and induced fit in the associations of p53 and 14-3-3 with their partners, *BMC Genomics 9*, S1.

138 Dawson, R., Muller, L., Dehner, A., Klein, C., Kessler, H., and Buchner, J. (2003) The N-terminal domain of p53 is natively unfolded, *J Mol Biol 332*, 1131–1141.

139 Romero, P., Obradovic, Z., Li, X., Garner, E. C., Brown, C. J., and Dunker, A. K. (2001) Sequence complexity of disordered protein, *Proteins 42*, 38–48.

140 Obradovic, Z., Peng, K., Vucetic, S., Radivojac, P., and Dunker, A. K. (2005) Exploiting heterogeneous sequence properties improves prediction of protein disorder, *Proteins 61*, Suppl 7, 176–182.

141 Peng, K., Radivojac, P., Vucetic, S., Dunker, A. K., and Obradovic, Z. (2006) Length-dependent prediction of protein intrinsic disorder, *BMC Bioinformatics 7*, 208.

142 Iakoucheva, L. M., Kimzey, A. L., Masselon, C. D., Bruce, J. E., Garner, E. C., Brown, C. J., Dunker, A. K., Smith, R. D., and Ackerman, E. J. (2001) Identification of intrinsic order and disorder in the DNA repair protein XPA, *Protein Sci 10*, 560–571.

143 Adkins, J. N. and Lumb, K. J. (2002) Intrinsic structural disorder and sequence features of the cell cycle inhibitor p57Kip2, *Proteins 46*, 1–7.

144 Dunker, A. K., Brown, C. J., Lawson, J. D., Iakoucheva, L. M., and Obradovic, Z. (2002) Intrinsic disorder and protein function, *Biochemistry 41*, 6573–6582.

145 Iakoucheva, L. M., Brown, C. J., Lawson, J. D., Obradovic, Z., and Dunker, A. K. (2002) Intrinsic disorder in cell-signaling and cancer-associated proteins, *J Mol Biol 323*, 573–584.

146 Iakoucheva, L. M., Radivojac, P., Brown, C. J., O'Connor, T. R., Sikes, J. G., Obradovic, Z., and Dunker, A. K. (2004) The importance of intrinsic disorder for protein phosphorylation, *Nucleic Acids Res 32*, 1037–1049.

147 Karlin, D., Ferron, F., Canard, B., and Longhi, S. (2003) Structural disorder and modular organization in Paramyxovirinae N and P, *J Gen Virol 84*, 3239–3252.

148 Munishkina, L. A., Fink, A. L., and Uversky, V. N. (2004) Conformational prerequisites for formation of amyloid fibrils from histones, *J Mol Biol 342*, 1305–1324.

149 Bandaru, V., Cooper, W., Wallace, S. S., and Doublie, S. (2004) Overproduction, crystallization and preliminary crystallographic analysis of a novel human DNA-repair enzyme that recognizes oxidative DNA damage, *Acta Crystallogr D Biol Crystallogr 60*, 1142–1144.

150 Oldfield, C. J., Ulrich, E. L., Cheng, Y., Dunker, A. K., and Markley, J. L. (2005) Addressing the intrinsic disorder bottleneck in structural proteomics, *Proteins 59*, 444–453.

151 Hansen, J. C., Lu, X., Ross, E. D., and Woody, R. W. (2006) Intrinsic protein disorder, amino acid composition, and histone terminal domains, *J Biol Chem 281*, 1853–1856.

152 Haag Breese, E., Uversky, V. N., Georgiadis, M. M., and Harrington, M. A. (2006) The disordered amino-terminus of SIMPL interacts with members of the 70-kDa heat-shock protein family, *DNA Cell Biol 25*, 704–714.

153 Radivojac, P., Vucetic, S., O'Connor, T. R., Uversky, V. N., Obradovic, Z., and Dunker, A. K. (2006) Calmodulin signaling: Analysis and prediction of a disorder-dependent molecular recognition, *Proteins 63*, 398–410.

154 Liu, J., Perumal, N. B., Oldfield, C. J., Su, E. W., Uversky, V. N., and Dunker, A. K. (2006) Intrinsic disorder in transcription factors, *Biochemistry 45*, 6873–6888.

155 Uversky, V. N., Roman, A., Oldfield, C. J., and Dunker, A. K. (2006) Protein intrinsic disorder and human papillomaviruses: Increased amount of disorder in E6 and E7 oncoproteins from high risk HPVs, *J Proteome Res 5*, 1829–1842.

156 Cheng, Y., LeGall, T., Oldfield, C. J., Dunker, A. K., and Uversky, V. N. (2006) Abundance of intrinsic disorder in protein associated with cardiovascular disease, *Biochemistry 45*, 10448–10460.

157 Sigalov, A. B., Aivazian, D. A., Uversky, V. N., and Stern, L. J. (2006) Lipid-binding activity of intrinsically unstructured cytoplasmic domains of multichain immune recognition receptor signaling subunits, *Biochemistry 45*, 15731–15739.

158 Singh, V. K., Zhou, Y., Marsh, J. A., Uversky, V. N., Forman-Kay, J. D., Liu, J., and Jia, Z. (2007) Synuclein-gamma targeting peptide inhibitor that enhances sensitivity of breast cancer cells to antimicrotubule drugs, *Cancer Res 67*, 626–633.

159 Ng, K. P., Potikyan, G., Savene, R. O., Denny, C. T., Uversky, V. N., and Lee, K. A. (2007) Multiple aromatic side chains within a disordered structure are critical for transcription and transforming activity of EWS family oncoproteins, *Proc Natl Acad Sci U S A 104*, 479–484.

160 Radivojac, P., Iakoucheva, L. M., Oldfield, C. J., Obradovic, Z., Uversky, V. N., and Dunker, A. K. (2007) Intrinsic disorder and functional proteomics, *Biophys J 92*, 1439–1456.

161 Vacic, V., Oldfield, C. J., Mohan, A., Radivojac, P., Cortese, M. S., Uversky, V. N., and Dunker, A. K. (2007) Characterization of molecular recognition features, MoRFs, and their binding partners, *J Proteome Res 6*, 2351–2366.

162 Kabsch, W. and Sander, C. (1983) Dictionary of protein secondary structure: Pattern recognition of hydrogen-bonded and geometrical features, *Biopolymers 22*, 2577–2637.

163 Minor, D. L., Jr. and Kim, P. S. (1996) Context-dependent secondary structure formation of a designed protein sequence, *Nature 380*, 730–734.

164 Jacoboni, I., Martelli, P. L., Fariselli, P., Compiani, M., and Casadio, R. (2000) Predictions of protein segments with the same amino acid sequence and different secondary structure: A benchmark for predictive methods, *Proteins 41*, 535–544.

165 Mezei, M. (1998) Chameleon sequences in the PDB, *Protein Eng 11*, 411–414.

166 Smith, C. A., Calabro, V., and Frankel, A. D. (2000) An RNA-binding chameleon, *Mol Cell 6*, 1067–1076.

167 Yoon, S. and Jung, H. (2006) Analysis of chameleon sequences by energy decomposition on a pairwise per-residue basis, *Protein J 25*, 361–368.

168 Guo, J. T., Jaromczyk, J. W., and Xu, Y. (2007) Analysis of chameleon sequences and their implications in biological processes, *Proteins 67*, 548–558.

169 Takano, K., Katagiri, Y., Mukaiyama, A., Chon, H., Matsumura, H., Koga, Y., and Kanaya, S. (2007) Conformational contagion in a protein: Structural properties of a chameleon sequence, *Proteins 68*, 617–625.

170 Uversky, V. N., Fuxreiter, M., Oldfield, C. J., Dunker, A. K., and Tompa, P. (2009) Intrinsically disordered proteins and their binding functions. In: Nussinov, R. and Schreiber, G. (eds.), *Computational protein-protein interactions*. Taylor & Francis Group, LLC, Boca Raton, pp. 223–252.

171 Anderson, C. W. and Appella, E. (2003) Signaling to the p53 tumor suppressor through pathways activated by genotoxic and nongenotoxic stress. In: Bradshaw, R. A. and Dennis, E. A. (eds.), *Handbook of cell signaling*. Academic Press, New York, pp. 237–247.

10

INTRINSIC FLEXIBILITY OF NUCLEIC ACID CHAPERONE PROTEINS FROM PATHOGENIC RNA VIRUSES

Roland Ivanyi-Nagy, Zuzanna Makowska, and Jean-Luc Darlix

10.1. INTRODUCTION

RNA is probably at the heart of life because it is endowed with essential functions from DNA maintenance to gene expression, translation, and regulation. However, long single-stranded RNAs such as hnRNAs (pre-mRNAs) and mRNAs are complex, flexible, and fragile macromolecules that can be trapped in misfolded conformations and are prone to degradation. In the virus world, single-stranded RNA is the genetic material of pathogenic viruses, notably retroviruses and flaviviruses (reviewed in Refs. [1–5]). In living cells and organisms RNAs are never found alone and are rather trafficking from one ribonucleoparticle (RNP) to another in a highly dynamic manner (reviewed in Refs. [6–8]). Indeed, there are many cellular and viral RNA-binding proteins that recognize RNA molecules with a broad sequence specificity, also named nonspecific nucleic acid binding proteins (NABPs). Some of these abundant NABPs are essential partners of RNA and as such are thought to protect, at least in part, the RNA from degradation or to resolve misfolding in order to reach the proper functional conformation (solving the "RNA folding problem") [6–11] and also to allow dynamic intermolecular RNA–RNA interactions, notably during

---

Protein and Peptide Folding, Misfolding, and Non-Folding, First Edition. Edited by Reinhard Schweitzer-Stenner.
© 2012 John Wiley & Sons, Inc. Published 2012 by John Wiley & Sons, Inc.

splicing (reviewed in Refs. [12–14]) and translational controls by micro-RNAs (reviewed in Refs. [15–17]).

The purpose of this review is to briefly present to the reader essential biochemical reactions chaperoned by viral nucleocapsid (NC) proteins that are required for the replication of widespread pathogenic viruses. Intrinsically unstructured regions confer numerous functional advantages to viral NC proteins. First, entropy changes associated with disorder-to-order transitions upon RNA binding may serve as a driving force for the ATP-independent nucleic acid melting activity of RNA chaperones ([18], reviewed in Refs. [19, 20]). In addition, their large accessible surface area and flexible conformation allows interactions with a plethora of protein, nucleic acid, and lipid partners [21], contributing to the important roles played by viral NC proteins in viral replication, persistence, and pathogenesis. Indeed, many NC/core proteins probably function as date hubs [22, 23] engaging in transient and dynamic interactions in "fuzzy" multiprotein and RNP complexes [24]. Finally, flexibility of the individual building blocks is essential for the proper assembly of the viral nucleocapsid shell [25].

Along this line we will discuss possible mechanisms involved in the interplay between viral RNA chaperones and RNA molecules, facilitated by the intrinsically unstructured regions present in the chaperone proteins [6, 18, 19]. We chose canonical viral RNA chaperones for major reasons such as their multifunctionality [26, 27], their intra- and intercellular trafficking in the form of specialized viral RNP structures and their involvement in many interactions with cellular and viral proteins [26–29], including other RNA chaperones necessary for viral replication, virus dissemination, and pathogenesis.

10.2. RETROVIRUSES AND RETROVIRAL NUCLEOCAPSID PROTEINS

Retroviruses comprise a large and diverse family of small enveloped plus strand RNA viruses with common structural, genetic, replicative, and pathological characteristics [1, 26]. The overall structure of the viral particle is globular with a mean diameter of 100–130 nm, decorated by envelope spikes, while the genomic RNA (gRNA) is in the interior of the viral capsid [1, 26]. The virion gRNA is dimeric [1, 26], with a genetic structure consisting of the Gag (structural proteins), Pol (the enzymes), and Env (the envelope glycoproteins) coding sequences flanked by long multifunctional 5' and 3' untranslated regions (UTRs). The general replicative properties of retroviruses comprise, in a timely fashion, cell infection, and reverse transcription of the single-stranded gRNA by the reverse transcriptase enzyme (RT) to generate a linear double-stranded DNA flanked by the long terminal repeats (LTRs) [[30–32], reviewed in [33, 34]], which is subsequently integrated into the host cell genome by the viral integrase (IN) [35, 36] to become an entirely new genetic entity called the provirus. In infected cells the provirus behaves as an active transcription unit leading to the synthesis of the full-length viral RNA, a fraction of which undergoes splicing. During the late steps of retrovirus replication, all the viral mRNAs are translated by the host cell ribosome machinery to generate large quantities of viral proteins [37], which comprise the Gag, Pol, and Env polyproteins and regulatory transcription and splicing factors such as Tat and Rev in HIV-1 infected cells [38]. The Gag, Pol, and Env proteins, as well as the gRNA, traffic to and assemble at the plasma membrane [28, 39, 40], where newly made virions leave the infected cell by budding, or else are transmitted to non-infected cells by means of cell-to-cell transmission [41].

Pathological consequences associated with retroviral infection and replication are diverse and involve immunodeficiencies such as AIDS in humans (HIV-1), certain forms of leukemia in rodents and human (Friend murine leukemia virus [MuLV] and HTLV-1, respectively), and neurological diseases (Cas-Br-MuLV) [1]. More recently, a new MuLV with a polytropic or xenotropic tropism provided by the viral envelope (XMRV) has been associated with prostate cancer [42, 43] and the chronic fatigue syndrome in humans [44, 45]. But the link between XMRV and disease is still a matter of debate [46].

10.2.1. Retroviral Nucleocapsid Proteins and Their Functions

From an historical point of view, NC was first isolated from nucleocapsid structures purified from viral particles of avian leukemia virus (ALV) and MuLV. These ribonucleoprotein (RNP) structures were able to support proviral DNA synthesis *in vitro*. In fact these viral RNP structures are composed of the gRNA in a dimeric form, coated by about 1500–2000 NC molecules, and contain molecules of the viral enzymes RT and IN and of cellular tRNAs [47]. In the decade from 1980 to 1990, two fundamental findings kickstarted the research on retroviral NC. First, the NC domain of Gag (GagNC) was found to drive the selection, dimerization, and packaging of the gRNA during assembly of the structural Gag polyprotein in infected cells [48, 49]. Second, NC was shown to direct gRNA dimerization and the annealing of the replication primer tRNA to the genomic primer binding site (PBS) in physiological conditions *in vitro* and in ALV and MuLV virions ([50–52], reviewed in Ref. [53]). This led to the discovery that NC from simple and complex retroviruses such as ALV, MuLV, feline immunodeficiency virus (FIV), and HIV-1 were viral proteins possessing nucleic acid chaperoning properties, which are absolutely required for proviral DNA synthesis, maintenance, and integration during the early phase of virus replication ([54–57], reviewed in Refs. [26, 58, 59]) and later for the control of reverse transcription during the late steps of virus assembly at the plasma membrane of infected cells [34, 59].

It should also be emphasized here that the interplay between NC molecules, the dimeric RNA genome, and the active RT enzyme is fueling frequent genetic recombinations by means of cDNA strand transfer reactions during viral DNA synthesis— also known as the forced and unforced copy choice recombinations [60]—which are at the source of the high genetic variability of retroviruses [1, 60–64]. In the case of HIV-1, such highly dynamic virus populations, which may differ from one organ to another [65, 66], allow the virus to resist individual immune defenses and to escape highly active antiretroviral therapies (HAARTs) by mounting multiple drug resistances such as those in the polymerase coding region, namely against anti-PR, anti-RT, and anti-IN drugs [67, 68]. These findings highlight the highly flexible genetic nature of HIV-1, the AIDS virus, and the multiple implications of the NC component in virus replication and variability.

10.2.2. Intrinsic Flexibility of the Nucleocapsid Protein

Retroviral NC proteins in their mature form are small basic proteins that originate from the protease-directed cleavage of the Gag structural polyprotein precursor in the course of the late steps of retrovirus assembly at the plasma membrane of infected cells (Fig. 10.1 HIV-1 and MoMuLV (Moloney murine leukemia virus) Gag, where protease cleavage sites are illustrated by scissors). NC proteins are characterized by one (MoMuLV) or two copies (HIV-1) of the highly conserved CCHC zinc-finger motif (ZnF), flanked by small regions rich in basic residues. In the absence of Zn^{2+} coordination, retroviral NC proteins have a completely disordered structure (not shown), while Zn^{2+} chelation by the CCHC residues causes ZnF to fold into a globular structure while the N- and C-terminal regions remained disordered, according to H1 nuclear magnetic resonance (NMR) analyses [69–72] (illustrated by dark segments in Fig. 10.1). In the case of HIV-1 NCp7, the two central ZnFs fold into a globular-like structure where the upper part forms a hydrophobic plateau constituted of hydrophobic and aromatic residues that specifically interact with the Psi packaging sequence of the gRNA (reviewed in Refs. [53, 58, 73, 74]). This specific retroviral Gag-NC–gRNA interaction is believed to represent the initial stage of virus assembly where the RNA acts as a scaffold promoting Gag oligomerization [58, 75, 76] via, at least in part, NC–NC interactions. More generally, such a specific recognition between Gag-NC and the gRNA appears to drive gRNA packaging in gamma-retroviruses (MoMuLV; Fig. 10.1) [53, 58, 77] and alpha-retroviruses (ASLV) (reviewed in Ref. [58]).

The intrinsic flexibility of NC goes far beyond retroviruses since it is also found in the ancient yeast retrotransposon TY3, where NCp9 was found to be totally unstructured (J. L. Darlix, unpublished results). Similarly, in the Gypsy retrotransposon of the fruit fly *Drosophila melanogaster*, the NC-like domain of Gag is totally unstructured (Fig. 10.1) [78]. Both TY3 NCp9 and the Gypsy NC-like were shown to be potent RNA chaperones *in vitro*, resisting heat denaturation. These unstructured retrotransposon NCs were also shown to direct the annealing of the homologous replication tRNA primer to the template RNA [78, 79].

10.2.3. Intrinsic Flexibility and Chaperoning Activity of Retroviral Nucleocapsid Proteins

As amply documented for HIV-1 NCp7, retroviral NC in its mature Zn^{2+} bound form is partially disordered, notably the N- and C-terminal regions [69–73]. Interestingly, NC in its apoform is totally disordered while maintaining its chaperoning activities *in vitro*, namely RNA annealing including primer tRNA annealing to the genomic PBS, strand transfer, and ribozyme activation [58, 80].

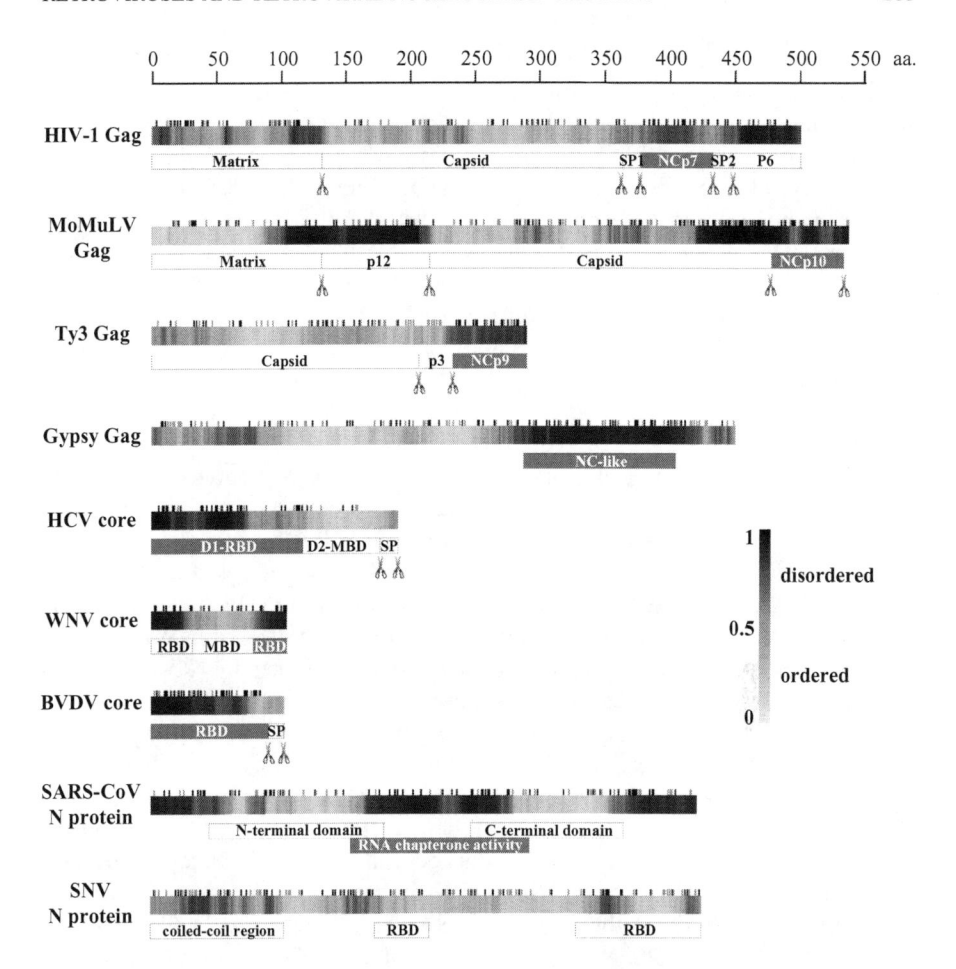

Figure 10.1. Domain organization and intrinsic disorder in viral nucleocapsid proteins with RNA chaperone activity. The heat map illustrates the predicted disorder in nucleocapsid proteins, with highly flexible segments in dark gray/black and well-folded domains in white/light gray. Computer prediction of disordered regions was obtained using the DisProt VL3-H predictor [205]. An amino acid with a disorder score greater than or equal to 0.5 is considered to be in a disordered environment, while below 0.5, in an ordered environment. Top: basic and acidic amino acid residues are indicated by black and gray vertical bars, respectively. Bottom: RBD stands for RNA-binding domain, while MBD for membrane-binding domain. Regions with known RNA chaperone activity are colored dark gray. Scissors indicate protease cleavage sites in HIV-1 and MoMuLV Gag precursors, the Ty3 yeast retrotransposon Gag precursor, and HCV and BVDV core. Note that the highly basic character and high flexibility of the RNA chaperone regions are conserved in the distinct viral groups. See color insert.

During viral DNA synthesis, which occurs soon after virus entry into the target cell, NC was shown to be an indispensable partner of the active RT enzyme throughout reverse transcription, from the initial step to the completion and maintenance of the viral double-stranded DNA [53–59]. It should be pointed out that this NC–RT interaction needs the completely folded ZnFs in the case of HIV-1 [81, 82]. The initial step corresponding to the annealing of the cellular primer tRNA to the genomic PBS is illustrated by simple schemes in Figure 10.2A. The contribution of NC in this specific annealing reaction can be separated into distinct but related steps [28, 73]: (1) binding of NC to the gRNA and to the primer tRNA under conditions where the NC to RNA ratio is in the order of 1 molecule per 8 nucleotide residues, which is a proportion similar to that existing in the viral particle; (2) this ensures a molecular crowding phenomenon whereby NC and RNA molecules are found highly concentrated in a large complex containing both ordered and disordered parts, thus corresponding to a "fuzzy globule"; (3) within such a large macromolecular complex, NC causes the intramolecular destabilization, or fraying, of complementary RNA sequences, notably that of the G6-U67 and T54-A58 regions within the primer tRNA^{Lys,3} structure [28, 73, 83–85]. NC-directed destabilization of the G6-U67 base pair promotes access by the PBS sequence to the weak bases at the four-way junction within the tRNA^{Lys,3} cloverleaf [85]; (4) then NC chaperones strand exchange at the level of the tRNA 3' acceptor stem; and (5) last, NC promotes, probably via its zinc fingers, the opening of the tRNA tertiary structure, enabling the complete annealing to the genomic PBS and primer activation signal (PAS) sequences (Fig. 10.2A). In this model structure, the RT enzyme can easily bind the tRNA anti-codon loop ACUUUUAA [52, 58], gaining access to the tRNA 3' end to start minus strand cDNA synthesis. Initiation of reverse transcription is further facilitated by NC–RT interactions [81, 82], as are the processivity and fidelity of the polymerization reaction [86, 87], since NC provides RT with some nucleotide excision–repair activity in vitro [86].

As simply illustrated and briefly discussed here, retroviral NC in the form of poorly defined oligomeric structures or globules is involved in a multitude of dynamic interactions, with the gRNA, the primer tRNA, the viral cDNA, and the RT enzyme once a retrovirus infects a target cell. In HIV-1, other interactions have been reported notably with the viral transactivator Tat and the viral infectivity factor Vif. Tat is another intrinsically disordered RNA chaperone protein, first shown to be indispensable for proviral DNA transcription in infected cells [88–90], and found to augment the initiation of reverse transcription [91]. Vif is also an RNA chaperone [92] that can interact with the NC–gRNA complex [93] and at the same time with the cellular restriction factor APOBEC 3G [94, 95]. This in turn prevents the incorporation of APOBEC3G into the nucleocapsid of newly formed virions [94, 95]. Thus, the interplay between Vif, NC, and APOBEC3G relieves the viral restriction imposed by the cellular factor that edits C to U during minus strand DNA synthesis in the reverse transcription complex [95].

During HIV assembly in infected cells, Gag-NC was shown to interact with other viral factors such as the viral VPR and the cellular ALIX and the prion protein, which is a cellular RNA chaperone with viral restriction activity [96–101]. We will not comment on these Gag-NC interactions since a review on this topic will soon be published [28].

How then could we describe the dynamic interactions between the gRNA, primer tRNA, NC molecules, and RT in the reverse transcription complexes, which pilot the synthesis and maintenance of a functional viral DNA? Such a large macromolecular complex can be considered as a "fuzzy globule" where the flexible gRNA

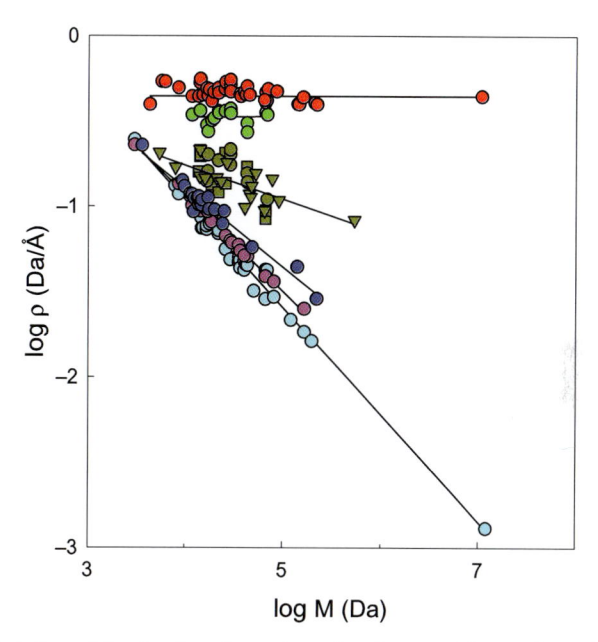

Figure 1.2. Variation of the density of protein molecules, ρ, with protein molecular weight, M, for ordered (red circles), molten globule (green circles), pre-molten globules (dark yellow symbols, where intermediates accumulated during the unfolding by urea or GdmCl are shown by circles); proteins with intact disulate bridges in 8 M urea or 6 M GdmCl are shown as squares; native pre-molten globules are shown as reversed triangles); native coils (blue circles); proteins without cross-links or with reduced cross-links unfolded in 8 M urea (pink circles); proteins without cross-links or with reduced cross-links unfolded in 6 M GdmCl (turquoise circles). The solid lines represent the best fit of the data.

Protein and Peptide Folding, Misfolding, and Non-Folding, First Edition. Edited by Reinhard Schweitzer-Stenner.
© 2012 John Wiley & Sons, Inc. Published 2012 by John Wiley & Sons, Inc.

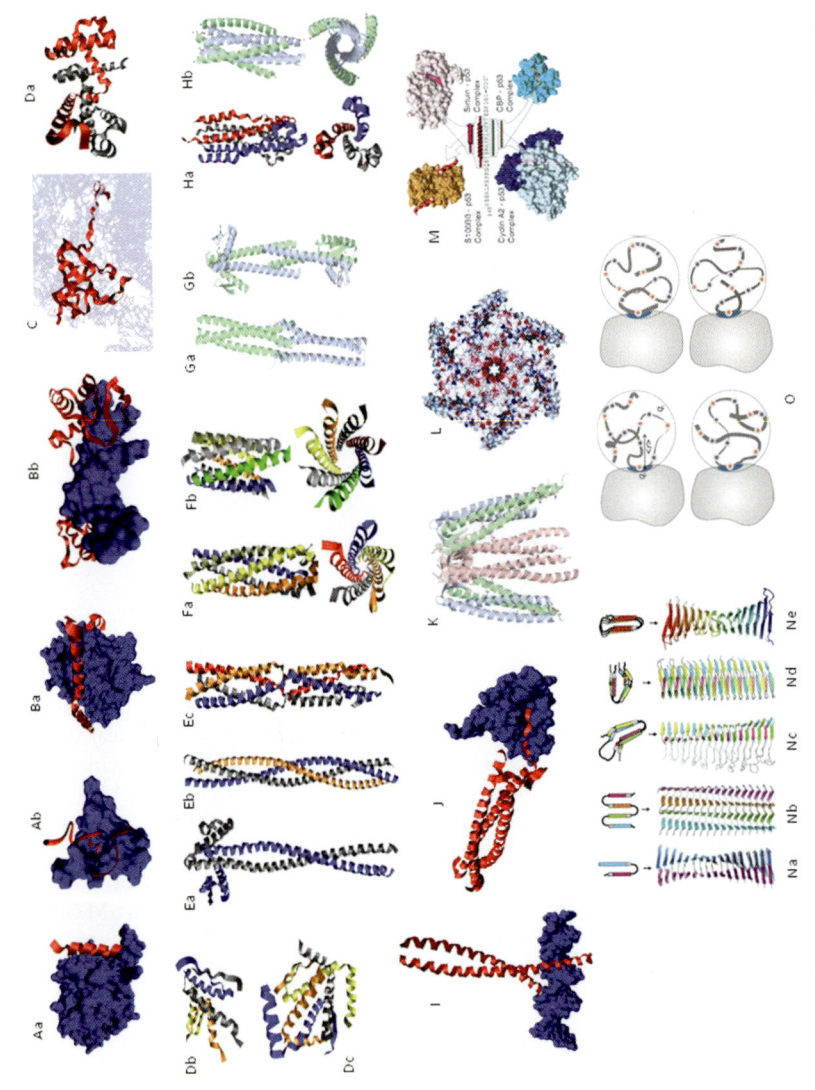

Figure 1.7. A portrait gallery of disorder-based complexes. Illustrative examples of various interaction modes of intrinsically disordered proteins are shown.

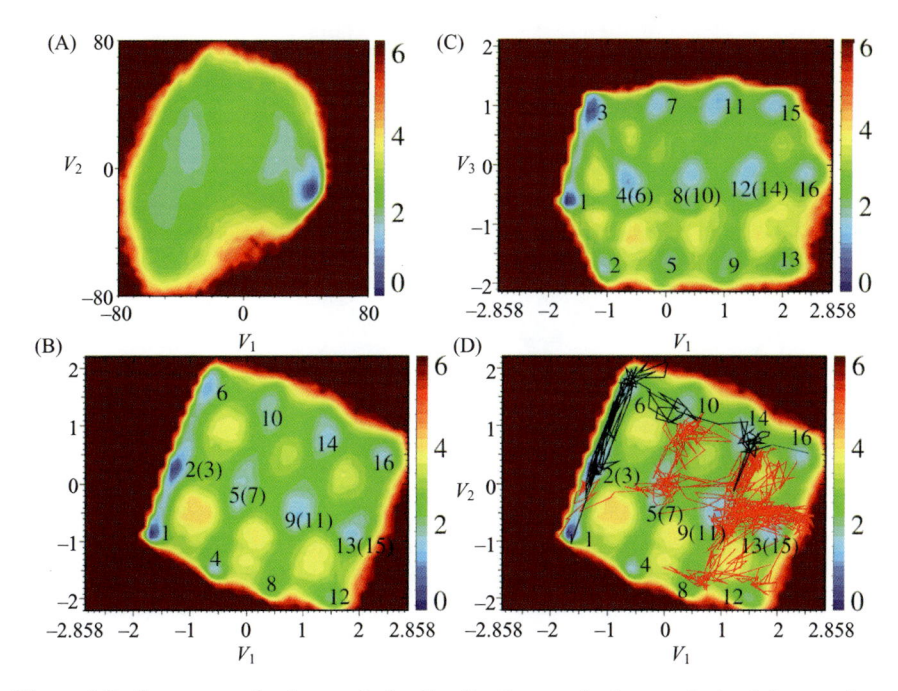

Figure 2.2. Free energy landscape (in kcal/mol) of penta-alanine as obtained from various PCAs of the 100 ns MD simulation. See text for further detail.

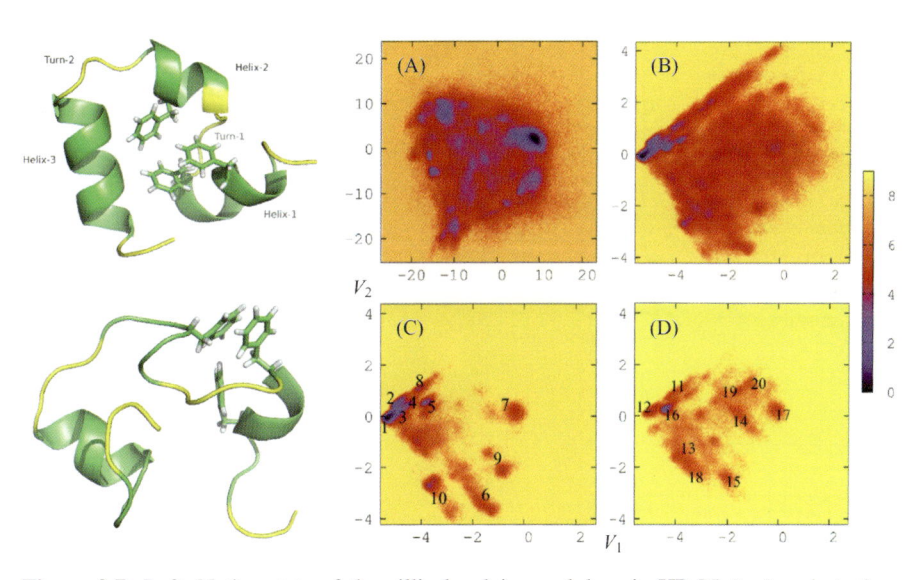

Figure 2.7. Left: Native state of the villin headpiece subdomain HP-35 (top) and starting structure of the folding MD simulations of HP-35 (bottom). Right: Two-dimensional representations of the free energy landscape $\Delta G(V_1, V_2)$ of HP-35 as obtained by the cPCA (A), dPCA (B), and pPCA (C and D). The latter shows the twenty most metastable conformational states (1–10 in A and 11–20 in B), using the eigenvectors of the overall dPCA in (B).

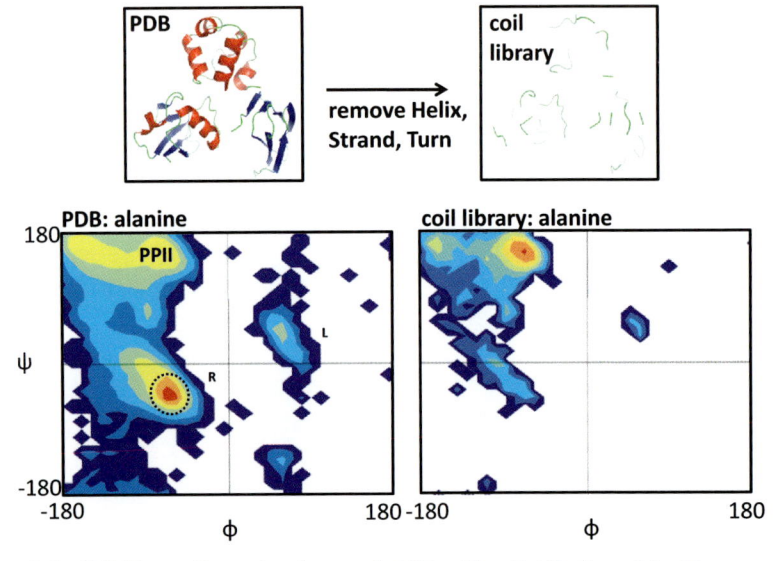

Figure 3.1. Coil library Ramachandran probability. The distribution of backbone φ, ψ torsional angles varies for the entire PDB and for a subset of the PDB that is restricted to coil regions, as depicted for alanine in each respective library. Structure images generated using the program PyMOL.

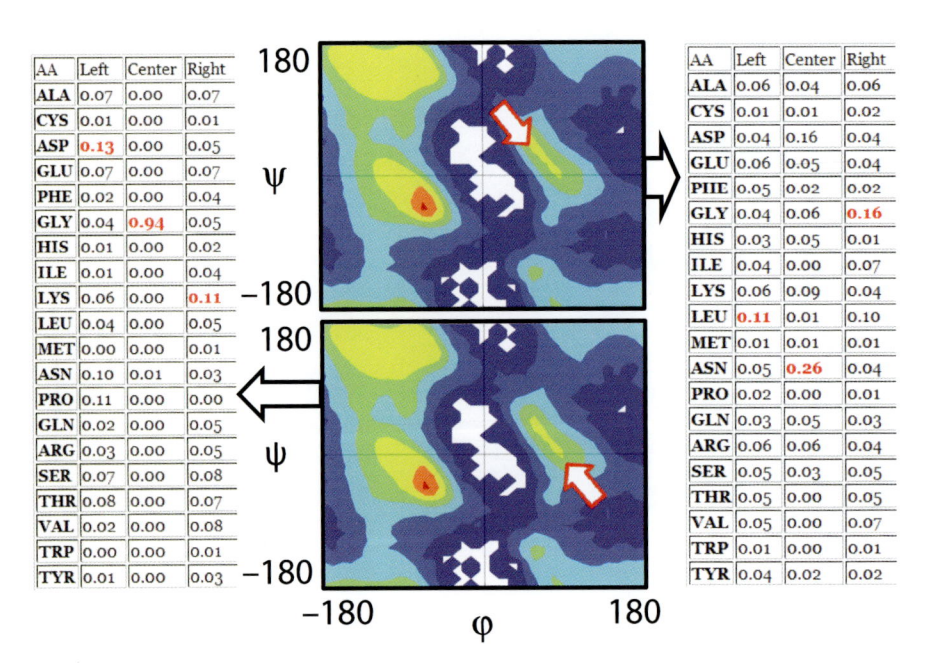

AA	Left	Center	Right
ALA	0.07	0.00	0.07
CYS	0.01	0.00	0.01
ASP	0.13	0.00	0.05
GLU	0.07	0.00	0.07
PHE	0.02	0.00	0.04
GLY	0.04	0.94	0.05
HIS	0.01	0.00	0.02
ILE	0.01	0.00	0.04
LYS	0.06	0.00	0.11
LEU	0.04	0.00	0.05
MET	0.00	0.00	0.01
ASN	0.10	0.01	0.03
PRO	0.11	0.00	0.00
GLN	0.02	0.00	0.05
ARG	0.03	0.00	0.05
SER	0.07	0.00	0.08
THR	0.08	0.00	0.07
VAL	0.02	0.00	0.08
TRP	0.00	0.00	0.01
TYR	0.01	0.00	0.03

AA	Left	Center	Right
ALA	0.06	0.04	0.06
CYS	0.01	0.01	0.02
ASP	0.04	0.16	0.04
GLU	0.06	0.05	0.04
PHE	0.05	0.02	0.02
GLY	0.04	0.06	0.16
HIS	0.03	0.05	0.01
ILE	0.04	0.00	0.07
LYS	0.06	0.09	0.04
LEU	0.11	0.01	0.10
MET	0.01	0.01	0.01
ASN	0.05	0.26	0.04
PRO	0.02	0.00	0.01
GLN	0.03	0.05	0.03
ARG	0.06	0.06	0.04
SER	0.05	0.03	0.05
THR	0.05	0.00	0.05
VAL	0.05	0.00	0.07
TRP	0.01	0.00	0.01
TYR	0.04	0.02	0.02

Figure 3.7. Identifying the preferred amino acid sequence for a particular native geometry. A web server presents the probability of finding each of the 20 amino acids for a given 10° × 10° position in the Ramachandran map. A left-handed helix prefers asparagines, whereas left-handed turns prefer glycine. The most probable neighbor amino acid identities also vary given the position in the map. Maps and tables are generated at http://godzilla.uchicago.edu/rama_AA.html.

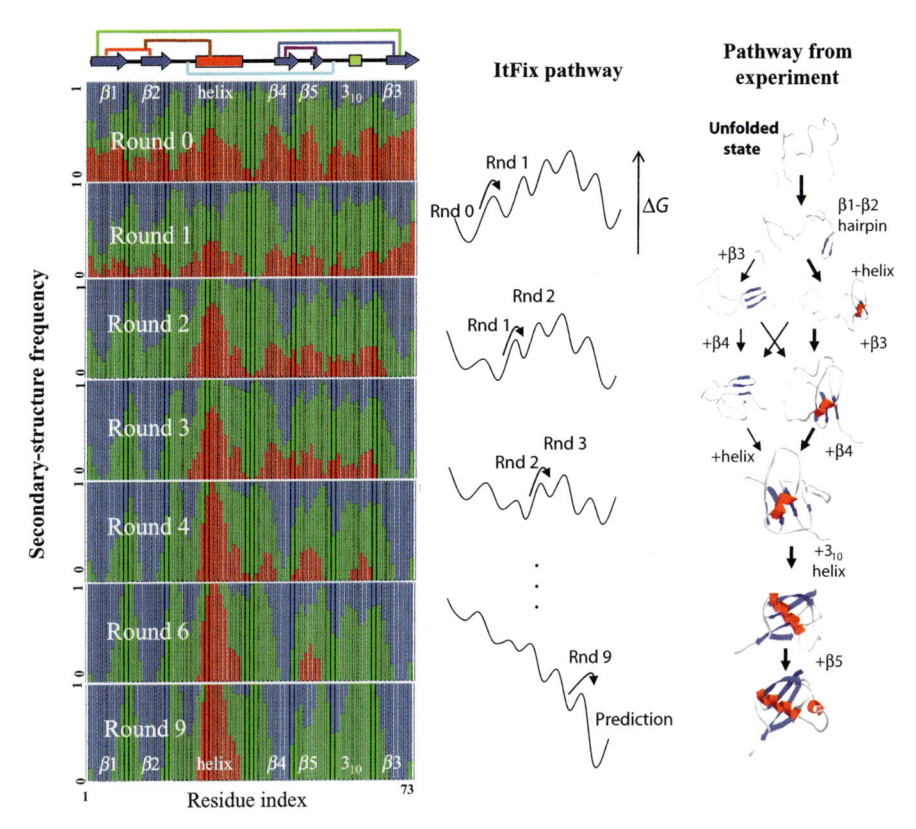

Figure 3.6. Computational and experimental folding pathways. The folding and prediction of protein structure using the ItFix algorithm begins with no secondary-structure assignments, but through subsequent rounds of folding incorporates tertiary-structure information in the formation of secondary-structure specifications for subsequent rounds. The position dependence of the secondary-structure frequencies at the end of each round, extended (blue), helix (red), and coil (green), are shown on the left. A single color bar represents a residue assigned to a single secondary-structure type (native secondary structure shown at top, along with long-range contacts). As the rounds progress, uncertainties in 2° structure diminish. This introduces a bias in the free energy surface which progressively tilts toward the native state through the folding rounds (middle diagram). The major steps in the proposed folding pathway [73, 81] are similar to the order of structure fixing over the multiple rounds: The hairpin forms, followed by the helix and $\beta 3$ strand, and then $\beta 4$. The final two events are the folding of the 3–10 helix and $\beta 5$. Their formation appears in some trajectories but not at a high enough frequency to be fixed.

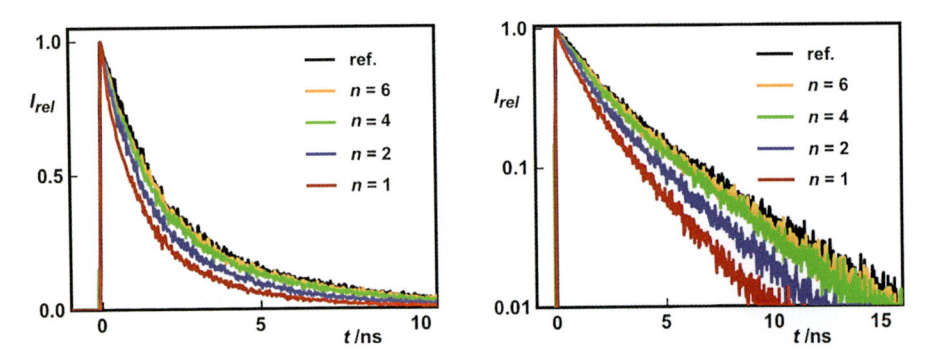

Figure 4.4. Time-resolved fluorescence decays of W–P$_n$–Dbo–NH$_2$ peptides in propylene glycol (normalized, linear intensity scale on the left and logarithmic one on the right) relative to the fluorescence decay of the W–P$_6$–NH$_2$ reference peptide lacking the acceptor (black trace). Modified from Ref. [31].

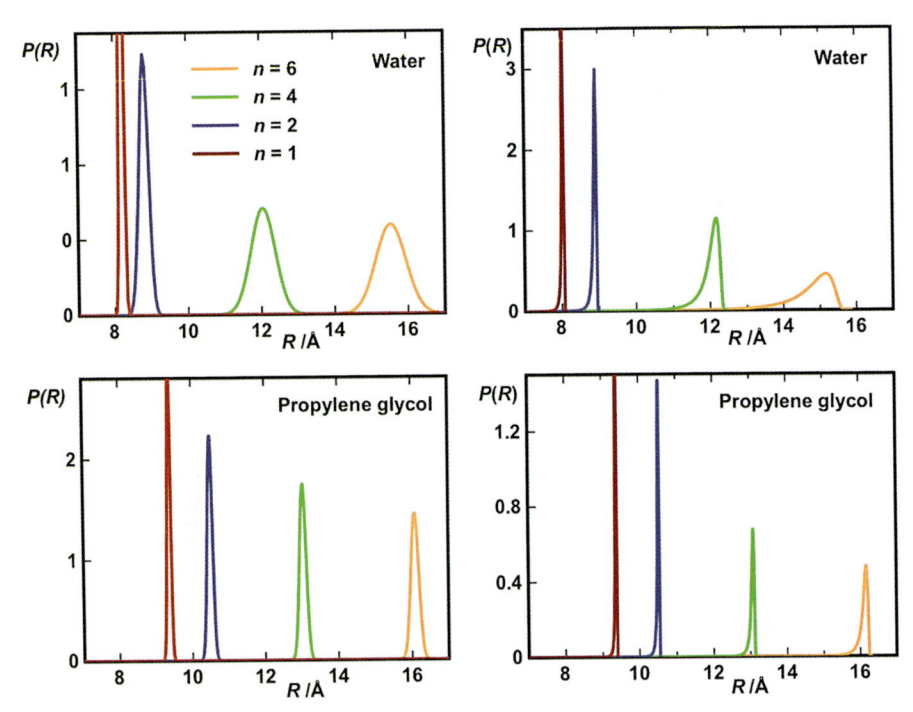

Figure 4.6. Probability distributions of donor–acceptor distances in W–P$_n$–Dbo peptides with $n = 1$ (red), 2 (blue), 4 (green), and 6 (yellow), in water (top row) and propylene glycol (bottom row) obtained from a Gaussian distribution function (left column) and a worm-like chain distribution function (right column) given by Equations 4.9 and 4.10. The area under each profile is one; the sharpest spike for $n = 1$ has been removed for clarity. Modified from Ref. [31].

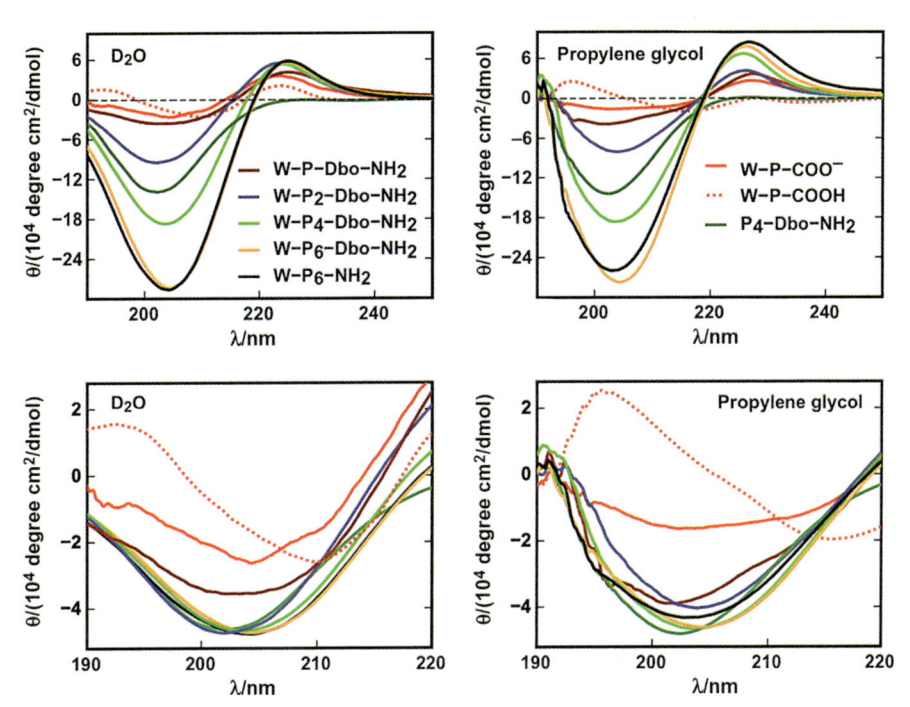

Figure 4.7. CD spectra of proline peptides in D$_2$O (left column) and propylene glycol (right column). The upper spectra show the molar ellipticity (per mole of peptide), and the lower spectra are normalized with respect to the number of proline residues in the peptide chain (per mole of proline residue).

β (ϕ = −120°, ψ = 120°)

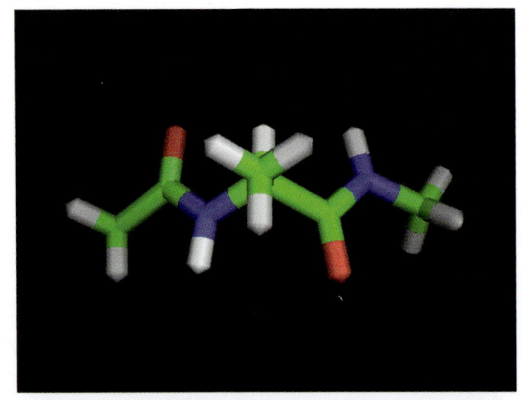

P$_{II}$ (ϕ = −75°, ψ = 145°)

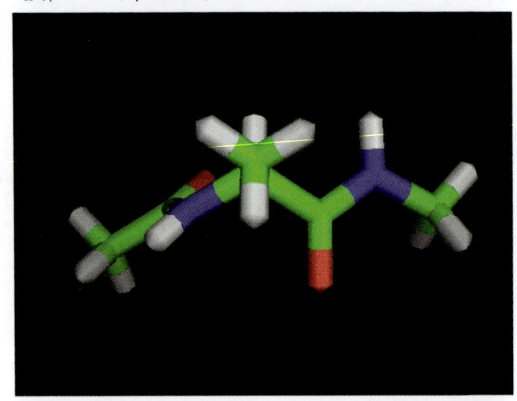

α$_R$ (ϕ = −60°, ψ = −40°)

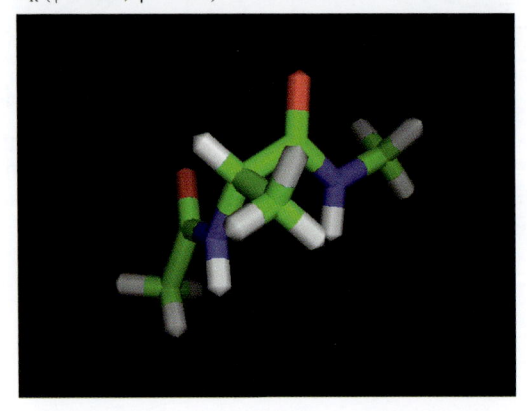

Figure 5.1. Alanine dipeptide in the β ($\varphi = -120°$, $\psi = 120°$), P$_{II}$ ($\varphi = -75°$, $\psi = 145°$), and α_R ($\varphi = -60°$, $\psi = -40°$) conformations.

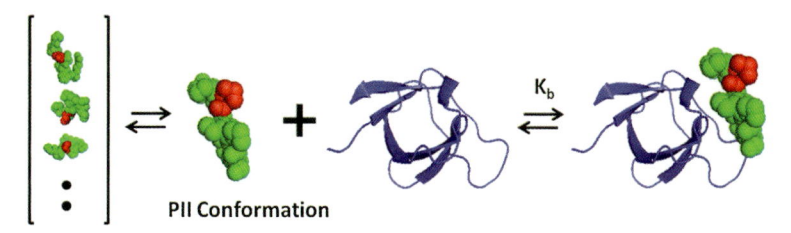

Figure 6.2. The SH3 : Sos model system employs a mutation strategy that allows calorimetric determination of a residue-specific PII propensity. Schematic of SH3 domain and SosY peptide (VPPPVPPRRRY) binding using the crystal structure (1SEM) of Lim et al. [38]. The mutation site in the SosY peptide, which is surface-exposed in the bound complex, is highlighted in red. The mutation is expected to perturb only the conformational equilibrium of the peptide between PII and other (binding incompetent) conformations. Images generated using the PyMOL Molecular Graphics System, Version 1.2r3pre, Schrödinger, LLC.

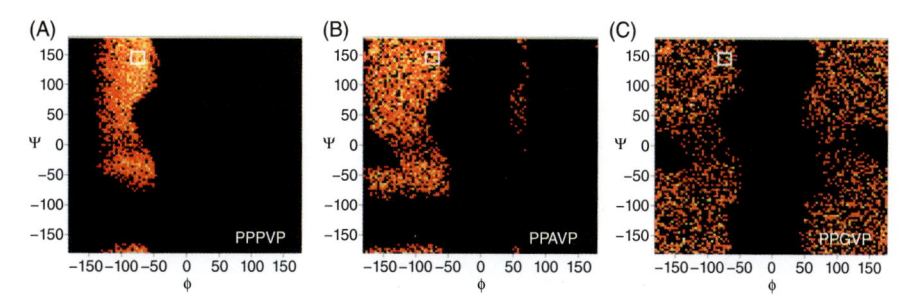

Figure 6.8. Ramachandran plot of dihedral angles adopted by the guest residue of PPXVP pentapeptides generated by MPMOD. (A) PPPVP, (B) PPAVP, and (C) PPGVP. Counts of the number of conformers in the PII region (−75 ± 10°, 145 ± 10°) boxed in white are given in Table 6.3. Conformers are not energy weighed.

(a)

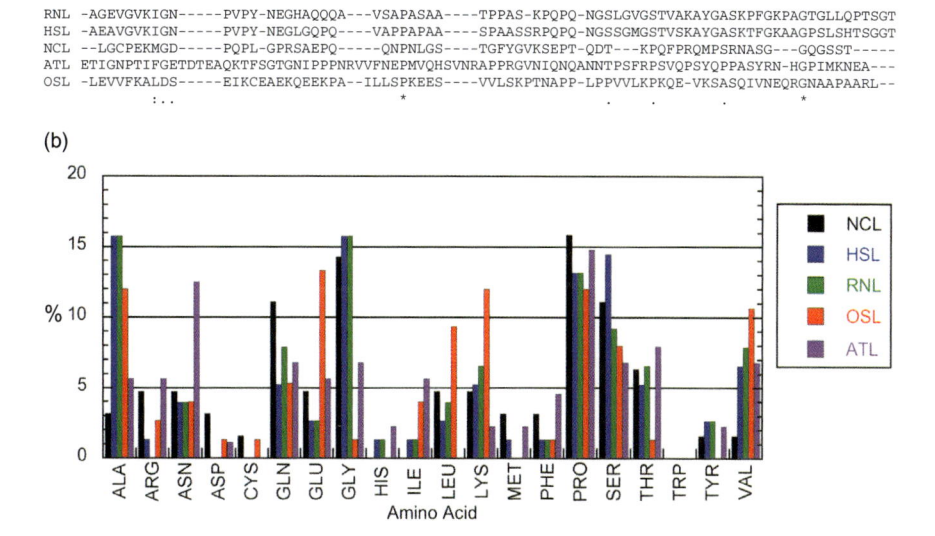

```
RNL -AGEVGVKIGN-----PVPY-NEGHAQQQA---VSAPASAA----TPPAS-KPQPQ-NGSLGVGSTVAKAYGASKPFGKPAGTGLLQPTSGT
HSL -AEAVGVKIGN-----PVPY-NEGLGQPQ----VAPPAPAA----SPAASSRPQPQ-NGSSGMGSTVSKAYGASKTFGKAAGPSLSHTSGGT
NCL --LGCPEKMGD-----PQPL-GPRSAEPQ-----QNPNLGS----TGFYGVKSEPT-QDT---KPQFPRQMPSRNASG---GQGSST-----
ATL ETIGNPTIFGETDTEAQKTFSGTGNIPPPNRVVFNEPMVQHSVNRAPPRGVNIQNQANNTPSFRPSVQPSYQPPASYRN-HGPIMKNEA---
OSL -LEVVFKALDS-----EIKCEAEKQEEKPA--ILLSPKEES----VVLSKPTNAPP-LPPVVLKPKQE-VKSASQIVNEQRGNAAPAARL--
                 :..                 *                        .  .     .        *
```

(b)

Figure 8.2. Amino acid sequence alignment of the five IULD homologues.

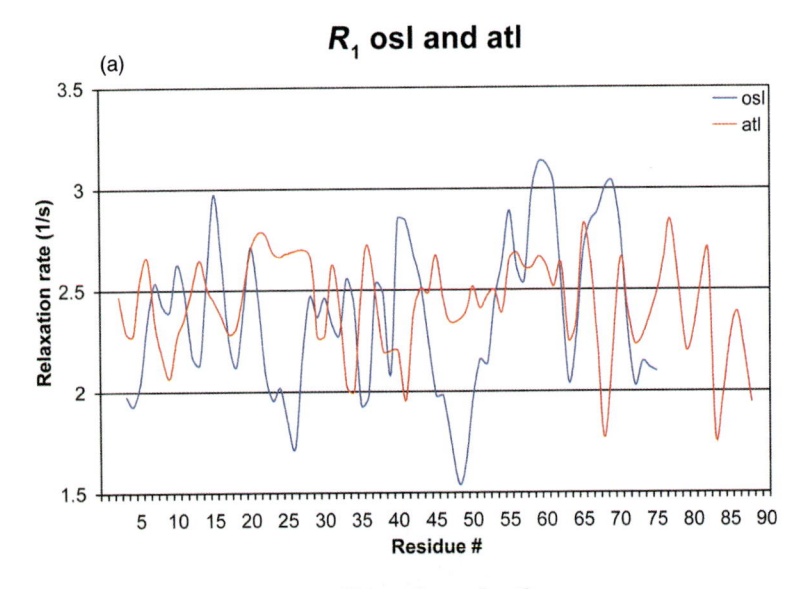

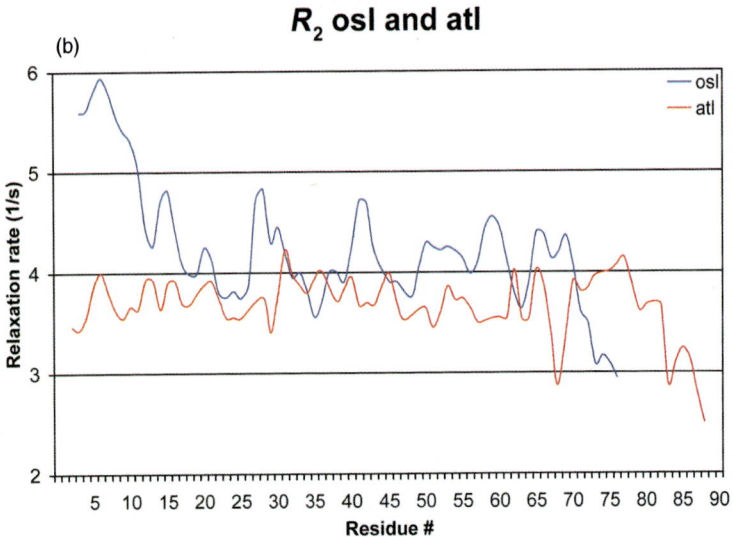

Figure 8.6. Plots showing (a) spin-lattice (R_1) and (b) spin-spin (R_2) relaxation rates for rnl (blue) and hsl (red). The relaxation rates are plotted on the vertical axis and residue number is plotted on the horizontal axis. Relaxation rates were averaged over a 2 residue window.

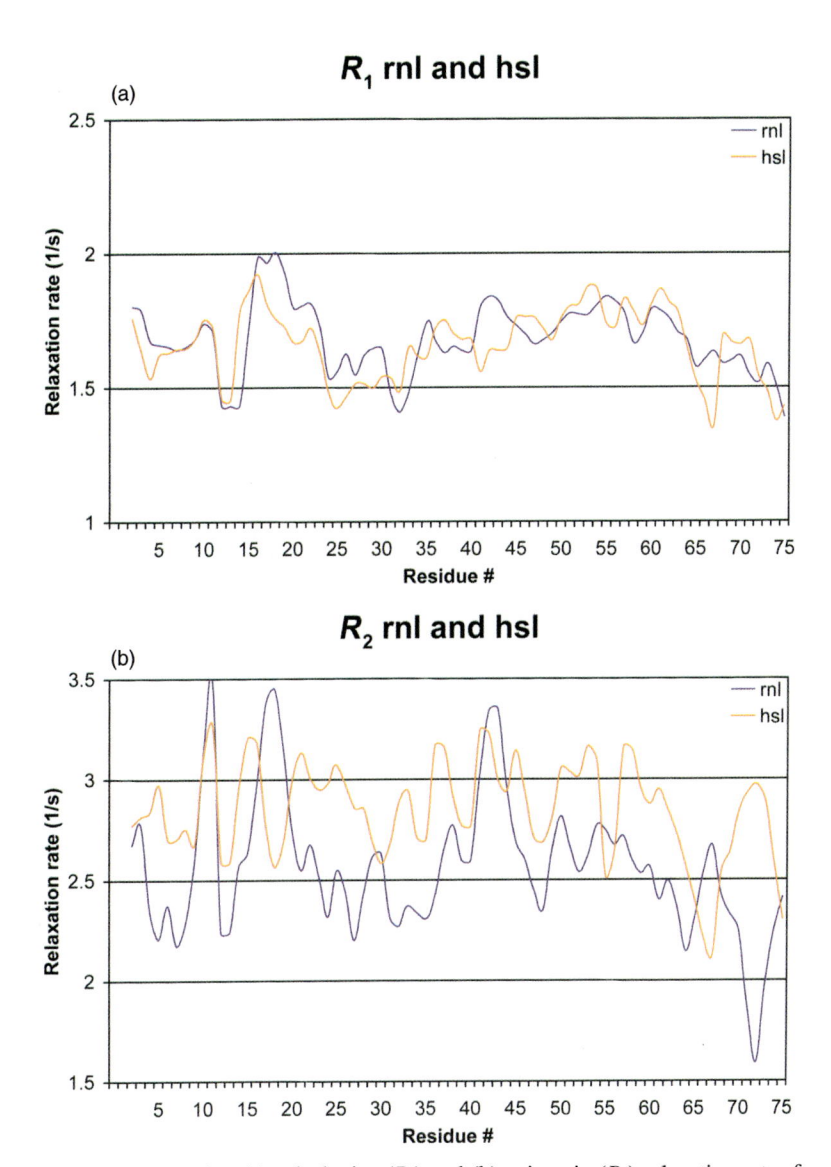

Figure 8.7. Plots showing (a) spin-lattice (R_1) and (b) spin-spin (R_2) relaxation rates for rnl (violet) and hsl (orange). The relaxation rates are plotted on the vertical axis and residue number is plotted on the horizontal axis. Relaxation rates were averaged over a two-residue window.

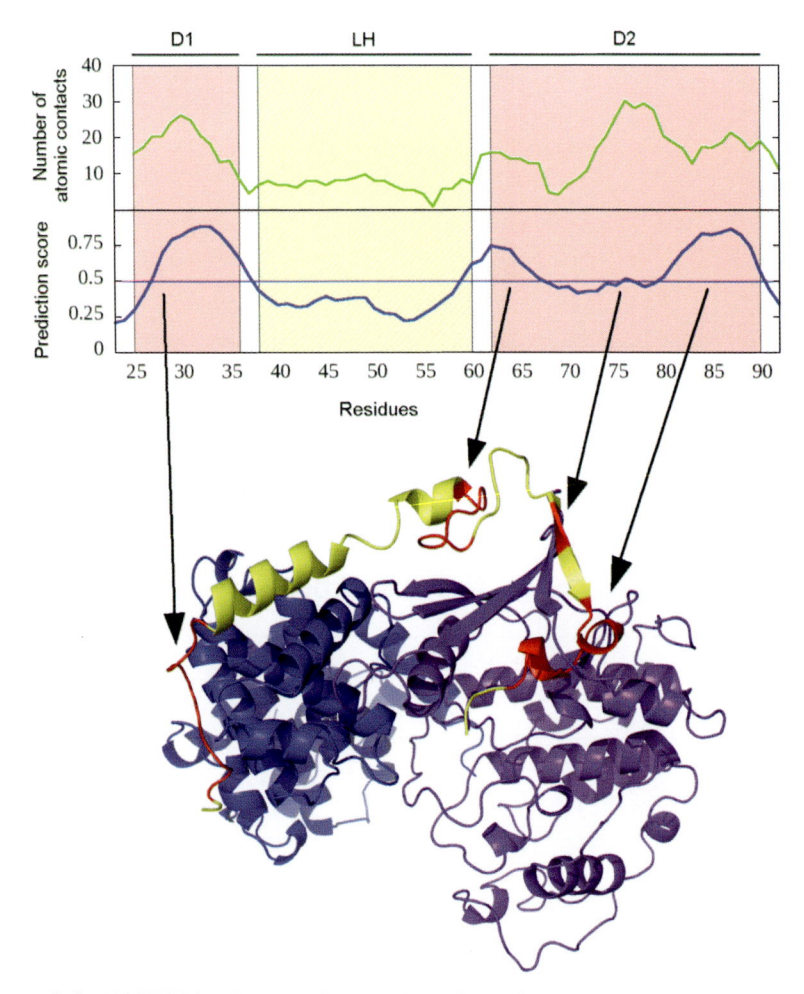

Figure 9.4. ANCHORing human p27. Number of atomic contacts (green) and prediction output (blue) and for the N-terminal binding region of human p27. "D1"and "D2" denote the two strongly interacting domains (red boxes) and "LH" denotes the weakly interacting linker domain between them (yellow box). Bottom: Crystal structure of human p27 (red and yellow) complexed with CDK2 (magenta) and cyclin A (blue) (PDB ID: 1jsu). Red parts denote regions that are predicted to bind by the predictor. These regions correspond to the experimentally verified strongly binding regions of p27. The figure was generated by PyMOL. Reproduced from Ref. [100].

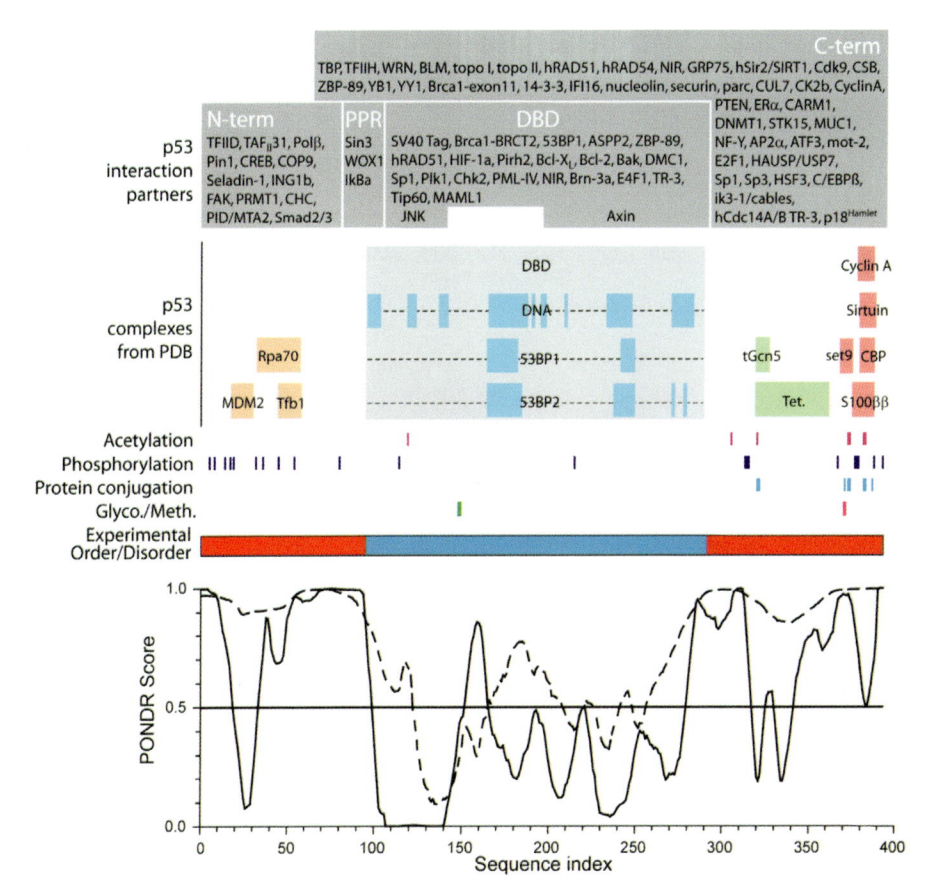

Figure 9.5. Summary of p53 interactions and structure. Gray boxes indicate the approximate binding regions of p53's known binding partners. The regions of p53 represented in structure complexes in PDB are represented by horizontal bars, labeled with the name of the binding partner. For the DBD, the extent of the globular domain is indicated by the tan box, where the internal horizontal bars indicate regions involved in binding to a particular partner. Posttranslational modifications sites are represented by vertical ticks. Experimentally characterized regions of disorder (red) and order (blue) are indicated by the horizontal bar. Finally, predictions of disorder (scores > 0.5) and order (scores < 0.5) are shown for two PONDR predictors: VLXT (solid line) and VSL2P (dashed line). All features are presented to scale, as indicated by the horizontal axis. The p53 interaction partners and posttranslational modification sites were adapted from Anderson and Appella [171]. Reproduced from Ref. [137].

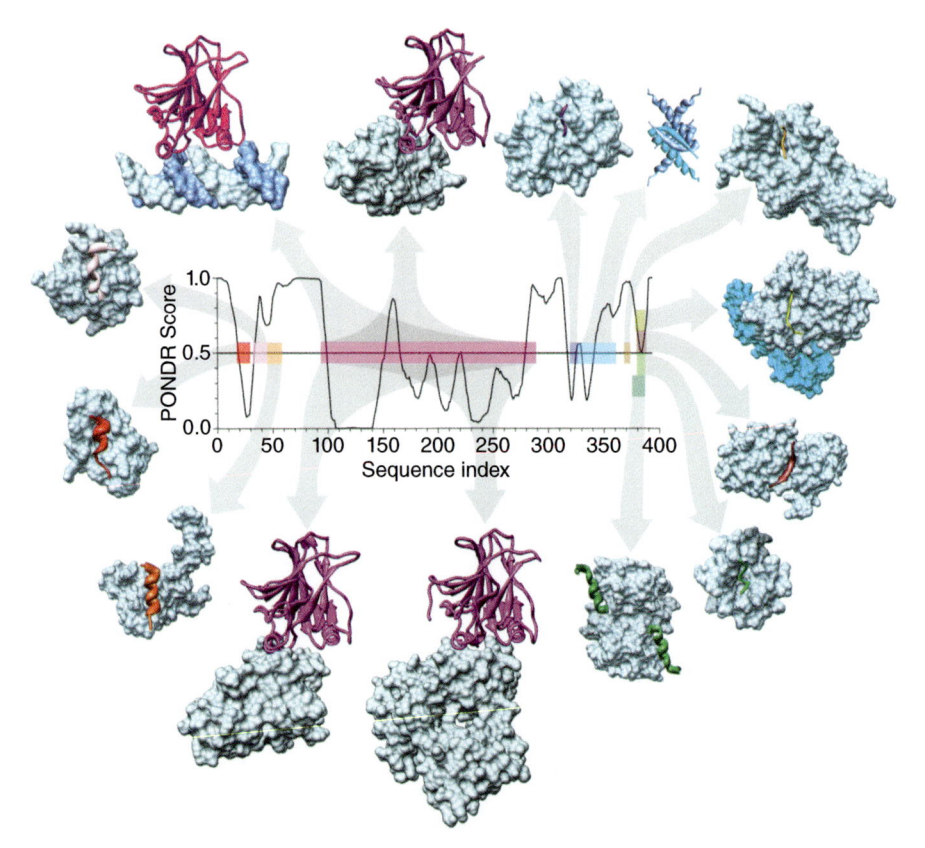

Figure 9.6. p53 interaction with different binding partners illustrate peculiarities of one-to-many signaling. (See text for full caption.)

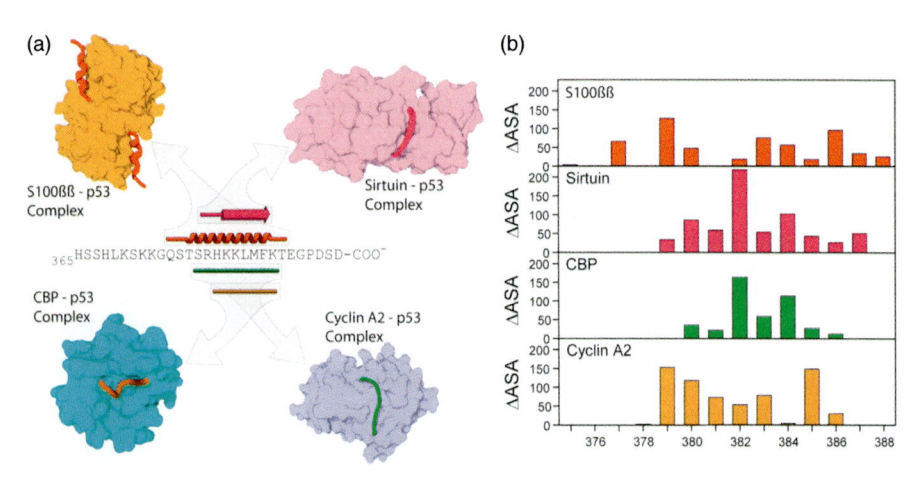

Figure 9.7. Sequence and structure comparison for the four overlapping complexes in the C-terminus of p53. (A) Primary, secondary, and quaternary structure of p53 complexes. (B) The ΔASA for rigid association between the components of complexes for each residue in the relevant sequence region of p53. Reproduced from Ref. [137].

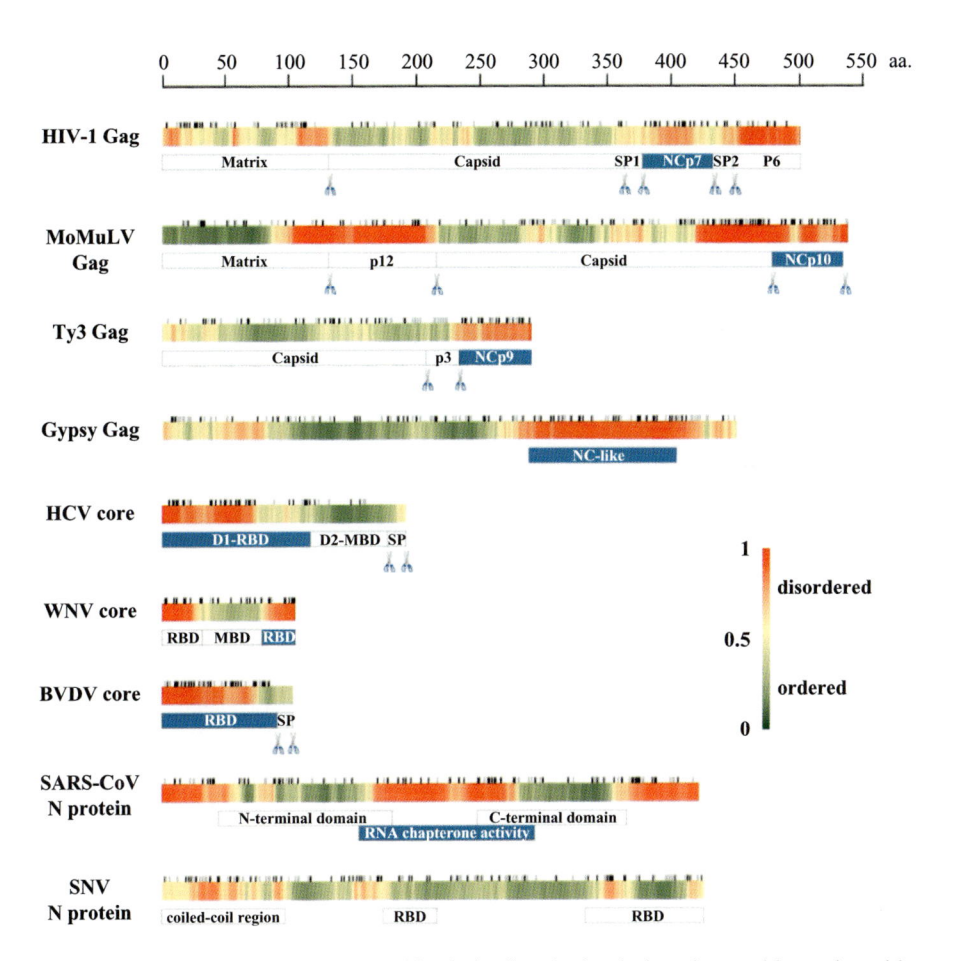

Figure 10.1. Domain organization and intrinsic disorder in viral nucleocapsid proteins with RNA chaperone activity.

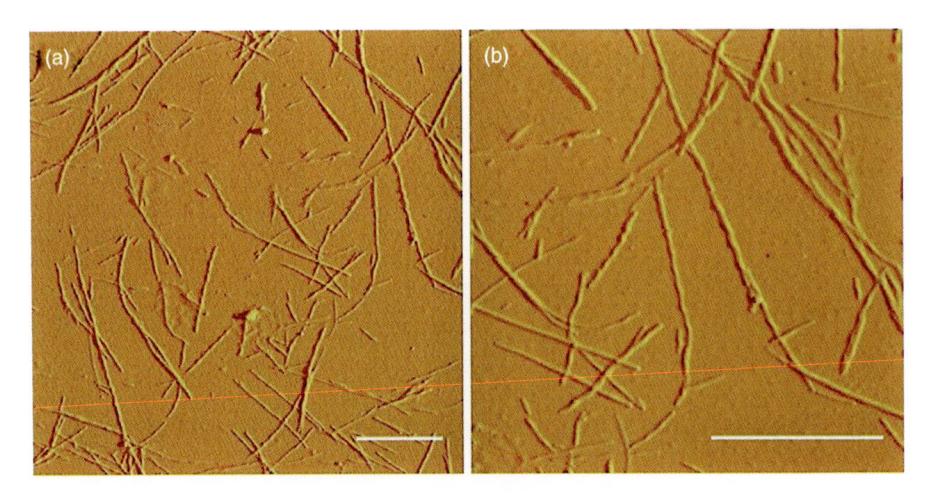

Figure 11.2. Amplitude AFM images of AKY8 fibrils. A mixture of twisted amyloid-like fibrils and more rod-like species can be discerned. The scale bar in each panel represents 1 μM.

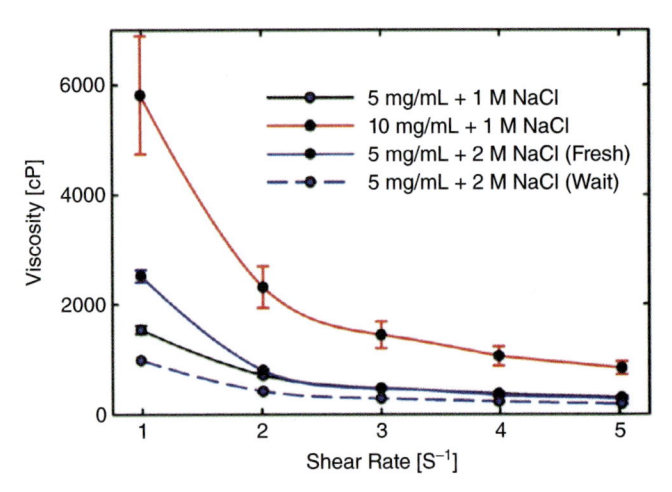

Figure 11.21. Viscosity of AK-16 hydrogels measured as a function of shear rate. Hydrogels were prepared with 5 mg/mL AK-16 and 1 M NaCl (black) and 10 mg/mL AK-16 and 1 M NaCl (red). To illustrate how β-sheet structure affects the viscosity of the resulting hydrogel, NaCl was added both to a freshly prepared portion of a 5 mg/mL AK-16 solution (blue, solid) and to a portion that was allowed to incubate overnight at room temperature (blue, dash) to yield a resulting salt concentration of 2 M. All samples were measured a total of three times.

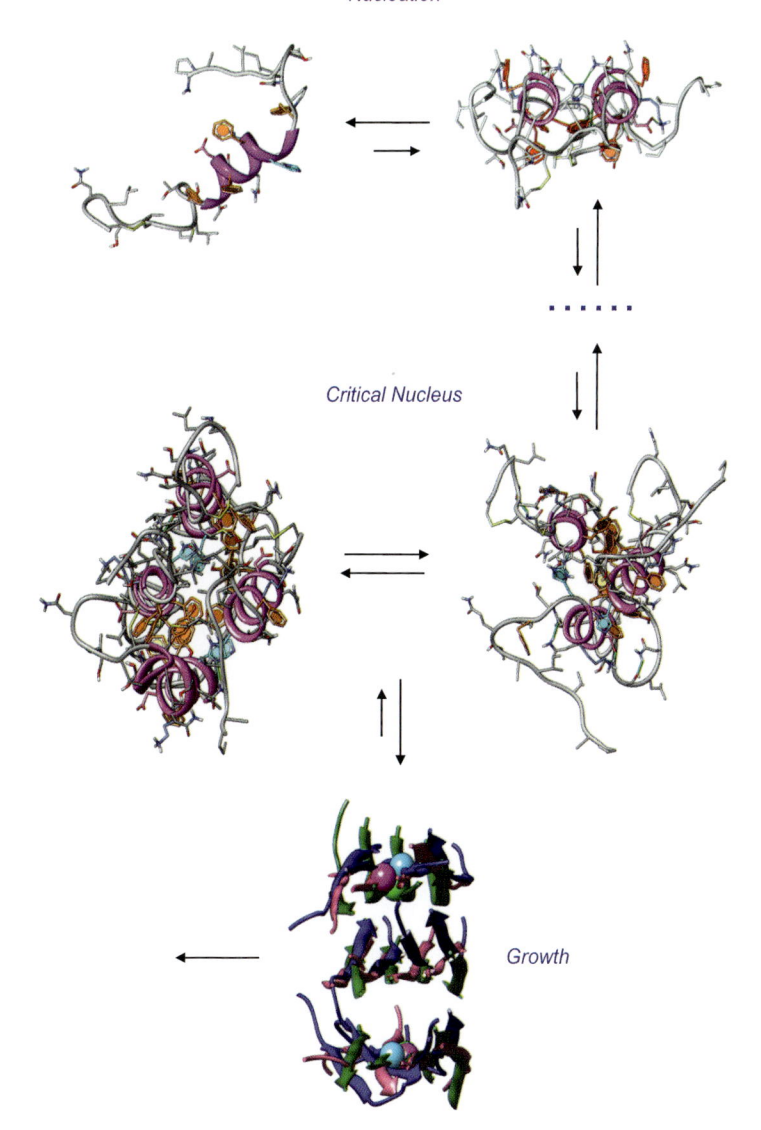

Figure 12.4. Double-nucleation fibrillation model for hCT. The first step is the homogeneous nucleation in which hCT monomers transiently interact, possibly forming dimers and aggregates of increasingly higher order. The dissociation back to monomers slows the formation of a critical nucleus, but when the stable nucleus is formed, dissociation back to monomers becomes negligible. After the critical nucleus is formed, the aggregate will grow "heterogeneously" in two directions to yield protofibrils. The protofibril growth and the assembly of the following complex fibrillar structures represent the heterogeneous second stage in the double-nucleation model. See text for further details.

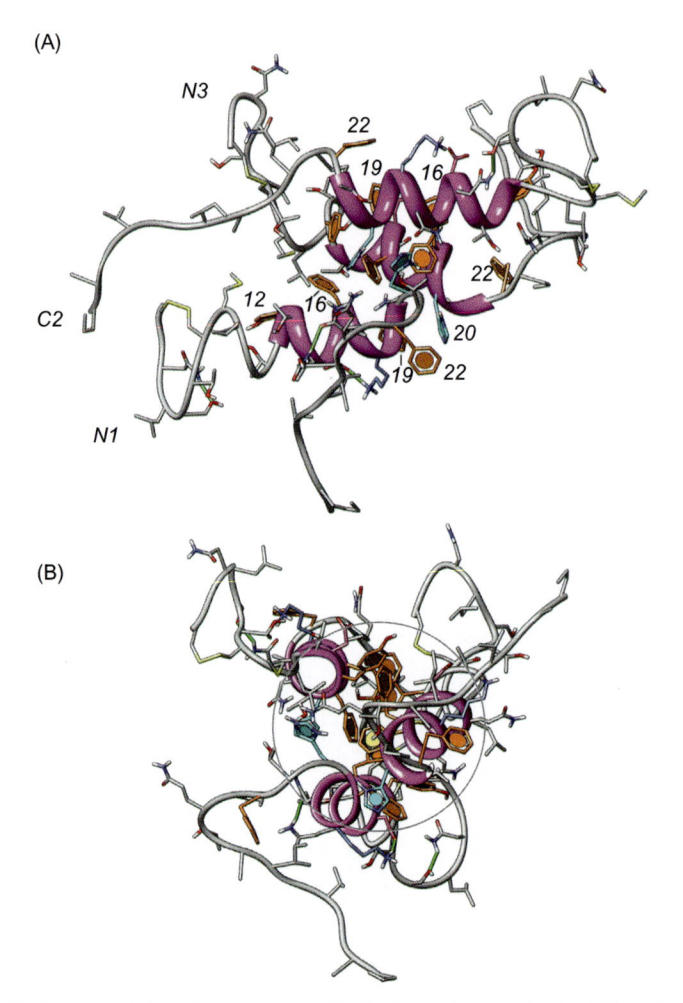

Figure 12.7. Representative structure of an hCT trimer. Perpendicular (A) and parallel (B) views to the helical axes. Sequence numbers label aromatic side chains, and *N1*, *C2*, and *N3* mark the N- and C-termini of the three monomers. In (B), the aromatic clusters are circled. For the color code, refer to the Figure 12.3 caption. Side-chains heavy atoms and polar hydrogens are shown as sticks, colored according to their atomic types, except for aromatic, His, basic, and acid residues, whose carbon atoms are painted orange, cyan, mid-blue, and pink, respectively. Aromatic rings are shown as disks colored according to the carbon atoms of the residue; side chain -side chain hydrogen bonds are depicted as green thin sticks. Ribbons are painted according to secondary structure: magenta for helix, gray for nonhelical, nonsheet regions. Where space permits, aromatic side chains are labeled with sequence numbers. *N1*, *C2*, and *N3* mark the N- and C-termini of the three monomers.

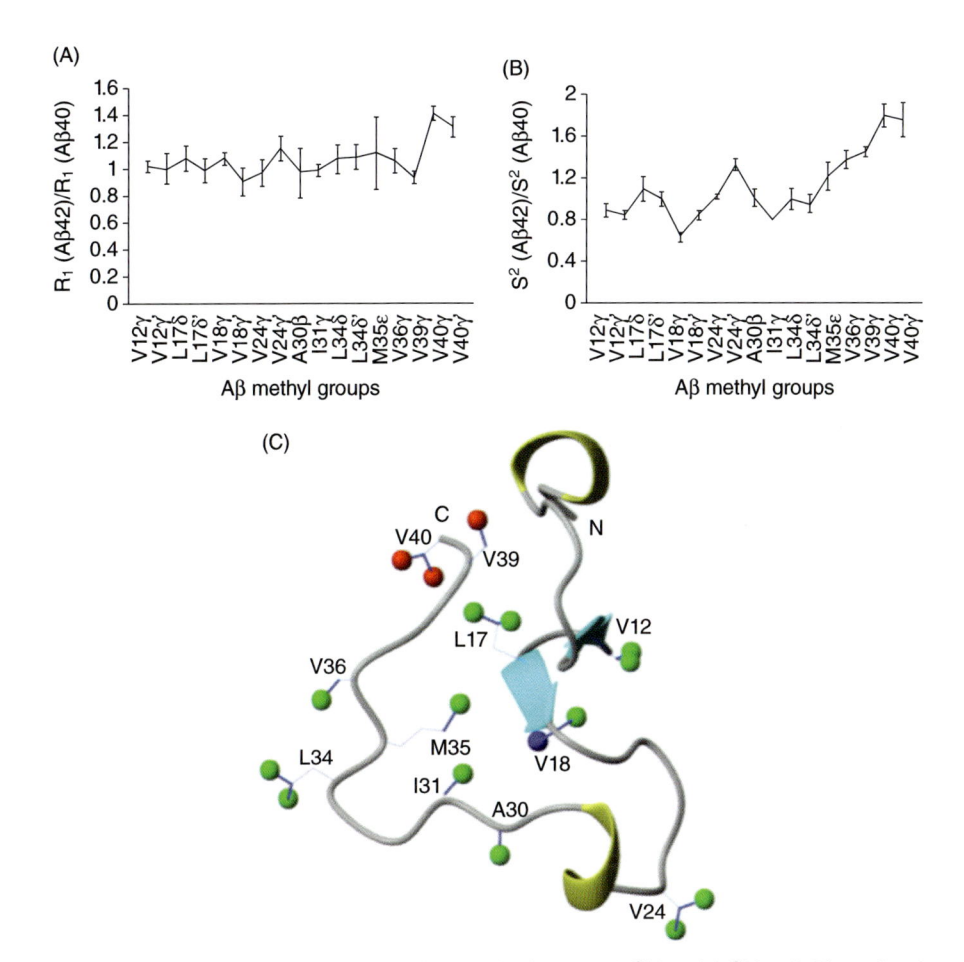

Figure 13.5. Comparison of the methyl dynamics between Aβ40 and Aβ42. (A) The ratio of R_1 of Aβ42 over Aβ40. (B) The ratio of order parameter S^2_{axis} of Aβ42 over Aβ40. (C). S^2_{axis} (Aβ42)/S^2_{axis} (Aβ40) values mapped onto the ribbon diagram of a simulated structure of Aβ40 monomer derived from MD simulation [58]. Methyl carbons are shown in a space-filling representation. Methyl groups are color-coded in red if the ratio is bigger than 1.4; green if $0.7 < S^2_{axis}$ (Aβ42)/S^2_{axis} (Aβ40) < 1.4; and blue if S^2_{axis} (Aβ42)/S^2_{axis} (Aβ40) < 0.7. From Yan et al. [73].

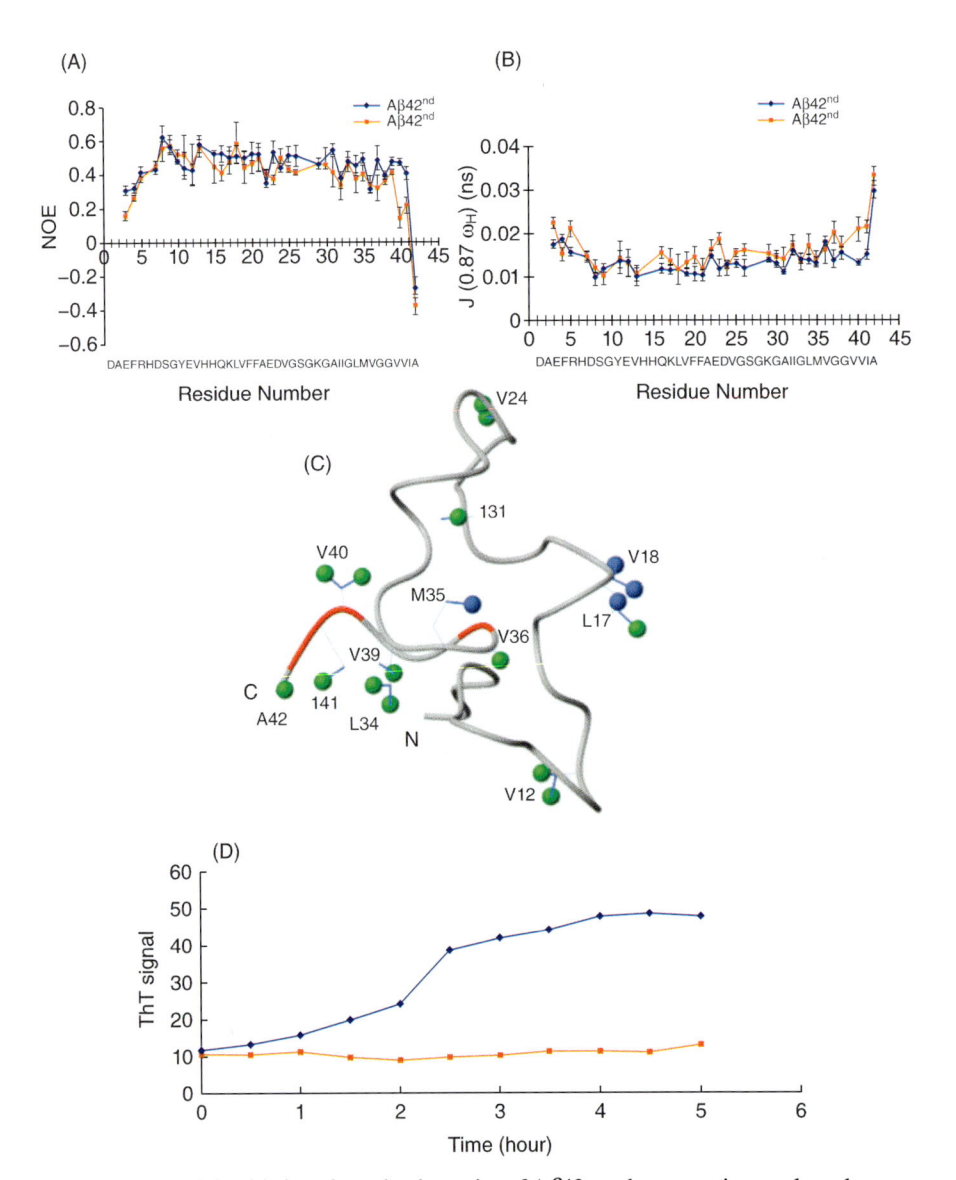

Figure 13.6. M35 oxidation alters the dynamics of Aβ42 on the ps–ns time scale and aggregation kinetics. Backbone ^{15}N NOE (A) and $J(0.87\omega_H)$ values (B) indicate increased mobility of the C-terminus of Aβ42 upon M35 oxidation. (C) Backbone ^{15}N NOE difference (NOE$_{Aβ42red}$ − NOE$_{Aβ42ox}$) and side-chain methyl groups $S^2_{Aβ42ox}/S^2_{Aβ42red}$ were mapped onto the ribbon diagram model of Aβ42 monomer from MD simulation [58]. Backbone is shown in red if the NOE difference (NOE$_{Aβ42red}$ − NOE$_{Aβ42ox}$) is bigger than 0.15; gray if −0.15 < NOE$_{Aβ42red}$ − NOE$_{Aβ42ox}$ < 0.15. Methyl groups are shown in blue if $S^2_{Aβ42ox}/S^2_{Aβ4red}$ is bigger than 1.2; green if 0.83 < $S^2_{Aβ42red}/S^2_{Aβ42ox}$ < 1.2. (D) M35 oxidation slows down Aβ42 aggregation, as shown by ThT assay (blue symbols: Aβ42^{red} and red: Aβ42^{ox}). Panels A–C from Yan et al. [57].

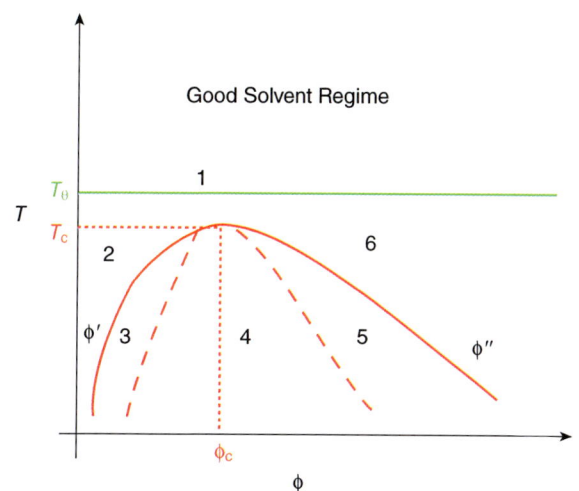

Figure 14.1. Archetypal phase diagram for polymer solutions. The ordinate denotes improving solvent quality expressed as temperature. At the theta-temperature, T_θ, and beyond (good solvent regime) no phase separation is observed. For $T < T_\theta$, a homogeneous mixed phase of polymer in solvent is formed in Region 2, and of solvent in polymer in Region 6. Conversely, phase separation is realized in Regions 3, 4, and 5. The solid red curve denotes the binodal, while the dashed red curve denotes the spinodal.

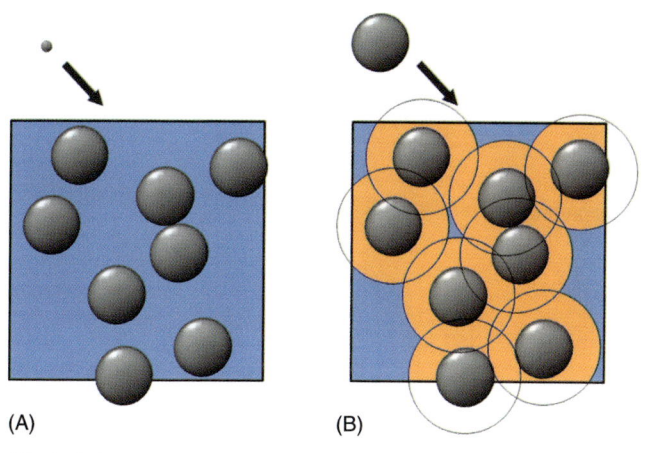

Figure 15.2. Excluded (orange and black) and available (blue) volume in a solution of spherical background macromolecules. (A) Volume available to a test molecule of infinitesimal size. (B) Volume available to a test molecule of a size comparable with background molecules [87].

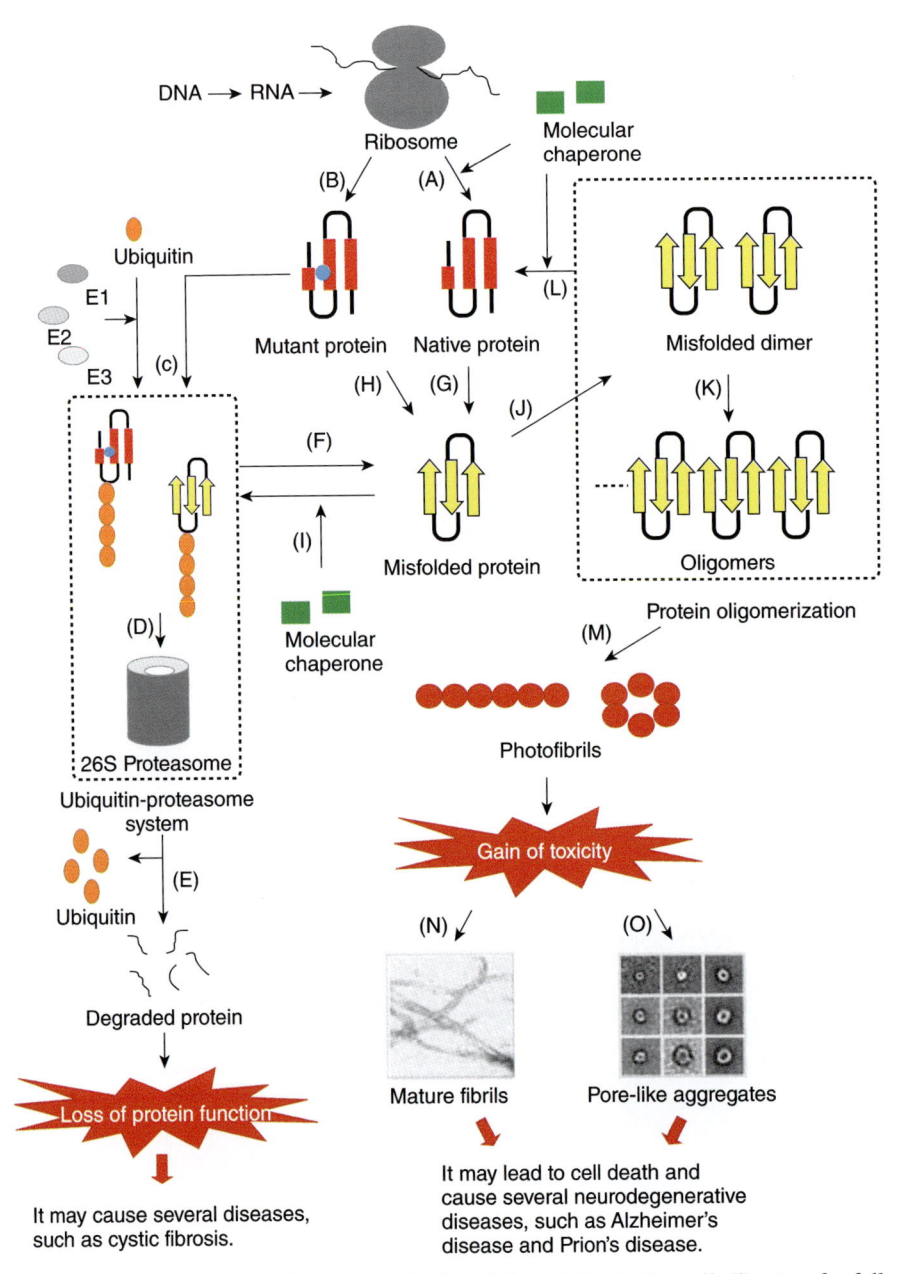

Figure 15.3. The pathway of protein synthesis and degradation in the cell. (See text for full caption.)

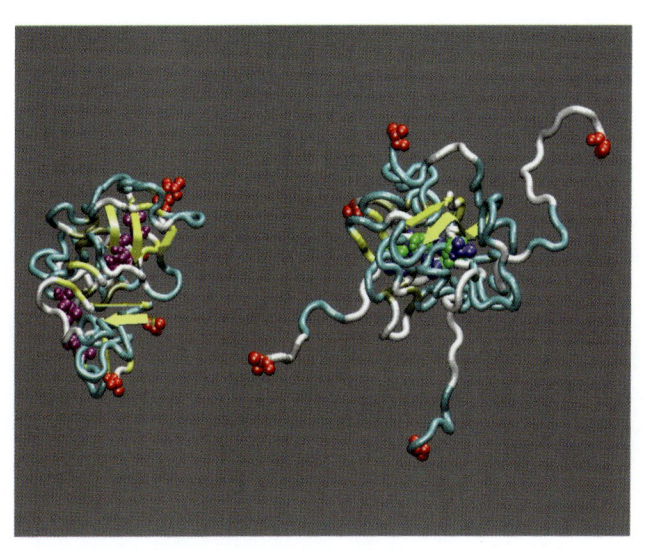

Figure 16.11. Structural features of pentamers of Aβ_{40} (left) and Aβ_{42} (right) obtained from DMD simulations, see Ref. [140]. The secondary structure of pentamers is shown as a silver tube (random coil-like structure), light-blue tube (turn), and yellow ribbon (β-strand). Red spheres in both (A) and (B) represent the N-terminal D1. (A) The C-terminal amino acids V39 and V40 are shown in purple. (B) The C-terminal amino acid I41 is shown in green, and A42 is shown in blue. Taken with permission from Urbanc, B., Cruz, L., Yun, S., Buldyrev, S. V., Bitan, G., Teplow, D. B., and Stanley, H. E (2004) In silico study of amyloid beta-protein folding and oligomerization, *Proc. Natl. Acad. Sci. U. S. A. 101*, 17345–17350. Copyright (2004) National Academy of Sciences, U.S.A.

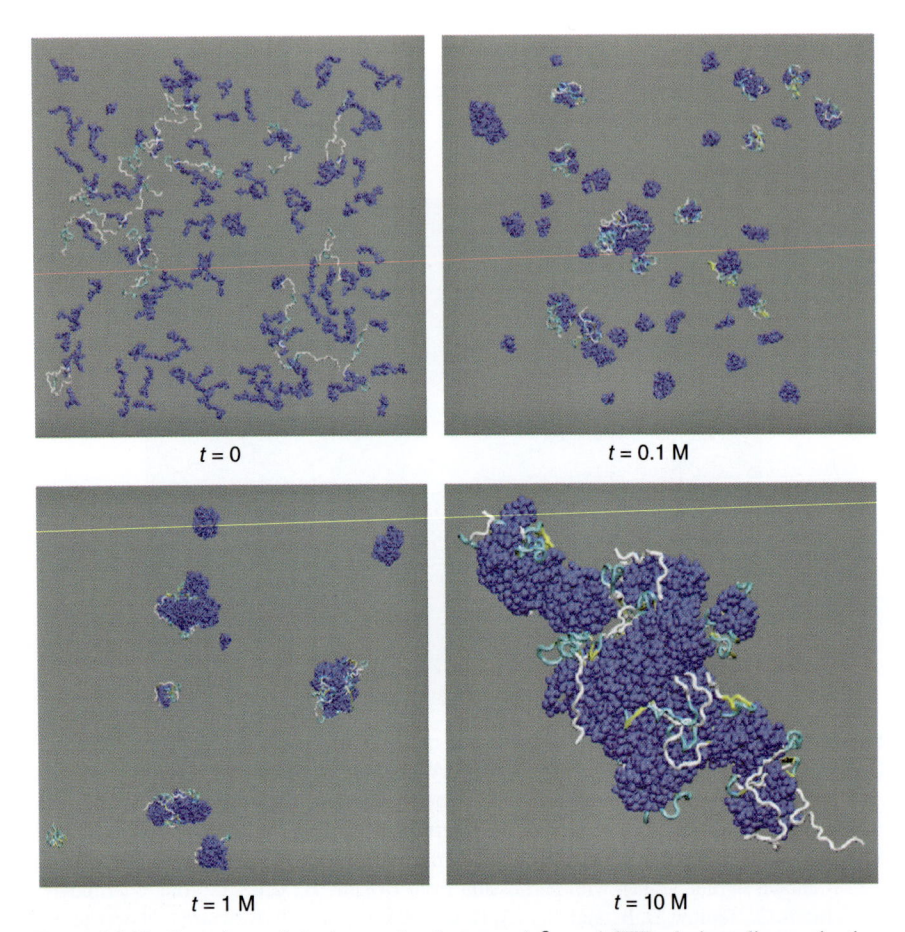

Figure 16.12. Snapshots of the interaction between $A\beta_{42}$ and CTFs during oligomerization. Configurations of 16 $A\beta_{42}$, and 128 $A\beta_{31-42}$ molecules at different time frames measured at given t simulation steps, see Ref. [126]. The dark blue is for CTFs, and $A\beta_{42}$ molecules are indicated by their secondary structure: yellow ribbons, β-strands; blue tubes, turns; silver tubes, random coil. Taken with permission from Fradinger, E. A., Monien, B. H., Urbanc, B., Lomakin, A., Tan, M., Li, H., Spring, S. M., Condron, M. M., Cruz, L., Xie, C. W., Benedek, G. B., and Bitan, G. (2008) C-terminal peptides coassemble into Abeta 42 oligomers and protect neurons against Abeta 42-induced neurotoxicity, *Proc. Natl. Acad. Sci. U.S.A. 105*, 14175–14180. Copyright (2008) National Academy of Sciences, U.S.A.

(A) HIV-1

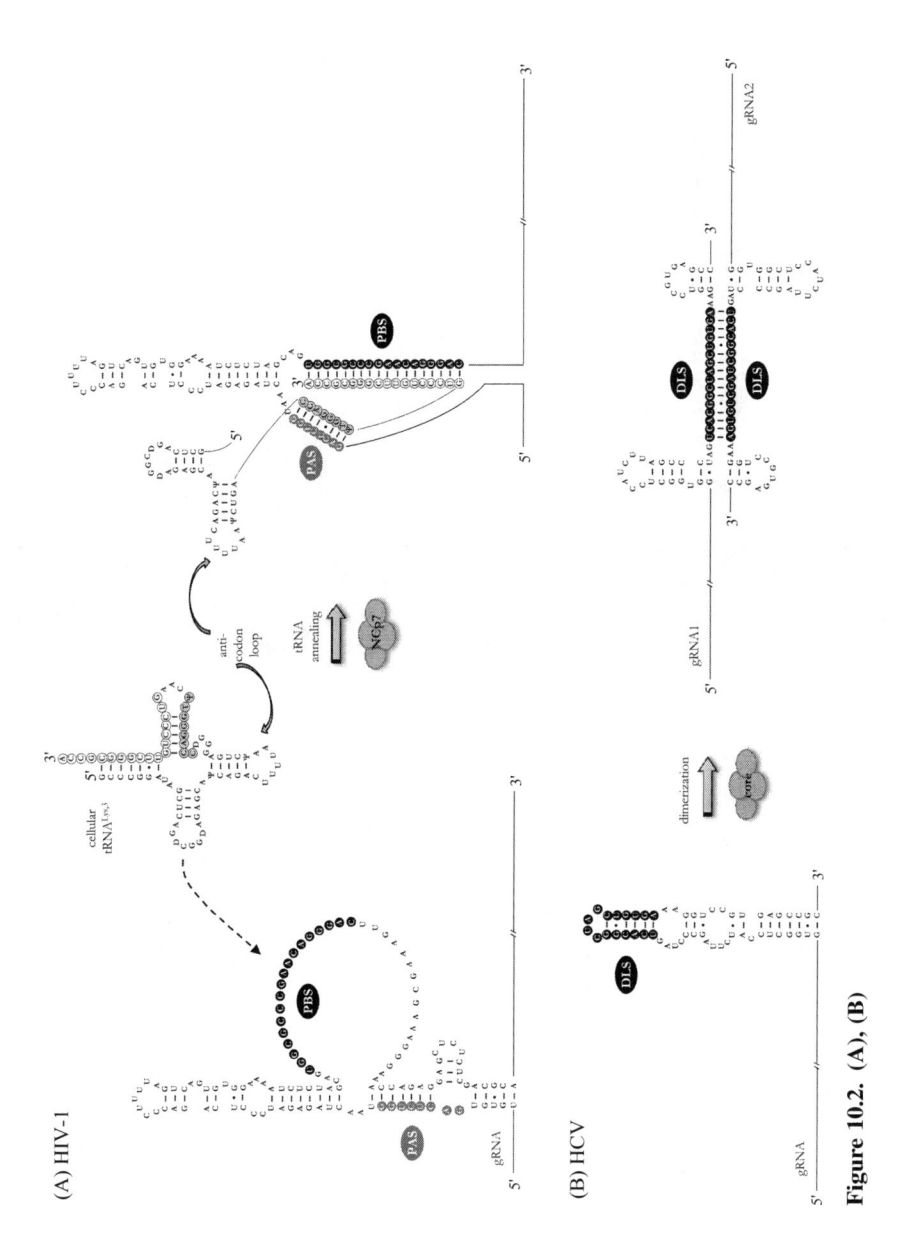

(B) HCV

Figure 10.2. (A), (B)

(C) WNV

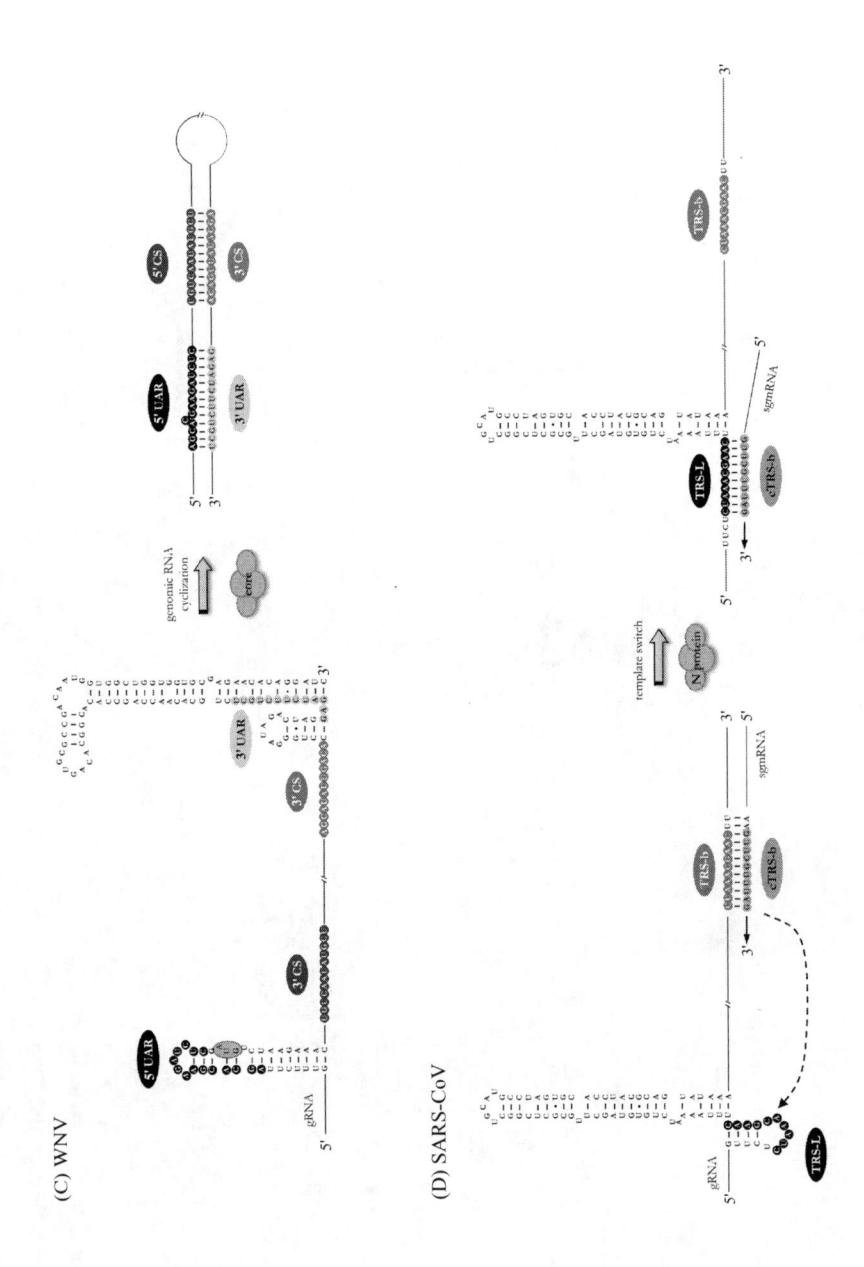

(D) SARS-CoV

Figure 10.2. Selected examples of inter- and intramolecular RNA structural rearrangements chaperoned by viral nucleocapsid proteins. (A) Chaperoning of cellular primer tRNA^{Lys,3} annealing to the HIV-1 genomic RNA (gRNA) by NCp7. This reaction can be arbitrarily considered as a multistep process: first, NCp7 molecules bind to both the primer tRNA of cellular origin and the gRNA, causing a molecular crowding phenomenon; second, NCp7 transiently unwinds the upper acceptor stem of primer tRNA (this is also known as fraying) [73], causing the annealing of the 3' last 18 nt of tRNA^{Lys,3} (open circles) to the genomic primer binding site (PBS) (black circles) in the 5' untranslated region (5' UTR) of the HIV-1 genome; third, the subsequent base pairing of tRNA^{Lys,3} to the primer activation signal (PAS) further stabilizes primer tRNA annealing and enhances the efficiency of reverse transcription initiation. In fact, in this model structure adapted from Ref. [206], the tRNA anti-codon stem-loop ACUUUAA appears to be free and thus prone to specifically interact with the viral DNA polymerase called reverse transcriptase (RT) [52], bringing the enzyme in close contact with the 3' terminal CCA of tRNA^{Lys,3} to initiate reverse transcription [34]. At the same time RT interacts with NCp7 [82], which augments the processivity and fidelity of viral cDNA synthesis [86, 87]. (B) Dimerization of hepatitis C virus (HCV) RNA by the viral core protein. *In vitro* dimerization of the 3' UTR of the gRNA is mediated by the palindromic, 16-nt-long dimer linkage sequence (DLS) and takes place in a reaction resembling that of primer tRNA annealing to the retroviral gRNA, that is, core binding, nucleic acid fraying, and annealing. The structural dimerization model is based on Ref. [111]. Note that the DLS sequence is completely conserved in all HCV strains [110] and required for virus replication [207]. (C) Facilitation of West Nile virus gRNA circularization by the core protein. Flavivirus gRNAs contain two complementary sequence pairs (UAR, upstream AUG region and CS, conserved sequence) in the 5' and 3' regions. The translation initiator AUG codon is indicated by a gray oval next to the 5' UAR. Genomic RNA cyclization takes place via the annealing of the UAR and CS sequences and was found to regulate translation, RNA synthesis, and virus replication. The structure is based on [208]. A similar structure applies for the dengue virus gRNA [209]. (D) Facilitation of template switching during SARS coronavirus mRNA synthesis by the viral nucleocapsid (N) protein. Discontinuous subgenomic RNA (sgmRNA) synthesis is achieved by template switching between transcription regulating sequences in the gRNA leader and body regions (TRS-L and TRS-b, respectively). This is achieved by TRS-b/cTRS unwinding by the viral N protein and repositioning of the transcription complex (arrow) [168]. The model structure is based on Ref. [210].

287

with its small secondary structures (see Fig. 10.2A) undergoes structural rearrangements by means of NC chaperoning. This might be accomplished by iterative loose–tight binding of NC molecules functioning at a time in a flexible and possibly elastic manner [73]. Such an NC mode of action might explain at the same time how the RT enzyme reads through the RNA template coated by NC molecules and be chaperoned by NC during the process of viral DNA synthesis [26, 34, 73, 82, 85–87]. In support of this view, mutating the CCHC ZnF of either HIV-1 or MoMuLV NC, which has little impact on viral particle production, suppresses the tight RNA binding determinant and in turn causes a defect in viral DNA synthesis and integrity, rendering the mutated virus completely defective [55, 57–59]. Along this line, the process of Gag trafficking and assembly is modified upon mutating the conserved CCHC residues of the retroviral ZnF [28, 102].

10.3. CORE PROTEINS IN THE *FLAVIVIRIDAE* FAMILY OF VIRUSES

The *Flaviviridae* family of single-stranded, positive sense RNA viruses is classified in three genera (hepaciviruses, flaviviruses, and pestiviruses), containing a number of pathogens of considerable medical and economic importance [3]. The hepatitis C virus (HCV), a hepacivirus, chronically infects ~2% of the world's population, and is a major cause of liver transplantation, due to the sequelae associated with progressive liver disease, including hepatic cirrhosis and hepatocellular carcinoma [103]. Medically important viruses in the flavivirus genus include the mosquito-borne West Nile virus (WNV), yellow fever virus (YFV), dengue viruses (DENV), and the tick-borne encephalitis virus complex (TBEV), many of which are currently emerging or re-emerging due to socio-biological impacts and climate change [104–106]. Pestiviruses, including bovine viral diarrhea virus (BVDV) of cattle, are important pathogens of livestock and wild ungulates.

Viruses in all three *Flaviviridae* genera encode a small (10–20 kDa), highly basic RNA-binding protein—the core protein—that serves as a nucleocapsid, coating and protecting the viral gRNA upon particle formation. *Flaviviridae* core proteins bind RNA and DNA with broad sequence specificity [107–109] and show potent RNA chaperone activities *in vitro*, facilitating a variety of nucleic acid annealing reactions, as well as ribozyme catalysis [109–111]. Despite their conserved function, core proteins show genus-specific domain organization, with markedly different secondary and tertiary structures and disorder content between the three genera [27] (Fig. 10.1).

HCV core consists of two structurally and functionally distinct domains (Fig. 10.1). The N-terminal domain 1 (D1) is characterized by the presence of three highly basic amino acid clusters, involved in RNA binding, chaperoning, and viral particle formation [107, 110–112]. Mapping of high-affinity core protein binding sites in the HCV genome identified target regions in the 5′ and 3′ UTRs [113, 114]. Core association with the 5′ UTR modulates viral translation [115–119], while binding to the 3′ UTR leads to RNA dimerization *in vitro* through a conserved palindromic sequence, called dimer linkage sequence (DLS), which is highly conserved in all HCV strains (Fig. 10.2B) [110, 111, 120].

The N-terminal region of HCV core was shown to be intrinsically unstructured by a variety of methods. Hypersensitivity to proteinase digestion [121], aberrant electrophoretic mobility [122], far-ultraviolet (UV) circular dichroism (CD) spectroscopy [109, 122, 123], and NMR spectroscopy [122] all suggest that D1 exists in a random coil-like conformation. Nevertheless, the isolated D1 domain is able to bind RNA and form virus-like particles in the presence of structured nucleic acids [107, 112, 124, 125]. The intrinsically unstructured nature of D1 might also play a role in the promiscuous interactions of core protein with cellular partners, allowing high-affinity/low-specificity binding to structurally diverse substrates [21]. By directly interacting with close to hundred cellular proteins [126], HCV core has a major effect on a variety of cellular processes, including immunomodulation, apoptosis, cellular transcription, and signal transduction (reviewed in Refs. [127, 128]). Perturbation of target cell metabolism by core protein probably plays an important role in viral pathogenesis, as evidenced by the characteristic symptoms of chronic HCV infection developed by core protein-transgenic mice, including insulin resistance, fatty liver (steatosis), and hepatocellular carcinoma [129–131].

The C-terminal D2 domain has a hydrophobic character and is responsible for the membrane association of HCV core [132, 133]. Core protein trafficking from the endoplasmic reticulum membrane to the surface of cellular lipid droplets—constituting the site of viral nucleocapsid assembly—is essential for infectious virus production [134, 135]. The C-terminal region adopts an α-helical conformation upon its interaction with cellular membranes [123]. Interestingly, folding of the D2 domain induces partial structure formation in the N-terminal region as well [123].

Flavivirus core proteins contain two independent RNA-binding regions at the N- and C-terminal extremities of the protein [136, 137], flanking an internal hydrophobic domain involved in homo-oligomerization and membrane association [138–140] (Fig. 10.1). Although the target regions for core protein binding in the viral gRNA are unknown, the RNA annealing activity of WNV core was shown to mimic genome cyclization *in vitro* by facilitating the interaction of two short, complementary RNA sequence pairs located at the 5′ and 3′ regions of the genome (R. I. N. and J. L.D., unpublished data) (Fig. 10.2C). Genome cyclization through these conserved sequences and a balance between linear and circular genome forms is essential for flavivirus RNA replication [141–144].

With the exception of the flexible N-terminal RNA-binding region, flavivirus core proteins contain stable, α-helical secondary structures and adopt a unique, conserved three-dimensional fold [145, 146]. Nevertheless, large deletions either in the terminal or in the central protein regions are tolerated [137, 147–149], suggesting that RNA-binding, membrane targeting, and particle formation do not require a rigid, well-defined structure. In agreement with this notion, the RNA binding and chaperoning activity of WNV core protein was found to be resistant to heat denaturation [109], a characteristic feature of intrinsically unstructured protein function.

Apart from the role they play in nucleocapsid formation, little is known about the functions of pestivirus core proteins. Bovine viral diarrhea virus (BVDV) core lacks any identifiable domain organization and was found to be completely unstructured by far-UV CD spectroscopy [108, 109]. In agreement with this, the distribution of basic residues over the entire protein length hints at the absence of a defined RNA-binding domain (RBD) [108], favoring a mechanism whereby the viral gRNA is bound through the mass action of basic amino acids in a disordered environment.

10.4. CORONAVIRUS NUCLEOCAPSID PROTEIN

Coronaviruses are enveloped, single-stranded, positive sense RNA viruses belonging to the Nidovirales order [150]. In humans, they are often associated with mild acute respiratory illness and enteric infections. With the recent emergence and worldwide spread of the coronavirus causing severe acute respiratory syndrome (SARS), the medical importance of the virus family has increased, entailing an urgent need for understanding the viral replication mechanism and protein functions [151–155].

Coronaviruses have by far the largest genome among RNA viruses (up to 32 kb in length), supported by a unique exoribonuclease activity endowing them with a proofreading mechanism required for high-fidelity replication [156, 157]. The packaging, encapsidation, and correct folding of this large genome is carried out by a viral RNA chaperone, the nucleocapsid (N) protein [29, 150]. Besides nucleocapsid formation, N protein is essential for efficient viral replication and interferes with a variety of cellular processes, including apoptosis and cell cycle progression (reviewed in Ref. [29]).

Bacterially expressed N protein binds RNA *in vitro* with a broad sequence specificity [158], but shows high-affinity, specific interaction with the 5′ leader sequence of the coronavirus genome in infected cells, a selectivity possibly due to posttranslational modifications [159–162]. A variety of *in vitro* assays, including facilitation of RNA annealing, ribozyme self-cleavage, and template switching in a heterologous system, attest to the potent RNA chaperone activities of the N protein [158, 163], demonstrated for two distantly related viruses, transmissible gastroenteritis virus (TGEV, a group 1 coronavirus) and SARS–coronavirus (SARS–CoV, a group 2b virus).

RNA rearrangements mediated by N protein play pivotal roles in viral transcription, possibly by facilitating template switching, similarly to the function of retroviral NC proteins in copy-choice recombination [163]. Transcription in coronaviruses proceeds in a discontinuous manner, resulting in the generation of a nested set of subgenomic mRNAs (sgmRNAs), which are 5′- and 3′-coterminal with the genome (reviewed in Refs. [150, 164]). The synthesis of sgmRNAs is regulated by a pair of short sequences separated in the RNA primary structure, named transcription regulating sequences (TRSs). TRS-L in the 5′ leader region serves as an acceptor for the template switch following pausing of the viral polymerase at one of the complementary TRSs (cTRS-body) preceding each gene, resulting in the fusion of a common leader sequence to the mRNAs (Fig. 10.2D). The stability of the duplex formed by base pairing between TRS-L and the cTRS region of the nascent negative-strand RNA determines the frequency of template switching and shows a good overall correlation with sgmRNA levels [165, 166]. Thus, the helix destabilizing and RNA annealing activities of N protein probably play an important role in increasing template switching frequency and regulating the relative amounts of the different mRNA species [159, 161, 163, 167, 168].

N protein sequences show a low level of sequence identity between coronaviruses belonging to different groups. Nevertheless, a conserved domain organization, based on the succession of ordered and disordered modules, has been proposed for N proteins [169]. Two structured domains account for less than half of the protein length, flanked and separated by long stretches of disordered segments (ID regions), best characterized for the SARS–CoV N protein [169] (Fig. 10.1).

The N-terminal structured domain (NTD) is involved in specific RNA binding [170, 171], while the C-terminal domain (CTD) is required for homo-oligomerization [172–174]. The disordered nature of the terminal and central ID regions was experimentally verified by NMR spectroscopy [169, 175]. In addition, small-angle X-ray scattering (SAXS) indicated that the central ID fragment exists in a partially extended conformation and serves as a flexible linker connecting the NTD and C-terminal structured domains [175]. Interestingly, all three disordered regions are involved in RNA binding independently [175], while the central ID segment is responsible for RNA chaperone activity [163, 168]. The intrinsically disordered and segmented nature of the RNA binding sites was suggested to contribute to high-affinity, cooperative interactions with RNA molecules, explained by the "fly-casting" and "coupled allostery" models [175–177].

10.5. HANTAVIRUS NUCLEOCAPSID PROTEIN

Hantaviruses are emerging pathogens belonging to the *Bunyaviridae* family of negative-strand, segmented RNA viruses. The tripartite genome of hantaviruses is composed of the S, M, and L RNA segments (vRNA), coding for the viral nucleocapsid (N) protein, the envelope glycoproteins, and the RNA-dependent RNA polymerase (RdRp), respectively [178]. Hantavirus infections in humans, acquired from rodent reservoirs through the inhalation of excreted virus in an aerosol form, cause outbreaks with substantial morbidity and high mortality rates [179]. Two distinct, serious diseases, namely hantavirus cardiopulmonary syndrome (HCPS) and hemorrhagic fever with renal syndrome (HFRS), are associated with New World and Old World hantaviruses, respectively [179].

N protein, encoded by the smallest genome segment, is a ~50-kDa protein with RNA binding and RNA chaperoning activities [180, 181]. Two distinct RNA-binding regions, exhibiting markedly different behaviors, were identified in the N protein [180, 182]. An unspecific RNA-interacting domain resides in the C-terminal region of the protein, while a central RBD—mapped between amino acids 175 and 217 in Hantaan virus—serves for the specific recognition of viral RNA, interacting with high-affinity with the 5′ ends of vRNA segments [182–188].

Facilitation of RNA–RNA annealing, as well as ATP-independent helix destabilization activities, has been demonstrated for the N protein of Sin Nombre hantavirus (SNV), the etiologic agent of HCPS [181, 189]. RNA binding and chaperoning play pivotal roles throughout the replication cycle of hantaviruses. The vRNA segments are encapsidated in an RNP complex with N protein (and the viral RdRp) in a circular conformation [190, 191], formed through base-pairing between short complementary sequences at the 5′ and 3′ ends of gRNAs ("panhandle formation"). N protein binding facilitates vRNA panhandle formation required for genome encapsidation, possibly by melting local RNA structures, hindering the formation of the globally stable long-range interaction [181].

Following virus infection, N protein ensures high-efficiency viral protein translation initiation by a truly unique mechanism among viruses [192]. Cap-dependent translation of cellular mRNAs requires the eIF4F complex, composed of three subunits (eIF4E, eIF4G, and eIF4A). eIF4E binds to the mRNA cap, eIF4G provides a link to the 43S pre-initiation complex, while eIF4A serves as a DEAD-box helicase in the scanning of the mRNA 5′ to the AUG start codon. A number of RNA viruses (e.g., hepatitis C virus, poliovirus, encephalomyocarditis virus) employ an alternative, cap-independent translation initiation mechanism that allows high levels of viral protein expression under conditions where cellular translation initiation is inhibited [193]. Uncoupling of translation from the eIF4F complex is achieved by complex RNA secondary structures (internal ribosome entry sites [IRESs]), which help to position the ribosome in the vicinity of the AUG codon and thus obviate the need for ribosomal scanning [193].

Although hantaviral mRNAs do not contain an IRES, they achieve the same eIF4F independence in an alternative manner. In an elegant series of experiments, Mir and Panganiban have shown that N protein alone is able to functionally replace the three independent activities associated with the eIF4F complex [192]. Direct binding of nucleocapsid to the mRNA cap and to the 43S complex mimic eIF4E and eIF4G activities, respectively. In addition, the ATP-dependent helicase activity of eIF4A is replaced by the ATP-independent RNA chaperone activity of nucleocapsid, mediating helix dissociation necessary for ribosome scanning [192].

Cap-binding by N protein is also important for viral replication. Hantaviruses employ a "cap-snatching"-mechanism, whereby short, capped RNAs generated from the 5′ end of host mRNA molecules serve as primers for the initiation of transcription by the viral RdRp. The N protein accumulates in cytoplasmic processing bodies (P bodies), protecting the 5′ mRNA caps from degradation [194]. For efficient transcription initiation, the helix-destabilizing activity of nucleocapsid unwinds the viral panhandle, followed by the annealing of the capped primer to the 3′ terminus of vRNAs [195].

N protein sequences within the hantavirus genus are relatively conserved and exhibit a common modular organization consisting of an N-terminal trimerization domain, followed by the central, viral genome-specific RBD and an unspecific RNA-binding region situated in the last hundred or so amino acids (Fig. 10.1). Structural information is only available for the N-terminal domain involved in trimerization, which forms a coiled-coil domain, as determined by NMR spectroscopy and in the crystal structure [196–200]. Bioinformatic analysis of the central RBD suggested a structured conformation with a high extent of flexibility [187], while the C-terminal

RNA-binding region is predicted to be partly unstructured (Fig. 10.1). This organization is reminiscent of that of the cellular RNA chaperone hnRNP A1 (heterogeneous nuclear RNP A1), which promotes nucleic acid annealing via its disordered C-terminal glycine-rich region [201, 202], while an N-terminal, folded RBD contacts RNA in a non-cooperative, sequence-specific manner [203, 204]. The role of the different regions of N protein in RNA chaperoning and the importance of disordered region(s) in this activity remain to be determined.

ACKNOWLEDGMENTS

Thanks are due to INSERM, ANRS, SIDACTION, and FINOVI (France) for their continuous support.

REFERENCES

1. Coffin, J. M., Hughes, S. H., and Varmus, H. E. (1997) *Retroviruses*, Cold Spring Harbor Laboratory Press, Cold Spring Harbor.
2. Penin, F., Dubuisson, J., Rey, F. A., Moradpour, D., and Pawlotsky, J. M. (2004) Structural biology of hepatitis C virus, *Hepatology 39*, 5–19.
3. Lindenbach, B., Thiel, H. J., and Rice, C. M. (2007) Flaviviridae: The viruses and their replication. In: Knipe, D. M. and Howley, P. M. (eds.), *Fields virology*. Lippincott-Raven Publishers, Philadelphia, pp. 1101–1152.
4. Beasley, D. W. (2005) Recent advances in the molecular biology of west Nile virus, *Curr Mol Med 5*, 835–850.
5. Holmes, E. C. (2009) RNA virus genomics: A world of possibilities, *J Clin Invest 119*, 2488–2495.
6. Cristofari, G. and Darlix, J. L. (2002) The ubiquitous nature of RNA chaperone proteins, *Prog Nucleic Acid Res Mol Biol 72*, 223–268.
7. Rajkowitsch, L., Chen, D., Stampfl, S., Semrad, K., Waldsich, C., Mayer, O., Jantsch, M. F., Konrat, R., Blasi, U., and Schroeder, R. (2007) RNA chaperones, RNA annealers and RNA helicases, *RNA Biol 4*, 118–130.
8. Schroeder, R., Barta, A., and Semrad, K. (2004) Strategies for RNA folding and assembly, *Nat Rev Mol Cell Biol 5*, 908–919.
9. Jankowsky, E. (2011) RNA helicases at work: Binding and rearranging, *Trends Biochem Sci 36*, 19–29.
10. Han, S. P., Tang, Y. H., and Smith, R. (2010) Functional diversity of the hnRNPs: Past, present and perspectives, *Biochem J 430*, 379–392.
11. Herschlag, D. (1995) RNA chaperones and the RNA folding problem, *J Biol Chem 270*, 20871–20874.
12. Kiss, T., Fayet-Lebaron, E., and Jady, B. E. (2010) Box H/ACA small ribonucleoproteins, *Mol Cell 37*, 597–606.
13. Newman, A. J. and Nagai, K. (2010) Structural studies of the spliceosome: Blind men and an elephant, *Curr Opin Struct Biol 20*, 82–89.

14. Hallegger, M., Llorian, M., and Smith, C. W. (2010) Alternative splicing: Global insights, *FEBS J 277*, 856–866.
15. Fabian, M. R., Sonenberg, N., and Filipowicz, W. (2010) Regulation of mRNA translation and stability by microRNAs, *Annu Rev Biochem 79*, 351–379.
16. Agami, R. (2010) microRNAs, RNA binding proteins and cancer, *Eur J Clin Invest 40*, 370–374.
17. Steitz, J. A. and Vasudevan, S. (2009) miRNPs: Versatile regulators of gene expression in vertebrate cells, *Biochem Soc Trans 37*, 931–935.
18. Tompa, P. and Csermely, P. (2004) The role of structural disorder in the function of RNA and protein chaperones, *FASEB J 18*, 1169–1175.
19. Ivanyi-Nagy, R., Davidovic, L., Khandjian, E. W., and Darlix, J. L. (2005) Disordered RNA chaperone proteins: From functions to disease, *Cell Mol Life Sci 62*, 1409–1417.
20. Zuniga, S., Sola, I., Cruz, J. L., and Enjuanes, L. (2009) Role of RNA chaperones in virus replication, *Virus Res 139*, 253–266.
21. Kriwacki, R. W., Hengst, L., Tennant, L., Reed, S. I., and Wright, P. E. (1996) Structural studies of p21Waf1/Cip1/Sdi1 in the free and Cdk2-bound state: Conformational disorder mediates binding diversity, *Proc Natl Acad Sci U S A 93*, 11504–11509.
22. Dosztanyi, Z., Chen, J., Dunker, A. K., Simon, I., and Tompa, P. (2006) Disorder and sequence repeats in hub proteins and their implications for network evolution, *J Proteome Res 5*, 2985–2995.
23. Dunker, A. K., Cortese, M. S., Romero, P., Iakoucheva, L. M., and Uversky, V. N. (2005) Flexible nets. The roles of intrinsic disorder in protein interaction networks, *FEBS J 272*, 5129–5148.
24. Tompa, P. and Fuxreiter, M. (2008) Fuzzy complexes: Polymorphism and structural disorder in protein-protein interactions, *Trends Biochem Sci 33*, 2–8.
25. Namba, K. (2001) Roles of partly unfolded conformations in macromolecular self-assembly, *Genes Cells 6*, 1–12.
26. Darlix, J. L., Garrido, J. L., Morellet, N., Mely, Y., and de Rocquigny, H. (2007) Properties, functions, and drug targeting of the multifunctional nucleocapsid protein of the human immunodeficiency virus, *Adv Pharmacol 55*, 299–346.
27. Ivanyi-Nagy, R. and Darlix, J. L. (2010) Intrinsic disorder in the core proteins of flaviviruses, *Protein Pept Lett 17*, 1019–1025.
28. Muriaux, D. and Darlix, J. L. (2010) Properties and functions of the nucleocapsid protein in virus assembly, *RNA Biol 7*, 744–753.
29. Surjit, M. and Lal, S. K. (2008) The SARS-CoV nucleocapsid protein: A protein with multifarious activities, *Infect Genet Evol 8*, 397–405.
30. Baltimore, D. (1970) RNA-dependent DNA polymerase in virions of RNA tumour viruses, *Nature 226*, 1209–1211.
31. Temin, H. M. and Mizutani, S. (1970) RNA-dependent DNA polymerase in virions of Rous sarcoma virus, *Nature 226*, 1211–1213.
32. Mizutani, S., Boettiger, D., and Temin, H. M. (1970) A DNA-dependent DNA polymerase and a DNA endonuclease in virions of Rous sarcoma virus, *Nature 228*, 424–427.
33. Gilboa, E., Mitra, S. W., Goff, S., and Baltimore, D. (1979) A detailed model of reverse transcription and tests of crucial aspects, *Cell 18*, 93–100.

REFERENCES

34. Mougel, M., Houzet, L., and Darlix, J. L. (2009) When is it time for reverse transcription to start and go? *Retrovirology 6*, 24.
35. Delelis, O., Carayon, K., Saib, A., Deprez, E., and Mouscadet, J. F. (2008) Integrase and integration: Biochemical activities of HIV-1 integrase, *Retrovirology 5*, 114.
36. Lewinski, M. K. and Bushman, F. D. (2005) Retroviral DNA integration—Mechanism and consequences, *Adv Genet 55*, 147–181.
37. Balvay, L., Lopez Lastra, M., Sargueil, B., Darlix, J. L., and Ohlmann, T. (2007) Translational control of retroviruses, *Nat Rev Microbiol 5*, 128–140.
38. Charnay, N., Ivanyi-Nagy, R., Soto-Rifo, R., Ohlmann, T., Lopez-Lastra, M., and Darlix, J. L. (2009) Mechanism of HIV-1 Tat RNA translation and its activation by the Tat protein, *Retrovirology 6*, 74.
39. Cimarelli, A. and Darlix, J. L. (2002) Assembling the human immunodeficiency virus type 1, *Cell Mol Life Sci 59*, 1166–1184.
40. Ono, A. (2010) Relationships between plasma membrane microdomains and HIV-1 assembly, *Biol Cell 102*, 335–350.
41. Martin, N. and Sattentau, Q. (2009) Cell-to-cell HIV-1 spread and its implications for immune evasion, *Curr Opin HIV AIDS 4*, 143–149.
42. Urisman, A., Molinaro, R. J., Fischer, N., Plummer, S. J., Casey, G., Klein, E. A., Malathi, K., Magi-Galluzzi, C., Tubbs, R. R., Ganem, D., Silverman, R. H., and DeRisi, J. L. (2006) Identification of a novel Gammaretrovirus in prostate tumors of patients homozygous for R462Q RNASEL variant, *PLoS Pathog 2*, e25.
43. Arnold, R. S., Makarova, N. V., Osunkoya, A. O., Suppiah, S., Scott, T. A., Johnson, N. A., Bhosle, S. M., Liotta, D., Hunter, E., Marshall, F. F., Ly, H., Molinaro, R. J., Blackwell, J. L., and Petros, J. A. (2010) XMRV infection in patients with prostate cancer: Novel serologic assay and correlation with PCR and FISH, *Urology 75*, 755–761.
44. Lombardi, V. C., Ruscetti, F. W., Das Gupta, J., Pfost, M. A., Hagen, K. S., Peterson, D. L., Ruscetti, S. K., Bagni, R. K., Petrow-Sadowski, C., Gold, B., Dean, M., Silverman, R. H., and Mikovits, J. A. (2009) Detection of an infectious retrovirus, XMRV, in blood cells of patients with chronic fatigue syndrome, *Science 326*, 585–589.
45. Lo, S. C., Pripuzova, N., Li, B., Komaroff, A. L., Hung, G. C., Wang, R., and Alter, H. J. (2010) Detection of MLV-related virus gene sequences in blood of patients with chronic fatigue syndrome and healthy blood donors, *Proc Natl Acad Sci U S A 107*, 15874–15879.
46. Weiss, R. A. (2010) A cautionary tale of virus and disease, *BMC Biol 8*, 124.
47. Chen, M., Garon, C. F., and Papas, T. S. (1980) Native ribonucleoprotein is an efficient transcriptional complex of avian myeloblastosis virus, *Proc Natl Acad Sci U S A 77*, 1296–1300.
48. Gorelick, R. J., Nigida, S. M., Jr., Bess, J. W., Jr., Arthur, L. O., Henderson, L. E., and Rein, A. (1990) Noninfectious human immunodeficiency virus type 1 mutants deficient in genomic RNA, *J Virol 64*, 3207–3211.
49. Meric, C., Gouilloud, E., and Spahr, P. F. (1988) Mutations in Rous sarcoma virus nucleocapsid protein p12 (NC): Deletions of Cys-His boxes, *J Virol 62*, 3328–3333.
50. Prats, A. C., Sarih, L., Gabus, C., Litvak, S., Keith, G., and Darlix, J. L. (1988) Small finger protein of avian and murine retroviruses has nucleic acid annealing activity and positions the replication primer tRNA onto genomic RNA, *EMBO J 7*, 1777–1783.

51. Prats, A. C., Housset, V., de Billy, G., Cornille, F., Prats, H., Roques, B., and Darlix, J. L. (1991) Viral RNA annealing activities of the nucleocapsid protein of Moloney murine leukemia virus are zinc independent, *Nucleic Acids Res 19*, 3533–3541.

52. Barat, C., Lullien, V., Schatz, O., Keith, G., Nugeyre, M. T., Gruninger-Leitch, F., Barre-Sinoussi, F., LeGrice, S. F., and Darlix, J. L. (1989) HIV-1 reverse transcriptase specifically interacts with the anticodon domain of its cognate primer tRNA, *EMBO J 8*, 3279–3285.

53. Levin, J. G., Guo, J., Rouzina, I., and Musier-Forsyth, K. (2005) Nucleic acid chaperone activity of HIV-1 nucleocapsid protein: Critical role in reverse transcription and molecular mechanism, *Prog Nucleic Acid Res Mol Biol 80*, 217–286.

54. Allain, B., Lapadat-Tapolsky, M., Berlioz, C., and Darlix, J. L. (1994) Transactivation of the minus-strand DNA transfer by nucleocapsid protein during reverse transcription of the retroviral genome, *EMBO J 13*, 973–981.

55. Buckman, J. S., Bosche, W. J., and Gorelick, R. J. (2003) Human immunodeficiency virus type 1 nucleocapsid zn(2+) fingers are required for efficient reverse transcription, initial integration processes, and protection of newly synthesized viral DNA, *J Virol 77*, 1469–1480.

56. Guo, J., Henderson, L. E., Bess, J., Kane, B., and Levin, J. G. (1997) Human immunodeficiency virus type 1 nucleocapsid protein promotes efficient strand transfer and specific viral DNA synthesis by inhibiting TAR-dependent self-priming from minus-strand strong-stop DNA, *J Virol 71*, 5178–5188.

57. Yu, Q. and Darlix, J. L. (1996) The zinc finger of nucleocapsid protein of Friend murine leukemia virus is critical for proviral DNA synthesis in vivo, *J Virol 70*, 5791–5798.

58. Darlix, J. L., Lapadat-Tapolsky, M., de Rocquigny, H., and Roques, B. P. (1995) First glimpses at structure-function relationships of the nucleocapsid protein of retroviruses, *J Mol Biol 254*, 523–537.

59. Thomas, J. A. and Gorelick, R. J. (2008) Nucleocapsid protein function in early infection processes, *Virus Res 134*, 39–63.

60. Coffin, J. M. (1979) Structure, replication, and recombination of retrovirus genomes: Some unifying hypotheses, *J Gen Virol 42*, 1–26.

61. Paillart, J. C., Shehu-Xhilaga, M., Marquet, R., and Mak, J. (2004) Dimerization of retroviral RNA genomes: An inseparable pair, *Nat Rev Microbiol 2*, 461–472.

62. Hu, W. S., Rhodes, T., Dang, Q., and Pathak, V. (2003) Retroviral recombination: Review of genetic analyses, *Front Biosci 8*, d143–d155.

63. Hu, W. S. and Temin, H. M. (1990) Retroviral recombination and reverse transcription, *Science 250*, 1227–1233.

64. Moore, M. D. and Hu, W. S. (2009) HIV-1 RNA dimerization: It takes two to tango, *AIDS Rev 11*, 91–102.

65. Galli, A., Kearney, M., Nikolaitchik, O. A., Yu, S., Chin, M. P., Maldarelli, F., Coffin, J. M., Pathak, V. K., and Hu, W. S. (2010) Patterns of Human Immunodeficiency Virus type 1 recombination ex vivo provide evidence for coadaptation of distant sites, resulting in purifying selection for intersubtype recombinants during replication, *J Virol 84*, 7651–7661.

66. Boutwell, C. L., Rolland, M. M., Herbeck, J. T., Mullins, J. I., and Allen, T. M. (2010) Viral evolution and escape during acute HIV-1 infection, *J Infect Dis 202*, Suppl 2, S309–S314.

REFERENCES

67 Wainberg, M. A., Brenner, B. G., and Turner, D. (2005) Changing patterns in the selection of viral mutations among patients receiving nucleoside and nucleotide drug combinations directed against human immunodeficiency virus type 1 reverse transcriptase, *Antimicrob Agents Chemother 49*, 1671–1678.

68 Turner, D., Roldan, A., Brenner, B., Moisi, D., Routy, J. P., and Wainberg, M. A. (2004) Variability in the PR and RT genes of HIV-1 isolated from recently infected subjects, *Antivir Chem Chemother 15*, 255–259.

69 Mely, Y., De Rocquigny, H., Morellet, N., Roques, B. P., and Gerad, D. (1996) Zinc binding to the HIV-1 nucleocapsid protein: A thermodynamic investigation by fluorescence spectroscopy, *Biochemistry 35*, 5175–5182.

70 Morellet, N., de Rocquigny, H., Mely, Y., Jullian, N., Demene, H., Ottmann, M., Gerard, D., Darlix, J. L., Fournie-Zaluski, M. C., and Roques, B. P. (1994) Conformational behaviour of the active and inactive forms of the nucleocapsid NCp7 of HIV-1 studied by 1H NMR, *J Mol Biol 235*, 287–301.

71 Summers, M. F., South, T. L., Kim, B., and Hare, D. R. (1990) High-resolution structure of an HIV zinc fingerlike domain via a new NMR-based distance geometry approach, *Biochemistry 29*, 329–340.

72 Bombarda, E., Grell, E., Roques, B. P., and Mely, Y. (2007) Molecular mechanism of the Zn2+-induced folding of the distal CCHC finger motif of the HIV-1 nucleocapsid protein, *Biophys J 93*, 208–217.

73 Godet, J. and Mely, Y. (2010) Biophysical studies of the nucleic acid chaperone properties of the HIV-1 nucleocapsid protein, *RNA Biol 7*, 48–60.

74 Rein, A. (2010) Nucleic acid chaperone activity of retroviral Gag proteins, *RNA Biol 7*, 61–66.

75 Muriaux, D., Mirro, J., Harvin, D., and Rein, A. (2001) RNA is a structural element in retrovirus particles, *Proc Natl Acad Sci U S A 98*, 5246–5251.

76 Muriaux, D., Costes, S., Nagashima, K., Mirro, J., Cho, E., Lockett, S., and Rein, A. (2004) Role of murine leukemia virus nucleocapsid protein in virus assembly, *J Virol 78*, 12378–12385.

77 D'Souza, V. and Summers, M. F. (2004) Structural basis for packaging the dimeric genome of Moloney murine leukaemia virus, *Nature 431*, 586–590.

78 Gabus, C., Ivanyi-Nagy, R., Depollier, J., Bucheton, A., Pelisson, A., and Darlix, J. L. (2006) Characterization of a nucleocapsid-like region and of two distinct primer tRNALys,2 binding sites in the endogenous retrovirus Gypsy, *Nucleic Acids Res 34*, 5764–5777.

79 Gabus, C., Ficheux, D., Rau, M., Keith, G., Sandmeyer, S., and Darlix, J. L. (1998) The yeast Ty3 retrotransposon contains a 5'-3' bipartite primer-binding site and encodes nucleocapsid protein NCp9 functionally homologous to HIV-1 NCp7, *EMBO J 17*, 4873–4880.

80 De Rocquigny, H., Gabus, C., Vincent, A., Fournie-Zaluski, M. C., Roques, B., and Darlix, J. L. (1992) Viral RNA annealing activities of human immunodeficiency virus type 1 nucleocapsid protein require only peptide domains outside the zinc fingers, *Proc Natl Acad Sci U S A 89*, 6472–6476.

81 Druillennec, S., Caneparo, A., de Rocquigny, H., and Roques, B. P. (1999) Evidence of interactions between the nucleocapsid protein NCp7 and the reverse transcriptase of HIV-1, *J Biol Chem 274*, 11283–11288.

82 Lener, D., Tanchou, V., Roques, B. P., Le Grice, S. F., and Darlix, J. L. (1998) Involvement of HIV-I nucleocapsid protein in the recruitment of reverse transcriptase into nucleoprotein complexes formed in vitro, *J Biol Chem 273*, 33781–33786.

83 Chan, B., Weidemaier, K., Yip, W. T., Barbara, P. F., and Musier-Forsyth, K. (1999) Intra-tRNA distance measurements for nucleocapsid protein-dependent tRNA unwinding during priming of HIV reverse transcription, *Proc Natl Acad Sci U S A 96*, 459–464.

84 Hargittai, M. R., Gorelick, R. J., Rouzina, I., and Musier-Forsyth, K. (2004) Mechanistic insights into the kinetics of HIV-1 nucleocapsid protein-facilitated tRNA annealing to the primer binding site, *J Mol Biol 337*, 951–968.

85 Tisne, C., Roques, B. P., and Dardel, F. (2004) The annealing mechanism of HIV-1 reverse transcription primer onto the viral genome, *J Biol Chem 279*, 3588–3595.

86 Bampi, C., Bibillo, A., Wendeler, M., Divita, G., Gorelick, R. J., Le Grice, S. F., and Darlix, J. L. (2006) Nucleotide excision repair and template-independent addition by HIV-1 reverse transcriptase in the presence of nucleocapsid protein, *J Biol Chem 281*, 11736–11743.

87 Grohmann, D., Godet, J., Mely, Y., Darlix, J. L., and Restle, T. (2008) HIV-1 nucleocapsid traps reverse transcriptase on nucleic acid substrates, *Biochemistry 47*, 12230–12240.

88 Gatignol, A. (2007) Transcription of HIV: Tat and cellular chromatin, *Adv Pharmacol 55*, 137–159.

89 Gatignol, A. and Jeang, K. T. (2000) Tat as a transcriptional activator and a potential therapeutic target for HIV-1, *Adv Pharmacol 48*, 209–227.

90 Kuciak, M., Gabus, C., Ivanyi-Nagy, R., Semrad, K., Storchak, R., Chaloin, O., Muller, S., Mely, Y., and Darlix, J. L. (2008) The HIV-1 transcriptional activator Tat has potent nucleic acid chaperoning activities in vitro, *Nucleic Acids Res 36*, 3389–3400.

91 Kameoka, M., Morgan, M., Binette, M., Russell, R. S., Rong, L., Guo, X., Mouland, A., Kleiman, L., Liang, C., and Wainberg, M. A. (2002) The Tat protein of human immunodeficiency virus type 1 (HIV-1) can promote placement of tRNA primer onto viral RNA and suppress later DNA polymerization in HIV-1 reverse transcription, *J Virol 76*, 3637–3645.

92 Henriet, S., Sinck, L., Bec, G., Gorelick, R. J., Marquet, R., and Paillart, J. C. (2007) Vif is a RNA chaperone that could temporally regulate RNA dimerization and the early steps of HIV-1 reverse transcription, *Nucleic Acids Res 35*, 5141–5153.

93 Bardy, M., Gay, B., Pebernard, S., Chazal, N., Courcoul, M., Vigne, R., Decroly, E., and Boulanger, P. (2001) Interaction of human immunodeficiency virus type 1 Vif with Gag and Gag-Pol precursors: Co-encapsidation and interference with viral protease-mediated Gag processing, *J Gen Virol 82*, 2719–2733.

94 Mangeat, B., Turelli, P., Caron, G., Friedli, M., Perrin, L., and Trono, D. (2003) Broad antiretroviral defence by human APOBEC3G through lethal editing of nascent reverse transcripts, *Nature 424*, 99–103.

95 Malim, M. H. and Emerman, M. (2008) HIV-1 accessory proteins—Ensuring viral survival in a hostile environment, *Cell Host Microbe 3*, 388–398.

96 Selig, L., Pages, J. C., Tanchou, V., Preveral, S., Berlioz-Torrent, C., Liu, L. X., Erdtmann, L., Darlix, J., Benarous, R., and Benichou, S. (1999) Interaction with the p6 domain of the gag precursor mediates incorporation into virions of Vpr and Vpx proteins from primate lentiviruses, *J Virol 73*, 592–600.

REFERENCES

97. Kondo, E., Mammano, F., Cohen, E. A., and Gottlinger, H. G. (1995) The p6gag domain of human immunodeficiency virus type 1 is sufficient for the incorporation of Vpr into heterologous viral particles, *J Virol 69*, 2759–2764.
98. de Rocquigny, H., Petitjean, P., Tanchou, V., Decimo, D., Drouot, L., Delaunay, T., Darlix, J. L., and Roques, B. P. (1997) The zinc fingers of HIV nucleocapsid protein NCp7 direct interactions with the viral regulatory protein Vpr, *J Biol Chem 272*, 30753–30759.
99. Popov, S., Popova, E., Inoue, M., and Gottlinger, H. G. (2008) Human immunodeficiency virus type 1 Gag engages the Bro1 domain of ALIX/AIP1 through the nucleocapsid, *J Virol 82*, 1389–1398.
100. Gabus, C., Derrington, E., Leblanc, P., Chnaiderman, J., Dormont, D., Swietnicki, W., Morillas, M., Surewicz, W. K., Marc, D., Nandi, P., and Darlix, J. L. (2001) The prion protein has RNA binding and chaperoning properties characteristic of nucleocapsid protein NCP7 of HIV-1, *J Biol Chem 276*, 19301–19309.
101. Leblanc, P., Baas, D., and Darlix, J. L. (2004) Analysis of the interactions between HIV-1 and the cellular prion protein in a human cell line, *J Mol Biol 337*, 1035–1051.
102. Grigorov, B., Decimo, D., Smagulova, F., Pechoux, C., Mougel, M., Muriaux, D., and Darlix, J. L. (2007) Intracellular HIV-1 Gag localization is impaired by mutations in the nucleocapsid zinc fingers, *Retrovirology 4*, 54.
103. Perz, J. F., Armstrong, G. L., Farrington, L. A., Hutin, Y. J., and Bell, B. P. (2006) The contributions of hepatitis B virus and hepatitis C virus infections to cirrhosis and primary liver cancer worldwide, *J Hepatol 45*, 529–538.
104. Gould, E., Gallian, P., De Lamballerie, X., and Charrel, R. (2010) First cases of autochthonous dengue fever and chikungunya fever in France: From bad dream to reality! *Clin Microbiol Infect 16*, 1702–1704.
105. Gould, E. A. and Solomon, T. (2008) Pathogenic flaviviruses, *Lancet 371*, 500–509.
106. Roehrig, J. T., Layton, M., Smith, P., Campbell, G. L., Nasci, R., and Lanciotti, R. S. (2002) The emergence of West Nile virus in North America: Ecology, epidemiology, and surveillance, *Curr Top Microbiol Immunol 267*, 223–240.
107. Santolini, E., Migliaccio, G., and La Monica, N. (1994) Biosynthesis and biochemical properties of the hepatitis C virus core protein, *J Virol 68*, 3631–3641.
108. Murray, C. L., Marcotrigiano, J., and Rice, C. M. (2008) Bovine viral diarrhea virus core is an intrinsically disordered protein that binds RNA, *J Virol 82*, 1294–1304.
109. Ivanyi-Nagy, R., Lavergne, J. P., Gabus, C., Ficheux, D., and Darlix, J. L. (2008) RNA chaperoning and intrinsic disorder in the core proteins of Flaviviridae, *Nucleic Acids Res 36*, 712–725.
110. Cristofari, G., Ivanyi-Nagy, R., Gabus, C., Boulant, S., Lavergne, J. P., Penin, F., and Darlix, J. L. (2004) The hepatitis C virus core protein is a potent nucleic acid chaperone that directs dimerization of the viral (+) strand RNA in vitro, *Nucleic Acids Res 32*, 2623–2631.
111. Ivanyi-Nagy, R., Kanevsky, I., Gabus, C., Lavergne, J. P., Ficheux, D., Penin, F., Fosse, P., and Darlix, J. L. (2006) Analysis of hepatitis C virus RNA dimerization and core-RNA interactions, *Nucleic Acids Res 34*, 2618–2633.
112. Majeau, N., Gagne, V., Boivin, A., Bolduc, M., Majeau, J. A., Ouellet, D., and Leclerc, D. (2004) The N-terminal half of the core protein of hepatitis C virus is sufficient for nucleocapsid formation, *J Gen Virol 85*, 971–981.

113 Yu, K. L., Jang, S. I., and You, J. C. (2009) Identification of in vivo interaction between Hepatitis C Virus core protein and 5' and 3' UTR RNA, *Virus Res 145*, 285–292.

114 Tanaka, Y., Shimoike, T., Ishii, K., Suzuki, R., Suzuki, T., Ushijima, H., Matsuura, Y., and Miyamura, T. (2000) Selective binding of hepatitis C virus core protein to synthetic oligonucleotides corresponding to the 5' untranslated region of the viral genome, *Virology 270*, 229–236.

115 Boni, S., Lavergne, J. P., Boulant, S., and Cahour, A. (2005) Hepatitis C virus core protein acts as a trans-modulating factor on internal translation initiation of the viral RNA, *J Biol Chem 280*, 17737–17748.

116 Lourenco, S., Costa, F., Debarges, B., Andrieu, T., and Cahour, A. (2008) Hepatitis C virus internal ribosome entry site-mediated translation is stimulated by cis-acting RNA elements and trans-acting viral factors, *FEBS J 275*, 4179–4197.

117 Zhang, J., Yamada, O., Yoshida, H., Iwai, T., and Araki, H. (2002) Autogenous translational inhibition of core protein: Implication for switch from translation to RNA replication in hepatitis C virus, *Virology 293*, 141–150.

118 Shimoike, T., Koyama, C., Murakami, K., Suzuki, R., Matsuura, Y., Miyamura, T., and Suzuki, T. (2006) Down-regulation of the internal ribosome entry site (IRES)-mediated translation of the hepatitis C virus: Critical role of binding of the stem-loop IIId domain of IRES and the viral core protein, *Virology 345*, 434–445.

119 Shimoike, T., Mimori, S., Tani, H., Matsuura, Y., and Miyamura, T. (1999) Interaction of hepatitis C virus core protein with viral sense RNA and suppression of its translation, *J Virol 73*, 9718–9725.

120 Shetty, S., Kim, S., Shimakami, T., Lemon, S. M., and Mihailescu, M. R. (2010) Hepatitis C virus genomic RNA dimerization is mediated via a kissing complex intermediate, *RNA 16*, 913–925.

121 Kunkel, M. and Watowich, S. J. (2002) Conformational changes accompanying self-assembly of the hepatitis C virus core protein, *Virology 294*, 239–245.

122 Duvignaud, J. B., Savard, C., Fromentin, R., Majeau, N., Leclerc, D., and Gagne, S. M. (2009) Structure and dynamics of the N-terminal half of hepatitis C virus core protein: An intrinsically unstructured protein, *Biochem Biophys Res Commun 378*, 27–31.

123 Boulant, S., Vanbelle, C., Ebel, C., Penin, F., and Lavergne, J. P. (2005) Hepatitis C virus core protein is a dimeric alpha-helical protein exhibiting membrane protein features, *J Virol 79*, 11353–11365.

124 Kunkel, M., Lorinczi, M., Rijnbrand, R., Lemon, S. M., and Watowich, S. J. (2001) Self-assembly of nucleocapsid-like particles from recombinant hepatitis C virus core protein, *J Virol 75*, 2119–2129.

125 Klein, K. C., Polyak, S. J., and Lingappa, J. R. (2004) Unique features of hepatitis C virus capsid formation revealed by de novo cell-free assembly, *J Virol 78*, 9257–9269.

126 de Chassey, B., Navratil, V., Tafforeau, L., Hiet, M. S., Aublin-Gex, A., Agaugue, S., Meiffren, G., Pradezynski, F., Faria, B. F., Chantier, T., Le Breton, M., Pellet, J., Davoust, N., Mangeot, P. E., Chaboud, A., Penin, F., Jacob, Y., Vidalain, P. O., Vidal, M., Andre, P., Rabourdin-Combe, C., and Lotteau, V. (2008) Hepatitis C virus infection protein network, *Mol Syst Biol 4*, 230.

127 McLauchlan, J. (2000) Properties of the hepatitis C virus core protein: A structural protein that modulates cellular processes, *J Viral Hepat 7*, 2–14.

REFERENCES

128 Ray, R. B. and Ray, R. (2001) Hepatitis C virus core protein: Intriguing properties and functional relevance, *FEMS Microbiol Lett 202*, 149–156.

129 Moriya, K., Fujie, H., Shintani, Y., Yotsuyanagi, H., Tsutsumi, T., Ishibashi, K., Matsuura, Y., Kimura, S., Miyamura, T., and Koike, K. (1998) The core protein of hepatitis C virus induces hepatocellular carcinoma in transgenic mice, *Nat Med 4*, 1065–1067.

130 Moriya, K., Yotsuyanagi, H., Shintani, Y., Fujie, H., Ishibashi, K., Matsuura, Y., Miyamura, T., and Koike, K. (1997) Hepatitis C virus core protein induces hepatic steatosis in transgenic mice, *J Gen Virol 78*, (Pt 7), 1527–1531.

131 Shintani, Y., Fujie, H., Miyoshi, H., Tsutsumi, T., Tsukamoto, K., Kimura, S., Moriya, K., and Koike, K. (2004) Hepatitis C virus infection and diabetes: Direct involvement of the virus in the development of insulin resistance, *Gastroenterology 126*, 840–848.

132 Hope, R. G. and McLauchlan, J. (2000) Sequence motifs required for lipid droplet association and protein stability are unique to the hepatitis C virus core protein, *J Gen Virol 81*, 1913–1925.

133 Hope, R. G., Murphy, D. J., and McLauchlan, J. (2002) The domains required to direct core proteins of hepatitis C virus and GB virus-B to lipid droplets share common features with plant oleosin proteins, *J Biol Chem 277*, 4261–4270.

134 Shavinskaya, A., Boulant, S., Penin, F., McLauchlan, J., and Bartenschlager, R. (2007) The lipid droplet binding domain of hepatitis C virus core protein is a major determinant for efficient virus assembly, *J Biol Chem 282*, 37158–37169.

135 Miyanari, Y., Atsuzawa, K., Usuda, N., Watashi, K., Hishiki, T., Zayas, M., Bartenschlager, R., Wakita, T., Hijikata, M., and Shimotohno, K. (2007) The lipid droplet is an important organelle for hepatitis C virus production, *Nat Cell Biol 9*, 1089–1097.

136 Khromykh, A. A. and Westaway, E. G. (1996) RNA binding properties of core protein of the flavivirus Kunjin, *Arch Virol 141*, 685–699.

137 Patkar, C. G., Jones, C. T., Chang, Y. H., Warrier, R., and Kuhn, R. J. (2007) Functional requirements of the yellow fever virus capsid protein, *J Virol 81*, 6471–6481.

138 Markoff, L., Falgout, B., and Chang, A. (1997) A conserved internal hydrophobic domain mediates the stable membrane integration of the dengue virus capsid protein, *Virology 233*, 105–117.

139 Wang, S. H., Syu, W. J., and Hu, S. T. (2004) Identification of the homotypic interaction domain of the core protein of dengue virus type 2, *J Gen Virol 85*, 2307–2314.

140 Samsa, M. M., Mondotte, J. A., Iglesias, N. G., Assuncao-Miranda, I., Barbosa-Lima, G., Da Poian, A. T., Bozza, P. T., and Gamarnik, A. V. (2009) Dengue virus capsid protein usurps lipid droplets for viral particle formation, *PLoS Pathog 5*, e1000632.

141 Khromykh, A. A., Meka, H., Guyatt, K. J., and Westaway, E. G. (2001) Essential role of cyclization sequences in flavivirus RNA replication, *J Virol 75*, 6719–6728.

142 Lo, M. K., Tilgner, M., Bernard, K. A., and Shi, P. Y. (2003) Functional analysis of mosquito-borne flavivirus conserved sequence elements within 3' untranslated region of West Nile virus by use of a reporting replicon that differentiates between viral translation and RNA replication, *J Virol 77*, 10004–10014.

143 Alvarez, D. E., Lodeiro, M. F., Luduena, S. J., Pietrasanta, L. I., and Gamarnik, A. V. (2005) Long-range RNA-RNA interactions circularize the dengue virus genome, *J Virol 79*, 6631–6643.

144 Villordo, S. M., Alvarez, D. E., and Gamarnik, A. V. (2010) A balance between circular and linear forms of the dengue virus genome is crucial for viral replication, *RNA 16*, 2325–2335.

145 Ma, L., Jones, C. T., Groesch, T. D., Kuhn, R. J., and Post, C. B. (2004) Solution structure of dengue virus capsid protein reveals another fold, *Proc Natl Acad Sci U S A 101*, 3414–3419.

146 Dokland, T., Walsh, M., Mackenzie, J. M., Khromykh, A. A., Ee, K. H., and Wang, S. (2004) West Nile virus core protein; tetramer structure and ribbon formation, *Structure 12*, 1157–1163.

147 Schlick, P., Taucher, C., Schittl, B., Tran, J. L., Kofler, R. M., Schueler, W., von Gabain, A., Meinke, A., and Mandl, C. W. (2009) Helices alpha2 and alpha3 of West Nile virus capsid protein are dispensable for assembly of infectious virions, *J Virol 83*, 5581–5591.

148 Kofler, R. M., Heinz, F. X., and Mandl, C. W. (2002) Capsid protein C of tick-borne encephalitis virus tolerates large internal deletions and is a favorable target for attenuation of virulence, *J Virol 76*, 3534–3543.

149 Kofler, R. M., Leitner, A., O'Riordain, G., Heinz, F. X., and Mandl, C. W. (2003) Spontaneous mutations restore the viability of tick-borne encephalitis virus mutants with large deletions in protein C, *J Virol 77*, 443–451.

150 Enjuanes, L., Almazan, F., Sola, I., and Zuniga, S. (2006) Biochemical aspects of coronavirus replication and virus-host interaction, *Annu Rev Microbiol 60*, 211–230.

151 Drosten, C., Gunther, S., Preiser, W., van der Werf, S., Brodt, H. R., Becker, S., Rabenau, H., Panning, M., Kolesnikova, L., Fouchier, R. A., Berger, A., Burguiere, A. M., Cinatl, J., Eickmann, M., Escriou, N., Grywna, K., Kramme, S., Manuguerra, J. C., Muller, S., Rickerts, V., Sturmer, M., Vieth, S., Klenk, H. D., Osterhaus, A. D., Schmitz, H., and Doerr, H. W. (2003) Identification of a novel coronavirus in patients with severe acute respiratory syndrome, *N Engl J Med 348*, 1967–1976.

152 Kuiken, T., Fouchier, R. A., Schutten, M., Rimmelzwaan, G. F., van Amerongen, G., van Riel, D., Laman, J. D., de Jong, T., van Doornum, G., Lim, W., Ling, A. E., Chan, P. K., Tam, J. S., Zambon, M. C., Gopal, R., Drosten, C., van der Werf, S., Escriou, N., Manuguerra, J. C., Stohr, K., Peiris, J. S., and Osterhaus, A. D. (2003) Newly discovered coronavirus as the primary cause of severe acute respiratory syndrome, *Lancet 362*, 263–270.

153 Rota, P. A., Oberste, M. S., Monroe, S. S., Nix, W. A., Campagnoli, R., Icenogle, J. P., Penaranda, S., Bankamp, B., Maher, K., Chen, M. H., Tong, S., Tamin, A., Lowe, L., Frace, M., DeRisi, J. L., Chen, Q., Wang, D., Erdman, D. D., Peret, T. C., Burns, C., Ksiazek, T. G., Rollin, P. E., Sanchez, A., Liffick, S., Holloway, B., Limor, J., McCaustland, K., Olsen-Rasmussen, M., Fouchier, R., Gunther, S., Osterhaus, A. D., Drosten, C., Pallansch, M. A., Anderson, L. J., and Bellini, W. J. (2003) Characterization of a novel coronavirus associated with severe acute respiratory syndrome, *Science 300*, 1394–1399.

154 Ksiazek, T. G., Erdman, D., Goldsmith, C. S., Zaki, S. R., Peret, T., Emery, S., Tong, S., Urbani, C., Comer, J. A., Lim, W., Rollin, P. E., Dowell, S. F., Ling, A. E., Humphrey, C. D., Shieh, W. J., Guarner, J., Paddock, C. D., Rota, P., Fields, B., DeRisi, J., Yang, J. Y., Cox, N., Hughes, J. M., LeDuc, J. W., Bellini, W. J., and Anderson, L. J. (2003) A novel coronavirus associated with severe acute respiratory syndrome, *N Engl J Med 348*, 1953–1966.

155 Peiris, J. S., Lai, S. T., Poon, L. L., Guan, Y., Yam, L. Y., Lim, W., Nicholls, J., Yee, W. K., Yan, W. W., Cheung, M. T., Cheng, V. C., Chan, K. H., Tsang, D. N., Yung, R. W., Ng, T. K., and Yuen, K. Y. (2003) Coronavirus as a possible cause of severe acute respiratory syndrome, *Lancet 361*, 1319–1325.

156 Gorbalenya, A. E., Enjuanes, L., Ziebuhr, J., and Snijder, E. J. (2006) Nidovirales: Evolving the largest RNA virus genome, *Virus Res 117*, 17–37.

157 Minskaia, E., Hertzig, T., Gorbalenya, A. E., Campanacci, V., Cambillau, C., Canard, B., and Ziebuhr, J. (2006) Discovery of an RNA virus 3'->5' exoribonuclease that is critically involved in coronavirus RNA synthesis, *Proc Natl Acad Sci U S A 103*, 5108–5113.

158 Zuniga, S., Sola, I., Moreno, J. L., Sabella, P., Plana-Duran, J., and Enjuanes, L. (2007) Coronavirus nucleocapsid protein is an RNA chaperone, *Virology 357*, 215–227.

159 Baric, R. S., Nelson, G. W., Fleming, J. O., Deans, R. J., Keck, J. G., Casteel, N., and Stohlman, S. A. (1988) Interactions between coronavirus nucleocapsid protein and viral RNAs: Implications for viral transcription, *J Virol 62*, 4280–4287.

160 Nelson, G. W., Stohlman, S. A., and Tahara, S. M. (2000) High affinity interaction between nucleocapsid protein and leader/intergenic sequence of mouse hepatitis virus RNA, *J Gen Virol 81*, 181–188.

161 Stohlman, S. A., Baric, R. S., Nelson, G. N., Soe, L. H., Welter, L. M., and Deans, R. J. (1988) Specific interaction between coronavirus leader RNA and nucleocapsid protein, *J Virol 62*, 4288–4295.

162 Chen, H., Gill, A., Dove, B. K., Emmett, S. R., Kemp, C. F., Ritchie, M. A., Dee, M., and Hiscox, J. A. (2005) Mass spectroscopic characterization of the coronavirus infectious bronchitis virus nucleoprotein and elucidation of the role of phosphorylation in RNA binding by using surface plasmon resonance, *J Virol 79*, 1164–1179.

163 Zuniga, S., Cruz, J. L., Sola, I., Mateos-Gomez, P. A., Palacio, L., and Enjuanes, L. (2010) Coronavirus nucleocapsid protein facilitates template switching and is required for efficient transcription, *J Virol 84*, 2169–2175.

164 Sawicki, S. G., Sawicki, D. L., and Siddell, S. G. (2007) A contemporary view of coronavirus transcription, *J Virol 81*, 20–29.

165 Zuniga, S., Sola, I., Alonso, S., and Enjuanes, L. (2004) Sequence motifs involved in the regulation of discontinuous coronavirus subgenomic RNA synthesis, *J Virol 78*, 980–994.

166 Sola, I., Moreno, J. L., Zuniga, S., Alonso, S., and Enjuanes, L. (2005) Role of nucleotides immediately flanking the transcription-regulating sequence core in coronavirus subgenomic mRNA synthesis, *J Virol 79*, 2506–2516.

167 Almazan, F., Galan, C., and Enjuanes, L. (2004) The nucleoprotein is required for efficient coronavirus genome replication, *J Virol 78*, 12683–12688.

168 Grossoehme, N. E., Li, L., Keane, S. C., Liu, P., Dann, C. E., 3rd, Leibowitz, J. L., and Giedroc, D. P. (2009) Coronavirus N protein N-terminal domain (NTD) specifically binds the transcriptional regulatory sequence (TRS) and melts TRS-cTRS RNA duplexes, *J Mol Biol 394*, 544–557.

169 Chang, C. K., Sue, S. C., Yu, T. H., Hsieh, C. M., Tsai, C. K., Chiang, Y. C., Lee, S. J., Hsiao, H. H., Wu, W. J., Chang, W. L., Lin, C. H., and Huang, T. H. (2006) Modular organization of SARS coronavirus nucleocapsid protein, *J Biomed Sci 13*, 59–72.

170 Huang, Q., Yu, L., Petros, A. M., Gunasekera, A., Liu, Z., Xu, N., Hajduk, P., Mack, J., Fesik, S. W., and Olejniczak, E. T. (2004) Structure of the N-terminal RNA-binding domain of the SARS CoV nucleocapsid protein, *Biochemistry 43*, 6059–6063.

171 Saikatendu, K. S., Joseph, J. S., Subramanian, V., Neuman, B. W., Buchmeier, M. J., Stevens, R. C., and Kuhn, P. (2007) Ribonucleocapsid formation of severe acute respiratory syndrome coronavirus through molecular action of the N-terminal domain of N protein, *J Virol 81*, 3913–3921.

172 Yu, I. M., Gustafson, C. L., Diao, J., Burgner, J. W., 2nd, Li, Z., Zhang, J., and Chen, J. (2005) Recombinant severe acute respiratory syndrome (SARS) coronavirus nucleocapsid protein forms a dimer through its C-terminal domain, *J Biol Chem 280*, 23280–23286.

173 Luo, H., Chen, J., Chen, K., Shen, X., and Jiang, H. (2006) Carboxyl terminus of severe acute respiratory syndrome coronavirus nucleocapsid protein: Self-association analysis and nucleic acid binding characterization, *Biochemistry 45*, 11827–11835.

174 Takeda, M., Chang, C. K., Ikeya, T., Guntert, P., Chang, Y. H., Hsu, Y. L., Huang, T. H., and Kainosho, M. (2008) Solution structure of the C-terminal dimerization domain of SARS coronavirus nucleocapsid protein solved by the SAIL-NMR method, *J Mol Biol 380*, 608–622.

175 Chang, C. K., Hsu, Y. L., Chang, Y. H., Chao, F. A., Wu, M. C., Huang, Y. S., Hu, C. K., and Huang, T. H. (2009) Multiple nucleic acid binding sites and intrinsic disorder of severe acute respiratory syndrome coronavirus nucleocapsid protein: Implications for ribonucleocapsid protein packaging, *J Virol 83*, 2255–2264.

176 Shoemaker, B. A., Portman, J. J., and Wolynes, P. G. (2000) Speeding molecular recognition by using the folding funnel: The fly-casting mechanism, *Proc Natl Acad Sci U S A 97*, 8868–8873.

177 Hilser, V. J. and Thompson, E. B. (2007) Intrinsic disorder as a mechanism to optimize allosteric coupling in proteins, *Proc Natl Acad Sci U S A 104*, 8311–8315.

178 Jonsson, C. B. and Schmaljohn, C. S. (2001) Replication of hantaviruses, *Curr Top Microbiol Immunol 256*, 15–32.

179 Jonsson, C. B., Figueiredo, L. T., and Vapalahti, O. (2010) A global perspective on hantavirus ecology, epidemiology, and disease, *Clin Microbiol Rev 23*, 412–441.

180 Gott, P., Stohwasser, R., Schnitzler, P., Darai, G., and Bautz, E. K. (1993) RNA binding of recombinant nucleocapsid proteins of hantaviruses, *Virology 194*, 332–337.

181 Mir, M. A. and Panganiban, A. T. (2006) The bunyavirus nucleocapsid protein is an RNA chaperone: Possible roles in viral RNA panhandle formation and genome replication, *RNA 12*, 272–282.

182 Xu, X., Severson, W., Villegas, N., Schmaljohn, C. S., and Jonsson, C. B. (2002) The RNA binding domain of the hantaan virus N protein maps to a central, conserved region, *J Virol 76*, 3301–3308.

183 Mir, M. A., Brown, B., Hjelle, B., Duran, W. A., and Panganiban, A. T. (2006) Hantavirus N protein exhibits genus-specific recognition of the viral RNA panhandle, *J Virol 80*, 11283–11292.

184 Mir, M. A. and Panganiban, A. T. (2004) Trimeric hantavirus nucleocapsid protein binds specifically to the viral RNA panhandle, *J Virol 78*, 8281–8288.

185 Mir, M. A. and Panganiban, A. T. (2005) The hantavirus nucleocapsid protein recognizes specific features of the viral RNA panhandle and is altered in conformation upon RNA binding, *J Virol 79*, 1824–1835.

186 Severson, W., Partin, L., Schmaljohn, C. S., and Jonsson, C. B. (1999) Characterization of the Hantaan nucleocapsid protein-ribonucleic acid interaction, *J Biol Chem 274*, 33732–33739.

187 Severson, W., Xu, X., Kuhn, M., Senutovitch, N., Thokala, M., Ferron, F., Longhi, S., Canard, B., and Jonsson, C. B. (2005) Essential amino acids of the hantaan virus N protein in its interaction with RNA, *J Virol 79*, 10032–10039.

188 Severson, W. E., Xu, X., and Jonsson, C. B. (2001) cis-Acting signals in encapsidation of Hantaan virus S-segment viral genomic RNA by its N protein, *J Virol 75*, 2646–2652.

189 Mir, M. A. and Panganiban, A. T. (2006) Characterization of the RNA chaperone activity of hantavirus nucleocapsid protein, *J Virol 80*, 6276–6285.

190 Hewlett, M. J., Pettersson, R. F., and Baltimore, D. (1977) Circular forms of Uukuniemi virion RNA: An electron microscopic study, *J Virol 21*, 1085–1093.

191 Pettersson, R. F. and von Bonsdorff, C. H. (1975) Ribonucleoproteins of Uukuniemi virus are circular, *J Virol 15*, 386–392.

192 Mir, M. A. and Panganiban, A. T. (2008) A protein that replaces the entire cellular eIF4F complex, *EMBO J 27*, 3129–3139.

193 Balvay, L., Soto Rifo, R., Ricci, E. P., Decimo, D., and Ohlmann, T. (2009) Structural and functional diversity of viral IRESes, *Biochim Biophys Acta 1789*, 542–557.

194 Mir, M. A., Duran, W. A., Hjelle, B. L., Ye, C., and Panganiban, A. T. (2008) Storage of cellular 5' mRNA caps in P bodies for viral cap-snatching, *Proc Natl Acad Sci U S A 105*, 19294–19299.

195 Mir, M. A., Sheema, S., Haseeb, A., and Haque, A. (2010) Hantavirus nucleocapsid protein has distinct m7G cap- and RNA-binding sites, *J Biol Chem 285*, 11357–11368.

196 Wang, Y., Boudreaux, D. M., Estrada, D. F., Egan, C. W., St Jeor, S. C., and De Guzman, R. N. (2008) NMR structure of the N-terminal coiled coil domain of the Andes hantavirus nucleocapsid protein, *J Biol Chem 283*, 28297–28304.

197 Boudko, S. P., Kuhn, R. J., and Rossmann, M. G. (2007) The coiled-coil domain structure of the Sin Nombre virus nucleocapsid protein, *J Mol Biol 366*, 1538–1544.

198 Alfadhli, A., Love, Z., Arvidson, B., Seeds, J., Willey, J., and Barklis, E. (2001) Hantavirus nucleocapsid protein oligomerization, *J Virol 75*, 2019–2023.

199 Alfadhli, A., Steel, E., Finlay, L., Bachinger, H. P., and Barklis, E. (2002) Hantavirus nucleocapsid protein coiled-coil domains, *J Biol Chem 277*, 27103–27108.

200 Alminaite, A., Halttunen, V., Kumar, V., Vaheri, A., Holm, L., and Plyusnin, A. (2006) Oligomerization of hantavirus nucleocapsid protein: Analysis of the N-terminal coiled-coil domain, *J Virol 80*, 9073–9081.

201 Kumar, A. and Wilson, S. H. (1990) Studies of the strand-annealing activity of mammalian hnRNP complex protein A1, *Biochemistry 29*, 10717–10722.

202 Pontius, B. W. and Berg, P. (1990) Renaturation of complementary DNA strands mediated by purified mammalian heterogeneous nuclear ribonucleoprotein A1 protein: Implications for a mechanism for rapid molecular assembly, *Proc Natl Acad Sci U S A 87*, 8403–8407.

203 Casas-Finet, J. R., Smith, J. D., Jr., Kumar, A., Kim, J. G., Wilson, S. H., and Karpel, R. L. (1993) Mammalian heterogeneous ribonucleoprotein A1 and its constituent

domains. Nucleic acid interaction, structural stability and self-association, *J Mol Biol* 229, 873–889.

204 Buvoli, M., Cobianchi, F., Biamonti, G., and Riva, S. (1990) Recombinant hnRNP protein A1 and its N-terminal domain show preferential affinity for oligodeoxynucleotides homologous to intron/exon acceptor sites, *Nucleic Acids Res 18*, 6595–6600.

205 Obradovic, Z., Peng, K., Vucetic, S., Radivojac, P., Brown, C. J., and Dunker, A. K. (2003) Predicting intrinsic disorder from amino acid sequence, *Proteins 53*, Suppl 6, 566–572.

206 Ooms, M., Cupac, D., Abbink, T. E., Huthoff, H., and Berkhout, B. (2007) The availability of the primer activation signal (PAS) affects the efficiency of HIV-1 reverse transcription initiation, *Nucleic Acids Res 35*, 1649–1659.

207 Yi, M. and Lemon, S. M. (2003) 3' nontranslated RNA signals required for replication of hepatitis C virus RNA, *J Virol 77*, 3557–3568.

208 Zhang, B., Dong, H., Stein, D. A., Iversen, P. L., and Shi, P. Y. (2008) West Nile virus genome cyclization and RNA replication require two pairs of long-distance RNA interactions, *Virology 373*, 1–13.

209 Friebe, P. and Harris, E. (2010) Interplay of RNA elements in the dengue virus 5' and 3' ends required for viral RNA replication, *J Virol 84*, 6103–6118.

210 Liu, P., Li, L., Millership, J. J., Kang, H., Leibowitz, J. L., and Giedroc, D. P. (2007) A U-turn motif-containing stem-loop in the coronavirus 5' untranslated region plays a functional role in replication, *RNA 13*, 763–780.

III

AGGREGATION OF DISORDERED PEPTIDES

11

SELF-ASSEMBLING ALANINE-RICH PEPTIDES OF BIOMEDICAL AND BIOTECHNOLOGICAL RELEVANCE

Thomas J. Measey and Reinhard Schweitzer-Stenner

11.1. BIOMOLECULAR SELF-ASSEMBLY

Molecular self-assembly is the autonomous and precise organization of constituent molecular components into supramolecular structures. The process of self-organization is ubiquitous throughout nature and technology [1–4]. While hydrophobic interactions and ionic and hydrogen bonding are inherently weak forces, such non-covalent interactions collectively govern the self-assembly process [4]. A defining feature of all living systems is the precise self-organization of its constituent components [5]. At the biomolecular level, self-assembly manifests itself as the characteristic tertiary fold of a protein, the intricate architecture of a cell membrane, and the organization of extracellular matrix proteins such as elastin and collagen, responsible for the elasticity and flexibility of most tissues and organs [6]. Biological self-assembly results in diverse architectures displaying various and often complex functions [3]. In addition to these manifestations of biological self-assembly necessary for the sustainability of living systems, other natural self-organizing processes result in the formation of toxic aggregates affiliated with various debilitating human pathologies [5, 7–9]. Moreover, if it is possible to exploit nature's ability to control

Protein and Peptide Folding, Misfolding, and Non-Folding, First Edition. Edited by Reinhard Schweitzer-Stenner.
© 2012 John Wiley & Sons, Inc. Published 2012 by John Wiley & Sons, Inc.

with precision and fidelity the self-assembly process, the design of various novel biomaterials can be envisaged with significant and possibly life-changing applications [10–13].

This review is essentially divided into two parts: the use of short alanine-rich peptides for (1) studying the mechanism of amyloid-like fibril formation and (2) exploitation of the self-assembly process for biotechnological applications.

11.2. MISFOLDING AND HUMAN DISEASE

As complete protein folding is often necessary for proper biological functioning, the inability to achieve or maintain a native fold is linked to various debilitating human pathologies [5, 7–9, 14]. The largest class of protein misfolding diseases results from the aggregation of specific misfolded or intrinsically disordered proteins (IDPs) into insoluble filamentous structures termed "amyloid fibrils" [5, 7, 15–17]. In particular, the formation of amyloid fibrils composed of misfolded proteins and/or IDPs and peptides is believed to be responsible for the onset of various neurodegenerative diseases, such as Alzheimer's, Huntington's, and Parkinson's disease, as well as the systemic amyloidoses and type II diabetes [5, 7, 8, 17–19].

Misfolded states are representative of the various local minima located either along or off the pathway (energy landscape) to the correctly folded state of a protein [17]. Owing to the complexity of the protein energy landscape, various states are often accessible in the cellular milieu, with the free energy of each state depending on the specific environmental conditions and various kinetic and thermodynamic factors [17, 20]. These various accessible states include aggregated states such as fibrils, which are the main constituents of proteinaceous buildup often described as a hallmark of the above-mentioned human diseases. A complete and detailed description of the protein energy landscape requires an understanding of the energetics accompanying the interconversion between the assortment of accessible states. Such a description is necessary for understanding amyloid formation for disease treatment and prevention.

A defining feature of the various pathologically diverse diseases that arise from protein misfolding and aggregation is the accumulation of cellular proteinaceous deposits, consisting primarily of fibrillar species. Amyloid-like plaques and Lewy bodies are formed from the accumulation of unbranched fibrils, which adopt a characteristic cross-β core, with the constituent β-strands of the intertwined protofilaments aligned perpendicular to the fibril axis [21]. Despite the sequence diversity of the various known fibril-forming polypeptides and proteins, and the diverse symptoms of the associated diseases, the architecture of the resulting fibrils is remarkably similar in all cases [21, 22]. The thickness of the fibrils is determined by the number of protofilaments, and typical amyloid-like fibrils are 6–12 nm in diameter, consisting of 2–6 protofilaments [18, 23]. Amyloid-like fibrils show well-defined diffraction patterns [21] and specifically bind the dyes Congo red and ThT [24, 25]. Even short peptides and globular proteins not affiliated with any known disease can form amyloid-like fibrils under certain conditions, leading to the suggestion that amyloid fibril formation is a general feature of all polypeptides [5, 7, 17]. Despite such generality, the often unique primary amino acid sequence and, thus, side-chain interactions, also play an important role, determined, for example, by the presence of charged side-chain functionalities and aromaticity [26]. It has even been hypothesized that the native state of proteins might, in fact, be a mesostable state [27] and that all peptides and proteins will form fibrils given an infinite amount of time [28].

In vitro fibril formation typical follows a nucleation mechanism, where a lag time often precedes an exponential growth of fibrillar aggregates [7, 27, 29]. Common biophysical methods employed to probe fibrillogenesis include fluorescence [25, 30] and turbidity assays [31–33], dynamic light scattering (DLS) [30, 34], atomic force [35], and electron microscopic [36, 37] imaging (AFM and EM, respectively), as well as vibrational spectroscopic methods such as Fourier transform infrared (FTIR) [38] and Raman [39] spectroscopies. Recently, vibrational circular dichroism (VCD) [40, 41] has been added to the list of biophysical techniques that can probe both fibrillogenesis and the chirality of the resulting supramolecular structures (T. Measey and R. Schweitzer-Stenner, unpublished results).

Various mechanisms exist for describing amyloid formation, as illustrated in Figure 11.1. These include templated assembly [42], monomer-direction conversion [43], nucleated polymerization [44–46], and nucleated-conformational conversion mechanisms [47]. For a review of these mechanisms, see Ref. [48]. A common feature of these mechanisms is the presence of a nucleation phase, which can consist of either monomers or oligomers as the nucleating species, depending on the mechanism. Moreover, models have been developed for describing the hierarchical self-assembly into fibrils and fibers. In particular, Aggeli et al. proposed a one-dimensional self-assembly model of chiral molecules, in which β-sheet tapes form two-layered ribbons. These ribbons in turn form protofilaments, and various protofilaments interact further to form amyloid-like fibrils [49, 50].

11.2.1. Short Peptides as Model Systems for Self-Assembly

Short peptides offer themselves as convenient model systems for studying protein aggregation [28, 50–52]. This is due, in part, to their inherent low cost and ease of experimental handling [28, 50]. In addition, various active fragments, so-called self-recognition elements (SREs) of amyloidogenic proteins, generally have fewer than 10 amino acid residues. It has been suggested that such short peptide segments form the core of amyloid fibrils [28, 53, 54]. The ability of hexa- [55], penta- [55, 56], tetra- [56, 57], tri- [58], and even dipeptides [59, 60] to form amyloid-like fibrils has been documented, and fibrils derived from such short amyloidogenic peptides have been shown to be cytotoxic [51, 61–63]. The interested reader should see the mini-review by Gazit [28]. Moreover, the kinetics of fibril formation of short peptides is similar to those of longer amyloid-forming proteins [51]. Thus, studying such short peptides may allow for the elucidation of the fundamental mechanism of fibril formation [51]. Also, their small size makes such peptides ideal candidates for *in silico* computational studies, for which the aggregation process of larger systems

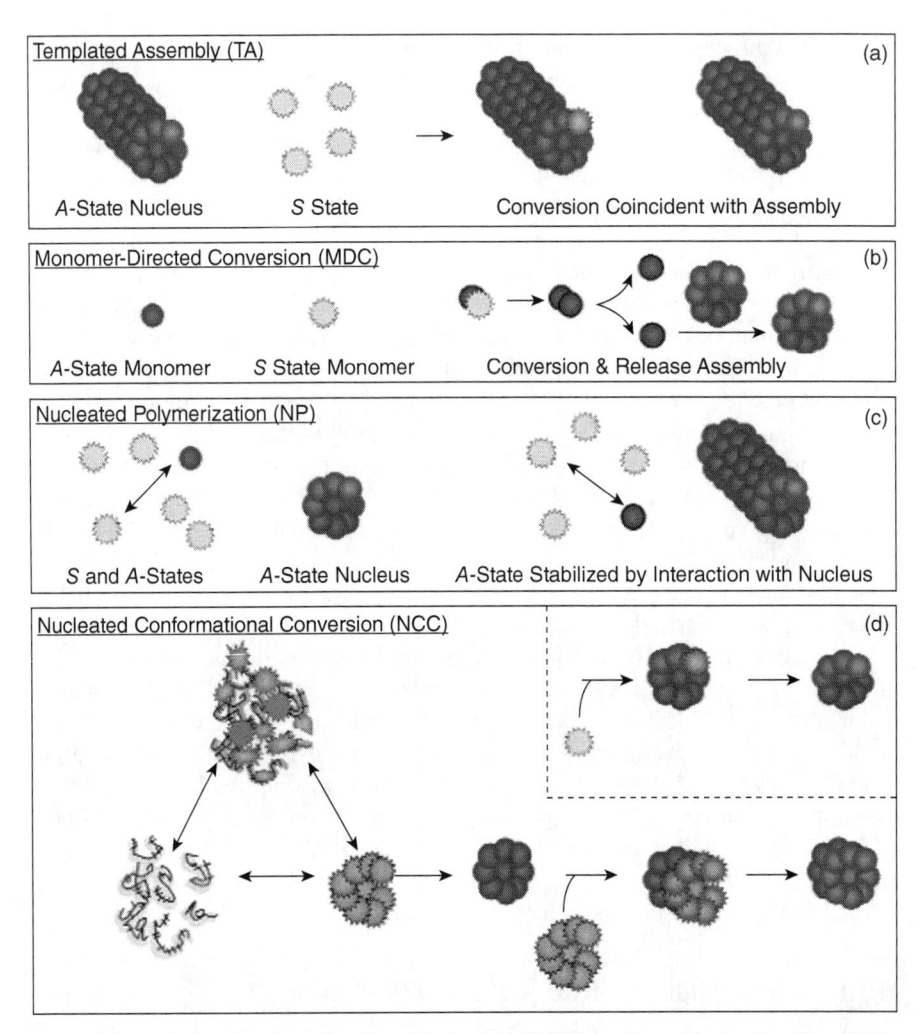

Figure 11.1. Mechanistic models of peptide/protein fibril formation: (a) templated assembly, (b) monomer-direction conversion, (c) nucleated polymerization, and (d) nucleated conformational conversion. This figure was taken from Ref. [48] and modified.

is impossible at present day [51, 64, 65]. Moreover, analyzing short peptide fragments is crucial for developing fibril inhibitors [28].

Studying short active segments or SREs can aid in elucidating the molecular recognition that plays a role in the self-assembly process, which has been suggested to originate in a short amino acid sequence [28]. Therefore, small model peptides are more suitable for investigating those sequence elements that favor aggregation [52]. Studies using short peptide models have provided valuable insight into the mechanism of amyloid-like fibril formation and paved the way toward understanding toxic amyloid fibril formation [28]. Also, since fibril formation represents a general feature of the polypeptide backbone, and short peptides are intrinsically disordered, they offer themselves as prime candidates for studying misfolding into fibrillar aggregates. An interesting hypothesis was recently put forward by Carny and Gazit, who suggested that assemblies formed from short peptides might have played a pivotal role in the origin of life, acting as templates for RNA polymerization under the harsh conditions existent on the primordial Earth [66].

11.2.1.1. Short Alanine-Rich Peptides Owing to the simplicity of the alanine side chain, and the abundance of alanine in nature, it would seem that short alanine-based peptides are ideal for short-peptide studies aimed at elucidating the mechanistic principles governing the peptide and protein self-assembly process. Moreover, the use of alanine as a base peptide has been implemented in a variety of *de novo*-designed peptides created and designed to self-assemble under well-defined conditions, for example, pH and temperature [67–70]. Examples of the latter will be discussed in the second half of this chapter. Short alanine-based peptides have been the subject of intense research activities since the late 1980s. This interest stemmed from the discovery by Marqusee et al. that short alanine-based peptides with at least 12 amino acid residues could adopt stable α-helices in aqueous solution if doped with a limited number of charged residues [71, 72]. Examples of the latter include Ac-(AEAAK)$_3$A-NH$_2$ and Ac-(AKAAE)$_3$A-NH$_2$ [73]. Since then, such peptides have been used to study helix-coil transitions of proteins [74–76]. Even shorter (<10 residues) alanine-rich peptides have been used extensively in recent years as convenient model systems to study the unfolded state of peptides and proteins [77–82]. Moreover, alanine-rich peptides of ≥8–16 residues in length have also been shown to aggregate under well-defined conditions, and polyalanines of >15 residues typically form β-sheet-rich aggregates in aqueous solution [83–85]. Short polyalanine peptides, such as deca-alanine (Ala$_{10}$), have also been used for studying peptide self-assembly in hydrophobic environments [86] which seek to mimic the cell membrane, as membrane interactions are believed to play a significant role in the aggregation and fibril formation of many amyloidogenic peptides and proteins.

Moreover, polyalanine stretches are implemented in the so-called polyalanine diseases [85, 87–90], which result from expansions of trinucleotide repeat sequences, which foster uninterrupted polyalanine tracts of >11 residues. Misfolding and aggregation appear to be characteristic features of such diseases. Thus, self-aggregating and amyloidogenic alanine-rich peptides offer themselves as convenient and necessary model systems for studying the mechanism of protein aggregation and fibril formation.

11.2.1.2. The Prion Fragment (ShPrP$_{(113-120)}$) As mentioned above, short, alanine-rich peptides are convenient models for studying the mechanism of toxic amyloid-like fibril formation, which is a hallmark of various neurodegenerative diseases, such a Alzheimer's, Parkinson's, Huntington's, and the prion diseases [5, 7, 9]. Prion diseases (or transmissible spongiform encephalopathies [TSEs]), are often fatal neurodegenerative diseases and include scrapie, mad cow disease, and Creutzfeldt–Jakob disease. Prion diseases result from the recruitment of normal, nontoxic monomeric proteins into toxic aggregates via aberrant misfolded conformations [91, 92]. Prion proteins (PrPs) are infectious agents that lack any genetic material. The infectious protein is often denoted as PrP^{Sc}, while the normal cellular protein is denoted as PrP^c. Aggregates of PrP^{Sc} deposit in various regions of the brain, ultimately resulting in neuronal death. In addition to the full-length fibril-forming PrPs of different species, many fragments of various PrPs can also form amyloid-like fibrils similar to those formed by the full-length proteins [93–95].

Gasset et al. investigated four peptides from the Syrian hampster PrP, ShPrP, which were predicted to adopt α-helical conformations [93]. In contrast to expectations, three of the four formed amyloid-like fibrils, with the most highly ayloidogenic of these found to be the alanine-rich palindrome sequence AGAAAAGA, corresponding to residues 113–120 of the ShPrP (ShPrP$_{113-120}$). This peptide is highly conserved among all species whose prion sequences have been determined [93]. The fragment is located in the hydrophobic core of the N-terminal segment, that is, the disordered segment believed to be responsible for misfolding and aggregation. AGAAAAGA is highly amyloidogenic and can form fibrils with remarkable similarity to the full-length PrP [93]. It has been shown by Norstrom et al. that this sequence is necessary for fibril formation of the full-length protein [96]. In particular, these authors showed that the protein lacking in the palindrome sequence did not adopt the proper conformation nor allow for proper interaction between the normal and aberrant PrP (PrP^{Sc} and PrP^C).

Ma and Nussinov carried out molecular dynamics (MD) simulations to try and shed light on the mechanism of fibril formation of AGAAAAGA [51]. They began their simulations using preformed oligomers (tri- , tetra-, hexa-, and octamers) to elucidate the effect that the oligomers have on the conformational change associated with the interaction of another monomer unit. The results of their simulation suggest that a minimum burial surface is required to stabilize oligomeric structures of AGAAAAGA [51]. In particular, they found that β-sheet structure in the oligomers becomes increasingly stable when $n = 6$ (where n is the number of β-strands in a sheet). This allows for buried methyl side chains of the alanine residues to interact, which would be absent in, for example, a tetramer, since they are all somewhat solvent-exposed in smaller oligomers. Moreover, the hexamer remained stable throughout the course of the simulation (3 ns, at 330 K).

Ma and Nussinov also investigated the *in silico* aggregation of AAAAAAAA (A8), where A8 tetramers were found to be as stable as tetramers formed from the plaindrome prion fragment, AGAAAAGA [51]. However, A8 octamers were found to be much more stable than octamers of AGAAAAGA. Ma and Nussinov put forth a hypothesis for fibril elongation by suggesting that monomer units add to an oligomer nucleus via hydrophobic interactions, where the ordered hydrophobic pattern of the nucleus acts as a conformational trap for an incoming monomer, binding perpendicular to the fibril axis. Stabilization of the aberrant monomer conformation then allows for the shift of the newly added unit to the end of the growing fibril, thus allowing for elongation. Such a scenario is in contrast to a conformational induction mechanism of an oligomeric nucleus on the incoming monomers, where the oliomeric "seed" catalyzes the conformational change of the incoming monomer [51].

11.2.1.3. AKY8 While the AGAAAAGA peptide fragment discussed above was originated from an amyloidogenic PrP, other peptides of similar length and amino acid sequence with no known disease affiliation can also form fibrillar structures with remarkable similarity to those toxic species formed during the course of disease [61–63]. As an example, our group recently reported the aggregation of an alanine-rich octapeptide, namely Ac-(A$_4$KA$_2$Y)-NH$_2$ (AKY8), into well-ordered amyloid-like fibrils. At first glance, the primary amino acid sequence of AKY8 suggests that the peptide should be in a monomeric, unfolded state and might be a suitable system for studying the unfolded state, as has been investigated thoroughly in the past decade. In fact, peptides of similar length and composition have been frequently used to explore the conformational distribution of unfolded peptides, and often adopt an ensemble of conformations in aqueous solution, with a predominant sampling of left-handed extended helices termed polyproline II (PPII) [77, 82, 97, 98].

In contrast to expectations, however, AKY8 forms amyloid-like fibrils upon incubation at room temperature in the presence of a small amount of added HCl [41]. This is surprising since the positive charge on the lysine side chain would be expected to prevent aggregation. The resulting fibrils are approximately 7 nm in diameter, some of which show a left-handed helical twist as depicted in Figure 11.2, characteristic of amyloid-like fibrils. The spontaneous and exponential growth of fibrils is preceded by a quiescent lag phase, consistent with a nucleation mechanism and also in line with typical amyloid fibril formation [7, 27, 99]. Thus, AKY8 undergoes a conformational transition from an unfolded ensemble of conformations into β-sheet-rich fibrillar aggregates.

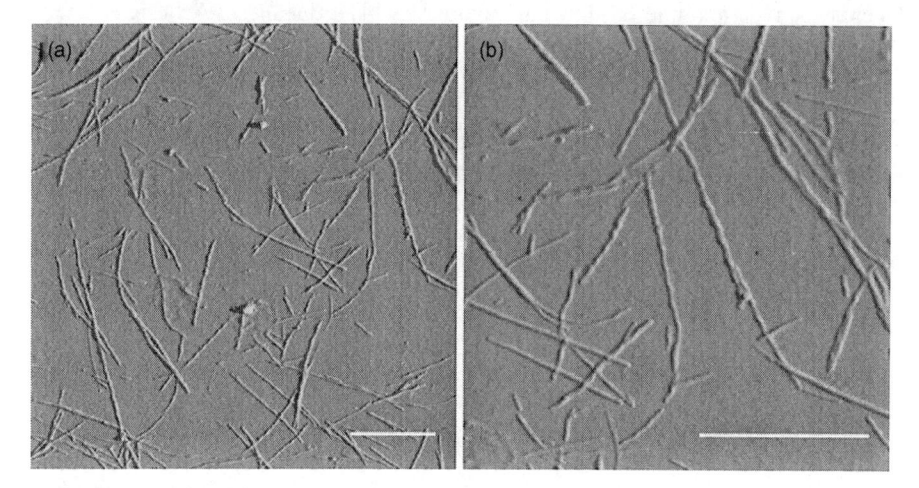

Figure 11.2. Amplitude AFM images of AKY8 fibrils. A mixture of twisted amyloid-like fibrils and more rod-like species can be discerned. The scale bar in each panel represents 1 μM. See color insert.

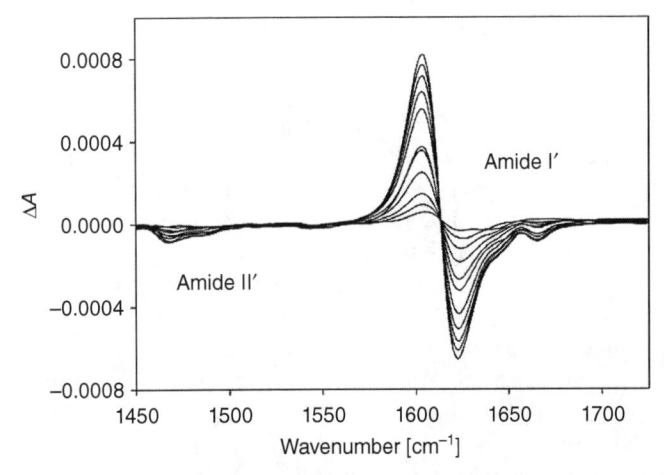

Figure 11.3. AKY8 fibril formation probed by VCD spectroscopy. Representative time-dependent VCD spectra of a 30 mM AKY8 solution, illustrating the increase in signals in the amide I' and amide II' regions attributed to AKY8 fibril formation. Starting with the spectrum displaying the lowest intensity in the amide I' couplet, these spectra were acquired at approximately 17, 18.5, 20, 22, 24, 26.5, 28.5, 30.5, 33.5, 37.5, and 44.5 hours, respectively, after the addition of HCl.

Measey et al. used VCD and FTIR spectroscopic methods to probe the structure and kinetics of AKY8 fibril formation from an unfolded ensemble to β-sheet-rich aggregates [41]. While various methods exist to characterize fibril formation, VCD is a novel and non-conventional method [40, 100]. The AKY8 fibril solution shows an enhanced couplet in the amide I' region of the VCD spectra, which increases with increasing reaction time, as shown in Figure 11.3 [41]. Enhanced VCD signals have also recently been observed for lysozyme and insulin fibrils [40]. Figure 11.4 shows the kinetic profile as probed by the positive maximum of the couplet in the amide I' region of the VCD spectra of AKY8. The kinetics exhibit a lag phase, which precedes an exponential growth of fibrils, consistent with a nucleated polymerization mechanism and characteristic of amyloid-like fibril formation. VCD can provide information on the local secondary structure, and can thus be considered a novel probe of fibril formation, especially when coupled with complementing methods.

Promotion of AKY8 fibril formation is attributed to the C-terminal tyrosine residue of AKY8, since the tyrosine-lacking analogue, namely Ac-AAAAKAA-NH$_2$ (AK7), does not form fibrils [41]. Measey et al. proposed that the pivotal role of tyrosine could reflect cation–π interactions between the lysine and tyrosine residues of neighboring strands and/or sheets, which can stabilize the final aggregated structure. A possible orientation of neighboring strands is illustrated in Figure 11.5. The role of aromatic residues in amyloid fibril formation is becoming increasingly important, and the fibril formation of other short peptides has also been suggested to be stabilized by aromatic interactions [101, 102], and C-terminal aromatic residues [103]. A growing number of experimental studies have highlighted the importance

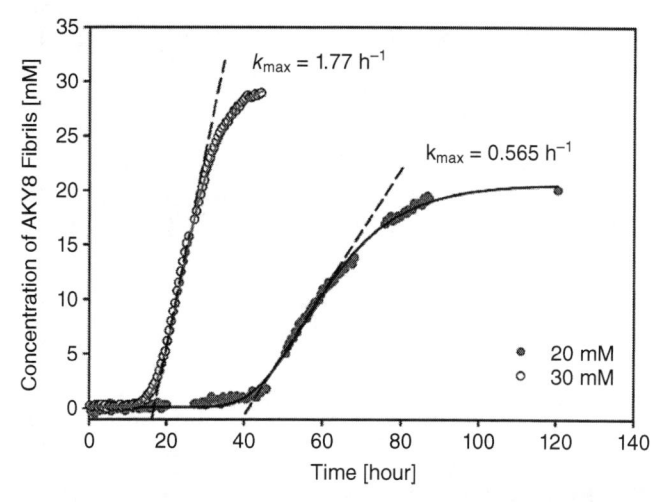

Figure 11.4. Kinetics of AKY8 fibril formation probed by the increase in the positive maximum of the enhanced amide I′ VCD couplet (1604 cm^{-1}), for two different concentrations of AKY8, namely, 20 and 30 mM.

of aromatic residues in promoting peptide and protein self-assembly and fibril formation [101, 103–107]. Many short sequences that are believed to be active in the fibril formation of various disease-related proteins containing aromatic residues have been documented [26], and the importance of aromatic interactions in inducing peptide self-assembly to form supramolecular hydrogels has been reported [101, 108, 109].

The formation of AKY8 fibrils offers further support for the notion of the generality of amyloid-like fibril formation, in the sense that a short peptide with no known disease affiliation can form structures with remarkable similarity to those amyloidogenic proteins linked to various human pathologies. The aggregation of AKY8 further suggests an important role for aromatic residues and cation–π interactions in such processes and also highlights the novel use of VCD to probe the fibril formation kinetics, which, when combined with other experimental and computational methods, may prove to be a useful probe of the chirality of the resulting fibril architecture and shed light on the mechanism of fibril formation [41].

11.2.2. Polyalanine Diseases

As illustrated in the preceding sections, alanine-rich peptide segments can be useful model systems for studying the mechanism of toxic amyloid-like fibril formation. Moreover, polyalanine expansions can lead to aberrant interactions, including misfolding and subsequent aggregation, and polyalanine stretches with more than 11 residues have been linked to the pathogenesis of the so-called polyalanine diseases, which result from trinucleotide expansions that encode for alanine. Trinucleotide expansions are also linked to amyloid diseases such as Huntington's disease, a neurodegenerative disease of which the protein, huntington, is mutated to contain long

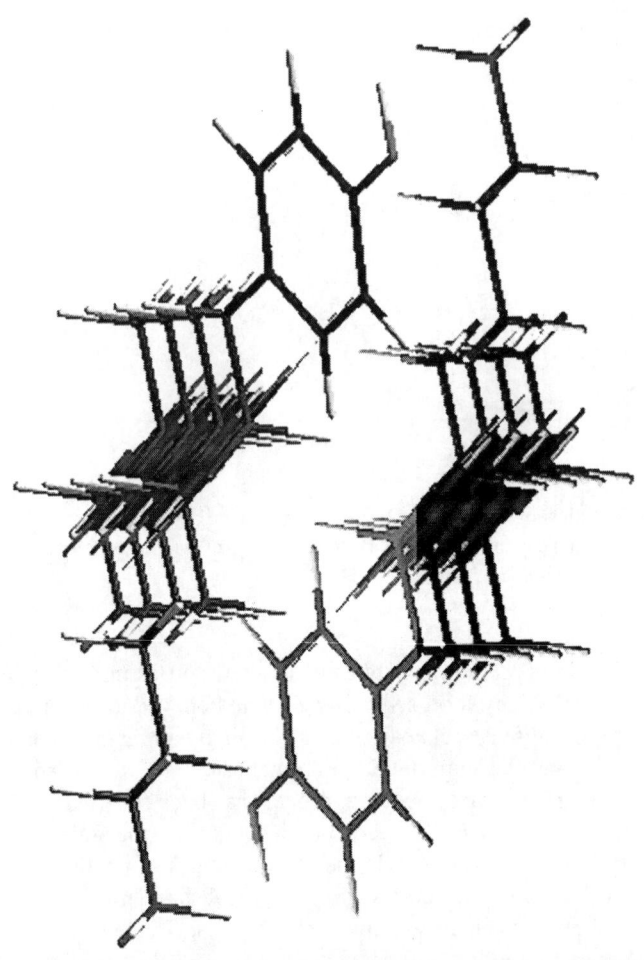

Figure 11.5. Schematic of an anti-parallel β-sheet arrangement (shown only for two strands) proposed for AKY8, in which every other strand is rotated 180° about the strand axis, so as to allow for favorable side-chain packing/interactions. The view is along the sheet axis.

stretches of glutamine (Gln) residues [110]. However, in contrast to the polyglutamine expansions associated with Huntington's disease, which can reach upwards of 180 residues [111], the polyalanine tracts associated with the various polyalanine diseases tend to be on the order of 10 residues or so.

Polyalanine diseases include various early developmental abnormalities [90] as well as the autosomal dominant disorder oculopharyngeal muscular dystrophy (OPMD), the latter of which is characterized by eyelid drooping, limb weakness, and difficulties in swallowing [87]. To date, 9 polyalanine diseases are known, and a recent list was compiled by Messaed and Rouleau [90]. Each disease results from transcription factors that exhibit aberrant polyalanine stretches and are detrimental to cells by causing either dysfunction or toxicity [90].

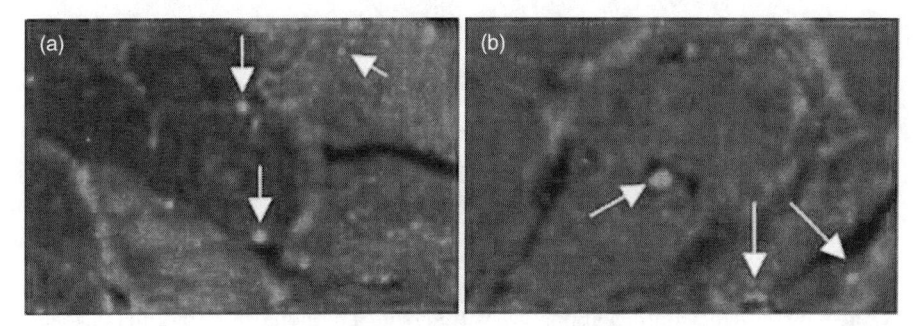

Figure 11.6. Intranuclear inclusions of PABPN1 found in the deltoid muscle of patients ((a), sample 1; (b) sample 2) with OPMD. The arrows show the positively stained inclusions. This figure was taken from Ref. [114] and modified.

It has been shown that the number of alanine residues in a tract correlates with the level of aggregation [90, 112]. It was found that cells expressing green fluorescent protein (GFP) fused to polyalanine stretches of 19–37 residues led to high levels of intracellular inclusions [113]. It has been demonstrated that polyalanine expansions are necessary for inducing protein aggregation in OPMD [87]. The alanine stretch in the normal protein increases from 10 to 12–17 in the mutant, disease-related protein [87]. The polyalanine expansions associated with the polyadenine-binding protein nuclear-1 (PABPN1) associated with OPMD result in misfolding and aggregation to yield intra-nuclear inclusions in skeletal muscle and hypothalamic neurons, as shown in Figure 11.6 [114–116]. These inclusions have been recognized as a hallmark of the disease [111, 115, 117], and are similar in structure (i.e., cross-β-strand) to those inclusion bodies composed of amyloid fibrils, which accompany various neurodegenerative diseases, such as Alzheimer's, Parkinson's, and the prion diseases, yet they do not bind the amyloid-specific dyes thioflavin T (ThT) or Congo red [85]. Moreover, like those of the amyloidogenic proteins and peptides, OPMD protein aggregates are cytotoxic [87].

Fan et al. concluded that the polyalanine stretch in PABPN1 is "necessary but not sufficient" to induce OPMD protein aggregation [87]. They suggested that other determinants must exist that might facilitate aggregation of PABPN1. Shinchuk et al. investigated *charge* versus *length* of short polyalanine peptides [85] to conclude that hydrophobicity was the driving force of aggregation. It was hypothesized by Fan et al. that longer alanine stretches result in misfolding, ultimately exposing the hydrophobic alanine tracts, which would otherwise remain buried, so that the longer the tract, the greater the exposure [87]. Fan et al. also demonstrated that inhibiting the oligomerization of PABPN1 prevents aggregation [87].

11.2.2.1. Short Polyalanines for Mechanistic Studies of Polyalanine Diseases The behavior of polyalanine stretches in the various polyalanine diseases is consistent with findings of Blondelle et al., Scheuermann et al., and Shinchuk et al. that polyalanines >15 residues typically form β-sheet-rich aggregates [83–85, 111]. The

Blondelle group undertook an attempt in the 1990s to design model β-sheet-forming peptides to study the mechanism of aggregate formation in relation to protein folding (hydrophobic collapse model) as well as amyloid formation [84]. Their initial attempts were inspired by the work of Zhang and associates, who had previously reported on the formation of a macroscopic membrane by an alanine-rich oligopeptide [67].

In an attempt to design peptides *de novo* that form soluble β-sheet structures in aqueous solution as model systems for studying β-sheet/fibril formation, Forood et al. reported the β-sheet formation of Ac-KA$_{14}$K-NH$_2$ (KAK) [84]. As the authors wanted to study the conformational transition from a monomeric coil or helical state to a β-sheet, they used alanine as the base amino acid because of the ability of short alanine-rich peptides to form stable α-helices in aqueous solution above a threshold number of residues [71, 118]. This peptide was found to adopt a stable β-sheet structure, even at concentrations as low as 1 μM, and was found to be protease-resistant and stable in 7 mM sodium dodecyl sulfate(SDS) and 80% trifluoroethanol (TFE). On the contrary, substitution of a central alanine residue with a proline (Ac-KA$_8$PA$_5$K-NH$_2$, KAPK) resulted in inhibition of the β-sheet structure, and partial α-helical structure in SDS and TFE solutions [84]. Hydrophobic packing was hypothesized to be the driving force for sheet formation of this peptide [84]. Work from this same group also involved the synthesis of tyrosine-containing alanine-rich peptides based on the Ac-YKA$_n$K-NH$_2$ (n = 3–25) motif to determine the minimal number of residues needed for β-sheet formation [83]. The tyrosine-containing peptide analogue, Ac-YKA$_{13}$K-NH$_2$, was found to exhibit conformational behavior similar to that of Ac-KA$_{14}$K-NH$_2$ [83]. As shown in Figure 11.7, the authors saw no indication of β-sheet secondary-structure formation for those peptides with $n \leq 9$.

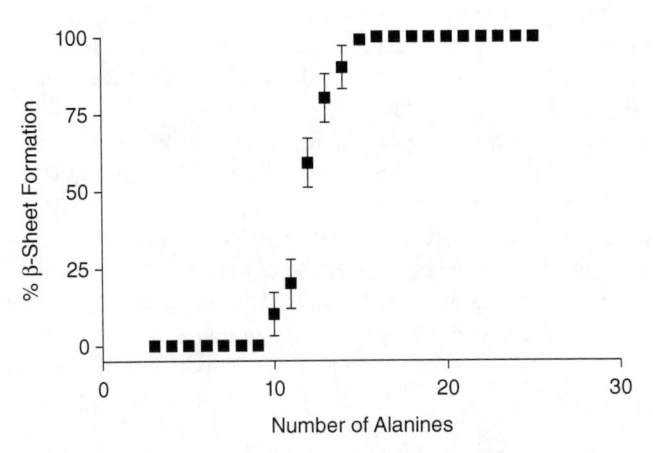

Figure 11.7. β-Sheet formation dependence on polyalanine length for peptides based on the Ac-YKA$_n$K-NH$_2$ (n = 3–25) motif. The percent of β-sheet formation at peptide concentrations of 1.4 mM was monitored by RP-HPLC after 24-hour incubation at 65°C. The figure was taken from Ref. [83] and modified.

For $10 \leq n \leq 14$, the authors found varying levels of β-sheet formation, while those with $n \geq 15$ showed nearly 100% β-sheet secondary structure [83].

To gauge the effects of the charged terminal lysine residues on the conformational behavior of polyalanine peptides, the Blondelle group also investigated peptides containing negatively charged glutamic acid residues (Ac-EA$_{13}$E-NH$_2$), as well as ion pairing, (Ac-KEA$_{13}$KE-NH$_2$), a histidine-containing analogue, that is, Ac-HYA$_{13}$H-NH$_2$, a peptide with a free N-terminal group, that is, KA$_{13}$K-NH$_2$, as well as a peptide containing two positively charged lysine residues, namely Ac-KKYA$_{13}$KK-NH$_2$. They found that the peptide containing the two neighboring positively charged lysines was the only one not able to undergo a conformational transition into a β-sheet, suggesting that the intermolecular hydrophobic interactions between alanine residues of neighboring strands are not sufficient to compensate for the repulsive electrostatic interactions of the neighboring positively charged lysine residues [83].

Giri et al. studied the pH-dependent conformational properties of a series of alanine-based peptides with the sequence Ac-KMA$_n$GY, where $n = 7$, 11, and 17, which have been referred to as 7-ala, 11-ala, and 17-ala, respectively [119]. Both of the longer peptides (11-ala and 17-ala) were found to adopt a substantial amount of β-sheet secondary structure under both acidic and alkaline conditions, and to a mild extent at neutral conditions. This is consistent with the findings of Blondelle et al. (see above and Fig. 11.7), who showed that above $n = 10$, the peptides, Ac-KYA$_n$K-NH$_2$, begin to adopt β-sheet secondary structure [83]. Giri et al. carried out a ThT binding assay to follow the fibril formation kinetics of 11-ala and 17-ala, as a function of peptide concentration, and found that the lag time decreases and the apparent rate constant increases upon increasing the peptide concentration [119]. As discussed above, this is typical of a nucleated polymerization mechanism, consistent with amyloid fibril formation of amyloidogenic peptides (c.f. AKY8 and Fig. 11.4).

What is interesting about the peptides studied by Giri et al. is that fibril solutions of 11-ala and 17-ala were able to form higher order organized structures with fractal-like patterns when deposited onto a glass substrate and allowed to air-dry. Optical micrographs of the structures formed by 11-ala and 17-ala are shown in Figure 11.8. Such structures are clearly distinct from the more fibrillar structures often observed and may be the result of off-pathway aggregation [119].

Shinchuck et al. showed that alanine-rich peptides with the sequence Ac-KYA$_n$K-NH$_2$ ($n = 3$–20) adopt varying amounts of β-sheet secondary structure if the number of residues exceeds 8 and are nearly 100% β-sheet when n > 15 [85]. In particular, peptides with $n = 11$, 13, and 20 form fibrils that, despite exhibiting a diffraction pattern indicative of a cross-β arrangement, do not bind the amyloid-specific dyes Congo red or ThT. This is in contrast to the peptides of similar sequence investigated by Giri, which indeed bind ThT [119]. These findings led Shinchuck et al. to suggest that PABPN1 misfolding and the proteinaceous intra-nuclear buildup is likely generated from the hydrophobic core of the alanine expansions [85]. Moreover, Baginska et al. showed that the peptide Ac-KA$_{11}$KGGY-NH$_2$ also predominantly adopts a β-sheet structure at concentrations of ~2 mM [121], and also forms amyloid-like fibrils at this concentration based on a ThT binding assay.

Figure 11.8. Images obtained with an optical microscope of (a) 11-ala and (b) 17-ala after incubation at pH = 11 for 20 days at room temperature. Figure taken from Ref. [120] and adapted.

The time scales of fibril formation can be as long as days and thus out of reach for most current computational capabilities, so that atomistic simulations can typically probe only oligomer formation [122–124]. Simplified models are necessary to follow the aggregation into larger fibrils. These include models in which individual protein monomers are considered as a single unit [125], or coarse-grained models, in which individual atoms and side-chain groups of a particular peptide residue are represented as beads [65, 126–129]. Such intermediate-resolution course-grained methods have been able to follow the formation of larger "fibril"-like species.

In attempts to explain the underlying physics of aggregation of short polyalanine peptides, Carol Hall and associates used intermediate-resolution protein models coupled with discontinuous molecular dynamics (DMD) to simulate the formation of short polyalanines [65, 126, 129–131]. These authors published a series of papers in which they utilized DMD simulations in conjunction with an intermediate-resolution protein model to study the conformational transitions of Ac-KA$_{14}$K-NH$_2$ in the monomeric state [130], as well as the aggregated and fibrillar states [65, 126, 129, 131]. DMD is a variant of more conventional MD techniques, but, when combined with the intermediate-resolution model, allows for the simulation of β-sheet and fibril formation, since it can allow for longer time scales and follow the trajectories of larger systems [65]. Otherwise, computer simulations of fibril formation are nearly impossible due to the large systems and time regimes that are typically involved. While such methods certainly have their limitations and downfalls, they can also provide details on the fibril formation process, which cannot be elucidated experimentally [65].

Hall and associates first investigated the thermodynamics of folding of a monomeric Ac-KA$_{14}$K-NH$_2$ peptide [130] under conditions that mimic the experimental conditions utilized by Blondelle and associates [83, 84]. Nguyen et al. examined the affect of solvent on the conformational transitions of Ac-KA$_{14}$K-NH$_2$ by varying both the hydrophobic interaction between nonpolar side chains as well as the temperature [130]. These authors found that at low temperatures and low hydrophobic interaction, the peptide adopts a mostly α-helical conformation due to favorable intramolecular interactions, that is, hydrogen bonding. The low hydrophobic interaction mimics the behavior of Ac-KA$_{14}$K-NH$_2$ in nonpolar solvents and is consistent with experimental results [83]. Under these artificial conditions, the helix decays into a random coil at higher temperatures, with the transition shifting to higher temperatures as the hydrophobic interaction is increased. At much higher hydrophobic interaction strengths, the peptide populates β-hairpin and β-sheet-like structures at lower temperatures, which decay into random coils at higher temperatures.

Nguyen and Hall used the same DMD technique and their protein intermediate-resolution model (PRIME) to follow the fibril formation of Ac-KA$_{14}$K-NH$_2$ [65, 126, 129]. Using their simplified PRIME model, which considers a peptide unit as consisting of three backbone spheres and one side-chain sphere as shown in Figure 11.9, these authors were able to sample a larger conformational space and longer time scales, which allowed them to follow the slow fibril formation of simulations carried out with 48–96 peptides. These authors investigated the fibril formation of Ac-KA$_{14}$K-NH$_2$ as a function of concentration and temperature. In contrast to the

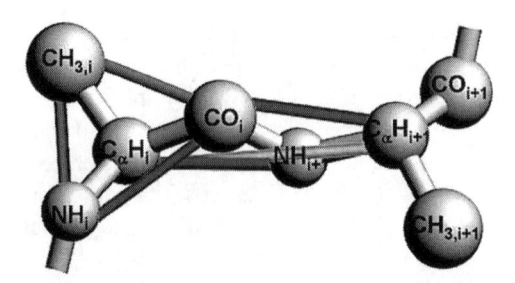

Figure 11.9. Illustration of the geometry of the protein intermediate-resolution model (PRIME) utilized by Nguyen and Hall for their DMD simulations of Ac-KA$_{14}$K-NH$_2$. Each peptide residue in this model is represented by three backbone spheres and one side-chain sphere. The figure was adapted from Ref. [65].

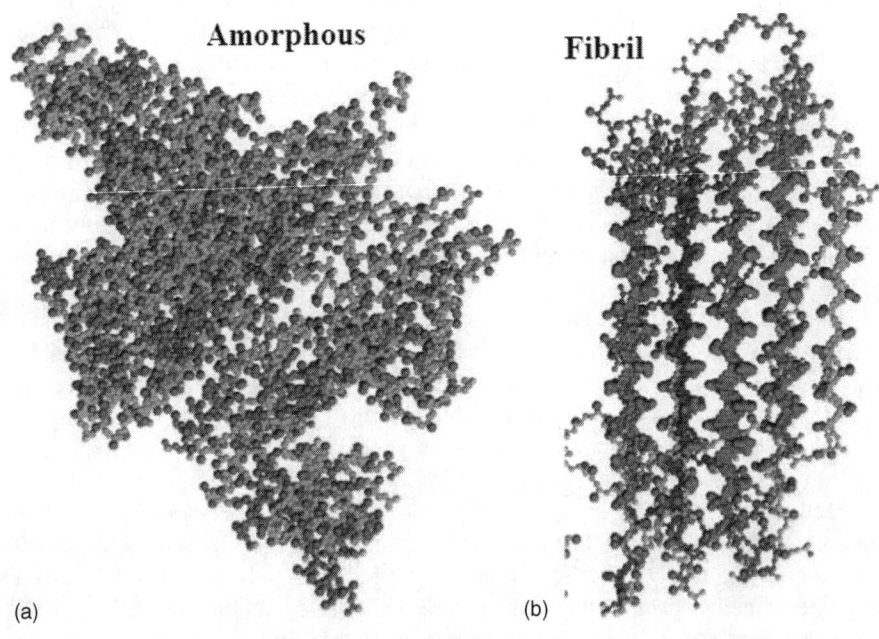

Figure 11.10. Early amorphous aggregate (a) and fibril-like structure (b) formed during the simulation of 96 Ac-KA$_{14}$-K-NH$_2$ peptides. The figure was taken from Ref. [131] and modified.

individual proposed mechanisms of peptide and protein aggregation (see the introduction of this chapter), these authors found that the *in silico* fibril formation of Ac-KA$_{14}$K-NH$_2$, instead, showed characteristics of each of the four models [126]. In particular, these authors found that small, amorphous aggregates preceded conformational conversion to form a nucleus. These amorphous aggregates are energetically favorable, but form more ordered fibril-like structures as the size of the oligomers increases [126]. Figure 11.10 depicts representative structures of the amorphous and fibril-like species observed by Nguyen and Hall and associates

MISFOLDING AND HUMAN DISEASE 325

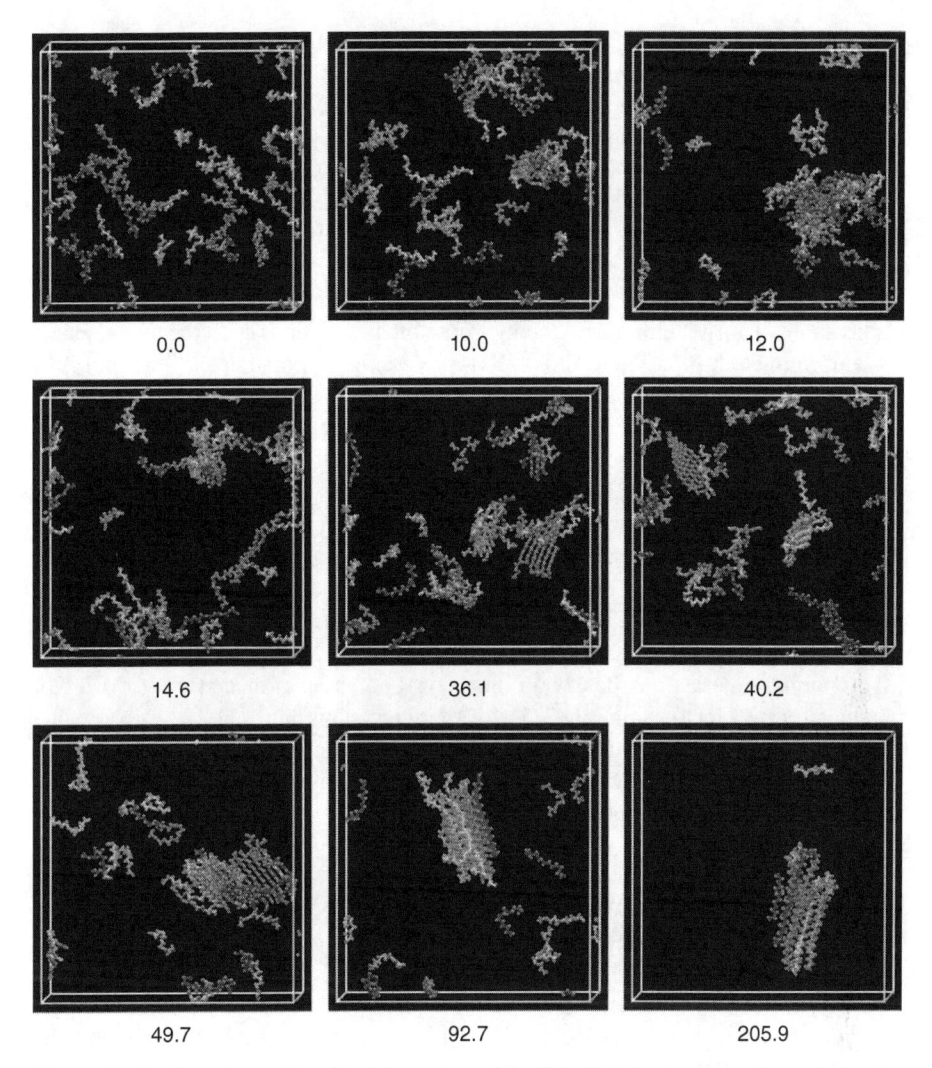

Figure 11.11. Snapshots of the fibril formation of Ac-KA$_{14}$Y-NH$_2$ at various times during the PRIME–DMD simulation of Nguyen and Hall. β-Sheet-rich fibril-like precursor structures begin to form at a reduced time of 36.1 (middle box). The figure was taken from Ref. [65] and modified.

during their *in silico* fibril formation study. The fibrils were found to elongate both in the direction of the individual sheets and laterally [65]. Figure 11.11 shows snapshots of the fibril formation simulation of Ac-KA$_{14}$K-NH$_2$, depicting the formation of small, β-sheet-rich structures beginning to form at a reduced time of 36.1, which then proceed to form fibril-like structures [65].

While the results of Hall and Nguyen and colleagues clearly exemplify the *in silico* behavior of polyalanines as a function of hydrophobic interactions (and, thus, implicit solvent), their studies do not capture the behavior in more realistic scenarios, for example, in membrane-like environments. Recently, Soto et al. used replica exchange molecular dynamics (REMD) to probe the dimerization of a 10-residue polyalanine peptide in a cyclohexane environment. REMD is one of the more efficient sampling schemes used for complex molecular systems, and it relies on the introduction, via trajectory exchange, of a structurally decorrelated conformer into a new "replica." The time interval between exchange is chosen so as to have the largest number of trajectory exchanges during a single simulation, while at the same time minimizing the correlation between consecutive exchanges [132].

Due to the competition between intra-peptide interactions, such as backbone–backbone hydrogen bonding and the hydrophobic interactions with the cyclohexane solvent, Soto et al. found that Ac-A_{10}-NH_2 preferentially populates a β-hairpin conformation in the monomeric state, with the most populated conformation corresponding to a type II' turn [86]. While other theoretical studies that have been carried out *in vacuo* suggest that monomeric polyalanines prefer α-helical conformations under conditions that mimic nonpolar solvents [133], such conformations were found to be unstable in the explicit cyclohexane environment used in the simulations of Soto et al. Above a critical concentration, which increases peptide–peptide recognition, two deca-alanine peptides were found to dimerize and form both helical and β-sheet dimers. While the former are favored by macrodipole interactions of the individual helices, the latter are favored by van der Waals and peptide–solvent (hydrophobic) interactions. The two-dimensional potential of mean force (PMF) plot obtained for two deca-alanines at 313 K is reproduced in Figure 11.12. The ordinate, $Q_{SecStruct}$, monitors the formation of secondary structure, while the abscissa, Q_{aggr}, monitors the aggregation (dimerization) state. The different dimeric substates populated under equilibrium conditions are shown.

11.3. EXPLOITATION OF PEPTIDE SELF-ASSEMBLY FOR BIOTECHNOLOGICAL APPLICATIONS

Aside from the unwanted and uncontrollable self-aggregation of peptides and proteins into toxic disease-related species of biomedical relevance, the protein self-assembly process has also been exploited in the past few decades in attempts to create novel nanostructured biomaterials with inherent biofunctionality [49]. For some recent reviews, see Refs. [4, 10, 13, 49]. These include nanofibrillar materials for industrial and medical applications [134, 135]. Along these lines, peptide hydrogels have shown potential to act as drug delivery systems [69, 70], and tissue-engineering scaffolds [68, 136], and to accelerate hemeostasis [137]. Peptide hydrogels are defined as a hydrophilic polypeptide network exhibiting the unique ability to swell in the presence of water [138, 139], and are thus able retain large amounts of fluids in their swollen states, which can affect different mechanical and surface properties, as well as permeability and biocompatibility [139].

De novo-designed oligopeptide systems have been of recent interest, since their small size can allow for careful sequence design and ease of control of the self-assembly process. Such oligopeptide systems typically rely on a sequence of

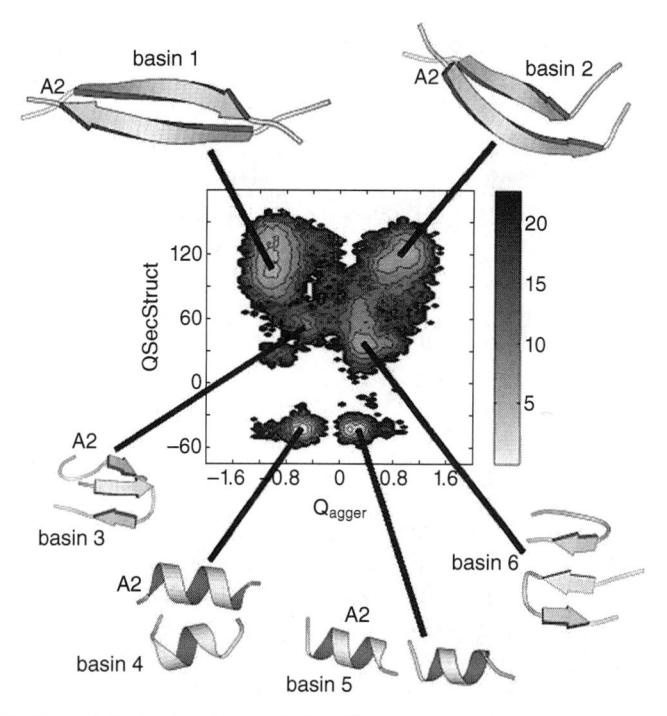

Figure 11.12. Potential of mean force (PMF) of deca-alanine in cyclohexane. Q_{aggr} monitors the aggregation state while $Q_{SecStruct}$ monitors the formation of secondary structure. The figure was taken from Ref. [86] and modified.

alternating hydrophobic and hydrophilic residues, containing complementary charge distributions alternating throughout [2]. A few examples of self-assembling oligopeptides include KLD-12 (AcN-(KLDL)$_4$-CNH$_2$) [140], RADA-16 (AcN-(RARADADA)$_2$-CNH$_2$) [69, 70], EAK-16 ((AEAEAKAKA)$_2$) [68], EMK16-II ((MEMEMKMK)$_2$) [141], and MAX-1 ((VK)$_4$V^{D}PPT(KV)$_4$-NH$_2$) [142]. Each of these peptides can form a polymeric network of nanofibers rich in β-sheet secondary structures under well-defined conditions, for example, pH and temperature. To advance the exploitation of peptide self-assembly in the creation of novel biomaterials, it is necessary to have a complete and detailed understanding of the rules that govern the self-assembly process.

Viscoelastic properties of peptide-based hydrogels and the density of the underlying nanofiber scaffold can be controlled by varying both peptide and salt concentrations, so that these properties can be tailored for specific applications [2, 137, 142]. Self-assembled gels typically exhibit shear-thinning behavior (i.e., drop in viscosity as a function of the applied shear) and are often quick in recovering their elastic properties once shearing has ceased [142]. The shear-thinning behavior results from the breaking of physical cross-links, for example, hydrogen bonds, by the applied strain [142].

11.3.1. Ala-Rich Peptides for Biotechnology

While the first half of this review focused on the spontaneous and aberrant conversion of otherwise monomeric soluble alanine-based peptides into toxic amyloid fibrils, self-assembly can also occur under very well-defined conditions, so that the process can be exploited for the creation of biomaterials with inherent biofunctionality, such as biocompatibility and biodegradability. In creating materials that mimic natural systems, insights gained from studying the natural assembly process is undoubtedly necessary [11]. In many self-assembling peptides and proteins, β-sheet formation dictates the properties of the resulting material, and such materials have shown application in wound healing and tissue engineering (including bone, cartilage, tendon, and ligament tissues) [143].

Fabrication of novel biomaterials requires the precise control of the self-assembly process, so that a detailed mechanistic understanding of such a process is necessary. This section will review some alanine-rich self-assembling peptides that have served as model systems, and display the potential to be used for various biomedical and biotechnological applications. These alanine-based systems and the insight gained from their studies have proved significant in designing novel biomaterials. These include systems designed to mimic the properties of spider silk, as well as alanine-rich *de novo*-designed oligopeptide hydrogels. While the constituent proteins of spider silk contain polyalanine-rich domains that control the crystallinity of the resulting silk fibers, *de novo*-designed oligopeptides often rely on an alanine-based system to provide simplicity yet hyrophobicity.

11.3.1.1. Spider Silk and Spidroin Proteins Spider silk is one of the toughest materials known [144], with a tensile strength greater than that of steel and similar to that of Kevlar [145]. Despite such robustness, these polymers are also elastic and can stretch upwards of 10% before breaking [144]. Polyalanine stretches play a pivotal role in the structure and properties of spider silk, where its has been attributed to the alanine-rich segments [146]. The constituent proteins of spider silk, namely the spidroin proteins, contain polyalanine stretches of up to 10 residues long [145, 147]. These proteins contain regions of high crystallinity separated by more amorphous regions. The ordered regions are rich in alanine content and β-sheet secondary structure, and contribute to the high tensile strength of spider silk, while the amorphous regions, on the other hand, impart elasticity and resiliency to silk [145, 148, 149]. Spider silk can be processed into fibers, gels, films, and sponges, as shown in Figure 11.13, with the resulting structure and function controlled by the crystalline

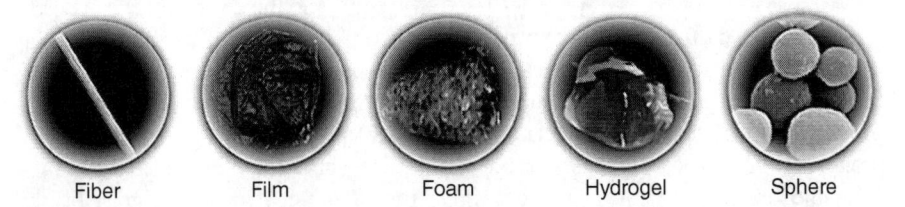

Figure 11.13. Variety of materials that spider silk can be processed into, including fibers, films, foams, hydrogels, and spheres. The figure was taken from Ref. [150] and adapted.

EXPLOITATION OF PEPTIDE SELF-ASSEMBLY 329

β-sheet, alanine-rich regions [143, 150]. Thus, designed materials that mimic the sequence and structure of the spidroin proteins can be envisaged to create materials for diverse biotechnological applications such as tendon implants and bullet-proof vests [151].

Solid-state nuclear magnetic resonance (NMR) has confirmed that the polyalanine stretches in the spider silk proteins adopt β-sheet conformations [146, 152]. These sheets are aligned with the fiber axis as probed by X-ray diffraction [153] and thus are similar to the cross-β structure observed for amyloid-like fibrils [21]. It has been suggested that the polyalanine stretches in the β-sheet regions of the spidroin proteins adopt anti-parallel sheet arrangements [154]. Further support that the alanine-rich regions predominantly adopt β-sheet conformations came from studies of the short peptides GAG-A_n-GGAG-GGY-NH_2 ($n = 4, 7$), which showed that both of these peptides adopt a β-sheet structure at concentrations less than 2 mM [155].

The self-assembly process of spider silk is driven by the hydrophobic alanine-rich domains [148], consistent with both experimental and computational results of short polyalanines [65, 83–85, 129]. The resulting silk fibrils and fibers form via the organization of a β-sheet rich liquid crystalline-like phase, as illustrated in Figure 11.14. The short hydrophobic side chain of alanine allows for tight packing of the neighboring sheets, which comprise the resulting silk fibers [143]. It was proposed over two decades ago that alanine and glycine are the two most abundant amino acids found in spider silk protein [156]. This was later confirmed by Holland et al., who found that alanine makes up approximately 82–86% of the amino acid content

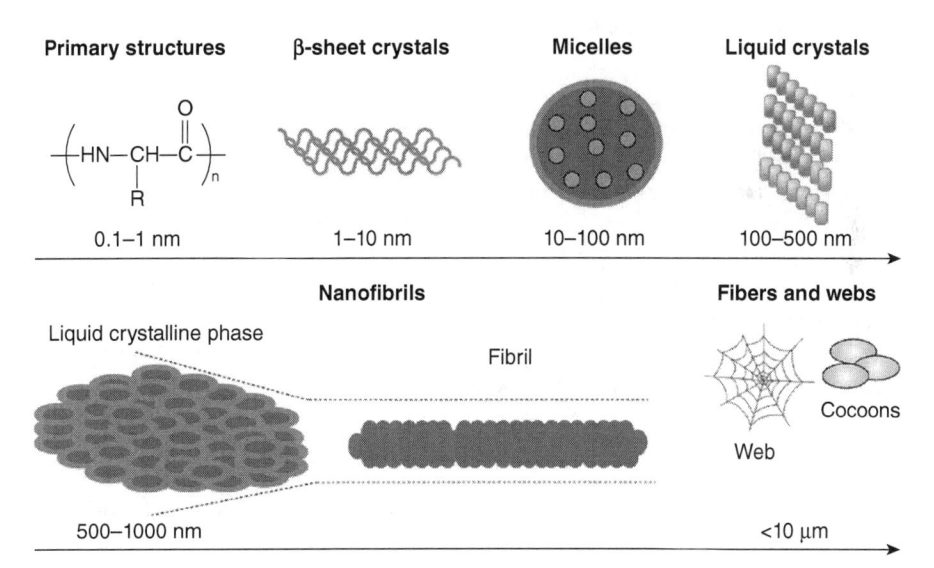

Figure 11.14. Hierarchical self-assembly of spidroin proteins into silk fibers. Monomeric silk proteins form β-sheets, which further assemble into micellar structures, and then into liquid crystalline structures. The liquid crystalline phase forms fibril-like structures, which compose the resulting spider silk. The figure was taken from Ref. [148] and modified.

MaSp1

```
1  QGAGAAAAAAGGAGQGGYGGLGGQGAGQGGYGGLGGQGAGQGAGAAAAAAAGGAGQGGYG
2  GLGSQGAGRGGQGAGAAAAAAGGAGQGGYGGLGSQGAGRGGLGGQGAGAAAAAAAGGAGQ
3  GGYGGLGNQGAGRGGQGAAAAAAGGAGQGGYGGLGSQGAGRGGLGGQGAGAAAAAAGGAG
4  QGGYGGLGGQGAGQGGYGGLGSQGAGRGGLGGQGAGAAAAAAAGGAGQGGLGGQGAGQGA
5  GASAAAAGGAGQGGYGGLGSQGAGRGGEGAGAAAAAAGGAGQGGYGGLGGQGAGQGGYGG
6  LGSQGAGRGGLGGQGAGAAAAGGAGQGGLGGQGAGQGAGAAAAAAGGAGQGGYGGLGSQG
7  AGRGGLGGQGAGAVAAAAAGGAGQGGYGGLGSQGAGRGGQGAGAAAAAAGGAGQRGYGGL
8  GNQGAGRGGLGGQGAGAAAAAAAGGAGQGGYGGLGNQGAGRGGQGAAAAAGGAGQGGYGG
9  LGSQGAGRGGQGAGAAAAAAVGAGQEGIRGQGAGQGGYGGLGSQGSGRGGLGGQGAGAAA
10 AAAGGAGQGGLGGQGAGQGAGAAAAAAGGVRQGGYGGLGSQGAGRGGQGAGAAAAAAGGA
11 GQGGYGGLGGQGVGRGGLGGQGAGAAAAGGAGQGGYGGVGSGASAASAAASRLSSPQASS
12 RLSSAVSNLVATGPTNSAALSSTISNVVSQIGASILVFLDVMSSFKLFSRLFLLLSRS
```

MaSp2

```
1  PGGYGPGQQGPGGYGPGQQGPSGPGSAAAAAAAAAAGPGGYGPGQQGPGGYGPGQQGPGR
2  YGPGQQGPSGPGSAAAAAAGSGQQGPGGYGPRQQGPGGYGQGQQGPSGPGSAAAASAAAS
3  AESGQQGPGGYGPGQQGPGGYGPGQQGPGGYGPGQQGPSGPGSAAAAAAAASGPGQQGPG
4  GYGPGQQGPGGYGPGQQGPSGPGSAAAAAAASGPGQQGPGGYGPGQQGPGGYGPGQQGL
5  SGPGSAAAAAAAGPGQQGPGGYGPGQQGPSGPGSAAAAAAAAGPGGYGPGQQGPGGYGP
6  GQQGPSGAGSAAAAAAAGPGQQGLGGYGPGQQGPGGYGPGQQGPGGYGPGSASAAAAAAG
7  PGQQGPGGYGPGQQGPSGPGSASAAAAAAAAGPGGYGPGQQGPGGYAPGQQGPSGPGSAS
8  AAAAAAAAGPGGYGPGQQGPGGYAPGQQGPSGPGSAAAAAAAAAGPGGYGPAQQGPSGPG
9  IAASAASAGPGGYGPAQQGPAGYGPGSAVAASAGAGSAGYGPGSQASAAASRLASPDSGA
10 RVASASVSNLVSSGPTSSAALSSVISNAVSQIGASNPGLSGCDVLIQALLEIVSACVTILS
11 SSSIGQVNYGAASQFAQVVGQSVLSAF
```

Figure 11.15. Primary amino acid sequence of major spidroin I and II spider silk proteins. The alanine residues are shown in red. This figure was taken from Ref. [158] and modified.

in β-sheet regions of spider silk [157, 158]. The primary amino acid sequences for the major spidroin I and II silk proteins are shown in Figure 11.15, clearly illustrating the abundance of alanine in the primary sequence. The remainder of the sequences of major spidroin I and II are made up mostly of glycine and serine [158]. Approximately 34% of the major silk proteins adopt β-sheet secondary structure [158].

Due to the spectacular tensile strength and elasticity of spider silk proteins, it is no surprise that their sequence architecture has been an inspiration for creating biomimetic polymers with silk-mimicking structures for use in biomaterials, and to better correlate biomaterial properties to structure [11, 144, 150, 159–161]. Potential applications of silk-mimicking materials include synthetic tendon implants and bullet-proof vests [151]. As one example of silk-inspired synthetic materials, Rathore et al. replaced the amorphous, non-β-sheet forming segments found in silk proteins with synthetic poly(ethylene glycol) (PEG) segments while retaining the polyalanine-rich segments [11]. They used polyalanines of varying lengths and found that by selectively replacing specific segments, they could obtain polymers whose properties mimic those of silk [11]. Rathore et al. retained those alanine-rich segments known to be essential for the mechanical properties of silk [11]. The polymers formed by

Rathore et al. were found to adopt anti-parallel β-sheets, with longer polyalanine segments, resulting in a larger amount of β-sheet secondary structure. A better understanding of the structural transition during silk self-assembly will be useful for creating novel silk-mimicking biomaterials [162].

11.3.2. *De Novo* Oligopeptide Hydrogels

Much attention has been focused on the *de novo* design of self-assembling oligopeptides. *De novo*-designed synthetic oligopeptides have shown the potential to self-assemble into macroscopic hydrogels with an underlying fibril-like network, rich in β-sheet content [50, 67, 69, 70, 136, 137, 163]. These studies have sought to design systems that can be easily synthesized and easily controlled to self-organize under well-defined conditions and in response to external factors such as pH and temperature. Oligopeptides offer themselves as convenient systems to exploit the hydrogel formation, as they can be easily fine-tuned to allow for specific functionality and conformational preference. Moreover, a large-scale production of such oligo-systems has clear processing advantages over production of more complex, larger systems [50].

Oligopeptide-based hydrogels have been demonstrated to be useful as drug delivery systems for controlled release [69, 70], as tissue-engineering scaffolds [164, 165], and have also found applications in wound healing [137]. Certain general guidelines have emerged in recent years for the formation of oligopeptide hydrogels [2]. In particular, oligopeptide-based hydrogels are typically formed from carefully designed sequences with alternating hydrophobic and hydrophilic residues, often containing complementary charge distributions recurring throughout [2, 69, 70]. In this regard, alanine-based sequences were the pioneering sequences and serve as the best model systems for creating such hydrogel-forming systems [67, 68, 166]. In particular, the oligopeptides AcN-(RARADADA)$_2$-CNH$_2$, RADA-16 [68, 70], and AcN-(AEAEKAKA)$_2$-NH$_2$, EAK-16 [67, 68] all possess charged side chains under neutral conditions, which allow for the stabilization of β-sheet secondary structures. Charge complementarity provides for favorable electrostatic interactions that can stabilize both intermolecular and intramolecular hydrogen bonding, ultimately resulting in hydrogel formation with an underlying filamentous architecture rich in β-sheet secondary structure [68–70, 140, 141].

11.3.2.1. EAK-16 Zhang and colleagues pioneered the field of the *de novo* design of alanine-based oligopeptides for the controlled self-assembly into macroscopic hydrogels. The first such oligopeptide they investigated was identified from the sequence of the yeast DNA binding protein, zuotin [167]. In particular, they designed the following sequence: AcN-(AEAEKAKA)$_2$-CNH$_2$, EAK-16 [67, 68]. EAK-16 can form a macroscopic membrane with an underlying filametous network architecture [67]. The resulting macroscopic membrane is resistant to high temperatures, extreme pH values, as well as chemical denaturants [67], and the peptide can form a hydrogel upon the addition of either buffer or physiological solutions. The resulting EAK-16 hydrogel exhibits the ability to foster mammalian cell attachment. This fact, along with the large water content and thus living-tissue-like properties, suggests that such systems may prove useful for tissue engineering and regeneration.

Zhang and colleagues proposed a model for the self-assembly of EAK-16 as reproduced in Figure 11.16, in which individual EAK-16 β-sheets stack via an ionic interface on one side and a hydrophobic interface on the other.

While electrostatic interactions are clearly important for the stabilization of EAK-16 into a β-sheet-rich macroscopic membrane, a recent study has demonstrated the importance of the hydrophobicity in the self-assembly process. Zou et al. showed that hydrogels formed by the EAK-16 analogue EMK-16, in which all of the alanine residues were replaced by methionine residues, could form at much lower critical concentrations [141]. This suggests that hydrophobic interactions are dominant in the self-assembly of such ionic self-complementary oligopeptides [141]. It is clear that to advance the exploitation of peptide self-assembly in the creation of novel biomaterials, it is necessary to have a complete and detailed understanding of the rules that govern the self-assembly process. This includes a detailed understanding of the self-assembly of ionic-complementary oligopeptides.

11.3.2.2. RADA-16 The charge complementarity of EAK-16 further inspired Zhang and colleagues to pursue the *de novo* design of similar alanine-based self-assembling peptides [68]. As is the case for EAK-16, the alternating complimentary charges of the designed alanine-rich oligopeptide AcN-(RARADADA)$_2$-CNH$_2$, (RADA-16) also allow for the stabilization of β-sheet structures and the formation of a macroscopic hydrogel, with an underlying nanofiber scaffold [69, 70]. Similar to the EAK-16 peptides, RADA-16 β-sheets exhibit both a hydrophobic and hydrophilic face, as depicted in Figure 11.17. The resulting RADA-16 hydrogel was found to contain >99.5% water, illustrating its similarity to living tissue [168]. RADA-16 hydrogels were found to reassemble after being fragmented by sonication, and the rigidity of the scaffold was found to increase with the nanofiber length [168]. Zhang and colleagues [168] proposed a sliding diffusion model for the formation of the macroscopic RADA-16 membrane, which is illustrated in Figure 11.18. The proposed model is similar to the diffusive sliding of nucleic acids exhibiting complementary DNA base pairs, for example, poly(A) and poly(U) [169, 170], as well as the translocation of repressor proteins on DNA segments [171, 172]. In the case of RADA-16 self-assembly, the nonspecific interactions of the all-alanine sheet faces result in diffusive sliding.

The "responsive" RADA-16 peptides form hydrogels under neutral conditions, which allow for favorable electrostatic interactions between the oppositely charged side chain residues. As was the case with EAK-16 hydrogels, the resulting RADA-16 hydrogel was found to act as scaffold for the attachment of a variety of mammalian cells [68, 136, 173], suggesting its possible role in tissue engineering and transplantation, as well as wound healing. RADA-16 hydrogels have also been shown to support neurite growth [136]. In particular, RADA-16 hydrogels were shown to exhibit the ability to support neuronal cell attachment and differentiation as well as extensive neurite outgrowth [136]. Moreover, the RADA-16 hydrogels do not elicit an immune response, which demonstrates their inherent low toxicity [136]. RADA-

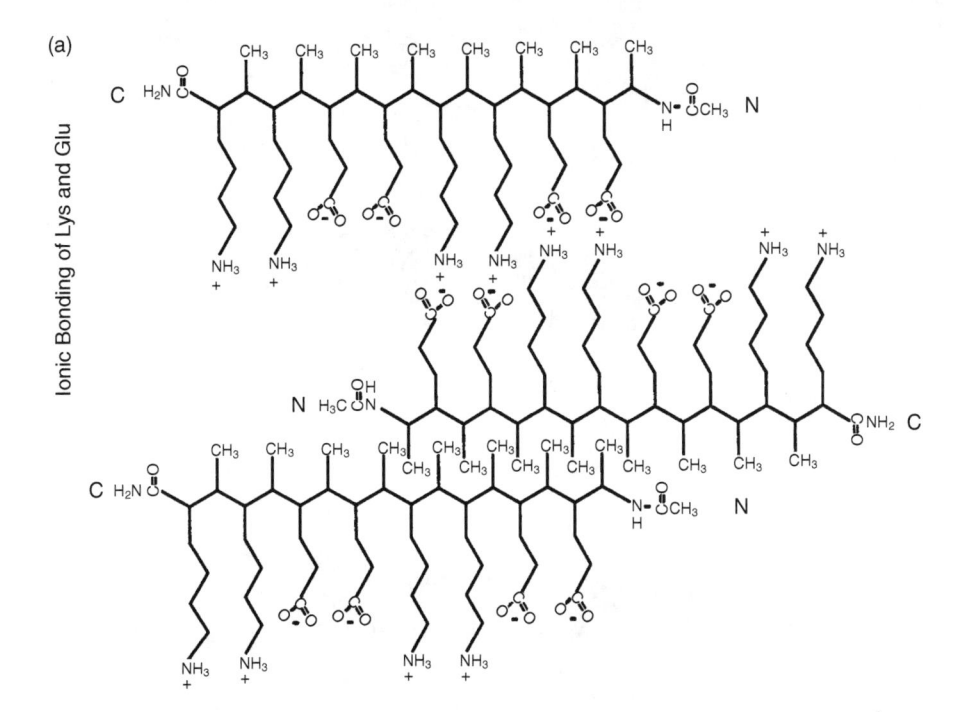

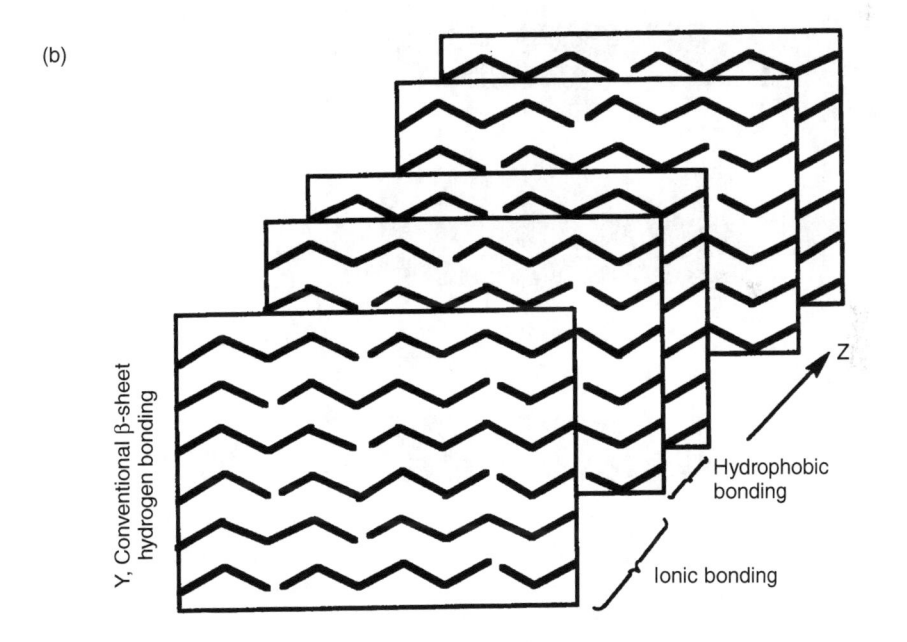

Figure 11.16. Model of β-sheet stacking and membrane formation of EAK-16. (a) Intra-sheet interactions showing both the ionic bonding between glutamic acid (E) and lysine (K) residues on one side and hydrophobic bonding of alanine side chains on the other side between neighboring strands. (b) Diagram of sheet stacking, depicting sheet stacking at both the ionic and hydrophobic interfaces. The figure was taken from Ref. [67] and modified.

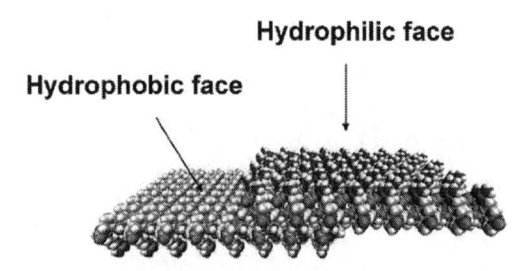

Figure 11.17. Model arrangement of stacked β-sheets in RADA-16 nanofibers, illustrating the hydrophilic (arginine, aspartic acid) and hydrophobic (alanine) faces. The figure was taken from Ref. [69] and modified.

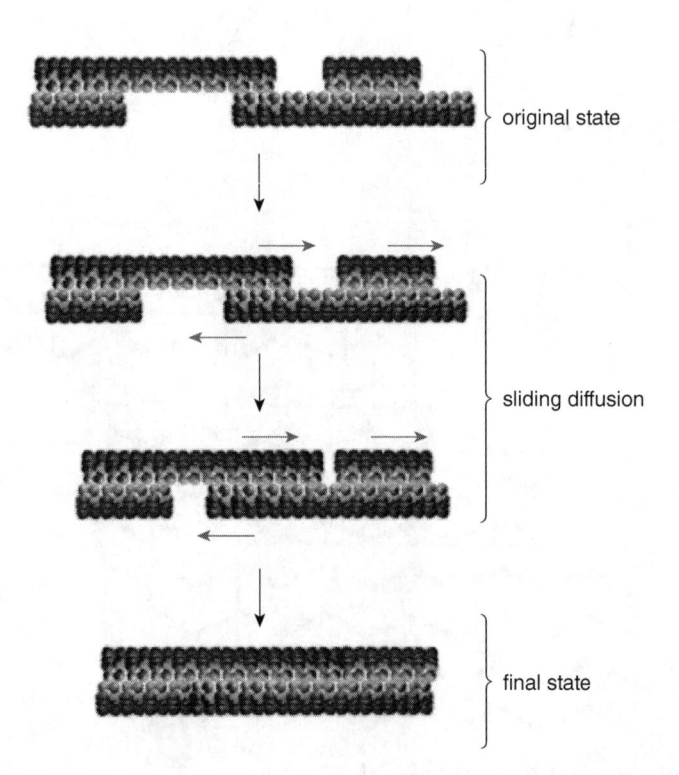

Figure 11.18. Sliding diffusion model proposed by Yokoi et al. to describe the reassembly of RADA-16 nanofibers after fragmentation. Nonspecific hydrophobic interactions permit the fiber to slide diffusively in either direction, eventually filling in the gaps. The figure was taken from Ref. [168] and modified.

16 hydrogels were observed to create an adequate environment for axonal regeneration, allowing for the hydrogel matrix to support repair and restoration of a severed optic tract in a hamster, which was able to regain its vision [173]. As another test of the tissue-regenerative capabilities of RADA-16 hydrogels, Horii coupled RADA-16 hydrogels with short biologically active motifs such as osteogenic peptides, showing that such a system supported pre-osteoblast cell proliferation and demonstrating the potential application of RADA-16 hydrogels in bone tissue regeneration [174].

RADA-16 hydrogels have also been shown to act as controlled-release devices [69, 70]. In particular, by tailoring the peptide concentration, Nagai et al. demonstrated the slow release of model dyes from RADA-16 hydrogel matrices [69] (see Fig. 11.19). The authors found that the diffusivity of the various dyes decreased with increasing hydrogel peptide concentration and were thus able to tailor the release profile of dyes by controlling the molecular interactions between the peptide nanofiber scaffold and the diffusing species [69]. Koutsopoulos et al. incorporated various proteins of differing hydrodynamic radii into the RADA-16 hydrogels and monitored the slow release of the proteins from the hydrogel matrix using fluorescence correlation spectroscopy (FCS) [70]. Using a variation of Fick's second law of diffusion, the authors were able to determine the diffusion coefficient of lysozyme, trypsin inhibitor, IgG, and bovine serum albumin (BSA) encapsulated in the RADA-16 hydrogel matrix [70]. Protein diffusion in gels depends primarily on protein size, and the diffusivity decreases with increasing hydrogel nanofiber density [70]. Thus, RADA-16 hydrogels can be used with a therapeutic protein for sustained-release applications *in vivo*.

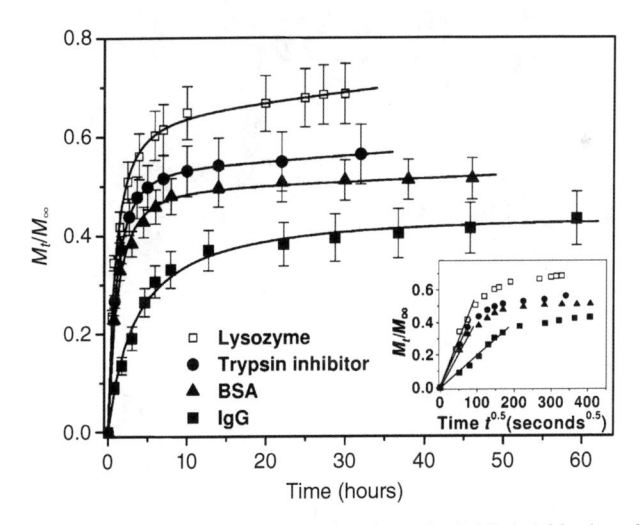

Figure 11.19. Release profiles of various proteins from the RADA-16 hydrogel matrix. The inset shows the linear fit to a model based on Fick's second law to determine the diffusivities of the encapsulated proteins as described by Koutsopoulos et al. [70]. The figure was taken from Ref. [70] and modified.

11.3.2.3. AK-16 In contrast to the *de novo*-designed oligopeptide sequences mentioned earlier, designed to form β-sheet structures under well-defined conditions via the inclusion of alternating complementary charges along the peptide backbone, Measey et al. recently reported the ability of a cationic alanine-based peptide, namely Ac-(AAKA)$_4$-NH$_2$ (AK-16), to form a self-supporting macroscopic hydrogel (see Fig. 11.20), rich in underlying β-sheet secondary structures [166, 175]. This finding is surprising for two reasons: (1) alanine-rich peptides of similar length and amino acid composition typically adopt α-helical conformations in aqueous solutions [71, 72] and (2) the positively charged lysine residues were expected to prevent self-aggregation. Despite such expectations, however, AK-16 forms thermodynamically unstable β-sheet structures, which decay into more extended PPII-like and turn-like conformations with both increasing time and temperature [166, 176]. In particular, AK-16 loses its β-sheet content over the course of days to months in the absence of any stabilizing factors, such as salts, which are required to shield the repulsive charged via electrostatic interactions, and ultimately result in hydrogel formation [166].

To gain a better understanding of the factors that contribute to the unusual conformational properties of AK-16, Measey et al. investigated various peptides in which the lysine residues of AK-16 were replaced by the charged amino acids, E (glutamic acid), R (arginine), and O (ornithine) [166]. Ornithine is essentially a

Figure 11.20. Self-supporting AK-16 hydrogel, formed by adding NaCl to a solution of 10 mg/mL AK-16 in H$_2$O solution to yield a final salt concentration of 2 M.

truncated lysine residue, containing one less methylene group. In addition, Measey et al. investigated the 12-residue, truncated version of AK-16, namely Ac-(AAKA)$_3$-NH$_2$ (AK-12), to see if sequence length played a role in the self-assembly [166]. While AR-12 was found to behave in a similar way to AK-16, neither AO-16, AE-16, nor AK-12 were found to form β-sheets or hydrogels under the conditions used for the AK-16 experiments. Thus, the unusual conformational properties of AK-16 were found to be the result of the combination of the sequence length and the length of the chosen charged side chain, that is, lysine [166].

The viscosity of the AK-16 hydrogels was measured as a function of shear rate for hydrogels formed using different peptide and NaCl concentrations. Measey et al. showed that the viscosity and physicochemical properties of the AK-16 hydrogels can by tuned by varying the type of salt and the concentrations of both salt and peptide, and, more importantly, by exploiting the conformational instability of the β-sheet aggregates. In particular, the viscosity of AK-16 hydrogels was found to decrease with both decreasing peptide and salt concentrations, as shown in Figure 11.21. Moreover, as the β-sheet content is necessary for AK-16 hydrogel formation, Measey et al. showed that by allowing a freshly prepared AK-16 solution to incubate overnight at room temperature, thus allowing for a loss of β-sheet content, the gel prepared from such a solution exhibits a much lower initial viscosity than the gel prepared from the fresh solution. These results illustrate the unique properties of a novel class of self-assembling cationic alanine-based peptides, where the viscosity of the resulting hydrogel can be tuned by exploiting the conformational instability

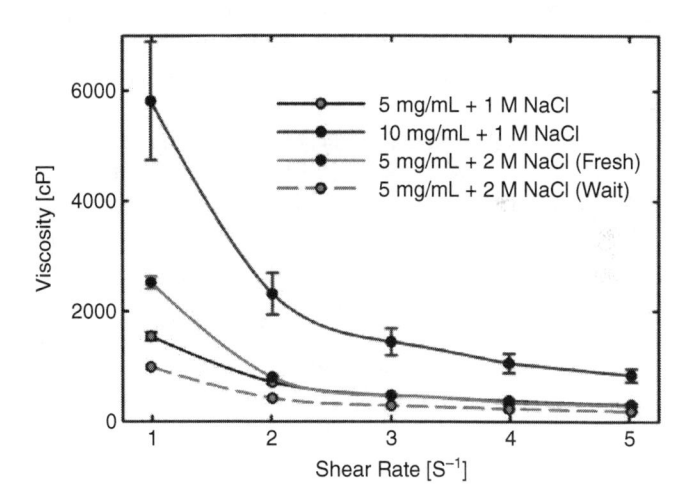

Figure 11.21. Viscosity of AK-16 hydrogels measured as a function of shear rate. Hydrogels were prepared with 5 mg/mL AK-16 and 1 M NaCl (black) and 10 mg/mL AK-16 and 1 M NaCl (red). To illustrate how β-sheet structure affects the viscosity of the resulting hydrogel, NaCl was added both to a freshly prepared portion of a 5 mg/mL AK-16 solution (blue, solid) and to a portion that was allowed to incubate overnight at room temperature (blue, dash) to yield a resulting salt concentration of 2 M. All samples were measured a total of three times. See color insert.

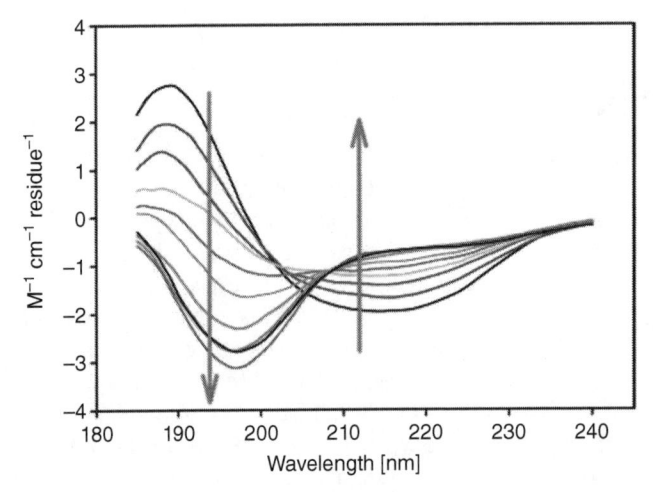

Figure 11.22. Temperature-dependent UV-CD spectra of AK-16 (3.5 mM, acidic pH). The arrows indicate increasing temperature. The figure was taken from Ref. [176] and modified.

of the β-sheet-rich structures. Moreover, these results provide insight into considerations for the future *de novo* design of self-assembling oligopeptides.

Alanine-based oligopeptide hydrogels have been shown to exhibit the ability to encapsulate and slowly release dyes and proteins, with the hopes of being used as slow-releasing drug delivery systems [69, 70]. Measey et al. demonstrated the ability of AK-16 hydrogels to encapulate and slowly release a model protein, namely cytochrome *c*, thus suggesting that AK-16 hydrogels might have biotechnological/biomedical applications [166].

As mentioned earlier, AK-16 loses its β-sheet content over time, as well as with increasing temperature. Far-ultraviolet circular dichroism (UV-CD) spectra of AK-16 as a function of temperature are shown in Figure 11.22. The loss of the negative maximum at ~215 nm is an indication of a loss of β-sheet secondary structure. In an attempt to gain a better understanding of the forces responsible for the aggregation and disaggregation of AK-16, Jang et al. used REMD in an all-atom study to investigate the dimer and trimer formation of AK-16 [176] *in silico*. Under conditions that mimic experimental conditions favoring AK-16 aggregation, Jang et al. detected the populations of various stable conformations on the energy landscape, as well as the thermal stability of AK-16 dimers and trimers. They found that the fraction of extended structures increased during the course of the simulation, which was run for a total of 64 ns. Moreover, the amount of trimers present was dependent on both the peptide concentration and the temperature, with only a small population of trimers present at low concentrations. At higher peptide concentration, the trimer population is high at lower temperatures and decreases sharply with increasing temperature, as shown in Figure 11.23, in accordance with experimental results [176]. Representative conformations from the six most populated clusters of trimers observed during the REMD simulation are shown in Figure 11.24.

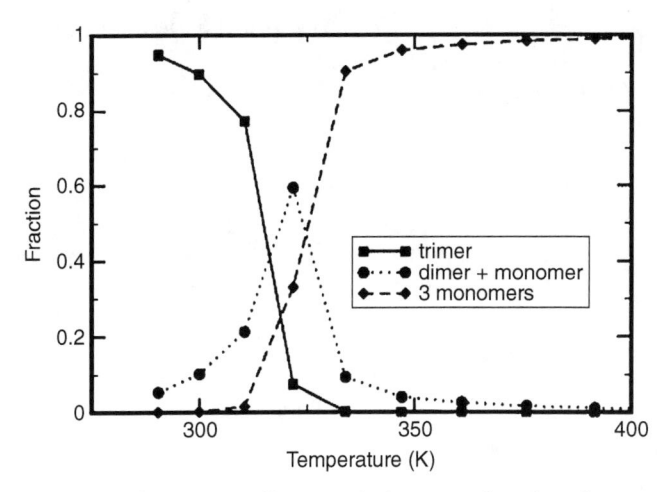

Figure 11.23. Fraction of monomers, dimers, and trimers as a function of temperature, which are present during the REMD simulation of AK-16. The figure was taken from Ref. [176] and modified.

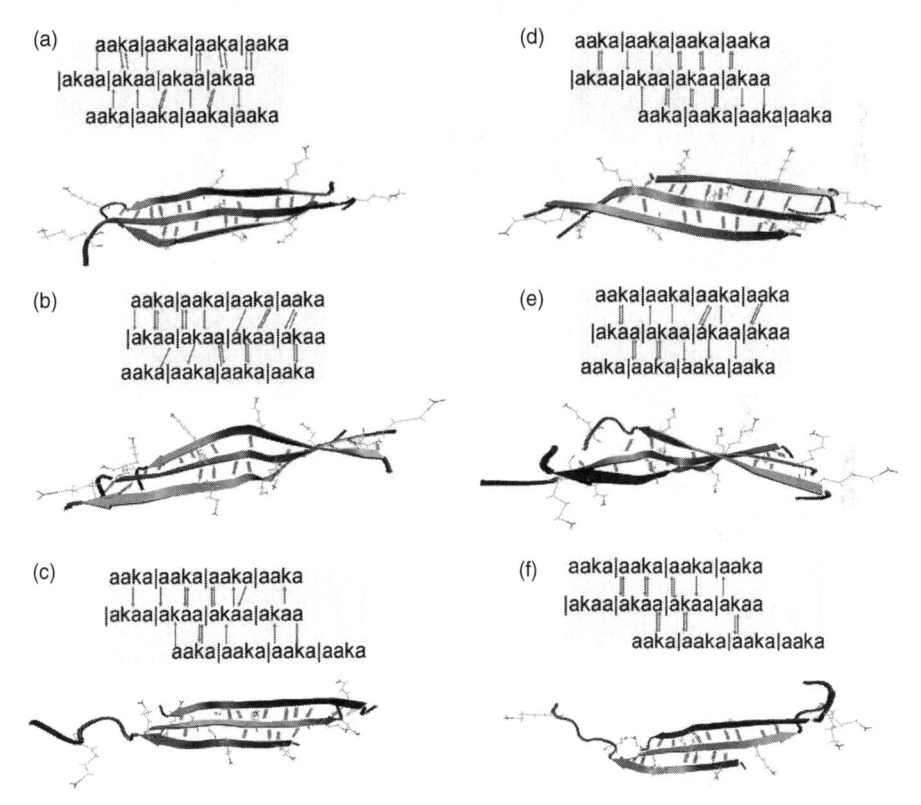

(a) aaka|aaka|aaka|aaka
    |akaa|akaa|akaa|akaa
    aaka|aaka|aaka|aaka

(b) aaka|aaka|aaka|aaka
    |akaa|akaa|akaa|akaa
    aaka|aaka|aaka|aaka

(c) aaka|aaka|aaka|aaka
    |akaa|akaa|akaa|akaa
    aaka|aaka|aaka|aaka

(d) aaka|aaka|aaka|aaka
    |akaa|akaa|akaa|akaa
    aaka|aaka|aaka|aaka

(e) aaka|aaka|aaka|aaka
    |akaa|akaa|akaa|akaa
    aaka|aaka|aaka|aaka

(f) aaka|aaka|aaka|aaka
    |akaa|akaa|akaa|akaa
    aaka|aaka|aaka|aaka

Figure 11.24. Representative clusters from the six most populated trimer clusters observed during the REMD simulation of Jang et al. The arrows represent backbone hydrogen bonding, where they point from the carbonyl oxygen to the amino hydrogen. (a)–(f) Six most populated clusters. The figure was taken from Ref. [176] and modified.

11.4. CONCLUDING REMARKS

Alanine is the most abundant amino acid found in nature. This fact, along with the simplicity and hydrophobicity of the methyl side chain of alanine, has resulted in the use of short alanine-based peptides as ideal model systems for studying peptide self-assembly. This includes self-assembly as it relates to toxic amyloid-fibril formation as well as to the formation of hydrogels with an underlying nanofiber scaffold. We have herein reviewed those simple systems that have been exploited to gain a better understanding of the mechanistic principles of self-assembly into well-ordered supramolecular structures. Knowledge of these mechanistic principles has provided, and is certain to further provide, information necessary to combat the various amyloid diseases and to better exploit the peptide self-assembly process for biotechnological applications. Moreover, short alanine-rich peptides are ideal for *in silico* studies.

Although other factors have been found to play significant roles in the self-assembly of alanine-based peptides, for example, aromaticity, sequence length, side-chain length of sequence-incorporated polar residues, and external conditions, one thing is clear: an alanine-based palette is a common theme among such aggregation-prone systems. Thus, one caveat protrudes: alanine-rich sequences are not sufficient to induce aggregation in all cases, but they are necessary in most cases. This fact is most likely due to their sterically less demanding and hydrophobic side chain.

For amyloidogenic peptide sequences, the identity of the primary amino acid sequence is becoming increasing less important as countless studies point to a general mechanism of fibril formation attributed to the polypeptide backbone, yet rules do exist for the *de novo* design of self-assembling and hydrogel-forming oligopeptides. Despite these rules, such as charge repetition and complementarity, as well as amphiphilicity, there also exist anomalous instances where self-assembly cannot be explained by such rules, for example, AK-16. Thus, there are still many fundamental details missing from a complete understanding of peptide and protein self-assembly, and short alanine-rich peptides are sure to continue to serve as ideal model systems for uncovering these mysteries.

ACKNOWLEDGMENTS

Part of the research reported herein was financially supported by a National Science Foundation grant (Chem 0804492) to R. S. S.

REFERENCES

1. Whitesides, G. M., Mathias, J. P., and Seto, C. T. (1991) Molecular self-assembly and nanochemistry: A chemical strategy for the synthesis of nanostructures, *Science 254*, 1312–1319.
2. Zhang, S. and Altman, M. (1999) Peptide self-assembly in functional polymer science and engineering, *React Funct Polym 41*, 91–102.

REFERENCES

3. Whitesides, G. M. (2002) Self-assembly at all scales, *Science 295*, 2418–2421.
4. Ma, P. X. (2008) Biomimetic materials for tissue engineering, *Adv Drug Deliv Rev 60*, 184–198.
5. Dobson, C. M. (2003) Protein folding and misfolding, *Nature 426*, 884–890.
6. Rosenbloom, J., Abrams, W. R., and Mecham, R. (1993) Extracellular matrix 4: The elastic fiber, *FASEB J 7*, 1208–1218.
7. Chiti, F. and Dobson, C. M. (2006) Protein misfolding, functional amyloid, and human disease, *Annu Rev Biochem 75*, 333–366.
8. Dobson, C. M. (1999) Protein misfolding, evolution and disease, *Trends Biochem Sci 24*, 329–332.
9. Dobson, C. M. (2009) Protein misfolding and disease: From the test tube to the organism, *J Neurochem 110*, 35.
10. Patterson, J., Martino, M. M., and Hubbell, J. A. (2010) Biomimetic materials in tissue engineering, *Mater Today 13*, 14–22.
11. Rathore, O. and Sogah, D. Y. (2001) Self-assembly of β-sheets into nanostructures by Poly(alanine) segments incorporated in multiblock copolymers inspired by spider silk, *J Am Chem Soc 123*, 5231–5239.
12. Mershin, A., Cook, B., Kaiser, L., and Zhang, S. (2005) A classic assembly of nanobiomaterials, *Nat Biotechnol 23*, 1379–1380.
13. Kyle, S., Aggeli, A., Ingham, E., and McPherson, M. J. (2009) Production of self-assembling biomaterials for tissue engineering, *Trends Biotechnol 27*, 423–433.
14. Dobson, C. M. (2004) Principles of protein folding, misfolding and aggregation, *Semin Cell Dev Biol 15*, 3–16.
15. Carrell, R. W. and Lomas, D. A. (1997) Conformational disease, *Lancet 350*, 134–138.
16. Uversky, V. N., Oldfield, C. J., and Dunker, A. K. (2008) Intrinsically disordered proteins in human diseases: Introducing the D2 concept, *Annu Rev Biophys 37*, 215–246.
17. Dobson, C. M. (2003) Protein folding and disease: A view from the first horizon symposium, *Nat Rev Drug Discov 2*, 154–160.
18. Stefani, M. (2008) Protein folding and misfolding, relevance to disease, and biological function. In: Smith, H. J., Simons, C., and Sewell, R. D. E. (eds.), *Protein misfolding in neurodegenerative diseases*, CRC Press, Boca Raton, FL, pp. 1–66.
19. Lee, H.-G., Zhu, X., Petersen, R. B., Perry, G., and Smith, M. A. (2003) Amyloids, aggregates and neuronal inclusions: Good or bad news for neurons? *Curr Med Chem Immunol Endocr Metab Agents 3*, 293–298.
20. Dobson, C. M. and Karplus, M. (1999) The fundamentals of protein folding: Bridging together theory and experiment, *Curr Opin Struct Biol 9*, 92–101.
21. Serpell, L. C., Fraser, P. E., and Sunde, M. (1999) X-ray fiber diffraction of amyloid fibrils, *Methods Enzymol 309*, 526–536.
22. Nelson, R., Sawaya, M. R., Balbirnie, M., Madsen, A. Ø., Riekel, C., Grothe, R., and Eisenberg, D. (2005) Structure of the cross-β spine of amyloid-like fibrils, *Nature 435*, 773–778.
23. Serpell, L. C. (2000) Alzheimer's amyoid fibrils: Structure and assembly, *Biochim Biophys Acta 1502*, 16–30.
24. Klunk, W. E., Jacob, R. F., and Mason, R. P. (1999) Quantifying amyloid by Congo red spectral shift assay, *Methods Enzymol 309*, 285–305.

25. LeVine, H., III (1993) Thioflavin T interaction with synthetic Alzheimer's disease β-amyloid peptides: Detection of amyloid aggregation in solution, *Protein Sci 2*, 404–410.
26. Gazit, E. (2007) Self-assembly of short aromatic peptides into amyloid fibrils and related nanostructures, *Prion 1*, 32–35.
27. Gazit, E. (2002) The "correctly folded" state of proteins: Is it a metastable state? *Angew Chem Int Ed Engl 41*, 257–259.
28. Gazit, E. (2005) Mechanisms of amyloid fibril self-assembly and inhibition: Model short peptides as a key research tool, *FEBS J 272*, 5971–5978.
29. Ferrone, F. (1999) Analysis of protein aggregation kinetics, *Methods Enzymol 309*, 256–274.
30. Modler, A. J., Gast, K., Lutsch, G., and Damaschun, G. (2003) Assembly of amyloid protofibrils via critical oligomers—A novel pathway of amyloid formation, *J Mol Biol 325*, 135–148.
31. Wood, S. J., Maleeff, B., Hart, T., and Wetzel, R. (1996) Physical, morphological and functional differences between pH 5.8 and 7.4 aggregates of the Alzheimer's amyloid peptide Aβ, *J Mol Biol 256*, 870–877.
32. Wood, S. J., Chan, W., and Wetzel, R. (1996) An ApoE-Aβ inhibition complex in Aβ fibril extension, *Chem Biol 3*, 949–956.
33. Jarrett, J. T. and Lansbury, P. T., Jr. (1992) Amyloid fibril formation requires a chemically discriminating nucleation event: Studies of an amyloidogenic sequence from the bacterial protein OsmB, *Biochemistry 31*, 12345–12352.
34. Walsh, D. M., Lomakin, A., Benedek, G. B., Condron, M. M., and Teplow, D. B. (1997) Amyloid β-protein fibrillogenesis, *J Biol Chem 272*, 22364–22372.
35. Harper, J. D., Wong, S. S., Lieber, C. M., and Lansbury, P. T., Jr. (1997) Observation of metastable Aβ amyloid protofibrils by atomic force microscopy, *Chem Biol 4*, 119–125.
36. Makin, O. S. and Serpell, L. C. (2004) Structural characterisation of islet amyloid polypeptide fibrils, *J Mol Biol 335*, 1279–1288.
37. Goldsbury, C. S., Cooper, G. J. S., Goldie, K. N., Müller, S. A., Saafi, E. L., Gruijters, W. T. M., Misur, M. P., Engel, A., Aebi, U., and Kistler, J. (1997) Polymorphic fibrillar assembly of human amylin, *J Struct Biol 119*, 17–27.
38. Zandomeneghi, G., Krebs, M. R. H., McCammon, M. G., and Fändrich, M. (2004) FTIR reveals structural differences between native β-sheet proteins and amyloid fibrils, *Protein Sci 13*, 3314–3321.
39. Shashilov, V. A. and Lednev, I. K. (2008) 2D correlation deep UV resonance Raman spectroscopy of early events of lysozyme fibrillation: Kinetic mechanism and potential interpretation pitfalls, *J Am Chem Soc 130*, 309–317.
40. Ma, S., Cao, X., Mak, M., Sadik, A., Walkner, C., Freedman, T. B., Lednev, I. K., Dukor, R. K., and Nafie, L. A. (2007) Vibrational circular dichroism shows unusual sensitivity to protein fibril formation and development in solution, *J Am Chem Soc 129*, 12364–12365.
41. Measey, T. J., Smith, K. B., Decatur, S. M., Zhao, L., Yang, G., and Schweitzer-Stenner, R. (2009) Self-aggregation of a polyalanine octamer promoted by its C-terminal tyrosine and probed by a strongly enhanced vibrational circular dichroism signal, *J Am Chem Soc 131*, 18218–18219.

REFERENCES

42. Uratani, Y., Asakura, S., and Imahori, K. (1972) A circular dichroism study of *Salmonella flagellin*: Evidence for conformational change on polymerization, *J Mol Biol 67*, 85–98.
43. Prusiner, S. B. (1982) Novel proteinaceous infectious particles cause scrapie, *Science 1982*, 136–144.
44. Hofrichter, J., Ross, P. D., and Eaton, W. A. (1974) Kinetics and mechanism of deoxyhemoglobin S gelation: A new approach to understanding sickle cell disease, *Proc Natl Acad Sci U S A 71*, 4864–4868.
45. Beaven, G. H., Gratzer, W. B., and Davies, H. G. (1969) Formation and structure of gels and fibrils from glucagon, *Eur J Biochem 11*, 37–42.
46. Jarrett, J. T. and Lansbury, P. T., Jr. (1993) Seeding "one-dimensional crystallization" of amyloid: A pathogenic mechanism in Alzheimer's disease and scrapie? *Cell 73*, 1055–1058.
47. Serio, T. R., Cashikar, A. G., Kowal, A. S., Sawicki, G. J., Moslehi, J. J., Serpell, L., Arnsdorf, M. F., and Lindquist, S. L. (2000) Nucleated conformational conversion and the replication of conformational information by a prion determinant, *Science 289*, 1317–1321.
48. Kelly, J. W. (2000) Mechanisms of amyloidogenesis, *Nat Struct Mol Biol 7*, 824–826.
49. Aggeli, A., Boden, N., and Zhang, S. (2001) *Self-assembling peptide systems in biology, medicine and engineering*, Kluwer Academic Publishers, Dordrecht.
50. Aggeli, A., Nyrkova, I. A., Bell, M., Harding, R., Carrick, L., McLeish, T. C. B., Semenov, A. N., and Boden, N. (2001) Hierarchical self-assembly of chiral rod-like molecules as a model for peptide β-sheet tapes, ribbons, fibrils, and fibers, *Proc Natl Acad Sci U S A 98*, 11857–11862.
51. Ma, B. and Nussinov, R. (2002) Molecular dynamics simulations of alanine rich β-sheet oligomers: Insight into amyloid formation, *Protein Sci 11*, 2335–2350.
52. de la Paz, M. L., Goldie, K., Zurdo, J., Lacroix, E., Dobson, C. M., Hoenger, A., and Serrano, L. (2002) De novo designed peptide-based amyloid fibrils, *Proc Natl Acad Sci U S A 99*, 16052–16057.
53. Tzotzos, S. and Doig, A. J. (2010) Amyloidogenic sequences in native protein structures, *Protein Sci 19*, 327–348.
54. Amijee, H., Madine, J., Middleton, D. A., and Doig, A. J. (2009) Inhibitors of protein aggregation and toxicity, *Biochem Soc Trans 37*, 692–696.
55. Tenidis, K., Waldner, M., Bernhagen, J., Fischle, W., Bergmann, M., Weber, M., Merkle, M.-L., Voelter, W., Brunner, H., and Kapurniotu, A. (2000) Identification of a penta- and hexapeptide of islet amyloid polypeptide (IAPP) with amyloidogenic and cytotoxic properties, *J Mol Biol 295*, 1055–1071.
56. Reches, M., Porat, Y., and Gazit, E. (2002) Amyloid fibril formation by pentapeptide and tetrapeptide fragments of human calcitonin, *J Biol Chem 277*, 35475–35480.
57. Tjernberg, L., Hosia, W., Bark, N., Thyberg, J., and Johansson, J. (2002) Charge attraction and β propensity are necessary for amyloido fibril formation from tetrapeptides, *J Biol Chem 277*, 43243–43246.
58. Maji, S. K., Drew, M. G. B., and Banerjee, A. (2001) First crystallographic signature of amyloid-like fibril forming b-sheet assemblage from a tripeptide with non-coded amino acids, *Chem Commun*, 1946–1947.

59 Reches, M. and Gazit, E. (2004) Formation of closed-cage nanostructures by self-assembly of aromatic dipeptides, *Nano Lett 4*, 581–585.

60 Reches, M. and Gazit, E. (2003) Casting metal nanowires within discrete self-assembled peptide nanotubes, *Science 300*, 625–627.

61 Guijarro, J. I., Sunde, M., Jones, J. A., Campbell, I. D., and Dobson, C. M. (1998) Amyloid fibril formation by an SH3 domain, *Proc Natl Acad Sci U S A 95*, 4224–4228.

62 Groβ, M., Wilkins, D. K., Pitkeathly, M. C., Chunk, E. W., Higham, C., Clark, A., and Dobson, C. M. (1999) Formation of amyloid fibrils by peptides derived from the bacterial cold shock protein CspB, *Protein Sci 8*, 1350–1357.

63 Chiti, F., Webster, P., Tadde, N., Clark, A., Stefani, M., Ramponi, G., and Dobson, C. M. (1999) Designing conditions for in vitro formation of amyloid protofilaments and fibrils, *Proc Natl Acad Sci U S A 96*, 3590–3594.

64 Gsponer, J., Haberthür, U., and Caflisch, A. (2003) The role of side-chain interactions in the early steps of aggregation: Molecular dynamics simulations of an amyliod-forming peptide from the yeast prion Sup35, *Proc Natl Acad Sci U S A 100*, 5154–5159.

65 Nguyen, H. D. and Hall, C. K. (2004) Molecular dynamics simulations of spontaneous fibril formation by random-coil peptides, *Proc Natl Acad Sci U S A 101*, 16180–16185.

66 Carny, O. and Gazit, E. (2005) A model for the role of short self-assembled peptides in the very early stages of the origin of life, *FASEB J 19*, 1051–1055.

67 Zhang, S., Holmes, T., Lockshin, C., and Rich, A. (1993) Spontaneous assembly of a self-complementary oligopeptide to form a stable macroscopic membrane, *Proc Natl Acad Sci U S A 90*, 3334–3338.

68 Zhang, S., Holmes, T. C., DiPersio, C. M., Hynes, R. O., Su, X., and Rich, A. (1995) Self-complementary oligopeptide matrices support mammalian cell attachment, *Biomaterials 16*, 1385–1393.

69 Nagai, Y., Unsworth, L. D., Koutsopoulos, S., and Zhang, S. (2006) Slow release of molecules in self-assembling peptide nanofiber scaffold, *J Control Release 115*, 18–25.

70 Koutsopoulos, S., Unsworth, L. D., Nagai, Y., and Zhang, S. (2009) Controlled release of functional proteins through designer self-assembling peptide nanofiber hydrogel scaffold, *Proc Natl Acad Sci U S A 106*, 4623–4628.

71 Marqusee, S., Robbins, V. H., and Baldwin, R. L. (1989) Unusually stable helix formation in short alanine-based peptides, *Proc Natl Acad Sci U S A 86*, 5286–5290.

72 Scholtz, J. M., Marqusee, S., Baldwin, R. L., Work, E. J., Stewart, J. M., Santoro, M., and Bolen, D. W. (1991) Calorimetric determination of the enthalpy change for the α-helix to coil transition of an alanine peptide in water, *Proc Natl Acad Sci U S A 88*, 2854–2858.

73 Marqusee, S. and Baldwin, R. L. (1987) Helix stabilization by Glu$^-$ ·· Lys$^+$ salt bridges in short peptides of de novo designed, *Proc Natl Acad Sci U S A 84*, 8898–8902.

74 Scholtz, J. M. and Baldwin, R. L. (1992) The mechanism of α-helix formation by peptides, *Annu Rev Biophys Biomol Struct 21*, 95–118.

75 Lednev, I. K., Karnoup, A. S., Sparrow, M. C., and Asher, S. A. (1999) α-Helix peptide folding and unfolding activation barriers: A nanosecond UV resonance Raman study, *J Am Chem Soc 121*, 8074–8086.

76 Huang, C.-Y., Klemke, J. W., Getahun, Z., DeGrado, W. F., and Gai, F. (2001) Temperature-dependent helix-coil transition of an alanine based peptide, *J Am Chem Soc 123*, 9235–9238.

77 Measey, T. and Schweitzer-Stenner, R. (2006) The conformations adopted by the octamer peptide (AAKA)$_2$ in aqueous solution probed by FTIR and polarized raman spectroscopy, *J Raman Spectrosc 37*, 248–254.

78 Woutersen, S. and Hamm, P. (2001) Isotope-edited two-dimensional vibrational spectroscopy of trialanine in aqueous solution, *J Chem Phys 114*, 2727–2737.

79 Shi, Z., Olson, C. A., Rose, G. D., Baldwin, R. L., and Kallenbach, N. R. (2002) Polyproline II structure in a sequence of seven alanine residues, *Proc Natl Acad Sci U S A 99*, 9190–9195.

80 McColl, I. H., Blanch, E. W., Hecht, L., Kallenbach, N. R., and Barron, L. D. (2004) Vibrational Raman optical activity characterization of Poly(L-proline) II helix in alanine oligopeptides, *J Am Chem Soc 126*, 5076–5077.

81 Eker, F., Griebenow, K., Cao, X., Nafie, L. A., and Schweitzer-Stenner, R. (2004) Preferred peptide backbone conformations in the unfolded state revealed by the structure analysis of alanine-based (AXA) tripeptides in aqueous solution, *Proc Natl Acad Sci U S A 101*, 10054–10059.

82 Schweitzer-Stenner, R. and Measey, T. J. (2007) The alanine-rich XAO peptide adopts a heterogeneous population, including turn-like and polyproline II conformations, *Proc Natl Acad Sci U S A 104*, 6649–6654.

83 Blondelle, S. E., Forood, B., Houghten, R. A., and Pérez-Payá, E. (1997) Polyalanine-based peptides as models for self-associated β-pleated-sheet complexes, *Biochemistry 36*, 8393–8400.

84 Forood, B., Pérez-Payá, E., Houghten, R. A., and Blondelle, S. E. (1995) Formation of an extremely stable polyalanine β-sheet macromolecule, *Biochem Biophys Res Commun 211*, 7–13.

85 Shinchuk, L. M., Sharma, D., Blondelle, S. E., Reixach, N., Inouye, H., and Kirschner, D. A. (2005) Poly-(L-Alanine) expansions from core β-sheets that nucleate amyloid assembly, *Proteins 61*, 579–589.

86 Soto, P., Baumketner, A., and Shea, J.-E. (2006) Aggregation of polyalanine in a hydrophobic environment, *J Chem Phys 124*, 134904-134901–134904-134907.

87 Fan, X., Dion, P., Laganiere, J., Brais, B., and Rouleau, G. A. (2001) Oligomerization of polyalanine expanded PABPN1 facilitates nuclear protein aggregation that is associated with cell death, *Hum Mol Genet 10*, 2341–2351.

88 Brais, B., Rouleau, G. A., Bouchard, J.-P., Fardeau, M., and Tomé, F. M. S. (1999) Oculopharyngeal muscular dystrophy, *Semin Neurol 19*, 59–66.

89 Toriumi, K., Oma, Y., Kino, Y., Futai, E., Sasagawa, N., and Ishiura, S. (2008) Expression of polyalanine stretches induces mitochondrial dysfunction, *J Neurosci Res 86*, 1529–1537.

90 Messaed, C. and Rouleau, G. A. (2009) Molecular mechanisms underlying polyalanine diseases, *Neurobiol Dis 34*, 397–405.

91 Miller, G. (2009) Could they all be prion diseases? *Science 326*, 1337–1339.

92 Geschwind, M. D. and Legname, G. (2008) Transmissible spongiform encephalopathis. In: Smith, H. J., Simons, C., and Sewell, R. D. E. (eds.), *Protein misfolding in neurodegenerative diseases*. CRC Press, Boca Raton, pp. 515–532.

93   Gasset, M., Baldwin, M. A., Lloyd, D. H., Gabriel, J.-M., Holtzman, D. M., Cohen, F., Fletterick, R., and Prusiner, S. B. (1992) Predicted α-helical regions of the prion protein when synthesized as peptides form amyloid, *Proc Natl Acad Sci U S A 89*, 10940–10944.

94   Satheeshkumar, K. S., Murali, J., and Jayakumar, R. (2004) Assemblages of prion fragments: Novel model systems for understanding amyloid toxicity, *J Struct Biol 148*, 176–193.

95   Balbirnie, M., Grothe, R., and Eisenberg, D. S. (2001) An amyloid-forming peptide from the yeast prion Sup35 reveals a dehydrated b-sheet structure for amyloid, *Proc Natl Acad Sci U S A 98*, 2375–2380.

96   Norstrom, E. M. and Mastrianni, J. A. (2005) The AGAAAAGA palindrome in PrP is required to generate a productive PrP^{Sc}-PrP^C complex that leads to prion propagation, *J Biol Chem 280*, 27236–27243.

97   Schweitzer-Stenner, R., Measey, T., Kakalis, L., Jordan, F., Pizzanelli, S., Forte, C., and Griebenow, K. (2007) Conformations of alanine-based peptides in water probed by FTIR, Raman, vibrational circular dichroism, electronic circular dichroism, and NMR spectroscopy, *Biochemistry 46*, 1587–1596.

98   Graf, J., Nguyen, P. H., Stock, G., and Schwalbe, H. (2007) Structure and dynamics of the homologous series of alanine peptides: A joint molecular dynamics/NMR study, *J Am Chem Soc 129*, 1179–1189.

99   Knowles, T. P. J., Waudby, C. A., Devlin, G. L., Cohen, S. I. A., Aguzzi, A., Vendruscolo, M., Terentjev, E. M., Welland, M. E., and Dobson, C. M. (2009) An analytical solution to the kinetics of breakable filament assembly, *Science 326*, 1533–1537.

100   Kurouski, D., Lombardi, R. A., Dukor, R. K., Lednev, I. K., and Nafie, L. A. (2010) Direct observation and pH control of reversed supramolecular chirality in insulin fibrils by vibrational circular dichroism, *Chem Commun 46*, 7154–7156.

101   Gazit, E. (2002) A possible role for π-stacking in the self-assembly of amyloid fibrils, *FASEB J 16*, 77–83.

102   Madine, J., Copland, A., Serpell, L. C., and Middleton, D. A. (2009) Cross-β-spine architecture of fibrils formed by the amyloidogenic segment NFGSVQFV of medin from sold-state NMR and X-ray fiber diffraction measurements, *Biochemistry 48*, 3089–3099.

103   Marshall, K. E., Hicks, M. R., Williams, T. L., Hoffman, S. V., Rodger, A., Dafforn, T. R., and Serpell, L. C. (2010) Characterizing the assembly of the Sup35 yeast prion fragment, GNNQQNY: Structural changes accompany a fiber-to-crystal switch, *Biophys J 98*, 330–338.

104   Tartaglia, G. G., Cavalli, A., Pellarin, R., and Caflisch, A. (2005) Prediction of aggregation rate and aggregation-prone segments in polypeptide sequences, *Protein Sci 14*, 2723–2734.

105   Makin, O. S., Atkins, E., Sikorski, P., Johansson, J., and Serpell, L. C. (2005) Molecular basis for amyloid fibril formation and stability, *Proc Natl Acad Sci U S A 102*, 315–320.

106   Azriel, R. and Gazit, E. (2001) Analysis of the minimal amyloid-forming fragment of the islet amyloid polypeptide, *J Biol Chem 276*, 34156–34161.

107   Tartaglia, G. G., Cavalli, A., Pellarin, R., and Caflisch, A. (2004) The role of aromaticity, exposed surface, and dipole moment in determining protein aggregation rates, *Protein Sci 13*, 1939–1941.

REFERENCES

108 Ma, M., Kuang, Y., Gao, Y., Zhang, Y., Gao, P., and Xu, B. (2010) Aromatic-aromatic interactions induce the self-assembly of pentapeptidic derivatives in water to form nanofibers and supramolecular hydrogels, *J Am Chem Soc 132*, 2719–2728.

109 Smith, A. M., Williams, R. J., Tang, C., Coppo, P., Collins, R. F., Turner, M. L., Saiani, A., and Ulijn, R. V. (2008) Fmoc-diphenylalanine self assembles to a hydrogel via a novel architecture based on π-π interlocked β-sheets, *Adv Mater 20*, 37–41.

110 Gutenkunst, C.-A. and Norflus, F. (2007) Huntington's disease. In: Smith, H. J., Simons, C., and Sewell, R. D. E. (eds.), *Protein misfolding in neurodegenerative diseases*. CRC Press, Boca Raton, pp. 447–465.

111 Scheuermann, T., Schulz, B., Blume, A., Wahle, E., Rudolph, R., and Schwarz, E. (2003) Trinucleotide expansions leading to an extended Poly-L-Alanine segment in the Poly (A) binding protein PABPN1 cause fibril formation, *Protein Sci 12*, 2685–2692.

112 Albrecht, A. N., Kornak, U., Böddrich, A., Süring, K., Robinson, P. N., Stiege, A. C., Lurz, R., Stricker, S., Wanker, E. E., and Mundlos, S. (2004) A molecular pathogenesis for transcription factor associated poly-alanine tract expansions, *Hum Mol Genet 13*, 2351–2359.

113 Rankin, J., Wyttenbach, A., and Rubinsztein, D. C. (2000) Intracellular green fluorescent protein-polyalanine aggregates are associated with cell death, *Biochem J 348*, 15–19.

114 Abu-Baker, A., Messaed, C., Laganiere, J., Gaspar, C., Brais, B., and Rouleau, G. A. (2003) Involvement of the ubiquitin-proteasome pathway and molecular chaperones in oculopharyngeal muscular dystrophy, *Hum Mol Genet 12*, 2609–2623.

115 Tomé, F. M. S. and Fardeau, M. (1980) Nuclear inclusion in oculopharyngeal dustrophy, *Acta Neuropathol 49*, 85–87.

116 Brais, B., Bouchard, J.-P., Xie, Y.-G., Rochefort, D. L., Chrétien, N., Tomé, F. M. S., Lafrenière, R. G., Rommens, J. M., Uyama, E., Nohira, O., Blumen, S., Korcyn, A. D., Heutink, P., Mathieu, J., Duranceau, A., Codère, F., Fardeau, M., and Rouleau, G. A. (1998) Short GCG expansions in the PABP2 gene cause oculopharyngeal muscular dystrophy, *Nat Genet 18*, 164–167.

117 Shanmugam, V., Dion, P., Rochefort, D., Laganière, J., Brais, B., and Rouleau, G. A. (2001) PABP2 polyalanine tract expansion causes intranuclear inclusions in oculopharyngeal muscular dystrophy, *Ann Neurol 48*, 798–802.

118 Chakrabartty, A., Kortemme, T., and Baldwin, R. L. (1994) Helix propensities of the amino acids measured in alanine-based peptides without helix-stabilizing side-chain interactions, *Protein Sci 3*, 843–852.

119 Giri, K., Bhattacharyya, N. P., and Basak, S. (2007) pH-dependent self-assembly of polyalanine peptides, *Biophys J 92*, 293–302.

120 Prince, J. T., McGrath, K. P., DiGirolamo, C. M., and Kaplan, D. L. (1995) Construction, cloning, and expression of synthetic genes encoding spider dragline silk, *Biochemistry 34*, 10879–10885.

121 Bagińska, K., Makowska, J., Wiczk, W., Kasprzykowski, F., and Chmurzyński, L. (2008) Conformational studies of alanine-rich peptide using CD and FTIR spectroscopy, *J Pept Sci 14*, 283–289.

122 Ma, B. and Nussinov, R. (2002) Stabilities and conformations of Alzheimer's β-amyloid peptide oligomers (Aβ_{16-22}, Aβ_{16-35}, and Aβ_{10-35}): Sequence effects, *Proc Natl Acad Sci U S A 99*, 14126–14131.

123 Wu, C., Lei, H., and Duan, Y. (2005) Elongation of ordered peptide aggregate of an amyloidogenic hexapeptide NFGAIL observed in molecular dynamics simulations with explicit solvent, *J Am Chem Soc 127*, 13530–13537.

124 Gnanakaran, S., Nussinov, R., and Garcia, A. E. (2006) Atomic-level description of amyloid β-dimer formation, *J Am Chem Soc 128*, 2158–2159.

125 Patro, S. Y. and Przybycien, T. M. (1996) Simulations of reversible protein aggregation and crystal structure, *Biophys J 70*, 2888–2902.

126 Nguyen, H. D. and Hall, C. K. (2005) Kinetics of fibril formation by polyalanine peptides, *J Biol Chem 280*, 9074–9082.

127 Bellesia, G. and Shea, J.-E. (2007) Self-assembly of β-sheet forming peptides into chiral fibrillar aggregates, *J Chem Phys 126*, 245104-245101–245104-245111.

128 Bellesia, G. and Shea, J.-E. (2009) Diversity of kinetic pathways in amyloid fibril formation, *J Chem Phys 131*, 111102-111101–111102-111104.

129 Nguyen, H. D. and Hall, C. K. (2006) Spontaneous fibril formation by polyalanines; discontinuous molecular dynamics simulations, *J Am Chem Soc 128*, 1890–1901.

130 Nguyen, H. D., Marchut, A. J., and Hall, C. K. (2004) Solvent effects on the conformational transition of a model polyalanine peptide, *Protein Sci 13*, 2909–2924.

131 Nguyen, H. D. and Hall, C. K. (2004) Phase diagrams describing fibrillization by polyalanine peptides, *Biophys J 87*, 4122–4134.

132 Nadler, W., Meinke, J. H., and Hansmann, U. H. E. (2008) Folding proteins by first-passage-times-optimized replica exchange, *Phys Rev E 78*, 061905-061901–061905-061904.

133 Vila, J. A., Ripoll, D. R., and Scheraga, H. A. (2000) Physical reasons for the unusual α-helix stabilization afforded by charged or neutral polar residues in alanine-rich peptides, *Proc Natl Acad Sci U S A 97*, 13075–13079.

134 Hamada, D., Yanagihara, I., and Tsumoto, K. (2004) Engineering amyloidogenicity towards the development of nanofibrillar materials, *Trends Biotechnol 22*, 93–97.

135 Rajagopal, K. and Schneider, J. P. (2004) Self-assembling peptides and proteins for nanotechnological applications, *Curr Opin Struct Biol 14*, 480–486.

136 Holmes, T. C., Lacalle, S. D., Su, X., Liu, G., Rich, A., and Zhang, S. (2000) Extensive neurite outgrowth and active synapse formation on self-assembling peptide scaffolds, *Proc Natl Acad Sci U S A 97*, 6728–6733.

137 Ruan, L., Zhang, H., Luo, H., Liu, J., Tang, F., Shi, Y.-K., and Zhao, X. (2009) Designed amphiphilic peptide forms stable nanoweb, slowly releases encapsulated hydrophobic drug, and accelerates animal hemostasis, *Proc Natl Acad Sci U S A 106*, 5105–5110.

138 Peppas, N. A., Huang, Y., Torres-Lugo, M., Ward, J. H., and Zhang, J. (2000) Physicochemical foundations and structural design of hydrogels in medicine and biology, *Annu Rev Biomed Eng 2*, 9–29.

139 Satish, C. S., Satish, K. P., and Shivakumar, H. G. (2006) Hydrogels as controlled drug delivery systems: Synthesis, crosslinking, water and drug transport mechanism, *Indian J Pharm Sci 68*, 133–140.

140 Kisiday, J., Jin, M., Kurz, B., Hung, H., Semino, C., Zhang, S., and Grodzinsky, A. J. (2002) Self-assemling peptide hydrogel fosters chondrocyte extracellular matrix production and cell division: Implications for cartilage tissue repair, *Proc Natl Acad Sci U S A 99*, 9996–10001.

141 Zou, D., Tie, Z., Lu, C., Qin, M., Lu, X., Wang, M., Wang, W., and Chen, P. (2010) Effects of hydrophobicity and anions on self-assembly of the peptide EMK16-II, *Biopolymers 93*, 318–329.

142 Schneider, J. P., Pochan, D. J., Ozbas, B., Rajagopal, K., Pakstis, L., and Kretsinger, J. (2002) Responsive hydrogels from the intramolecular folding and self-assembly of a designed peptide, *J Am Chem Soc 124*, 15030–15037.

143 Vepari, C. and Kaplan, D. L. (2007) Silk as a biomaterial, *Prog Polym Sci 32*, 991–1007.

144 Lewis, R. V. (2006) Spider silk: Ancient ideas for new biomaterials, *Chem Rev 106*, 3762–3774.

145 Xu, M. and Lewis, R. V. (1990) Structure of a protein superfiber: Spider dragline silk, *Proc Natl Acad Sci U S A 87*, 7120–7124.

146 Liivak, O., Flores, A., Lewis, R., and Jelinski, L. W. (1997) Conformation of the polyalanine repeats in minor ampullate gland silk of the spider nephila clavipes, *Macromolecules 30*, 7127–7130.

147 Hinman, M. B. and Lewis, R. V. (1992) Isolation of a clone encoding a second dragline silk fibroin, *J Biol Chem 267*, 19320–19324.

148 Kluge, J. A., Rabotyagova, O., Leisk, G. G., and Kaplan, D. L. (2008) Spider silks and their applications, *Trends Biotechnol 26*, 244–251.

149 Bini, E., Knight, D. P., and Kaplan, D. L. (2004) Mapping domain structures in silks from insects and spiders related to protein assembly, *J Mol Biol 335*, 27–40.

150 Hardy, J. G. and Scheibel, T. R. (2009) Silk-inspired polymers and proteins, *Biochem Soc Trans 37*, 677–681.

151 Swanson, B. O., Blackledge, T. A., Beltrán, J., and Hayashi, C. Y. (2006) Variation in the material properties of spider dragline silk across species, *Appl Phys A 82*, 213–218.

152 Simmons, A., Ray, E., and Jelinski, L. W. (1994) Solid-state ^{13}C NMR of nephila clavipes dragline silk establishes structure and identity of crystalline regions, *Macromolecules 27*, 5235–5237.

153 Parkhe, A. D., Seeley, S. K., Gardner, K., Thompson, L., and Lewis, R. V. (1997) Structural studies of spider silk proteins in the fiber, *J Mol Recognit 10*, 1–6.

154 Yamauchi, K., Horiguchi, K., Okonogi, M., and Asakura, T. (2008) Structural analysis of (Ala)$_n$ in crystalline region of silk fibroins using solid-state NMR, *Polym Preprints 49*, 732–733.

155 Spek, E. J., Wu, H.-C., and Kallenbach, N. R. (1997) The role of alanine sequences in forming β-sheets of spider dragline silk, *J Am Chem Soc 119*, 5053–5054.

156 Work, R. W. and Young, C. T. (1987) The amino acid compositions of major and minor ampullate silks of certain orb-web-building spiders (Araneae, Araneidae), *J Arachnol 15*, 65–80.

157 Holland, G. P., Jenkins, J. E., Creager, M. S., Lewis, R. V., and Yarger, J. L. (2007) Quantifying the fraction of glycine and alanine in β-sheet and helical conformations in spider dragline silk using solid-state NMR, *Chem Commun*, 5568–5570.

158 Jenkins, J. E., Creager, M. S., Lewis, R. V., Holland, G. P., and Yarger, J. L. (2010) Quantitative correlation between the protein primary sequences and secondary structures in spider dragline silks, *Biomacromolecules 11*, 192–200.

159 Rathore, O. and Sogah, D. Y. (2001) Nanostructure formation through β-sheet self-assembly in silk-based materials, *Macromolecules 34*, 1477–1486.

160 Liu, H., Xu, W., Zhao, S., Huang, J., Yang, H., Wang, Y., and Ouyang, C. (2010) Silk-inspired polyurethane containing GlyAlaGlyAla tetrapeptide. I. Synthesis and primary structure, *J Appl Polym Sci 117*, 235–242.

161 Winningham, M. J. and Sogah, D. Y. (1997) A modular approach to polymer architecture control via catenation of prefabricated biomolecular segments: Polymers containing parallel β-sheets templated by a phenoxathiin-based reverse turn mimic, *Macromolecules 30*, 862–876.

162 Rabotyagova, O. S., Cebe, P., and Kaplan, D. L. (2010) Role of polyalanine domains in β-sheet formation in spider silk block copolymers, *Macromol Biosci 10*, 49–59.

163 Hamley, I. W. (2007) Peptide fibrillization, *Angew Chem Int Ed Engl 46*, 8128–8147.

164 Semino, C. E., Merok, J. R., Crane, G. G., Panagiotakos, G., and Zhang, S. (2003) Functional differentiation of hepatocyte-like spheroid structures from putative live propenitor cells in three-dimensional peptide scaffolds, *Differentiation 71*, 262–270.

165 Sun, J. and Zheng, Q. (2009) Experimental study on self-assembly of KLD-12 peptide hydrogel and 3-D culture of MSC encapsulated with hydrogel in vitro, *J Huazhong Univ Sci Technol Med Sci 29*, 512–516.

166 Measey, T. J., Schweitzer-Stenner, R., Sa, V., and Kornev, K. (2010) Anomalous conformational instability and hydrogel formation of a cationic class of self-assembling oligopeptides, *Macromolecules 43*, 7800–7806.

167 Zhang, S., Lockshin, C., Herbet, A., Winter, E., and Rich, A. (1992) Zuotin, a putative Z-DNA binding protein in *Saccharomyces cerevisiae*, *EMBO J 11*, 3787–3796.

168 Yokoi, H., Kinoshita, T., and Zhang, S. (2005) Dynamic reassembly of peptide RADA16 nanofiber scaffold, *Proc Natl Acad Sci U S A 102*, 8414–8419.

169 Rich, A. and Davies, D. R. (1956) New two stranded helical structure: Polyadenylic acid and polyuridylic acid, *J Am Chem Soc 78*, 3548–3549.

170 Felsenfeld, G., Davies, D. R., and Rich, A. (1957) Formation of a three-stranded polynucleotide molecule, *J Am Chem Soc 79*, 2023–2024.

171 Richter, P. H. and Eigen, M. (1974) Diffusion controlled reaction rates in spheroidal geometry. Application to repressor-operator association and membrane bound enzymes, *Biophys Chem 2*, 255–263.

172 Berg, O. G., Winter, R. B., and Hippel, P. H. V. (1981) Diffusion-driven mechanisms of protein translocation on nucleic acids. 1. Models and theory, *Biochemistry 20*, 6929–6948.

173 Ellis-Behnke, R. G., Liang, Y.-X., You, S.-W., Tay, D. K. C., Zhang, S., So, K.-F., and Schneider, G. E. (2006) Nano neuro knitting: Peptide nanofiber scaffold for brain repair and axon regeneration with functional return of vision, *Proc Natl Acad Sci U S A 103*, 5054–5059.

174 Horii, A., Wang, Z., Gelain, F., and Zhang, S. (2007) Biological designer self-assembling peptide nanofiber scaffolds significantly enhance osteoblast proliferation, differentiation and 3-D migration, *PLoS ONE 2*, 1–9.

175 Measey, T. J. and Schweitzer-Stenner, R. (2006) Aggregation of the amphipathic peptides $(AAKA)_n$ into antiparallel β-sheets, *J Am Chem Soc 128*, 13324–13325.

176 Jang, S., Yuan, J.-M., Shin, J., Measey, T. J., Schweitzer-Stenner, R., and Li, F.-Y. (2009) Energy landscapes associated with the self-aggregation of an alanine-based oligopeptide $(AAKA)_4$, *J Phys Chem B 113*, 6054–6061.

12

STRUCTURAL ELEMENTS REGULATING INTERACTIONS IN THE EARLY STAGES OF FIBRILLOGENESIS: A HUMAN CALCITONIN MODEL SYSTEM

Rosa Maria Vitale, Giuseppina Andreotti, Pietro Amodeo, and Andrea Motta

12.1. STATING THE PROBLEM

Nowadays, at least 27 human and 9 animal diseases of unrelated etiology are associated with protein misfolding, loss of specific conformational function, gain of toxicity, and/or protein aggregation [1]. Although different proteins are involved, they are often referred to as protein-misfolding diseases to emphasize the common molecular mechanisms of their origin, but many factors, intrinsic and extrinsic to polypeptide, acting independently, additively or synergistically, can increase propensity to misfold and/or aggregate [2]. Intrinsic factors involve charge [3–5], hydrophobicity [6–8], patterns of polar and nonpolar residues [9], and the propensities to adopt different secondary-structure motifs [3, 8, 10, 11]. Extrinsic factors comprise the interaction with cellular components such as molecular chaperones [12], proteases that generate or process the amyloidogenic precursors [13], and the effectiveness of quality control mechanisms, such as the ubiquitin–proteasome system [14, 15]. They also include solution physicochemical parameters defining the environment of the polypeptides, such as pH, temperature, ionic strength, and concentration [16–19].

Protein and Peptide Folding, Misfolding, and Non-Folding, First Edition. Edited by Reinhard Schweitzer-Stenner.
© 2012 John Wiley & Sons, Inc. Published 2012 by John Wiley & Sons, Inc.

Protein-misfolding diseases can affect a single organ or be spread through multiple tissues. For example, numerous amyloidoses and various neurodegenerative disorders originate from the conversion of soluble functional states of proteins into stable, highly ordered amyloid fibrils, and from their deposition in organs and tissues. Tables 12.1 and 12.2 list the currently known human and animal proteins identified as the main fibril components in deposits with classical amyloid properties [1]. Amyloid is defined as an "*in-vivo* deposited material, which can be distinguished from non-amyloid deposits by characteristic fibrillar electron microscopic appearance, typical X-ray diffraction pattern and histological staining reactions, particularly affinity for the dye Congo red with resulting green birefringence" [1]. Although there is one fibril protein in each type of amyloidosis, the deposits, together with the polypeptide fibril, may contain other components such as glycosaminoglycans, apolipoprotein E, and serum amyloid P-component. In earlier studies, amyloid only referred to extracellular material. However, there is increasing evidence that many types of amyloid may start intracellularly, showing inclusions with typical amyloid structure. Therefore, nowadays all deposits fulfilling the above definition are called amyloid irrespective of their appearance intra- or extracellularly [1]. On the other hand, an extracellular or intracellular deposit lacking affinity for Congo red should not be called amyloid.

Amyloid fibril deposition goes through increasingly larger assemblies, and during the early phase of the process (referred to as lag phase) monomers interact through thermodynamically unfavorable interactions forming transient and unstable intermediates, until a critical mass nucleus is formed (nucleation time). This is followed by a rapid growth phase, during which molecules interact with the growing nucleus through thermodynamically favorable interactions [16, 17, 20]. The nucleus formation is a limiting step of the process as the presence of preformed nuclei or "seeds" can shorten or bypass the nucleation time [16, 17, 20, 21].

The molecular mechanism of self-assembly into amyloid fibrils is still poorly understood, although information at atomic level on the aggregates has been obtained [22–26]. According to the nucleation–elongation polymerization model of proteins [27], the nucleus is in thermodynamic equilibrium with a monomer, and the fibril mass is proportional to t^2 at the beginning of the reaction (t being the experiment time). However, such a model does not include the lag-phase evolution and lacks generality. Studies on the hemoglobin S (HbS) polymerization [28] allowed an extension of the model, defining a heterogeneous-nucleation mechanism [29, 30]. The first step is the formation of nuclei of critical size starting from monomers (homogeneous nucleation); then the aggregates catalyze the development of additional fibrils. A power-law dependence of the concentration on HbS lag-phase length was observed, and kinetic analysis confirmed that the lag-phase length is proportional to $C^{n/2}$ (where C is the monomer concentration, and n is the nucleus size). Therefore, the nucleus size can be obtained from an apparent reaction order if the kinetics progress curve is consistent with the existence of a homogenous mechanism. Through the years, several studies have indicated that the kinetic profile of amyloid fibril formation can be described by heterogeneous nucleation mechanism [20, 31–33].

STATING THE PROBLEM

TABLE 12.1. Amyloid Fibril Proteins and Their Precursors in Human[a]

Amyloid Protein	Precursor	Systemic (S) or Localized, Organ Restricted (L)	Syndrome or Involved Tissues
AL	Immunoglobulin light chain	S, L	Primary Myeloma-associated
AH	Immunoglobulin heavy chain	S, L	Primary
$A\beta_2M$	β_2-microglobulin	S, L?	Hemodialysis-associated Joints
ATTR	Transthyretin	S, L?	Familial Senile systemic Tenosynovium
AA	(Apo)serum AA	S	Secondary, reactive
AApoAI	Apolipoprotein AI	S, L	Familial Aorta, meniscus
AApoAII	Apolipoprotein AII	S	Familial
AApoAIV	Apolipoprotein AIV	S	Sporadic, associated with aging
AGel	Gelsolin	S	Familial, Finnish
ALys	Lysozyme	S	Familial
AFib	Fibrinogen α-chain	S	Familial
ACys	Cystatin C	S	Familial
ABri	ABriPP	S	Familial dementia, British
ADan[b]	ADanPP	L	Familial dementia, Danish
$A\beta$	$A\beta$ protein precursor ($A\beta$PP)	L	Alzheimer's disease, aging
APrP	Prion protein	L	Spongiform encephalopathies
ACal	(Pro)calcitonin	L	C-cell thyroid tumors
AIAPP	Islet amyloid polypeptide[c]	L	Islets of Langerhans insulinomas
AANF	Atrial natriuretic factor	L	Cardiac atria
APro	Prolactin	L	Aging pituitary Prolactinomas
AIns	Insulin	L	Iatrogenic
AMed	Lactadherin	L	Senile aortic, arterial media
AKer	Kerato-epithelin	L	Cornea, familial
ALac	Lactoferrin	L	Cornea
AOaap	Odontogenic Ameloblast-associated protein	L	Odontogenic tumors
ASemI	Semenogelin I	L	Vesicula seminalis
ATau	Tau	L	Alzheimer's disease, fronto-temporal dementia, aging, other cerebral

[a] Adapted from Ref. [1]. When possible, proteins are grouped according to relationship. Thus, apolipoproteins and polypeptide hormones are listed together.
[b] ADan and ABri originate from the same gene.
[c] Also called "amylin."

TABLE 12.2. Amyloid Fibril Proteins and Their Precursors in Animals[a]

Amyloid Protein	Precursor	Systemic (S) or Localized (L)	Syndrome or Involved Tissues	Species
AL	Immunoglobulin light chain	L	Plasmocytoma	Horse
AA	(Apo)serum AA	S	Secondary, reactive	Mouse, guinea pig, cat, dog, cow, duck, etc.
AApoAI	Apolipoprotein AI	S	Age-related	Dog
AApoAII	Apolipoprotein AII	S	Age-related	Mouse
Aβ	Aβ protein precursor	L	Age-related	Dog, sheep, wolverine
AIAPP	Islet amyloid polypeptide	L	Islets of Langerhans Insulinoma	Cats, apes, raccoon
AIns	Insulin	L	Islets of Langerhans	*Octodon degus*
ACas	α-S2C casein	L	Mammary gland	Cow
ATau	Tau	L	Brain	Polar bear, wolverine, apes, etc.

[a] Adapted from Ref. [1].

The mechanism of amyloid formation *in vivo* may be quite different from that observed *in vitro*. In fact, besides the contribution of specific structure elements and amino acidic sequences to amyloidogenesis, there is growing evidence that interactions with lipid and membrane play significant roles in the progression and toxicity of amyloid diseases, since amylodogenic molecules act in a heterogeneous environment ([34, 35] and references therein). Furthermore, the conformational solution state of amyloidogenic polypeptides and proteins before aggregation takes place is also important. They can be broadly divided into "globular" and "natively disordered," the latter including those biomolecules in a "mixed globular/disordered" state. Globular systems show compact structure with a definite fold, while intrinsically disordered systems, including those slightly structured, are flexible, and often with a low tendency to aggregate. However, the possible presence of conformational constraints can act as a possible seed favoring their aggregation [36–38]. In fact, many unstructured polypeptides, including model peptides, do form amyloid *in vivo*, and some pathologically important (ABri through AANF) are reported in Table 12.1.

12.2. AGGREGATION MODELS: THE STATE OF THE ART

Regarding the conformation taken up before aggregation occurs, amyloidogenic polypeptides and proteins are frequently divided into "globular" and "natively disordered," the latter including "mixed globular/disordered" state. A globular conformation presents a definite secondary/tertiary fold, which has to unfold partially or totally before aggregation can take place. Intrinsically disordered systems present a flexible structure and require the presence of local structural (turn, helix-like, β-sheet-like) elements that can trigger their aggregation.

One of the major prerequisites for aggregation and fibrillogenesis in natively disordered polypeptides is the presence of hydrophobic stretches in the primary polypeptide structure [39–43]. For example, hydrophobic residues in the central and C-terminal regions of the amyloid β (Aβ) peptide with 42 amino acids (Aβ42) have been reported to promote aggregation [40–42]. Their role can be twofold: (1) the hydrophobic side chains mediate specific interactions that direct the self-assembly of Aβ42; and (2) the characteristics of the nonpolar side chains are irrelevant, and hydrophobicity per se promotes aggregation. Support for the first possibility ("specific" nonpolar side chains promote aggregation) comes from crystal structures of model amyloidogenic peptides [44]. These systems are characterized by well-packed structures with specific side-chain interactions, forming highly ordered "steric zippers." Support for the second possibility ("generic" hydrophobic key positions in the amino acidic sequence are sufficient to promote aggregation) comes from protein design, where patterning of polar and hydrophobic residues, but not the exact identities of these residues, is sufficient to design *de novo* libraries of proteins ([45, 46] and references therein). To distinguish between the possibilities, a library of mutants was constructed and characterized, in which 12 hydrophobic residues in the central and C-terminal stretches of Aβ42 were replaced by a stochastic mixture of the nonpolar residues leucine, isoleucine, valine, phenylalanine, and methionine. The results indicated that the positioning of the hydrophobic residues is more important than the exact identities of the hydrophobic side chains. However, the mutated Aβ42 sequences aggregate with altered kinetics [45]. In particular, an Aβ42 mutant completely lacking Phe amino acids aggregated somewhat slower than either wild-type Aβ42 or the other nonpolar → nonpolar mutants, and was more soluble than the other hydrophobic variants. The slower aggregation rate of the no-Phe Aβ42 variant was attributed to the lack of aromatic residues in its C-terminal half. In fact, electron microscopy studies of the mutated Aβ42 peptides showed that the variants containing four Phe residues in their C-terminal halves aggregate faster than wild-type Aβ42 or no-Phe, which contain two or zero Phe residues, respectively, in their C-terminal halves [45]. The results reported for the mutants of Aβ42 showed that although sequences devoid of aromatics can form amyloid structures, they do so more slowly, therefore suggesting a specific role for aromatic residues and their side-chain packing.

Many globular proteins that form aggregates under physiological conditions preserve their well-defined secondary and tertiary structures [47]. Some of them are included in Table 12.1 and discussed in Ref. [48]. When folded, the propensity of the proteins to aggregate is generally very low, but under solution conditions (low pH, high temperature, high pressure, and in the presence of co-solvents) that favor their partial unfolding they are able to readily form amyloid fibril *in vitro* [49–56]. Furthermore, many of the mutations associated with hereditary forms of amyloidoses destabilize the globular native fold and promote aggregation *in vitro*, suggesting that the disease is a consequence of the reduced conformational stability [57–60].

According to these results, a conformational change hypothesis has been put forward. It relies on the idea that a global or partial unfolding is required to initiate the aggregation of a globular protein that is normally cooperatively folded [61, 62]. If a mutation or a change of solution conditions decreases the ΔG between the native fold (N) and the fully unfolded (U) or a partially unfolded (I) states, the non-native U and I states at equilibrium will have a higher population. Moreover, if the transition state is not destabilized, or is destabilized to a lesser extent with respect to the native state, the U and I states will also be kinetically accessible. In such states, many of the hydrophobic and backbone moieties, normally buried in the interior of the protein fold, become solvent-exposed and thus accessible for intermolecular interactions, bringing about a greatly enhanced propensity for aggregation.

The above model involves a transition from the N state of a protein across the major free energy barrier to reach the totally and partially unfolded states U and I. However, recent findings suggest that unfolding processes of this magnitude are not essential to generate precursor states that favor amyloid formation. Conformational states thermodynamically distinct from the native state, but structurally similar to it, can be accessed directly from the native state through thermal fluctuations. These conformational states, termed N*, are separated from N by a relatively low energy barrier; although only transiently populated under physiological conditions, they can be sampled more frequently than states such as I and U [63]. An experimental indication of structural fluctuation can certainly be inferred from the amide hydrogens buried in the interior of a native protein, which often exchange with the solvent more rapidly than global unfolding [64–66]. Therefore, such states provide a general mechanism for pathological amyloid formation *in vivo* that does not require the existence of the relatively drastic *in vitro* conditions. It has been proposed [47] that local unfolded states (occurring through thermal fluctuations of the native folded state) may not only be populated under conditions close to physiological, but also favor amyloid formation. However, ligands and antibodies that bind the native structure of such proteins significantly decrease the aggregation propensity not only by preventing the global unfolding but also by hampering the thermal fluctuations of the native conformation [67].

Some proteins preserve their native structure in the initial aggregates only, and afterward reorganize the global fold to mature into fibrils, as does insulin, which converts from completely helical initial aggregates to fibrils that have an almost completely β-sheet structure [68]. Some other systems show aggregates that retain their native-like fold also in the mature fibrillar state. This occurs more naturally for proteins that contain only or mainly a β-sheet structure (for example, S134N mutant of copper–zinc superoxide dismutase 1, transthyretin, and β_2-microglobulin) [47].

12.3. HUMAN CALCITONIN HCT AS A MODEL SYSTEM FOR SELF-ASSEMBLY

A number of low-molecular-mass inhibitors have been identified by screening compound libraries as well as rational design strategies. For the Aβ peptide they include chemically diverse compounds such as curcumin, inositol, and nicotine [69], as well as peptide and protein mimetics [70]. This strategy has two major drawbacks. First, these compounds lack structurally defined targets, and some of them may work in a nonspecific manner; second, they block the later stages of fibril formation and do not affect the formation of protofibrillar oligomeric species formed in the early stages of aggregation. A growing body of *in vitro* and *in vivo* experimental studies indicates that soluble oligomers often undetectable, rather than mature full-length fibrils, are responsible for amyloid-induced cytotoxicity [71–76]. This suggests that successful inhibitors of amyloidoses should be able to interfere with the early stages of aggregation. When protofibril formation is mediated by unstable α-helix or turn-like polypeptide stretches, targeting those regions offers a novel approach to finding specific inhibitors that are effective *in vivo*, since a molecule that stabilizes, even modestly, the α-helical/turn-like structure causes a delay of the subsequent β-sheet oligomerization and aggregation [77]. This strategy also includes the formation of "mixed oligomers": that is, a mixture of an amyloidogenic polypeptide with an aggregation-resistant analogue in which a "mixed" protofibril is formed *before* the critical mass nucleus. Therefore, the aggregation-resistant analogue works as an efficient separator that prevents maturation of the fibril [36, 78, 79].

Here we investigate the early stages of amyloid fibril formation by hCT, aiming at defining a possible procedure to explore the formation of early aggregates in fibrillogenic biopolymers. hCT is a 32-residue hormone synthesized and secreted by the C cells of the thyroid, involved in calcium regulation and bone dynamics [80]. In its common form, hCT presents an N-terminal disulfide bridge between positions 1 and 7, and an amidated C-terminal proline residue (Fig. 12.1).

Only eight residues are common to all species so far studied, and these are clustered at the two ends of the molecule. The salmon variant (sCT) is widely used in the treatment of osteoporosis and Paget's disease, as well as malignancy-caused hypercalcemia and musculoskeletal pain [81, 82]. This is because hCT shows an extremely high tendency to form amyloid fibrils both *in vivo* in patients with medullar carcinoma of the thyroid [83], and *in vitro* in preparations designed for patient administration [31]. Therefore, aggregation constitutes a serious problem that leads to a significant decrease in the hormone activity [84]. Moreover, aggregation stimulates undesirable immune responses resulting in resistance or allergic reactions in patients [85, 86], and drug-induced cytotoxicity [87]. On the contrary, sCT shows a

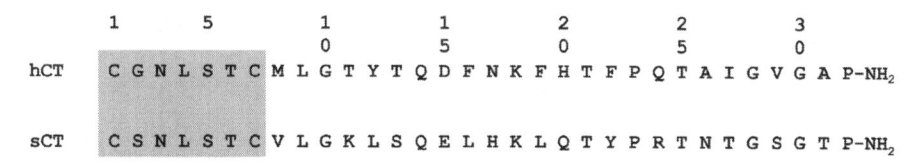

```
           1         5         1         1         2         2         3
                               0         5         0         5         0
hCT        C G N L S T C M L G T Y T Q D F N K F H T F P Q T A I G V G A P-NH₂

sCT        C S N L S T C V L G K L S Q E L H K L Q T Y P R T N T G S G T P-NH₂
```

Figure 12.1. Primary structure of human (hCT) and salmon (sCT) calcitonins. Amino acids are represented by the single-letter code. The N-terminal disulfide bridge between positions 1 and 7 is shaded.

higher potency combined with a longer *in vivo* half-life when compared with hCT, and substantially lower propensity to aggregate [31], but it has been shown to develop side effects such as anorexia and vomiting [88, 89]. When aggregation is prevented by drastic chemical conditions, hCT shows a much higher potency than sCT [90]; however, those conditions are difficult to implement during the production, storage, and administration to patients, explaining why hCT has never been extensively used as a therapeutic.

It has been proposed that the first step of hCT fibril formation is a homogeneous association of α-helices to form the nucleus of a fibril [91], followed by an autocatalytic heterogeneous fibrillation of β-sheets to form a mature fibril [31, 92]. A morphological and structural TEM investigation has clearly shown how the protofibril aggregation generates polymorphism of fibrillar supramolecular assemblies [93]. Notwithstanding the pH- and, therefore, the charge-dependent structural diversity [92, 94], all of hCT fibril types are formed according to the two-step (nucleation and maturation) mechanism [91, 92]. Furthermore, an increase in α and β secondary-structure components is observed [95], with hydrophobic interactions and charged amino acids respectively favoring the formation of α-helical bundles and β-sheet association [91, 92]. In addition, π–π aromatic interactions have been hypothesized to stabilize the structure of hCT fibrils at neutral pH [91, 92, 96]. These findings are in agreement with the concept that factors intrinsic to the amino acidic polypeptide chain (e.g., hydrophobicity, charge, and the propensity of the polypeptide chain to adopt α-helical or β-sheet structure) affect amyloidosis ([32, 97] and references therein, [98]).

The fibrillation of sCT is much slower than hCT, because at 1 mg/mL and pH 7.2 sCT requires more than 8 months while hCT fibrillates in only 21 minutes [31]. Furthermore, the physiological pH accelerates hCT fibrillation with respect to low pH [31]. Considering that the proposed fibrillation mechanism involves a homogeneous aggregation of α-helices, it is surprising that sCT does not fibrillate, despite having a helical propensity higher than hCT [99, 100]. However, sCT mature fibrils, grown in 10 mg/mL solution at high temperature, after 9 days are characterized by a β-structure analogous to that of hCT [101].

hCT may represent a useful model for characterizing the structural determinants that favor the formation of intermediates at the beginning of the amyloidogenic process. In general, understanding those factors would be very useful in hindering the formation of malignant aggregates and delaying their progression toward mature fibrils. As such, the search for successful pharmacological inhibitors should also benefit from such studies, as efficient inhibition should target the early stages of the nucleation process. Finally, clarifying the mechanism of amyloid formation by hCT and controlling this process would improve the therapeutic use of calcitonin.

12.4. THE "PREFIBRILLAR" STATE OF HCT

The double-nucleation model [28–30] indicates that the first step requires the formation of a critical nucleus, which may transform into a stable aggregate with negligible back dissociation. Before formation of the critical nucleus, the association of molecules is homogeneous (i.e., there is no preferential direction for aggregation). When critical nuclei are formed, possible changes on the molecular surface limit the number of potential sites for the binding of new molecules. The aggregate will now grow bi-directionally to yield protofibrils, which interact to generate a variety of higher order assemblies: protofibril-ribbons, fibrils, multistranded fibril-ribbons, tubes, transition bundles, and multistranded cables [93]. The protofibril growth and the subsequent building of complex fibrillar structures represent the heterogeneous fibrillation, the second step in the double-nucleation model. Several questions arise: First, does hCT exist in a "prefibrillar" state? Second, how many hCT molecules are required to make up a critical nucleus in the early stages of fibrillogenesis? Third, are prefibrillar helical aggregates of hCT stable? Answering these and related questions will likely clarify the structural determinants that are involved in hCT fibrillogenesis.

The first question can be tackled by investigating the possible presence of the preliminary assemblies of hCT by size exclusion chromatography at pH 3.0 and 7.2 and by comparing the results with the aggregation-resistant analogue sCT.

The apparent molecular weights of hCT and sCT at non-fibrillating concentrations [31] were measured by gel filtration (Fig. 12.2) on a Sephadex G-50 column, using as eluents 20 mM phosphate containing 100 mM NaCl (pH 7.2), and 20 mM acetate containing 100 mM NaCl (pH 3.0). At acidic pH (Fig. 12.2A), sCT (circles) eluted with an apparent molecular mass of 6.6 kDa (as compared with bovine pancreatic trypsin inhibitor (BPTI), 6.6 kDa, squares) instead of 3.4 kDa, as expected from its amino acid composition. On the contrary, hCT (lozenges) eluted with an apparent molecular mass corresponding to that of the insulin B-chain (3.5 kDa, triangles). The above data indicate that the two analogues behave differently: sCT elutes as a dimer while hCT appears as a monomer. As for the two standards, both elution patterns are symmetric, implying the presence of a single species in solution.

At pH 7.2 (Fig. 12.2B), sCT shows an elution profile (circles) with a predominant component corresponding to the molecular weight of a dimer, with a slight tail due to the presence of a second minor component corresponding to sCT monomer. Such an asymmetry of the eluted peak is characteristic of a dissociating system [102], suggesting an equilibrium between the two forms strongly favoring the dimer. The elution pattern of hCT (lozenges) is asymmetric, with a sharper front edge between monomer and dimer, and a tail region showing a distinct second component corresponding to hCT monomer, also suggestive of a slow equilibrium between the two forms. These results well reproduce those reported at slightly different pH values [103]. By using guanidine hydrochloride unfolding experiments, we demonstrated that hydrophobic interactions stabilize the dimeric structure of both hormones [103]. Analogously, guanidine hydrochloride denaturation profiles for both hCT and sCT dimers (not shown) here confirm the relevance of hydrophobic interactions for the formation of the dimer at both acidic and physiological pH values. Recently, an elution size exclusion chromatographic profile corresponding to a dimeric assembly has been observed at pH 7.2 [104].

Taken together, the above results indicate that the hCT prefibrillar state is monomeric at pH 3.0, and in a dimer-to-monomer equilibrium at pH 7.2, while sCT is

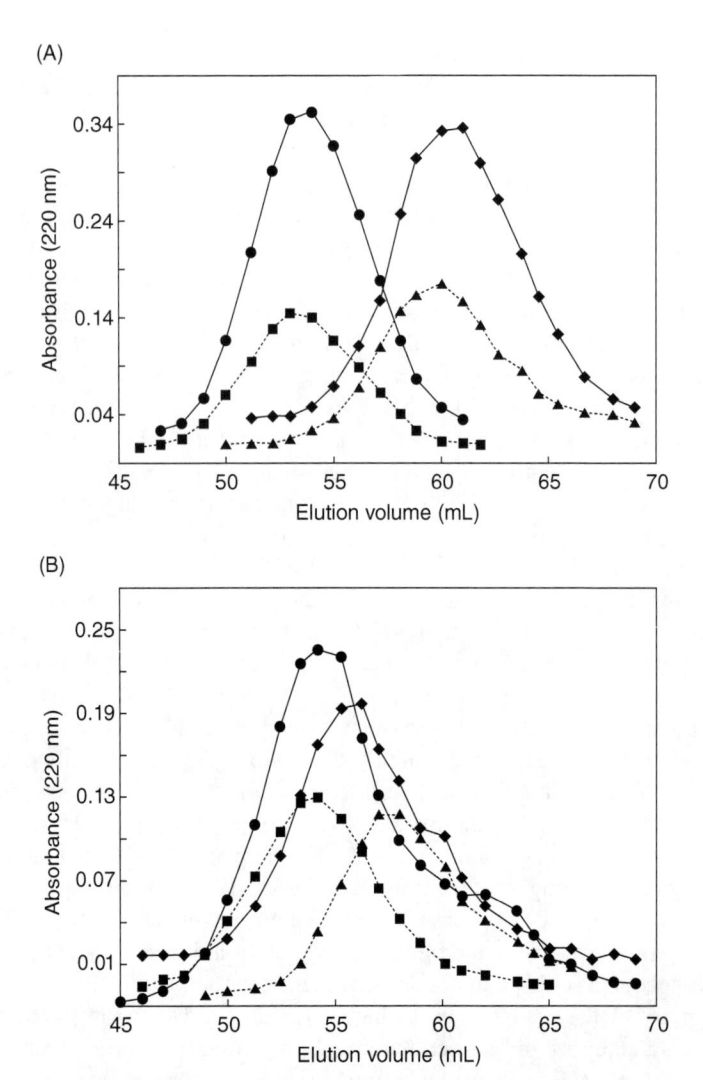

Figure 12.2. Molecular sizes by gel filtration of hCT and sCT. (A) Elution profiles for hCT (lozenges, 0.040 mM, 0.13 mg/mL) and sCT (circles, 0.42 mM, 1.43 mg/mL) in 20 mM acetate, 100 mM NaCl, pH 3.0. (B) Elution profiles for hCT (lozenges, 0.016 mM, 0.054 mg/mL) and sCT (circles, 0.42 mM, 1.43 mg/mL) in 20 mM phosphate, 100 mM NaCl, pH 7.2. Insulin B-chain (triangles, 3.5 kDa) and BPTI (squares, 6.6 kDa), used as molecular size standards, are also reported. Measurements were carried out at room temperature, using a 1.5 × 50-cm Sephadex G-50 fine column at a flow rate of 0.3 mL/min. Peptide concentrations were determined by ultraviolet absorption spectroscopy using coefficients at 275 nm of 1531 and 1515 cm^{-1} M^{-1} for hCT and sCT, respectively.

dimeric at both pH values. According to guanidine hydrochloride denaturation profiles, hydrophobic interactions stabilize the dimeric structure of both hormones. Furthermore, the presence of a monomer–dimer equilibrium for hCT at physiological pH, when fibrillation is faster [31], suggests a dynamic evolution of hCT not observed for sCT.

12.5. HOW MANY MOLECULES FOR THE CRITICAL NUCLEUS?

The kinetics of hCT fibrillation can be followed by fluorescence measurements in a thioflavin T (ThT) binding assay. The lag time, also known as fibrillation time t_f (the time during which the solution remains clear), depends upon several parameters, including temperature, pH, ionic strength, and polypeptide concentration [93, 105]. We have characterized the dependence of hCT fibrillation time on the initial concentration (C_i) of the hormone for values ranging from 1 (3.2×10^{-3} mg/mL) to 40 (12.8×10^{-2} mg/mL) μM. The use of such low concentrations was found to extend the lag-phase period, therefore enabling an accurate characterization of the kinetics of the early stages of the fibrillation process. Each concentration was examined by monitoring three independent fluorescence measurements, and the results are reported in Figure 12.3A, which shows the ThT fluorescence binding values as a function of the time. At low concentrations, virtually no fibrils formed; upon concentration increase, fibrils were produced more rapidly, implying that the hCT concentration affects the length of the lag time. The time length of lag-phase (t_f) for each concentration was derived as in Figure 12.3B, namely by a linear interpolation of the initial slope of the ThT-versus-time curves (Fig. 12.3A), which represents the growth phase in the fibrillation process [31, 104]. The dependence of the fibrillation time t_f on the initial hCT concentration is depicted in Figure 12.3C. As expected, t_f decreased from 280 to 0.8 hours upon increasing the hCT concentration from 1 to 40 μM (Table 12.3). In a double-nucleation model [28, 31, 104, 105], the fibrillation time–concentration dependence should be fitted with a straight line in a logarithmic representation. In a log-log plot we did observe a linear dependence between the decimal logarithm of the reciprocal lag-time length and the decimal logarithm of the corresponding hCT concentration, therefore confirming that the hCT fibrillation process can be explained by a heterogeneous nucleation mechanism. According to the model, the critical nucleus size equals twice the slope of the fitting linear function [29, 30]. For our system, we obtained the following equation $y = 1.47x - 5.68$ ($R^2 = 0.98$), and from the calculated slope of 1.47 we derived that the critical nucleus contains $2 \times 1.47 = 2.94$ (~3) hCT monomers. Following the same procedure on a smaller data set, a value of 1.3 has been reported [104], which again hints at the presence of three hCT monomers in the fibril nucleus.

From turbidity measurements of higher concentration (0.5–40 mg/mL) solutions, it was derived that the critical nucleus consists instead of four molecules [32, 105]. Electron microscopy photographs of hCT solutions at 0.5 mg/mL concentration indirectly confirmed this finding, revealing the predominant presence of filaments with an apparent width of ca. 4 nm as the smallest ever observed [93]. In addition,

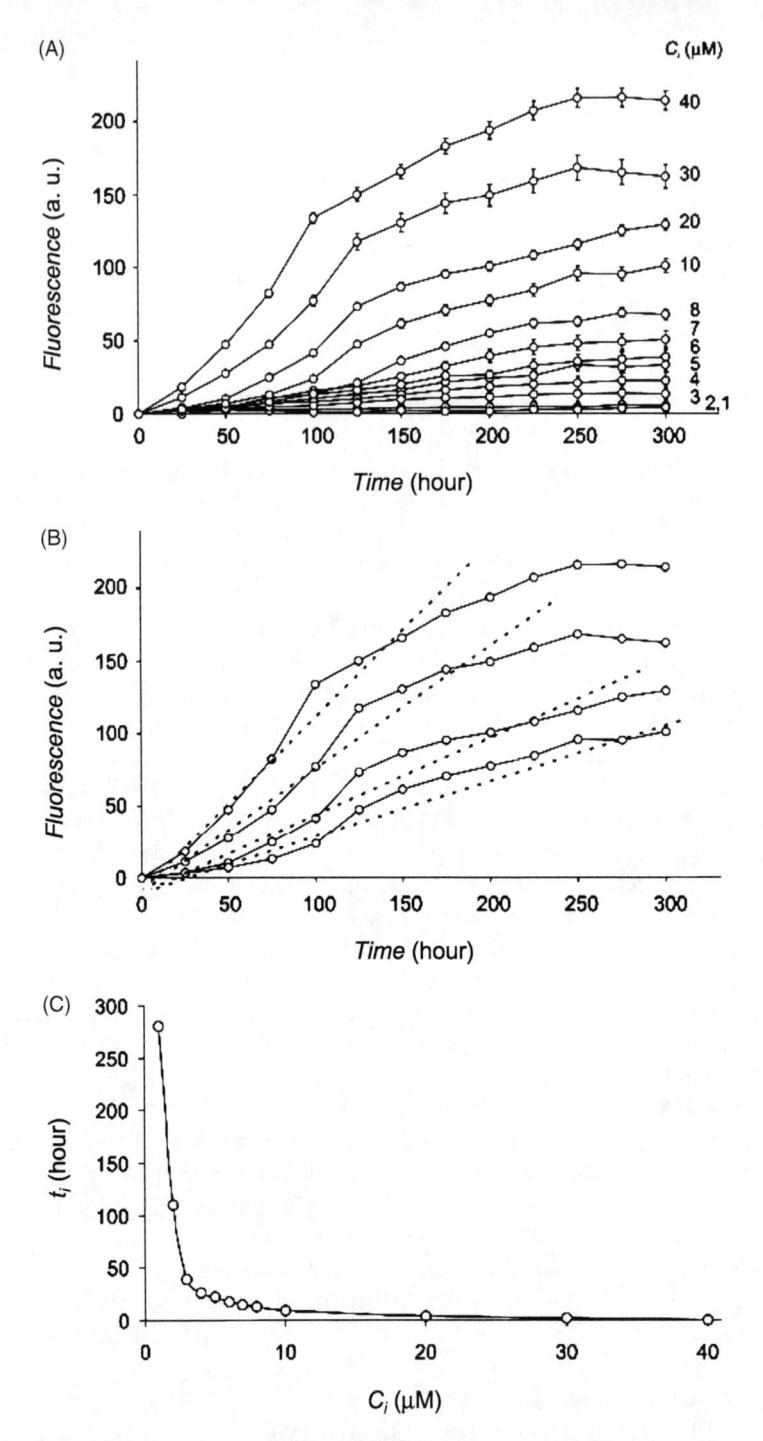

Figure 12.3. Kinetic characterization of hCT fibrillation. (A) A ThT binding assay was performed in the range of 1–40 μM. Each point represents the average value of three independent experiments; standard deviations are also reported. The initial concentration values (C_i) are reported on the right side of each curve. (B) The time length of the lag phase at each concentration was obtained from the intercept of the linear fitting of the initial slope (representing the fibrillation growth phase) of each curve with the time axis. (C) Dependence of the fibrillation time (t_f) on the initial concentration (C_i) of hCT. t_f at each concentration was determined as in (B). Fluorescence values were measured using a Fluoromax 3 Fluorometer (Jobin Yvon Horiba, Tokyo, Japan), excitation at 450 nm, and emission at 480 nm.

TABLE 12.3. Length of the Lag Phase (t_f) at Different Initial Concentration (C_i) of hCT

t_f (hour)	C_i (μM)
280	1
110	2
39	3
26	4
22	5
18	6
15	7
13	8
9	10
4	20
2	30
0.8	40

a plot of the pitch versus the diameter of the fibrils gives an intercept value of 4.1 nm, which corresponds to the minimum width observed for the protofibrils. Therefore, the 4-nm protofibrils always form first and can be considered the thinnest and stable structural building block of all polymorphic hCT assemblies. As a conclusion, it was proposed that the initial aggregate may contain four hCT monomers per cross-section [93, 105], although accommodation of three molecules is also possible [93]. The experimental data described above indicate that the critical nucleus of hCT may contain either a trimer or a tetramer, depending on the initial concentration of the solutions under investigation.

According to the above considerations, the double-nucleation model for the hCT fibrillation process can be illustrated as follows (Fig. 12.4). The first step is the homogeneous nucleation in which hCT monomers transiently interact, possibly forming dimers and aggregates of increasingly higher order. They may mature into a critical nucleus that, depending on the initial hormone concentration, may be composed of trimers and/or tetramers. Since dissociation back to monomers is the principal reaction occurring in solution, several reaction steps may take place until a stable nucleus is formed; but when the stable nucleus is formed, dissociation back to monomers becomes negligible. The transitory molecular association is "homogeneous" in that aggregation is random, without any preferential direction. After the critical nucleus is formed, the aggregate will grow "heterogeneously" in two directions to yield protofibrils, which, upon lateral interactions, form fibrils, cables, and bundles [93]. The protofibril growth and the assembly of the following complex fibrillar structures represent the heterogeneous second stage in the double-nucleation model.

Since the critical nucleus is made of a trimer and/or a tetramer, it is therefore reasonable to assume a dynamic situation in which more than one species is present in solution. This was also suggested by an investigation of small, metastable

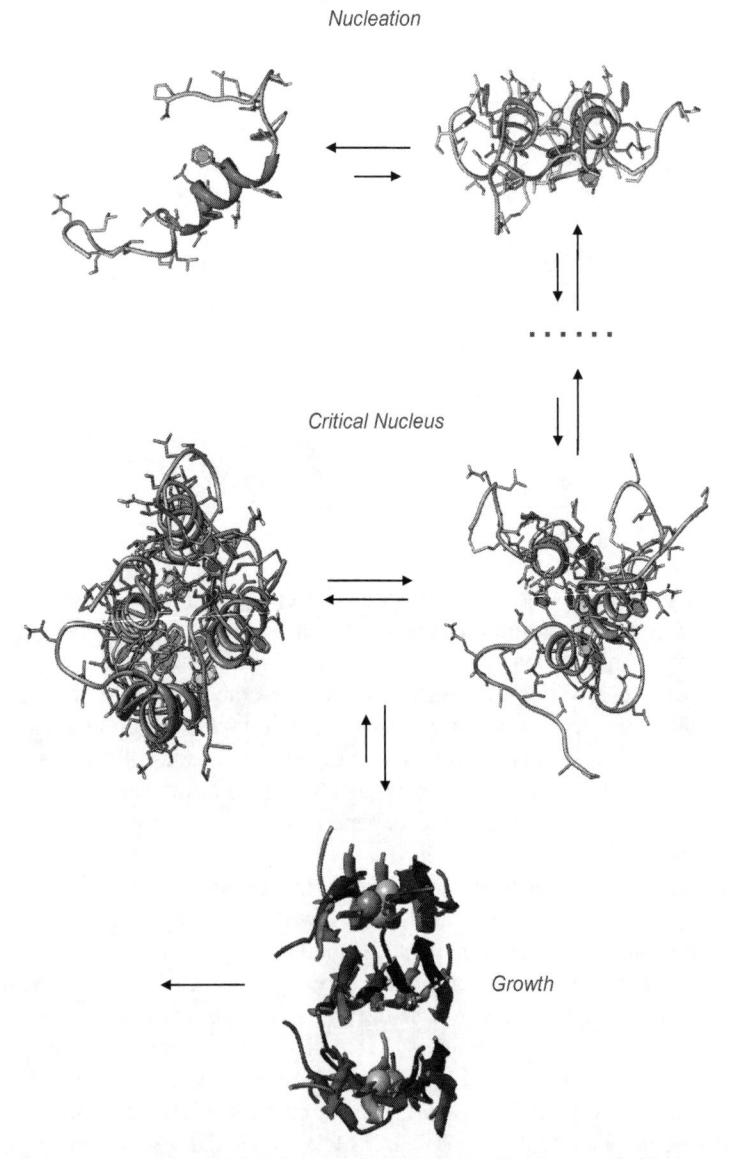

Figure 12.4. Double-nucleation fibrillation model for hCT. The first step is the homogeneous nucleation in which hCT monomers transiently interact, possibly forming dimers and aggregates of increasingly higher order. The dissociation back to monomers slows the formation of a critical nucleus, but when the stable nucleus is formed, dissociation back to monomers becomes negligible. After the critical nucleus is formed, the aggregate will grow "heterogeneously" in two directions to yield protofibrils. The protofibril growth and the assembly of the following complex fibrillar structures represent the heterogeneous second stage in the double-nucleation model. See text for further details. See color insert.

oligomers of hCT trapped by rapid photochemical cross-linking, which revealed the presence of a monomer-to-hexamer species distribution with percentages of ca. 24%, 17%, 29%, 18%, 9%, and 2%, respectively [106]. Although the cross-linked trimer is the most populated oligomer, a safe interpretation of the data does imply that at least the dimer-to-tetramer species (which account for 89% of the distribution) have comparable relevance in the aggregation process. Dynamic light scattering data also indicated that many types of intermediates are formed and coexist during the early stages of the hCT fibrillation process, although data analysis pointed toward the predominance of 9-kDa molecules, corresponding to an hCT trimer [104].

The distribution of hCT oligomers can be analyzed by a simple mathematical model (a "toy model"), previously applied to describe the metastable aggregates of Aβ protein [106]. Assuming that the molecules behave like monomeric spheres interacting through random elastic collisions, we can obtain the distribution of assemblies at any time point during the oligomerization process [106, 107]. If only an irreversible dimerization reaction takes place, aggregation may progress when dimers add monomers to yield trimers, or other dimers to form tetramers, and conceivably aggregates of increasing order by a continuous addition of the basic monomeric or dimeric unit. Since third- or higher order reactions are rarely occurring [108], only second-order reactions will be considered, and, consequently, all of the products are formed through bimolecular reactions. For such a system, the concentration of the nth-order oligomer O_n changes with time according to Ref. [108]:

$$\frac{d[O_n]}{dt} = \sum_{i=1}^{n-1} k_{i,n-1}[O_i][O_{n-1}] - \sum_{i=1}^{\infty} k_{i,n}[O_i][O_n].$$

If we consider that monomers are the only species present at the beginning of the reaction, the distribution of oligomers O_n at a given time point is a function of the initial monomer concentration and the binding constants $k_{i,n}$. By using an identical constant for all reactions occurring during the lag-phase, integration of the above equation yields a series of theoretical distributions of aggregates in the prefibrillar state [106]. The approximation on the rate constant is based on the fact that the formation of larger aggregates is a balance between the enhanced binding efficiency of the increasing oligomer size (more binding sites are present on the molecular surface) and the reduced diffusion coefficients (the reduced molecular motions due to increased molecular size make intermolecular interactions less probable) [106]. As a result, the counteracting actions significantly affect the rate constants, and the calculated distribution values have to be considered with care. Evaluation of the nucleation reaction rate for hCT was carried out according to a reported kinetic fibrillogenesis theory [31, 92, 109], and the result was $1.86 \pm 0.07 \times 10^{-6} S^{-1}$. With this value, we obtained a distribution of species covering a monomer-to-hexamer interval with the following percentages: 28%, 15%, 35%, 14%, 5%, and 3%, which well correlate with the distribution of oligomers trapped by rapid photochemical cross-linking of hCT [106].

The above considerations strongly suggest that the early stage of hCT fibrillation is characterized by the presence of a population from monomer through hexamer species, with the monomer-to-tetramer accounting for 92% of the oligomers. By size exclusion chromatography, we have found that hCT is in a monomer–dimer equilibrium, and from fluorescence measurements in ThT binding assays, we and others [104] derived a trimeric assembly for the critical nucleus. At the used concentrations (10^{-3} to 10^{-2} mg/mL), we were unable to experimentally detect the hCT tetramer, and this could imply a slow rate for the trimer-to-tetramer conversion, and/or the instability of the tetrameric assembly. However, the finding of the tetrameric species at a higher concentration (mg/mL) [93, 105] indicates that the rate constant for the trimer-to-tetramer reaction is smaller than those of the monomer-to-dimer and the dimer-to-trimer aggregation steps. Analogously, the experimental absence of higher order oligomers (pentamers, hexamers, etc.) was linked to a concentration effect and the increased molecular size of the aggregates, because the lag-phase length is proportional to $C^{n/2}$, where C is the monomer concentration and n is the nucleus size [29, 30].

The above considerations suggest that dimer-to-tetramer assemblies essentially represent the prefibrillar aggregates.

12.6. MODELING PREFIBRILLAR AGGREGATES

A model for fibril formation suggests that the early step of the process is characterized by the existence of micelles in rapid equilibrium with monomers [16, 103]. Such micellar aggregates provide domains of high local concentration in which fibril nuclei can form. hCT micelles are most likely to originate from an α-helix bundle [91] in which an α-helix or a helical-like region can act as a "seed" for protofibrillar aggregates formation [36, 37, 103]. To gain insights into the early aggregates of hCT, we modeled the structures of the dimer-to-hexamer oligomers and assessed their stability by using molecular dynamics (MD) simulations. Pre-association of hCT and the final fibril structure has been reported to depend upon the pH of starting solution [92, 94, 103] through favorable electrostatic interactions of the Asp^{15}, Lys^{18}, and His^{20} side chains. However, the absence of a negative charge at Asp^{15} does not alter the ability of fibril formation, while the positively charged side chains of Lys^{18} and His^{20} only delay fibril maturation [110]. This suggests that some other local factor or factors are involved in the early-stage assembly. Since in the aggregation-resistant sCT Lys^{18} is conserved and Asp^{15} is homologously mutated into Glu^{15}, we concentrated on His^{20} (substituted for Gln in sCT), investigating its role in π-stacking with other aromatic side chains. Therefore, MD simulation was used to evaluate the role of aromatic residues in providing a structural framework for the interaction of critical core residues.

12.7. HCT HELICAL OLIGOMERS

Early stages of hCT aggregation involve conformations with an appreciable α-helical content [31, 94, 95, 105], in line with the conformational tendencies exhibited by hCT [99, 100, 111]. Accordingly, we tested the stability of the helical dimer-to-hexamer oligomers at pH 7.2, when the fibrillation process is faster than at pH 3.0 [31]. All assemblies used the monomeric hCT helical structure previously reported

[111] and were built by starting from an anti-parallel dimer, because such an assembly, stabilized by Leu hydrophobic interactions, was found for sCT [103]. A plausible starting model for oligomerization was based on the following considerations. The most prominent hCT species detected by rapid photochemical cross-linking [106] and confirmed by our calculated species distribution are monomers and trimers, while a dimer and a trimer were experimentally detected at low concentration; a tetramer was observed at sufficiently high concentration. Therefore, we considered a progressive addition of helical monomers to the basic dimer, to yield a trimer, a tetramer, a pentamer, and a hexamer. We also considered the addition of a dimer to a dimer for the tetramer, and a further dimer for the hexamer. No difference in the final structures of the aggregates was observed with the procedure of the single monomer addition. In all models, the helix content varied during the simulations, and, on average, we observed a lower helical content than that detected in sodium dodecyl sulfate (SDS) [100, 111] or methanol [112]. Such a structural fluctuation is linked to the flexibility of hCT for which we found evidence of a structure interconverting between an extended chain and a sequence of turns located in the central region [103]. Indeed the angular order parameter (S) of ϕ and ψ dihedral angles indicates that hCT exhibits a higher flexibility with respect to sCT, more uniformly distributed along the polypeptide chain, including the helix, and is associated with a rather continuous distribution of conformers with similar energy [100].

The dimer (Fig. 12.5) exhibits an interhelical angle of 157.6°, with the helix formed in the region 14–22 (monomer *N1*, panel A) and 11–22 (monomer *C2*, panel A). For all simulated oligomers, we observed an average helix length comprising residues 11–22, with one monomer often unfolding the first helical turn in the N-terminal region (monomer *N1*, panel A). The interhelix core is occupied by the two His[20] side chains (Fig. 12.5A, side chains labeled with sequence number), which are involved in a reciprocal intermolecular aromatic stacking interaction. This is further stabilized by intermolecular hydrogen bonds with Asn[17] side chains and by a π-hydrogen bond with Phe[16] side chain for each His residue. Therefore, an interhelix aromatic pattern, Phe[16](*N1*)-His[20](*C2*)-His[20](*N1*)-Phe[16](*C2*), is observed (Fig. 12.5B), with the four aromatic side chains (labeled *16* and *20*) that face the internal side of the helix, and His–His and Phe–Phe intercalating in a parallel-displaced geometry. Interestingly, clustering of side chains forms a hydrophobic face (Fig. 12.6A) involving the helix–helix interface, where aromatic stacking (Fig. 12.6A) plays an essential role. The N- and C-terminal regions also interact, forming a cluster of polar residues (Thr[6], Gln[2], and Thr[25]) stabilized by side chain–side chain or side chain–backbone hydrogen bonds. However, these interactions fall in substantially unstructured regions, and, as such, they appear to fluctuate dynamically in bound–unbound states.

The above hydrophobic surface is a natural candidate for further addition of a monomer to yield a trimer. Accordingly, the most stable structure corresponded to the addition of an hCT molecule on this face (*N3* shaded monomer, Fig. 12.6B), with the aromatic side chains pointing toward the ring cluster of the dimer. Although insertion of the side chains of the third monomer determines a partial rearrangement of the interhelix interactions, the hydrophobic surface described in Figure 12.6A remains essentially unchanged. The trimer (Fig. 12.7) forms interhelix angles of

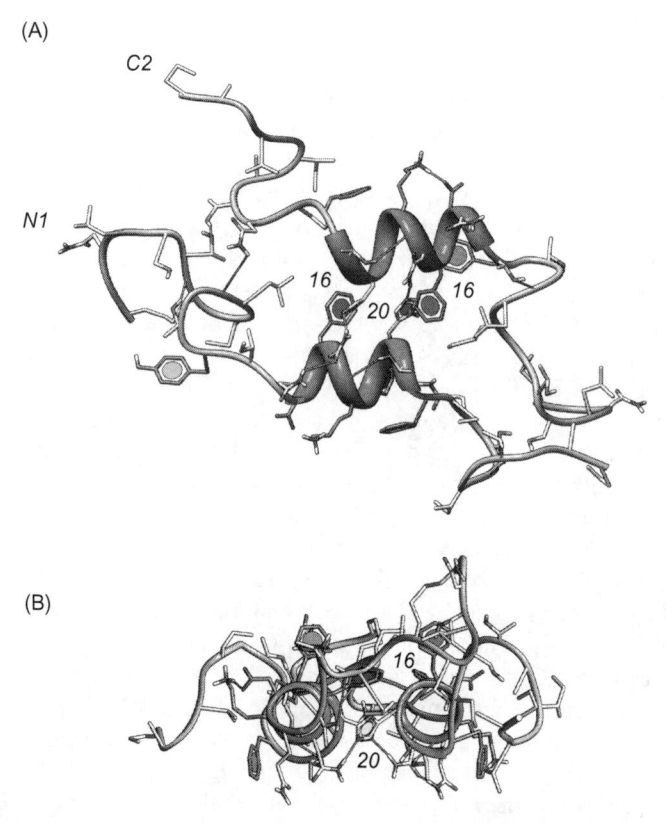

Figure 12.5. Representative structure of an hCT dimer as obtained from molecular dynamics simulation. A ribbon plus side chain representation of perpendicular (A) and parallel (B) views to the helical axes are drawn. Side-chains heavy atoms are shown as sticks, and aromatic rings are shown as disks. Where space permits, aromatic side chains are labeled with sequence numbers. *N1* and *C2* mark the N- and C-termini, respectively, of the two monomers. Peptide assemblies were manually built by using the MOLMOL program [155]. Each multimer underwent energy minimization with NAMD 2.7b4 [156] package using Charmm22 force field [157], and after being subjected to molecular dynamics simulations in solvent. Molecular graphics images were produced using the UCSF Chimera package [158] from the Resource for Biocomputing, Visualization, and Informatics at the University of California, San Francisco (supported by NIH P41 RR001081).

160.2°, 156.9°, and 28.6°. Surprisingly, the trimer preserves the central aromatic core observed in the dimer; it is further reinforced by the additive aromatic stacking due to the third monomer, and such interactions are responsible for the overall stability of the aggregate. In particular, the aromatic core (circled in Fig. 12.7B) is formed by two out of three His[20] residues, three Phe[16], two Phe[19], and one Tyr[12] (monomer *C2*). The third His[20] (monomer *N3*) and the third Phe[19] (monomer *C2*) form a more isolated intermolecular aromatic stack (lower part of the circled region), while one of the remaining two Tyr[12] (opposite view of Fig. 12.7B, monomer *N1*) is also partly interacting with the aromatic core. As found for the dimer, the formation of hydrogen

HCT HELICAL OLIGOMERS 369

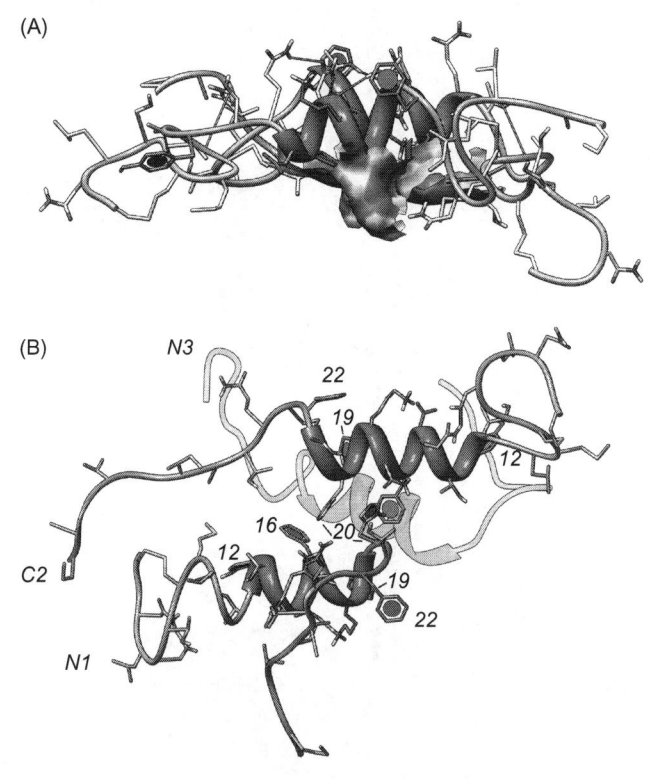

Figure 12.6. Modeling the structure of an hCT trimer by adding a monomer to the simulated dimer. Perpendicular view (A) of the previous dimer depicting the aromatic cluster surface of the helix–helix interface; perpendicular view (B) of the trimer in which the monomer added face-to-face to the hydrophobic surface depicted in (A) is shaded. Sequence numbers label aromatic side chains. *N1*, *C2*, and *N3* mark the N- and C-termini of the three monomers.

bond of each His[20] imidazole ring with the corresponding Asn[17] residue, and a π-hydrogen bond with the related Phe[16] side chain, is also present in the trimer. However, compared with the dimer, these interactions are weaker and fluctuate more freely during the MD simulations.

Addition of a monomer to the lowest energy trimeric structure forms a tetramer (Fig. 12.8), with interhelix angles of 174.7°, 157.8°, 173.9°, 166.4°, 9.3°, and 17.3°. The structure preserves the interhelical interactions observed for isolated dimers, suggesting that the tetramer can also be considered a "dimer of a dimer." His[20] rings (drawn in *magenta* in Fig. 12.8B) take up a pivotal role for the aromatic stacking, which is reinforced by the formation of hydrogen bond between each imidazole and the corresponding Asn[17]. Such interactions occur between monomers *N1* and *C2*, and *N3* and *C4*; furthermore, Phe[19] of a monomer faces Phe[19] and Phe[16] of the other in the pair *C2* and *N3*, as well as in the pair *N1* and *C4*. A stacking between Phe[16] residues belonging to monomers *N1* and *N3* also stabilizes the hydrophobic core of the tetramer, while all Tyr[12] side chains, located at the beginning of each helix, act as a cap on both sides of the aromatic core, being oriented toward the internal channel of the tetramer.

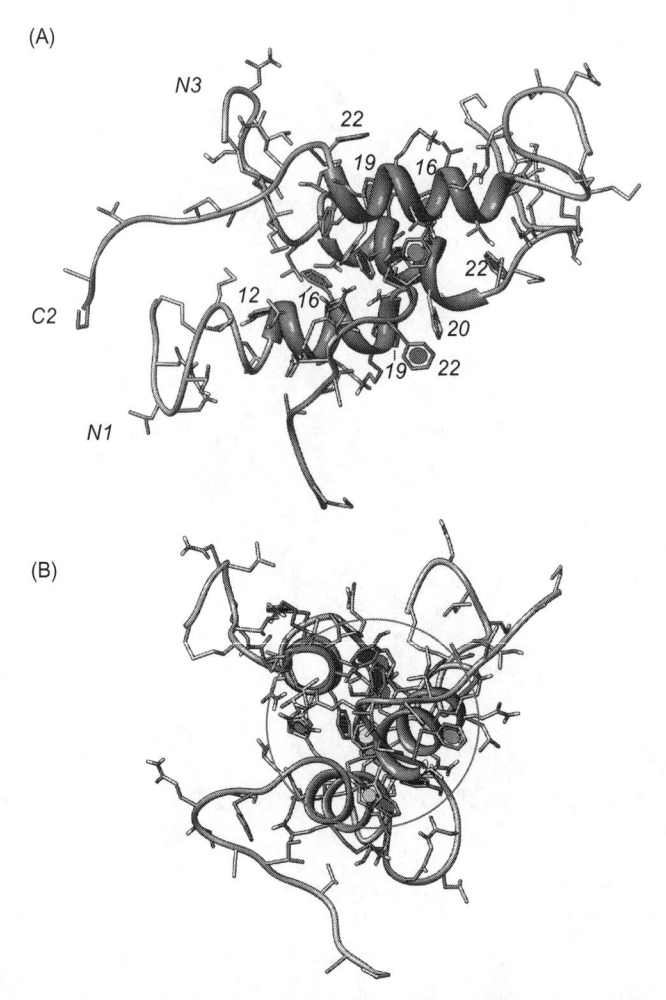

Figure 12.7. Representative structure of an hCT trimer. Perpendicular (A) and parallel (B) views to the helical axes. Sequence numbers label aromatic side chains, and *N1*, *C2*, and *N3* mark the N- and C-termini of the three monomers. In (B), the aromatic clusters are circled. Side-chains heavy atoms and polar hydrogens are shown as sticks, colored according to their atomic types, except for aromatic, His, basic, and acid residues, whose carbon atoms are painted orange, cyan, mid-blue, and pink, respectively. Aromatic rings are shown as disks colored according to the carbon atoms of the residue; side chain-side chain hydrogen bonds are depicted as green thin sticks. Ribbons are painted according to secondary structure: magenta for helix, gray for nonhelical, nonsheet regions. Where space permits, aromatic side chains are labeled with sequence numbers. *N1*, *C2*, and *N3* mark the N- and C-termini of the three monomers. See color insert.

HCT HELICAL OLIGOMERS 371

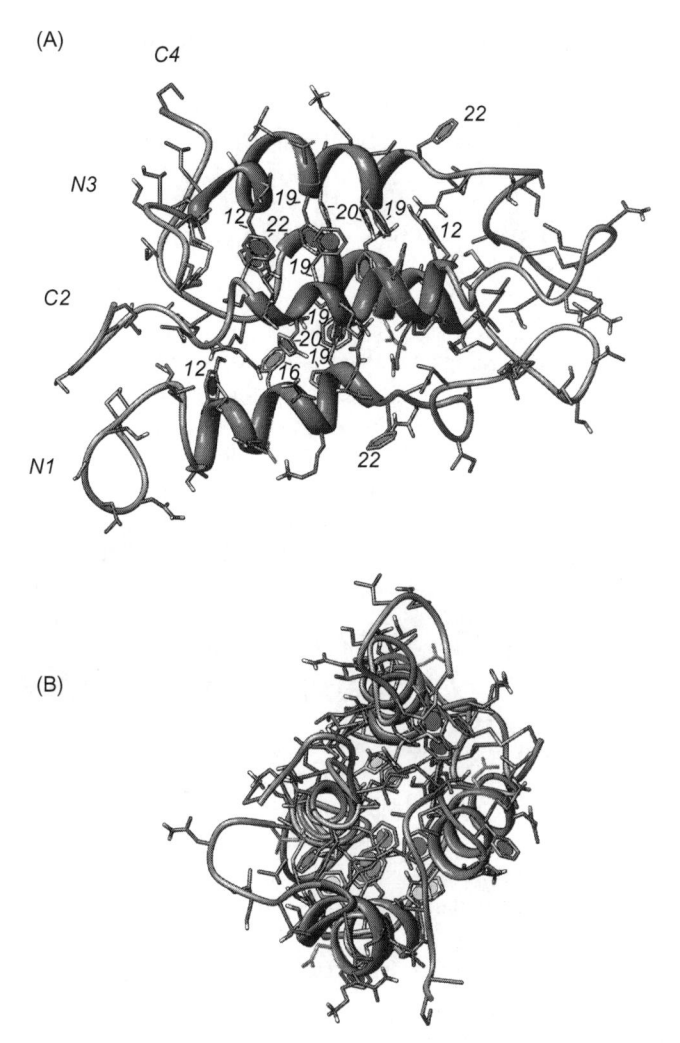

Figure 12.8. Representative structure of an hCT tetramer. Perpendicular (A) and parallel (B) views to the helical axes. Where space permits, sequence numbers label aromatic side chains, and *N1*, *C2*, *N3*, and *C4* mark the N- and C-termini of the four monomers.

The final structures of all modeled hCT aggregates suggest that His[20] can efficiently drive the aggregation of the sampled oligomers, forming a His–His ring stacking that somehow orients the assembly. The oligomer is further stabilized by intermolecular hydrogen bonds formed by the N3 atoms, which are electrophile and acceptor for hydrogen bonding in a nonionized His side chain (observed at physiological pH). In addition, the other aromatic residues form large clusters of stacked rings, which generate a stable hydrophobic core in the aggregates. The complementarity of aromatic residues is optimal in the tetramer, where parallel-displaced stacking occurs all along the central channel of the oligomer, but such a packing hampers the access of the aromatics of other monomers to build pentamers and hexamers. The prevalence of aromatic side chains in hCT over Leu amino acids found in sCT explains the differences in aggregation properties of the two hormones.

12.8. THE ROLE OF AROMATIC RESIDUES IN THE EARLY STAGES OF AMYLOID FORMATION

Amyloid fibrils are found in many different fatal diseases, each characterized by a specific polypeptide that aggregates into insoluble amyloid fibrils. They form toxic deposits in tissues, where they manifest the pathological effects of the disease. Although the different polypeptides do not share any obvious sequence homology, functional fragments of amyloid-forming peptides do show a remarkable regularity of aromatic residues, which are thought to play a primary role in fibrillogenesis [96]. Their importance in fibril stability has been recognized: aromatic side chains, especially Phe, are closely interlocked in mature fibrils [26, 94] by taking up a parallel-displaced ring geometry that dominates interactions between adjacent stands, whereas a T-shape arrangement is observed for aromatic ring interactions between sheets [26, 113]. Similar studies on small synthetic peptides that self-assemble and form amyloid fibrils *in vitro* [3, 114–117] confirmed that aromatic residues clearly play an important role in aggregation.

Do aromatic rings play a role also in the early stages of amyloid formation? Data on the islet amyloid polypeptide [118, 119] have suggested that Phe residues may accelerate aggregation, which would imply that ring interactions should take place before aggregation, and afterward stabilize the mature fibril. Stacking of aromatic rings immediately before the appearance of insoluble fibrils was indeed documented by near-ultraviolet (UV) circular dichroism (CD) spectra of islet amyloid polypeptide [120].

hCT presents aromatic residues at sites Tyr^{12}, Phe^{16}, Phe^{19}, His^{20}, and Phe^{22}, which become Leu^{12}, Leu^{16}, Leu^{19}, Gln^{20}, and Tyr^{22} in sCT (Fig. 12.1). Formation of transient complexes in hCT is characterized by the presence of α-helices [31, 94, 95, 105], and our MD simulations of the hCT aggregates did confirm that the region between residues Thr^{11} and Phe^{22} is a helix comprising all the aromatic residues. A possible mechanism for helix–helix interaction is coiled-coiling of adjacent strands [121, 122]. In coiled-coils, the bundle motif is exemplified by the heptad repeat $(abcdefg)_n$ containing hydrophobic residues at positions *a* and *d*, and polar residues generally elsewhere. Leu (33%) is the predominant residue in the *a* and *d* positions of the motif [121, 123], while Phe and Tyr have low and moderate occurrence, respectively [124]. Applying the COILS software to sCT and hCT amino acidic sequences (http://www.ch.embnet.org/software/COILS_form.html) [125, 126], we found that sCT has an 85% probability of forming a coiled-coil in the region Leu^9–Thr^{21} through intermolecular Leu–Leu contacts, as experimentally found [103]. On the contrary, no hCT region has the potential of forming a coiled-coil [103], because, except for Leu^9 (an *a* site), the other *a* and *d* positions are occupied by aromatic residues (Fig. 12.1). Therefore, helical interactions in hCT most likely rely upon Phe–Phe interactions, reported to be the most common in proteins [127], and responsible for amyloidosis of short hCT peptides based upon the 15–19 region, which form fibrils if Phe^{16} and Phe^{19} are preserved [115].

12.9. THE FOLDING OF HCT BEFORE AGGREGATION

Peptides that form helices in solution go through a complex mixture of conformers, frequently taking up central helices with frayed ends [128]. Accordingly, their folding can be described by using the so-called nucleation–propagation models [129–131]. They assume a nucleation-growth mechanism in which the first step is the formation of a helical turn representing an entropically unfavorable, slow nucleation reaction, to be balanced by favorable enthalpic contributions. Therefore, since three or four amino acids decrease simultaneously their conformational entropy, the nucleation process must overcome a large free energy barrier. On the contrary, the propagation steps are energetically favorable because the loss of the conformational entropy of a single residue is balanced by the energy gained from the formation of one extra hydrogen bond [132–134].

Nuclear magnetic resonance (NMR) spectroscopy and calculations indicate that in structure-promoting media the helical region of hCT is endowed with high mobility and often resembles a sequence of turns [100]. Following the thermal unfolding of hCT in water by NMR, we have observed transient chemical shifts of Phe^{16} and Phe^{19} in the range expected for residues being in a helical conformation. Estimation of the secondary structure from CD spectra in water of hCT by the k2d neural network algorithm [135] suggests a helix percentage of 12% [36], which corresponds to a single helical turn (a regular α-helix has 3.6 residues per turn). The central turn, most likely including residues of Phe^{16} to Phe^{19}, represents the most probable site for helix nucleation, which then propagates toward the two terminal ends. The stabilizing side chain–side chain interactions among Tyr^{12}, Phe^{16,19,22}, and His^{20}, observed in the simulated aggregates, could also play a crucial role in further stabilizing and enlarging the nucleating turn of the central helix. A strict relationship between the helix content and the distance between the Trp and His side chains has been reported for a 21-residue α-helical heteropeptide [136].

Notably, the presence of a helical or turn-like segment within a monomeric polypeptide significantly limits the number of possible chain conformations and hence would provide an effective structural framework for the interaction of critical core residues in the early phases of fibrillation. We therefore may speculate that stable dimer formation of hCT involves the interaction of two helical trigger sites at some stage in the fibrillation pathway. As illustrated schematically in Figure 12.4, such a mechanism ideally would align the dimer in an anti-parallel register, with the interhelical aromatic interactions that participate in the chain recognition and alignment process. Collisions of short helical stretches could be sufficient to bring together local determinants, such as aromatic side chains, which help oligomerization.

Such a mechanism is supported by computer simulations on possible folding pathways of GCN4 leucine zipper peptide [137], which suggested that dimer formation starts from the collision of short helical stretches [138]. Finally, interacting helices then "zip up" along the molecule to form a stable dimeric structure (Fig. 12.5).

Aromatic residues can therefore act as a local factor, affording an energetic contribution that may modulate the preliminary self-assembly process of fibril formation [96, 139]. Therefore, the side-chain interactions result in specific directionality, order, and orientation, which can later mature into a highly ordered, stable fibrils rather than amorphous aggregates. The relative spatial displacement of the two aromatic side chains is critical in determining the preferred angle between the ring planes, and interactions with other side chains can interfere with the π–π stacking [127]. We have suggested [103] that interstrand recognition in amyloid peptides bearing aromatic residues may occur through significant interaction between a hydrogen bond donor and the center of an aromatic ring, which acts as a hydrogen bond acceptor [140]. This interaction, which is about half as strong as a normal hydrogen bond, contributes ca. 13 kJ/mol to the stability. Intermolecular hydrogen bonds have been reported to play an important role in the association of the hCT molecules [91].

12.10. MODEL EXPLAINS THE DIFFERENCES IN AGGREGATION PROPERTIES BETWEEN HCT AND SCT

The formation in hCT oligomers of a hydrophobic core essentially based on aromatic stacking explains the differences in aggregation properties between hCT and sCT. We have observed that sCT is able to dimerize via the hydrophobic face of an amphipathic helix formed by leucines at sites 12, 16, and 19 [103]. In hCT, leucines are substituted for Tyr^{12} and Phe^{16,19} residues (Fig. 12.1). As found in the above simulations, together with His^{20} and Phe^{22}, they drive a specific orientation through aromatic hydrogen bond and π-stacking, and form a homogeneous distribution of aromatic rings along the central region of the aggregates. In sCT, the presence of Leu residues in the central region allows for a different kind of helix–helix packing that largely favors dimeric aggregates versus higher states of oligomerization. Therefore, leucines (but not aromatic residues) in the central region of sCT favor a stable prefibrillar helical dimer that prevents amyloid maturation.

In all the modeled hCT oligomers, MD simulations have clearly shown the role of His^{20} in the π–π aromatic stacking with other His^{20} residues, together with intermolecular hydrogen bonds formed by the His N3 atoms, which at pH 7.2 are electrophile and hydrogen bond acceptors. The other aromatic residues (Tyr^{12}, Phe^{16}, Phe^{19}, and Phe^{22}) participate in the cluster of stacked rings, and the optimal packing is observed in the tetramer, where the stacking occurs symmetrically all along the central tunnel of the oligomer (Fig. 12.8B). Interactions of aromatic residues at the interface increasingly stabilizes dimer-to-tetramer assemblies, while insertion of further hCT molecules within the tetramer to generate pentamers and hexamers appears to be limited by the compactness of the aggregate, and only significant structural fluctuations can allow accommodation of other aromatic side chains. As built, the MD-simulated oligomers, based on progressive addition of monomers to generate hCT dimers, trimers, and tetramers, do interpret the experimental results found for the early stages of hCT fibrillation. Monomers and trimers are the most prominent hCT species detected by rapid photochemical cross-linking [106], and, at the used concentration (10^{-3} to 10^{-2} mg/mL), we experimentally detected for hCT a monomer–dimer equilibrium, a dimer, and a trimer, while at a higher concentration (mg/ml) a tetramer was detected [93, 105]. As found for human apolipoprotein C-II, the presence of higher order oligomers depends on concentration, because an increase in the fractional volume occupancy of macromolecules in a physiological fluid can nonspecifically accelerate the formation of amyloid fibrils [141]. Therefore, evaluation of the hCT fibrillar species present in the early stages should also consider the hormone *in vivo* concentration, which can physiologically increase because of the loss of cellular and tissue water due to advancing age. The effect is a reduction of the "cellular space" with a corresponding increase in the fraction of cellular or tissue volume occupied by solutes [142, 143], and such a physiological macromolecular crowding is expected to correspondingly accelerate fibril maturation [141]. Therefore, the "idyllic" model proposed above (a "basic" dimer that evolves by progressive addition of hCT monomers) is obviously complicated by many factors. However, interaction of aromatic side chains in driving formation of hCT intermediates should be a prerequisite at the beginning of the amyloidogenic process. Therefore, all chemical factors that hamper $\pi-\pi$ stacking would delay the formation of malignant aggregates and delay their progression toward mature fibrils. We have proposed [103] that hCT fibrillation could be blocked by stabilizing a Leu-based helical dimer, which can be achieved by substituting aromatic amino acids for leucine residues at sites 12, 16, and 19, the pivotal sites for dimeric formation in sCT. More recently [36], inhibition of hCT amyloidosis has been obtained by mixing an aggregation-resistant hCT mutant and the wild-type hCT, the analogue being obtained by structure homology with the non-fibrillating sCT. Such a strategy seems to rely on the formation of mixed hCT–mutant oligomers, in which fibrillating intermolecular interactions are hindered [78, 79].

12.11. HCT FIBRIL MATURATION

A proposed model for hCT fibril formation is depicted in Figure 12.9 for pH values 3.3 and 7.5 [92]. Depending on the pH value, intermolecular interactions (mainly electrostatic by the positively and negatively charged side chains) between the hCT molecules in solution (Fig. 12.9A) may favor hCT assembling in an anti-parallel way at pH 7.5, where the positively charged side chains associate in both parallel and anti-parallel ways at pH 3.3 to form α-helical bundles (micelle) (Fig. 12.9B). Subsequently, the α-helical bundles undergo a conformational change to take up an oligomeric β-sheet as the first nucleation step (Fig. 12.9C). The maturation process (Fig. 12.9C,D) induces the formation of long fibrils in the direction of the red arrows

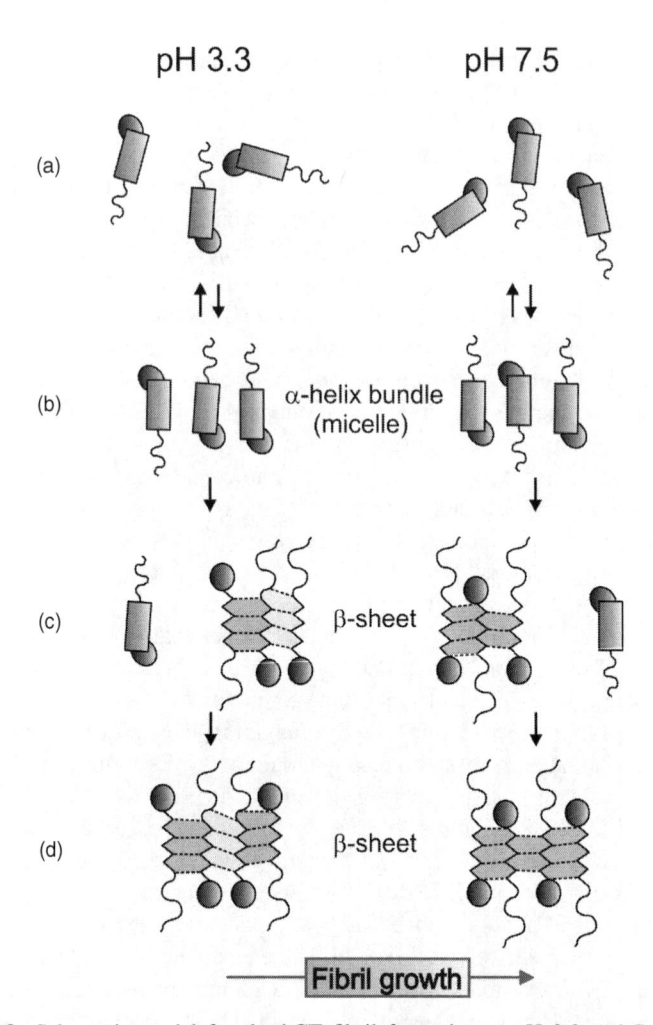

Figure 12.9. Schematic model for the hCT fibril formation at pH 3.3 and 7.5. The hCT monomers (A) associate homogeneously to form the α-helical bundle (micelle) (B). (C) A homogeneous nucleation process favors the β-sheet formation and gives rise to a heterogeneous associating process. Parallel and anti-parallel β-sheets are in yellow and cyan, respectively. (D) Large fibrils are formed through a heterogeneous fibrillation process along the red arrow. The figure has been redrawn and modified from Ref. [93].

indicated at the bottom of Figure 12.9. Based on high-resolution solid-state ^{13}C NMR spectroscopy data, it was proposed [92] that, at pH 7.5, hCT forms a homogeneous fibril with an anti-parallel β-sheet conformation in the central region and a random coil in the C-terminus region (Fig. 12.9C), while at pH 3.3 a mixture of parallel and anti-parallel β-sheet is formed. According to the model, all molecules forming the helical bundle have to simultaneously change the conformation from an α-helix to a β-sheet in the first nucleation process, while in the second heterogeneous fibrillation process, only one α-helix has to be converted to a β-sheet [92]. This explains why the second process is much faster than the first. Such a mechanism has been invoked for the formation of amyloid deposits in Alzheimer's disease through a conversion from an α-helix to a β-sheet [144].

Using ^{13}C-labeled hCT molecules, solid-state NMR spectroscopy revealed that the conformational change from an α-helix to a β-sheet in the process of fibril formation should involve amino acidic stretches around Gly^{10} and Phe^{22} [92]. In particular, in the D15N-hCT mutant, the region converted into the β-sheet structure was located around Phe^{16} [110]. Intermolecular contacts between β-strands are mediated by π–π interactions, as an hCT mutant, whose Tyr^{12} and Phe^{16,19} are replaced by Leu residues, showed a dramatic reduction of the fibrillation rate [145]. This suggests that aromatic residues most likely align to stabilize the hCT parallel/anti-parallel β-sheets by π–π interaction. Phe^{16} and Phe^{19} have been reported to favor fibril formation and stabilization in the model of the region Asp^{15}-to-Phe^{19} [146–148], which is the shortest hCT fragment to form amyloid fibril [114]. Because of the small size of the hCT peptide, an intramolecular parallel β-sheet structure can be excluded, even though a short double-stranded anti-parallel β-sheet made by residues 16–21 and connected by a two-residue hairpin loop formed by residues 18 and 19 was observed in the central region of hCT [149]. However, such a structure was observed in a 15% dimethyl sulfoxide–85% water (v/v) cryoprotective mixture at 278 K. The presence of a β-sheet in such a structure-inducing solvent indirectly confirms the tendency of hCT to assume a β-sheet in its central region.

12.12. α-HELIX →β-SHEET CONFORMATIONAL TRANSITION AND HCT FIBRILLATION

Conformational equilibria of a polypeptide can be elucidated by using principal coordinate analysis [150]. This method has been successfully applied to describe the conformational transitions of the Aβ with 40 amino acids (Aβ40) and Aβ42 peptides [151]. The free energy surfaces obtained for both peptides are characterized by: (1) two large basins, one dominated by extended or β-sheet conformations ("β-basin") and the other dominated by α-helical structures ("α-basin"); and (2) a relatively high number (~20) of minima with comparable free energies, which are separated by low barriers. Consequently, conformational conversions corresponding to migration among these minima frequently take place. On average, the Aβ42 local minima in both basins share comparable free energies, while the local minima in the Aβ40 α-basin are lower than those in the β-basin [151]. In fact, the maximum percentage of α-helix observed for Aβ40 and Aβ42 was 32% and 19%, respectively, with Aβ42 assuming the α-helix percentage appreciably faster than did Aβ40 [152]. During fibril formation of the Aβ peptide, a conformational transition going through a statistical coil → α-helix → β-sheet is observed, and Aβ42 migrates through the path faster than does Aβ40 to fibrillate rapidly. Accordingly, an efficient stabilization that locks the molecule in the so-called α-basin avoids the migration toward the "β-basin," therefore hampering fibril formation [153].

We are currently extending this approach to hCT and sCT, and preliminary results indicate that, as found for Aβ42, the "α-basin" and the "β-basin" are characterized by low barriers for α-helix $\rightarrow$ β-sheet transitions, meaning that a rapid progression toward the β-sheet is possible. On the contrary, a deeper α-helix basin is observed for sCT, meaning that sCT conformers will populate these regions more frequently than those in the "β-basin." In parallel with Aβ40 and Aβ42, the different stability of α-helix-containing conformers of sCT and hCT explains the recognized divergence between the fibril-forming property of the two hormones, and the inherent major flexibility of hCT polypeptide chain [100].

The α-helix $\rightarrow$ β-sheet conversion going through progressively looser helical segments can be hampered by targeting either the intact α-helix with ligand molecules [77] or the structurally independent folding units (turn-like structures) present in the peptide [151, 154].

12.13. CONCLUDING REMARKS

The observations reported here reveal the importance of understanding the early stages of hCT fibrillogenesis to shed light on the structure and dynamics of precursor species and the mechanisms by which they self-assemble. The idea is that targeting such species with inhibitors would significantly hamper amyloidosis by interfering with the early aggregates. Experimental data, both *in vitro* and *in vivo*, indicate that soluble aggregates often undetectable, rather than mature full-length fibrils, are responsible for amyloid-induced cytotoxicity. If protofibril formation is mediated by unstable α-helix or turn-like polypeptide stretches, targeting those regions offers a novel approach to finding specific inhibitors that might be effective *in vivo*. This strategy also includes the use of a mixture of an amyloidogenic polypeptide with an aggregation-resistant analogue, the latter working as an efficient separator that prevents maturation of the early aggregates into fibrils.

An important point would be to understand the interactions between "natively unfolded" peptides and the molecular chaperones and control processes that protect the cell from peptide/protein aggregation. Such processes safely target unfolded or misfolded species, avoiding their aggregation and favoring their folding or degradation. However, for systems like hCT and Aβ peptide, such mechanisms appear to fail. The comprehension of these mechanisms will not only clarify how these polypeptides function both normally in health and abnormally in disease, but will also prompt the design of therapeutic strategies against amyloidosis.

ACKNOWLEDGMENTS

We thank Dominique Melck for the excellent laboratory assistance, as well as Emilio P. Castelluccio and Salvatore Donadio for the skillful computer system maintenance.

REFERENCES

1. Westermark, P., Benson, M. D., Buxbaum, J. N., Cohen, A. S., Frangione, B., Ikeda, S.-I., Masters, C. L., Merlini, G., Saraiva, M. J., and Sipe, J. D. (2007) A primer of amyloid nomenclature, *Amyloid 14*, 179–183.

2. Uversky, V. N., Oldfield, C. J., Midic, U., Xie, H., Xue, B., Vucetic, S., Iakoucheva, L. M., Obradovic, Z., and Dunker, A. K. (2009) Unfoldomics of human diseases: Linking protein intrinsic disorder with diseases, *BMC Genomics 10*, S7. doi:10.1186/1471-2164-10-S1-S7.

3. Tjernberg, L., Hosia, W., Bark, N., Thyberg, J., and Johansson, J. (2002) Charge attraction and beta propensity are necessary for amyloid fibril formation from tetrapeptides, *J Biol Chem 277*, 43243–43246.

4. Chiti, F., Calamai, M., Taddei, N., Stefani, M., Ramponi, G., and Dobson, C. M. (2002) Studies of the aggregation of mutant proteins in vitro provide insights into the genetics of amyloid diseases, *Proc Natl Acad Sci U S A 99*, 16419–16426.

5. de la Paz, M. L., Goldie, K., Zurdo, J., Lacroix, E., Dobson, C. M., Hoenger, A., and Serrano, L. (2002) De novo designed peptide-based amyloid fibrils, *Proc Natl Acad Sci U S A 99*, 16052–16057.

6. Otzen, D. E., Kristensen, O., and Oliveberg, M. (2000) Designed protein tetramer zipped together with a hydrophobic Alzheimer homology: A structural clue to amyloid assembly, *Proc Natl Acad Sci U S A 97*, 9907–9912.

7. Schwartz, R., Istrail, S., and King, J. (2001) Frequencies of amino acid strings in globular protein sequences indicate suppression of blocks of consecutive hydrophobic residues, *Protein Sci 10*, 1023–1031.

8. Chiti, F., Taddei, N., Baroni, F., Capanni, C., Stefani, M., Ramponi, G., and Dobson, C. M. (2002) Kinetic partitioning of protein folding and aggregation, *Nat Struct Biol 9*, 137–143.

9. West, M. W., Wang, W. X., Patterson, J., Mancias, J. D., Beasley, J. R., and Hecht, M. H. (1999) De novo amyloid proteins from designed combinatorial libraries, *Proc Natl Acad Sci U S A 96*, 11211–11216.

10. Villegas, V., Zurdo, J., Filimonov, V. V., Aviles, F. X., Dobson, C. M., and Serrano, L. (2000) Protein engineering as a strategy to avoid formation of amyloid fibrils, *Protein Sci 9*, 1700–1708.

11. Kallberg, Y., Gustafsson, M., Persson, B., Thyberg, J., and Johansson, J. (2001) Prediction of amyloid fibril-forming proteins, *J Biol Chem 276*, 12945–12950.

12. Muchowski, P. J. (2002) Protein misfolding, amyloid formation, and neurodegeneration: A critical role for molecular chaperones? *Neuron 35*, 9–12.

13. Citron, M., Westaway, D., Xia, W. M., Carlson, G., Diehl, T., Levesque, G., Johnson-Wood, K., Lee, M., Seubert, P., Davis, A., Kholodenko, D., Motter, R., Sherrington, R., Perry, B., Yao, H., Strome, R., Lieberburg, I., Rommens, J., Kim, S., Schenk, D., Fraser, P., St. George Hyslop, P., and Selkoe, D. J. (1997) Mutant presenilins of Alzheimer's disease increase production of 42-residue amyloid beta-protein in both transfected cells and transgenic mice, *Nat Med 3*, 67–72.

14. Bence, N. F., Sampat, R. M., and Kopito, R. R. (2001) Impairment of the ubiquitin-proteasome system by protein aggregation, *Science 292*, 1552–1555.

15 Tofaris, G. K., Razzaq, A., Ghetti, B., Lilley, K. S., and Spillantini, M. G. (2003) Ubiquitination of alpha-synuclein in Lewy bodies is a pathological event not associated with impairment of proteasome function, *J Biol Chem 278*, 44405–44411.

16 Lomakin, A., Chung, D. S., Benedek, G. B., Kirschner, D. A., and Teplow, D. B. (1996) On the nucleation and growth of amyloid β-protein fibrils: Detection of nuclei and quantitation of rate constants, *Proc Natl Acad Sci U S A 93*, 1125–1129.

17 Harper, J. D. and Lansbury, P. T. (1997) Models of amyloid seeding in Alzheimer's disease and scrapie: Mechanistic truths and physiological consequences of the time-dependent solubility of amyloid proteins, *Annu Rev Biochem 66*, 385–407.

18 Kusumoto, Y., Lomakin, A., Teplow, D. B., and Benedek, G. B. (1998) Temperature dependence of amyloid beta-protein fibrillization, *Proc Natl Acad Sci U S A 95*, 12277–12282.

19 Zurdo, J., Guijarro, J. I., Jimenez, J. L., Saibil, H. R., and Dobson, C. M. (2001) Dependence on solution conditions of aggregation and amyloid formation by an SH3 domain, *J Mol Biol 311*, 325–340.

20 Jarrett, J. T. and Lansbury, P. T., Jr. (1993) Seeding "one-dimensional crystallization" of amyloid: A pathogenic mechanism in Alzheimer's disease and scrapie? *Cell 73*, 1055–1058.

21 Gazit, E. (2002) The "correctly-folded" state of proteins: Is it a metastable state? *Angew Chem Int Ed Engl 41*, 257–259.

22 Sawaya, M. R., Sambashivan, S., Nelson, R., Ivanova, M. I., Sievers, S. A., Apostol, M. I., Thompson, M. J., Balbirnie, M., Wiltzius, J. J., McFarlane, H. T., Madsen, A. Ø., Riekel, C., and Eisenberg, D. (2007) Atomic structures of amyloid cross-beta spines reveal varied steric zippers, *Nature 447*, 453–457.

23 Wasmer, C., Lange, A., Van Melckebeke, H., Siemer, A. B., Riek, R., and Meier, B. H. (2008) Amyloid fibrils of the HET-s(218–289) prion form a beta solenoid with a triangular hydrophobic core, *Science 319*, 1523–1526.

24 Margittai, M. and Langen, R. (2008) Fibrils with parallel in-register structure constitute a major class of amyloid fibrils: Molecular insights from electron paramagnetic resonance spectroscopy, *Q Rev Biophys 41*, 265–297.

25 Tycko, R. (2006) Molecular structure of amyloid fibrils: Insights from solid-state NMR, *Q Rev Biophys 39*, 1–55.

26 Makin, O. S., Atkins, E., Sikorski, P., Johansson, J., and Serpell, L. C. (2005) Molecular basis for amyloid fibril formation and stability, *Proc Natl Acad Sci U S A 102*, 315–320.

27 Oosawa, F. and Asakura, S. (1975) *Thermodynamics of the polymerization of protein*, Academic Press, New York, pp. 47–55.

28 Hofrichter, J., Ross, P. D., and Eaton, W. A. (1974) Kinetics and mechanism of deoxyhemoglobin S gelation: A new approach to understanding sickle cell disease, *Proc Natl Acad Sci U S A 71*, 4864–4868.

29 Ferrone, F. A., Hofrichter, J., Sunshine, H. R., and Eaton, W. A. (1980) Kinetic studies on photolysis-induced gelation of sickle cell hemoglobin suggest a new mechanism, *Biophys J 32*, 361–380.

30 Ferrone, F. A., Hofrichter, J., and Eaton, W. A. (1985) Kinetics of sickle hemoglobin polymerization. II. A double nucleation mechanism, *J Mol Biol 183*, 611–631.

REFERENCES 381

31 Arvinte, T., Cudd, A., and Drake, A. F. (1993) The structure and mechanism of formation of human calcitonin fibrils, *J Biol Chem 268*, 6415–6422.

32 Padrick, S. B. and Miranker, A. D. (2002) Islet amyloid: Phase partitioning and secondary nucleation are central to the mechanism of fibrillogenesis, *Biochemistry 41*, 4694–4703.

33 Librizzi, F. and Rischel, C. (2005) The kinetic behavior of insulin fibrillation is determined by heterogeneous nucleation pathways, *Protein Sci 14*, 3129–3134.

34 Yip, C. M. and McLaurin, J. (2001) Amyloid-beta peptide assembly: A critical step in fibrillogenesis and membrane disruption, *Biophys J 80*, 1359–1371.

35 Shtainfeld, A., Sheynis, T., and Jelinek, R. (2010) Specific mutations alter fibrillation kinetics, fiber morphologies, and membrane interactions of pentapeptides derived from human calcitonin, *Biochemistry 49*, 5299–5307.

36 Andreotti, G., Vitale, R. M., Avidan-Shpalter, C., Amodeo, P., Gazit, E., and Motta, A. (2011) Converting the highly amyloidogenic human calcitonin into a powerful fibril inhibitor by 3D structure homology with a non-amyloidogenic analogue, *J Biol Chem 286*, 2707–2718.

37 Abedini, A. and Raleigh, D. P. (2009) A role for helical intermediates in amyloid formation by natively unfolded polypeptides? *Phys Biol 6*, 015005.

38 Abedini, A. and Raleigh, D. P. (2009) A critical assessment of the role of helical intermediates in amyloid formation by natively unfolded proteins and polypeptides, *Protein Eng Des Sel 22*, 453–459.

39 Chiti, F., Stefani, M., Taddei, N., Ramponi, G., and Dobson, C. M. (2003) Rationalization of the effects of mutations on peptide and protein aggregation rates, *Nature 424*, 805–808.

40 Williams, A. D., Portelius, E., Kheterpal, I., Guo, J. T., Cook, K. D., Xu, Y., and Wetzel, R. (2004) Mapping Aβ amyloid fibril secondary structure using scanning proline mutagenesis, *J Mol Biol 335*, 833–842.

41 Wurth, C., Guimard, N. K., and Hecht, M. H. (2002) Mutations that reduce aggregation of the Alzheimer's Aβ42 peptide: An unbiased search for the sequence determinants of Aβ amyloidogenesis, *J Mol Biol 319*, 1279–1290.

42 Kheterpal, I., Williams, A., Murphy, C., Bledsoe, B., and Wetzel, R. (2001) Structural features of the Aβ amyloid fibril elucidated by limited proteolysis, *Biochemistry 40*, 11757–11767.

43 Morimoto, A., Irie, K., Murakami, K., Masuda, Y., Ohigashi, H., Nagao, M., Fukuda, H., Shimizu, T., and Shirasawa, T. (2004) Analysis of the secondary structure of β-amyloid (Aβ42) fibrils by systematic proline replacement, *J Biol Chem 279*, 52781–52788.

44 Nelson, R., Sawaya, M. R., Balbirnie, M., Madsen, A. O., Riekel, C., Grothe, R., and Eisenberg, D. (2005) Structure of the cross-β spine of amyloid-like fibrils, *Nature 435*, 773–778.

45 Kim, W. and Hecht, M. H. (2006) Generic hydrophobic residues are sufficient to promote aggregation of the Alzheimer's Aβ42 peptide, *Proc Natl Acad Sci U S A 103*, 15824–15829.

46 Kim, W. and Hecht, M. H. (2008) Mutations enhance the aggregation propensity of the Alzheimer's Aβ peptide, *J Mol Biol 377*, 565–574.

47 Chiti, F. and Dobson, C. M. (2009) Amyloid formation by globular proteins under native conditions, *Nat Chem Biol 5*, 15–22.
48 Chiti, F. and Dobson, C. M. (2006) Protein misfolding, functional amyloid, and human disease, *Annu Rev Biochem 75*, 333–366.
49 Guijarro, J. I., Sunde, M., Jones, J. A., Campbell, I. D., and Dobson, C. M. (1998) Amyloid fibril formation by an SH3 domain, *Proc Natl Acad Sci U S A 95*, 4224–4228.
50 Litvinovich, S. V., Brew, S. A., Aota, S., Akiyama, S. K., Haudenschild, C., and Ingham, K. C. (1998) Formation of amyloid-like fibrils by self-association of a partially unfolded fibronectin type III module, *J Mol Biol 280*, 245–258.
51 Chiti, F., Webster, P., Taddei, N., Clark, A., Stefani, M., Ramponi, G., and Dobson, C. M. (1999) Designing conditions for in vitro formation of amyloid protofilaments and fibrils, *Proc Natl Acad Sci U S A 96*, 3590–3594.
52 Ferrão-Gonzales, A. D., Souto, S. O., Silva, J. L., and Foguel, D. (2000) The preaggregated state of an amyloidogenic protein: Hydrostatic pressure converts native transthyretin into the amyloidogenic state, *Proc Natl Acad Sci U S A 97*, 6445–6450.
53 McParland, V. J., Kad, N. M., Kalverda, A. P., Brown, A., Kirwin-Jones, P., Hunter, M. G., Sunde, M., and Radford, S. E. (2000) Partially unfolded states of β_2-microglobulin and amyloid formation in vitro, *Biochemistry 39*, 8735–8746.
54 Fändrich, M., Fletcher, M. A., and Dobson, C. M. (2001) Amyloid fibrils from muscle myoglobin, *Nature 410*, 165–166.
55 Schmittschmitt, J. P. and Scholtz, J. M. (2003) The role of protein stability, solubility, and net charge in amyloid fibril formation, *Protein Sci 12*, 2374–2378.
56 De Felice, F. G., Vieira, M. N., Meirelles, M. N., Morozova-Roche, L. A., Dobson, C. M., and Ferreira, S. T. (2004) Formation of amyloid aggregates from human lysozyme and its disease-associated variants using hydrostatic pressure, *FASEB J 18*, 1099–1101.
57 Booth, D. R., Sunde, M., Bellotti, V., Robinson, C. V., Hutchinson, W. L., Fraser, P. E., Hawkins, P. N., Dobson, C. M., Radford, S. E., Blake, C. C., and Pepys, M. B. (1997) Instability, unfolding and aggregation of human lysozyme variants underlying amyloid fibrillogenesis, *Nature 385*, 787–793.
58 Raffen, R., Dieckman, L. J., Szpunar, M., Wunschl, C., Pokkuluri, P. R., Dave, P., Wilkins Stevens, P., Cai, X., Schiffer, M., and Stevens, F. J. (1999) Physicochemical consequences of amino acid variations that contribute to fibril formation by immunoglobulin light chains, *Protein Sci 8*, 509–517.
59 Stathopulos, P. B., Rumfeldt, J. A., Scholz, G. A., Irani, R. A., Frey, H. E., Hallewell, R. A., Lepock, J. R., and Meiering, E. M. (2003) Cu/Zn superoxide dismutase mutants associated with amyotrophic lateral sclerosis show enhanced formation of aggregates in vitro, *Proc Natl Acad Sci U S A 100*, 7021–7026.
60 Sekijima, Y., Wiseman, R. L., Matteson, J., Hammarström, P., Miller, S. R., Sawkar, A. R., Balch, W. E., and Kelly, J. W. (2005) The biological and chemical basis for tissue-selective amyloid disease, *Cell 121*, 73–85.
61 Kelly, J. W. (1996) Alternative conformations of amyloidogenic proteins govern their behavior, *Curr Opin Struct Biol 6*, 11–17.
62 Dobson, C. M. (1999) Protein misfolding, evolution and disease, *Trends Biochem Sci 24*, 329–332.

63 Karplus, M. and McCammon, J. A. (2002) Molecular dynamics simulations of biomolecules, *Nat Struct Biol 9*, 646–652.
64 Woodward, C., Simon, I., and Tüchsen, E. (1982) Hydrogen exchange and the dynamic structure of proteins, *Mol Cell Biochem 48*, 135–160.
65 Englander, S. W. and Kallenbach, N. R. (1984) Hydrogen exchange and structural dynamics of proteins and nucleic acids, *Q Rev Biophys 16*, 521–655.
66 Roder, H., Elove, G. A., and Shastry, M. C. R. (2000) In: Pain, R. (ed.), *Mechanisms of protein folding*, 2nd ed. Oxford University Press, Oxford, pp. 70–72.
67 Dumoulin, M., Last, A. M., Desmyter, A., Decanniere, K., Canet, D., Larsson, G., Spencer, A., Archer, D. B., Sasse, J., Muyldermans, S., Wyns, L., Redfield, C., Matagne, A., Robinson, C. V., and Dobson, C. M. (2003) A camelid antibody fragment inhibits the formation of amyloid fibrils by human lysozyme, *Nature 424*, 783–788.
68 Bouchard, M., Zurdo, J., Nettleton, E. J., Dobson, C. M., and Robinson, C. V. (2000) Formation of insulin amyloid fibrils followed by FTIR simultaneously with CD and electron microscopy, *Protein Sci 9*, 1960–1967.
69 Mason, J. M., Kokkoni, N., Stott, K., and Doig, J. (2003) Design strategies for antiamyloid agents, *Curr Opin Struct Biol 13*, 526–532.
70 Takahashi, T. and Mihara, H. (2008) Peptide and protein mimetics inhibiting amyloid beta-peptide aggregation, *Acc Chem Res 41*, 1309–1318.
71 Cohen, F. E. and Kelly, J. W. (2003) Therapeutic approaches to protein-misfolding diseases, *Nature 426*, 905–909.
72 Bucciantini, M., Giannoni, E., Chiti, F., Baroni, F., Formigli, L., Zurdo, J., Taddei, N., Ramponi, G., Dobson, C. M., and Stefani, M. (2002) Inherent toxicity of aggregates implies a common mechanism for protein misfolding diseases, *Nature 416*, 507–511.
73 Caughey, B. and Lansbury, P. T. (2003) Protofibrils, pores, fibrils, and neurodegeneration: Separating the responsible protein aggregates from the innocent bystanders, *Annu Rev Neurosci 26*, 267–298.
74 Kayed, R., Head, E., Thompson, J. L., McIntire, T. M., Milton, S. C., Cotman, C. W., and Glabe, C. G. (2003) Common structure of soluble amyloid oligomers implies common mechanism of pathogenesis, *Science 300*, 486–489.
75 Lesné, S., Koh, M. T., Kotilinek, L., Kayed, R., Glabe, C. G., Yang, A., Gallagher, M., and Ashe, K. H. (2006) A specific amyloid-beta protein assembly in the brain impairs memory, *Nature 440*, 352–357.
76 Cheng, I. H., Scearce-Levie, K., Legleiter, J., Palop, J. J., Gerstein, H., Bien-Ly, N., Puolivali, J., Lesné, S., Ashe, K. H., Muchowski, P. J., and Mucke, L. (2007) Accelerating amyloid-beta fibrillization reduces oligomer levels and functional deficits in Alzheimer disease mouse models, *J Biol Chem 282*, 23818–23828.
77 Nerelius, C., Sandegren, A., Sargsyan, H., Raunak, R., Leijonmarck, H., Chatterjee, U., Fisahn, A., Imarisio, S., Lomas, D. A., Crowther, D. C., Strömberg, R., and Johansson, J. (2009) Alpha-helix targeting reduces amyloid-beta peptide toxicity, *Proc Natl Acad Sci U S A 106*, 9191–9196.
78 Abedini, A., Meng, F., and Raleigh, D. P. (2007) A single-point mutation converts the highly amyloidogenic human islet amyloid polypeptide into a potent fibrillization inhibitor, *J Am Chem Soc 129*, 11300–11301.
79 Murray, M. M., Bernstein, S. L., Nyugen, V., Condron, M. M., Teplow, D. B., and Bowers, M. T. (2009) Amyloid beta protein: Aβ40 inhibits Aβ42 oligomerization, *J Am Chem Soc 131*, 6316–6317.

80. Sexton, P. M., Findlay, D. M., and Martin, T. J. (1999) Calcitonin, *Curr Med Chem 6*, 1067–1093.
81. Zaidi, M., Inzerillo, A. M., Troen, B., Moonga, B. S., Abe, E., and Burckhardt, P. (2002) In: Bilezikian, J. P., Raisz, L. G., and Rodan, G. A. (eds.), *Principles of bone biology*, Academic Press, San Diego, pp. 1423–1440.
82. Manicourt, D. H., Devogelaer, J. P., Azria, M., and Silverman, S. (2005) Rationale for the potential use of calcitonin in osteoarthritis, *J Musculoskelet Neuronal Interact 5*, 285–293.
83. Sletten, K., Westermark, P., and Natvig, J. B. (1973) Characterization of amyloid fibril proteins from medullary carcinoma of the thyroid, *J Exp Med 143*, 993–998.
84. Zaidi, M., Inzerillo, A. M., Moonga, B. S., Bevis, P. J., and Huang, C. L. (2002) Forty years of calcitonin—Where are we now? A tribute to the work of Iain Macintyre, FRS, *Bone 30*, 655–663.
85. Singer, F. R., Aldred, J. P., Neer, R. M., Krane, S. M., Potts, J. T., Jr., and Bloch, K. J. (1972) An evaluation of antibodies and clinical resistance to salmon calcitonin, *J Clin Investig 51*, 2331–2338.
86. Wada, S., Martin, T. J., and Findlay, D. M. (1995) Homologous regulation of the calcitonin receptor in mouse osteoclast-like cells and human breast cancer T47D cells, *Endocrinology 136*, 2611–2621.
87. Rymer, D. L. and Good, T. A. (2001) The role of G protein activation in the toxicity of amyloidogenic Aβ-(1–40), Aβ-(25–35), and bovine calcitonin, *J Biol Chem 276*, 2523–2530.
88. Feletti, C. and Bonomini, V. (1979) Effect of calcitonin on bone lesions in chronic dialysis patients, *Nephron 24*, 85–88.
89. Yamamoto, Y., Nakamuta, H., Koida, M., Seyler, J. K., and Orlowski, R. C. (1982) Calcitonin-induced anorexia in rats: A structure-activity study by intraventricular injections, *Jpn J Pharmacol 32*, 1013–1017.
90. Cudd, A., Arvinte, T., Das, R. E., Chinni, C., and MacIntyre, I. (1995) Enhanced potency of human calcitonin when fibrillation is avoided, *J Pharm Sci 84*, 717–719.
91. Kanaori, K. and Nosaka, A. Y. (1995) Study of human calcitonin fibrillation by proton nuclear magnetic resonance spectroscopy, *Biochemistry 34*, 12138–12143.
92. Kamihira, M., Naito, A., Tuzi, S., Nosaka, Y. A., and Saitô, H. (2000) Conformational transitions and fibrillation mechanism of human calcitonin as studied by high-resolution solid-state ^{13}C NMR, *Protein Sci 9*, 867–877.
93. Bauer, H. H., Aebi, U., Haner, M., Hermann, R., Muller, M., Arvinte, T., and Merkle, H. P. (1995) Architecture and polymorphism of fibrillar supramolecular assemblies produced by in vitro aggregation of human calcitonin, *J Struct Biol 115*, 1–15.
94. Naito, A., Kamihira, M., Inoue, R., and Saitô, H. (2004) Structural diversity of amyloid fibril formed in human calcitonin as revealed by site-directed ^{13}C solid-state NMR spectroscopy, *Magn Reson Chem 42*, 247–257.
95. Bauer, H. H., Muller, M., Goette, J., Merkle, H. P., and Fringeli, U. P. (1994) Interfacial adsorption and aggregation associated changes in secondary structure of human calcitonin monitored by ATR-FTIR spectroscopy, *Biochemistry 33*, 12276–12282.
96. Gazit, E. (2002) A possible role for π-stacking in the self-assembly of amyloid fibrils, *FASEB J 16*, 77–83.

REFERENCES

385

97 DuBay, K. F., Pawar, A. P., Chiti, F., Zurdo, J., Dobson, C. M., and Vendruscolo, M. (2004) Prediction of the absolute aggregation rates of amyloidogenic polypeptide chains, *J Mol Biol 341*, 1317–1326.

98 Routledge, K. E., Tartaglia, G. G., Platt, G. W., Vendruscolo, M., and Radford, S. E. (2009) Competition between intramolecular and intermolecular interactions in an amyloid-forming protein, *J Mol Biol 389*, 776–786.

99 Siligardi, G., Samorì, B., Melandri, S., Visconti, M., and Drake, A. F. (1994) Correlations between biological activities and conformational properties for human, salmon, eel, porcine calcitonins and Elcatonin elucidated by CD spectroscopy, *Eur J Biochem 221*, 1117–1125.

100 Amodeo, P., Motta, A., Strazzullo, G., and Castiglione Morelli, M. A. (1999) Conformational flexibility in calcitonin: The dynamic properties of human and salmon calcitonin in solution, *J Biomol NMR 13*, 161–174.

101 Gilchrist, P. J. and Bradshow, J. P. (1993) Amyloid formation by salmon calcitonin, *Biochim Biophys Acta 1182*, 111–114.

102 Cunningham, B. C., Mulkerrin, M. G., and Wells, J. A. (1991) Dimerization of human growth hormone by zinc, *Science 253*, 545–548.

103 Andreotti, G. and Motta, A. (2004) Modulating calcitonin fibrillogenesis: An antiparallel alpha-helical dimer inhibits fibrillation of salmon calcitonin, *J Biol Chem 279*, 6364–6370.

104 Avidan-Shpalter, C. and Gazit, E. (2006) The early stages of amyloid formation: Biophysical and structural characterization of calcitonin prefibrillar assemblies, *Amyloid 13*, 216–225.

105 Arvinte, T. (1996) Human calcitonin fibrillogenesis. In: Bock, G.R. and Goode, J.A. (eds.), *"The nature and origin of amyloid fibrils,"* Ciba Foundation Symposium 199. J. Wiley & Sons Ltd, Chichester, pp. 90–103.

106 Bitan, G., Lomakin, A., and Teplow, D. B. (2001) Amyloid beta-protein oligomerization: Prenucleation interactions revealed by photoinduced cross-linking of unmodified proteins, *J Biol Chem 276*, 35176–35184.

107 Hall, D., Hirota, N., and Dobson, C. M. (2005) A toy model for predicting the rate of amyloid formation from unfolded protein, *J Mol Biol 195*, 195–205.

108 Capellos, C. and Bielski, B. H. J. (1972) *Kinetic systems: Mathematical description of chemical kinetics in solution*, J. Wiley & Sons Ltd, New York.

109 Lomakin, A., Teplow, D. B., Kirschner, D. A., and Benedek, G. B. (1997) Kinetics theory of fibrillogenesis of amyloid β-protein, *Proc Natl Acad Sci U S A 94*, 7942–7947.

110 Kamihira, M., Oshiro, Y., Tuzi, S., Nosaka, A. Y., Saitô, H., and Naito, A. (2003) Effect of electrostatic interaction on fibril formation of human calcitonin as studied by high resolution solid state ^{13}C NMR, *J Biol Chem 278*, 2859–2865.

111 Motta, A., Andreotti, G., Amodeo, P., Strazzullo, G., and Castiglione Morelli, M. A. (1998) Solution structure of human calcitonin in membrane-mimetic environment: The role of the amphipathic helix, *Proteins 32*, 314–323.

112 Meadows, R. P., Nikonowicz, E. P., Jones, C. R., Bastian, J. W., and Gorenstein, D. G. (1991) Two-dimensional NMR and structure determination of salmon calcitonin in methanol, *Biochemistry 30*, 1247–1254.

113 McGaughey, G. B., Gagné, M., and Rappé, A. K. (1998) π-Stacking interactions. Alive and well in proteins, *J Biol Chem 273*, 15458–15463.

114 Azriel, R. and Gazit, E. (2001) Analysis of the minimal amyloid-forming fragment of the islet amyloid polypeptide. An experimental support for the key role of the phenylalanine residue in amyloid formation, *J Biol Chem 276*, 34156–34161.

115 Reches, M., Porat, Y., and Gazit, E. (2002) Amyloid fibril formation by pentapeptide and tetrapeptide fragments of human calcitonin, *J Biol Chem 277*, 35475–35480.

116 Zanuy, D., Porat, Y., Gazit, E., and Nussinov, R. (2004) Peptide sequence and amyloid formation; molecular simulations and experimental study of a human islet amyloid polypeptide fragment and its analogs, *Structure 12*, 439–455.

117 Bellesia, G. and Shea, J.-E. (2009) What determines the structure and stability of KFFE monomers, dimers, and protofibrils? *Biophys J 96*, 875–886.

118 Tracz, S. M., Abedini, A., Driscoll, M., and Raleigh, D. P. (2004) Role of aromatic interactions in amyloid formation by peptides derived from human amylin, *Biochemistry 43*, 15901–15908.

119 Marek, P., Abedini, A., Song, B., Kanungo, M., Johnson, M. E., Gupta, R., Zaman, W., Wong, S. S., and Raleigh, D. P. (2007) Aromatic interactions are not required for amyloid fibril formation by islet amyloid polypeptide but do influence the rate of fibril formation and fibril morphology, *Biochemistry 46*, 3255–3261.

120 Kayed, R., Bernhagen, J., Greenfield, N., Sweimeh, K., Brummer, H., Voelter, W., and Kapurniotu, A. (1999) Conformational transitions of islet amyloid polypeptide (IAPP) in amyloid formation in vitro, *J Mol Biol 287*, 781–796.

121 Cohen, C. and Parry, D. A. D. (1990) Alpha-helical coiled coils and bundles: How to design an alpha-helical protein, *Proteins 7*, 1–15.

122 Parry, D. A. D., Fraser, R. D. B., and Squire, J. M. (2008) Fifty years of coiled-coils and α-helical bundles: A close relationship between sequence and structure, *J Struct Biol 163*, 258–269.

123 Eilers, M., Patel, A. B., Liu, W., and Smith, S. O. (2002) Comparison of helix interactions in membrane and soluble alpha-bundle proteins, *Biophys J 82*, 2720–2736.

124 Langosch, D. and Heringa, J. (1998) Interaction of transmembrane helices by a knobs-into-holes packing characteristic of soluble coiled coils, *Proteins 31*, 150–159.

125 Lupas, A., Van Dyke, M., and Stock, J. (1991) Predicting coiled coils from protein sequences, *Science 252*, 1162–1164.

126 Lupas, A. (1996) Prediction and analysis of coiled-coil structures, *Methods Enzymol 266*, 513–525.

127 Singh, J. and Thornton, J. M. (1985) The interaction between phenilalanine rings in proteins, *FEBS Lett 191*, 1–6.

128 Fierz, B., Reiner, A., and Kiefhaber, B. (2009) Local conformational dynamics in α-helices measured by fast triplet transfer, *Proc Natl Acad Sci U S A 106*, 1057–1062.

129 Zimm, B. H. and Bragg, J. K. (1959) Theory of the phase transition between helix and random coil in polypeptide chains, *J Chem Phys 31*, 526–535.

130 Lifson, S. and Roig, A. (1961) On the theory of helix-coil transition in polypeptides, *J Chem Phys 34*, 1963–1974.

131 Streicher, W. W. and Makhatadze, G. I. (2006) Calorimetric evidence for a two-state unfolding of the beta-hairpin peptide trpzip4, *J Am Chem Soc 128*, 30–31.

132 Huang, C.-Y., Getahun, Z., Zhu, Y., Klemke, J. W., DeGrado, W. F., and Gai, F. (2002) Helix formation via conformation diffusion search, *Proc Natl Acad Sci U S A 99*, 2788–2793.

133 Werner, J. H., Dyer, B. R., Fesinmeyer, M. R., and Andersen, N. H. (2002) Dynamics of the primary processes of protein folding: Helix nucleation, *J Phys Chem B 106*, 487–494.

134 Ihalainen, J. A., Paoli, B., Muff, S., Backus, E. H. G., Bredenbeck, J., Woolley, G. A., Caflisch, A., and Hamm, P. (2008) α-Helix folding in the presence of structural constraints, *Proc Natl Acad Sci U S A 105*, 9588–9593.

135 Andrade, M. A., Chacón, P., Merelo, J. J., and Morán, F. (1993) Evaluation of secondary structure of proteins from UV circular dichroism spectra using an unsupervised learning neural network, *Protein Eng 6*, 383–390.

136 Jas, G. S. and Kuczera, K. (2004) Equilibrium structure and folding of a helix-forming peptide: Circular dichroism measurements and replica-exchange molecular dynamics simulations, *Biophys J 87*, 3786–3798.

137 Kammerer, R. A., Schulthess, T., Landwehr, R., Lustig, A., Engel, J., Aebi, U., and Steinmetz, M. O. (1998) An autonomous folding unit mediates the assembly of two-stranded coiled coils, *Proc Natl Acad Sci U S A 95*, 13419–13424.

138 Vieth, M., Kolinski, A., Brooks, C. L., III, and Skolnick, J. (1994) Prediction of the folding pathways and structure of the GCN4 leucine zipper, *J Mol Biol 237*, 361–367.

139 Hunter, C. A. (1993) The role of aromatic interactions in molecular recognition, *Angew Chem Int Ed Engl 32*, 1584–1586.

140 Levitt, M. and Perutz, M. F. (1988) Aromatic rings act as hydrogen bond acceptors, *J Mol Biol 201*, 751–754.

141 Hatters, D. M., Minton, A. P., and Howlett, G. J. (2002) Macromolecular crowding accelerates amyloid formation by human apolipoprotein C-II, *J Biol Chem 277*, 7824–7830.

142 Dobson, C. M. (2004) Chemical space and biology, *Nature 432*, 824–828.

143 Lipinski, C. and Hopkins, A. (2004) Navigating chemical space for biology and medicine, *Nature 432*, 855–861.

144 Zagorski, M. G. and Barrow, C. J. (1992) NMR studies of amyloid β-peptides: Proton assignments, secondary structure, and mechanism of an α-helix $\rightarrow$ β-sheet conversion for a homologous, 28-residue, N-terminal fragment, *Biochemistry 31*, 5621–5631.

145 Naito, A. and Kawamura, I. (2007) Solid-state NMR as a method to reveal structure and membrane-interaction of amyloidogenic proteins and peptides, *Biochim Biophys Acta 1768*, 1900–1912.

146 Tsai, H.-H., Zanuy, D., Haspel, N., Gunasekaran, K., Ma, B., Tsai, C. J., and Nussinov, R. (2004) The stability and dynamics of the human calcitonin amyloid peptide DFNKF, *Biophys J 87*, 146–158.

147 Tsai, H.-H., Reches, M., Tsai, C. J., Gunasekaran, K., Gazit, E., and Nussinov, R. (2005) Energy landscape of amyloidogenic peptide oligomerization by parallel-tempering molecular dynamics simulation: Significant role of Asn ladder, *Proc Natl Acad Sci U S A 102*, 8174–8179.

148 Haspel, N., Zanuy, D., Ma, B., Wolfson, H., and Nussinov, R. (2005) A comparative study of amyloid fibril formation by residues 15–19 of the human calcitonin hormone:

A single beta-sheet model with a small hydrophobic core, *J Mol Biol 345*, 1213–1227.
149 Motta, A., Temussi, P. A., Wünsch, E., and Bovermann, G. (1991) A ¹H NMR study of human calcitonin in solution, *Biochemistry 30*, 2364–2371.
150 Becker, O. M. (1998) Principal coordinate maps of molecular potential energy surfaces, *J Comput Chem 19*, 1255–1267.
151 Yang, M. and Teplow, D. B. (2008) Amyloid β-protein monomer folding: Free-energy surfaces reveal alloform-specific differences, *J Mol Biol 384*, 450–464.
152 Kirkitadze, M. D., Condron, M. M., and Teplow, D. B. (2001) Identification and characterization of key kinetic intermediates in amyloid β-protein fibrillogenesis, *J Mol Biol 312*, 1103–1119.
153 Fezoui, Y. and Teplow, D. B. (2002) Kinetic studies of amyloid β-protein fibril assembly. Differential effects of α-helix stabilization, *J Biol Chem 277*, 36948–36954.
154 Maji, S. K., Ogorzalek Loo, R. R., Inayathullah, M., Spring, S. M., Vollers, S. S., Condron, M. M., Bitan, G., Loo, J. A., and Teplow, D. B. (2009) Amino acid position-specific contributions to amyloid β-protein oligomerization, *J Biol Chem 284*, 23580–23591.
155 Koradi, R., Billeter, M., and Wüthrich, K. (1996) MOLMOL: A program for display and analysis of macromolecular structures, *J Mol Graph 14*, 51–55.
156 Phillips, J. C., Braun, R., Wang, W., Gumbart, J., Tajkhorshid, E., Villa, E., Chipot, C., Skeel, R. D., Kale, L., and Schulten, K. (2005) Scalable molecular dynamics with NAMD, *J Comput Chem 26*, 1781–1802.
157 MacKerell, A. D., Jr., Bashford, D., Bellott, M., Dunbrack, R. L., Jr., Evanseck, J. D., Field, M. J., Fischer, S., Gao, J., Guo, H., Ha, S., Joseph-McCarthy, D., Kuchnir, L., Kuczera, K., Lau, F. T. K., Mattos, C., Michnick, S., Ngo, T., Nguyen, D. T., Prodhom, B., Reiher, W. E., III, Roux, B., Schlenkrich, M., Smith, J. C., Stote, R., Straub, J., Watanabe, M., Wiorkiewicz-Kuczera, J., Yin, D., and Karplus, M. (1998) All-atom empirical potential for molecular modeling and dynamics studies of proteins, *J Phys Chem B 102*, 3586–3616.
158 Pettersen, E. F., Goddard, T. D., Huang, C. C., Couch, G. S., Greenblatt, D. M., Meng, E. C., and Ferrin, T. E. (2004) UCSF chimera—A visualization system for exploratory research and analysis, *J Comput Chem 25*, 1605–1612.

13

SOLUTION NMR STUDIES OF Aβ MONOMERS AND OLIGOMERS

Chunyu Wang

13.1. INTRODUCTION

Numerous experimental and clinical data support the hypothesis that amyloid β-peptide (Aβ) is a major causative factor in Alzheimer's disease (AD) (for reviews, see Hardy [1–4]). Aβ fibrils and oligomers are toxic to neurons and can activate a cascade of pathological events culminating in neuronal and memory loss. It is therefore imperative to understand the structure and dynamics of Aβ and their relationship with Aβ aggregation and toxicity. Solution nuclear magnetic resonance (NMR) is uniquely suited for the study of soluble forms of Aβ, namely Aβ monomers and oligomers. Aβ monomers are at the starting gate of the aggregation pathway, and their conformational and motional properties are important for aggregation kinetics and thermodynamics. In recent years, mounting evidence has indicated that Aβ oligomers may play an even more prominent role than Aβ fibrils in causing Aβ toxicity and dementia. Neither Aβ oligomers nor Aβ monomers are amenable to crystallization due to the intrinsic flexibility of Aβ. Solution NMR is one of the few techniques that can study Aβ monomer and oligomers at atomic resolution. Compared with solid-state NMR, Fourier transform infrared (FTIR), Raman spectroscopy, and fluorescence techniques, which require site-specific labeling, solution NMR can report on almost all Aβ atoms at once with atomic resolution, providing a detailed picture of

---

Protein and Peptide Folding, Misfolding, and Non-Folding, First Edition. Edited by Reinhard Schweitzer-Stenner.
© 2012 John Wiley & Sons, Inc. Published 2012 by John Wiley & Sons, Inc.

Aβ structure and dynamics. In addition, solution NMR is one of the most powerful experimental techniques for studying protein dynamics, which can be a crucial determinant in the mechanism of Aβ aggregation and possibly in Aβ toxicity [5].

I will review the numerous methods for recombinant production of Aβ in *Escherichia coli* cells for NMR sample preparation. The data on NMR characterization of Aβ monomers in aqueous solution and detergent micelles will be discussed. Then I will review the NMR relaxation data, and their relationship with the mechanism of Aβ aggregation. An example combining NMR J-coupling measurement and molecular dynamics simulation is presented which reveals conformational differences between Aβ40 and Aβ42. Finally, NMR structure data on Aβ oligomers are summarized, as well as saturation transfer experiments to study monomer–oligomer equilibrium.

13.2. OVEREXPRESSION AND PURIFICATION OF RECOMBINANT Aβ

Although Aβ can be obtained by peptide synthesis, for solution NMR studies uniformly ^{13}C- and/or ^{15}N-labeled Aβ is usually preferred to maximize the information content of NMR data. To tackle the challenge of large molecular weights (MWs) of certain Aβ oligomers, deuteration of Aβ to enhance NMR sensitivity is highly desirable, which can be achieved at a reasonable cost by recombinant overexpression. Recombinant overexpression also offers the possibility of unique labeling patterns, for example, those achieved with 1-^{13}C glucose, 2-^{13}C glucose, and fluorine-labeled amino acid. Recently, it has been shown that recombinant Aβ is purer than synthetic Aβ, leading to faster aggregation and enhanced toxicity of recombinant Aβ [6].

Because of the rapid aggregation, the small size and the toxicity of Aβ, recombinant overexpression in *E. coli* and the subsequent purification of Aβ have proved difficult. However, in recent years, many elegant and efficient solutions to this challenge have appeared in the literature.

13.2.1. Overexpression without Enhancement Tags

To overexpress Aβ without any fusion partner, an N-terminal methionine has to be added to the Aβ sequence for transcription initiation. In this review, such forms of Aβ, Aβ40, and Aβ42 with an extra N-terminal methionine are called MAβ, MAβ40, and MAβ42, respectively.

When expressed alone in *E. coli*, MAβ is produced at very high levels and goes into inclusion bodies. MAβ40 and MAβ42 can be purified by solubilization of the inclusion body in 8 M urea, anion exchange chromatography, and centrifugal filtration [7]. Interestingly, Walsh et al. noted anion exchange in batch mode worked much better than in column mode, presumably due to high concentration and aggregation of MAβ during elution in column. MAβ can also be obtained by solubilization in dimethylsulfoxide (DMSO) and purification by HPLC [8]. Walsh et al. have shown that MAβ has indistinguishable kinetics in fibril formation by thioflavin T (ThT) assay and similar toxicity on hippocampal neurons as native Aβ sequence [7].

OVEREXPRESSION AND PURIFICATION OF RECOMBINANT Aβ   391

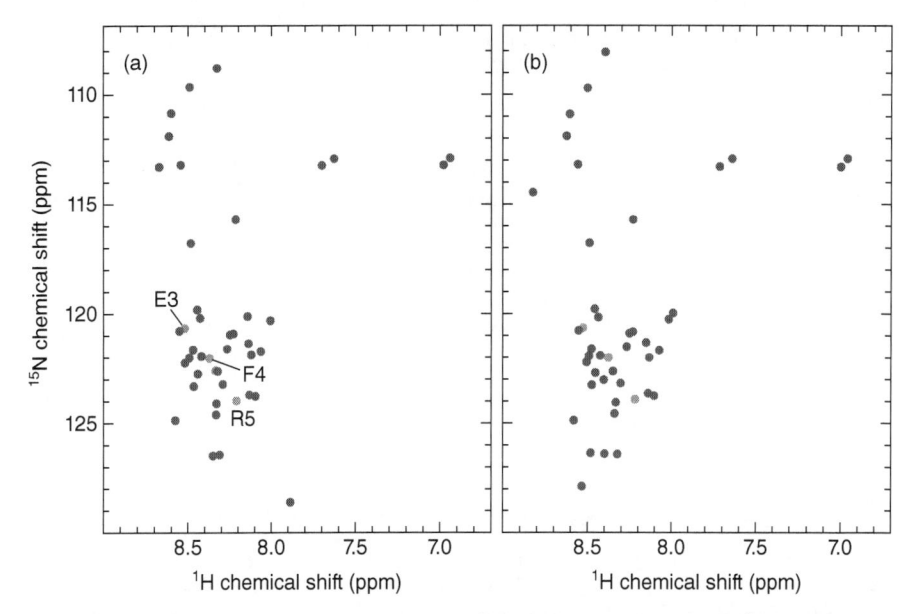

Figure 13.1. HSQC spectra of recombinant MAβ40 (dark gray) (A) and MAβ42 (dark gray) (B), overlaid with native Aβ40 (light gray) (A) and Aβ42 (light gray) (B). Only three residues at the N-terminus have significant chemical shift changes. From Ref. [9].

MAβ can be prevented from going into inclusion body by the co-expression of an affibody, $Z_{Aβ3}$ [9]. Affibodies are small helical proteins engineered to bind target molecules with affinity similar to antibodies. $Z_{Aβ3}$ is selected from phage display of a library of three-helix bundle proteins and binds to Aβ with nanomolar affinity, trapping Aβ13–37 in a β-hairpin conformation [10]. The complex of MAβ and His-tagged $Z_{Aβ3}$ can be purified from the cell lysate by nickel-nitrilotriacetic acid (NTA) affinity chromatography. MAβ is then separated from the affibody by denaturation of the complex and purified by size exclusion chromatography. The HSQC of MAβ is identical to Aβ, except for residues E3, F4, and R5, while more N-terminal residues are invisible due to fast solvent exchange (Fig. 13.1). Thus, the extra methionine leads to small changes in the N-terminus of Aβ.

Aβ42 fused with a His-tag and a thrombin recognition site has been expressed into inclusion body and purified via nickel-NTA affinity chromatography [11]. The on-column digestion by thrombin had to be adjusted carefully to prevent nonspecific cleavage within Aβ. Another drawback of this method is that the thrombin cleavage leaves non-Aβ GSHM residues at the N-terminus.

13.2.2. Overexpression with Enhancement Tags

Protein fusions can enhance expression level and solubility, and inhibit the degradation and toxicity of Aβ in *E. coli*. However, an extra cleavage step is required to release the native or near-native Aβ sequence from the fusion partner.

In 1995, the first overexpression of $A\beta$ in *E. coli* with a fusion protein was achieved by the pioneering work of Döbeli et al. [12] using a surface protein from malaria, 19 repeats of the tetrapeptide sequence NANP. A His-tagged is present at the N-terminus to facilitate purification and a methionine residue separates the fusion partner from the $A\beta42$ sequence. Cleavage of $A\beta$ from the fusion protein is carried out with cyanogen bromide, which cleaves at methionine residues. To avoid cleavage at methionine 35 (M35) within the $A\beta$ sequence, CBr cleavage is either carried out with $A\beta$ immobilized on a reverse-phase (RP) high-performance liquid chromatography (HPLC) column to prevent accessibility to M35 or using $A\beta$ with an M35L mutation. Subsequently, the fibril structure of $A\beta42$ is solved using $A\beta42$ M35L [13]. Recently, this $(NANP)_{19}$ fusion was combined with a TEV site to produce intact, native $A\beta42$ sequence [6]. Fandrich et al. first used TEV protease to generate intact $A\beta$ sequence from a fusion with maltose-binding protein [14]. A variant site, ENLYFQ/D, where D is the first residue in $A\beta$ and the "/" represents the site of protease cleavage, can be recognized by TEV protease as efficiently as the canonical ENLYFQ/G [14]. TEV protease can be readily prepared in the lab, thus this strategy represents an economical way of obtaining native $A\beta$.

Lee et al. took advantage of a de-ubiquitinating enzyme, which recognizes the ubiquitin fold instead of a specific sequence, to obtain intact $A\beta42$ sequence [15]. A fusion of His-tagged ubiquitin and $A\beta42$ is mostly expressed into the inclusion body and can be purified by urea solubilization, nickel-NTA affinity chromatography, yeast ubiquitin hydrolase-1 (YUH-1) digestion and RP HPLC [15]. Ubiquitin–$A\beta42$ can also be overexpressed in soluble fraction by an additional fusion to a heat-stable *E. coli* protein trigger factor (TF) and then purified in a similar fashion using a de-ubiquitinating enzyme, Usp2-c [16].

$A\beta$ has been recently overexpressed with a fatty acid-binding protein (FABP) [17]. FABP is cleaved off from $A\beta$ by factor X_a, which also generates intact $A\beta$ sequence. The unique feature of the purification protocol is a deliberate aggregation step, which requires Zn^{2+} for $A\beta40$ but not for $A\beta42$. $A\beta42$ has also been successfully overexpressed and purified from a fusion protein with glutathione-S-transferase (GST) [18].

Meredith et al. have proposed an ingenious semi-synthetic route for native $A\beta40$ production that involves intein cleavage, CBr cleavage, protein ligation, and selective desulfuration [19]. Here $A\beta1-29$ is overexpressed with a C-terminal intein fusion in *E. coli*, while $A\beta30-40$ is chemically synthesized with an A30C mutation. With no methionine residue in the $A\beta1-29$ sequence, the methionine at the N-terminus of the overexpressed $MA\beta1-29$ can be cleanly removed by CBr cleavage. The full $A\beta$ sequence is joined by protein ligation, taking advantage of the reactive thiol generated by intein and C30 in $A\beta30-40$/A30C. C30 is reverted to A30 by selective desulfuration. This offers a paradigm for segmental labeling in $A\beta$, and site-specific labeling in the C-terminal part of $A\beta$.

High-quality recombinant $A\beta$ with uniform ^{15}N and/or ^{13}C labeling (Fig. 13.2) is available commercially from rpeptide (http://www.rpeptide.com), made with proprietary and undisclosed methods.

Aβ MONOMERS

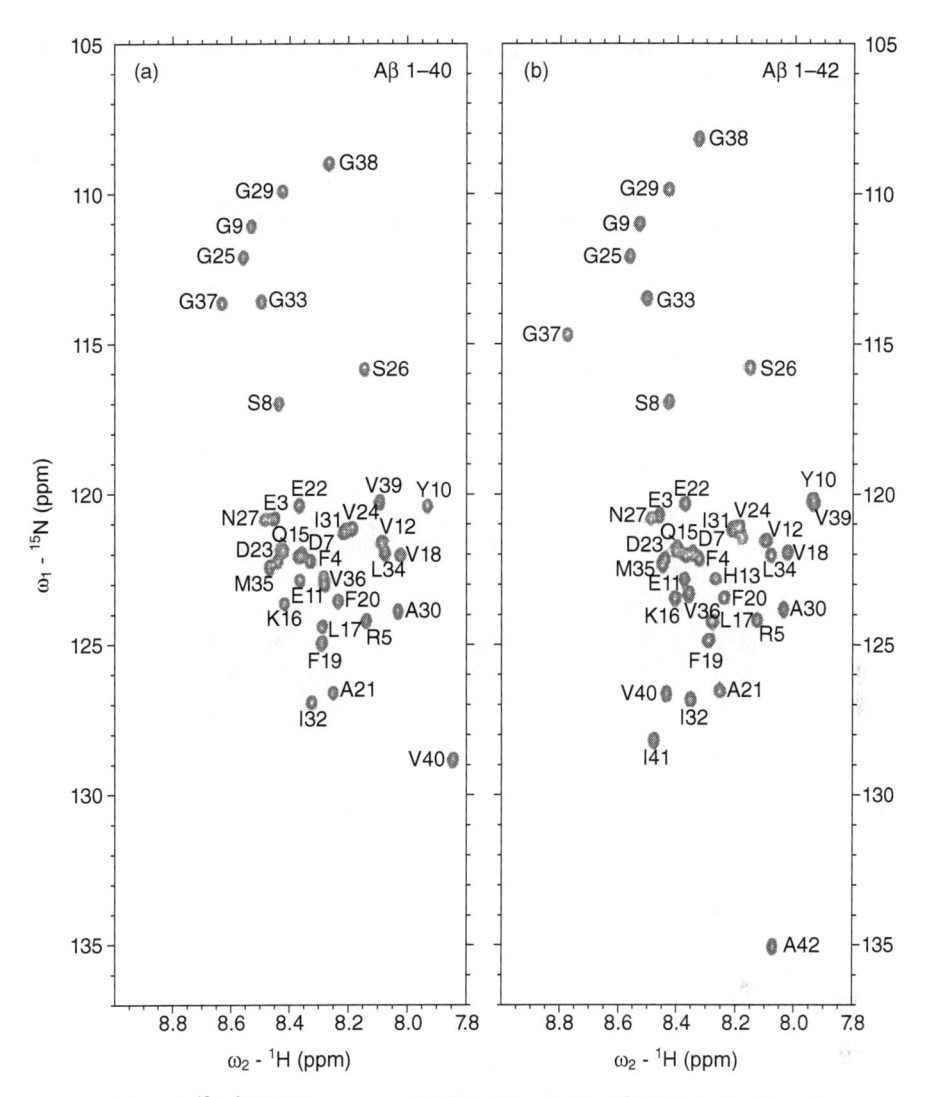

Figure 13.2. 2-D ^{15}N-^{1}H NMR spectra of Aβ40 (A) and (B) Aβ42 labeled with assignment, respectively [5]. The peptide was purchased from rPeptide (Bogart, GA) and the NMR sample was prepared according to Zagorski et al. [20].

13.3. Aβ MONOMERS

Aβ monomers are at the "starting gate" of the Aβ aggregation pathway, thus the structure and dynamics of Aβ monomers will influence the subsequent aggregation mechanism, such as the thermodynamics and kinetics of aggregation.

13.3.1. NMR Signals Originate from Aβ Monomers in Aqueous Solution after Proper Disaggregation

Excellent one-dimensional (1-D) nuclear magnetic resonance (NMR) and two-dimensional (2-D) HSQC spectra of Aβ have been consistently obtained by many groups after careful disaggregation, such as NaOH treatment and HFIP treatment [20, 21]. However, it is crucial to ascertain whether such NMR signals are from monomers or from a mixture of Aβ species. The following data demonstrate that NMR signals come from Aβ monomers in properly disaggregated samples:

13.3.1.1. Analytical Ultracentrifugation Analytical ultracentrifugation (AUC) has been regarded as the gold standard for determining the oligomeric state of proteins. Zagorski et al. carried out AUC studies of Aβ samples prepared with NaOH treatment [20, 22] (Fig. 13.3A). An MW of 4800 ± 100 was obtained for Aβ40, with an expected MW of 4330 for a monomer and 8660 for a dimer [20]. The AUC data were well fit with a monomer–dimer dissociation constant of 1.9 mM, while the fit with monomer–trimer or monomer–tetramer equilibrium was not successful. It was concluded that with the employed conditions at 0.15 mM Aβ40 (10 times less than the K_d), at least ~90% of the Aβ40 in the NMR sample are monomers. Considering higher MW species contribute little to the observed NMR signal compared with the monomer, it is likely that the observed NMR signal originates exclusively from Aβ40 monomers.

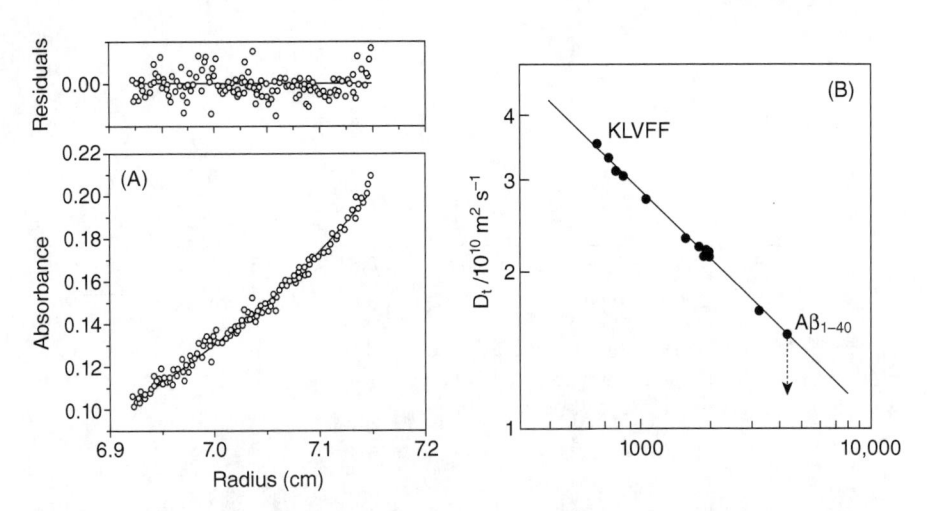

Figure 13.3. Analytical ultracentrifugation (AUC) of Aβ40 (A) and translational diffusion coefficient of Aβ fragments and full-length Aβ40 (B). (A) AUC data fitting showed Aβ40 has a MW of 4700 ± 100 Da, indicating that at least 90% of Aβ in the NMR sample is in monomeric form [20]. Figure from Hou et al. (B) The translational diffusion coefficient of Aβ40 is in line with other monomer Aβ, demonstrating that Aβ40 monomers give rise to solution NMR signals. Arrow points to the coefficient expected for Aβ dimer according to scaling law. Figure from Danielsson et al. [23].

13.3.1.2. Translational Diffusion Coefficient (Fig. 13.3B) NMR pulse-field gradient measurement directly addresses the question of whether NMR signals originate from monomers, because the diffusion coefficient represents the mobility of the molecules that give rise to NMR signals. In contrast, data from AUC may include contribution from Aβ species that do not give rise to NMR signal, such as large-MW aggregates, thereby overestimating the contribution from larger MW species in NMR. Gräslund et al. have measured the translational diffusion coefficient of Aβ fragments ranging from 5-residue peptide to full-length Aβ40, at 25°C and at physiological pH [23]. These coefficients, including those of monomer fragments, can be well fit to a scaling law as a function of the MW (Fig. 13.3B). Therefore, the NMR signals of the full-length Aβ40 are from monomers.

13.3.1.3. A Single Set of NMR Resonances (Fig. 13.2) In either ^{13}C-^{1}H or ^{15}N-^{1}H HSQC spectrum, a single set of resonances are observed for chemical groups in Aβ, again demonstrating NMR signals are from monomers. Symmetric oligomers, giving rising also to a single set of resonances, had to be highly structured; but the chemical shifts of Aβ clearly point to a mostly unstructured peptide in aqueous solution [20]. The single set of resonances could also be the results of fast chemical exchange between monomer and a small-MW oligomer (e.g., dimer, trimer, and tetramer), but the small-MW species have to be a very minor population, based on AUC data above and paramagnetic relaxation enhancement (PRE) data below.

13.3.1.4. Lack of PRE When spin-labeled Aβ40 is added to Aβ40 without spin label [24]. Fawzi and Clore mixed two types of Aβ40: (1) ^{15}N-labeled Aβ40 without spin label; and (2) natural abundance Aβ40 with spin label. If NMR signals from ^{15}N-labeled Aβ40 formed oligomers with Aβ40 with spin label, significant PRE should have been observed in ^{15}N spin relaxation rates. Two types of spin-labeled Aβ40 was used, one with an extra cysteine at the N-terminus and the other with a F20C mutation. For Cys-Aβ40, PRE effects are below reliably detection limit while for F20C Aβ40 there is essentially no PRE effect at all. The lack of PRE effects in the mixing experiment conclusively demonstrates that the observed NMR signals arise from monomers only.

Although most of the evidence cited is based on Aβ40, it is clear that NMR signals for Aβ42 are from monomers as well, based on the similarities in HSQC spectra (Fig. 13.2) and relaxation rates between these two peptides [5].

13.3.2. Aβ Monomers Are Mostly Unstructured in Aqueous Solution at Physiological pH and Adopts Helical Conformation in Micelles

13.3.2.1. Full-Length Aβ in Aqueous Solution Michael Zagorski pioneered the NMR studies of Aβ monomers in aqueous solutions with 2-D NMR [25]. A thorough study using triple-resonance experiments yielded the vast majority of chemical shifts of Aβ40 and Aβ42 monomers and established that these peptides are largely in random, extended conformation, based on NOE, chemical shift indices, and temperature coefficients of amide protons. This agrees with earlier work by Riek et al. [26] on M35-oxidized Aβ. The chemical shift index of HA and CB demonstrate a

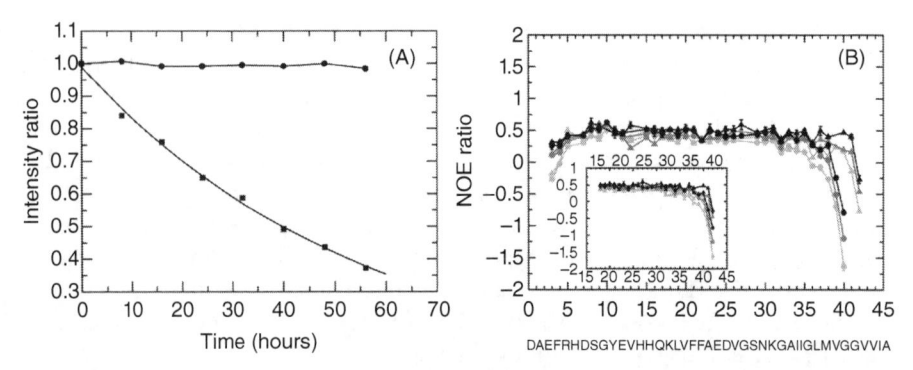

Figure 13.4. Correlation between aggregation kinetics of Aβ and protein dynamics. (A) Aβ aggregation monitored by the monomer signal of Aβ40 (•) and Aβ42 (▲) over time. (B) Temperature dependence of NOE at 273.3 K (black), 280.4 K (gray) and 287.6 K (light gray) for Aβ40 (•) and Aβ42 (▲). Higher ^{1}H-^{15}N steady-state NOE values indicate more rigidity. Aβ42 aggregates faster than Aβ40, and Aβ42 has a more rigid C-terminus, indicated by higher NOE values at three temperatures. From Yan and Wang [5].

β-propensity for C-terminal residues for Aβ42 while such a propensity is absent in Aβ40. A turn or bend like structure was also identified at D7-E11 and F20-S26. These studies have also established that Aβ42 and Aβ40 monomers seem to have similar structural properties in solution, even though Aβ42 aggregates much faster than Aβ40 (Fig. 13.4).

13.3.2.2. Aβ Fragments in Aqueous Solution

A large number of NMR studies on Aβ fragment have been carried out, partly due to the ease in sample preparation by peptide synthesis. Only a select few works on Aβ fragments will be reviewed here.

Teplow et al. identified a protease-resistant core of Aβ, Aβ21-30 [27], and the structure of this peptide was characterized by both NMR [27] and replica exchange molecular dynamics (REMD) [28], demonstrating a β-turn structure stabilized by hydrophobic interaction between the side chains of V24 and K28, and the electrostatic interaction between K28 and either E22 or D23. The lower stability of turn in FAD mutants, characterized by decreased Nuclear Overhauser Effect (NOE) intensity between A30 NH and E22 HA and decreased solvent protection of K28 side-chain amide, was correlated with the increased propensity to form oligomers [29]. Interestingly, recent papers by Härd et al. show that A21 and A30 are close to each other in a structure of Aβ–affibody complex [10]. In contrast to Teplow et al.'s idea that the turn in Aβ21-30 may be a deterrent to aggregation, a disulfide bond-linked Aβ42 through A21C and A30C mutation stabilizes toxic Aβ42 oligomer [30]. A recent combined MD and NMR studies of this fragment strongly argue for a more diverse conformation for Aβ21-30, with 60% of the conformations in unstructured state, while a minority of conformations are in a β-turn stabilized by the K28-D23 salt bridge [31]. Impressive agreement was found in experimental and predicted values of $^3J_{HNHA}$, ^{13}C chemical shifts and ^{13}C relaxation rates [31]. The Rotating

Frame Overhauser Effect (ROE) between A30 NH and E22 HA was not observed in the data acquired at 900 MHz and may have been a mis-assignment of ROE between A30NH and K28 HA [31].

Lee et al. calculated NMR structures of Aβ10-35 fragment in aqueous solution and found that regular secondary structures are absent but that there was a collapsed core at the central hydrophobic cluster region, residue 17–21 LVFFA [32].

Gräslund et al. characterized the Aβ12-28 fragment [33] and found evidence that this fragment is in a temperature-dependent equilibrium between a left-handed 3_{10} helix and a random coil conformation, based on chemical shifts, translational diffusion, and hydrogen-deuterium (HD) exchange measurements.

13.3.2.3. Aβ in Non-aqeous Solution   When Aβ is released by γ-secretase cleavage from C99, the C-terminal half of Aβ is embedded in the membrane. Thus, the conformation of Aβ in a membrane-like environment may have important implications for Aβ aggregation [34, 35].

Sodium Dodecyl Sulfate (SDS)   Craik et al. solved the structure of Aβ40 in 100 mM SDS micelle, composed of two helices at residues 15–24 (helix 1) and 28–36 (helix 2) [36]. The rest of the peptide is in random coil conformation. Similar results were obtained by Zagorski et al [37] in SDS micelle. M35 oxidation causes a local and selective disruption of helix 2 [38]. Graslund et al. probed the orientation of Aβ40 within the micelle with paramagnetic 5-Doxyl stearic acid and Mn^{2+} [39]. Helix 1 was found to be located in the SDS headgroup region while helix 2 is buried inside the micelle. The termini and the loop connecting the two helices were found to be exposed to solvent. Gräslund et al. further investigated the effect of detergents at sub-micellar concentrations and found that LiDS or SDS induces Aβ aggregation and β-conformation [40]. DPC has also been shown to induce β-conformaton in Aβ40 [41].

Organic Solvent   In organic solvent such as TFE and HFIP, Aβ generally has two helical segments and the location and extent of the helices depend on the solvent condition [42, 43], although β-conformation can be induced at low concentrations of organic solvent [44].

13.3.3. Protein Dynamics and the Mechanism of Aβ Aggregation

13.3.3.1. Backbone Dynamics of Aβ40 and Aβ42 Monomers ^{15}N relaxation rates, R_1, R_2, and steady-state ^1H-^{15}N NOE (ssNOE) have become standard experiments for probing ns–ps time scale motion in proteins. These relaxation rates are intimately related to molecular motion at ps to ns time scale, which is faster than the global tumbling motion of proteins in solution. ssNOE is particularly sensitive to such fast time scale motion, which decreases ssNOE values. ssNOE of Aβ backbone amides was first characterized by Riek et al. on M35-oxidized Aβ [26], and it was found that Aβ42 has a larger NOE than Aβ40 in the C-terminus, indicating that Aβ42 has a more rigid C-terminus. This was later confirmed by full characterization of ^{15}N R_1,

R_2, and NOE, spectral density mapping, and order parameter comparison on reduced Aβ by Yan et al. (Fig. 13.4) [5]. The effect of aggregation was removed from relaxation rates by an interleaved technique in NMR experiments [5]. Lim et al. also observed such a dynamic difference in the C-terminus between Aβ42 and Aβ40 [45]. These data suggest that although Aβ42 and Aβ40 are similar in their averaged conformational properties in aqueous solution, the C-terminal rigidity are remarkably different between Aβ40 and A42 and may contribute to the mechanism of Aβ aggregation.

13.3.3.2. Methyl Dynamics of Aβ40 and Aβ42 Monomers Dynamics of protein side chains are of great interest [46, 47] due to their essential role in protein interaction and protein aggregation [48, 49]. The two C-terminal residues of Aβ42, I41 and A42, which are responsible for its enhanced amyloidogenecity [50], have three methyl groups. Thus, methyl groups may play an important role in Aβ aggregation. Yan et al. measured ^{13}C relaxation rate R_1 and dipole–dipole cross-correlated relaxation rates using the pulse sequences developed by Yang et al. [51]. From these rates, an order parameter S_{axis}^2 can be derived to describe the amplitude of motion of the methyl C-C axis. S_{axis}^2 ranges from 0 to 1, with 0 describing complete disorder and 1 describing complete rigidity of the C-C bond. The ratio of the order parameter values between Aβ40 and Aβ42 shows that the C-terminal residues of Aβ42 display strikingly larger S_{axis}^2 values than those of Aβ40 (Fig. 13.5B). Thus, the data from methyl dynamics corroborated the findings from backbone ^{15}N dynamics and again demonstrated that the C-terminus of Aβ42 is more rigid than Aβ40. Interestingly, two methyl groups of V18 in Aβ42 have significantly smaller order parameter values than those in Aβ40 with the ratios $S^2(A\beta 42)/S^2(A\beta 40)$ of 0.6 and 0.8, respectively (Fig. 13.5C), suggesting that the V18 side chain is more mobile in Aβ42 than in Aβ40. V18, located in the central hydrophobic cluster (CHC), is an important residue for Aβ aggregation [52, 53].

13.3.3.3. M35 Oxidation Correlates Dynamics with Aggregation The side chain of M35 can be oxidized to sulfoxide and the oxidized form (Aβ^{ox}) comprises 10–50% of total Aβ in post-mortem senile plaques in brains inflicted with AD [54]. Recent studies have shown delayed aggregation of Aβ42^{ox} compared with Aβ42^{red} [22]. Oligomer assembly of Aβ42 also becomes indistinguishable from that of Aβ40 after M35 oxidation based on both photoinduced cross-linking of unmodified proteins (PICCUPs) [55] and ion mobility coupled mass spectroscopy [56]. However, it is not clear how M35 oxidation can cause such profound changes in Aβ42 aggregation.

Yan et al. characterized the effect of M35 oxidation on Aβ dynamics [57]. Aβ42^{ox} and Aβ42^{red} have similar backbone ^{15}N R_1 and R_2 values, indicating similar global motions. NOE and $J(0.87\omega_H)$ values demonstrate that the C-terminus of Aβ42 becomes more dynamic upon M35 oxidation (Fig. 13.6A,B). As expected, the extra oxygen led to the reduced motion of the side-chain methyl group of M35 (Fig. 13.6C). M35 oxidation increased the side-chain mobility toward the C-terminus as indicated by the decrease in order parameter values by ~10% (Fig. 13.6B). More

Aβ MONOMERS 399

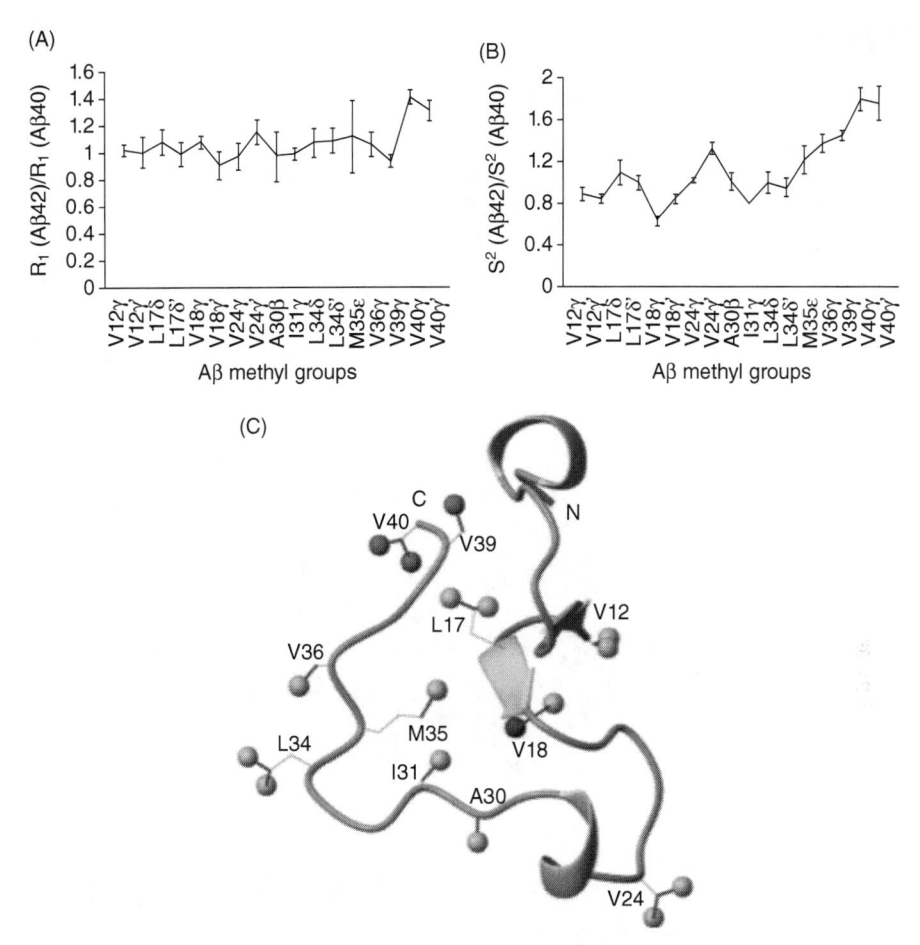

Figure 13.5. Comparison of the methyl dynamics between Aβ40 and Aβ42. (A) The ratio of R_1 of Aβ42 over Aβ40. (B) The ratio of order parameter S^2_{axis} of Aβ42 over Aβ40. (C). S^2_{axis} (Aβ42)/S^2_{axis} (Aβ40) values mapped onto the ribbon diagram of a simulated structure of Aβ40 monomer derived from MD simulation [58]. Methyl carbons are shown in a space-filling representation. Methyl groups are color-coded in red if the ratio is bigger than 1.4; green if $0.7 < S^2_{axis}$ (Aβ42)/S^2_{axis} (Aβ40) < 1.4; and blue if S^2_{axis} (Aβ42)/S^2_{axis} (Aβ40) < 0.7. From Yan et al. [73]. See color insert.

interestingly, the order parameters of methyl groups in the central hydrophobic region (L17 and V18) increased significantly compared with those in Aβ42^{red} (Fig. 13.6B).

The dynamic differences between reduced Aβ40 and reduced Aβ42 are characterized by a more rigid C-terminus and more mobile methyl groups in CHC in Aβ42 than in Aβ40. M35 oxidation enhances the mobility of Aβ42 C-terminus and reduces the rigidity of CHC methyl groups (Fig. 13.6C). Thus, M35 caused Aβ40-like changes in Aβ42 dynamics, while at the same time M35 oxidation reduced Aβ42 aggregation

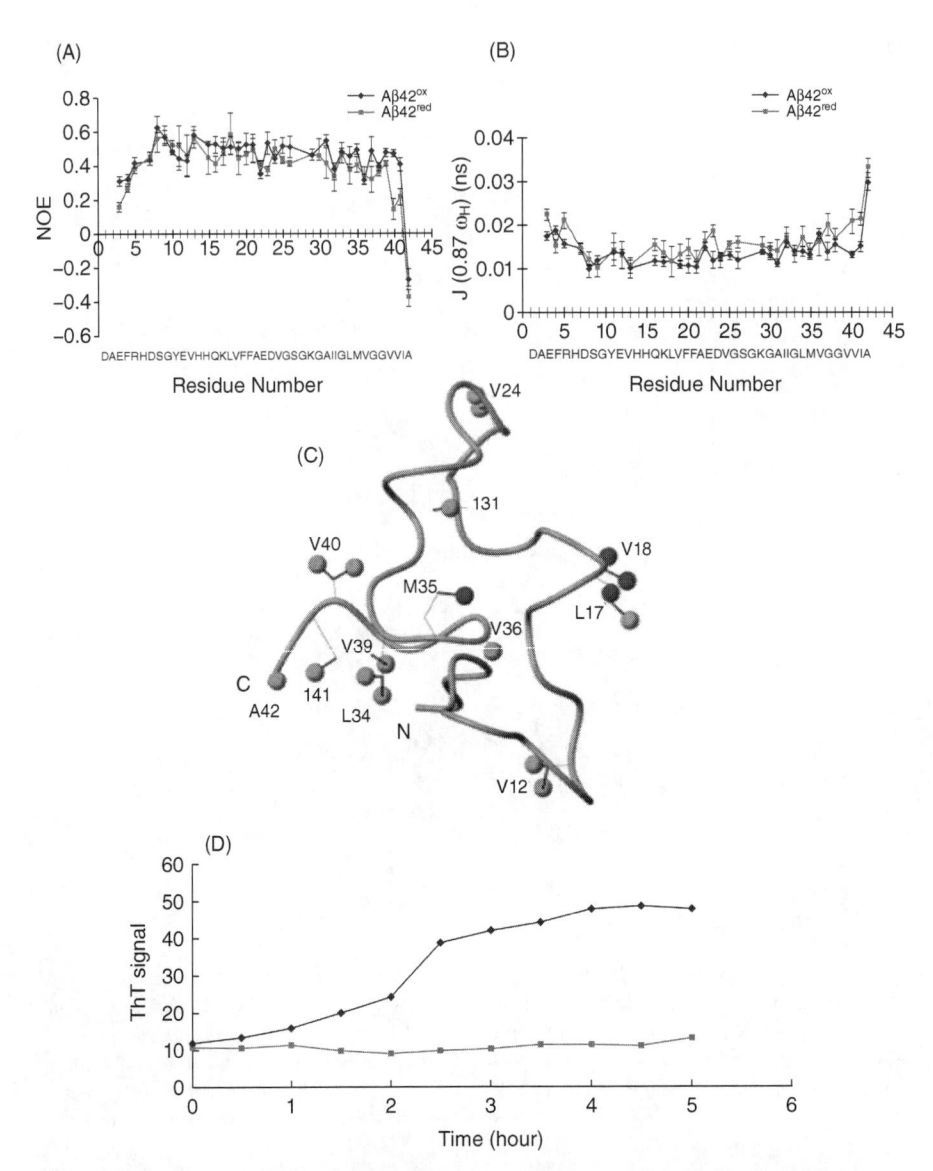

Figure 13.6. M35 oxidation alters the dynamics of Aβ42 on the ps–ns time scale and aggregation kinetics. Backbone ^{15}N NOE (A) and J(0.87ω_H) values (B) indicate increased mobility of the C-terminus of Aβ42 upon M35 oxidation. (C) Backbone ^{15}N NOE difference (NOE$_{Aβ42red}$ − NOE$_{Aβ42ox}$) and side-chain methyl groups $S^2_{Aβ42ox}/S^2_{Aβ42red}$ were mapped onto the ribbon diagram model of Aβ42 monomer from MD simulation [58]. Backbone is shown in red if the NOE difference (NOE$_{Aβ42red}$ − NOE$_{Aβ42ox}$) is bigger than 0.15; gray if −0.15 < NOE$_{Aβ42red}$ − NOE$_{Aβ42ox}$ < 0.15. Methyl groups are shown in blue if $S^2_{Aβ42ox}/S^2_{Aβ4red}$ is bigger than 1.2; green if 0.83 < $S^2_{Aβ42red}$ < 1.2. (D) M35 oxidation slows down Aβ42 aggregation, as shown by ThT assay (blue symbols: Aβ42^{red} and red: Aβ42^{ox}). Panels A–C from Yan et al. [57]. See color insert.

(Fig. 13.6D). This suggests that reduced aggregation of $A\beta 42^{ox}$ may come from changes in protein dynamics. These data strongly support the idea that motional properties in $A\beta$ monomers contribute to the mechanism of $A\beta$ aggregation.

13.3.3.4. MD Simulation Links Structure and Dynamics in Aβ40 and Aβ42 Monomers

Garcia et al. have carried out all-atom simulation with explicit water of both $A\beta 40$ and $A\beta 42$ monomers on the microsecond time scale [58]. A variety of force fields were tested through direct comparison with $^3J_{HNH\alpha}$. It was found that the OPLS, GROMOS, and FF99SB force fields reproduce the J-coupling data well, while OPLS and FF99SB reached the best agreement (Fig. 13.7).

$A\beta 40$ and $A\beta 42$ monomers display distinct structural features (Fig. 13.8). The simulated conformations of $A\beta 40$ can be grouped in two families according to the

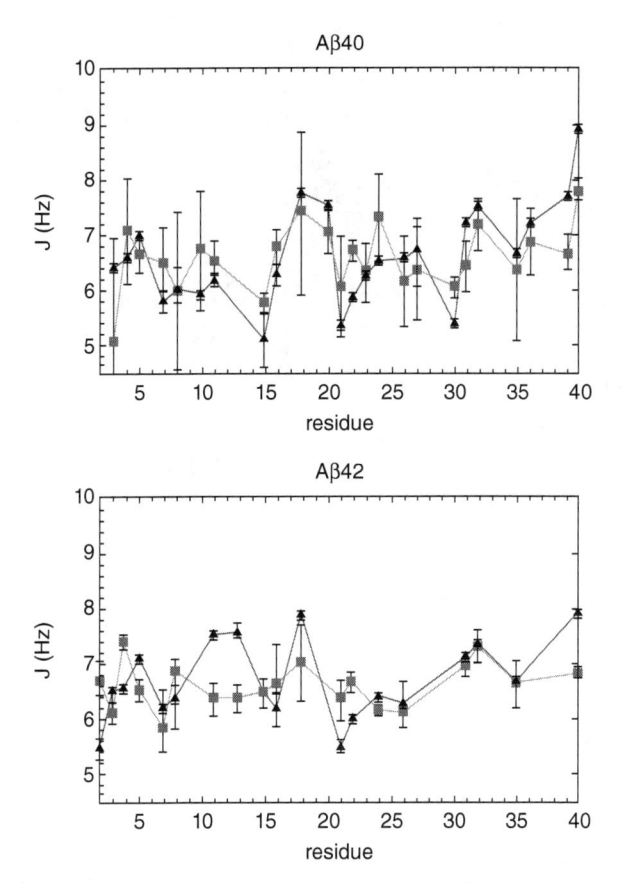

Figure 13.7. Agreement between experimental (gray) and predicted (black) three-bond $J_{HNH\alpha}$ values, plotted against the residue number of $A\beta 40$ and $A\beta 42$. Pearson correlation coefficient is 0.66 for $A\beta 40$ and 0.44 for $A\beta 42$. Both the average value of J and variations along the sequence of the two peptides agree with predicted values from MD using OPLS force field. Adapted from Sgourakis et al. [58].

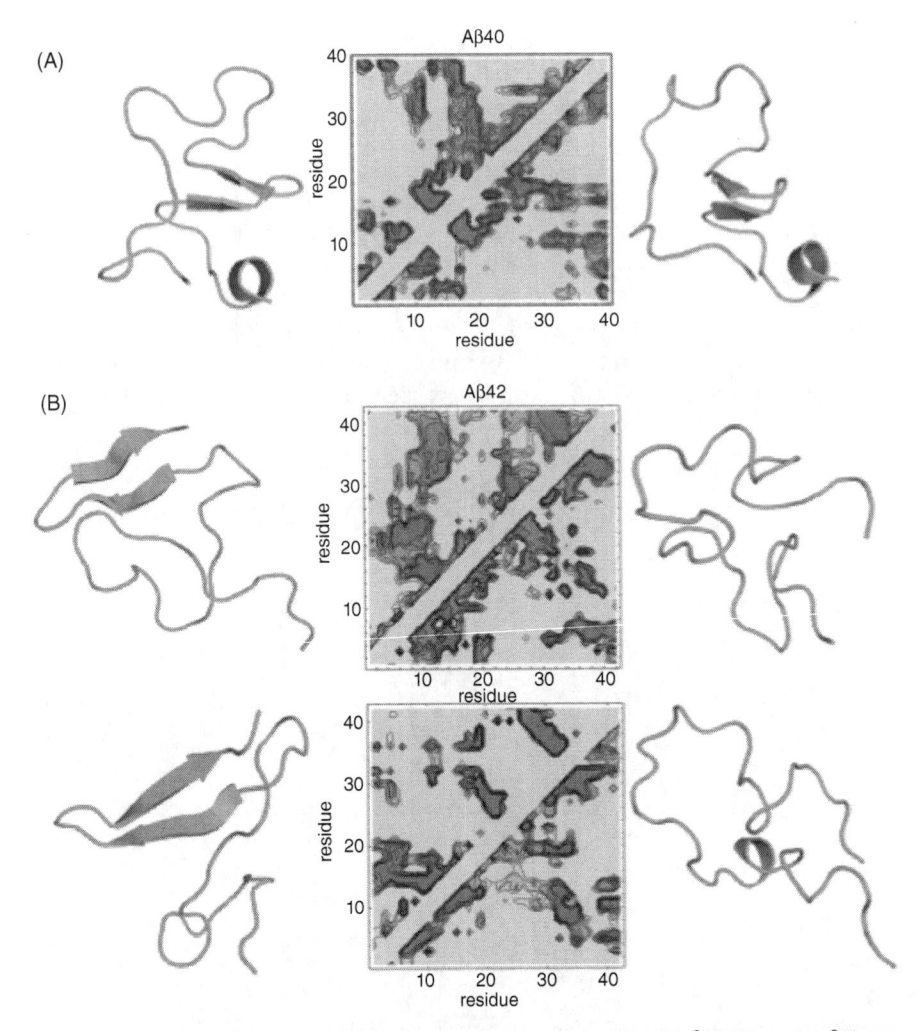

Figure 13.8. Representative conformations within the ensemble of Aβ40 (A) and Aβ42 (B), simulated with the OPLS force field. (A) Central structures from the two dominant clusters are displayed next to the contact maps corresponding to all members of these clusters where a darker color denotes a high probability of contact. They consist of collapsed structures with a short helical region in the N-terminus of the peptide, a short γ-hairpin toward the center of the structure and a disordered C-terminus that can form either a right-handed (left) or a left-handed conformation (right) of the backbone trail. The sizes of the clusters displayed here cover 21% and 11% of the total population, respectively. (B) Central structures from the four dominant clusters of Aβ42 conformations. A diversity of topologies was observed for the backbone of the peptide; however, the C-terminus is usually trapped in a β-hairpin or an extended loop. Clusters are presented in a decreasing size order, from the upper left to the lower right of the figure. Their sizes cover 21%, 5.9%, 5.8%, and 5.6% on a total sample of 21,120 conformations. In all images, the conformations are presented with the N-terminus down and the C-terminus up. From Sgourakis et al. [58].

orientation of the flexible C-terminal backbone (Fig. 13.8A). However, conformations of Aβ42 are more diverse and the C-terminus of Aβ42 is more structured, through the formation of a β-hairpin (Fig. 13.8B). These simulation results provide a framework for interpreting the NMR dynamics data of Aβ40 and Aβ42. The C-terminus of Aβ42 is more rigid than that of Aβ40, probably because the major conformation of the Aβ42 C-terminus is characterized by a β-hairpin while the Aβ40 C-terminus is unstructured (Fig. 13.8). The methyl groups in the CHC of Aβ40 is more rigid because of a γ-hairpin, which is absent in Aβ42 (Fig. 13.8).

13.4. Aβ OLIGOMERS AND MONOMER–OLIGOMER INTERACTION

In recent years, Aβ oligomers have become widely regarded as the more toxic Aβ species than Aβ fibrils. Indeed, Aβ oligomers have been shown to have better correlation with memory loss [59, 60]. Different types of Aβ oligomers abound in the literature, such as Aβ* [61], dimer [62], tetramers [63, 64], pentamers [65], globulomers [66], and Aβ-derived diffusible ligand (ADDL) [67]; however, it is not clear which form is indeed the toxic species in the brain of AD patients or if a range of oligomers with different sizes are responsible for Aβ toxicity. Structural and dynamic data on Aβ oligomer at atomic resolution is urgently needed for understanding AD pathogenesis and structure-based drug discovery.

Olejniczak et al. at Abbot Laboratories provided the first in-depth structural characterization of an Aβ oligomer by solution NMR [8]. MAβ42 was used in the preparation of NMR samples of a toxic "globulomer" (globular oligomer) formed in SDS [66]. Based on HD exchange, inter- and intramolecular NOEs, a novel conformation of Aβ dimer unit within the globulomer, was revealed, containing both parallel, intermolecular and anti-parallel, intramolecular β-strands (Fig. 13.9). This surprising arrangement is confirmed by an L17C/L34C double mutant that should form a disulfide bond that can constrain Aβ into the proposed NMR structure. The disulfide bond-constrained oligomer formed by MAβ42L17C/L34C was shown to have a similar affinity to a globulomer-specific antibody and an HD exchange pattern similar to wild-type globulomer. HD exchange experiments are usually carried out by monitoring how the amide peaks intensity decrease over time using ^{15}N-^{1}H HSQC. However, for HD exchange of fibrils and oligomers, the extent of deuterium exchange is instead read out with monomers after disaggregating the oligomer, due to the low sensitivity and resolution of NMR of the Ab aggregates. In this study, the oligomer sample in D$_2$O is flash-frozen at various time points and then disaggregated in DMSO, and the extent of deuterium exchange is determined by amide signals of Aβ monomers in DMSO. The authors argue that such large structural differences between Aβ oligomer and fibril (Fig. 13.9) are consistent with the idea that Aβ oligomers and fibrils must aggregate through different pathways from the monomer.

Two solid-state (ss) NMR groups have also provided crucial structural insights into Aβ oligomers. Ishii et al. studied a different oligomer, I$_\beta$, which shows enhanced toxicity compared with fibrils in PC12 cells [68]. The oligomer is formed with synthetic Aβ40, which has a diameter of 15–35 nm, much bigger than that of ADDL

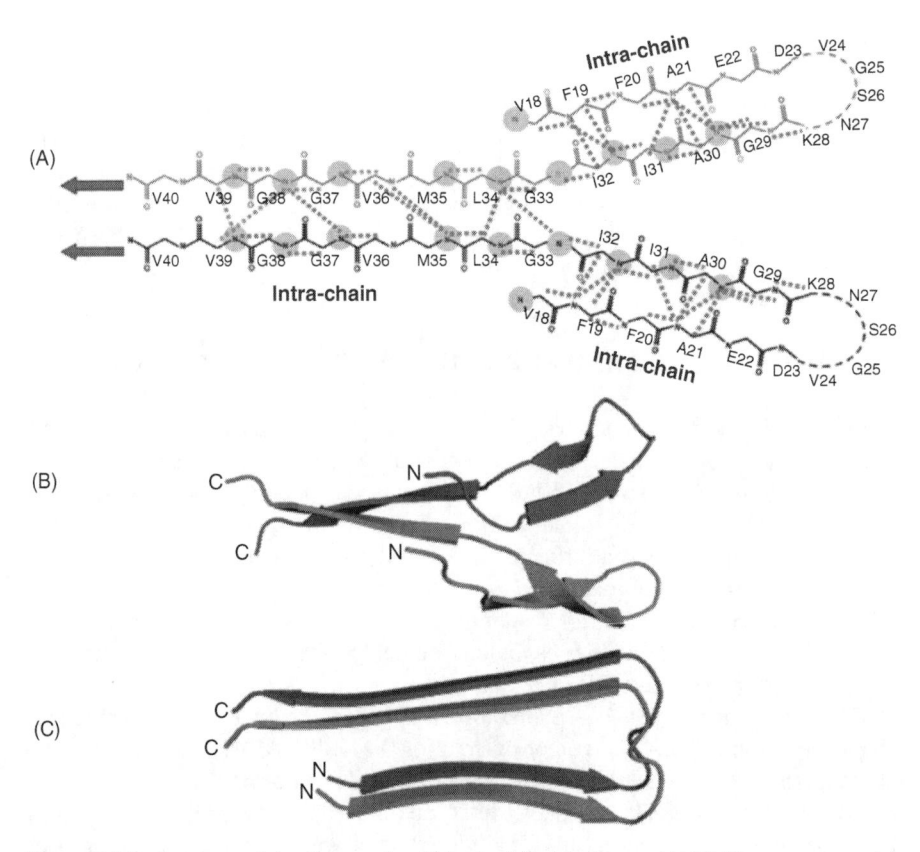

Figure 13.9. Structure of the dimer unit within the Aβ globulomer. (A) NMR structure model of the dimer composed of both inter- and intramolecular β-strands. Dashed lines represent observed NOEs. Yellow sphere represents protected amide groups. Ribbon diagram of the dimer unit within the globulomer (B) and Aβ fibrils (C). From Yu et al. [8].

(~5 nm). It was found that Aβ oligomers and fibrils have the similar ssNMR chemical shift, linewidth, dihedral angle predicted using chemical shifts, and interstrand distances. Thus, I_β predefines Aβ fibril structure and is on the pathway to fibril formation. The differences between this study and the work by Ishii et al. may be reconciled by the differences in the aggregation pathway in the presence (globulomer) and absence (I_β) of detergents. In contrast, recently work by Smith et al. showed that a highly toxic *Aβ42* pentamer has different HD exchange patterns, different carbon–carbon distances from those found in fibrils [65]. Based on these measurements, the oligomer structure is proposed to have three turns at His13-Q15, G25-G29, and G37–G38, while Aβ42 fibrils only have one turn at G25–G29 (Fig. 13.10). The author proposed that the oligomer must convert from loose-packed β-strands to a highly ordered structure in fibrils with completely different β-strand arrangement.

Interactions between Aβ monomers and oligomers can be characterized by the saturation transfer experiment. In this experiment, the large-MW oligomer is selec-

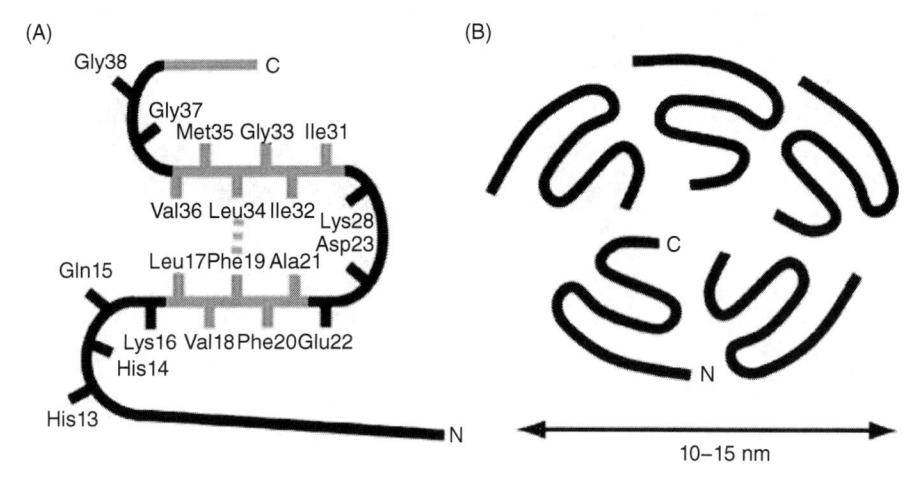

Figure 13.10. Aβ42 pentamer structure based on ssNMR distance measurement and HD exchange. (A) Monomer structure model based on distance constraint between L34 and F19 (dashed line) and solvent protection data. Gray represents amides protected from exchange in the oligomer. (B) Pentamer model. The dimensions are based on atomic force microscopy (AFM), gel filtration, while the C-terminus arrangement is based on HD exchange. From Ahmed et al. [65].

tively saturated by using a resonance frequency specific for the oligomer such as upfield shifted methyl groups. Then the saturation will be transferred to the monomer through the monomer–oligomer complex. Reif et al. found that 0.5 mM Aβ40 in physiological aqueous solution (pH 6.9, 50 mM sodium phosphate) exists in an equilibrium between a monomer and an oligomer [69]. The oligomer has a broad peak at about −0.1 ppm and has an MW of more than 100 kD, as determined by the measurement of translational diffusion coefficient with pulse-field gradient. The interaction sites were determined by the saturation transfer difference experiment and found to involve the central hydrophobic core region, residues 15–24. A similar approach was used for Aβ fragments by Melacini et al. [70–72].

Fawzi and Clore recently quantified the interaction between Aβ monomer and oligomer by combining spin relaxation and saturation transfer [24]. The transverse relaxation rates (R_2) of amide nitrogens and protons increase with concentration from 60 μM, 150 μM to 300 μM. The increase in R_2 is largely uniform along the Aβ sequence, independent of magnetic field strength and nucleus type (Fig. 13.11). Such R_2 enhancement with sample concentration must be due to the exchange with an NMR-invisible Aβ oligomer, "the dark state." This exchange must be in the slow exchange on the NMR time scale, otherwise the rate enhancement would have shown a field dependence. In the slow exchange regime, the R_2 rate increase can be interpreted as the apparent on-rate for monomer conversion to the NMR-invisible oligomer, estimated as 3.1/s at 300 μM. Proton saturation transfer experiment was carried out to probe "the dark state" at a different resonance offset and two field strengths.

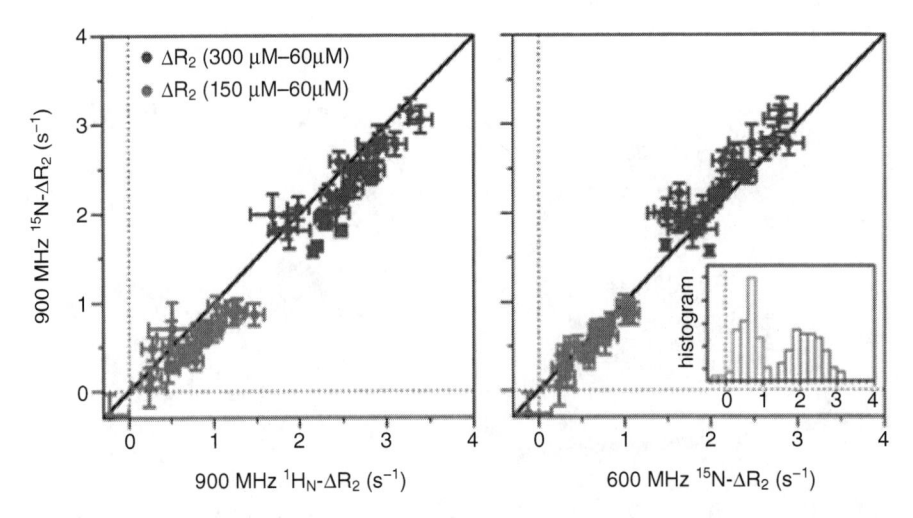

Figure 13.11. Increase in transverse relaxation rate with concentration is independent of nucleus type (left panel) and magnetic field strength (right panel), suggesting that this is due to a slow chemical exchange. From Fawzi et al. [24].

Analysis of the saturation yielded an off-rate of 73/s, while the fraction of peptide within the oligomer "dark state" was about 3%. Variations in R_2 rates in the monomer may be explained by distinct relaxation rates for different regions of $A\beta$ in the oligomer.

13.5. CONCLUSION

In recent years, tremendous progress has been made in the recombinant expression and purification of $A\beta$, NMR sample preparation, and structure and dynamic of $A\beta$ monomers and oligomers. In the next a few years, solution NMR is expected to make major contributions to more in-depth characterization of $A\beta$ structure, dynamics, and aggregation, in particular the structure of $A\beta$ oligomers and protein dynamics in $A\beta$.

REFERENCES

1. Hardy, J. A. and Higgins, G. A. (1992) Alzheimer's disease: The amyloid cascade hypothesis, *Science 256*, 184–185.
2. Hardy, J. (1999) The shorter amyloid cascade hypothesis, *Neurobiol Aging 20*, 85, discussion 87.
3. 3journalCit>Hardy, J. and Selkoe, D. J. (2002) The amyloid hypothesis of Alzheimer's disease: Progress and problems on the road to therapeutics, *Science 297*, 353–356.
4. Hardy, J. (2009) The amyloid hypothesis for Alzheimer's disease: A critical reappraisal, *J Neurochem 110*, 1129–1134.

5. Yan, Y. and Wang, C. (2006) Abeta42 is more rigid than Abeta40 at the C terminus: Implications for Abeta aggregation and toxicity, *J Mol Biol 364*, 853–862.
6. Finder, V. H., Vodopivec, I., Nitsch, R. M., and Glockshuber, R. (2010) The recombinant amyloid-beta peptide Abeta1-42 aggregates faster and is more neurotoxic than synthetic Abeta1-42, *J Mol Biol 396*, 9–18.
7. Walsh, D. M., Thulin, E., Minogue, A. M., Gustavsson, N., Pang, E., Teplow, D. B., and Linse, S. (2009) A facile method for expression and purification of the Alzheimer's disease-associated amyloid β-peptide, *FEBS J 276*, 1266–1281.
8. Yu, L., Edalji, R., Harlan, J. E., Holzman, T. F., Lopez, A. P., Labkovsky, B., Hillen, H., Barghorn, S., Ebert, U., Richardson, P. L., Miesbauer, L., Solomon, L., Bartley, D., Walter, K., Johnson, R. W., Hajduk, P. J., and Olejniczak, E. T. (2009) Structural characterization of a soluble amyloid beta-peptide oligomer, *Biochemistry 48*, 1870–1877.
9. Macao, B., Hoyer, W., Sandberg, A., Brorsson, A.-C., Dobson, C. M., and Härd, T. (2008) Recombinant amyloid beta-peptide production by coexpression with an affibody ligand, *BMC Biotechnol 8*, 82.
10. Hoyer, W., Gronwall, C., Jonsson, A., Stahl, S., and Hard, T. (2008) Stabilization of a beta-hairpin in monomeric Alzheimer's amyloid-beta peptide inhibits amyloid formation, *Proc Natl Acad Sci U S A 105*, 5099–5104.
11. Wiesehan, K., Funke, S. A., Fries, M., and Willbold, D. (2007) Purification of recombinantly expressed and cytotoxic human amyloid-beta peptide 1–42, *J Chromatogr B Analyt Technol Biomed Life Sci 856*, 229–233.
12. Dobeli, H., Draeger, N., Huber, G., Jakob, P., Schmidt, D., Seilheimer, B., Stuber, D., Wipf, B., and Zulauf, M. (1995) A biotechnological method provides access to aggregation competent monomeric Alzheimer's 1–42 residue amyloid peptide, *Biotechnology (N Y) 13*, 988–993.
13. Luhrs, T., Ritter, C., Adrian, M., Riek-Loher, D., Bohrmann, B., Dobeli, H., Schubert, D., and Riek, R. (2005) 3D structure of Alzheimer's amyloid-beta(1-42) fibrils, *Proc Natl Acad Sci U S A 102*, 17342–17347.
14. Hortschansky, P., Schroeckh, V., Christopeit, T., Zandomeneghi, G., and Fändrich, M. (2005) The aggregation kinetics of Alzheimer's β-amyloid peptide is controlled by stochastic nucleation, *Protein Sci 14*, 1753–1759.
15. Lee, E., Hwang, J., Shin, D., Kim, D., and Yoo, Y. (2005) Production of recombinant amyloid-beta peptide 42 as an ubiquitin extension, *Protein Expr Purif 40*, 183–189.
16. Thapa, A., Shahnawaz, M., Karki, P., Raj Dahal, G., Golam Sharoar, M., Yub Shin, S., Sup Lee, J., Cho, B., and Park, I.-S. (2008) Purification of inclusion body—Forming peptides and proteins in soluble form by fusion to *Escherichia coli* thermostable proteins, *BioTechniques 44*, 787–796.
17. Garai, K., Crick, S. L., Mustafi, S. M., and Frieden, C. (2009) Expression and purification of amyloid-beta peptides from *Escherichia coli*, *Protein Expr Purif 66*, 107–112.
18. Zhang, L., Yu, H., Song, C., Lin, X., Chen, B., Tan, C., Cao, G., and Wang, Z. (2009) Expression, purification, and characterization of recombinant human beta-amyloid42 peptide in *Escherichia coli*, *Protein Expr Purif 64*, 55–62.
19. Bockhorn, J. J., Lazar, K. L., Gasser, A. J., Luther, L. M., Qahwash, I. M., Chopra, N., and Meredith, S. C. (2010) Novel semisynthetic method for generating full length beta-amyloid peptides, *Biopolymers 94*, 511–520.

20 Hou, L., Shao, H., Zhang, Y., Li, H., Menon, N. K., Neuhaus, E. B., Brewer, J. M., Byeon, I. J., Ray, D. G., Vitek, M. P., Iwashita, T., Makula, R. A., Przybyla, A. B., and Zagorski, M. G. (2004) Solution NMR studies of the A beta(1–40) and A beta(1-42) peptides establish that the Met35 oxidation state affects the mechanism of amyloid formation, *J Am Chem Soc 126*, 1992–2005.

21 Bitan, G. and Teplow, D. B. (2005) Preparation of aggregate-free, low molecular weight amyloid-beta for assembly and toxicity assays, *Methods Mol Biol 299*, 3–9.

22 Hou, L., Kang, I., Marchant, R. E., and Zagorski, M. G. (2002) Methionine 35 oxidation reduces fibril assembly of the amyloid abeta-(1-42) peptide of Alzheimer's disease, *J Biol Chem 277*, 40173–40176.

23 Danielsson, J., Jarvet, J., Damberg, P., and Graslund, A. (2002) Translational diffusion measured by PFG-NMR on full length and fragments of the Alzheimer A beta(1–40) peptide. Determination of hydrodynamic radii of random coil peptides of varying length, *Magn Reson Chem 40*, S89–S97.

24 Fawzi, N. L., Ying, J., Torchia, D. A., and Clore, G. M. (2010) Kinetics of amyloid beta monomer-to-oligomer exchange by NMR relaxation, *J Am Chem Soc 132*, 9948–9951.

25 Barrow, C. J. and Zagorski, M. G. (1991) Solution structures of beta peptide and its constituent fragments: Relation to amyloid deposition, *Science 253*, 179–182.

26 Riek, R., Guntert, P., Dobeli, H., Wipf, B., and Wuthrich, K. (2001) NMR studies in aqueous solution fail to identify significant conformational differences between the monomeric forms of two Alzheimer peptides with widely different plaque-competence, A beta(1–40)(ox) and A beta(1-42)(ox), *Eur J Biochem 268*, 5930–5936.

27 Lazo, N. D., Grant, M. A., Condron, M. C., Rigby, A. C., and Teplow, D. B. (2005) On the nucleation of amyloid beta-protein monomer folding, *Protein Sci 14*, 1581–1596.

28 Baumketner, A., Bernstein, S. L., Wyttenbach, T., Lazo, N. D., Teplow, D. B., Bowers, M. T., and Shea, J. E. (2006) Structure of the 21–30 fragment of amyloid beta-protein, *Protein Sci 15*, 1239–1247.

29 Grant, M. A., Lazo, N. D., Lomakin, A., Condron, M. M., Arai, H., Yamin, G., Rigby, A. C., and Teplow, D. B. (2007) Familial Alzheimer's disease mutations alter the stability of the amyloid beta-protein monomer folding nucleus, *Proc Natl Acad Sci U S A 104*, 16522–16527.

30 Sandberg, A., Luheshi, L. M., Sollvander, S., Pereira de Barros, T., Macao, B., Knowles, T. P. J., Biverstal, H., Lendel, C., Ekholm-Petterson, F., Dubnovitsky, A., Lannfelt, L., Dobson, C. M., and Hard, T. (2010) Stabilization of neurotoxic Alzheimer amyloid-oligomers by protein engineering, *Proc Natl Acad Sci U S A 107*, 15595–15600.

31 Fawzi, N. L., Yap, E. H., Okabe, Y., Kohlstedt, K. L., Brown, S. P., and Head-Gordon, T. (2008) Contrasting disease and nondisease protein aggregation by molecular simulation, *Acc Chem Res 41*, 1037–1047.

32 Zhang, S., Iwata, K., Lachenmann, M. J., Peng, J. W., Li, S., Stimson, E. R., Lu, Y.-A., Felix, A. M., Maggio, J. E., and Lee, J. P. (2000) The Alzheimer's peptide Aβ adopts a collapsed coil structure in water, *J Struct Biol 130*, 130–141.

33 Jarvet, J., Damberg, P., Danielsson, J., Johansson, I., Eriksson, L. E., and Graslund, A. (2003) A left-handed 3(1) helical conformation in the Alzheimer Abeta(12-28) peptide, *FEBS Lett 555*, 371–374.

34 Bystrom, R., Aisenbrey, C., Borowik, T., Bokvist, M., Lindstrom, F., Sani, M. A., Olofsson, A., and Grobner, G. (2008) Disordered proteins: Biological membranes as two-dimensional aggregation matrices, *Cell Biochem Biophys 52*, 175–189.

35 Aisenbrey, C., Borowik, T., Bystrom, R., Bokvist, M., Lindstrom, F., Misiak, H., Sani, M. A., and Grobner, G. (2008) How is protein aggregation in amyloidogenic diseases modulated by biological membranes? *Eur Biophys J 37*, 247–255.

36 Coles, M., Bicknell, W., Watson, A. A., Fairlie, D. P., and Craik, D. J. (1998) Solution structure of amyloid beta-peptide(1-40) in a water-micelle environment. Is the membrane-spanning domain where we think it is? *Biochemistry 37*, 11064–11077.

37 Shao, H., Jao, S., Ma, K., and Zagorski, M. G. (1999) Solution structures of micelle-bound amyloid beta-(1-40) and beta-(1-42) peptides of Alzheimer's disease, *J Mol Biol 285*, 755–773.

38 Watson, A. A., Fairlie, D. P., and Craik, D. J. (1998) Solution structure of methionine-oxidized amyloid beta-peptide (1–40). Does oxidation affect conformational switching? *Biochemistry 37*, 12700–12706.

39 Jarvet, J., Danielsson, J., Damberg, P., Oleszczuk, M., and Graslund, A. (2007) Positioning of the Alzheimer Abeta(1-40) peptide in SDS micelles using NMR and paramagnetic probes, *J Biomol NMR 39*, 63–72.

40 Wahlstrom, A., Hugonin, L., Peralvarez-Marin, A., Jarvet, J., and Graslund, A. (2008) Secondary structure conversions of Alzheimer's Abeta(1-40) peptide induced by membrane-mimicking detergents, *FEBS J 275*, 5117–5128.

41 Mandal, P. K. and Pettegrew, J. W. (2004) Alzheimer's disease: Soluble oligomeric Abeta(1-40) peptide in membrane mimic environment from solution NMR and circular dichroism studies, *Neurochem Res 29*, 2267–2272.

42 Sticht, H., Bayer, P., Willbold, D., Dames, S., Hilbich, C., Beyreuther, K., Frank, R. W., and Rosch, P. (1995) Structure of amyloid A4-(1-40)-peptide of Alzheimer's disease, *Eur J Biochem 233*, 293–298.

43 Crescenzi, O., Tomaselli, S., Guerrini, R., Salvadori, S., D'Ursi, A. M., Temussi, P. A., and Picone, D. (2002) Solution structure of the Alzheimer amyloid beta-peptide (1–42) in an apolar microenvironment. Similarity with a virus fusion domain, *Eur J Biochem 269*, 5642–5648.

44 Tomaselli, S., Esposito, V., Vangone, P., van Nuland, N. A., Bonvin, A. M., Guerrini, R., Tancredi, T., Temussi, P. A., and Picone, D. (2006) The alpha-to-beta conformational transition of Alzheimer's Abeta-(1-42) peptide in aqueous media is reversible: A step by step conformational analysis suggests the location of beta conformation seeding, *Chembiochem 7*, 257–267.

45 Lim, K. H., Henderson, G. L., Jha, A., and Louhivuori, M. (2007) Structural, dynamic properties of key residues in Aβ amyloidogenesis: Implications of an important role of nanosecond timescale dynamics, *Chembiochem 8*, 1251–1254.

46 Zheng, Y. and Yang, D. (2004) Measurement of dipolar cross-correlation in methylene groups in uniformly 13C-, 15N-labeled proteins, *J Biomol NMR 28*, 103–116.

47 Houben, K. and Boelens, R. (2004) Side chain dynamics monitored by 13C-13C cross-relaxation, *J Biomol NMR 29*, 151–166.

48 Lee, A. L., Kinnear, S. A., and Wand, A. J. (2000) Redistribution and loss of side chain entropy upon formation of a calmodulin-peptide complex, *Nat Struct Biol 7*, 72–77.

49 Ishima, R. and Torchia, D. A. (2000) Protein dynamics from NMR, *Nat Struct Biol 7*, 740–743.

50 Kim, W. and Hecht, M. H. (2005) Sequence determinants of enhanced amyloidogenicity of Alzheimer A{beta}42 peptide relative to A{beta}40, *J Biol Chem 280*, 35069–35076.

51. Zhang, X., Sui, X., and Yang, D. (2006) Probing methyl dynamics from 13C autocorrelated and cross-correlated relaxation, *J Am Chem Soc 128*, 5073–5081.
52. Soto, C., Castano, E. M., Frangione, B., and Inestrosa, N. C. (1995) The alpha-helical to beta-strand transition in the amino-terminal fragment of the amyloid beta-peptide modulates amyloid formation, *J Biol Chem 270*, 3063–3067.
53. Christopeit, T., Hortschansky, P., Schroeckh, V., Guhrs, K., Zandomeneghi, G., and Fandrich, M. (2005) Mutagenic analysis of the nucleation propensity of oxidized Alzheimer's beta-amyloid peptide, *Protein Sci 14*, 2125–2131.
54. Kuo, Y. M., Kokjohn, T. A., Beach, T. G., Sue, L. I., Brune, D., Lopez, J. C., Kalback, W. M., Abramowski, D., Sturchler-Pierrat, C., Staufenbiel, M., and Roher, A. E. (2001) Comparative analysis of amyloid-beta chemical structure and amyloid plaque morphology of transgenic mouse and Alzheimer's disease brains, *J Biol Chem 276*, 12991–12998.
55. Bitan, G., Tarus, B., Vollers, S. S., Lashuel, H. A., Condron, M. M., Straub, J. E., and Teplow, D. B. (2003) A molecular switch in amyloid assembly: Met35 and amyloid beta-protein oligomerization, *J Am Chem Soc 125*, 15359–15365.
56. Bernstein, S. L., Dupuis, N. F., Lazo, N. D., Wyttenbach, T., Condron, M. M., Bitan, G., Teplow, D. B., Shea, J.-E., Ruotolo, B. T., Robinson, C. V., and Bowers, M. T. (2009) Amyloid-beta protein oligomerization and the importance of tetramers and dodecamers in the aetiology of Alzheimer's disease, *Nat Chem 1*, 326–331.
57. Yan, Y., McCallum, S. A., and Wang, C. (2008) M35 oxidation induces Abeta40-like structural and dynamical changes in Abeta42, *J Am Chem Soc 130*, 5394–5395.
58. Sgourakis, N. G., Yan, Y., McCallum, S. A., Wang, C., and Garcia, A. E. (2007) The Alzheimer's peptides Abeta40 and 42 adopt distinct conformations in water: A combined MD/NMR study, *J Mol Biol 368*, 1448–1457.
59. Kuo, Y. M., Emmerling, M. R., Vigo-Pelfrey, C., Kasunic, T. C., Kirkpatrick, J. B., Murdoch, G. H., Ball, M. J., and Roher, A. E. (1996) Water-soluble Abeta (N-40, N-42) oligomers in normal and Alzheimer disease brains, *J Biol Chem 271*, 4077–4081.
60. McLean, C. A., Cherny, R. A., Fraser, F. W., Fuller, S. J., Smith, M. J., Beyreuther, K., Bush, A. I., and Masters, C. L. (1999) Soluble pool of Abeta amyloid as a determinant of severity of neurodegeneration in Alzheimer's disease, *Ann Neurol 46*, 860–866.
61. Lesne, S., Koh, M. T., Kotilinek, L., Kayed, R., Glabe, C. G., Yang, A., Gallagher, M., and Ashe, K. H. (2006) A specific amyloid-beta protein assembly in the brain impairs memory, *Nature 440*, 352–357.
62. Shankar, G. M., Li, S., Mehta, T. H., Garcia-Munoz, A., Shepardson, N. E., Smith, I., Brett, F. M., Farrell, M. A., Rowan, M. J., Lemere, C. A., Regan, C. M., Walsh, D. M., Sabatini, B. L., and Selkoe, D. J. (2008) Amyloid-beta protein dimers isolated directly from Alzheimer's brains impair synaptic plasticity and memory, *Nat Med 14*, 837–842.
63. Chen, Y. R. and Glabe, C. G. (2006) Distinct early folding and aggregation properties of Alzheimer amyloid-beta peptides Abeta40 and Abeta42: Stable trimer or tetramer formation by Abeta42, *J Biol Chem 281*, 24414–24422.
64. Bitan, G., Lomakin, A., and Teplow, D. B. (2001) Amyloid beta-protein oligomerization: Prenucleation interactions revealed by photo-induced cross-linking of unmodified proteins, *J Biol Chem 276*, 35176–35184.
65. Ahmed, M., Davis, J., Aucoin, D., Sato, T., Ahuja, S., Aimoto, S., Elliott, J. I., Van Nostrand, W. E., and Smith, S. O. (2010) Structural conversion of neurotoxic amyloid-beta(1-42) oligomers to fibrils, *Nat Struct Mol Biol 17*, 561–567.

66 Barghorn, S., Nimmrich, V., Striebinger, A., Krantz, C., Keller, P., Janson, B., Bahr, M., Schmidt, M., Bitner, R. S., Harlan, J., Barlow, E., Ebert, U., and Hillen, H. (2005) Globular amyloid beta-peptide1-42 oligomer—A homogenous and stable neuropathological protein in Alzheimer's disease, *J Neurochem 95*, 834–847.

67 Kayed, R., Head, E., Thompson, J. L., McIntire, T. M., Milton, S. C., Cotman, C. W., and Glabe, C. G. (2003) Common structure of soluble amyloid oligomers implies common mechanism of pathogenesis, *Science 300*, 486–489.

68 Chimon, S., Shaibat, M. A., Jones, C. R., Calero, D. C., Aizezi, B., and Ishii, Y. (2007) Evidence of fibril-like beta-sheet structures in a neurotoxic amyloid intermediate of Alzheimer's beta-amyloid, *Nat Struct Mol Biol 14*, 1157–1164.

69 Narayanan, S. and Reif, B. (2005) Characterization of chemical exchange between soluble and aggregated states of beta-amyloid by solution-state NMR upon variation of salt conditions, *Biochemistry 44*, 1444–1452.

70 Huang, H., Milojevic, J., and Melacini, G. (2008) Analysis and optimization of saturation transfer difference NMR experiments designed to map early self-association events in amyloidogenic peptides, *J Phys Chem B 112*, 5795–5802.

71 Milojevic, J., Esposito, V., Das, R., and Melacini, G. (2007) Understanding the molecular basis for the inhibition of the Alzheimer's Abeta-peptide oligomerization by human serum albumin using saturation transfer difference and off-resonance relaxation NMR spectroscopy, *J Am Chem Soc 129*, 4282–4290.

72 Milojevic, J., Raditsis, A., and Melacini, G. (2009) Human serum albumin inhibits Abeta fibrillization through a "monomer-competitor" mechanism, *Biophys J 97*, 2585–2594.

73 Yan, Y., Liu, J., McCallum, S. A., Yang, D., and Wang, C. (2007) Methyl dynamics of the amyloid-beta peptides Abeta40 and Abeta42, *Biochem Biophys Res Commun 362*, 410–414.

14

THERMODYNAMIC AND KINETIC MODELS FOR AGGREGATION OF INTRINSICALLY DISORDERED PROTEINS

Scott L. Crick and Rohit V. Pappu

14.1. INTRODUCTION

Like most generic carbon-based polymers, proteins are only marginally soluble in aqueous solvents. For most proteins solubility becomes an issue when protein concentrations are in the micromolar range or higher. Experiments have shown that protein solubility decreases as the stability of the folded state decreases vis-à-vis partially unfolded states [1]. This suggests that stable folded states lead to homogeneous protein solutions with protein molecules dispersed in aqueous solvents. The alternative is protein aggregation, which is the process that leads to separation of homogeneous protein solutions into a dilute phase of soluble, isolated proteins, and a protein-rich phase characterized by solvent exclusion and significant intermolecular interactions between proteins. Avoiding phase separation and keeping proteins soluble is essential for realizing the rich set of functional possibilities that can be derived from interactions with protein surfaces [2–5]. Approximately 70% of proteins in eukaryotic proteomes fold into well-defined compact three-dimensional structures [2–4]. Protein folding may be viewed as a strategy to keep proteins

Protein and Peptide Folding, Misfolding, and Non-Folding, First Edition. Edited by Reinhard Schweitzer-Stenner.
© 2012 John Wiley & Sons, Inc. Published 2012 by John Wiley & Sons, Inc.

soluble, and in this regard sequences that fold are viewed as being special when compared with random sequences of similar length and composition.

In contrast to proteins with stable folded states, roughly 30% of proteins in eukaryotic proteomes are intrinsically disordered proteins (IDPs) [2–4]. These are a class of proteins that purportedly fail to adopt stable three-dimensional structures in aqueous solutions [2, 4, 6]. IDPs are characterized by lower overall hydrophobicity than their foldable counterparts [7, 8]. Recent work has shown that IDPs either form disordered globules, namely ensembles of disparate structures of equivalent compactness and stabilities or swollen random coils [9–17]. The specific ensemble type is determined by the net charge per residue [16, 17]. Unlike for their foldable counterparts, solubility in aqueous solvents has to be a major problem for IDPs, specifically for those IDPs that are disordered globules,[1] because these sequences cannot fold unless they are part of protein–protein or protein–nucleic acid complexes. The question of how IDPs maintain their functional relevance in regulation and signaling without creating problems of aggregation and insolubility is of considerable importance. Several hypotheses have been put forth to answer this question. These include: (1) the co-evolution of "gatekeeper" flanking sequences that protect IDPs from aggregation by increasing their lifetime, that is, by making IDPs kinetically stable [18]; (2) co-evolution of chaperones to protect aggregation-prone IDPs from dysfunctional heterotypic and homotypic interactions [19, 20]; (3) regulation of IDP expression levels and half-lives through synergistic control of IDP transcription and degradation [21]; (4) increased solubility through posttranslational modifications that alter properties such as the net charge per residue [22]; (5) expressing IDPs as fusions with stable protein domains and releasing the disordered regions through proteolysis when necessary and following this up by degradation through the ubiquitin–proteasome system [5]. All of these strategies resemble those pursued by engineers and scientists who work with colloids, where the goal is to ensure against phase separation.

Colloids are particles or molecules with diameters between one nanometer and tens of microns that are dispersed in solution [23–27]. Colloidal dispersions are thermodynamically unstable (or more precisely metastable) because the globally stable ground state for a colloidal dispersion is the phase-separated state [23–25, 27]. However, this ground state is undesirable because colloids lose their natural functions. For example, pigments found in paint are colloids and if these pigments form insoluble aggregates, the paint ceases to be useful. A major focus in colloid science is keeping colloidal dispersions stable, which of course means maintaining their kinetic stability [23–25, 27]. If colloidal particles can be kept from aggregating on time scales that are relevant for their intended use, then the colloids are considered stable [27]. Of course proteins are different from colloidal particles because they have internal degrees of freedom; they are also different from generic polymers because they are considerably shorter and their sequences have higher complexity when

---

[1] Analysis of IDP sequences suggests that a majority have low values for the net charge per residue and fit the description of being disordered globules (A. H. Mao, A. Spencer, and R. V. Pappu, unpublished data).

compared with simple homopolymers or block copolymers. Despite these obvious differences, the coarse-grain phase behavior of proteins, specifically IDPs, can be described using theories borrowed from colloidal science and the physics of flexible block copolymers—a fact that remains underappreciated in the protein literature.

Recent reviews have suggested that the phase-separated state might be the globally stable state for proteins [1, 28, 29]. This does not contradict Anfinsen's thermodynamic hypothesis because the issue of stability cannot be decoupled from protein concentration, although this has not precluded the depiction of confusing cartoons of free energy landscapes that disregard the effects of protein concentration. In dilute solutions of foldable proteins, the folded state must minimize the free energy of the protein plus solvent system, thus giving rise to a homogeneously mixed protein solution. For IDPs, the situation is complicated. One can envisage applicability of either Anfinsen's thermodynamic hypothesis or the colloidal scenario. In the former, the free energy of the protein plus solvent system is minimized when the IDP is part of either a binary or multi-molecular complex. In the colloidal scenario, IDPs dispersed in aqueous solvents represent a kinetically stable situation. Irrespective of how or why stability is realized in dilute solutions, the situation is be fundamentally different when protein concentrations increase, because the effects of homotypic intermolecular interactions will eventually lead to the thermodynamically stable phase-separated state. This issue becomes particularly relevant when we consider the crowded environments in which IDPs have to perform their natural functions.

The remainder of this chapter is organized as follows: We start with a summary of a thermodynamic description for phase separation. We then contrast this description with more traditional descriptions based on microstate partitioning that is common in the protein aggregation literature [30–37]. From here, we segue into models for the kinetics of protein aggregation, which is then followed by an overview of the weaknesses. Prior to presenting the concluding section and an outlook for future work, we discuss the salient features of protein aggregation kinetics from the vantage point of colloidal science.

14.2. THERMODYNAMICS OF PROTEIN AGGREGATION— THE PHASE DIAGRAM APPROACH

It is worth reiterating to remember that proteins are polymers and aggregation is a phase separation process. Casting aggregation of IDPs in terms of a polymer phase separation process allows us to touch base with the rich theoretical framework of polymer physics [38–50]. The properties of polymers are determined by the nature of the subunits comprising the polymer chain and the nature of the solvent and for proteins, the subunits are amino acids and the solvent is generally aqueous. For a given polymer–solvent combination, the conformational and phase behavior of a polymer can be understood from the interplay between chain–chain interactions, chain–solvent interactions, and solvent–solvent interactions [38, 47]. In solutions where the chain–chain interactions are favored over chain–solvent interactions, polymers collapse to adopt compact, globular conformations to minimize the polymer–solvent interface [38, 47]. Under such conditions, the polymer is said to be in a poor solvent. In solutions where the chain–solvent interactions are preferred to chain–chain interactions, polymers prefer the swollen coil state whereby the polymer–solvent interface is maximized and the polymer and solvent mix on all length scales [38, 47]. Under such conditions the polymer is said to be in a good solvent. Intrinsically flexible polymers transition between globule and coil states as the solvent quality is altered either through changes in thermodynamic parameters such as temperature and hydrostatic pressure or by altering solution conditions through the addition of cosolutes, salts, or changes in pH [51, 52].

The driving forces that cause a polymer to collapse in a poor solvent are similar to the forces that cause polymers to self-associate and eventually phase separate. A generic phase diagram for a polymer with an upper critical solution temperature (UCST)[2] is shown in Figure 14.1 [47]. Here the temperature controls solvent quality. As T increases, the solvent quality improves until eventually a temperature is reached above which the solvent is good for all polymer concentrations. T_θ is the crossover or theta temperature and represents the temperature at which the chain–chain interactions exactly balance the chain–solvent interactions. The abscissa, labeled φ, denotes the volume fraction of polymer in a solution. Low values of φ represent dilute solutions.

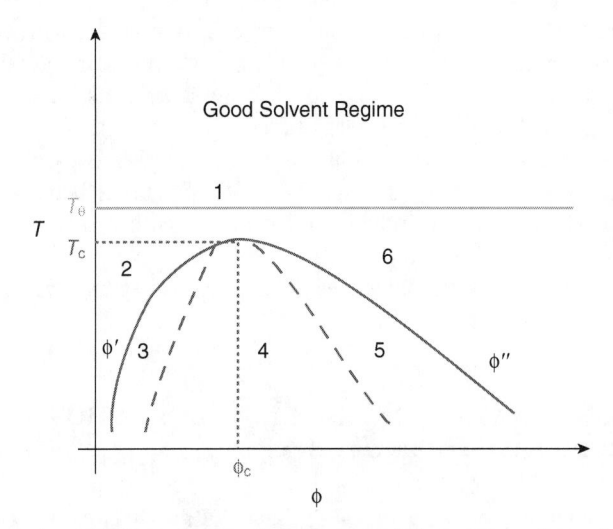

Figure 14.1. Archetypal phase diagram for polymer solutions. The ordinate denotes improving solvent quality expressed as temperature. At the theta-temperature, T_θ, and beyond (good solvent regime) no phase separation is observed. For $T < T_\theta$, a homogeneous mixed phase of polymer in solvent is formed in Region 2, and of solvent in polymer in Region 6. Conversely, phase separation is realized in Regions 3, 4, and 5. The solid red curve denotes the binodal, while the dashed red curve denotes the spinodal. See color insert.

---

[2] The assumption of a UCST is reasonable for polypeptides because they are variant polyamides with secondary amides as the main repeating units and side chains of diverse chemistries.

There are six distinct regions in Figure 14.1. Region 1 corresponds to the good solvent regime. In this region, aggregation does not occur because chain–solvent interactions are favored over chain–chain interactions. The solid red curve is the binodal or the phase separation boundary and is identical to the coexistence curve for a binary mixture. Let $\Delta G_{mix}(\varphi,T)$ denote the free energy of mixing for a binary mixture of polymers in a low-molecular-weight solvent. For each temperature, T, the binodal represents the envelope of points φ' and φ'' that satisfy the condition:

$$\left.\frac{\partial \Delta G_{mix}}{\partial \varphi}\right|_{\varphi=\varphi'} = \left.\frac{\partial \Delta G_{mix}}{\partial \varphi}\right|_{\varphi=\varphi''}.$$

If the combination of T and φ place the polymer + solvent system below the binodal, then the system undergoes spontaneous phase separation. The kinetic mechanism of spontaneous phase separation will depend on the precise location below the binodal, and this location is referred to as the quench depth. The dashed curve in Figure 14.1 is the spinodal or the stability boundary, which is the envelope of points for which

$$\frac{\partial^2 \Delta G_{mix}(\varphi,T)}{\partial \varphi^2} = 0.$$

Below the spinodal line,

$$\frac{\partial^2 \Delta G_{mix}(\varphi,T)}{\partial \varphi^2} < 0,$$

and the homogeneously mixed state is unstable, whereas between the binodal and spinodal,

$$\frac{\partial^2 \Delta G_{mix}(\varphi,T)}{\partial \varphi^2} > 0,$$

and the homogeneous mixed state is metastable.

For all points below the binodal, the concentration of polymer that remains in solution at equilibrium with the precipitate is determined by the location of the binodal [47]. This concentration is the saturation concentration, φ_s, and can be determined using cloud point measurements that assess the concentration (for a given T) beyond which the solution becomes cloudy [53]. In the protein aggregation literature, the saturation concentration is often incorrectly referred to as the critical concentration [32, 33, 54, 55]. The critical point (φ_c, T_c) on the phase diagram is the point where the binodal and spinodal coincide and is characterized by conformational and concentration fluctuations on all length scales [47]. For homopolymers, the chain length determines the location of the binodal and spinodal; for block copolymers and heteropolymers (such as proteins) the ratio of solvophilic to solvophobic groups is an additional parameter that contributes to the location of the binodal and spinodal [46].

Region 2 lies to the left of the binodal region. In this sub-saturated region, $\varphi < \varphi_s$ and the solution is therefore too dilute for phase separation to be thermodynamically favored. Instead, homogeneously mixed globules or mesoglobules[3] characterize the solution. If the concentration is increased beyond the binodal, the solution is in either Region 3 or Region 4 and is supersaturated. In Region 3, between the binodal and spinodal, the solution is metastable. This is the nucleation–elongation regime where, for a given temperature, the concentration fluctuations must satisfy the condition $(\langle\varphi^2\rangle - \langle\varphi\rangle^2)^{1/2} > |\Delta\varphi|$, where $|\Delta\varphi|$ is magnitude of the gap between the binodal and the spinodal. These concentration fluctuations are needed to form the nucleus or nuclei that have the same composition as the new thermodynamically favored phase-separated state so this new phase can grow within the old, homogeneously mixed metastable state. Region 4 lies below the spinodal, and within this region there is no barrier to phase separation. As a result, phase separation in this region is thermodynamically downhill and is kinetically limited by the diffusion of individual chains or clusters of chains [56]. In Region 5, the solution is concentrated and a barrier exists for growing the soluble phase inside the polymer aggregate. Finally, Region 6 represents a stable, polymer-rich phase with low-molecular-weight solvent dispersed in it.

The preceding discussion is important because it emphasizes the extent of information one can glean from full knowledge of a polymer's phase diagram. This demonstration defies the anecdotal view that thermodynamic descriptions do not provide insights into the mechanisms of polymer aggregation. On the contrary, knowledge of the phase diagram or even parts of the phase diagram, such as the low-concentration arm of the binodal and spinodal or the binodal alone, will narrow the range of mechanisms that are applicable to the aggregation of specific polymer + solvent systems. Such information can be gathered using combinations of osmotic and scattering measurements and is available for a variety of polymer + solvent combinations [47]. Indeed, these types of characterization were the purview of classical biophysics measurements on proteins. However, to our knowledge, phase diagrams are unavailable for any of the important aggregation-prone IDPs. This shortcoming leads to the development of a range of cartoon-based models for IDP aggregation, the tenets of which have not been tested in any quantitative manner. Alternatively, one uses rather elaborate models, which we refer to as microstate partitioning models (MPMs), to fit rather sparse kinetic and thermodynamic data [30–37]. The next two sections will provide a summary of these models or schemes. It will become readily obvious that quantitative modeling of the detailed schemes requires knowledge or fitting of large numbers of parameters. However, most, if not all experiments, tend to be silent about the large number of microscopic equilibria and rate constants [33]. In our opinion, MPMs are inherently inferior to simpler descriptions based on phase diagrams because the number of measurements needed is rather overwhelming and generally inaccessible, whereas knowledge of a phase diagram or even regions of the phase diagram helps narrow down the family of conceivable mechanisms.

---

[3] Mesoglobules are clusters or oligomers of globules and may be viewed as globules of globules.

Before delving into mechanisms based on MPMs for phase separation of IDPs, it is important to clarify some criteria for this process. These criteria are derived from the inspection of a polymer phase diagram and are as follows:

1. There exists a saturation concentration (C_s) below which phase separation will not occur. For $C > C_s$, phase separation is thermodynamically favored. C_s can be quantified as the concentration of soluble material that is in equilibrium with the precipitate.

2. For a given polymer, many variables, including the solvent type, temperature, pressure, pH, salts, and the presence of macromolecular cosolutes or small molecule cosolutes such as osmolytes, determine the saturation concentration.

3. If the combination of φ and solution conditions places the polymer solution between the binodal and spinodal (Region 3 in Fig. 14.1), phase separation requires the formation of one or more barrier-limited species, referred to as the nucleus (if the process is homogeneous) or nuclei (if the distribution of barriers is heterogeneous). Conversely, if the combination of φ and solution conditions places the polymer solution below the spinodal, then phase separation is limited only by inter- and intramolecular diffusion and the mechanism is referred to as spinodal decomposition because even the smallest perturbation in φ causes phase separation.

4. It is important to distinguish between aggregation and phase separation, which are often used interchangeably. Aggregation refers to the formation of intermolecular clusters that are stabilized either by confinement (as in the aggregation and packing of hard spheres) or through favorable intermolecular interactions (as is the case with colloidal particles). These clusters/aggregates are characterized by an aggregate number n_A (or degree of polymerization if one is referring to linear aggregates) where $n_A = 2$ is the smallest aggregate. Whether an aggregate of size n_A is dispersed in solvent or part of the phase-separated state will depend on the precise balance between chain–chain and chain–solvent interactions. The phase-separated state will be characterized by two separate values for $\langle n_A \rangle$ to denote the average aggregate number $\langle n_A^s \rangle$ for the dispersed phase in solution that is in equilibrium with the polymer-rich phase or precipitate characterized by $\langle n_A^p \rangle$. Typically, one makes the simplifying assumption that $\langle n_A^s \rangle \to 1$ and $\langle n_A^p \rangle \to \infty$; that is, the soluble material is assumed to be purely monomeric. However, there is no a priori justification for making this stringent assumption and there is growing evidence that $\langle n_A^s \rangle > 1$ for aggregation-prone IDPs. Additionally, the distribution of aggregate numbers under conditions where $\varphi < \varphi_s$ need not be similar to the distributions of aggregate numbers within the soluble material that is in equilibrium with the polymer-rich precipitate when $\varphi > \varphi_s$.

5. In the literature, one sees implicit terminological conflation between aggregation and phase separation. Indeed, the latter term is seldom used. To maintain consistency with the protein literature, we will co-opt the conflation for the remainder of this document, with the caveat that this conflation is conceptually inaccurate. This conflation originates, at least partially, in the formal structure of the MPMs.

14.3. THERMODYNAMICS OF IDP AGGREGATION (PHASE SEPARATION)—MPM DESCRIPTION

Aggregation of IDPs typically results in the formation of two types of structures in the polymer-rich phase: amorphous aggregates and ordered aggregates such as amyloid fibrils [57, 58]. Historically, models of IDP aggregation have been focused on the formation of amyloid fibrils and these models have largely discounted the formation of amorphous aggregates. The two most likely reasons are that these structures were originally implicated as a possible cause for the disease and that the formation of these structures is easily followed with simple techniques such as fluorescence enhancement of thioflavin T [59–62]. MPMs for IDP aggregation are built on a wealth of literature describing the formation of fibrillar structures such as F-actin and sickle cell hemoglobin [63–67]. Our discussion of MPMs is not meant to be a comprehensive review of these models because comprehensive coverage is already available in the literature [30–37, 68]. We will instead provide a critical overview of basic mechanisms and evaluate these MPMs in terms of their predictive power and ability to reproduce known characteristics of IDP aggregation.

MPMs for thermodynamics of linear aggregation: We start with the most intuitive mechanism for the formation of a linear aggregate, which involves growth via monomer addition. This mechanism is shown in Scheme 14.1. With knowledge of the rate constants of each step and the starting concentration, this simple mechanism can describe an aggregation mechanism with no off-pathway intermediates (where off-pathway means that these species are non-productive toward the formation of the final product). It is impractical to enumerate every elementary step required to form high-molecular-weight species such as fibrils. Even if such an enumeration were practical, one could never measure the concentration of each species along the pathway, which is a requirement for determining the associated rate constants for each step. As a result, researchers are forced to make simplifying assumptions to this model.

$$M_1 + M_1 \underset{k_{-1}}{\overset{k_1}{\rightleftharpoons}} M_2; \quad K_2 = \frac{[M_2]}{[M_1][M_1]}$$

$$M_2 + M_1 \underset{k_{-2}}{\overset{k_2}{\rightleftharpoons}} M_3; \quad K_3 = \frac{[M_3]}{[M_2][M_1]} \qquad \text{(Scheme 14.1)}$$

$$\vdots$$

$$M_i + M_1 \underset{k_{-2}}{\overset{k_2}{\rightleftharpoons}} M_{i+1}; \quad K_{i+1} = \frac{[M_{i+1}]}{[M_{i+1}][M_1]}.$$

Isodesmic aggregation: One simplification is to assume that every step in the reaction has the same equilibrium constant (K) [69, 70]. This is known as isodesmic polymerization and the consequences of this simplification are shown in the equations for isodesmic aggregation. Using this assumption, it can be shown that the concentration of the monomeric species M_1, and therefore any species along the

THERMODYNAMICS OF IDP AGGREGATION (PHASE SEPARATION)

reaction coordinate, is determined simply by the initial monomer concentration C_0 and the equilibrium constant K. Assuming that one can measure this monomer concentration M_1 at equilibrium, it is simple to determine K from a concentration dependence of M_1 as a function of C_0. Figure 14.2A shows a plot of DP, the degree of aggregation, as a function of the dimensionless parameter KC_0. This reveals that extensive aggregation cannot be realized unless $KC_0 \gg 1$, requiring that for typical concentrations ($C_0 \ll 1$ M), K must be large. This is illustrated in Figure 14.2B.

Equations for isodesmic aggregation

$$K_i = K \text{ for } 2 \leq i < \infty$$

$$[M_2] = K[M_1]^2$$

$$[M_3] = K[M_2][M_1] = K^2[M_1]^3$$

$$\vdots$$

$$[M_{i+1}] = K^i[M_1]^{i+1}$$

For $K[M_1] < 1$

$$C_o = \sum_{i=1}^{\infty} i[M_i] = \frac{[M_1]}{(1-K[M_1])^2} \text{ is the initial concentration} \quad \text{(Scheme 14.2)}$$

Solving for $[M_1]$ yields: $[M_1] = \dfrac{\left(2K + \dfrac{1}{C_o}\right) - \sqrt{\left(2K + \dfrac{1}{C_o}\right)^2 - 4K^2}}{2K^2}$

$$C_P = \sum_{i=1}^{\infty}[M_i] = \frac{[M_1]}{(1-K[M_1])^2} \text{ is the partition function}$$

$$\langle i \rangle = DP = \frac{C_o}{C_P} = \frac{1}{1-K[M_1]}.$$

Isodesmic aggregation is the simplest possible mechanism and that remains its strength. However, it fails to capture several key features of the presumed linear aggregation of many proteins. The weaknesses are multifold: the concept of a saturation concentration is not defined nor is the possibility of seeding a reaction using preformed aggregates. As can be seen in Figure 14.2A, the concentration of higher order aggregates will increase with the initial monomer concentration C_0 for a given equilibrium constant K. Addition of preformed aggregates to a reaction containing a monomer concentration above the saturation concentration will cause growth of the aggregates and a decrease in the monomer concentration [71–74]. In isodesmic aggregation, the monomers of the preformed aggregates would repartition into other species and a new equilibrium would be established whereby the concentration of monomer would actually increase.

Linked isodesmic processes: To explain the cooperativity observed in actin polymerization, Oosawa and coworkers considered two isodesmic processes that are

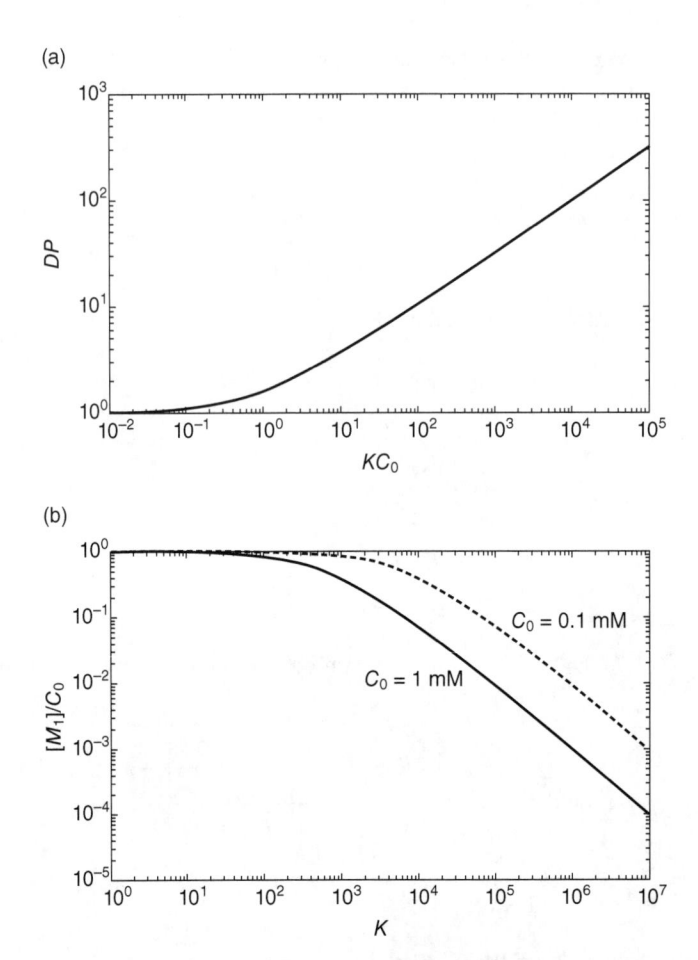

Figure 14.2. (A) The degree of polymerization (*DP*) as a function of the initial monomer concentration (C_0) and the equilibrium constant (K) for an isodesmic polymerization mechanism. (B) The concentration of monomer [M_1] relative to the initial monomer concentration (C_0) remaining in solution at equilibrium in an isodesmic polymerization mechanism as a function of the equilibrium constant (K). Two different initial monomer concentrations are shown: 1 mM (solid line) and 0.1 mM (dashed line). The position of the 1 mM curve relative to the 0.1 mM curve is a reflection of the increased monomer incorporation into polymer at a given K. This can also be seen from Figure 14.2A where, for a given K, the higher C_0 will lead to a higher degree of polymerization.

linked such that the first process models the consequences of a free energy profile as a function of aggregate size for which $G(n) - G(n-1) > 0$ for $n < n^*$ and a second process for which $G(n) - G(n-1) < 0$ for $n > n^*$ [63, 64, 68]. This species of size n^* is the nucleus and represents a peak on the free energy profile of G versus n. Oosawa's mechanism, depicted in Scheme 14.3, is referred to as homogeneous nucleation or the nucleation–elongation mechanism.

$$M_1 + M_1 \underset{k_{-1}}{\overset{k_1}{\rightleftharpoons}} M_2; K_2 = \frac{[M_2]}{[M_1][M_1]}; [M_2] = K[M_1]^2$$

$$M_2 + M_1 \underset{k_{-2}}{\overset{k_2}{\rightleftharpoons}} M_3; K_3 = \frac{[M_3]}{[M_2][M_1]}; [M_3] = K^2[M_1]^3$$

$$\vdots$$

$$M_{n*-1} + M_1 \underset{k_{-n*}}{\overset{k_{n*}}{\rightleftharpoons}} M_{n*}; K_c = \frac{[M_{n*}]}{[M_{n*-1}][M_1]}; [M_{n*}] = K_c K^{n*-2}[M_1]^{n*}$$

$$\vdots$$

$$M_i + M_1 \underset{k_{-i}}{\overset{k_i}{\rightleftharpoons}} M_{i+1}; K_{i+1} = \frac{[M_{i+1}]}{[M_i][M_1]}; [M_{i+1}] = (K_c)^{i+1-n*} K^{n*-2}[M_1]^i \text{ for } i \geq n*$$

$$\text{or } [M_{i+1}] = \sigma(K_c)^{i-1}[M_1]^i \text{ for } i \geq n*$$

$$\text{where } \sigma = \left(\frac{K}{K_c}\right)^{n*-2}$$

(Scheme 14.3)

By introducing a set of unfavorable steps (followed by favorable steps) into the reaction described in the equations for isodesmic aggregation, one models two linked isodesmic processes through the cooperativity parameter σ. Here, K is less than unity and K_c is greater than unity and as a result σ is <1. We can obtain the expression for the total monomer concentration to demonstrate that this leads to the appearance of a saturation concentration C_s. Since

$$C_o = [M_1] + \sum_{i=2}^{\infty} i\sigma K_c^{-1}(K_c[M_1])^i,$$

which for $\sigma \ll 1$ and $K < K_c$ yields

$$C_o \approx [M_1] + \frac{\sigma[M_1]}{(1 - K_c[M_1])^2}.$$

That the Oosawa model supports the existence of a saturation concentration is demonstrated in Figure 14.3A. For the purpose of illustration, we set $n* = 4$ and $K_c = 10^5/M$ with $K = 1/M$. If $C_o < K_c^{-1}$, then the model yields nearly a one-to-one correspondence between the monomer pool ($[M_1]$) and C_o. For $C_o > K_c^{-1}$, the monomer concentration shows a plateau value, and this becomes the fraction of C_o that does not get incorporated into higher molecular weight species. In the Oosawa model K_c^{-1} is the saturation concentration because it fits both criteria outlined in the previous section for the definition of C_s. Essentially no higher order aggregates form until the C_s is crossed. This is clear from a plot of the degree of polymerization:

$$DP = \frac{1}{1 - K_c[M_1]}.$$

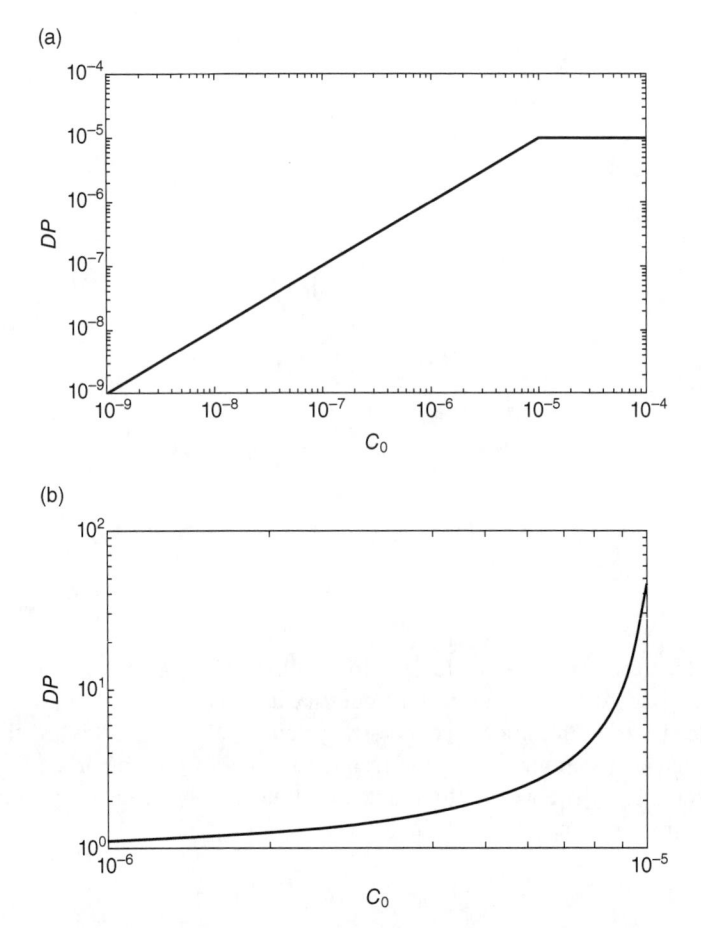

Figure 14.3. (A) The concentration, in molar, of monomer [M_1] remaining in solution at equilibrium in a nucleation–elongation polymerization mechanism with a nucleus size of 4, a prenucleation equilibrium constant (K) of 1/M, and a post-nucleation equilibrium constant (K_c) of 10^5/M. This leads to a predicted saturation concentration of 10^{-5} M, which is evident by the plateau in [M_1] beginning at the saturation concentration. (B) The degree of polymerization as a function of the initial monomer concentration (C_0) for the polymerization mechanism and conditions described in (A). It is expected that the degree of polymerization will be low at values of C_0 less than the saturation concentration, and will start to increase dramatically at concentrations near the saturation concentration and higher. This is due to the fact that, because the concentration of [M_1] remains constant at equilibrium at C_0 higher than the saturation concentration, all additional monomer must be incorporated into the high-order polymers as C_0 is increased.

as a function of the initial monomer concentration, shown in Figure 14.3B.

Oosawa's model provided one of the original descriptions of the thermodynamics of aggregation (polymerization in his parlance) in the MPM framework. This framework can be generalized to include additional steps such as condensation of fibrils, fragmentation of higher molecular weight species, and the effects of conformational changes [30, 75]. Roberts and coworkers have included the effects of conformational changes through the so-called Lumry–Eyring nucleated polymerization (LENP) model [76, 77]. All of these nuances require the inclusion of additional microstates and steps to describe the conversion between these states. At a minimum, each additional microstate requires three extra parameters, namely two rate constants and the activity (concentration) of the new species. This poses severe challenges because one cannot use standard data sets to determine the requisite parameters.

14.4. KINETICS OF HOMOGENEOUS NUCLEATION AND ELONGATION USING MPMS

In the previous section we considered the equilibrium behavior of nucleation–elongation processes. Although this is useful for illustrating the concentration dependence of aggregation, it ignores the rich time-dependent evolution of nucleation–elongation mechanisms. One feature of nucleation–elongation processes is the presence of a lag phase associated with the formation of species larger than the nucleus. A second feature is shortening of the lag phase by seeding the reaction with preformed aggregates [33, 65, 66]. To simplify the modeling of the time dependence of a nucleation–elongation process, two assumptions are generally made. These are: (1) the monomer (M) is in rapid equilibrium with the nucleus (N) with equilibrium constant (K_{PN}), and (2) the addition of the monomer to the nucleus or any species larger than the nucleus leads to the formation of a fibril and is irreversible [34, 35, 37, 68]. Applying these simplifying assumptions yields the pre-equilibrated nucleus model shown in Scheme 14.4.

$$n * M \xrightleftharpoons{K_{PN}} N$$
this is the pre-equilibration step for forming the nucleus
$$M + N \xrightarrow{k_{1PN}} F$$
this step refers to irreversible elongation of the nucleus
$$M + F \xrightarrow{k_{2PN}} F$$
irreversible loss of monomer to the fibril
$$[N] = K_{PN}[M]^{n*}$$
concentration of the nucleus in the pre-equilibrium model
$$\frac{d[M]}{dt} = -k_{1PN}K_{PN}[M]^{n*+1} - k_{2PN}[M][F]$$
rate of loss of monomer
$$\frac{d[F]}{dt} = k_{1PN}K_{PN}[M]^{n*+1}$$
rate of growth of fibril.

(Scheme 14.4)

In this mechanism, n^* is the number of monomers in the nucleus, M denotes the concentration of monomers, N is the nucleus, and F is the fibril. These equations are co-dependent on the concentration of the monomer and, consequently, monomer loss kinetics alone cannot be used to determine both the pre-equilibrium constant (K_{PN}) and the rate constant for elongation of the nucleus, k_{1PN}. Instead, one must follow both monomer loss and the formation of fibril. These equations can be numerically integrated with $[M] = [M]_0$ and $[F] = 0$ at time $t = 0$ for an unseeded reaction.

The monomer-loss and fibril growth kinetics using a pre-equilibrated nucleus model with a monomeric nucleus are shown in Figure 14.4. The equilibrium and

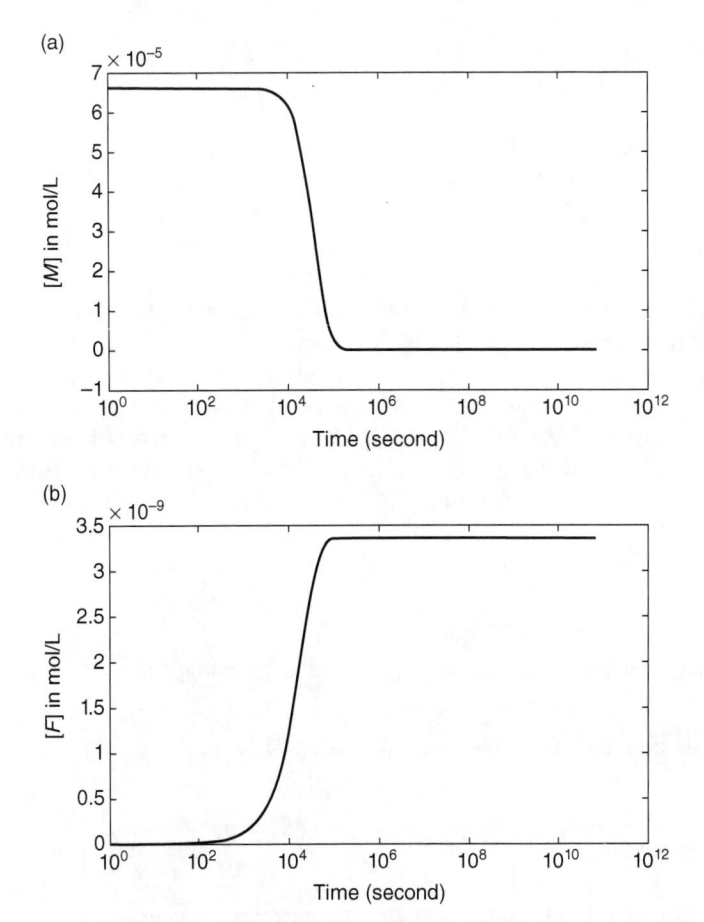

Figure 14.4. (A) The monomer concentration [M] as a function of time in a pre-equilibrated nucleus model for nucleation elongation. The pre-equilibrium constant (K_{PN}), rate constants, and nucleus size (n^*) used were those determined by Wetzel and coworkers for the aggregation of polyglutamine ($K_{PN} = 2.6 \times 10^{-9}$, $k_{1PN} = k_{2PN} = 11,400/s$ M, $n^* = 1$). The starting concentration was 66 μM. There is a pronounced lag phase before the monomer concentration begins to decrease. (B) The fibril concentration [F] as a function of time in the same model and conditions described in (A). As with the monomer concentration in (A), there is a pronounced lag phase before the fibril concentration begins to increase. The fibril growth is concomitant with the monomer loss.

rate constants used were those determined by Wetzel and coworkers for the aggregation of polyglutamine ($K_{PN} = 2.6 \times 10^{-9}$ and $k_{1PN} = k_{2PN} = 11,400/s\ M$) [54, 55]. As expected for a nucleation–elongation process, both the monomer-loss and fibril growth kinetics show a pronounced lag phase before aggregation is readily apparent. For a given set of equilibrium constants and rate constants, the rate of monomer loss in the pre-equilibrated nucleus model will decrease with increasing nucleus size and increase with increasing monomer concentration. This is clearly illustrated in Figures 14.5A,B. Also, the initial rate of fibril formation is strongly dependent on the nucleus size. This is illustrated in Figure 14.5C. Another feature of a nucleation–elongation reaction is that the lag time can be significantly shortened with increasing amounts of preformed aggregate. This effect is illustrated in Figure 14.5D. As the starting percentage of preformed aggregate increases, the lag time is shortened dramatically.

Going beyond pre-equilibration models: There are a number of other mechanisms proposed for fitting kinetic data that give similar kinetic behavior to the nucleation–elongation mechanism. Many of these have been described and evaluated by Bernacki and Murphy [34]. Even though these mechanisms are differently formulated, they share the feature that there is a rate-limiting step prior to the formation of the fibrillar species. More importantly, because many different mechanisms can fit the same kinetic data equally well, this means that the ability of a mechanism to fit kinetic data does not mean that the mechanism represents reality.

The nucleation–elongation mechanism can capture many features typically observed in an aggregation reaction of an IDP: there exists a saturation concentration; there is a lag phase in both the formation of fibrils and the loss of soluble material; and the lag phase can be shortened by the presence of preformed aggregate at the start of the reaction. However, a major flaw in this mechanism and other mechanisms of the MPM flavor is their inability to account for the accretion of stable, soluble oligomers. Indeed, recent data show that many different IDPs form stable oligomers prior to the formation of the final fibrillar species [78–86]. Even though these reactions show the characteristics of nucleation elongation, it is clear from the equations presented in Schemes 14.3 and 14.4 that any species in the reaction that is larger than the nucleus will form the final product and species that are smaller than the nucleus are unfavorable and therefore unstable. Another problem with MPMs is their inability to account for the context under which aggregation occurs. For instance, there is no way to predict from these mechanisms what the effect of changes in salt concentrations or pH might be—unless of course one were willing to float more empirical parameters into the modeling. Colloidal science provides an alternative for modeling the effects of solution conditions on IDP aggregation. In the following section we will provide the rationale for borrowing ideas from colloidal science and conclude with a brief overview of the main ideas.

14.5. CONCEPTS FROM COLLOIDAL SCIENCE

An important feature of aggregates of colloids is that their structure is determined by the aggregation mechanism [87–91]. Therefore, careful investigation of aggregate

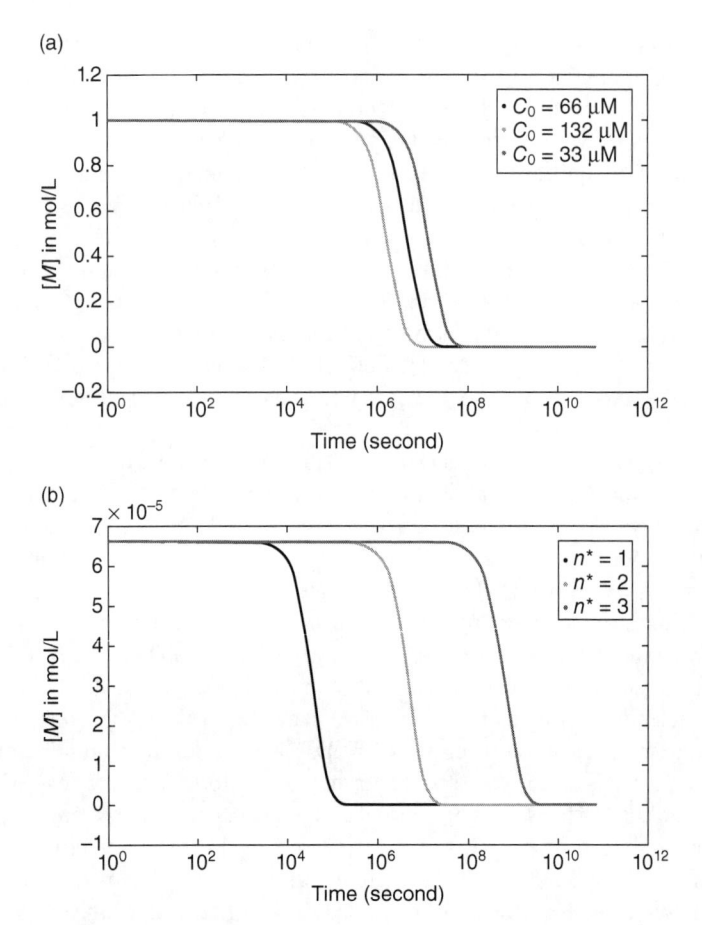

Figure 14.5. (A) The relative monomer concentration $[M]/C_0$ as a function of time in a pre-equilibrated nucleus model for nucleation elongation shows a dependence on the starting monomer concentration. In this figure, $K_{PN} = 2.6 \times 10^{-9}$/M, $k_{1PN} = k_{2PN} = 11,400$/s M, $n^* = 2$. The lag phase decreases with increasing C_0. (B) An investigation of the effect of the nucleus size on the monomer concentration as a function of time for the conditions shown in (A) with $C_0 = 66$ μM. An increasing nucleus size greatly increases the lag phase. For $n^* = 1, 2$, and 3, K_{PN} was set to be $K_{PN} = 2.6 \times 10^{-9}$, 2.6×10^{-9}/M, 2.6×10^{-9}/M^2, respectively. (C) An investigation of the effect of the nucleus size on the fibril concentration as a function of time for the conditions shown in (B). The rate of fibril growth is fastest at early times. The concentration of fibril remaining at the end of the polymerization is dependent on the nucleus size, with smaller nuclei leading to higher concentrations of fibrils. (D) The effect of adding various concentration of preformed fibril [F] on the monomer concentration as a function of time for the conditions shown in (A) with $C_0 = 66$ μM. The addition of preformed fibril to the beginning of a polymerization reaction reduces the lag phase. In the pre-equilibrated nucleus model, it is not possible to completely eliminate the lag phase with seeding.

CONCEPTS FROM COLLOIDAL SCIENCE 429

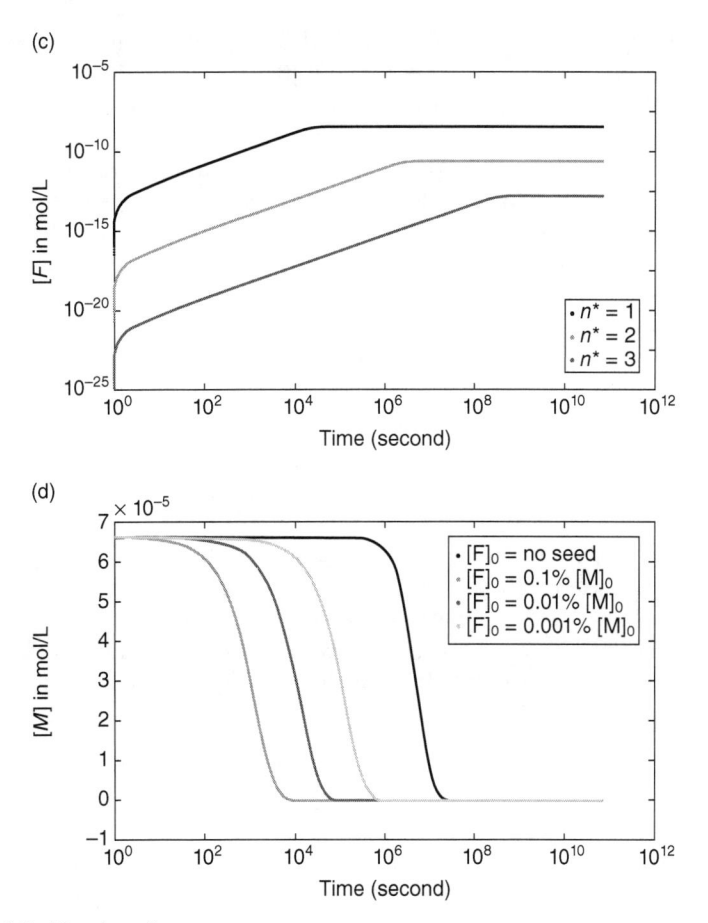

(c)

- $n^* = 1$
- $n^* = 2$
- $n^* = 3$

[F] in mol/L vs Time (second)

(d)

- $[F]_0$ = no seed
- $[F]_0 = 0.1\% [M]_0$
- $[F]_0 = 0.01\% [M]_0$
- $[F]_0 = 0.001\% [M]_0$

[M] in mol/L vs Time (second)

Figure 14.5. (Continued)

growth and final morphology is very informative for understanding aggregation mechanism. For our purposes, it is important to note that proteins are colloids and the framework of colloidal science can be used to understand protein aggregation. Recent work has shown that aggregation-prone IDPs form compact, disordered globules in aqueous solvents [9–15]. Additionally, the time scales for conformational relaxation are predicted to be either on the same time scale as bimolecular diffusion, or even slower [13, 56]. Indeed, there is considerable evidence that IDP globules initially associate to form spherical, molten oligomers that are metastable [83, 92–97]. Further conversion of oligomers to linear, fibrillar aggregates appears to be templated by rate-limiting conformational conversions within molten oligomers or mesoglobules. Of course, this process might have to compete with diffusion-limited cluster growth, which leads to precipitation through the formation of amorphous aggregates [87, 91]. The spherical oligomers appear have characteristics of spherical polymer micelles, with solvophobic cores and solvophilic surfaces [92, 93]. In so far as this mapping is meaningful, there exist interesting thermodynamic theories describing the conversion from spherical micelles to cylindrical micelles, the so-called sphere-to-rod transition, which might be directly applicable to nucleated conformational conversion from molten oligomers to fibrils as first postulated by Lindquist and coworkers [98]. The framework that we discuss below would be applicable to the formation of molten oligomers/mesoglobules, which is precisely what is missing in the MPMs.

The aggregation of colloids is a competition between long-range electrostatic repulsions and short-range attractions [23–27]. The Derjaguin–Landau–Verwey–Overbeek (DLVO) theory provides a simple framework for describing these interactions [23, 25–27]. The colloidal pair-potential (V_T) describes the potential of mean force for the interaction between two colloids, and is a sum of the repulsive potential (V_R) and the attractive potential (V_A). An illustration of how these potentials behave as a function of the separation h between two colloids is shown in Figure 14.6. The primary maximum of the total interaction potential is considered to be the rate-determining barrier in colloid aggregation [27]. We will first describe the interaction potentials and then we will describe how the aggregation rates can be modulated by the potentials.

In the DLVO theory, the attractive potential energy is dominated by van der Waals (vdW) interactions [27]. Because colloids have molecular dimensions (rather than atomic dimensions), vdW interactions between colloidal particles have a longer range than small molecules [99]. For a colloid of radius r, the attraction potential has the following form:

$$V_A = -\frac{A}{12}\left[\frac{1}{x(x+2)} + \frac{1}{(x+1)^2} + 2\ln\frac{x(x+2)}{(x+1)^2}\right],$$

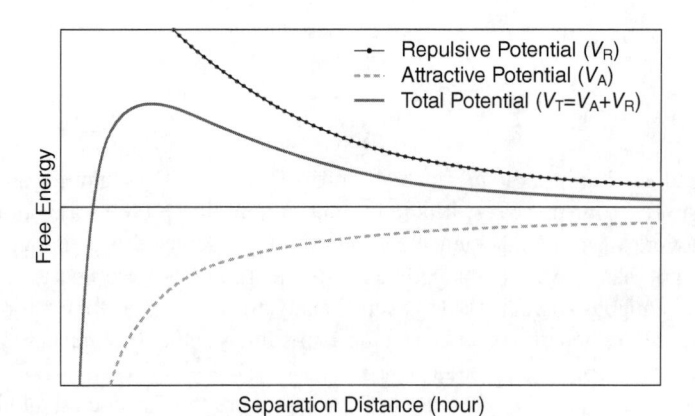

Figure 14.6. An illustration of the forces affecting interactions between a pair of colloidal particles in the DLVO theory. The repulsive potential is long-range and arises from electrostatics. The attractive potential works over shorter separation distances and is determined by van der Waals interactions. The total interaction potential is a sum of the attractive and repulsive potentials. The maximum of the total interaction potential is a critical determinant of association kinetics.

where $x = h/2r$ [27, 99]. Here, A is known as the Hamaker constant, which reflects the relative strength of attraction between two colloids. It is a function of the electronic polarizability and density of a given material. For a given colloidal dispersion, the Hamaker constant is a geometric mean of the Hamaker constant of the colloidal particle and the dispersion medium [27]. A common way of stabilizing colloids is to increase the net charge on each colloid. This increases the surface potential of each colloid and the repulsion between colloids. The long-range electrostatic repulsion potential (V_R) between identical colloids is

$$V_R = \frac{\varepsilon r \psi^2}{2} \ln(1 + \exp(-\kappa h)),$$

where $\kappa = \sqrt{\left(\dfrac{2I}{\varepsilon \varepsilon_o k_B T}\right)}$ [27].

Here, ε is the dielectric constant of the medium, ε_o is the relative permittivity of free space, κ is the screening parameter and is the reciprocal of the Debye screening length, r is the particle radius, h is the inter-particle separation, k_B is the Boltzmann constant, T is temperature in Kelvin, I is the ionic strength of the solution, and ψ is the surface potential of the colloid given as

$$\psi(r) = \frac{Nq}{\varepsilon r (1 + \kappa r)} \text{ [100]}.$$

Here, N is the number of particles with net charge q, comprising the colloid of radius r. Using this form for the surface potential, we obtain

$$V_R = \frac{(Nq)^2}{2\varepsilon r (1 + \kappa r)^2} \ln(1 + \exp(-\kappa h)).$$

The net charge and number of particles in the colloid dominate the magnitude of the repulsive potential. The distance dependence of the repulsive potential is modulated by the screening parameter, which is most sensitive to changes in the ionic strength. Increasing the ionic strength of the solution promotes aggregation by lowering the barrier in the total interaction potential.

The total interaction potential determines the aggregation rate and mechanism. If the primary maximum is lower than $k_B T$, there is no barrier to association and aggregation will only be limited by diffusion [27, 91]. This is known as diffusion-limited aggregation (DLA). This assumes that every particle collision will lead to irreversible association. In this case, the collision rate constant between two molecules is the same as the aggregation rate constant. This rate constant (k_D) is directly related to the diffusion coefficient (D) of the colloid and is given as $k_D = 8\pi D r$ [27].

When the height of the barrier (V_{max}) of the total interaction potential is larger than $k_B T$, the aggregation rate will be slower than the diffusion limit [27, 89, 91]. This rate is called the reaction-limited rate, and aggregation that occurs in this regime is called reaction-limited aggregation (RLA). Fuchs defined a stability ratio (W), which is defined as the ratio of the diffusion-limited rate constant to the reaction-limited rate constant (k_R), $W = k_D/k_R$. A reasonable approximation to this stability ratio is

$$W = \frac{1}{2\kappa r} \exp\left(\frac{V_{max}}{k_B T}\right),$$

which yields the reaction-limited rate constant to be

$$k_R \approx 16\pi\kappa D r^2 \exp\left(-\frac{V_{max}}{k_B T}\right) [27].$$

This expression for k_R shows that the reaction-limited rate is most sensitive to the barrier height. As a result, one common way to slow the aggregation of colloids dramatically is to increase their surface charge. Likewise, increasing the salt concentration leads to more rapid aggregation [27].

The height of the repulsive barrier will also increase as the square of the number of colloids (N) in an aggregate. The aggregation rate constant will decrease as the aggregates grow due to the higher charge repulsion toward approaching monomers. The aggregation rate is a product of the rate constant and the concentration of monomers in solution. At low concentrations, such as those found in sub-saturated solutions or after phase separation has occurred, the aggregation rate will tend to zero, leaving metastable aggregates remaining in solution. Additionally, the size of the soluble aggregates will depend on the net charge and ionic strength, with smaller aggregates remaining at a high net charge and/or low ionic strength and vice versa. Maiti and coworkers have demonstrated this for aggregates of several different proteins [78].

Inferring mechanisms from morphologies: An important feature of colloid aggregation is that the morphology of aggregates is determined by the kinetics of aggregation, and this is independent of the nature of the colloids [89, 91]. As mentioned previously, aggregation of colloids falls into two regimes: DLA and RLA. In both regimes, the structure of the aggregates is fractal and the mass of the aggregate follows a power-law behavior such that

$$M \propto \left(\frac{R_g}{r}\right)^{d_f}$$

where R_g is the radius of gyration of the aggregate, r is the radius of the individual colloids in the aggregate, and d_f is the fractal dimension of the aggregate [89, 91, 101]. The fractal dimension is $d_f \approx 1.8$ for aggregates formed via DLA and $d_f \approx 2.1$ for those formed via RLA [89, 91]. The fractal dimension is experimentally accessible using static light scattering and therefore one can readily infer the mechanism of aggregation by comparing with theoretical predictions.

Light scattering depends not only on the size of the scattering particle, but also on the scattering vector,

$$q = \left(\frac{2\pi n}{\lambda}\right) \sin\left(\frac{\theta}{2}\right) [101].$$

Here, n is the refractive index of the medium, λ is the wavelength of incident light, and θ is the scattering angle. At high values of qR_g, the total intensity of scattered light scales linearly with the mass (M) of the scattering object. As a result, the total intensity of scattered light scales is q^{-d_f}, allowing easy determination of the fractal dimension [89, 91, 101].

Knowledge of the fractal dimension of aggregates is important for inferring the mechanism of aggregation. If the fractal dimension matches that predicted by DLA and RLA, the mechanism can be understood using simple DLVO theory. More important, if the fractal dimension deviates from the prediction, it provides evidence that more complex events are taking place. If the time-dependent evolution of the fractal dimension is followed, deviations may provide clues as to when and over what length scales these more complex events manifest themselves. This is of major importance in the aggregation of proteins where many aggregates of IDPs have fractal dimensions larger than what is predicted for both DLA and RLA, likely as a result of restructuring of the aggregate [91, 102, 103].

14.6. CONCLUSIONS

We have provided an overview of a different blend of concepts to understand the phase behavior of IDPs. We sandwiched the classical MPM-based descriptions using concepts borrowed from the classical theories for phase separation and colloidal science. We believe that these theories, as opposed to MPMs, provide the best prospects for a more complete understanding of both driving forces and mechanisms of IDP aggregation/phase separation. Indeed these concepts provide the optimal way of connecting theory, simulation, and experimental data. Preliminary examples of these efforts are available in the IDP aggregation literature [9, 78, 104, 105]. The successes of these preliminary efforts emphasize the need for adapting and refining phase diagram and colloidal science-based approaches. The importance is further underscored by the growing recognition that insoluble inclusions may play a protective role vis-à-vis cell toxicity, suggesting that soluble oligomers, that is, the species that play no role in MPMs, might be the main contributors to toxicity [94, 106–110].

ACKNOWLEDGMENTS

We are grateful to Andreas Vitalis and Evan Powers, who have helped us immensely through many stimulating conversations about phase separation and aggregation.

This work was supported by grant 5RO1 NS56114 from the National Institutes of Health.

REFERENCES

1. Chiti, F. and Dobson, C. M. (2006) Protein misfolding, functional amyloid, and human disease, *Annu Rev Biochem 75*, 333–366.
2. Dunker, A. K., Brown, C. J., Lawson, J. D., Iakoucheva, L. M., and Obradovic, Z. (2002) Intrinsic disorder and protein function, *Biochemistry 41*, 6573–6582.
3. Dunker, A. K., Cortese, M. S., Romero, P., Iakoucheva, L. M., and Uversky, V. N. (2005) Flexible nets. The roles of intrinsic disorder in protein interaction networks, *FEBS J 272*, 5129–5148.
4. Dunker, A. K., Silman, I., Uversky, V. N., and Sussman, J. L. (2008) Function and structure of inherently disordered proteins, *Curr Opin Struct Biol 18*, 756–764.
5. Dyson, H. J. and Wright, P. E. (2005) Intrinsically unstructured proteins and their functions, *Nat Rev Mol Cell Biol 6*, 197–208.
6. Wright, P. E. and Dyson, H. J. (1999) Intrinsically unstructured proteins: Re-assessing the protein structure-function paradigm, *J Mol Biol 293*, 321–331.
7. Uversky, V. N., Gillespie, J. R., and Fink, A. L. (2000) Why are "natively unfolded" proteins unstructured under physiologic conditions? *Proteins 41*, 415–427.
8. Weathers, E. A., Paulaitis, M. E., Woolf, T. B., and Hoh, J. H. (2004) Reduced amino acid alphabet is sufficient to accurately recognize intrinsically disordered protein, *FEBS Lett 576*, 348–352.
9. Crick, S. L., Jayaraman, M., Frieden, C., Wetzel, R., and Pappu, R. V. (2006) Fluorescence correlation spectroscopy shows that monomeric polyglutamine molecules form collapsed structures in aqueous solutions, *Proc Natl Acad Sci U S A 103*, 16764–16769.
10. Dougan, L., Li, J., Badilla, C. L., Berne, B. J., and Fernandez, J. M. (2009) Single homopolypeptide chains collapse into mechanically rigid conformations, *Proc Natl Acad Sci U S A 106*, 12605–12610.
11. Moglich, A., Joder, K., and Kiefhaber, T. (2006) End-to-end distance distributions and intrachain diffusion constants in unfolded polypeptide chains indicate intramolecular hydrogen bond formation, *Proc Natl Acad Sci U S A 103*, 12394–12399.
12. Mukhopadhyay, S., Krishnan, R., Lemke, E. A., Lindquist, S., and Deniz, A. A. (2007) A natively unfolded yeast prion monomer adopts an ensemble of collapsed and rapidly fluctuating structures, *Proc Natl Acad Sci U S A 104*, 2649–2654.
13. Vitalis, A., Wang, X., and Pappu, R. V. (2007) Quantitative characterization of intrinsic disorder in polyglutamine: Insights from analysis based on polymer theories, *Biophys J 93*, 1923–1937.
14. Walters, R. H. and Murphy, R. M. (2009) Examining polyglutamine peptide length: A connection between collapsed conformations and increased aggregation, *J Mol Biol 393*, 978–992.
15. Wang, X., Vitalis, A., Wyczalkowski, M. A., and Pappu, R. V. (2006) Characterizing the conformational ensemble of monomeric polyglutamine, *Proteins 63*, 297–311.
16. Mao, A. H., Crick, S. L., Vitalis, A., Chicoine, C. L., and Pappu, R. V. (2010) Net charge per residue modulates conformational ensembles of intrinsically disordered proteins, *Proc Natl Acad Sci U S A 107*, 8183–8188.

17 Muller-Spath, S., Soranno, A., Hirschfeld, V., Hofmann, H., Ruegger, S., Reymond, L., Nettels, D., and Schuler, B. (2010) From the Cover: Charge interactions can dominate the dimensions of intrinsically disordered proteins, *Proc Natl Acad Sci U S A 107*, 14609–14614.

18 Rousseau, F., Schymkowitz, J., and Serrano, L. (2006) Protein aggregation and amyloidosis: Confusion of the kinds? *Curr Opin Struct Biol 16*, 118–126.

19 Balch, W. E., Morimoto, R. I., Dillin, A., and Kelly, J. W. (2008) Adapting proteostasis for disease intervention, *Science 319*, 916–919.

20 Powers, E. T., Morimoto, R. I., Dillin, A., Kelly, J. W., and Balch, W. E. (2009) Biological and chemical approaches to diseases of proteostasis deficiency, *Annu Rev Biochem 78*, 959–991.

21 Gsponer, J., Futschik, M. E., Teichmann, S. A., and Babu, M. M. (2008) Tight regulation of unstructured proteins: From transcript synthesis to protein degradation, *Science 322*, 1365–1368.

22 Uversky, V. N. (2002) Natively unfolded proteins: A point where biology waits for physics, *Protein Sci 11*, 739–756.

23 Kruyt, H. R. (1959) *Colloid science*, Vol. 1, Elsevier, New York.

24 Derjaguin, B. and Landau, L. (1941) Theory of the stability of strongly charged lyophobic sols and of the adhesion of strongly charged particles in solutions of electrolytes, *Acta Physico Chemica URSS 14*, 633–662.

25 Levine, S. (1939) Problems of stability in hydrophobic colloidal solutions I. On the interaction of two colloidal metallic particles. General discussion and applications, *Proc R Soc Lond A 170*, 145–165.

26 Verwey, E. J. W. and Overbeek, J. T. G. (1948) *Theory of the stability of lyophobic colloids*, Elsevier, Amsterdam.

27 Cosgrove, T. (2010) *Colloid science: Principles, methods, and applications*, 2nd ed., John Wiley & Sons Ltd, Hoboken, NJ.

28 Perczel, A., Hudaky, P., and Palfi, V. K. (2007) Dead-end street of protein folding: Thermodynamic rationale of amyloid fibril formation, *J Am Chem Soc 129*, 14959–14965.

29 Chiti, F. and Dobson, C. M. (2009) Amyloid formation by globular proteins under native conditions, *Nat Chem Biol 5*, 15–22.

30 Roberts, C. J. (2007) Non-native protein aggregation kinetics, *Biotechnol Bioeng 98*, 927–938.

31 Powers, E. T. and Powers, D. L. (2006) The kinetics of nucleated polymerizations at high concentrations: Amyloid fibril formation near and above the "supercritical concentration," *Biophys J 91*, 122–132.

32 Harper, J. D. and Lansbury, P. T., Jr. (1997) Models of amyloid seeding in Alzheimer's disease and scrapie: Mechanistic truths and physiological consequences of the time-dependent solubility of amyloid proteins, *Annu Rev Biochem 66*, 385–407.

33 Frieden, C. (2007) Protein aggregation processes: In search of the mechanism, *Protein Sci 16*, 2334–2344.

34 Bernacki, J. P. and Murphy, R. M. (2009) Model discrimination and mechanistic interpretation of kinetic data in protein aggregation studies, *Biophys J 96*, 2871–2887.

35 Morris, A. M., Watzky, M. A., and Finke, R. G. (2009) Protein aggregation kinetics, mechanism, and curve-fitting: A review of the literature, *Biochim Biophys Acta 1794*, 375–397.

36 Morris, A. M., Watzky, M. A., Agar, J. N., and Finke, R. G. (2008) Fitting neurological protein aggregation kinetic data via a 2-step, minimal/"Ockham's razor" model: The Finke-Watzky mechanism of nucleation followed by autocatalytic surface growth, *Biochemistry 47*, 2413–2427.

37 Ferrone, F. (1999) Analysis of protein aggregation kinetics, *Methods Enzymol 309*, 256–274.

38 Flory, P. J. (1953) *Principles of polymer chemistry*, Cornell University Press, Ithaca and London.

39 Ganazzoli, F., Raos, G., and Allegra, G. (1999) Polymer association in poor solvents: From monomolecular micelles to clusters of chains and phase separation, *Macromol Theory Simul 8*, 65–84.

40 Grosberg, A. Y. and Khokhlov, A. R. (1997) *Statistical physics of macromolecules (AIP series in polymers and complex materials)*, 1st ed., AIP Press, New York.

41 Grosberg, A. Y. and Kuznetsov, D. V. (1992) Phase-separation of polymer-solutions and interactions of globules, *J Phys II 2*, 1327–1339.

42 Grosberg, A. Y. and Kuznetsov, D. V. (1992) Quantitative theory of the globule-to-coil transition. 4. Comparison of theoretical results with experimental-data, *Macromolecules 25*, 1996–2003.

43 Grosberg, A. Y. and Kuznetsov, D. V. (1992) Quantitative theory of the globule-to-coil transition. 3. Globule–globule interaction and polymer-solution binodal and spinodal curves in the globular range, *Macromolecules 25*, 1991–1995.

44 Grosberg, A. Y. and Kuznetsov, D. V. (1992) Quantitative theory of the globule-to-coil transition. 2. Density–density correlation in a globule and the hydrodynamic radius of a macromolecule, *Macromolecules 25*, 1980–1990.

45 Grosberg, A. Y. and Kuznetsov, D. V. (1992) Quantitative theory of the globule-to-coil transition. 1. Link density distribution in a globule and its radius of gyration, *Macromolecules 25*, 1970–1979.

46 Pappu, R. V., Wang, X., Vitalis, A., and Crick, S. L. (2008) A polymer physics perspective on driving forces and mechanisms for protein aggregation, *Arch Biochem Biophys 469*, 132–141.

47 Rubinstein, M. and Colby, R. H. (2003) *Polymer physics*, Oxford University Press, Oxford and New York.

48 Raos, G. and Allegra, G. (1996) Chain interactions in poor-solvent polymer solutions: Equilibrium and nonequilibrium aspects, *Macromolecules 29*, 6663–6670.

49 Raos, G. and Allegra, G. (1996) Chain collapse and phase separation in poor-solvent polymer solutions: A unified molecular description, *J Chem Phys 104*, 1626–1645.

50 Raos, G. and Allegra, G. (1997) Macromolecular clusters in poor-solvent polymer solutions, *J Chem Phys 107*, 6479–6490.

51 Bondos, S. E. (2006) Methods for measuring protein aggregation, *Curr Anal Chem 2*, 157–170.

52 Vernaglia, B. A., Huang, J., and Clark, E. D. (2004) Guanidine hydrochloride can induce amyloid fibril formation from hen egg-white lysozyme, *Biomacromolecules 5*, 1362–1370.

53 Lu, J., Carpenter, K., Li, R.-J., Wang, X.-J., and Ching, C.-B. (2004) Cloud-point temperature and liquid-liquid phase separation of supersaturated lysozyme solution, *Biophysical Chemistry 109*, 105–112.

54 Bhattacharyya, A. M., Thakur, A. K., and Wetzel, R. (2005) Polyglutamine aggregation nucleation: Thermodynamics of a highly unfavorable protein folding reaction, *Proc Natl Acad Sci U S A 102*, 15400–15405.

55 Chen, S., Ferrone, F. A., and Wetzel, R. (2002) Huntington's disease age-of-onset linked to polyglutamine aggregation nucleation, *Proc Natl Acad Sci U S A 99*, 11884–11889.

56 Chuang, J., Grosberg, A. Y., and Tanaka, T. (2000) Topological repulsion between polymer globules, *J Chem Phys 112*, 6434–6442.

57 Sipe, J. D. and Cohen, A. S. (2000) Review: History of the amyloid fibril, *J Struct Biol 130*, 88–98.

58 Krebs, M. R., Domike, K. R., and Donald, A. M. (2009) Protein aggregation: More than just fibrils, *Biochem Soc Trans 37*, 682–686.

59 Novitskaya, V., Bocharova, O. V., Bronstein, I., and Baskakov, I. V. (2006) Amyloid fibrils of mammalian prion protein are highly toxic to cultured cells and primary neurons, *J Biol Chem 281*, 13828–13836.

60 Lorenzo, A. and Yankner, B. A. (1994) Beta-amyloid neurotoxicity requires fibril formation and is inhibited by Congo red, *Proc Natl Acad Sci U S A 91*, 12243–12247.

61 Lorenzo, A., Razzaboni, B., Weir, G. C., and Yankner, B. A. (1994) Pancreatic islet cell toxicity of amylin associated with type-2 diabetes mellitus, *Nature 368*, 756–760.

62 LeVine, H., 3rd (1993) Thioflavine T interaction with synthetic Alzheimer's disease beta-amyloid peptides: Detection of amyloid aggregation in solution, *Protein Sci 2*, 404–410.

63 Oosawa, F. and Kasai, M. (1962) A theory of linear and helical aggregations of macromolecules, *J Mol Biol 4*, 10–21.

64 Kasai, M., Asakura, S., and Oosawa, F. (1962) The cooperative nature of G-F transformation of actin, *Biochim Biophys Acta 57*, 22–31.

65 Frieden, C. and Goddette, D. W. (1983) Polymerization of actin and actin-like systems: Evaluation of the time course of polymerization in relation to the mechanism, *Biochemistry 22*, 5836–5843.

66 Frieden, C. (1985) Actin and tubulin polymerization: The use of kinetic methods to determine mechanism, *Annu Rev Biophys Biophys Chem 14*, 189–210.

67 Ferrone, F. A. (1993) The polymerization of sickle hemoglobin in solutions and cells, *Experientia 49*, 110–117.

68 Oosawa, F. and Asakura, S. (1975) *Thermodynamics of the polymerization of protein*, Academic Press, Waltham, MA.

69 De Greef, T. F., Smulders, M. M., Wolffs, M., Schenning, A. P., Sijbesma, R. P., and Meijer, E. W. (2009) Supramolecular polymerization, *Chem Rev 109*, 5687–5754.

70 Ciferri, A. (2005) *Supramolecular polymers*, CRC Press, Boca Raton, FL.

71 O'Nuallain, B., Williams, A. D., Westermark, P., and Wetzel, R. (2004) Seeding specificity in amyloid growth induced by heterologous fibrils, *J Biol Chem 279*, 17490–17499.

72 Krebs, M. R., Wilkins, D. K., Chung, E. W., Pitkeathly, M. C., Chamberlain, A. K., Zurdo, J., Robinson, C. V., and Dobson, C. M. (2000) Formation and seeding of amyloid fibrils from wild-type hen lysozyme and a peptide fragment from the beta-domain, *J Mol Biol 300*, 541–549.

73 Jarrett, J. T. and Lansbury, P. T., Jr. (1993) Seeding "one-dimensional crystallization" of amyloid: A pathogenic mechanism in Alzheimer's disease and scrapie? *Cell 73*, 1055–1058.

74. Come, J. H., Fraser, P. E., and Lansbury, P. T., Jr. (1993) A kinetic model for amyloid formation in the prion diseases: Importance of seeding, *Proc Natl Acad Sci U S A 90*, 5959–5963.
75. Xue, W. F., Homans, S. W., and Radford, S. E. (2008) Systematic analysis of nucleation-dependent polymerization reveals new insights into the mechanism of amyloid self-assembly, *Proc Natl Acad Sci U S A 105*, 8926–8931.
76. Li, Y. and Roberts, C. J. (2009) Lumry-Eyring nucleated-polymerization model of protein aggregation kinetics. 2. Competing growth via condensation and chain polymerization, *J Phys Chem B 113*, 7020–7032.
77. Andrews, J. M. and Roberts, C. J. (2007) A Lumry-Eyring nucleated polymerization model of protein aggregation kinetics: 1. Aggregation with pre-equilibrated unfolding, *J Phys Chem B 111*, 7897–7913.
78. Sahoo, B., Nag, S., Sengupta, P., and Maiti, S. (2009) On the stability of the soluble amyloid aggregates, *Biophys J 97*, 1454–1460.
79. Garai, K., Sengupta, P., Sahoo, B., and Maiti, S. (2006) Selective destabilization of soluble amyloid beta oligomers by divalent metal ions, *Biochem Biophys Res Commun 345*, 210–215.
80. Shin, T. M., Isas, J. M., Hsieh, C. L., Kayed, R., Glabe, C. G., Langen, R., and Chen, J. (2008) Formation of soluble amyloid oligomers and amyloid fibrils by the multifunctional protein vitronectin, *Mol Neurodegener 3*, 16.
81. Shekhawat, G. S., Lambert, M. P., Sharma, S., Velasco, P. T., Viola, K. L., Klein, W. L., and Dravid, V. P. (2009) Soluble state high resolution atomic force microscopy study of Alzheimer's beta-amyloid oligomers, *Appl Phys Lett 95*, 183701.
82. Martins, S. M., Frosoni, D. J., Martinez, A. M., De Felice, F. G., and Ferreira, S. T. (2006) Formation of soluble oligomers and amyloid fibrils with physical properties of the scrapie isoform of the prion protein from the C-terminal domain of recombinant murine prion protein mPrP-(121-231), *J Biol Chem 281*, 26121–26128.
83. Kayed, R., Head, E., Thompson, J. L., McIntire, T. M., Milton, S. C., Cotman, C. W., and Glabe, C. G. (2003) Common structure of soluble amyloid oligomers implies common mechanism of pathogenesis, *Science 300*, 486–489.
84. Huang, T. H., Yang, D. S., Plaskos, N. P., Go, S., Yip, C. M., Fraser, P. E., and Chakrabartty, A. (2000) Structural studies of soluble oligomers of the Alzheimer beta-amyloid peptide, *J Mol Biol 297*, 73–87.
85. Haass, C. and Selkoe, D. J. (2007) Soluble protein oligomers in neurodegeneration: Lessons from the Alzheimer's amyloid beta-peptide, *Nat Rev Mol Cell Biol 8*, 101–112.
86. Ferreira, S. T., Vieira, M. N., and De Felice, F. G. (2007) Soluble protein oligomers as emerging toxins in Alzheimer's and other amyloid diseases, *IUBMB Life 59*, 332–345.
87. Asnaghi, D., Carpineti, M., Giglio, M., and Sozzi, M. (1992) Coagulation kinetics and aggregate morphology in the intermediate regimes between diffusion-limited and reaction-limited cluster aggregation, *Phys Rev A 45*, 1018.
88. Anderson, V. J. and Lekkerkerker, H. N. W. (2002) Insights into phase transition kinetics from colloid science, *Nature 416*, 811–815.
89. Lin, M. Y., Lindsay, H. M., Weitz, D. A., Ball, R. C., Klein, R., and Meakin, P. (1990) Universal reaction-limited colloid aggregation, *Phys Rev A 41*, 2005–2020.

REFERENCES

90. Ball, R. C., Weitz, D. A., Witten, T. A., and Leyvraz, F. (1987) Universal kinetics in reaction-limited aggregation, *Phys Rev Lett 58*, 274.
91. Lin, M. Y., Lindsay, H. M., Weitz, D. A., Ball, R. C., Klein, R., and Meakin, P. (1989) Universality in colloid aggregation, *Nature 339*, 360–362.
92. Rhoades, E. and Gafni, A. (2003) Micelle formation by a fragment of human islet amyloid polypeptide, *Biophys J 84*, 3480–3487.
93. Yong, W., Lomakin, A., Kirkitadze, M. D., Teplow, D. B., Chen, S. H., and Benedek, G. B. (2002) Structure determination of micelle-like intermediates in amyloid beta-protein fibril assembly by using small angle neutron scattering, *Proc Natl Acad Sci U S A 99*, 150–154.
94. Glabe, C. G. and Kayed, R. (2006) Common structure and toxic function of amyloid oligomers implies a common mechanism of pathogenesis, *Neurology 66*, S74–S78.
95. Meier, J. J., Kayed, R., Lin, C. Y., Gurlo, T., Haataja, L., Jayasinghe, S., Langen, R., Glabe, C. G., and Butler, P. C. (2006) Inhibition of human IAPP fibril formation does not prevent beta-cell death: Evidence for distinct actions of oligomers and fibrils of human IAPP, *Am J Physiol Endocrinol Metab 291*, E1317–E1324.
96. Smith, A. M., Jahn, T. R., Ashcroft, A. E., and Radford, S. E. (2006) Direct observation of oligomeric species formed in the early stages of amyloid fibril formation using electrospray ionisation mass spectrometry, *J Mol Biol 364*, 9–19.
97. Wacker, J. L., Zareie, M. H., Fong, H., Sarikaya, M., and Muchowski, P. J. (2004) Hsp70 and Hsp40 attenuate formation of spherical and annular polyglutamine oligomers by partitioning monomer, *Nat Struct Mol Biol 11*, 1215–1222.
98. Krishnan, R. and Lindquist, S. L. (2005) Structural insights into a yeast prion illuminate nucleation and strain diversity, *Nature 435*, 765–772.
99. Hamaker, H. C. (1937) The London—van der Waals attraction between spherical particles, *Physica 4*, 1058–1072.
100. Nagasawa, S. A. R. M. (1961) *Polyelectrolyte solutions*, Academic Press, London.
101. Martin, J. E. and Ackerson, B. J. (1985) Static and dynamic scattering from fractals, *Phys Rev A 31*, 1180.
102. Harada, S., Tanaka, R., Nogami, H., and Sawada, M. (2006) Dependence of fragmentation behavior of colloidal aggregates on their fractal structure, *J Colloid Interface Sci 301*, 123–129.
103. Kumagai, H., Matsunaga, T., and Hagiwara, T. (1999) Effect of salt addition on the fractal structure of aggregates formed by heating dilute BSA solutions, *Biosci Biotechnol Biochem 63*, 223–225.
104. Xu, S. (2009) Cross-beta-sheet structure in amyloid fiber formation, *J Phys Chem B 113*, 12447–12455.
105. Sahin, E., Grillo, A. O., Perkins, M. D., and Roberts, C. J. (2010) Comparative effects of pH and ionic strength on protein-protein interactions, unfolding, and aggregation for IgG1 antibodies, *J Pharm Sci 99*, 4830–4848.
106. Ceru, S., Kokalj, S. J., Rabzelj, S., Skarabot, M., Gutierrez-Aguirre, I., Kopitar-Jerala, N., Anderluh, G., Turk, D., Turk, V., and Zerovnik, E. (2008) Size and morphology of toxic oligomers of amyloidogenic proteins: A case study of human stefin B, *Amyloid 15*, 147–159.
107. Glabe, C. G. (2008) Structural classification of toxic amyloid oligomers, *J Biol Chem 283*, 29639–29643.

108 Lin, C. Y., Gurlo, T., Kayed, R., Butler, A. E., Haataja, L., Glabe, C. G., and Butler, P. C. (2007) Toxic human islet amyloid polypeptide (h-IAPP) oligomers are intracellular, and vaccination to induce anti-toxic oligomer antibodies does not prevent h-IAPP-induced beta-cell apoptosis in h-IAPP transgenic mice, *Diabetes 56*, 1324–1332.

109 Congdon, E. E. and Duff, K. E. (2008) Is tau aggregation toxic or protective? *J Alzheimers Dis 14*, 453–457.

110 Gotz, J., Ittner, L. M., Fandrich, M., and Schonrock, N. (2008) Is tau aggregation toxic or protective: A sensible question in the absence of sensitive methods? *J Alzheimers Dis 14*, 423–429.

15

MODIFIERS OF PROTEIN AGGREGATION—FROM NONSPECIFIC TO SPECIFIC INTERACTIONS

Michal Levy-Sakin,[1] Roni Scherzer-Attali,[1] and Ehud Gazit

15.1. INTRODUCTION

Proteins are more likely to stay soluble in aqueous solutions while they are in their native state. Loss of the native state may result in the exposure of the hydrophobic core followed by protein aggregation owing to hydrophobic interactions between protein monomers. Electrostatic interactions may also induce protein aggregation. For example, if a protein is at its isoelectric point it has a zero net charge. Consequently, protein monomers will not repulse each other, leading to aggregation and precipitation [1]. A broad review of the protein aggregation mechanism can be found in Philo and Arakawa and others [1–3].

Protein aggregation can be controlled by physical properties such as temperature and pressure, by changing the pH of the solution or by adding different solutes to the solution [4]. Solution additives that modify the protein's aggregation tendency can interact with a protein in a specific or a nonspecific manner. Specific modifiers interact through binding sites on the protein's surface. Examples of specific modifiers are ligands, inhibitors, pharmacological chaperones, agonists, and antagonists [5]. The second class of compounds alters protein aggregation through nonspecific,

---

[1] These two authors contributed equally to this work.

Protein and Peptide Folding, Misfolding, and Non-Folding, First Edition. Edited by Reinhard Schweitzer-Stenner.
© 2012 John Wiley & Sons, Inc. Published 2012 by John Wiley & Sons, Inc.

weak interactions. Accordingly, these compounds will influence protein stability and the protein's tendency to aggregate only at high concentrations, in the range of millimolar to molar concentrations.

In the following chapter we will discuss these two modifying protein aggregation strategies. Compounds affecting protein aggregation in a nonspecific manner are more relevant in different processes and areas of interest than compounds that interact specifically with a given protein. Nonspecific modifiers are fundamental when studying protein structure *in vitro* and *in vivo* and in enzymology studies. Such modifiers occur naturally in all organisms and are very dominant in extremophiles [6]. Specific modifiers, on the other hand, have been the focus of many pharmacological and clinical studies.

The two main factors that influence protein aggregation are (1) protein stability and misfolding and (2) protein–protein interactions. Henceforth, we will review the influence of modifiers on these two aspects of protein aggregation.

In recent years there has been a growing interest in amyloid-forming proteins and in their aggregation behavior. Hence, we will give them special attention in our discussion. An in-depth discussion of amyloid proteins' pathological role and potential treatments can be found in the second part of this chapter (specific modifiers of protein aggregation).

15.2. NONSPECIFIC MODIFIERS

15.2.1. Salts and Hofmeister's Series

The first milestone in the research on protein aggregation modifiers was achieved by Franz Hofmeister and coworkers in the 1880s and 1890s. A series of seven papers was published under the title *About the Science of the Effect of Salts* (*Zur Lehre von der Wirkung der Salze*) [7–13]; one of the seven is Hofmeister's famous work [9]. The paper was originally written in German, but in 2004 an English translation was published by Kunz and Ninham, making this fundamental work more available [14].

In his classical work Hofmeister studied the tendency of different ions to precipitate a whole egg white protein mixture. The ability of the ions to precipitate proteins was as follows:

Anions: $SO_4^{2-} > HPO_4^{2-} > CH_3COO^- > Cl^- > NO_3^-$

Cations: $Mg^{2+} > Li^+ > Na^+ = K^+ > NH_4^+$

Hofmeister hypothesized that the differential ability of the salts to precipitate proteins and other macromolecules derives from their water-attracting capacity. This list of ions is now termed "the Hofmeister series" and their influence on water and on macromolecules is termed "the Hofmeister effect."

Nowadays, more than 100 years later, it is known that many other processes function according to the Hofmeister series; these include enzyme activity, protein folding, colloidal assembly, protein crystallization, and more [15, 16]. However, a full explanation on the Hofmeister effect regarding protein precipitation and other processes is still lacking.

Water interactions mediate the influence of ions on protein solubility and aggregation. The simple and basic explanation that is commonly given for protein precipitation by salt ions is competition for water molecule interaction. When ions with a high tendency to interact with water are introduced to a protein solution, fewer water molecules are free to interact with the protein and therefore the solute's relative concentration increases and precipitation occurs. Ions can be divided into two groups, based on their ability to interact with water and to affect water activity: (1) ions that interact with water more strongly than water molecules themselves, which are termed kosmotropes, and (2) ions that form weaker interactions with water than water itself, which are termed chaotropes. Kosmotropic ions are small or have high charge densities. They induce stabilization of water–water interactions and accordingly stabilize intermolecular interactions of proteins and other macromolecules. Proteins in kosmotropic ion solutions are stabilized and at high salt concentrations they may precipitate. Chaotropic ions are structure breakers and in their presence a protein will have a higher tendency to lose its structure. Proteins that aggregate in the presence of chaotropes aggregate as denatured species [15, 17, 18]. Figure 15.1 summarizes some major characteristics of chaotropic (in red) and kosmotropic (in green) ions [17].

In addition, ions influence protein stability and aggregation by interacting directly with proteins. Electrostatic interactions stabilize intramolecular and intermolecular binding of charged groups. Ions compete with the electrostatic interactions and consequently may cause destabilization and aggregation of proteins [19]. They can also interact with the peptide's backbone owing to their dipole moment (partial positive charge on the amino group and partial negative charge on the carboxyl group) [19, 20]. The effect of salt ions on protein structure is greatly dependent on ion concentrations. At low concentrations salt ions reduce electrostatic interactions owing to charge shielding. At higher concentrations, preferential binding also takes place, in addition to the charge shielding effect [19, 21].

Hofmeister series

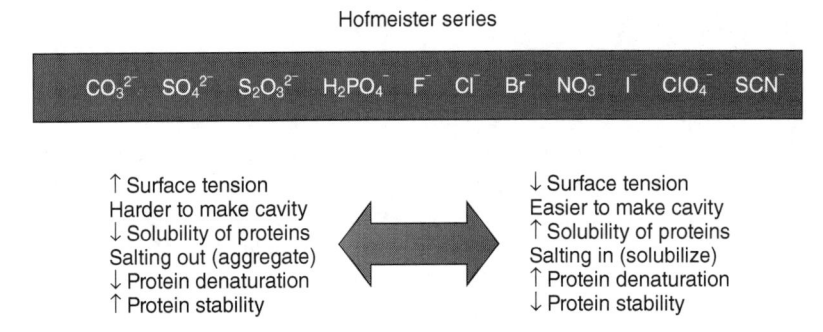

CO_3^{2-} SO_4^{2-} $S_2O_3^{2-}$ $H_2PO_4^-$ F^- Cl^- Br^- NO_3^- I^- ClO_4^- SCN^-

↑ Surface tension
Harder to make cavity
↓ Solubility of proteins
Salting out (aggregate)
↓ Protein denaturation
↑ Protein stability

↓ Surface tension
Easier to make cavity
↑ Solubility of proteins
Salting in (solubilize)
↑ Protein denaturation
↓ Protein stability

Figure 15.1. Chaotropes and and kosmotropes (in dark and light gray, respectively). From Zhang and Cremer [17], modified from http://tinyurl.com/ed5gj.

15.2.1.1. The Influence of Chaotropes and Kosmotropes on the Assembly of Amyloid Fibrils
Amyloid-forming proteins tend to misfold and self-assemble into well-ordered amyloid fibril aggregates. Several research works focused on the Hofmeister effect and the amyloid fibril formation process. The use of different sodium salts showed that chaotropes and kosmotropes have opposite effects on the fibrillization of the prion Sup35NM. Order maker ions promoted fibril formation, whereas structure breaker ions inhibited Sup35NM fibril formation [22]. Similar results were achieved with other amyloid-forming proteins, such as immunoglobulin light chain protein and amyloid beta $(A\beta)_{1-40}$ [23, 24]. However, in another case, the effect of salts on mouse prion fibrillization did not follow the Hofmeister series. Here the effect of the tested salts on protein self-assembly was in correlation with the electroselectivity of anion binding to the protein [25].

15.2.2. Ionic Liquids

An ionic liquid is a salt in the liquid state. In contrast to water, which is predominantly composed of electrically neutral molecules, ionic liquids are largely made of ions and short-lived ion pairs. The ionic bond is usually stronger than the van der Waals forces between the molecules of ordinary liquids.

The field of ionic liquids research in the context of protein structure and function is rather new. A fundamental work by Summers and Flowers [26] demonstrated protein aggregation suppression by an ionic liquid, giving rise to a new approach for modifying protein aggregation. They showed that the liquid organic salt, ethylammonium nitrate, prevented aggregation of denatured lysozyme. Inspired by their work, other groups investigated additional ionic liquids. For example, the effect of a series of ionic liquids of N′-alkyl and N′-(ω-hydroxyalkyl)-methylimidazolium chlorides on lysozyme and on a single chain antibody fragment was studied [27]. All the ionic liquids that had been tested were found to decrease protein aggregation, allowing renaturation of the unfolded proteins. With regard to native folded proteins, the ionic liquid salts had a destabilizing effect. Their destabilizing and aggregation-suppressing properties were found to correlate with the hydrophobicity of the substituted imidazolium cations. For further examples of the effect of ionic liquids on protein solubility and stability, see Fujita et al. and Buchfink et al. [28–30].

15.2.3. Protein Stabilizers

15.2.3.1. Terminology and Background
Protein stabilizers include compounds such as protective osmolytes and chemical chaperones. Osmolytes are small organic solutes that accumulate in different organisms because of osmotoic pressure or other cellular stress states [31–34]. Cells use organic osmolytes but not inorganic ions (unless for minor and small tuning of osmotic pressure) since inorganic ions at high concentrations can cause DNA breakage and interfere with protein structure and function. Usually, organic osmolytes do not perturb macromolecules even at high concentrations; hence, they do not hamper cell activities. In these cases they are

often referred to as compatible solutes [35–37]. Moreover, sometimes they have a protective effect on macromolecules, specifically on proteins, in which case they are termed chemical chaperones or osmoprotectants. Urea is an exceptional example of a non-compatible osmolyte, although it does accumulate under stress conditions; it interrupts cellular processes and macromolecule stability.

The ability of small organic molecules to stabilize proteins and protect them from thermal denaturation was already observed in the 1970s [38]. However, Yancey and coworkers' contributions had a great impact on this field [34–36, 39–43]. In particular, their studies on osmoregulation in marine animals and in other systems greatly advanced our understanding of the *in vivo* significance of osmolytes in maintaining protein structure and solubility. In their famous work Yancey et al. pinpointed the existence of the denaturant urea as the main osmolyte that balances seawater and trimethylamine-*N*-oxide (TMAO) as the second most abundant solute. In many animals the *in vivo* concentrations of these two solutes were found to be at a constant ratio of 1:2 TMAO:urea [35, 41]. It was suggested that TMAO and other methylamines accumulate in that ratio in order to counteract the destabilizing effect of urea on protein structure. Indeed, this ratio was also shown *in vitro* and *in silico* as an optimum ratio for TMAO to counteract the effect of urea [44, 45].

As mentioned above, chemical chaperon is a term used to describe small molecules that stabilize protein structure. However, it is used also for describing small organic molecules that do not occur naturally, although these kinds of molecules are rarely found. An example of a synthetic chemical chaperone is 4-phenylbutyric acid (PBA), which was found to stabilize proteins both *in vitro* and *in vivo* [46]. The term "chemical chaperone" was coined in the mid 1990s by Welch and coworkers because of the similarity to the actions of molecular chaperones [47].

Osmolytes influence protein conformation and stability mainly by their tendency to interact with the protein backbone. Osmolytes with unfavorable interactions with the protein backbone shift the equilibrium toward the folded state of the protein, since the backbone is exposed when a protein is unfolded. Favoring of the folded state usually suppresses protein aggregation. However, we will discuss both cases in which chemical chaperones inhibit and accelerate protein aggregation.

15.2.3.2. Origin of Chemical Chaperones   Naturally occurring solutes that stabilize protein structures against different stresses can be categorized according to four different groups (Table 15.1): (1) carbohydrates and polyols, such as trehalose, inositol, glycerol, and sorbitol, (2) amino-acids, such as glycine, proline, taurine, and β-alanine, (3) methylamines, such as TMAO and betaine, and (4) methylsulfonium compounds such as Choline-*O*-sulfate and β-dimethylsulfoniopropionate [34]. Osmolytes accumulate in organisms from all kingdoms when stress (e.g., heat, osmotic, freezing, and hydrostatic stress) resistance is needed [40, 41].

15.2.3.3. Mechanism of Protein Stabilization Used by Osmoprotectants and Chemical Chaperones   The mechanism underlying protein stabilization that is used by chemical chaperones is not fully understood; however, several non-mutually exclusive theories exist:

TABLE 15.1. Chemical Structure of Some Stabilizing Osmolytes

Compound	Structure	Type
Trehalose		Carbohydrate
Myo-inositol		Polyol
Glycerol		Polyol
Sorbitol		Polyol
Glycine		Amino acid
Proline		Amino acid
Taurine		Amino acid
β-Alanine		Amino acid
TMAO		Methylamine
Betaine		Methylamine
Choline-O-sulfate		Methylsulfonium
β-Dimethylsulfoniopropionate		Methylsulfonium

1. Preferential exclusion—It was suggested that chemical chaperones and other solutes are preferentially excluded from the protein surface. Owing to their tendency to be excluded from the protein surface, the hydration shell of the protein is increased, and consequently tighter packing with lower relative surface is induced [48–50].
2. Influence on water structure—Similar to kosmotrope ions, it was postulated that chemical chaperones bind strongly to water molecules; thus they maintain the hydration shell around the protein [32, 51]. Indeed, molecular dynamics simulations showed that TMAO influences the protein structure through indirect interactions; it was demonstrated that TMAO ordered and stabilized the water structure and therefore helped maintain the folded conformation of the protein [44].
3. Steric effect—One mechanism underlying protein unfolding is through water perturbation into the protein core, for example, under conditions of high hydrostatic pressure [52]. It was proposed that chemical chaperones decrease the proteins' accessible volume, and since the chaperones are larger than water molecules they protect the proteins from water penetration [32].
4. Replacing water molecules—This mechanism differs in its nature from the above-mentioned mechanisms. More specifically, it is relevant in protecting proteins from freezing and drying stresses. Chemical chaperones, such as trehalose, can bind proteins through hydrogen bonding in a manner similar to water binding. The binding maintains the proteins in their conformation without being damaged due to loss of water molecules [43].

Similar to salts, osmolytes are also assumed to change water activity and thus to affect the protein hydration shell. Unlike salt ions, osmolytes usually do not have a net charge [32, 33].

15.2.3.4. Aggregation Suppression by Protective Solutes It is often reported that chemical chaperones can inhibit protein aggregation, as detailed in the following examples: TMAO was shown to inhibit homoserine *trans*-succinylase aggregation both *in vitro* and *in vivo* [53]; glycerol suppressed aggregation of wild-type (WT) and mutant rabbit muscle creatine kinase [54]; and trehalose was found to inhibit bovine serum albumin (BSA) aggregation [55]. Inhibition of protein aggregation, facilitated by compatible osmolytes, is generally attributed to their stabilizing effect on the protein's native structure. However, in other cases, the destabilizing and solubilizing effect of the solutes was thought to play a role in aggregation inhibition: Proline was found to inhibit the aggregation of the cellular retinoic acid-binding protein (CRABP). Surprisingly, there was no evidence for a proline stabilizing effect on the native structure of CRABP. Moreover, it was suggested that a destabilizing effect on partially folded species and on small aggregates of the protein are the key to proline's inhibitory effect on CRABP aggregation [56].

15.2.3.5. The Influence of Stabilizing Osmolytes on Amyloid Aggregation
Compatible solutes were shown both to accelerate and to inhibit amyloid fibril

formation. Belfort and coworkers studied the influence of sugars on insulin fibril formation. Sugars have the ability to inhibit insulin fibril formation in the following order: tri > di > monosacharids [57]. In contrast, in another study, McLaurin and coworkers showed that whereas fructose inhibits Aβ fibrillization, glucose accelerates its nucleation and both galactose and mannose enhance its self-assembly into mature fibrils. Since sugars of the same molecular weight, with similar hydrodynamic volumes, and similar hard sphere radii were examined, it was suggested that the hydrogen bonding capacity of the saccharides is the characteristic that influences their tendency to accelerate or inhibit fibril formation: Hydrogen-bonded water molecules are important for protein stability. In addition of carbohydrates to the solution they can replace the hydrogen-bonded water molecules. A high number of hydrogen bonds between a protein and sugars increase native protein stability and therefore inhibit fibrillogenesis [58]. Induction of amyloid fibril formation by carbohydrates was also demonstrated with other proteins, such as the immunoglobulin light chains [59].

Methylamines were also shown to stabilize the native fold of amyloidogenic proteins and to reduce fibrillization. For example, betaine inhibited immunoglobulin light chain aggregation [59].

The influence of different osmolytes on amyloid fibril formation is presented in Table 15.2. No simple rules are reflected from this list, implying that osmolyte–amyloid interactions are complex. According to our current knowledge, no model allows us to predict whether an osmolyte will inhibit or accelerate certain amyloidogenic protein fibrillization.

TABLE 15.2. The Influence of Stabilizing Osmolytes on Amyloid Aggregation

Protein	Cosolute	Effect	Reference
Aβ_{1-40} and Aβ_{1-42}	Trehalose	Inhibits fibrillization. Adducts of fibril–trehalose were formed.	[199, 200]
Aβ_{1-40} and Aβ_{1-42}	Ducrose	Inhibits fibrillization. Adducts of fibril–trehalose were formed.	[199, 201]
Aβ_{1-40} and Aβ_{1-42}	TMAO	Inhibits fibrillization in the presence of lipids and does not inhibit it without lipids.	[199]
Aβ_{1-42}	α-D-mannosylglycerate	Inhibits fibrillization	[202]
Aβ_{1-42}	α-D-mannosylglyceramide	No effect	[202]
Aβ_{1-42}	Mannose	No effect	[202]
Aβ_{1-42}	Methylmannoside	No effect	[202]
Aβ_{1-42}	Glycerol	Inhibits fibrillization	[202]
Aβ_{1-42}	Ectoine	Inhibits fibrillization	[203]
Aβ_{1-42}	Hydroxyectoine	Inhibits fibrillization	[203]
Aβ_{1-40} and Aβ_{1-42}	Glucose	Promotes nucleation	[58]

TABLE 15.2. (Continued)

Protein	Cosolute	Effect	Reference
$A\beta_{1-40}$ and $A\beta_{1-42}$	Galactose	Promotes fibrillization	[58]
$A\beta_{1-40}$ and $A\beta_{1-42}$	Mannose	Promotes fibrillization	[58]
Insulin	Glucose	Inhibits fibrillization	[57]
Insulin	Fructose	Inhibits fibrillization	[57]
Insulin	Maltose	Inhibits fibrillization	[57]
Insulin	Sucrose	Inhibits fibrillization	[57, 204]
Insulin	Trehalose	Inhibits fibrillization	[57, 205, 206]
Insulin	Raffinose	Inhibits fibrillization	[57]
Insulin	Melezitose	Inhibits fibrillization	[57]
Insulin	TMAO	Inhibits fibrillization	[204]
Insulin	Betaine	Inhibits fibrillization	[205]
Insulin	Ectoine	Inhibits fibrillization	[205]
Insulin	Citrulline	Inhibits fibrillization	[205]
1SS-α-lac	TMAO	Promotes fibrillization	[207]
1SS-α-lac	Sucrose	Promotes fibrillization	[207]
1SS-α-lac	Trehalose	Promotes fibrillization	[207]
α-Syn	Sarcosine	Inhibits fibrillization	[208]
α-Syn	Betaine	Inhibits fibrillization	[208]
α-Syn	TMAO	Inhibits fibrillization at high TMAO concentrations and promotes it at lower concentrations	[208, 209]
α-Syn	Taurine	Promotes fibrillization	[208]
α-Syn	Glycerol	Promotes fibrillization	[208]
P39A tetra-Cys CRABP	Proline	Inhibits fibrillization	[56]
Tetra-Cys Htt53	Proline	Inhibits fibrillization	[56]
W7FW14F apomyoglobin	Trehalose	Inhibits fibrillization	[206]
MB-Gln35	Trehalose	Inhibits fibrillization	[210]
α-PrP	TMAO	Inhibits fibrillization	[211]
Immunoglobulin light chain	Sorbitol	Inhibits fibrillization	[59]
Immunoglobulin light chain	Betaine	Inhibits fibrillization	[59]

15.2.4. Denaturants

In contrast to protein stabilizers, which interact with a protein backbone in an unfavorable manner, denaturants tend to interact favorably with a protein backbone [60]. Thus, denaturants induce protein unfolding, which causes the protein backbone to be exposed. However, interactions of denaturants with protein side chains were reported as well. The binding of denaturants to hydrophobic moieties helps them solubilize proteins and consequently prevents aggregation, as reviewed by Shiraki and coworkers [4]. The ability of denaturants to solubilize proteins makes them very useful in refolding proteins, for example, from inclusion bodies [61]. In other cases, owing to protein unfolding, denaturants may cause aggregation.

The most widely used and studied denaturants are guanidine hydrochloride (GdnHCl) and urea [4]. The basis for the interactions of urea and GdnHCl with proteins and their denaturating effect has been under debate for several years [62]. In molecular dynamics simulations, GdnHCl and urea were found to interact with protein via hydrogen bonding with the carbonyl groups of the peptide backbone, as well as with charged residues [63]. In contrast, nuclear magnetic resonance (NMR) hydrogen-deuterium exchange studies concluded that GdnHCl does not form hydrogen bonds with the peptide backbone group [64].

15.2.4.1. Protein Aggregation Suppression and Induction by Denaturants As previously mentioned, denaturants can solubilize proteins, thus preventing protein aggregation. This phenomenon usually occurs at high denaturant concentrations. Nevertheless, lower concentrations of denaturants may be sufficient for protein unfolding but not for solubilizing; consequently, aggregation may be induced. For example, the effect of GdnHCl on recombinant human interferon gamma (rhIFNγ) was studied. It was shown that below 2 M, GdnHCl induced protein aggregation and at higher concentrations, above 3 M, GdnHCl could dissolve rhIFNγ aggregates [65].

15.2.4.2. The Effect of Denaturants on Amyloid Aggregation Belfort and coworkers, who studied the effect of protecting osmolytes on insulin fibrillization, studied the effect of urea and GdnHCl on this aggregation process as well. Both urea and GdnHCl accelerated the fibrillization process [57]. In contrast, when urea was incubated with Aβ, the self-assembly process was reduced [66]. For more details on the effect of denaturants on the amyloid fibril formation process, see Table 15.3.

15.2.5. Solvents

Research on the effect of organic solvents on protein structure, function, and aggregation revealed different aggregation phenomena [67, 68]. For example, an interest-

TABLE 15.3. The Influence of Denaturants on Amyloid Aggregation

Protein	Denaturant	Effect	Reference
Insulin	Urea	Promotes fibrillization	[57, 204]
Insulin	GdnHCl	Promotes fibrillization	[57]
β-Lactoglobulin	Urea	Promotes fibrillization	[212]
Immunoglobulin light chain	Urea	Promotes fibrillization	[59]
Aβ_{1-40} and Aβ_{1-42}	Urea	Inhibits fibrillization	[66]
Prion	GdnHCl	Promotes fibrillization	[213]
Lysozyme	GdnHCl	Promotes fibrillization	[214]

ing observation was made when gelatin was incubated with ethanol: aggregation was induced and when the aggregates were analyzed by electron microscopy, fractal-shaped aggregates were observed [69].

Alcohols, especially trifluoroethanol (TFE), destabilize proteins and induce α-helical conformation [70]. The effect of TFE on protein structure is mediated by weakening protein hydrophobic interactions and strengthening intramolecular bonding. TFE, similar to other organic solvents, changes the dielectric constant of the solvent surrounding the protein [67].

Usually when protein aggregation is associated with structural changes, the β-sheet conformation becomes more dominant (e.g., in the case of amyloid aggregation). However, it was reported that in the presence of TFE, rich α-helical aggregates can be observed. Thus, protein aggregation was investigated in relation to various alcohols. Briefly, BSA was incubated at high pHs with various alcohols. Methanol, ethanol, and 2-propanol induced secondary and tertiary structures, rich in β-sheet structure. The amount of the induced structure was dependent on the solvent and was in the following order: 2-propanol > ethanol > methanol. TFE had the greatest effect on secondary- and tertiary-structure formation, but it induced rich α-helical structures rather than β-sheet structures [71].

The effect of hexafluoroisopropanol (HFIP) and TFE on enzyme stability, aggregation, and activity was studied on the creatine kinase enzyme. It was shown that low concentrations of HFIP (below 5%) and TFE (below 10%) induced enzyme inactivation, whereas high concentrations of HFIP and TFE induced protein aggregation [67].

In another study, poly-L-lysine peptide aggregation was studied. The effect of ethanol and the organic solvent dimethylsulfoxide (DMSO) on peptide aggregation was compared. In ethanol a higher peptide concentration is needed in order to aggregate, in comparison with the peptide concentration required for aggregation in DMSO solution. In addition, ethanol had a disaggregative effect, attributed to its ability to directly bind poly-L-lysine peptides [72].

15.2.5.1. The Effect of Solvents on Amyloid-Fibril Formation Various solvents promote amyloid-fibril formation. Two examples are lysozyme, which forms amyloid fibrils in the presence of 90% ethanol [73], and insulin, which forms amyloid fibrils with different properties in TFE and ethanol solutions [74]. However, under other conditions ethanol inhibited insulin fibril formation [75].

Solvents are often used for dissolving amyloidogenic peptides, before diluting them in buffer solutions. The choice of the solvent can influence significantly the fibrillation rate. In the case of Aβ it was shown that the fibrillization rate after dilution to buffer was as follows: 100% DMSO < 0.1% trifluoroacetic acid (TFA) < 10% DMSO < 35% acetonitrile/0.1% TFA [76]. Organic solvents, such as DMSO, TFE, and HFIP, were also shown to dissolve preformed amyloid fibrils [77].

15.2.6. Detergents and Surfactants

To date, we focused on protein aggregation at the solvent level. However, the surface with which the protein interacts can significantly affect protein folding and aggregation. Surfactants are surface active agents due to their amphiphilic characteristics. Surfactants alter the surface properties of water at the water–air or water–solid interfaces, whereas their hydrophilic moieties, oriented to the water and the hydrophobic moieties are transformed to the air or the solid phase. Similar to surfactants, proteins are also surface active. Surface tension forces can cause protein perturbation and thus protein misfolding, followed by aggregation or by protein absorption to the surface [19, 78, 79]. Surfactant and detergents decrease surface tension and are often used as solution additives to prevent protein aggregation while shipping, purifying, and during long-term storage of proteins [19, 80].

Similar to denaturants, detergents can solubilize proteins and thus are very useful in refolding proteins from inclusion bodies [61]. However, whereas urea and GdnHCl unfold the protein structure into a highly flexible structure, detergents form a more structured conformation [61]. Sodium dodecyl sulfate (SDS) is frequently used with proteins. It binds to hydrophobic patches on the protein, causing the protein to denaturate and to become negatively charged. In contrast to other compounds that were already discussed, SDS–protein interactions are not weak, SDS monomers can bind the protein very tightly, and dialysis and dilution are insufficient for complete removal of the detergent. Sulfobetaines can act as non-detergent surfactants, they can help in protein solubilization and can prevent aggregation, but they do not bind proteins as tightly as SDS and hence can be removed [4].

15.2.6.1. The Effect of Detergents and Surfactants on Amyloid Fibrillization and Oligomerization Various independent studies showed that SDS and other detergents can interact with amyloidogenic proteins and alter their aggregation pathways. Otzen and coworkers reported that SDS induced flexible worm-like structures, which are transformed into classical straight fibrils only after agitation [81].

In another study the influence of SDS on lysozyme was studied at acidic conditions. The effect of SDS was shown to be concentration dependent. At low SDS concentrations (up to 0.1 mM SDS) amyloid fibrils were formed in similar to the control samples (without SDS) and at high concentrations (0.25–20 mM) amyloid fibril formation was inhibited [82]. SDS was also shown to play a key role in forming $A\beta$ toxic oligomers *in vitro* [83].

Gemini surfactants consist of two conventional surfactant molecules chemically bonded together by a spacer. They self-assemble at lower concentrations and are superior in surface activity in comparison to other surfactants. The influence of a cationic gemini surfactant on $A\beta_{1-40}$ was studied [84]. At low surfactant concentrations the process of $A\beta_{1-40}$ fibril formation was promoted, while at higher concentrations the surfactant micelles inhibited the fibrillization process. In another study, insulin fibrillization was inhibited in the presence of amphiphilic surfactants [85].

15.2.7. Molecular Crowders

Usually, *in vitro* studies of protein folding are performed in low-concentration solutions. However, in the cellular context, proteins are exposed to a very crowded environment. It is estimated that the concentration of macromolecules in the cyto-

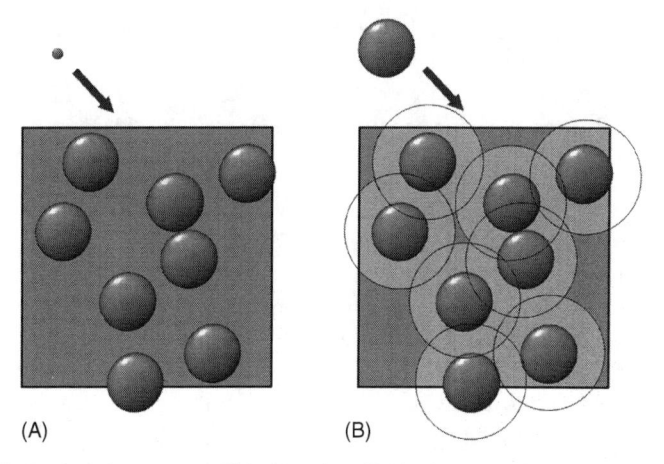

Figure 15.2. Excluded (orange and black) and available (blue) volume in a solution of spherical background macromolecules. (A) Volume available to a test molecule of infinitesimal size. (B) Volume available to a test molecule of a size comparable with background molecules [87]. See color insert.

plasm is in the 80–400 mg/mL range. In recent years new evidence has indicated that molecular crowding can influence protein folding and aggregation [86].

It has been estimated that 20–30% of the cell volume is occupied by macromolecules, making it sterically unavailable. In practice, the available volume differs for each molecule, depending on its size. Figure 15.2 presents a simple demonstration of this principle [87]. This crowded environment induces a constant nonspecific repulsion termed "the excluded-volume effect." It affects protein folding and aggregation differently, mainly depending on protein size and its tendency to fold [88]. In crowded surroundings the more compact forms of proteins are thermodynamically preferred. On the one hand, the folded state of a protein will be favored, implying that proteins will tend less to aggregate. On the other hand, the association of molecules, including protein association, decreases the total occupied volume, implying that protein aggregation will be favored in crowded solutions [88]. Indeed, both cases in which protein aggregation is induced or diminished by a crowding effect are observed. In general, for a nascent polypeptide chain, the native folded state will be favored when the protein tends to fold rapidly and aggregation will be favored when the folding process is slower.

According to dynamic simulation studies, Kinjo and Takada indicated that molecular crowding influences protein aggregation by three different pathways: (1) stabilization of the native fold before aggregation occurs, (2) decreased lag time before aggregation begins and accelerated aggregate growth at the beginning of the aggregation process, and (3) slowing down the aggregation process at late aggregation stages [89]. Using *in silico* tools they studied the effect of molecular crowding both on a folded protein and on a denatured protein. Similar studies were carried out *in vitro* by van den Berg et al., with both reduced and oxidized lysozyme. The reduced form refolded very poorly, but the oxidized form refolded into an active enzyme very efficiently [90, 91]. These studies were in agreement, concluding that (1) for fast-folding proteins the folding rate increases as the crowding is higher, and (2) for slow-folding proteins the aggregation rate increases as the crowder's concentration increases.

15.2.7.1. Molecular Crowding and Amyloid Fibril Formation The tendency of macromolecules to associate in the presence of high molecular crowding was well demonstrated in a study of amyloid fibril formation. Briefly, α-synuclein (α-syn) fibrillization was accelerated in the presence of various crowding agents (PEG 200, PEG 400, PEG 600, PEG 3350, PEG 6000, PEG 20000, Dextran 70, Dextran 138, Ficoll 70, Ficoll 400, lysozyme, and BSA) [92]. Similar results were observed with the amyloid-forming protein human apolipoprotein C-II and the Dextran T-10 crowder [93]. Consequently, it was suggested that changes in neuronal cytoplasm crowding may play a key role in amyloid-related disease etiology. In general, amyloid diseases are age-dependent, with a higher propensity in elderly individuals. Elevated crowding levels with age can be explained both by dehydration of tissue and cellular environment and by reduced protein degradation. Thus, increased molecular crowding with age may be an important factor in the development and progression of amyloid diseases [92, 93]. Further studies demonstrated that crowding specifically accelerates the nucleation step in the formation of amyloid fibrils [94]. However, different results were observed when lysozyme amyloid formation was studied. Here, BSA alone or in a mixed solution with Ficoll 70 inhibited lysozyme fibril formation [95].

15.3. SPECIFIC MODIFIERS

Thus far we have discussed various protein aggregation processes and general ways to modify them. In this section we will focus on specific protein modifiers. These are highly relevant in clinical and pharmacological studies. In accordance, we will extend our discussion on the role of protein aggregation in human diseases.

Protein aggregation is associated with many different proteins. The incorrect folding and aggregation of a protein is normally prevented by complex cellular quality control mechanisms. Yet, under certain conditions, several proteins are able to aggregate within or around cells, resulting in several protein-aggregation-associated diseases. Usually, protein misfolding takes place because of an undesirable mutation in the polypeptide chain, an unfavorable physiological environment, or, in a few cases, some lesser known reasons [96]. Although there is no homology in the amino acid sequence between different aggregated proteins and peptides associated with these different diseases, most of these proteins share a common mechanism of aggregation. In the last decades, a great effort has been directed toward studying and better understanding different protein-aggregation-associated diseases, also termed conformational diseases. This is because an increasing number of diseases have been found to be associated with protein conformational abnormalities and protein aggregation.

The largest class of conformational diseases is termed amyloid diseases. In this class all peptides or proteins seem to adopt a similar, insoluble, highly ordered structure when aggregated, known as the cross-β spine. Under extreme conditions this protein misfolding and aggregation is beyond the influence of intracellular quality-control systems. The histology of the resulting protein deposits was discovered more than 150 years ago and was termed amyloid [97, 98]. Amyloids are formed by the self-assembly of β-sheet-rich monomers into large fibrilar structures. Several diseases, including Alzheimer's disease (AD), Parkinson's disease (PD), type II diabetes, and several forms of Prion disorders, such as bovine spongiform encephalopathy and Creutzfeldt–Jakob disease, are associated with the formation of large amyloid deposits [99].

However, several conformation-associated diseases are not amyloid related, such as p53 and ubiquitin enzyme (E1)-related cancer, cystic fibrosis (CF), Gaucher's disease, emphysema and liver disease, and nephrogenic diabetes insipidus.

A great effort has been directed toward the prevention and the treatment of these diseases. Good prevention and therapeutic candidates are compounds that will specifically interact with the different proteins and peptides associated with these diseases, which in turn will prevent their aggregation. In the following section we will discuss several compounds that have been found to prevent protein aggregation via one or more mechanisms: (1) steric hindrance of protein–protein interaction, which can consequently inhibit the self-assembly process, especially in the case of amyloids, (2) shielding regions of exposed hydophobicity, or (3) in the case of amyloid fibrillogenesis, minimizing structural changes leading to β-sheet formation [100]. Compounds that can accelerate protein aggregation by specific binding will be discussed next.

15.3.1. Protein Aggregation Inhibitors (Self-Assembly): Small Molecules and Peptides

As previously mentioned, molecules that are able to prevent aggregation of peptides and proteins associated with conformational diseases have been intensively investigated. Two main strategies have been followed: (1) random screening of chemical libraries for small molecular inhibitors, and (2) rational design of short peptide sequences that inhibit aggregation [101]. Through these approaches several peptides and small molecules have been found to inhibit aggregate formation, particularly in the case of amyloid self-assembly.

Small molecules, mostly small aromatic molecules, have been extensively studied in relation to amyloidogenic peptides and proteins associated with different diseases. The small molecule inhibitors approach was initially based on early findings that demonstrated that small aromatic molecules such as Congo red and thioflavin T interact specifically with amyloid fibrils and inhibit their formation [102–104]. In the past few years, accumulating reports have described small molecule inhibitors of amyloid fibril formation. A partial list of these reports shows that aromatic-rich compounds dramatically inhibit cell death in cell culture amyloidogenic cytotoxicity assays. Several of these compounds were also described as efficient *in vitro* inhibitors of amyloid fibril formation.

15.3.1.1. Polyphenols One example of these small aromatic inhibitors are polyphenolic small molecular compounds [105]. Polyphenols are a large group of natural and synthetic small molecules that are composed of one or more aromatic phenolic rings. Natural polyphenols are usually found in high concentrations in wine, tea, nuts, berries, cocoa, and a wide variety of other plants. Several hundereds of polyphenols have been identified as well as their natural function, usually associated with protection of plants. Natural polyphenols can be grouped into several categories: vitamins (e.g., β-carotene and α-tocopherol), phenolic acids (e.g., benzoicacid and phenylacetic acid), flavonoids (e.g., flavanone and isoflavone), and other miscellaneous polyphenols (ellagic acid, sesamol, eugenol, thymol, etc.). Synthetic polyphenols are commonly used as pH indicators in cell culture media and synthetic food additives (e.g., phenolsulfonphthaleine, butylated hydroxylanisole [BHA], and butylated hydroxytoluene [BHT]) [105]. In the past few years, several polyphenols have been found to inhibit the self-assembly process of several proteins associated with amyloidogenic diseases.

For example, tannic acid, resveratrol, and catechin [105] were shown to inhibit the self-assembly process of Aβ (Table 15.4). Aβ is the major peptide associated with one of the most studied neurodegenerative disorders, AD.

AD pathology is characterized by the formation of two types of protein aggregates in the brain: (1) amyloid plaques, which form extracellular lesions composed of the Aβ peptide; and (2) intracellular neurofibrillary tangles, which are composed of hyperphosphorylated filaments of the microtubule-associated protein tau [98]. Several of these polyphenols, such as purpurogallin, exifon, hypericin, myricetin, gossypetin, pentahydroxybenzophenon, epicatechin gallate, and 2,3,4-trihydroxybenzophenone (THBP) have also been shown to inhibit the self-assembly of tau-protein into neurofibrillar tangles with an IC$_{50}$ in the low micromolar range (Table 15.4).

Several polyphenols such as apomorphine, baicalein, and dobutamine can also inhibit amyloid formation by other proteins and peptides associated with other amyloidogenic diseases, such as the major protein associated with PD, α-syn (Table 15.3). Accumulation of fibrillar forms of α-syn is found in Lewy bodies, a defining pathological characteristic of PD and dementia with Lewy bodies [106, 107]. In addition, phenolsulfonphthalein [108] and epigallocatechin gallate [Levy, unpublished work] [109], were shown to inhibit the self-assembly process of islet amyloid polypeptide (IAPP, also termed amylin). IAPP is a putative polypeptide hormone, which is produced by pancreatic β-cells, and is the major constituent of amyloid deposits seen mainly in islets of type II diabetic humans and diabetic cats [110]. Another amyloidogenic protein inhibited by polyphenols such as tannic acid and epigallocatechin gallate (Table 15.4) is the infectious prion protein (PrP), which is mainly associated with different forms of prion diseases. PrP is normally present in its native conformation (PrP^c); however, in all prion diseases, the protein is present

TABLE 15.4. Several Polyphenols and Their Inhibition Toward Several Different Amyloidogenic Protein, as Modified from Porat et al. [105]

Polyphenol	Studied Protein
NDGA	$A\beta_{1-40}$ and $A\beta_{1-42}$
Curcumin	$A\beta_{1-40}$ and $A\beta_{1-42}$
Rosmarinic acid	$A\beta_{1-40}$ and $A\beta_{1-42}$
Reservetol	$A\beta_{1-40}$, $A\beta_{1-42}$ and $A\beta_{23-35}$
Dobutamin	α-Syn
Myricetin	$A\beta_{1-40}$ and $A\beta_{1-42}$
Morin	$A\beta_{1-40}$ and $A\beta_{1-42}$
Quercetin	$A\beta_{1-40}$ and $A\beta_{1-42}$
Exifone	$A\beta_{1-40}$ and Tau_{412}
Kaempferol	$A\beta_{1-40}$ and $A\beta_{1-42}$
2,3,4,2',4'-Pentahydroxybenzophenone	$A\beta_{1-40}$ and Tau_{412}
Gossypetin	$A\beta_{1-40}$ and Tau_{412}
Purpurogallin	$A\beta_{1-40}$ and Tau_{412}
THBP	$A\beta_{1-40}$ and Tau_{412}
	α-syn
Biacalein	α-syn
Catechin	$A\beta_{1-40}$ and $A\beta_{1-42}$
Epicatechin	$A\beta_{1-40}$, $A\beta_{1-42}$ and Tau_{412}
Phenolsulfonphthaleine	IAPP
	Insulin
	$A\beta_{1-40}$
Epicatechin gallate	$A\beta_{1-40}$ and Tau_{412}
Epigallocatechin gallate	$A\beta_{1-40}$ and Insulin
	Calcitonin
	Prion PrP^{sc}
Hypericin	$A\beta_{1-40}$ and Tau412
Tannic acid	Prion PrP^{sc}
	$A\beta_{1-40}$ and $A\beta_{1-42}$

in an abnormal conformation (PrP^{sc}) that accumulates and forms amyloids around neurons [111].

15.3.1.2. Quinones Quinones are another group of small molecules that act as inhibitors of the self-assembly process of amyloidogenic proteins associated with different diseases.

Quinones have long been known to inhibit various metabolic pathways in the cell, and to have anti-bacterial, anti-viral, and anti-cancer properties [112, 113]. More recently several quinones have been shown to effectively inhibit the aggregation of some amyloidogenic proteins. One example of this is coenzyme Q (CoQ) and *p*-benzoquinone, which were reported to inhibit amyloid self-assembly of both the Aβ peptide in AD [114] and the IAPP peptide in type II diabetes [115] as well as to inhibit amyloid fibril formation by hen egg-white lysozymes [116]. Another example is a group of quinone-bearing polyamines, such as memoquin, which have displayed anti-aggregation activity toward Aβ. These quinone-bearing polyamines act as multi-target-directed ligands toward AD pathology, exhibiting therapeutic ability toward other non-amyloid-related aspects of the disease [117]. Likewise, anthraquinones, a tricyclic quinone, was found to be an effective inhibitor of tau-protein aggregation [118] as well as Aβ_{1-42} aggregation [119]. More recently, naphthoquinones, dicyclic quinones, were shown to effectively inhibit Aβ_{42} aggregation. This can occur by inhibiting *in vitro* oligomerization of Aβ, which was mediated by 1,2-naphthoquinone [120], or by inhibiting both oligomerization and fibrillization of Aβ by a tryptophan–naphthoquinone hybrid [121].

15.3.1.3. Indoles Another group of small molecules widely researched for their ability to inhibit amyloid formation is the indole group. Indoles are aromatic heterocyclic compounds, natural and synthetic, with a wide range of therapeutic targets, such as anti-inflammatories, phosphodiesterase inhibitors, 5-hydroxytryptamine receptor agonists and antagonists, cannabinoid receptors and agonists, and hydroxy-3-methyl-glutaryl-CoA reductase (HMG-CoA reductase) inhibitors [122]. Indole derivatives were found to very effectively inhibit amyloid formation of various amyloidogenic peptides. For example, three hydroxylindole derivatives: indole-3 carbinol, 3-hydroxyindole, and 4-hydroxyindole (Table 15.5) were the most effective inhibitors of amyloid formation by Aβ out of a screen of 29 indole derivatives, with a low micromolar IC$_{50}$ [123, 124]. Inhibition of α-syn fibrillization is also apparent with a large indole derivative, melatonin (Table 15.5). In addition, α-syn aggregation inhibition is apparent with dopamine derivatives that have been indole substituted. Three examples of these indole–dopamine derivatives are dopaminochrome, 5,6-dihydroxyindole, and indol-5,6-quinone (Table 15.5) [124]. All three molecules have been shown to bind non-covalently to α-syn and to successfully inhibit its fibrillization *in vitro*.

15.3.1.4. Peptide Inhibitors Not only have small molecular inhibitors of protein aggregation and self-assembly been extensively studied, the anti-aggregation ability

TABLE 15.5. Several Indoles and Their Inhibition Toward Several Different Amyloidogenic Proteins

Indole	Studied Protein
Indole-3 carbinol	Aβ_{1-40} and Aβ_{1-42}
3-Hydroxyindole	Aβ_{1-40} and Aβ_{1-42}
4-Hydroxyindole	Aβ_{1-40} and Aβ_{1-42}
Melatonin	α-Syn
Dopaminochrome	α-Syn
5,6-Dihydroxyindole	α-Syn
Indol-5,6-quinone	α-Syn

of small peptides in several conformational diseases have also been studied. Several approaches have been put forth toward inhibiting protein aggregation using small peptides. One approach uses small peptide fragments derived from the amyloidogenic protein itself to inhibit its own aggregation or the aggregation of other amyloidogenic proteins. Another approach uses natural existing peptide inhibitors such as β-synuclein (β-syn), which is known to inhibit the formation of α-syn aggregates *in vitro* in a dose-dependent manner [125].

Relying on the discovery that β-syn inhibits the aggregation of α-syn, Windisch and colleagues conducted experiments indicating that the N-terminal amino acid residues 1–15 are important for interaction with α-syn. Based on this finding, they initiated a drug development program, creating a peptide library containing all the different variations of amino acid compositions derived from this sequence of β-syn, with the specific aim of finding short, metabolically stable sequences that can be used for therapeutic application immediately or can serve as a basis for developing peptidomimetic small molecules [126].

Since a key feature in the process of amyloid formation is the transition into a rich β-sheet structure [127–131], preventing the ability of amyloidogenic proteins to adopt a β-sheet conformation appears to be a useful way to interfere with the amyloid assembly process. Several studies have demonstrated the use of β-sheet breaker peptides as agents for inhibiting amyloid formation. These short synthetic peptides are capable of binding the protein but are unable to become part of the β-sheet structure. Hence, they can destabilize the amyloidogenic protein conformer and preclude amyloid formation. A milestone in proving this concept was provided by Soto et al. [132]. Using β-breaker elements incorporated into short peptides composed of the recognition sequence of the amyloidogenic protein, Soto and others were able to inhibit amyloid formation [130, 132, 133]. The proline residue is widely used as the natural amino acid with the highest β-breakage potential.

Another approach is based on N-methylation of peptide inhibitors that prevent β-sheet stacking by interfering with the intermolecular backbone NH to CO hydrogen bonds composing the structure [134]. An example of this is the non-natural amino acid α-aminoisobutyric acid (Aib), which is a methylated alanine and is known to have a high β-sheet breaker potential, even higher than proline. Small peptide inhibitors derived from the full IAPP peptide that incorporated the Aib amino acid exhibited a strong inhibition potential toward the fibrillization of IAPP [135]. Importantly, a small dipeptide, which includes tryptophan and Aib, completely reduced Aβ aggregation both *in vitro* and *in vivo* [136]. Also, naturally existing peptides that incorporate both aromatic moieties and beta-breaking elements such as endomorphines have been shown to inhibit Aβ aggregation.

15.3.2. Protein Aggregation Inhibition by Refolding and Misfolding Misfolded Proteins: Pharmacological Chaperones

As previously mentioned, misfolded protein may lead to harmful effects. These latter can be divided into two categories (see Fig. 15.3 for details): (1) gain of function as

DNA → RNA → Ribosome

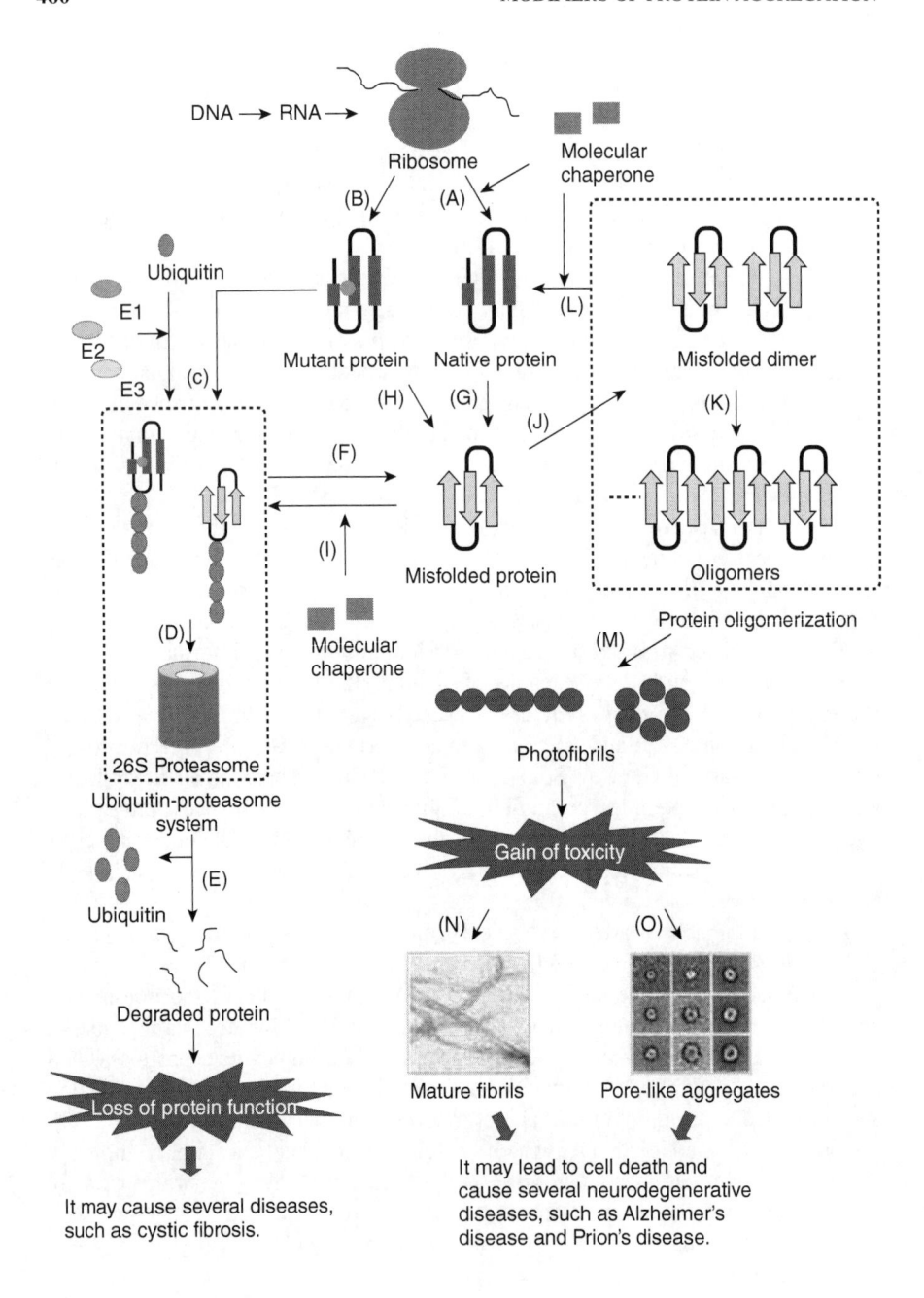

Molecular chaperone

(B) (A)

Ubiquitin
E1
E2
E3
(c)

Mutant protein   Native protein   Misfolded dimer

(H) (G) (J) (K)

(F)

(I)

Misfolded protein   Oligomers

Molecular chaperone

(D)

Protein oligomerization

(M)

Photofibrils

26S Proteasome
Ubiquitin-proteasome system

(E)

Ubiquitin

Gain of toxicity

(N)   (O)

Degraded protein

Mature fibrils   Pore-like aggregates

Loss of protein function

It may cause several diseases, such as cystic fibrosis.

It may lead to cell death and cause several neurodegenerative diseases, such as Alzheimer's disease and Prion's disease.

Figure 15.3. The pathway of protein synthesis and degradation in the cell. (A) Nascent polypeptide chain is converted into its native folded protein with the help of molecular chaperones. (B) Nascent polypeptide chain with a mutation (blue ball) folds into its native-like protein (or partial unfolded protein). (C) Mutant (or partial unfolded) protein may be re-recognized as imperfect proteins, leading to ubiquitination by E1 (ubiquitin-activating enzyme), E2 (ubiquitin-conjugating enzyme), and E3 (ubiquitin ligase). (D) Misfolded protein enters into the proteasome system with the help of the ubiquitin complex. (E) Misfolded protein is degraded into small peptide by proteasome and ubiquitin is regenerated. (F) Impaired proteasome system could not degrade the misfolded protein. (G) Native protein molecule is converted into a misfolded structure, which is caused by destabilization of the α-helical structure and the simultaneous formation of the β-sheet structure. (H) Mutations accelerate protein misfolding. (I) Molecular chaperones facilitate directing the misfolded proteins to the proteasomal pathway. (J) Misfolded monomers aggregate into dimer as initial building blocks for the formation of amyloid fibrils. (K) These building blocks further polymerize to form oligomers. (L) Molecular chaperones disaggregate the compact aggregates and develop native folded monomers. (M) Oligomers further form photofibrils. (N) Amyloid hypothesis. (O) Channel hypothesis. [96] See color insert.

observed in AD, Huntington's disease, PD, and Prion's disease (as discussed above), and (2) loss of function as with CF and α_1-antitrypsin deficiency [96].

In the former category and discussed above, the misfolded proteins may further aggregate to form amyloids, which cause cell toxicity and eventually death. Aggregation, which is usually accelerated with mutant proteins, can also occur with proteins in their native conformation. In other words, mutations are not an absolute requirement for protein misfolding and for treating diseases [96]. Moreover, mutations can lead to defects in folding, which can, in turn, lead to the accumulation of a misfolded protein in the endoplasmic reticulum (ER), which can result in increased ER stress [137]. Moreover, accumulation of the misfolded protein can lead to an absence of the correctly folded protein, which can result in the loss of physiologically important functions in the native protein.

As mentioned, many WT proteins fold inefficiently in the ER, even with the aid of the heat shock protein 40/70/90 (hsp40, hsp70, and hsp90) chaperone system that facilitates the refolding or retrotranslocation of misfolded proteins back into the cytoplasm; this can be due to protein mutation or other reasons. For this reason small molecule ligands, sometimes termed pharmacological chaperones, which bind to the native state of these mutant proteins, can stabilize the native state or destabilize the transition state to compensate for the influence of the mutation [138].

15.3.2.1. CF One example of non-amyloidogenic disease associated with protein misfolding that results in a loss of protein function is CF. The human cystic fibrosis transmembrane conductance regulator (CFTR) gene encodes an integral membrane glycoprotein of 1480 amino acid residues with two N-linked glycosylation sites [139]. The CFTR is a cAMP-regulated chloride (Cl-) channel localized at the apical membrane of secretory epithelia. Mutations in this channel cause CF, a disease characterized by the inability of epithelial cells to secrete chloride. This results in the production of thick and viscous mucus that causes severe functional obstruction of lungs and pancreas. A majority of CF patients have a deletion of a phenylalanine residue at position 508, resulting in an F508del-CFTR protein [140]. The clinical importance of this mutation becomes evident because it accounts for 90% of patients diagnosed with CF [141]. This mutation results in a misfolded channel retained in the ER in an immature state; this is rapidly degraded by a process involving the ubiquitin-dependent proteasomal system [139]. Thus, very little of this protein can reach the cellular membrane, resulting in the loss of function phenotype.

Rescue of misfolded trafficking-defective mutant proteins by pharmacological chaperones emerged some years ago with the finding that ligands (agonists and antagonists) increased the efficiency of receptor maturation and restored the function of these proteins [142–144]. In particular, numerous efforts aimed at developing small molecule pharmacotherapy for CF focused on identifying compounds that can either stabilize the tertiary structure of the F508del-CFTR protein or can modify the interactions of the mutant protein with ER chaperones. By preventing these interactions, the newly synthesized, misfolded but functional F508del-CFTR protein might escape recognition by mechanisms responsible for its retention and its ultimate degradation [145]. A limited number of molecules have been shown to restore partial function in F508del-CFTR mutant cells: curcumin [146–148], CFTRcor-325 [149], and the 1,2-glucosidase inhibitor miglustat [150].

In addition, benzo[c]quinolizinium (MPB) derivatives have been shown to activate WT CFTR, along with several CF mutants such as the F508del mutation [151–154]. Two derivatives, MPB-07 and MPB-91, have also been described as correctors of F508del-CFTR trafficking [152, 155] and as inhibitors of the first cytoplasmic-domain degradation of F508del-CFTR [156]. More recently, Gly622 amino acid was identified as part of the putative site interfering with the corrective effect of MPB, and it was shown that the rescue of F508del-CFTR by MPB is due to prevention of proteasomal degradation.

15.3.2.2. Gaucher's Disease Another conformational, non-amyloidogenic disease associated with protein aggregation is Gaucher's disease [157]. Gaucher's disease is the most prevalent lysosomal storage disease and is caused by deficient activity elicited by several mutated forms of the enzyme glucocerebrosidase (β-Glu), the β-glucosidase that hydrolyzes glucosylceramide into glucose and ceramide [158]. The most prominent mutation (N370S) and other β-Glu mutations are thought to alter active-site residues [138]. Accumulation of the substrate (glucosylceramide) leads to hepatomegaly, splenomegaly, bone crisis, anemia, and central nervous system (CNS) involvement [159].

Cellular levels of the mutated, misfolded enzyme are abnormally low because of its premature degradation by specific cytosolic endoproteases in the ER. Several therapeutic strategies for Gaucher's disease have been developed over the past years [160]. Among them, use of pharmacological chaperones, competitive inhibitors of the target enzyme, which assist the proper folding of the defective protein at subinhibitory concentrations [5, 161, 162], is an active field of research [163].

These inhibitors seem to provide a template that complements the active site of β-Glu by stabilizing the native state in the ER, allowing β-Glu to leave the ER and to be trafficked to the lysosome. The high concentration of the glycolipid substrate in the lysosomes of β-Glu-deficient cells can competitively displace the inhibitor, allowing the enzyme to function. Kelly and colleagues [159] showed that a subset of N-alkylated deoxynorjirimycin and simpler six-membered ring N-heterocycles increase the activity of β-Glu within human cells [159]. Also, several other iminosugars and aminocyclitols have been reported in the literature [5, 158, 160]. Di'az and coworkers reported new N-alkylaminocyclitols bearing a 1,2,3-triazole system at different positions of the alkyl chain [164].

15.3.3. Protein Aggregation Acceleration by Metals

Although a large number of specific interactions of different molecules and aggregative proteins focus on inhibiting and modifying protein aggregation, many are also directed toward specific interactions with these proteins, which enhance aggregation. For example, a large effort has been made in studying the effect of metals on protein aggregation. Metal ions have been widely found to be implicated as potential risk cofactors in several neurodegenerative disorders, such as AD, PD, Creutzfeldt–Jacob disease, and amyotrophic lateral sclerosis (ALS) [165–167]. Specifically related to AD, several recent studies reported that some metals can accelerate the dynamics of Aβ aggregation, thus increasing the neurotoxic effects on neuronal cells as a consequence of marked biophysical alteration properties of the peptide [168–172]. The influence of metal ions such as Fe^{2+}, Cu^{2+}, and Zn^{2+} in stimulating Aβ aggregation as well as their interaction modalities with the Aβ peptide have been widely studied *in vitro* [173, 174]. Of these metals, zinc has been found to induce rapid precipitation of Aβ and also to increase the protease-resistant aggregates under different conditions, such as acidic and alkaline [169]. Copper and iron have also been shown to increase aggregation but only at acidic pH [175, 176]. Aluminum has also been implicated in AD pathogenesis and has been thought to induce Aβ aggregation [177, 178]. It has been controversially suggested that aluminum-contaminated drinking water can be a risk factor for AD [178], although the role for aluminum as a risk factor for AD and its direct influence on the aggregation of Aβ have been debated in the literature. A detailed characterization of the *in vitro* conformational and aggregational changes stimulated by aluminum on different Aβ fragments and, in particular, on human $Aβ_{1-42}$, has recently been published [171]. These recent studies have demonstrated that among the various metal ions that have been implicated in increasing aggregation, aluminum seems to be the cation that most efficiently promotes Aβ aggregation *in vitro* while dramatically increasing cellular neurotoxicity [171, 179, 180]. Aluminum has also been implicated in promoting the spontaneous increase of $Aβ_{1-42}$ surface hydrophobicity, compared with Aβ alone, by converting the peptide into partially folded conformations with solvent-exposed hydrophobic patches [171].

The involvement of heavy metals in PD was first suggested based on the considerable increase in total iron, zinc, and aluminum content observed in the parkinsonian substantia nigra when compared with control tissues [181, 182]. However, the exact mechanism by which metals promote the onset of PD and their exact role in the mechanism leading to degeneration of dopaminergic neurons in PD patients remains controversial [183]. One hypothesis is their possible role in promoting α-syn aggregation and fibrillization in the cells of PD patients. Several mechanisms for metal aggregation stimulation have been proposed, for example, simple direct interactions between α-syn and the metal, leading to structural changes in α-syn that may enhance its propensity to form fibrils [184, 185]. However, the physiological concentration of these metals as well as the cytosol composition and redox states accessible to different metal ions may lead to more complex interactions than would the simple coordination chemistry between α-syn and the different metals, which in turn leads to aggregation stimulation [183]. It has been shown that polyvalent, rather than monovalent [185, 186] ions, can directly promote α-syn oligomerization [184] as well as fibrillization [184, 185]. In the presence of $AlCl_3$, $FeCl_3$, $CoCl_3$, or $CuCl_2$ at what can be considered high metal ion concentrations (2 mM, approximately three orders of magnitude higher than the expected concentration in cells), the rate of fibril formation was > 50-fold faster than in their absence [183].

One polyvalent metal whose homeostasis is one of the factors to be considered in discussing the mechanism underlying neurodegeneration in PD is iron. There is evidence of a sizable increase in iron content in the parkinsonian substantia nigra when compared with other metals, together with reduced ferritin levels [187]. The high levels of reactive iron in dopaminergic neurons have also been associated with the observation that the amount of iron bound to neuromelanin is 50 times higher than that of copper [188]. The interplay between iron and α-syn may occur through different mechanisms, one being the direct binding of either Fe(II) or Fe(III) to the protein. This was shown by intrinsic tyrosine fluorescence from which a Ki value of 173 mM was obtained [189]. More recent studies, using intrinsic Tyr fluorescence and an isothermal titration calorimeter (ITC), confirmed the presence in α-syn of a single binding site for iron with similar binding properties [190]. In addition, cell culture data suggested that α-syn interacts with iron [183]. Recently, the role of iron in oligomer formation was also investigated. This was carried out using single-molecule fluorescence techniques and atomic force microscopy (AFM). α-Syn oligomerization was first triggered by organic solvents, and then Fe(III) at micromolar concentrations was added. Interestingly, the addition of Fe(III), but not Fe(II), significantly increased the oligomerization of α-syn [191].

A large body of epidemiological, histological, and biochemical data are reported in the literature, generated in the process of unraveling the role of copper in PD. Among the first observations is the finding of high copper levels in the cerebrospinal fluid of PD patients [192]. The specificity of Cu(II), relative to all other transition metal ions, to bind to monomeric α-syn indicates that Cu(II) has a specific binding mode. Also, among the several metal ions analyzed, Cu(II) is the only one that is effective in accelerating α-syn aggregation at physiologically relevant concentrations [193]. Results obtained by different investigators often contradicted each other; however, the body of evidence now available may lead to a clearer picture. Several studies have aimed toward unveiling the exact interactions between copper and α-syn. Recent evidence, using a combination of low- and high-resolution spectroscopic techniques [193–196], suggests that the high-affinity copper-binding site of α-syn is located in the N-terminus of the protein. However, the mechanisms by which the interactions between α-syn and Cu(II) promote the aggregation events remain to be unraveled. Much effort has been put forth into elucidating the exact underlying mechanism. For example, experimental evidence by Rasia et al. indicated that Cu(II) promotes the nucleation, but not the growth phase, suggesting that the Cu(II)-bound form of α-syn is more prone to nucleate than is the uncomplexed protein [193]. It has also been suggested that copper binding to the high-affinity site of α-syn might be the critical step in rendering the protein a target for destabilization. This process *in vivo* might lead to a cascade of subsequent structural alterations promoting the generation of a pool of α-syn molecules that are more prone to aggregation [186].

Several other metal ions such as manganese, cobalt, and nickel have been shown to bind α-syn, but with low affinity [186]. Elevated concentrations of these metals were required to observe any promoting effect in the aggregation process of α-syn. This therefore rules out the relevance of these metal ion complexes of α-syn in the pathology of PD.

Metal ions were also shown to play a key role in the aggregation of the prion-associated protein PrP. Ricchelli and coworkers investigated the role of Cu^{2+}, Mn^{2+}, Zn^{2+}, and Al^{3+} in inducing defective conformational rearrangements of the recombinant human prion protein (hPrP), which triggers aggregation and fibrillogenesis. Aluminum strongly stimulated the conversion of native hPrP into the altered conformation, and its potency in inducing aggregation was very high. However, zinc was more efficient than aluminum in promoting the organization of hPrP aggregates into well-structured, amyloid-like fibrillar filaments, whereas manganese delayed and copper prevented this process [197].

15.3.4. Protein Aggregation Acceleration by Small Molecules

Not only have metals been shown to enhance aggregation—several other small molecules have enhanced protein aggregation, especially in the context of amyloidogenic proteins. For example, Levy et al. [197], using small molecules such as suramin, demonstrated dose-dependent inhibition of IAPP fibrillization at a 2–4 mM range; however, at a concentration higher than 5 mM, suramin induced a marked increase in fluorescence, which indicated a significant dose-dependent increase in IAPP fibrillization.

Another small molecule that had an accelerating effect on protein fibrillization, in this case on $A\beta$ fibrillization, is methylene blue. This small molecule had a strong inhibitory effect on $A\beta$ oligomer formation while promoting $A\beta$ fibrillization. Methylene blue-mediated promotion of fiber formation occurred via a dose-dependent decrease in the lag time and an increase in the fibrillization rate, consistent with promotion of both filament nucleation and elongation. Addition of methylene blue to preformed oligomers led to oligomer loss and promotion of fibrillization [198].

ACKNOWLEDGMENTS

We thank members of the Gazit laboratory for helpful discussions. We thank the Deutsch-Israelische Projektkooperation and the Levi Eshkol Fellowship from the Ministry of Science and Technology of Israel for financial support.

REFERENCES

1. Cromwell, M. E., Hilario, E., and Jacobson, F. (2006) Protein aggregation and bioprocessing, *AAPS J 8*, E572–E579.
2. Philo, J. S. and Arakawa, T. (2009) Mechanisms of protein aggregation, *Curr Pharm Biotechnol 10*, 348–351.
3. Dobson, C. M. (2004) Principles of protein folding, misfolding and aggregation, *Semin Cell Dev Biol 15*, 3–16.
4. Hamada, H., Arakawa, T., and Shiraki, K. (2009) Effect of additives on protein aggregation, *Curr Pharm Biotechnol 10*, 400–407.
5. Morello, J. P., Petaja-Repo, U. E., Bichet, D. G., and Bouvier, M. (2000) Pharmacological chaperones: A new twist on receptor folding, *Trends Pharmacol Sci 21*, 466–469.
6. Lentzen, G. and Schwartz, T. (2006) Extremolytes: Natural compounds from extremophiles for versatile applications, *Appl Microbiol Biotechnol 72*, 12.
7. Hofmeister, F. (1891) About the science of the effect of salts: The contribution of dissolved components to swelling processes, *Arch Exp Pathol Pharmakol XXVIII*, 29.
8. Lewith, S. (1887) About the science of the effect of salts: The behaviour of the proteins in the blood serum in the presence of salts, *Arch Exp Pathol Pharmakol XXIV*, 16.
9. Hofmeister, F. (1887) About the science of the effect of salts: About regulatrities in the protein precipitating effects of salts and the relation of these effects with the physiological behaviour of salts, *Arch Exp Pathol Pharmakol XXIV*, 14.
10. Hofmeister, F. (1888) About the science of the effect of salts: About the water withdrawing effect of the salts, *Arch Exp Pathol Pharmakol XXV*, 30.
11. Limbeck, R. V. (1888) About the science of the effect of salts: About the diuretic effect of salts, *Arch Exp Pathol Pharmakol XXV*, 18.
12. Hofmeister, F. (1890) About the science of the effect of salts: Investigations about the swelling process, *Arch Exp Pathol Pharmakol XXVII*, 18.
13. Münzer, E. (1898) About the science of the effect of salts: The general effect of salts, *Arch Exp Pathol Pharmakol XLI*, 23.
14. Kunz, W. and Ninham, B. W. (2004) 'Zur Lehre von der Wirkung der Salze' (about the science of the effect of salts): Franz Hofmeister's historical papers, *Curr Opin Colloid Interface Sci 9*, 19.
15. Collins, K. D. (2006) Ion hydration: Implications for cellular function, polyelectrolytes, and protein crystallization, *Biophys Chem 119*, 271–281.
16. Zhang, Y. and Cremer, P. S. (2010) Chemistry of Hofmeister anions and osmolytes, *Annu Rev Phys Chem 61*, 63–83.
17. Zhang, Y. and Cremer, P. S. (2006) Interactions between macromolecules and ions: The Hofmeister series, *Curr Opin Chem Biol 10*, 658–663.
18. Broering, J. M. and Bommarius, A. S. (2005) Evaluation of Hofmeister effects on the kinetic stability of proteins, *J Phys Chem B 109*, 20612–20619.
19. Chi, E. Y., Krishnan, S., Randolph, T. W., and Carpenter, J. F. (2003) Physical stability of proteins in aqueous solution: Mechanism and driving forces in nonnative protein aggregation, *Pharm Res 20*, 1325–1336.
20. Curtis, R. A., Ulrich, J., Montaser, A., Prausnitz, J. M., and Blanch, H. W. (2002) Protein-protein interactions in concentrated electrolyte solutions, *Biotechnol Bioeng 79*, 367–380.

REFERENCES

21. Arakawa, T. and Timasheff, S. N. (1984) Mechanism of protein salting in and salting out by divalent cation salts: Balance between hydration and salt binding, *Biochemistry 23*, 5912–5923.
22. Yeh, V., Broering, J. M., Romanyuk, A., Chen, B., Chernoff, Y. O., and Bommarius, A. S. (2010) The Hofmeister effect on amyloid formation using yeast prion protein, *Protein Sci 19*, 47–56.
23. Sikkink, L. A. and Ramirez-Alvarado, M. (2008) Salts enhance both protein stability and amyloid formation of an immunoglobulin light chain, *Biophys Chem 135*, 25–31.
24. Klement, K., Wieligmann, K., Meinhardt, J., Hortschansky, P., Richter, W., and Fandrich, M. (2007) Effect of different salt ions on the propensity of aggregation and on the structure of Alzheimer's abeta(1-40) amyloid fibrils, *J Mol Biol 373*, 1321–1333.
25. Jain, S. and Udgaonkar, J. B. (2010) Salt-induced modulation of the pathway of amyloid fibril formation by the mouse prion protein, *Biochemistry 49*, 7615–7624.
26. Summers, C. A. and Flowers, R. A., 2nd (2000) Protein renaturation by the liquid organic salt ethylammonium nitrate, *Protein Sci 9*, 2001–2008.
27. Lange, C., Patil, G., and Rudolph, R. (2005) Ionic liquids as refolding additives: N'-alkyl and N'-(omega-hydroxyalkyl) N-methylimidazolium chlorides, *Protein Sci 14*, 2693–2701.
28. Fujita, K., MacFarlane, D. R., and Forsyth, M. (2005) Protein solubilising and stabilising ionic liquids, *Chem Commun (Camb) Issue 38*, 4804–4806.
29. Fujita, K., Forsyth, M., MacFarlane, D. R., Reid, R. W., and Elliott, G. D. (2006) Unexpected improvement in stability and utility of cytochrome c by solution in biocompatible ionic liquids, *Biotechnol Bioeng 94*, 1209–1213.
30. Buchfink, R., Tischer, A., Patil, G., Rudolph, R., and Lange, C. (2010) Ionic liquids as refolding additives: Variation of the anion, *J Biotechnol 150*, 64–72.
31. Holthauzen, L. M. and Bolen, D. W. (2007) Mixed osmolytes: The degree to which one osmolyte affects the protein stabilizing ability of another, *Protein Sci 16*, 293–298.
32. Roberts, M. F. (2005) Organic compatible solutes of halotolerant and halophilic microorganisms, *Saline Systems 1*, 5.
33. Harries, D. and Rosgen, J. (2008) A practical guide on how osmolytes modulate macromolecular properties, *Methods Cell Biol 84*, 679–735.
34. Yancey, P. H. (2005) Organic osmolytes as compatible, metabolic and counteracting cytoprotectants in high osmolarity and other stresses, *J Exp Biol 208*, 2819–2830.
35. Yancey, P. H., Blake, W. R., and Conley, J. (2002) Unusual organic osmolytes in deep-sea animals: Adaptations to hydrostatic pressure and other perturbants, *Comp Biochem Physiol A Mol Integr Physiol 133*, 667–676.
36. Yancey, P. H. (2004) Compatible and counteracting solutes: Protecting cells from the Dead Sea to the deep sea, *Sci Prog 87*, 1–24.
37. Brown, A. D. and Simpson, J. R. (1972) Water relations of sugar-tolerant yeasts: The role of intracellular polyols, *J Gen Microbiol 72*, 589–591.
38. Gerlsma, S. Y. and Stuur, E. R. (1972) The effect of polyhydric and monohydric alcohols on the heat-induced reversible denaturation of lysozyme and ribonuclease, *Int J Pept Protein Res 4*, 377–383.
39. Yancey, P. H. and Burg, M. B. (1990) Counteracting effects of urea and betaine in mammalian cells in culture, *Am J Physiol 258*, R198–R204.

40 Yancey, P. H., Clark, M. E., Hand, S. C., Bowlus, R. D., and Somero, G. N. (1982) Living with water stress: Evolution of osmolyte systems, *Science 217*, 1214–1222.

41 Yancey, P. H., Fyfe-Johnson, A. L., Kelly, R. H., Walker, V. P., and Aunon, M. T. (2001) Trimethylamine oxide counteracts effects of hydrostatic pressure on proteins of deep-sea teleosts, *J Exp Zool 289*, 172–176.

42 Yancey, P. H., Heppenstall, M., Ly, S., Andrell, R. M., Gates, R. D., Carter, V. L., and Hagedorn, M. (2010) Betaines and dimethylsulfoniopropionate as major osmolytes in cnidaria with endosymbiotic dinoflagellates, *Physiol Biochem Zool 83*, 167–173.

43 Yancey, P. H., Rhea, M. D., Kemp, K. M., and Bailey, D. M. (2004) Trimethylamine oxide, betaine and other osmolytes in deep-sea animals: Depth trends and effects on enzymes under hydrostatic pressure, *Cell Mol Biol (Noisy-le-grand) 50*, 371–376.

44 Bennion, B. J. and Daggett, V. (2004) Counteraction of urea-induced protein denaturation by trimethylamine N-oxide: A chemical chaperone at atomic resolution, *Proc Natl Acad Sci U S A 101*, 6433–6438.

45 Lin, T. Y. and Timasheff, S. N. (1994) Why do some organisms use a urea-methylamine mixture as osmolyte? Thermodynamic compensation of urea and trimethylamine N-oxide interactions with protein, *Biochemistry 33*, 12695–12701.

46 Perlmutter, D. H. (2002) Chemical chaperones: A pharmacological strategy for disorders of protein folding and trafficking, *Pediatr Res 52*, 832–836.

47 Tatzelt, J., Prusiner, S. B., and Welch, W. J. (1996) Chemical chaperones interfere with the formation of scrapie prion protein, *EMBO J 15*, 6363–6373.

48 Welch, W. J. and Brown, C. R. (1996) Influence of molecular and chemical chaperones on protein folding, *Cell Stress Chaperones 1*, 109–115.

49 Gekko, K. and Timasheff, S. N. (1981) Mechanism of protein stabilization by glycerol: Preferential hydration in glycerol-water mixtures, *Biochemistry 20*, 4667–4676.

50 Gekko, K. and Timasheff, S. N. (1981) Thermodynamic and kinetic examination of protein stabilization by glycerol, *Biochemistry 20*, 4677–4686.

51 Baldwin, R. L. (1996) How Hofmeister ion interactions affect protein stability, *Biophys J 71*, 2056–2063.

52 Hummer, G., Garde, S., Garcia, A. E., Paulaitis, M. E., and Pratt, L. R. (1998) The pressure dependence of hydrophobic interactions is consistent with the observed pressure denaturation of proteins, *Proc Natl Acad Sci U S A 95*, 1552–1555.

53 Gur, E., Biran, D., Gazit, E., and Ron, E. Z. (2002) In vivo aggregation of a single enzyme limits growth of *Escherichia coli* at elevated temperatures, *Mol Microbiol 46*, 1391–1397.

54 Feng, S. and Yan, Y. B. (2008) Effects of glycerol on the compaction and stability of the wild type and mutated rabbit muscle creatine kinase, *Proteins 71*, 844–854.

55 Barreca, D., Lagana, G., Ficarra, S., Tellone, E., Leuzzi, U., Magazu, S., Galtieri, A., and Bellocco, E. (2010) Anti-aggregation properties of trehalose on heat-induced secondary structure and conformation changes of bovine serum albumin, *Biophys Chem 147*, 146–152.

56 Ignatova, Z. and Gierasch, L. M. (2006) Inhibition of protein aggregation in vitro and in vivo by a natural osmoprotectant, *Proc Natl Acad Sci U S A 103*, 13357–13361.

57 Nayak, A., Lee, C. C., McRae, G. J., and Belfort, G. (2009) Osmolyte controlled fibrillation kinetics of insulin: New insight into fibrillation using the preferential exclusion principle, *Biotechnol Prog 25*, 1508–1514.

REFERENCES

58 Fung, J., Darabie, A. A., and McLaurin, J. (2005) Contribution of simple saccharides to the stabilization of amyloid structure, *Biochem Biophys Res Commun 328*, 1067–1072.

59 Kim, Y. S., Cape, S. P., Chi, E., Raffen, R., Wilkins-Stevens, P., Stevens, F. J., Manning, M. C., Randolph, T. W., Solomon, A., and Carpenter, J. F. (2001) Counteracting effects of renal solutes on amyloid fibril formation by immunoglobulin light chains, *J Biol Chem 276*, 1626–1633.

60 De Young, L. R., Dill, K. A., and Fink, A. L. (1993) Aggregation and denaturation of apomyoglobin in aqueous urea solutions, *Biochemistry 32*, 3877–3886.

61 Tsumoto, K., Ejima, D., Kumagai, I., and Arakawa, T. (2003) Practical considerations in refolding proteins from inclusion bodies, *Protein Expr Purif 28*, 1–8.

62 Dunbar, J., Yennawar, H. P., Banerjee, S., Luo, J., and Farber, G. K. (1997) The effect of denaturants on protein structure, *Protein Sci 6*, 1727–1733.

63 O'Brien, E. P., Dima, R. I., Brooks, B., and Thirumalai, D. (2007) Interactions between hydrophobic and ionic solutes in aqueous guanidinium chloride and urea solutions: Lessons for protein denaturation mechanism, *J Am Chem Soc 129*, 7346–7353.

64 Lim, W. K., Rosgen, J., and Englander, S. W. (2009) Urea, but not guanidinium, destabilizes proteins by forming hydrogen bonds to the peptide group, *Proc Natl Acad Sci U S A 106*, 2595–2600.

65 Kendrick, B. S., Cleland, J. L., Lam, X., Nguyen, T., Randolph, T. W., Manning, M. C., and Carpenter, J. F. (1998) Aggregation of recombinant human interferon gamma: Kinetics and structural transitions, *J Pharm Sci 87*, 1069–1076.

66 Kim, J. R., Muresan, A., Lee, K. Y., and Murphy, R. M. (2004) Urea modulation of beta-amyloid fibril growth: Experimental studies and kinetic models, *Protein Sci 13*, 2888–2898.

67 Wang, X. Y., Meng, F. G., and Zhou, H. M. (2003) Inactivation and conformational changes of creatine kinase at low concentrations of hexafluoroisopropanol solutions, *Biochem Cell Biol 81*, 327–333.

68 Yang, H. P. and Zhou, H. M. (1997) Conformational changes of creatine kinase in trifluoroethanol solutions, *Biochem Mol Biol Int 43*, 1297–1304.

69 Bohidar, H. B. and Mohanty, B. (2004) Anomalous self-assembly of gelatin in ethanol-water marginal solvent, *Phys Rev E 69*, 021902-1–021902-9.

70 Roccatano, D., Colombo, G., Fioroni, M., and Mark, A. E. (2002) Mechanism by which 2,2,2-trifluoroethanol/water mixtures stabilize secondary-structure formation in peptides: A molecular dynamics study, *Proc Natl Acad Sci U S A 99*, 12179–12184.

71 Sen, P., Ahmad, B., Rabbani, G., and Khan, R. H. (2010) 2,2,2-Trifluroethanol induces simultaneous increase in alpha-helicity and aggregation in alkaline unfolded state of bovine serum albumin, *Int J Biol Macromol 46*, 250–254.

72 Sabate, R. and Estelrich, J. (2003) Disaggregating effects of ethanol at low concentration on beta-poly-L-lysines, *Int J Biol Macromol 32*, 10–16.

73 Cao, A., Hu, D., and Lai, L. (2004) Formation of amyloid fibrils from fully reduced hen egg white lysozyme, *Protein Sci 13*, 319–324.

74 Dzwolak, W., Grudzielanek, S., Smirnovas, V., Ravindra, R., Nicolini, C., Jansen, R., Loksztejn, A., Porowski, S., and Winter, R. (2005) Ethanol-perturbed amyloidogenic self-assembly of insulin: Looking for origins of amyloid strains, *Biochemistry 44*, 8948–8958.

75 Grudzielanek, S., Velkova, A., Shukla, A., Smirnovas, V., Tatarek-Nossol, M., Rehage, H., Kapurniotu, A., and Winter, R. (2007) Cytotoxicity of insulin within its self-assembly and amyloidogenic pathways, *J Mol Biol 370*, 372–384.

76 Shen, C. L. and Murphy, R. M. (1995) Solvent effects on self-assembly of beta-amyloid peptide, *Biophys J 69*, 640–651.

77 Hirota-Nakaoka, N., Hasegawa, K., Naiki, H., and Goto, Y. (2003) Dissolution of beta2-microglobulin amyloid fibrils by dimethylsulfoxide, *J Biochem 134*, 159–164.

78 Sonesson, A. W., Blom, H., Hassler, K., Elofsson, U. M., Callisen, T. H., Widengren, J., and Brismar, H. (2008) Protein-surfactant interactions at hydrophobic interfaces studied with total internal reflection fluorescence correlation spectroscopy (TIR-FCS), *J Colloid Interface Sci 317*, 449–457.

79 Chou, D. K., Krishnamurthy, R., Randolph, T. W., Carpenter, J. F., and Manning, M. C. (2005) Effects of Tween 20 and Tween 80 on the stability of Albutropin during agitation, *J Pharm Sci 94*, 1368–1381.

80 Randolph, T. W. and Jones, L. S. (2002) Surfactant-protein interactions, *Pharm Biotechnol 13*, 159–175.

81 Giehm, L., Oliveira, C. L., Christiansen, G., Pedersen, J. S., and Otzen, D. E. (2010) SDS-induced fibrillation of alpha-synuclein: An alternative fibrillation pathway, *J Mol Biol 401*, 115–133.

82 Hung, Y. T., Lin, M. S., Chen, W. Y., and Wang, S. S. (2010) Investigating the effects of sodium dodecyl sulfate on the aggregative behavior of hen egg-white lysozyme at acidic pH, *Colloids Surf B Biointerfaces 81*, 141–151.

83 Barghorn, S., Nimmrich, V., Striebinger, A., Krantz, C., Keller, P., Janson, B., Bahr, M., Schmidt, M., Bitner, R. S., Harlan, J., Barlow, E., Ebert, U., and Hillen, H. (2005) Globular amyloid beta-peptide oligomer—A homogenous and stable neuropathological protein in Alzheimer's disease, *J Neurochem 95*, 834–847.

84 Cao, M., Han, Y., Wang, J., and Wang, Y. (2007) Modulation of fibrillogenesis of amyloid beta(1-40) peptide with cationic gemini surfactant, *J Phys Chem B 111*, 13436–13443.

85 Wang, S. S., Liu, K. N., and Han, T. C. (2010) Amyloid fibrillation and cytotoxicity of insulin are inhibited by the amphiphilic surfactants, *Biochim Biophys Acta 1802*, 519–530.

86 Samiotakis, A., Wittung-Stafshede, P., and Cheung, M. S. (2009) Folding, stability and shape of proteins in crowded environments: Experimental and computational approaches, *Int J Mol Sci 10*, 572–588.

87 Minton, A. P. (2001) The influence of macromolecular crowding and macromolecular confinement on biochemical reactions in physiological media, *J Biol Chem 276*, 10577–10580.

88 Ellis, R. J. (2001) Macromolecular crowding: Obvious but underappreciated, *Trends Biochem Sci 26*, 597–604.

89 Kinjo, A. R. and Takada, S. (2002) Effects of macromolecular crowding on protein folding and aggregation studied by density functional theory: Dynamics, *Phys Rev E Stat Nonlin Soft Matter Phys 66*, 051902.

90 van den Berg, B., Ellis, R. J., and Dobson, C. M. (1999) Effects of macromolecular crowding on protein folding and aggregation, *EMBO J 18*, 6927–6933.

91 van den Berg, B., Wain, R., Dobson, C. M., and Ellis, R. J. (2000) Macromolecular crowding perturbs protein refolding kinetics: Implications for folding inside the cell, *EMBO J 19*, 3870–3875.

REFERENCES

92. Shtilerman, M. D., Ding, T. T., and Lansbury, P. T., Jr. (2002) Molecular crowding accelerates fibrillization of alpha-synuclein: Could an increase in the cytoplasmic protein concentration induce Parkinson's disease? *Biochemistry 41*, 3855–3860.

93. Hatters, D. M., Minton, A. P., and Howlett, G. J. (2002) Macromolecular crowding accelerates amyloid formation by human apolipoprotein C-II, *J Biol Chem 277*, 7824–7830.

94. Zhou, Z., Fan, J. B., Zhu, H. L., Shewmaker, F., Yan, X., Chen, X., Chen, J., Xiao, G. F., Guo, L., and Liang, Y. (2009) Crowded cell-like environment accelerates the nucleation step of amyloidogenic protein misfolding, *J Biol Chem 284*, 30148–30158.

95. Zhou, B. R., Zhou, Z., Hu, Q. L., Chen, J., and Liang, Y. (2008) Mixed macromolecular crowding inhibits amyloid formation of hen egg white lysozyme, *Biochim Biophys Acta 1784*, 472–480.

96. Zhao, J. H., Liu, H. L., Lin, H. Y., Huang, C. H., Fang, H. W., Chen, S. S., Ho, Y., Tsai, W. B., and Chen, W. Y. (2007) Chemical chaperone and inhibitor discovery: Potential treatments for protein conformational diseases, *Perspect Med Chem 1*, 39–48.

97. Virchow, R. (1853) Über eine im gehirn und rückenmark des menschen aufgefundene substanz mit der chemischen reaktion der cellulose, *Virchows Arch 6*, 135–138.

98. Aguzzi, A. and O'Conno, T. (2010) Protein aggregation diseases: Pathogenicity and therapeutic perspectives, *Nat Rev 9*, 237–248.

99. Amijee, H., Madine, J., Middletona, D. A., and Doig, A. J. (2009) Inhibitors of protein aggregation and toxicity, *Biochem Soc Trans 37*, 692–696.

100. Wyatt, A. R., Yerbury, J. J., Poon, S., and Wilson, M. R. (2009) Therapeutic targets in extracellular protein deposition diseases, *Curr Med Chem 16*, 2855–2866.

101. Smith, H. J., Simons, C., and Sewell, R. D. E. (2008) *Protein misfolding in neurodegenerative diseases: Mechanisms and therapeutic strategies, illustrated ed.*, CRC Press, Boca Raton, FL.

102. Lorenzo, A. and Yankner, B. A. (1994) Beta-amyloid neurotoxicity requires fibril formation and is inhibited by Congo red, *Proc Natl Acad Sci U S A 91*, 12243–12247.

103. Lee, V. M. (2002) Amyloid binding ligands as Alzheimer's disease therapies, *Neurobiol Aging 23*, 1039–1042.

104. Poli, G., Ponti, W., Carcassola, G., Ceciliani, F., Colombo, L., Dall'Ara, P., Gervasoni, M., Giannino, M. L., Martino, P. A., Pollera, C., Villa, S., and Salmona, M. (2003) In vitro evaluation of the anti-prionic activity of newly synthesized Congo red derivatives, *Arzneimittelforschung 53*, 875–888.

105. Porat, Y., Abramowitz, A., and Gazit, E. (2006) Inhibition of amyloid fibril formation by polyphenols: Structural similarity and aromatic interactions as a common inhibition mechanism, *Chem Biol Drug Des 67*, 27–37.

106. Conway, K. A., Rochet, J. C., Bieganski, R. M., and Lansbury, P. T., Jr. (2001) Kinetic stabilization of the α-synuclein protofibril by a dopamine-α-synuclein adduct, *Science 294*, 1346–1349.

107. Spillantini, M. G., Schmidt, M. L., Lee, V. M. Y., Trojanowski, J. Q., Jakes, R., and Goedert, M. (1997) α-Synuclein in Lewy bodies, *Nature 388*, 839–840.

108. Levy, M., Porat, Y., Bacharach, E., Shalev, D. E., and Gazit, E. (2008) Phenolsulfonphthalein, but not phenolphthalein, inhibits amyloid fibril formation: Implications for the modulation of amyloid self-assembly, *Biochemistry 47*, 5896–5904.

109. Ehrnhoefer, D. E., Bieschke, J., Boeddrich, A., Herbst, M., Masino, L., Lurz, R., Engemann, S., Pastore, A., and Wanker, E. E. (2008) EGCG redirects amyloidogenic

polypeptides into unstructured, off-pathway oligomers, *Nat Struct Mol Biol 15*, 558–566.
110 Westermark, P., Engstrom, U., Johnson, K. H., Westermark, G. T., and Betsholtz, C. (1990) Islet amyloid polypeptide: Pinpointing amino acid residues linked to amyloid fibril formation, *Proc Natl Acad Sci U S A 87*, 5036–5040.
111 Szegedi, V., Juhasz, G., Rozsa, E., Juhasz-Vedres, G., Datki, Z., Fulop, L., Bozso, Z., Lakatos, A., Laczko, I., Farkas, T., Kis, Z., Toth, G., Soos, K., Zarandi, M., Budai, D., Toldi, J., and Penke, B. (2006) Endomorphin-2, an endogenous tetrapeptide, protects against Abeta1-42 in vitro and in vivo, *FASEB J 20*, 1191–1193.
112 Frew, T., Powis, G., Berggren, M., Gallegos, A., Abraham, R. T., Ashendel, C. L., Zalkow, L. H., Hudson, C., Gruszecka-Kowalik, E., and Burgess, E. M. (1995) Novel quinone antiproliferative inhibitors of phosphatidylinositol-3-kinase, *Anticancer Drug Des 10*, 347–359.
113 Gulielmo, B. J. and MacDougall, C. (2004) Pharmacokinetics of valaciclovir, *J Antimicrob Chemother 6*, 899–901.
114 Ono, K., Hasegawa, K., Naiki, H., and Yamada, M. (2005) Preformed β-amyloid fibrils are destabilized by coenzyme Q10 in vitro, *Biochem Biophys Res Commun 330*, 111–116.
115 Tomiyama, T., Kaneko, H., Kataoka, K., Asano, S., and Endo, N. (1997) Rifampicin inhibits the toxicity of pre-aggregated amyloid peptides by binding to peptide fibrils and preventing amyloid-cell interaction, *Biochem J 322*, 859–865.
116 Lieu, V. H., Wu, J. W., Wang, S. S., and Wu, C. H. (2007) Inhibition of amyloid fibrillization of hen egg-white lysozymes by rifampicin and p-benzoquinone, *Biotechnol Prog 23*, 698–706.
117 Bolognesi, M. L., Banzi, R., Bartolini, M., Cavalli, A., Tarozzi, A., Andrisano, V., Minarini, A., Rosini, M., Tumiatti, V., Bergamini, C., Fato, R., Lenaz, G., Hrelia, P., Cattaneo, A., Recanatini, M., and Melchiorre, C. (2007) Novel class of quinone-bearing polyamines as multi-target-directed ligands to combat Alzheimer's disease, *J Med Chem 50*, 4882–4897.
118 Pickhardt, M., Gazova, Z., von Bergen, M., Khlistunova, I., Wang, Y., Hascher, A., Mandelkow, E. M., Biernat, J., and Mandelkow, E. (2005) Anthraquinones inhibit tau aggregation and dissolve Alzheimer's paired helical filaments in vitro and in cells, *J Biol Chem 280*, 3628–3635.
119 Convertino, M., Pellarin, R., Catto, M., Carotti, A., and Caflisch, A. (2009) 9,10-Anthraquinone hinders β-aggregation: How does a small molecule interfere with Aβ-peptide amyloid fibrillation? *Prot Sci 18*, 792–800.
120 Necula, M., Kayed, R., Milton, S., and Glabe, C. G. (2007) Small molecule inhibitors of aggregation indicate that amyloid β oligomerization and fibrillization Pathways are independent and distinct, *J Biol Chem 282*, 10311–10324.
121 Scherzer-Attali, R., Pellarin, R., Convertino, M., Frydman-Marom, A., Egoz-Matia, N., Peled, S., Levy-Sakin, M., Shalev, D. E., Caflisch, A., Gazit, E., and Segal, D. (2010) Complete phenotypic recovery of an Alzheimer's disease model by a quinone-tryptophan hybrid aggregation inhibitor, *PLoS ONE 5*, e11101.
122 de Sá Alves, F. R., Barreiro, E. J., and Fraga, C. A. (2009) From nature to drug discovery: The indole scaffold as a "privileged structure," *Mini Rev Med Chem 7*, 782–793.
123 Cohen, T., Frydman-Marom, A., Rechter, M., and Gazit, E. (2006) Inhibition of amyloid fibril formation and cytotoxicity by hydroxyindole derivatives, *Biochemistry 45*, 4727–4735.

REFERENCES

124 Latawiec, D., Herrera, F., Bek, A., Losasso, V., Candotti, M., Benetti, F., Carlino, E., Kranjc, A., Lazzarino, M., Gustincich, S., Carloni, P., and Legname, G. (2010) Modulation of alpha-synuclein aggregation by dopamine analogs, *PLoS ONE 5*, e9234.

125 Jensen, P. H., Sorensen, E. S., Petersen, T. E., Gliemann, J., and Rasmussen, L. K. (1995) Residues in the synuclein consensus motif of the alpha-synuclein fragment, NAC, participate in transglutaminase-catalysed cross-linking to Alzheimer-disease amyloid beta A4 peptide, *Biochem J 310*, (Pt 1), 91–94.

126 Windisch, M., Hutter-Paier, B., Schreiner, E., and Wronski, R. (2004) Beta-Synuclein-derived peptides with neuroprotective activity: An alternative treatment of neurodegenerative disorders? *J Mol Neurosci 24*, 155–165.

127 Sunde, M. and Blake, C. C. F. (1997) The structure of amyloid fibrils by electron microscopy and X-ray diffraction, *Adv Protein Chem 50*, 123–159.

128 Rochet, J. C. and Lansbury, P. T. (2000) Amyloid fibrillogenesis: Themes and variations, *Curr Opin Struct Biol 10*, 60–68.

129 Stefani, M. and Dobson, C. M. (2003) Protein aggregation and aggregate toxicity: New insights into protein folding, misfolding diseases and biological evolution, *J Mol Med 81*, 678–699.

130 Soto, C. (2003) Unfolding the role of protein misfolding in neurodegenerative diseases, *Nat Rev Neurosci 4*, 49–60.

131 Gazit, E. (2002) A possible role for p-stacking in self-assembly of amyloid fibrils, *FASEB J 16*, 77–83.

132 Soto, C., Sigurdsson, E. M., Morelli, L., Kumar, R. A., Castaño, E. M., and Frangione, B. (1998) Beta-sheet breaker peptides inhibit fibrillogenesis in a rat brain model of amyloidosis: Implications for Alzheimer's therapy, *Nat Med 4*, 822–826.

133 Findeis, M. A., Musso, G. M., Arico-Muendel, C. C., Benjamin, H. W., Hundal, A. M., Lee, J. J., Chin, J., Kelley, M., Wakefield, J., Hayward, N. J., and Molineaux, S. M. (1999) Modified-peptide inhibitors of amyloid-beta-peptide polymerization, *Biochemistry 38*, 6791–6800.

134 Kapurniotu, A., Schmauder, A., and Tenidis, K. (2002) Structure-based design and study of non-amyloidogenic, double N-methylated IAPP amyloid core sequences as inhibitors of IAPP amyloid formation and cytotoxicity, *J Mol Biol 315*, 339–350.

135 Gilead, S. and Gazit, E. (2004) Inhibition of amyloid fibril formation by peptide analogues modified with α-aminoisobutyric acid, *Angew Chem Int Ed Engl 43*, 4041–4044.

136 Frydman-Marom, A., Rechter, M., Shefler, I., Bram, Y., Shalev, D. E., and Gazit, E. (2009) Cognitive-performance recovery of Alzheimer's disease model mice by modulation of early soluble amyloidal assemblies, *Angew Chem Int Ed Engl 48*, 1981–1986.

137 Kaufman, R. J. (1999) Stress signaling from the lumen of the endoplasmic reticulum: Coordination of gene transcriptional and translational controls, *Genes Dev 13*, 1211–1233.

138 Cohen, F. E. and Kelly, J. W. (2003) Therapeutic approaches to protein misfolding diseases, *Nature 426*, 905–909.

139 Kopito, R. R. (1999) Biosynthesis and degradation of CFTR, *Physiol Rev 79*, S167–S173.

140 Kerem, B. S., Rommens, J. M., Buchanan, J. A., Markiewicz, D., Cox, T. K., Chakravarti, A., Buchwald, M., and Tsui, L. C. (1989) Identification of the cystic fibrosis gene: Genetic analysis, *Science 245*, 1073–1080.

141. Amaral, M. D. (2006) Therapy through chaperones: Sense or antisense? Cystic fibrosis as a model disease, *J Inherit Metab Dis 29*, 477–487.
142. Robert, J., Auzan, C., Ventura, M. A., and Clauser, E. (2005) Mechanisms of cell-surface rerouting of an endoplasmic reticulum-retained mutant of the vasopressin V1b/V3 receptor by a pharmacological chaperone, *J Biol Chem 280*, 42198–42206.
143. Ulloa-Aguirre, A., Janovick, J. A., Brothers, S. P., and Conn, P. M. (2004) Pharmacologic rescue of conformationally-defective proteins: Implications for the treatment of human disease, *Traffic 5*, 821–837.
144. Gong, Q., Jones, M. A., and Zhou, Z. (2006) Mechanisms of pharmacological rescue of trafficking-defective hERG mutant channels in human long QT syndrome, *J Biol Chem 281*, 4069–4074.
145. Powell, K. and Zeitlin, P. L. (2002) Therapeutic approaches to repair defects in deltaF508 CFTR folding and cellular targeting, *Adv Drug Deliv Rev 54*, 1395–1408.
146. Egan, M. E., Glockner-Pagel, J., Ambrose, C., Cahill, P. A., Pappoe, L., Balamuth, N., Cho, E., Canny, S., Wagner, C. A., Geibel, J., and Caplan, M. J. (2002) Calcium-pump inhibitors induce functional surface expression of Delta F508-CFTR protein in cystic fibrosis epithelial cells, *Nat Med 8*, 485–492.
147. Norez, C., Heda, G. D., Jensen, T., Kogan, I., Hughes, L. K., Auzanneau, C., Derand, R., Bulteaux-Pignoux, L., Li, C., Ramjeesingh, M., Sheppard, D. N., Bear, C. E., Riordan, J. R., and Becq, F. (2004) Determination of CFTR chloride channel activity and pharmacology using radiotracer flux methods, *J Cyst Fibros 3*, Suppl 2, 119–121.
148. Norez, C., Antigny, F., Becq, F., and Vandebrouck, C. (2006) Maintaining low Ca2+ level in the endoplasmic reticulum restores abnormal endogenous F508del-CFTR trafficking in airway epithelial cells, *Traffic 7*, 562–573.
149. Wang, Y., Bartlett, M. C., Loo, T. W., and Clarke, D. M. (2006) Specific rescue of cystic fibrosis transmembrane conductance regulator processing mutants using pharmacological chaperones, *Mol Pharmacol 70*, 297–302.
150. Norez, C., Noel, S., Wilke, M., Bijvelds, M., Jorna, H., Melin, P., DeJonge, H., and Becq, F. (2006) Rescue of functional delF508-CFTR channels in cystic fibrosis epithelial cells by the alpha-glucosidase inhibitor miglustat, *FEBS Lett 580*, 2081–2086.
151. Becq, F., Mettey, Y., Gray, M. A., Galietta, L. J., Dormer, R. L., Merten, M., Metaye, T., Chappe, V., Marvingt-Mounir, C., Zegarra-Moran, O., Tarran, R., Bulteau, L., Dérand, R., Pereira, M. M., McPherson, M. A., Rogier, C., Joffre, M., Argent, B. E., Sarrouilhe, D., Kammouni, W., Figarella, C., Verrier, B., Gola, M., and Vierfond, J. M. (1999) Development of substituted benzo[c]quinolizinium compounds as novel activators of the cystic fibrosis chloride channel, *J Biol Chem 274*, 27415–27425.
152. Dormer, R. L., Derand, R., McNeilly, C. M., Mettey, Y., Bulteau-Pignoux, L., Metaye, T., Vierfond, J. M., Gray, M. A., Galietta, L. J. V., Morris, M. R., Pereira, M. M., Doull, I. J., Becq, F., and McPherson, M. A. (2001) Correction of delF508-CFTR activity with benzo(c)quinolizinium compounds through facilitation of its processing in cystic fibrosis airway cells, *J Cell Sci 114*, 4073–4081.
153. Marivingt-Mounir, C., Norez, C., Derand, R., Bulteau-Pignoux, L., Nguyen-Huy, D., Viossat, B., Morgant, G., Becq, F., Vierfond, J. M., and Mettey, Y. (2004) Synthesis, SAR, crystal structure, and biological evaluation of benzoquinoliziniums as activators of wild-type and mutant cystic fibrosis transmembrane conductance regulator channels, *J Med Chem 47*, 962–972.

REFERENCES

154 Melin, P., Thoreau, V., Norez, C., Bilan, F., Kitzis, A., and Becq, F. (2004) The cystic fibrosis mutation G1349D within the signature motif LSHGH of NBD2 abolishes the activation of CFTR chloride channels by genistein, *Biochem Pharmacol 67*, 2187–2196.

155 Dormer, R. L., McNeilly, C. M., Morris, M. R., Pereira, M. M., Doull, I. J., Becq, F., Mettey, Y., Vierfond, J. M., and McPherson, M. A. (2001) Localisation of wild-type and DeltaF508-CFTR in nasal epithelial cells, *Pflugers Arch 443*, S117–S120.

156 Stratford, F. L., Pereira, M. M., Becq, F., McPherson, M. A., and Dormer, R. L. (2003) Benzo-(c)quinolizinium drugs inhibit degradation of Delta F508-CFTR cytoplasmic domain, *Biochem Biophys Res Commun 300*, 524–530.

157 Nowak, R. J., Cuny, G. D., Choi, S., Lansbury, P. T., and Ray, S. S. (2010) Improving binding specificity of pharmacological chaperones that target mutant superoxide dismutase-1 linked to familial amyotrophic lateral sclerosis using computational methods, *J Med Chem 53*, 2709–2718.

158 Brady, R. O. (1997) Gaucher's disease: Past, present and future, *Baillieres Clin Haematol 10*, 621–634.

159 Sawkar, A. R., Cheng, W. C., Beutler, E., Wong, C. H., Balch, W. E., and Kelly, J. W. (2002) Chemical chaperones increase the cellular activity of N370S beta-glucosidase: A therapeutic strategy for Gaucher disease, *Proc Natl Acad Sci U S A 99*, 15428–15433.

160 Sawkar, A. R., D'Haeze, W., and Kelly, J. W. (2006) Therapeutic strategies to ameliorate lysosomal storage disorders; a focus on Gaucher disease, *Cell Mol Life Sci 63*, 1179–1192.

161 Fan, J.-Q. (2003) A contradictory treatment for lysosomal storage disorders: Inhibitors enhance mutant enzyme activity, *Trends Pharmacol Sci 24*, 355–360.

162 Butters, T. D. (2007) Pharmacotherapeutic strategies using small molecules for the treatment of glycolipid lysosomal storage disorders, *Expert Opin Pharmacother 8*, 427–435.

163 Grabowski, G. A. (2008) Treatment perspectives for the lysosomal storage diseases, *Expert Opin Emerg Drugs 13*, 197–211.

164 Di'az, L., Bujons, J., Casas, J., Llebaria, A., and Delgado, A. (2010) Click chemistry approach to new N-substituted aminocyclitols as potential pharmacological chaperones for Gaucher disease, *J Med Chem 53*, 5248–5255.

165 Zatta, P., Lucchini, R., Van Rensburg, S. J., and Taylor, A. (2003) The role of metals in neurodegenerative processes: Aluminum, manganese and zinc, *Brain Res Bull 62*, 15–28.

166 Zatta, P. (2003) *Metal ions and neurodegenerative disorders*, World Scientific, Singapore and London.

167 Liu, G., Huang, W., Moir, R., Vanderburg, C. R., Lai, B., Peng, Z., Tanzi, R. E., Rogers, J. T., and Huang, X. (2006) Metal exposure and Alzheimer's pathogenesis, *J Struct Biol 155*, 45–51.

168 Morgan, D. M., Dong, J., Jacob, J., Lu, K., Apkarian, R. P., Thiyagarajan, P., and Lynn, D. G. (2002) Metal switch for amyloid formation: Insight into the structure of the nucleus, *J Am Chem Soc 124*, 12644–12645.

169 Bush, A. I. (2003) The metallobiology of Alzheimer's disease, *Trends Neurosci 26*, 207–214.

170 House, E., Collingwod, J., Khan, A., Korchazkina, O., Berthon, G., and Exley, C. (2004) Aluminium, iron, zinc and copper influence the in vitro formation of amyloid fibrils of Abeta42 in a manner which may have consequences for metal chelation therapy in Alzheimer's disease, *J Alzheimer Dis 6*, 291–301.

171 Ricchelli, F., Drago, D., Filippi, B., Tognon, G., and Zatta, P. (2005) Aluminum triggered structural modifications and aggregation of betaamyloids, *Cell Mol Life Sci 62*, 1724–1733.

172 Maynard, C. J., Bush, A. I., Masters, C. L., Cappai, R., and Li, Q. X. (2005) Metals and amyloid-β in Alzheimer's disease, *Int J Exp Pathol 86*, 147–159.

173 Yoshiike, Y., Tanemura, K., Murayama, O., Akagi, T., Murayama, M., Sato, S., Sun, X., Tanaka, N., and Takashima, A. (2001) New insights on how metals disrupt amyloid beta-aggregation and their effects on amyloid-beta cytotoxicity, *J Biol Chem 276*, 32293–32299.

174 Huang, X., Atwood, C. S., Moir, R., Hartshorn, M. A., Tanzi, R. E., and Bush, A. I. (2004) Trace elements contamination initiates the apparent autoaggregation, amyloidosis, and oligomerization of Alzheimer's Abeta peptides, *J Biol Inorg Chem 9*, 954–960.

175 Mantyh, P. W., Ghilardi, J. R., Rogers, S., DeMaster, E., Allen, C. J., Stimson, E. R., and Maggio, J. E. (1993) Aluminum, iron, and zinc ions promote aggregation of physiological concentrations of beta-amyloid peptide, *J Neurochem 61*, 1171–1174.

176 Atwood, C. S., Moir, R. D., Huang, X., Scarpa, R. C., Bacarra, N. M., Romano, D. M., Hartshorn, M. A., Tanzi, R. E., and Bush, A. I. (1998) Dramatic aggregation of Alzheimer abeta by Cu(II) is induced by conditions representing physiological acidosis, *J Biol Chem 273*, 12817–12826.

177 Beauchemin, D. and Kisilevsky, R. (1998) A method based on ICP-MS for the analysis of Alzheimer's amyloid plaques, *Anal Chem 70*, 1026–1029.

178 Drago, D., Bolognin, S., and Zatta, P. (2008) Role of metal ions in the Aβ oligomerization in Alzheimer's disease and in other neurological disorders, *Curr Alzheimer Res 5*, 500–507.

179 Kawahara, M., Kato, M., and Kuroda, Y. (2001) Effects of aluminium on the neurotoxicity of primary cultured neurons and on the aggregation of beta-amyloid protein, *Brain Res Bull 55*, 211–217.

180 Kawahara, M. (2005) Effects of aluminum on the nervous system and its possible link with neurodegenerative diseases, *J Alzheimer Dis 8*, 171–182.

181 Dexter, D. T., Wells, F. R., Lees, A. J., Agid, F., Agid, Y., Jenner, P., and Marsden, C. D. (1989) Increased nigral iron content and alterations in other metal ions occurring in brain in Parkinson's disease, *J Neurochem 52*, 1830–1836.

182 Riederer, P., Sofic, E., Rausch, W. D., Schmidt, B., Reynolds, G. P., Jellinger, K., and Youdim, M. B. (1989) Transition metals, ferritin, glutathione, and ascorbic acid in parkinsonian brains, *J Neurochem 52*, 515–520.

183 Bisaglia, M., Tessari, I., Mammi, S., and Bubacco, L. (2009) Interaction between α-synuclein and metal ions, still looking for a role in the pathogenesis of Parkinson's disease, *Neuromol Med 11*, 239–251.

184 Paik, S. R., Shin, H. J., Lee, J. H., Chang, C. S., and Kim, J. (1999) Copper(II)-induced self-oligomerization of alpha-synuclein, *Biochem J 340*, 821–828.

185 Uversky, V. N., Li, J., and Fink, A. L. (2001) Metal-triggered structural transformations, aggregation, and fibrillation of human alpha-synuclein. A possible molecular NK between Parkinson's disease and heavy metal exposure, *J Biol Chem 276*, 44284–44296.

REFERENCES

186 Binolfi, A., Rasia, R. M., Bertoncini, C. W., Ceolin, M., Zweckstetter, M., Griesinger, C., Jovin, T. M., and Fernández, C. O. (2006) Interaction of alpha-synuclein with divalent metal ions reveals key differences: A link between structure, binding specificity and fibrillation enhancement, *J Am Chem Soc 128*, 9893–9901.

187 Dexter, D. T., Carayon, A., Javoy-Agid, F., Agid, Y., Wells, F. R., Daniel, S. E., Lees, A. J., Jenner, P., and Marsden, C. D. (1991) Alterations in the levels of iron, ferritin and other trace metals in Parkinson's disease and other neurodegenerative diseases affecting the basal ganglia, *Brain 114*, 1953–1975.

188 Zecca, L., Stroppolo, A., Gatti, A., Tampellini, D., Toscani, M., Gallorini, M., Giaveri, G., Arosio, P., Santambrogio, P., Fariello, R. G., Karatekin, E., Kleinman, M. H., Turro, N., Hornykiewicz, O., and Zucca, F. A. (2004) The role of iron and copper molecules in the neuronal vulnerability of locus coeruleus and substantia nigra during aging, *Proc Natl Acad Sci U S A 101*, 9843–9848.

189 Golts, N., Snyder, H., Frasier, M., Theisler, C., Choi, P., and Wolozin, B. (2002) Magnesium inhibits spontaneous and iron-induced aggregation of alpha-synuclein, *J Biol Chem 277*, 16116–16123.

190 Bharathi, K. S. J. Rao (2007) Thermodynamics imprinting reveals differential binding of metals to alpha-synuclein: Relevance to Parkinson's disease, *Biochem Biophys Res Commun 359*, 115–120.

191 Kostka, M., Hogen, T., Danzer, K. M., Levin, J., Habeck, M., Wirth, A., Wagner, R., Glabe, C. G., Finger, S., Heinzelmann, U., Garidel, P., Duan, W., Ross, C. A., Kretzschmar, H., and Giese, A. (2008) Single particle characterization of iron-induced pore-forming alpha-synuclein oligomers, *J Biol Chem 283*, 10992–11003.

192 Pall, H. S., Williams, A. C., Blake, D. R., Lunec, J., Gutteridge, J. M., Hall, M., and Taylor, A. (1987) Raised cerebrospinal-fluid copper concentration in Parkinson's disease, *Lancet 2*, 238–241.

193 Rasia, R. M., Bertoncini, C. W., Marsh, D., Hoyer, W., Cherny, D., Zweckstetter, M., Griesinger, C., Jovin, T. M., and Fernández, C. O. (2005) Structural characterization of copper(II) binding to alpha-synuclein: Insights into the bioinorganic chemistry of Parkinson's disease, *Proc Natl Acad Sci U S A 102*, 4294–4299.

194 Brown, D. R., Qin, K., Herms, J. W., Madlung, A., Manson, J., Strome, R., Fraser, P. E., Kruck, T., von Bohlen, A., Schulz-Schaeffer, W., Giese, A., Westaway, D., and Kretzschmar, H. (1997) The cellular prion protein binds copper in vivo, *Nature 390*, 684–687.

195 Sung, Y. H., Rospigliosi, C., and Eliezer, D. (2006) NMR mapping of copper binding sites in alpha-synuclein, *Biochim Biophys Acta 1764*, 5–12.

196 Binolfi, A., Lamberto, G. R., Duran, R., Quintanar, L., Bertoncini, C. W., Souza, J. M., Cerveñansky, C., Zweckstetter, M., Griesinger, C., and Fernández, C. O. (2008) Site-specific interactions of Cu(II) with alpha and beta-synuclein: Bridging the molecular gap between metal binding and aggregation, *J Am Chem Soc 130*, 11801–11812.

197 Ricchelli, F., Buggio, R., Drago, D., Forloni, G., Negro, A., Tognon, G., and Zatta, P. (2006) Aggregation/fibrillogenesis of recombinant human prion protein and Gestmann-Straussler-Scheinker disease peptides in the presence of metal ions, *Biochemistry 45*, 6724–6732.

198 Mihaela, N., Breydo, L., Milton, S., Kayed, R., van der Veer, W. E., Tone, P., and Glabe, C. G. (2007) Methylene blue inhibits amyloid Aβ oligomerization by promoting fibrillization, *Biochemistry 46*, 8850–8860.

199 Qi, W., Zhang, A., Good, T. A., and Fernandez, E. J. (2009) Two disaccharides and trimethylamine N-oxide affect Abeta aggregation differently, but all attenuate oligomer-induced membrane permeability, *Biochemistry 48*, 8908–8919.

200 Liu, R., Barkhordarian, H., Emadi, S., Park, C. B., and Sierks, M. R. (2005) Trehalose differentially inhibits aggregation and neurotoxicity of beta-amyloid 40 and 42, *Neurobiol Dis 20*, 74–81.

201 Ueda, T., Nagata, M., Monji, A., Yoshida, I., Tashiro, N., and Imoto, T. (2002) Effect of sucrose on formation of the beta-amyloid fibrils and D-aspartic acids in Abeta 1–42, *Biol Pharm Bull 25*, 375–378.

202 Ryu, J., Kanapathipillai, M., Lentzen, G., and Park, C. B. (2008) Inhibition of beta-amyloid peptide aggregation and neurotoxicity by alpha-d-mannosylglycerate, a natural extremolyte, *Peptides 29*, 578–584.

203 Kanapathipillai, M., Lentzen, G., Sierks, M., and Park, C. B. (2005) Ectoine and hydroxyectoine inhibit aggregation and neurotoxicity of Alzheimer's beta-amyloid, *FEBS Lett 579*, 4775–4780.

204 Nielsen, L., Khurana, R., Coats, A., Frokjaer, S., Brange, J., Vyas, S., Uversky, V. N., and Fink, A. L. (2001) Effect of environmental factors on the kinetics of insulin fibril formation: Elucidation of the molecular mechanism, *Biochemistry 40*, 6036–6046.

205 Arora, A., Ha, C., and Park, C. B. (2004) Inhibition of insulin amyloid formation by small stress molecules, *FEBS Lett 564*, 121–125.

206 Vilasi, S., Iannuzzi, C., Portaccio, M., Irace, G., and Sirangelo, I. (2008) Effect of trehalose on W7FW14F apomyoglobin and insulin fibrillization: New insight into inhibition activity, *Biochemistry 47*, 1789–1796.

207 Bomhoff, G., Sloan, K., McLain, C., Gogol, E. P., and Fisher, M. T. (2006) The effects of the flavonoid baicalein and osmolytes on the Mg2+ accelerated aggregation/fibrillation of carboxymethylated bovine 1SS-alpha-lactalbumin, *Arch Biochem Biophys 453*, 75–86.

208 Hegde, M. L. and Rao, K. S. (2007) DNA induces folding in alpha-synuclein: Understanding the mechanism using chaperone property of osmolytes, *Arch Biochem Biophys 464*, 57–69.

209 Uversky, V. N., Li, J., and Fink, A. L. (2001) Trimethylamine-N-oxide-induced folding of alpha-synuclein, *FEBS Lett 509*, 31–35.

210 Tanaka, M., Machida, Y., Niu, S., Ikeda, T., Jana, N. R., Doi, H., Kurosawa, M., Nekooki, M., and Nukina, N. (2004) Trehalose alleviates polyglutamine-mediated pathology in a mouse model of Huntington disease, *Nat Med 10*, 148–154.

211 Nandi, P. K., Bera, A., and Sizaret, P. Y. (2006) Osmolyte trimethylamine N-oxide converts recombinant alpha-helical prion protein to its soluble beta-structured form at high temperature, *J Mol Biol 362*, 810–820.

212 Hamada, D. and Dobson, C. M. (2002) A kinetic study of beta-lactoglobulin amyloid fibril formation promoted by urea, *Protein Sci 11*, 2417–2426.

213 Polano, M., Bek, A., Benetti, F., Lazzarino, M., and Legname, G. (2009) Structural insights into alternate aggregated prion protein forms, *J Mol Biol 393*, 1033–1042.

214 Vernaglia, B. A., Huang, J., and Clark, E. D. (2004) Guanidine hydrochloride can induce amyloid fibril formation from hen egg-white lysozyme, *Biomacromolecules 5*, 1362–1370.

16

COMPUTATIONAL STUDIES OF FOLDING AND ASSEMBLY OF AMYLOIDOGENIC PROTEINS

J. Srinivasa Rao, Brigita Urbanc, and Luis Cruz

16.1. INTRODUCTION

During the last few decades intense experimental and theoretical efforts have made it clear that a number of amyloid-forming proteins are responsible for triggering the pathology in a vast number of neurodegenerative diseases, such as Alzheimer's disease (AD), Parkinson's disease (PD), type II diabetes mellitus, Huntington's disease (HD), amyotrophic lateral sclerosis (ALS), and prion diseases, to name a few [1–5]. The term "amyloid" refers to the ability of these proteins to form fibrils with a characteristic cross β-sheet structure. Although the identities of these proteins are known, the mechanisms through which amyloid proteins assemble into fibrils and how the fibrils grow are still not known. Equally obscure are interactions between the amyloid protein assemblies and the cells that presumably cause toxicity. A general confounding fact is that these proteins form amyloid fibrils while displaying apparently unrelated amino acid sequences. This apparent contradiction suggests that some basic principles play the key role in folding and assembly. On the other hand, naturally occurring mutations of amyloid proteins often exhibit a more radical form and/or earlier onset of the diseases. An additional obstacle lies in the fact that

---

Protein and Peptide Folding, Misfolding, and Non-Folding, First Edition. Edited by Reinhard Schweitzer-Stenner.
© 2012 John Wiley & Sons, Inc. Published 2012 by John Wiley & Sons, Inc.

although amyloid fibrils can be and mostly are well characterized experimentally, the pathway (possibly multiple) from the monomeric state to fibril is also unknown. Yet it is not the insoluble fibrillar form of the proteins that is the proximate neurotoxic species. Much smaller, soluble, and metastable oligomeric forms (comprising a small number of non-covalently bonded protein molecules) are implicated as the most toxic structures that trigger the onset of the associated disease. In trying to circumvent experimental difficulties, computer studies have taken to heart the elucidation of the precise mechanisms of folding and assembly as well as the relevance of the resulting structures to their toxicity. This chapter is an attempt to introduce the reader to computational work aimed at understanding, at the atomic and molecular levels, folding, dynamics, and assembly of amyloidogenic proteins. As such, this is not an extensive compendium of work, but an introductory sampling of the field. The oligomer and amyloid-forming general topic covers many diseases, and by the shear size of the literature only a fraction of the literature will be reviewed here. Although many diseases will be presented, an emphasis will be given to AD because of its larger prevalence among the elderly.

The topics in this chapter are organized as follows. First, a discussion on the amyloid proteins and their relationship to the various diseases will be presented. Second, an overview of studies that aim to understand folding of protein fragments as well as their full-length counterparts will be given in the context of folding mechanisms and early stages of assembly that precede fibril formation. Finally, studies that encompass several levels of the assembly process will be reviewed. As a whole, this chapter will take the reader through the entire process from the monomer state through oligomeric and protofibrillar intermediates to amyloid fibrils.

16.2. AMYLOIDS

Amyloids are insoluble, intra- or extracellular fibrillar protein deposits. Each fibril is composed of several filaments and is about 10 Å in diameter. Fibril lengths vary from 0.1 to 10 μm [6, 7]. One characteristic feature of amyloids is that irrespective of the particular amino acid sequence and native fold of the underlying protein, amyloid filaments share a common β-sheet core structure, the cross-β-structure. In this structure, the individual protein β-strands are aligned perpendicular to the fibril axis while forming intermolecular hydrogen bonds to form long β-sheets parallel to the fibril axis [8–10].

The association of these self-assembled β-sheet structures in many human neurodegenerative diseases makes it imperative to understand the mechanisms of formation. This level of importance is transformed to one of great concern by the fact that besides this cross-β-sheet structure, little else is known about the high-resolution structure of amyloid fibrils or its smaller, yet toxic, aggregates. A big part of the problem is that fibrils are insoluble and thus it is very difficult to obtain the structure using conventional biophysical techniques such as X-ray crystallography and solution nuclear magnetic resonance (NMR) [11–14].

AMYLOIDS 481

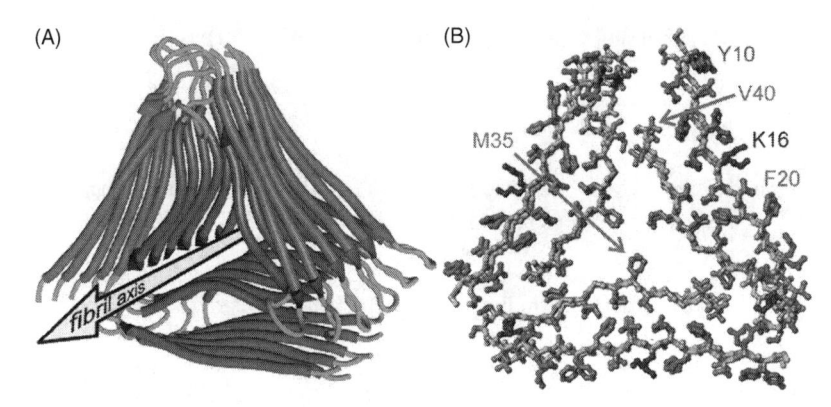

Figure 16.1. Structure of amyloid fibrils based on experimental and modeling data, see Ref. [18]. (A) Shows a triangular model with ribbon representation, and (B) shows the atomic representation of the same viewed down the fibril axis. Hydrophobic, polar, negatively charged, and positively charged amino acid side chains are green, magenta, red, and blue, respectively. Backbone nitrogen and carbonyl oxygen atoms are cyan and pink. Taken with permission from Paravastu, A. K., Leapman, R. D., Yau, W. M., and Tycko, R (2008) Molecular structural basis for polymorphism in Alzheimer's beta-amyloid fibrils, *Proc. Natl. Acad. Sci. U.S.A.* 105, 18349–18354. Copyright (2008) National Academy of Sciences, U.S.A.

While the three-dimensional structure of amyloid fibrils (see Fig. 16.1) can be well characterized by X-ray diffraction [15] and solid-state NMR [16], the *in vitro* characterization of folded and oligomeric structures is hindered by their lack of ordered structure, heterogeneity, and their typically transient nature [17]. Even though the cross-β-structure is a robust feature of amyloid fibrils, the molecular details may depend on the external conditions, resulting in polymorphism of amyloid fibrils [18].

16.2.1. Relation to Disease

The proteins that form amyloids associated with neurodegenerative diseases in humans include immunoglobulin light chains [19], serum amyloid A protein [20], transthyretin [21], gelsolin [22], amyloid β-protein (Aβ) [23], β2-microglobulin [24], immunoglobulin [25], and prion protein (PrP) [26], among others. Common features of the amyloid include the now familiar β-pleated sheet secondary structure, forming insoluble aggregates, exhibiting green birefringence after dye staining, and possessing a characteristic fibrillar structure as observed under electron microscopy [4, 27–31]. At present, at least 21 different forms of amyloidogenic proteins are known, some of which are rare and some of which play a central role in the pathogenesis of the disease that is affecting millions of patients worldwide [32]. The study of amyloid as a major causative agent of pathology has become an enterprise by itself. Table 16.1 shows the names of the most widespread diseases characterized by

TABLE 16.1. Examples of Diseases Associated with Amyloid Deposition

Clinical Syndrome	Fibrillar Component	Proteinaceous Deposit
Alzheimer's disease	Aβ peptide 1–40, 1–42	Extracellular amyloid
Parkinson's disease	α-Synuclein	Lewy body
Amyotrophic lateral sclerosis	Insoluble SOD1	Intraneuronal inclusions
Transmissable spongiform encephalopathies	Full-length prion protein (PrP) or fragments	Extracellular plaques
Fronto-temporal dementia	Fibrillar tau	Neurofibrillary tangles
Diffuse Lewy body disease	Fibrillar α-synuclein	Cortical Lewy bodies
Huntington's disease	Fibrillar Htn	Intranuclear neuronal inclusions
Primary systemic amyloidosis	Intact light chain or fragments	Extracellular deposits of amyloid
Secondary systemic amyloidosis	76-residue fragment of amyloid A protein	Extracellular deposits of amyloid
Familial amyloidotic poly neuropathy I	Transthyretin variants and fragments	Extracellular deposits of transthyretin
Senile systemic amyloidosis	Wild-type transthyretin and fragments	Extracellular deposits of transthyretin
Hereditary cerebral amyloid angiopathy	Fragment of cystatin-C	Extracellular cerebral amyloid deposits
Hemodialysis-related amyloidosis	β2-Microglobulin	
Familial amyloidotic polyneuropathy II	Fragments of apolipoprotein A-I	Extracellular deposits of transthyretin
Finnish hereditary amyloidosis	71-residue fragment of gelsolin	
Type II diabetes	Islet-associated polypeptide (IAPP)	Intracellular amyloid-like deposits
Medullary carcinoma of the thyroid	Calcitonin	Extra- or intracellular lumps of amorphous sub- stance with amyloid criteria
Atrial amyloidosis	Atrial natriuretic factor	Extracellular deposits of amyloid
Lysozyme amyloidosis	Full-length lysozyme variants	Fibrils deposition in tissues
Insulin-related amyloidosis	Full-length insulin	Intracellular amyloid
Fibrinogen α-chain amyloidosis	Fibrinogen α-chain variants	Extracellular deposition of amyloid fibrils

amyloid formation, the name of the specific protein, and deposit associated with the diseases.

Substantial evidence shows that amyloid-forming proteins might adopt their most toxic structures not at the fibril level, but prior to fibril formation, at the early stages of assembly into oligomers [33]. In AD, for example, the characteristic senile plaques composed of the fibrillar aggregates of Aβ were for several decades considered the proximate neurotoxic agent. Recently, these amyloid deposits lost their prominence due to the discovery of the more toxic small soluble Aβ oligomers [34–36].

To put the focus of this chapter—computational studies—in a broader context of neurodegenerative diseases, the next few subsections offer a brief introduction to the most prevalent diseases that are hypothesized to be triggered by protein misfolding and aberrant protein assembly.

16.2.1.1. Alzheimer's Disease AD is an incurable progressive neurological disorder believed to be caused by Aβ assembly. It has the potential to become an epidemic among the elderly at about 13% occurrence for people over 65 years old in the United States, and up to 40% occurrence for people over 85 years old. Among the symptoms of AD are a progressive loss of memory, starting from a short-term memory loss and escalating to deficits in identifying people and objects, and ultimately having problems with speech and even more basic tasks. In addition to a substantial neuronal loss, the hallmarks of AD are senile plaques and neurofibrillary tangles, both a result of protein aggregation. Senile plaques are extracellular inclusions while neurofibrillary tangles occur inside of cells. The major constituents of senile plaques are the 40 or 42 amino acids long Aβ peptides derived from the proteolytic cleavages of the amyloid precursor protein (APP) [3, 34, 37]. NMR studies on the Aβ_{10-35} fragment, which comprises amino acids from positions 10 to 35 of the full-length peptides, showed that fibrils in AD formed in-register parallel β-sheets that ran along the fibril axis [38]. This finding was later confirmed for the full length Aβ_{1-40} [39].

AD is most prevalent in its sporadic form where no preconditions or salient risk factors have been found. There are also more radical and devastating forms of AD that can be traced to certain small groups of individuals from the same family (familial AD [FAD]). Many of these forms have been traced to single-point mutations of Aβ, in which an amino acid either at position 22 or 23 is substituted by another amino acid, resulting in the Dutch (E22Q), Iowa (D23N), Arctic (E22G), and Italian (E22K) familial forms [40–44], among others.

Until recently, senile plaques were believed to be the causative agent of the pathology in AD. This is known as the amyloid hypothesis [34, 45–47]. Evidence from multiple sources, however, resulted in a paradigm shift [48] that revised the original view to include pre-fibrillar, low-molecular-weight structures, composed of two or more Aβ peptides with no substantial ordered structure, called oligomers, as the proximate neurotoxins in AD and, as such, implicated in neuron death [49–51]. These oligomers, which are sometimes referred to as Aβ-derived diffusible ligands (ADDLs) [49] or globulomers [52], and paranuclei [53] exist for short times in metastable states in various orders, including monomeric states, which complicate structural studies (see Fig. 16.2).

16.2.1.2. Huntington's Disease HD is a progressive disorder associated with significant neurodegeneration specifically in the striatum [54]. With this deterioration

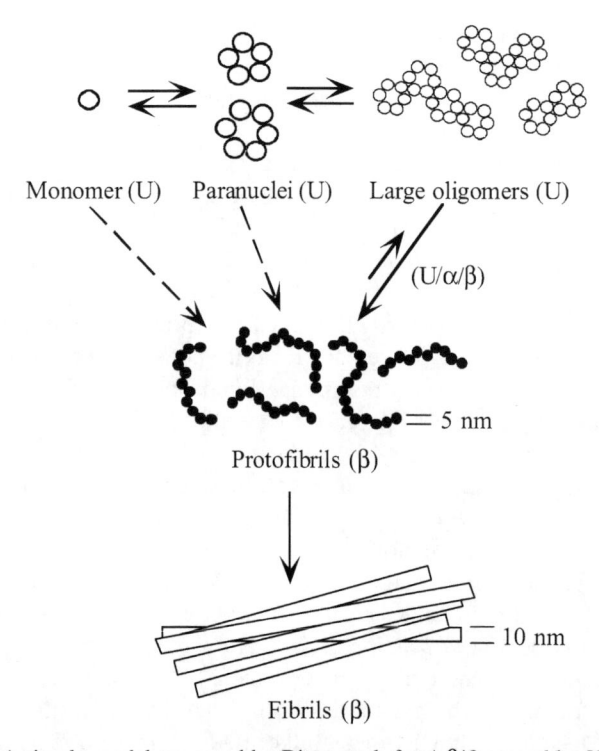

Figure 16.2. A simple model proposed by Bitan et al. for Aβ42 assembly. [53]. This model shows that the oligomerization of monomers is a rapid process and leads to paranuclei, which again oligomerize to form larger, beaded structures. These larger oligomers can convert into protofibrils and further into amyloid fibrils. Taken with permission from Bitan, G., Kirkitadze, M. D., Lomakin, A., Vollers, S. S., Benedek, G. B., and Teplow, D. B (2003) Amyloid β-protein (Aβ) assembly: Aβ40 and Aβ42 oligomerize through distinct pathways, *Proc. Natl. Acad. Sci. U. S. A. 100*, 330–335. Copyright (2003) National Academy of Sciences, U.S.A.

comes the onset of abnormal involuntary movements, dystonia, behavioral difficulties, failing coordination, and cognitive decline [55]. HD is caused by a single-gene mutation that causes a polyglutamine [poly(Q)] repeat expansion in the first exon of the huntingtin protein. The Huntington protein is in charge of delivering small "packages" (vesicles containing important molecules) to the outside of the cell [56, 57]. In general, the coding region of this gene contains multiple repetitions of the DNA sequence CAG. People with HD have an abnormally high number of these CAG triplets, approximately 40 or more [58] compared with the normal number of CAG triplets in the coding region of DNA ranging between 10 and 26. This overabundance of the triplet is believed to disrupt the function of the gene's protein production, but the process by which the expansion of the CAG repeat causes disease is not yet known.

16.2.1.3. Prion Diseases Transmissible spongiform encephalopathies (TSEs), also known as prion diseases, are a group of neurodegenerative diseases. The common symptoms of prion diseases include personality changes, psychiatric problems such as depression, and the lack of coordination. The progression of prion diseases is rapid: after the initial onset of symptoms death normally ensues within 1 to 3 years. The most common form of TSE is the Creutzfeldt–Jakob disease (CJD), which is caused by conformational changes in the PrP [26, 59–65]. The normal cellular protein known as PrP^c (for cellular) is a normal host protein encoded by a single axon of a single copy gene [66]. This protein is found predominantly on the surface of neurons attached by a glycoinositol phospholipid anchor. A widely accepted mechanism suggests that the soluble protease-sensitive PrP^c is converted into protease-resistant PrP^{Sc} by a process in which a portion of its α-helical and coil structure are converted into a β-sheet conformation, which has a high propensity to form the aggregates [67]. The exact mechanism by which this happens and how it propagates the conformational change is unknown.

16.2.1.4. Parkinson's Disease   PD, a motor system disorder, arises from a loss of dopamine-producing brain cells. PD is both a chronic and a progressive disorder, in which genetic and environmental factors play important roles. The main symptoms of PD include muscular stiffness, slow movements, and tremor. Most of the experimental studies show that α-synuclein, a small highly conserved presynaptic protein with as yet undetermined function, is responsible for both familial and sporadic PD [68, 69]. The α-synuclein is a major component of the amyloid fibrils forming the characteristic intracellular deposits found in PD, namely the Lewy bodies. The α-synuclein protein has two closely related homologues, called β-synuclein and γ-synuclein, located in different regions of the brain [70]. Several studies have reported that α-synuclein acquires various conformations that include partially folded states, different oligomeric forms, as well as amorphous aggregates [71–74]. The aggregation of α-synuclein has been found to be tunable, that is, accelerated or inhibited, by different intrinsic and extrinsic factors [74].

16.2.1.5. Amyotrophic Lateral Sclerosis   ALS is a progressive neurological disease that directly affects motor neurons located in the spinal cord and the part of the brain that is connected to the spinal cord (the brainstem) that control muscle movement [75–80]. The symptoms of ALS disease are not common for all people; some of the symptoms include muscle weakness, slurred speech, and uncontrolled emotions. The most prevalent form of ALS is the sporadic or non-inherited type. The nature of the occurrence of the sporadic form of ALS is not clear, but it is believed that a combination of genetic and environmental factors are probable causes [75, 81]. Inherited forms of ALS, the familial forms, account for about 10% of the ALS cases.

16.3. COMPUTER SIMULATIONS

16.3.1. Importance

The fibril insolubility and a high sensitivity of these naturally disordered proteins to small changes in external conditions limit the ability to experimentally track the conformational changes involved at every step of assembly from monomers through oligomers to fibrils. Several factors controlling the assembly of these disordered proteins into amyloid fibrils have been identified by different experimental techniques [82–84]; however, the exact pathway(s) of assembly are still unknown. Light, neutron, and X-ray scattering methods have only provided an average ensemble picture of fibril formation [11]. Experiments understanding amyloid fibrils at the molecular level are relatively recent using techniques such as atomic force microscopy (AFM) and transmission electron microscopy (TEM) [85, 86]. However, to date no experimental procedure can elucidate the entire aggregation process.

Computer simulations have served as a viable alternative to investigate mechanisms and highlight molecular interactions that help bridge the gap between microscopic processes and experimental observables by exploring the atomic details of early formation of oligomers and other metastable intermediates. However, protein aggregation has proven to be a complex problem presenting computational studies with challenges and difficulties that for its most part reside in the limitations of current computer hardware. Aggregation of these amyloid proteins is a process that typically takes from seconds to days in *in vitro* experiments while all-atom molecular dynamics (MD) simulations containing the most detail and atomic resolution can consider only up to tens or only hundreds of nanoseconds per trajectory. These limitations impose restrictions on the number of atoms that can be considered in a simulation and also on the total simulation.

Because the atomistic description of the process of assembly is greatly restricted by the available computational capabilities, thus limiting accessible time scales, computational approaches that consider simplified models of the system at hand are needed to be able to consider biologically relevant time scales. Simplifications can take multiple shapes: from replacing the solvent molecules by an effective "mean-field" medium, using a coarse-grain description of individual amino acids by modeling groups of atoms rather than individual atoms, replacing standard MD by discrete molecular dynamics (DMD), to using kinetic models. Nonetheless, MD simulations with the most atomic detail have been applied to study fragments of the amyloidogenic proteins in the study of monomer dynamics and folding, and is increasingly being used in larger systems, such as full-length proteins in the presence of membranes, with the help of the ever-increasing computer power, and the use of new algorithmic formalisms.

16.3.2. Structure, Dynamics, and Stability of Monomers

The exact mechanism of amyloid formation that results in a fibril structure is poorly understood. A fibril can grow by an addition of monomers that first dock to the end and then lock to it by forming the intermolecular hydrogen bonds [87] (see Fig. 16.3) or through formation of elongated protofibril-like intermediates that assemble from quasi-spherical oligomers [30]. One of the main thrusts in studies of aggregation is to understand these pre-fibrillar assemblies as either an important stepping stone toward the fibril formation or as the structures that mediate toxicity in diseases such as AD and are significantly more toxic to cells than fibrils [49, 88–92]. Structural

COMPUTER SIMULATIONS

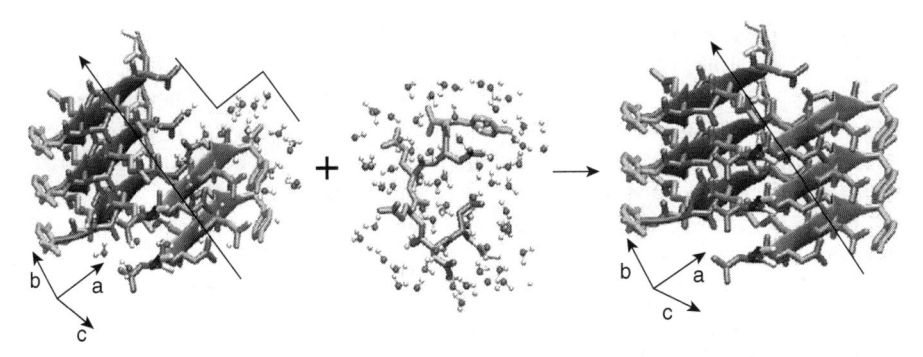

Figure 16.3. Addition of monomer to fibrils through dock–lock mechanism for Aβ microcrystal [87]. This mechanism was probed using all-atom MD simulations. Arrows in the figure represent the fibril axis. In the left-hand side structure the silver sheet has two monomers creating a vacant position in the crystal for the incoming monomer. The addition of an unstructured solvated monomer leads to the fibril on the right side. Taken with permission from Reddy, G., Straubb, J. E., and Thirumalai, D. (2009) Dynamics of locking of peptides onto growing amyloid fibrils, *Proc. Natl. Acad. Sci. U. S. A. 106*, 11948–11953.

characterization and understanding of the dynamics of these intermediate species formed along the pathway of amyloid formation are clearly important for the rational design of inhibitors of amyloid formation. To understand assembly, however, we have to take a step back and understand the early events in the process of assembly from monomers to oligomers that are key to elucidating the initial stages of the folding of these toxic species [93, 94] but are also poorly understood. Common themes emerge in the monomer dynamics, such as the importance of β-sheet conformations in, for example, AD [95–97], prions [98], and early-onset PD [99].

16.3.2.1. Monomers: Fragments Because of the complexity involved even in the folding of a single peptide, experimental and computational studies have concentrated their focus on fragments of particular amyloid proteins. This approach has been successful in many ways, especially when many fragments have been shown experimentally to form amyloid fibrils similar to those of their parent proteins [100]. Fragments are also a very convenient system for explicit solvent MD computational studies (the most detailed but most computationally demanding) because they can be fully considered and still be within the intrinsic limitations of computing hardware that usually limits the length of the peptide as well as the size of the solvent–peptide system that can be simulated. In addition, fragments have proven key in studies that aim to inhibit assembly or induce alternative assembly pathways, as discussed below.

Aβ_{10-35} In AD, studies of fragments usually consider sequences of amino acids that include the central hydrophobic cluster (CHC from L17 to A21) of the Aβ. The CHC is believed to play an important role in early folding events of the full-length

Aβ. Experimentally, Zhang et al. demonstrated that Aβ_{10-35} adopts a collapsed coil in aqueous environments [101].

Using an atomistic description of Aβ_{10-35} and an overdamped diffusive dynamics framework called the MaxFlux algorithm, an initial theoretical study explored likely pathways for global macromolecular conformational transitions among a collapsed coil, α-helical, folded β-sheet, and extended conformations [102]. The results demonstrated that these conformational states were separated by significant energy barriers between them [102].

An MD study of Aβ_{10-35} folding in explicit solvent found that the CHC region exhibited a stable turn [103]. In a follow-up, the authors compared the dynamics of the Aβ_{10-35} fragment in both, the wild-type (WT) and the more aggressive Dutch (E22Q) mutant using multiple nanosecond MD trajectories in explicit solvent. The Dutch mutation is pathologically characterized by an increased deposition rate in the walls of blood vessels, linked to the hereditary cerebral hemorrhage with amyloidosis [104]. Contrary to expectations, the main conclusions of this study indicated that both the WT and the Dutch mutant were mostly found in a collapsed-coil conformation. However, the Dutch mutant peptide exhibited increased structural fluctuations relative to the WT. Also, this study reported a change in the solvation of the water around the Dutch mutant in the region of the peptide's "hydrophobic patch," suggesting an increased solvation of the mutant. The authors suggested that the decreased desolvation barrier as well as increased structural fluctuations in the Dutch mutant was directly linked to an experimentally observed increased aggregation propensity and thus to an increased deposition rate of the Dutch mutant relative to the WT [105, 106].

A similar conclusion was found in another study set to test the hypothesis that the fibril elongation rate of the Dutch mutant was higher compared with the WT. This increased elongation rate was experimentally attributed to a larger propensity for the formation of β-sheet structure in the monomeric E22Q mutant peptide in solution [107]. The authors tested this hypothesis using multiple nanosecond MD simulations of the WT and Dutch mutation of the Aβ_{10-35} peptide in aqueous solution. Their results indicated that the higher fibril elongation rate of the Dutch mutant was not due to a larger propensity for β-sheet formation, but rather due to a greater stability of the WT folded structure in aqueous solvent, which decreased the fibril growth relative to the Dutch mutant. It was argued that the differential stabilities of the two peptides were a result of the different charge states of these two peptides [108].

Other all-atom MD studies considering the same Aβ_{10-35} fragment found that a strand–loop–strand (SLS) structure formed with the loop located in the D23-K28 region, thus resembling the structure found in fibrils. They also found that mutants [E22Q]Aβ_{10-35} and [D23N]Aβ_{10-35} had an increased propensity to form the SLS structure relative to the WT. If the stable SLS structure represents an obligatory intermediate in the process of fibril formation, they argued that the more populated the SLS structure, the higher the aggregation rate should be, providing a plausible explanation for the experimentally observed enhanced aggregation rates of the mutant peptides [109]. A similar study, however, indicated a more dynamic and less structured conformation for the two mutant peptides [110]. The dynamical structure of $A\beta_{10-35}$ was found to be associated with multiple minima in the free energy landscape that shared a common structural motif, that is, a turn–loop structure in the V24-N27 region, and differed by the absence or presence of the 23–28 salt bridge. The authors argued that if aggregation was induced by the folding nucleus in the region 24–27 and if the 23–28 salt bridge played a lesser role, then the intermolecular interactions as well as crowding played important roles in the aggregation process [111]. Another group applied the replica exchange MD (REMD) to study $A\beta_{10-35}$ folding, confirming that this fragment exists as a mixture of collapsed globular states that remain in rapid dynamic equilibrium with each other dominated by a collapsed coil and bend structures but without a significant secondary structure. They argued that the high propensity for the salt-bridge D23–K28 formation in $A\beta_{10-35}$ monomer conformations stabilized the structural motif that was consistent with the fibril structure, resulting in increased aggregation propensity [110].

$A\beta_{21-30}$ This fragment was recently hypothesized to nucleate folding of the full-length $A\beta$ based on results of proteolysis and solution state NMR performed on both the full-length $A\beta$ and $A\beta_{21-30}$ [112]. These data showed that in both cases the peptides were protease-resistant in the same 21–30 region, suggesting the existence of a folding nucleus involving the region V24-K28 [112].

To elucidate the structure and dynamics of $A\beta_{21-30}$ folding, a DMD study with a united-atom protein model, which accounts for all atoms in the protein except hydrogens, was conducted using an implicit solvent [113]. This study confirmed that the effective hydrophobic interactions between V24 and the butyl portion of K28 facilitated $A\beta_{21-30}$ folding into a loop-like structure (see Fig. 16.4). By varying the strength of the electrostatic interactions (EIS) the authors found that at intermediate EIS strengths unfolded conformations disappeared, while at EIS strengths comparable to the strengths in the interior of proteins, the packing between the side chains of V24 and K28 was destabilized. Additional transient EIS between the negatively charged E22 or D23 and positively charged K28 at intermediate strengths of EIS stabilized the loop conformation [113].

In a subsequent study using all-atom MD simulations in explicit solvent, the $A\beta_{21-30}$ and its Dutch mutant [E22Q]$A\beta_{21-30}$ exhibited a loop stabilized by the packing of the side chains of V24 and K28 [114]. The Dutch mutant was shown to exhibit a more open structure than the WT with the salt-bridge D23-K28 stabilizing the fold. This study also addressed the existence of ions in the solvent, which further stabilized the salt bridges. Reducing the water density was used to mimic a membrane environment, resulting in an increased propensity to form π-helix conformation [114].

The existence of the loop conformation in the $A\beta_{21-30}$ fragment was confirmed by an alternative computational methodology [115]. Using the activation-relaxation technique (ART nouveau) coupled with a coarse-grained force field, the $A\beta_{21-30}$ was found to lack any secondary structure and to populate three structural families with a loop spanning residues V24-K28 [115].

To locate the global minimum conformations of this fragment, computer simulations were carried out using the REMD technique [116]. The results indicated that

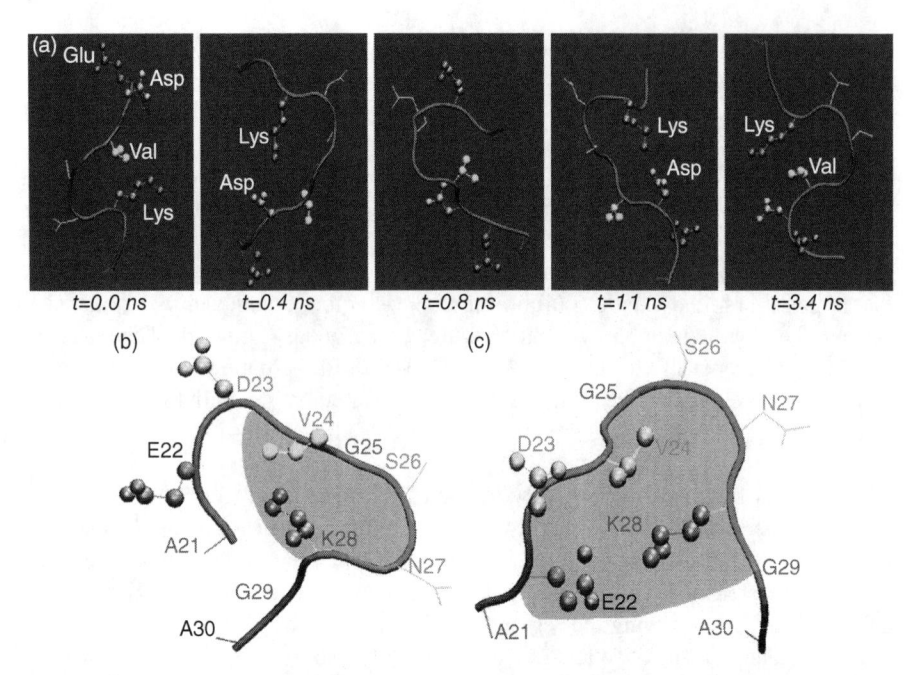

Figure 16.4. Representative $A\beta_{21-30}$ conformations obtained from DMD simulations from Ref. [113]. In (A) various structural forms that have V24-K28 unpacking/packing events due to formation of transient D23-K28 contact are shown. (B and C) Show loop conformations in which the first one has K28 side-chain points opposite to the normal vector of the loop plane, and in the second conformation the E22 is close to K28 and K28 points along the normal vector to the loop plane. Taken with permission from Borreguero, J. M., Urbanc, B., Lazo, N. D., Buldyrev, S. V., Teplow, D. B., and Stanley, H. E (2005). Folding events in the 21–30 region of amyloid-beta-protein (Abeta) studied in silico, *Proc. Natl. Acad. Sci. U. S. A. 102*, 6015–6020. Copyright (2005) National Academy of Sciences, U.S.A.

the global minimum structure contained a bend motif spanning the V24-K28 region of this fragment (the two most stable conformations are shown in Fig. 16.5). The results of the study showed that this bend is stabilized by a network of hydrogen bonds involving the side chain of residue D23 and the amide hydrogens of adjacent residues G25, S26, N27, and K28, as well as by a salt bridge formed between side chains of K28 and E22.

To determine in more detail the stable $A\beta_{21-30}$ conformations, REMD simulations addressed the effects of Dutch (E22Q), Arctic (E22G), and Italian (E22K) mutations at position 22, as well as the effect of the Iowa (D23N) mutation at position 23 [117]. The study found that in all mutant peptides the loop in the A21-A30 region was preserved. The mutations at position 22 resulted in a bend motif while the Iowa mutation resulted in a turn motif. The authors argued that these results indicated that the pathways of folding and assembly could be different for these two classes of mutations [117]. On another front, along with an experimental study of $A\beta_{21-30}$

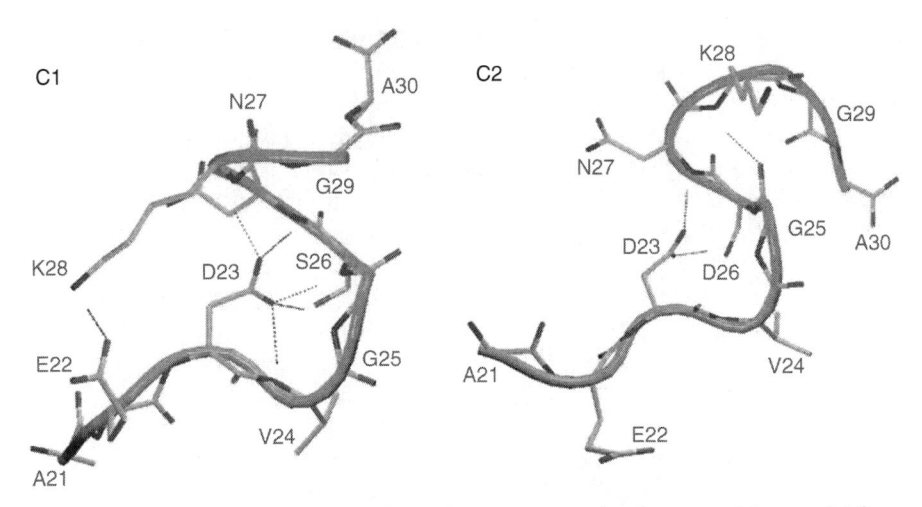

Figure 16.5. The centroid conformations of two most populated structural forms of $A\beta_{21-30}$ obtained from REMD simulations, from Ref. [116]. Reprinted with permission from Baumketner, A., Bernstein, S. L., Wyttenbach, T., Lazo, N. D., Teplow, D. B., Bowers, M. T., and Shea, J.-E. (2006). Structure of the 21–30 fragment of amyloid β-protein, *Protein Sci 15*, 1239–1247. Copyright (2006) Elsevier.

folding using gas-phase conditions with no solvent, the corresponding simulations of $A\beta_{21-30}$ and its mutants (E22Q, E22G, E22K, and D23N), the WT peptide decapeptide adopted a turn controlled by D23N hydrogen bonding. Mutations at position 22 led to only subtle variations to the overall folded structure of $A\beta_{21-30}$. Comparison of the gas-phase results with those obtained in the solution phase revealed that $A\beta_{21-30}$ also folded in the absence of solvent and suggested the importance of the D23N hydrogen bonding network, observed in both gas phase and solvent, in the stabilization of the $A\beta_{21-30}$ folded conformation [118].

Another combined experimental and computational study of the $A\beta_{21-30}$, in which the solution NMR data were compared with REMD simulations results using a variety of force fields and water models, showed that the $A\beta_{21-30}$ fragment existed mostly in unstructured conformations, lacking any secondary structure or persistent hydrogen-bonding networks [119]. Intriguingly, the remaining minority population of conformations contained conformers with a β-turn centered at V24 and G25, as well as evidence of the D23-K28 salt bridge formation that is known to contribute to stabilty of the Aβ fibril structure [119].

Other Aβ Fragments The ubiquitous bend structures as well as β-hairpins are also present in studies considering other Aβ fragments, especially those including or in the proximity of the CHC. For example, a study on the $A\beta_{16-22}$ fragment, experimentally shown to form fibrils [120], showed that the monomer was mostly found in a β-sheet conformation at high concentrations of urea, whereas in water it existed mostly in a random coil conformation. This was attributed to the increased solvation of the peptide backbone by the urea that formed hydrogen bonds between the urea and the backbone, thus promoting the β-sheet conformation over the random coil [121].

In a study of long time scale MD simulations in explicit water of the $A\beta_{12-28}$ fragment, a transition from the α-helical to β-hairpin conformation was also observed [122]. This transition was also observed in the H1 peptide of the PrP. The main conclusions from this study highlighted the unfolding of α-helices, followed by the formation of a bend conformations with a final convergence to ordered in-register β-hairpin conformations. The β-hairpins observed, despite different sequences, exhibited a common dynamic behavior and the presence of a peculiar pattern of the hydrophobic side chains, in particular in the region of the turns. These observations emphasized the possible common aggregation mechanism for the onset of different amyloid diseases and a plausible common mechanism in the transition to the β-hairpin structures. Additional simulations using an organic fluorinated co-solvent, a 2,2,2-trifluoroethanol (TFE)–water mixture, evinced the stabilization of the α-helical conformations, thus giving credence to the hypothesis that the TFE molecules, by providing a peptide coating, were responsible for the stabilization of the soluble helical conformation [122].

Qualitatively similar results were obtained in a later study on the $A\beta_{12-28}$ fragment using REMD with explicit solvent [123]. This study reported a mostly collapsed coil with some non-negligible β-turn structure (but less than 10%) at physiological temperatures. The Dutch mutant of the $A\beta_{15-28}$ fragment exhibited a weakened interaction between the CHC and a bend at E22-K28 with respect to the WT, and resulted in a β-strand conformation in the CHC region that could seed the fibril formation, thereby providing a plausible explanation of the experimentally observed increased aggregation of the Dutch mutant relative to the WT [123].

A similar study on the $A\beta_{12-28}$ fragment using REMD simulatons at physiological temperatures showed that this peptide mostly existed in a collapsed-coil structure. Other structural forms such as hairpins with a β-turn at position V18-F19 (less than 10%), and 10% of hairpin-like conformations possessing a bend rather than a turn in the central VFFA positions were also observed. The most representative structures obtained from this study are shown in Figure 16.6 [124].

In the $A\beta_{25-35}$ fragment, a REMD study in explicit water and also hexafluoroisopropanol (HFIP)–water co-solvent showed that this fragment preferentially populated a helical structure in apolar organic solvent, while in pure water it adopted a collapsed-coil conformation and to a lesser extent β-hairpin conformations [125]. The β-hairpin was characterized by a type II' β-turn involving residues G29 and A30 and two short β-strands involving residues N27, K28, I31, and I32. Furthermore, the hairpin was found to be stabilized by backbone hydrogen-bonding interactions between residues K28 and I31, and S26 and G33, and by side-chain-to-side-chain interactions between N27 and I32 [125].

Recently, C-terminal fragments (CTFs) from the $A\beta$ peptide have garnered much attention as possible inhibitors of $A\beta_{1-42}$-induced toxicity [126]. An explicit solvent REMD study examined folding of the CTFs of different lengths [127]. It was shown that the fragments of the $A\beta_{x-42}$ ($x = 29$–$31, 39$) type adopted metastable β-structures,

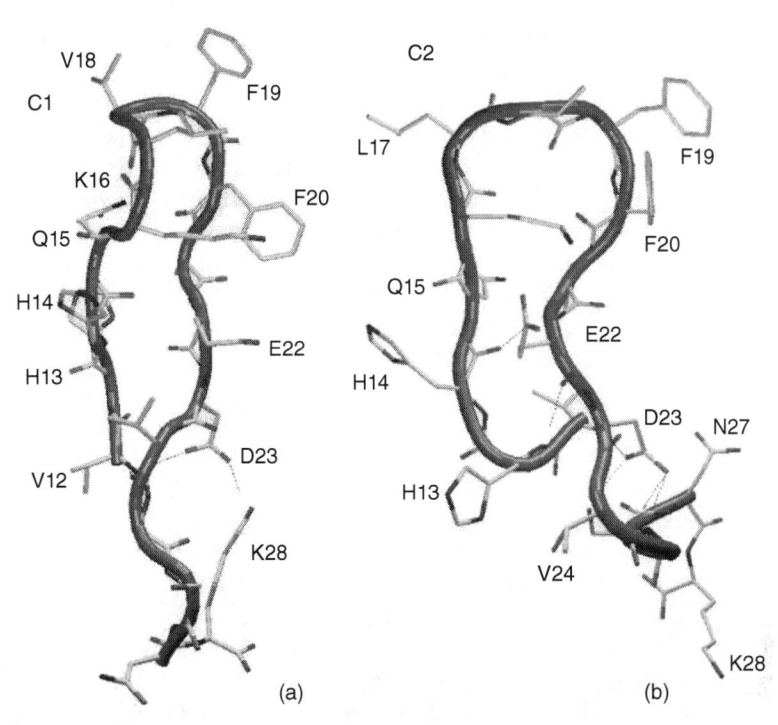

Figure 16.6. Centroids of the two most representative conformations of the two most populated clusters (A) C1 and (B) C2, obtained from REMD simulations of Aβ_{12-28} fragment, from Ref. [124]. Reprinted from Baumketner, A. and Shea, J.-E (2006) Folding landscapes of the Alzheimer amyloid-β_{12-28} peptide, *J Mol Biol* 362, 567–579. Copyright (2006), with permission from Elsevier.

namely a β-hairpin for Aβ_{x-42} (= 29–31) and an extended β-strand for Aβ_{39-42}. Interestingly, when the last two hydrophobic residues were removed from the Aβ_{30-42} fragment in particular, it was then mostly found in a turn-coil conformation. This result highlighted the importance of these last two hydrophobic residues in the aggregation process, as the difference between the Aβ_{1-40} and the more toxic Aβ_{1-42} resides in precisely these two amino acids. Comparison of structures with solvent and without solvent indicated that the hydrophobic interactions were critical for the formation of the β-hairpin [127].

Similar efforts were carried out on proteins from other diseases; for example, in prion disease some antiprion compounds have been shown to interfere with the pathological conversion of the PrP into its misfolded isoform. In an all-atom MD simulation with explicit solvent, different conformations of PrP with and without ligand binding were considered to clarify the role of a typical antiprion compound termed GN8. In this approach, urea-driven unfolding simulations were employed to assay whether GN8 prevented the denaturation of the PrP. Results showed that urea mediated a partial unfolding at helix B of the PrP, suggesting a transition into the intermediate states of the pathological conversion. However, GN8 efficiently suppressed local fluctuations by binding to flexible spots on helix B and prevented its urea-induced denaturation. The authors concluded that GN8 inhibited the pathological conversion by suppressing the level of expression of the intermediate, thus giving evidence supporting the chemical chaperone hypothesis, which states that GN8 acts as a chaperone to stabilize the normal form of the PrP and thus constitutes a possible target as a therapeutic compound in drug design for prion diseases [128].

In a coarse-grained implicit solvent MD simulation study of the $PrP_{125-228}$ protein fragment along with its CJD mutation T183A, the mutant showed a decrease in its thermodynamic stability. This loss of stability was related to a destabilization of a two-helix subdomain. The authors suggested that the difference between the WT and the CJD laid in the helices but not in their β-sheet content or conversions from the existing α- to β-sheet structure, even in a dimer examined in the same work [129].

16.3.2.2. Monomers: Full Length

The previous studies make it clear that understanding the folding dynamics of protein fragments is important for identification of the protein regions that are key to folding and assembly of the full-length protein. Because only the naturally occurring full-length pathological protein and alloforms are directly relevant to the disease, folding studies of the full-length proteins are critical to understanding the early protein misfolding events. Clues are offered by experimental studies which indicate that the full-length $A\beta$ in AD is largely helical in a membrane or membrane-mimicking environment [130–133], whereas it is characterized by a collapsed-coil structure with a small amount (10–20%) of the β-strand secondary structure in aqueous solutions [101, 134, 135]. Structural studies based on solid state NMR and electron microscopy of the full-length fibrils suggest that $A\beta_{1-40}$ comprised two β-strand regions (12–24 and 30–40) connected by a bend at 25–29. This bend would bring the two β-strands together to form a parallel β-sheet (see Fig. 16.7) through side chain–side chain interactions [136, 137]. Examination

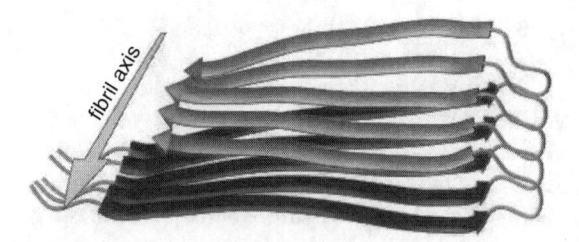

Figure 16.7. Structural model of $A\beta_{1-40}$ fibrils, based on NMR and modeling studies, from Ref. [136]. Figure shows a ribbon representation of single molecular layer or cross-β-unit. The yellow arrow represents the fibril axis. As shown in the figure the cross-β-unit has a double-layered structure, with in-register parallel β-sheets formed by residues 12–24 (orange ribbons) and 30–40 (blue ribbons). Taken with permission from Petkova, A. T., Ishii, Y., Balbach, J. J., Antzutkin, O. N., Leapman, R. D., Delaglio, F., and Tycko, R (2002) A structural model for Alzheimer's beta-amyloid fibrils based on experimental constraints from solid state NMR, *Proc. Natl. Acad. Sci. U. S. A.* 99, 16742–16747. Copyright (2002) National Academy of Sciences, U.S.A.

of the $A\beta$ fibril structure at the molecular level led to the discovery of the molecular polymorphism in fibril formation, suggesting an existence of multiple fibrillization pathways [138, 139].

To gain a deeper understanding, at the atomic level, of the nature and formation dynamics of these secondary structures believed to be intermediate steps toward aggregation, computational studies are again invoked. Recent advances in computing hardware as well as innovative algorithms have permitted the exploration of the dynamics of the full-length amyloid peptide, specifically those targeted toward $A\beta$ in AD. Although some studies use all-atom MD, many studies use other simplifying assumptions to be able to explore longer simulation times and an increased region of the energy landscape. Typical techniques involve using REMD with implicit solvent and DMD with coarse-grained models and implicit solvent.

On that note, we start our discussion with a study that considered the folding events of $A\beta_{1-40}$ and $A\beta_{1-42}$ using DMD with a four-bead protein model with backbone hydrogen bonding and effective amino acid specific hydropathy [140]. In this study, both peptides were found to fold first at the C-terminus followed by the central region with the subsequent formation of the contacts involving the N-terminus. The authors found that the folded structures of $A\beta_{1-40}$ and $A\beta_{1-42}$ differed at the C-terminus. $A\beta_{1-42}$ but not $A\beta_{1-40}$ was characterized by a turn centered at G37-G38. This structural difference was later confirmed by several independent experimental studies [112, 141–143]. In addition, the $A\beta_{1-40}$ and $A\beta_{1-42}$ folded structures differed also at the N-terminal region with a short β-strand at the N-terminus of $A\beta_{1-40}$, not present in $A\beta_{1-42}$ [140].

Xu et al. applied all-atom MD to study $A\beta_{1-40}$ folding using an explicit water model [144]. The aim of this study was to structurally characterize the dynamic behavior of $A\beta_{1-40}$ and to examine the effect of mutations of glycines at positions 25, 29, 33, and 37 to alanines, in both aqueous and a biomembrane environment. In aqueous solution, an α-helix to β-sheet conformational transition was observed, and a complete helix to coil transition was observed. Structures with β-sheet structure occurred as intermediates in the helix to coil transition pathway. The $A\beta_{1-40}$ mutant, in which four glycines were replaced by alanines, showed almost no β-sheet structure and instead showed an increased α-helix propensity. In the membrane environment, $A\beta_{1-40}$ structure was mostly α-helical, consistent with experimentally determined folded structure. $A\beta_{1-40}$ tended to exit the membrane environment and lie down on the surface of the lipid bilayer [144].

Studies trying to elucidate stable conformations of the full-length $A\beta$ from AD used REMD to explore the phase space of the protein [145]. But again because of the size of the system, the peptide had to be considered in implicit solvent. Results showed conformations exhibiting mostly loops and turns with small helical structures near the C-terminal hydrophobic tail, a region with high propensity for forming helices in apolar media. These studies also performed simulations of the peptide without any solvent showing collapsed-coil conformations with hydrophobic regions residing in regions geometrically opposite to those found in the implicit solvent simulations. Fits of these structures to accompanying experimental results using ion mobility mass spectroscopy suggested that the experimental results could be replicated by considering a mixture of the computationally determined conformations [145].

In a combined experimental and computational study, REMD simulations in explicit water were combined with NMR spectroscopy to investigate possible conformations of the $A\beta_{1-40}$ and $A\beta_{1-42}$ peptides [143]. This study confirmed the existence of structured regions and suggested that the C-terminus of $A\beta_{1-42}$ was more structured than the $A\beta_{1-40}$, in agreement with the earlier study by Urbanc et al. [140]. Furthermore, the formation of a β-hairpin in the 31–42 amino acid sequence involving short strands at residues 31–34 and 38–41 were interpreted as being able to reduce the C-terminal flexibility of the $A\beta_{1-42}$. The authors suggested that this loss of flexibility may be responsible for the higher propensity of the $A\beta_{1-42}$ peptide to form amyloids [143]. In contrast to the above studies, all-atom MD in explicit solvent showed that the C-terminus of the $A\beta_{1-42}$ was more unfolded than that of $A\beta_{1-40}$ [146].

Other MD studies of $A\beta_{1-40}$, $A\beta_{1-42}$, and their mutants also identified regions of the full-length peptides with significant structures. Specifically, studies showed that initial α-helices gave way to β-rich stable substructures that could seed aggregation in all-atom MD [147]. Microsecond time scale REMD simulations of both, $A\beta_{1-40}$ and $A\beta_{1-42}$ in aqueous solvent demonstrated that there were five independent folding structural units, with residues 1–5, 10–13, 17–22, 28–37, and 39–42, interconnected by four turn structures. The authors also report that the additional two amino acids in $A\beta_{1-42}$ at the C-terminus increased the contacts within the C-terminus and the CHC, making the $A\beta_{1-42}$ favor the β-structure more than the $A\beta_{1-40}$ [148].

A Monte Carlo study of folding of $A\beta_{1-42}$ and its mutants showed a qualitatively similar variety of β-sheet structures with distinct turns. It was shown that the WT preferentially populated two major classes of conformations, either extended with a high β-sheet content or more compact with a lower β-sheet content. Three mutants (E22G, F20E, E22G/I31E) included in this study altered the balance between these classes by affecting structures based in a region centered at residues 23–26, where $A\beta_{1-42}$ tended to form a turn. The aggregation-accelerating E22G mutation made this turn region conformationally more diverse, whereas the aggregation-decelerating F20E mutation had the reverse effect, and the E22G/I31E mutation reduced the turn population [149]. However, in stark contrast to the previous two studies, other implicit solvent REMD simulations showed a lack for any β-structures [150].

In an implicit solvent DMD simulations using a four-bead protein model, the temperature dependence of the β-structure was studied in both $A\beta_{1-40}$ and $A\beta_{1-42}$, and their Arctic (G22) mutants [151]. Both $A\beta_{1-40}$ and $A\beta_{1-42}$ were shown to adopt a collapsed-coil conformation with several β-strands. The amount of the β-structure increased with temperature (see Fig. 16.8), in agreement with experimental observations [135, 152]. Specifically, the $A\beta_{1-42}$ showed higher β-strand propensity than the $A\beta_{1-40}$ at just above physiological temperature, and for both alloforms it was observed that a turn structure was formed in the $A\beta_{21-30}$ region. The Arctic mutation was observed to change the structure of the $A\beta_{1-40}$ and make it more similar to the more toxic $A\beta_{1-42}$, thus suggesting that both Arctic $A\beta$ peptides might assemble into structures similar to the toxic $A\beta_{1-42}$ oligomers [151].

COMPUTER SIMULATIONS 497

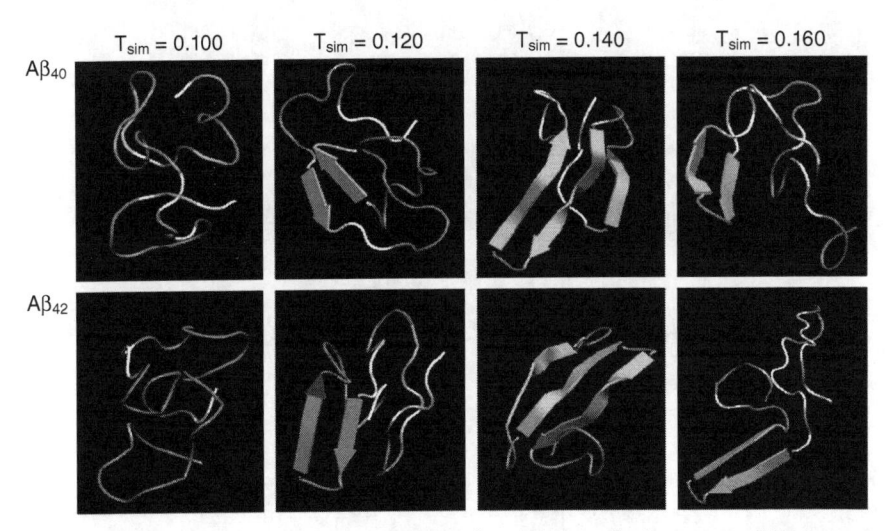

Figure 16.8. Representative conformations of $A\beta_{40}$ (top) and $A\beta_{42}$ (bottom) at different temperature, from Ref. [151]. Adapted with permission from *J Am Chem Soc* 130, 17413–17422. Copyright (2008) American Chemical Society.

16.3.2.3. Monomers: Interactions with Surfaces Computational studies usually consider the folding and aggregation of amyloid-forming proteins in bulk water. This is not the actual *in vivo* environment in which these proteins exist and exert their deleterious effects on cells, but is a necessary simplification to a very complex problem. The main tenet of the bulk water consideration is that the main mechanisms of folding and aggregation should show up and be intrinsic to the protein. However, other experimental and computational studies have shown that the particular composition of the solvent and external conditions are crucially significant to, for example, $A\beta$ of AD [114, 130–133].

In an all-atom MD study of the full-length $A\beta_{1-40}$ protein of AD, the relationship between insertion depth in a membrane lipid bilayer, protonation state of key residues, and ionic strength of the protein were investigated [153]. These simulations showed that $A\beta_{1-40}$ was not expelled from the membrane, but that, depending on its protonation state, it would stay embedded in the membrane at differing depths. Because many elements of secondary structures, in particular β-structures, were observed in the 21–30 region, it was suggested that membranes could provide a template for aggregation [153].

More recent all-atom MD studies of the $A\beta_{29-40}$ fragment revealed that in aqueous solution the peptide exhibited a collapsed-coil conformation, but in the presence of trehalose—a simple disaccharide—the α-helical conformation was preferred [154]. This fragment also formed an α-helix upon insertion into a membrane. Their results suggested that because trehalose promoted the α-helix formation, its proximity to $A\beta_{29-40}$ in water while close to a membrane should favor the insertion of $A\beta_{29-40}$ into the membrane more readily, thus suggesting that trehalose promotes the insertion of $A\beta_{29-40}$ into biological membranes [154].

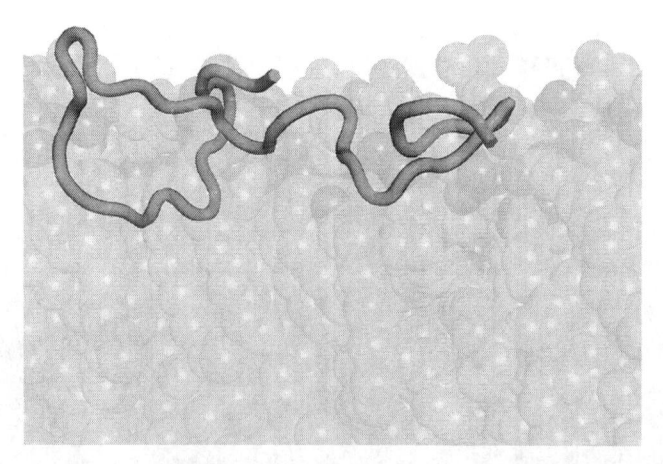

Figure 16.9. The structure of the $A\beta_{1-42}$ monomer used for the simulations on DPPC bilayer, see ref. [155]. Adapted with permission from *J Phys Chem B* 113, 14480–14486. Copyright (2009) American Chemical Society.

Other studies using REMD with umbrella sampling investigated the role that the membrane could have on $A\beta_{1-42}$ aggregation [155, 156]. By studying the effects of the membrane on the $A\beta_{1-42}$ folding, it was concluded that $A\beta_{1-42}$ did not adopt a stable structure at the bilayer surface and that the interaction with the bilayer likely precluded $A\beta_{1-42}$ from adopting a stable folded structure (see Fig. 16.9). Their results suggest that the observation of the structural conversion during aggregation likely required extensive protein–protein interactions rather than protein–membrane interactions. The authors proposed that the interaction between the $A\beta_{1-42}$ and the membrane could facilitate aggregation by reducing the peptide diffusion from a three-dimensional to a two-dimensional process, locally increasing the $A\beta$ concentration on the bilayer surface due to the highly favorable free energy of binding, and decreasing the local pH on the bilayer surface, resulting in $A\beta$ disordered conformations that would be amenable to protein–protein interactions and would promote oligomer and fibril formation [155, 156].

16.3.3. Process of Assembly

The studies presented above have brought us far in the understanding of monomer stability and dynamics but do not yet present a complete picture of the process. Despite this lack of a full understanding of the monomer, which poses a big barrier to the full understanding of aggregation as a framework for the pathogenesis of the neurodegenerative diseases targeted in this review, other studies have embarked on the more ambitious task of understanding the initial stages of assembly. Naturally, the processes of oligomer assembly and amyloid formation represent the most hotly debated research topics in the last two decades, but they still remain unsolved problems [157].

There are clues, nonetheless, on the nature of amyloid fibrils. One is given by the hypothesis of a common molecular organization for fibrils, supported by the finding that amyloid fibrils formed by many different proteins (each associated with its own clinical syndrome) showed similar cross-β diffraction patterns. This degree of similarity points to a common core molecular structure [158] as if the formation of fibrillar amyloid was a generic property of polypeptide chains [30]. Other advances in the understanding of amyloid formation have been recently achieved by using synthetic peptides whose solid aggregates exhibit most of the features of amyloid fibrils [159, 160], and even proteins with no apparent link to disease but that form fibrillar aggregates *in vitro* [33]. On another front, other clues related to the process of formation of the pre-amyloid assemblies or oligomers are emerging, such as the evidence that a common antibody recognizes oligomers formed by proteins with different sequences with no sequence homology [161], suggesting that there are common processes involved in their formation [162].

Similarly to the case of the theoretical monomer folding studies reviewed in the previous section, it is useful to explore the mechanistic details of assembly by considering short peptide sequences that adopt a native-like fold and can be converted to pre-amyloid and amyloid fibrils. Because of the high computational demand of the system of multiple peptide chains, studies of protein fragments were carried out using all-atom descriptions, whereas the other studies using full-length peptides were performed using simplified, but powerful, computational approaches. Some of the work reviewed below addresses the stability of small assemblies, given an initially preformed structure taken from the well-characterized fibrillar models. These stability studies, although not technically studies of assembly, provide the much needed input on the structures that are more likely to be the end result of the fibril formation processes.

16.3.4. Protein Assembly in Aqueous Solution

16.3.4.1. Fragment Assembly: AD

$A\beta_{16-22}$ An initial all-atom MD study using explicit solvent considered fibril-like assembly structures of the Aβ fragments 16–22, 16–35, and 10–35 [163]. In their results, the simulations indicated that an antiparallel β-sheet assembly was the most stable for the Aβ_{16-22}, in agreement with the solid-state NMR-based model [120]. A model with 24 Aβ_{16-22} strands indicated a highly twisted fibril. Whereas the short Aβ_{16-22} and Aβ_{24-36} may exist in fully extended form, the linear parallel β-sheets for Aβ_{16-35} appeared impossible, mainly because of the polar region in the middle of the 16–35 sequence. However, the authors observed that a bent double layered hook-like structure with the polar residue in the exposed loop–turn region formed parallel β-sheets. An intra-strand salt bridge (D23-K28) stabilized the hook-like structure. The bent double-β-sheet model for the Aβ_{10-35} similarly resulted in stability of shorter pieces of a fibril [163].

In another all-atom MD simulations in explicit water meant to probe the assembly mechanism of the Aβ_{16-22} peptides, the authors showed that the assembly into low-energy structures, in which the peptides formed anti-parallel β-sheets, occurred by multiple pathways with the formation of an obligatory α-helical intermediate [164]. This observation and the experimental results on fibril formation of $A\beta_{1-40}$ and $A\beta_{1-42}$ suggest that the assembly mechanism (random coil → α-helix → β-strand) was universal for this class of peptides. Their results also showed that both interpeptide hydrophobic and EIS were critical in the formation of the anti-parallel β-sheet structure. They also found that mutations of either hydrophobic or charged residues destabilized the assembly, suggesting that the Arctic (E22G), Dutch (E22Q), and Italian (E22K) mutants of $A\beta_{16-22}$ were less likely to form ordered fibrils [164].

Another all-atom MD study of dimer formation of five β-sheet-forming peptides using implicit solvent did not find any preferred β-sheet structure [165]. The peptides used were the $A\beta_{16-22}$, its Arctic mutant, and three designed sequences, each 6 to 8 amino acids long, typically used in other studies of fibril formation. The resulting dimers exhibited all possible combinations of β-sheets, with an overall preference for anti-parallel arrangements. Through statistical analysis of 1000 1-ns-long dimerization trajectories, the authors showed that the observed distribution of dimer configurations was kinetically determined, whereas effective hydrophobic interactions oriented the peptides to minimize the solvent-accessible surface area. Consequently, the dimer structures were trapped in energetically unfavorable conformations. In addition, once the hydrophobic contacts were formed, the backbone hydrogen bonds formed rapidly by a zipper-like mechanism. Their results indicated that the relaxation time of dimers was longer than the time for diffusional encounters with other oligomers at typical concentrations. Their results suggest that kinetic trapping could play a role in the structural evolution of early assembly of amyloid proteins [165].

$A\beta_{16-22}$: Molecules in Solution All-atom MD with explicit solvent was also used to probe the stability of $A\beta_{16-22}$ assemblies in aqueous urea solution [121]. The high concentration of urea promoted the formation of β-strand structures in monomers, whereas in water they adopted largely collapsed coil structures. A tripeptide system, which formed stable anti-parallel β-sheet structures in water, was shown to be destabilized in the urea solution. Klimov et al.'s results showed that the enhancement of the β-strand content in the monomers and the disruption of assembly structures occurred largely due to a direct interaction of the urea with the peptide backbone, suggesting that the assembly was affected by two opposing effects, namely by the increased propensity of monomers to form β-strands and the rapid disruption of the assembly by hydrogen bonding of individual peptides to the urea. The authors also hypothesized that a high urea concentration should destabilize assemblies formed by other amyloidogenic peptides [121].

The molecular mechanisms underlying the *in vitro* finding that a disaccharide molecule, trehalose, inhibited Aβ aggregation were explored by all-atom MD in explicit solvent [166]. The $A\beta_{16-22}$ assembly was studied at different trehalose concentrations. These simulations confirmed that trehalose prevented $A\beta_{16-22}$ aggregation in a dose-dependent manner. The preferential excluded-volume effect of trehalose was found to be the origin of its aggregation inhibition. Namely, there was a preferential hydration on the peptide surface within the 3-Å-thick layer, and trehalose molecules clustered around the peptides at a larger distance of 4–5 Å. At high trehalose concentrations, the preferential exclusion of trehalose led to three inhibition mechanisms: (1) the secondary structure of $A\beta_{16-22}$ monomers was stabilized in a turn, bend, or coil conformation, so that the β-sheet-rich structures prone to assembly were inhibited, (2) the thin hydration layer and trehalose clusters weakened the effective hydrophobic interactions that drive protein assembly, and (3) more hydrogen bonds between trehalose and $A\beta_{16-22}$ formed, which suppressed the intermolecular hydrogen bonding among $A\beta_{16-22}$ peptides [166].

On the D23–K28 Salt Bridge An all-atom MD study in explicit water assessed the stability of preformed fibril-like hexamers of $A\beta_{16-35}$ and mutant peptides relevant to familial form of AD [167]. These cross-β $A\beta_{16-35}$ hexamer structures were shown to be destabilized by the solvation of the two charged residues D23 and K28, which were initially buried in the interior of the hexamer structure, suggesting that the desolvation of charged residues may contribute to the kinetic barrier needed to initiate the process of assembly into fibrils. On the other hand, the E22Q/D23N, D23N/K28Q, and E22Q/D23N/K28Q mutants had residues with neutral amide side chains that made hydration less significant, suggesting that this relative insensitivity of the structure to hydration contributed to an increased stability of fibril-like hexamers of the mutant peptides, reduced the desolvation barrier, and thus increased the aggregation rate [167].

The existence of the lactam bridge at D23-K28 in $A\beta_{1-40}$ was experimentally shown to increase the rate of $A\beta_{1-40}$(D23-K28) fibril formation by several orders of magnitude relative to the WT peptide [168]. The dynamics of $A\beta_{10-35}$(D23-K28) monomers and dimers was probed by all-atom MD in explicit solvent [169]. To model the lactam bridge, D23-K28 was harmonically constrained. The reduction in the free energy barrier to fibril formation in $A\beta_{10-35}$(D23-K28) relative to the WT was mainly due to the entropic restriction provided by the restrained bridge. An entropy decrease of the unfolded state and a reduced conformational reorganization energy needed for salt-bridge formation resulted in a reduction in the kinetic barrier in $A\beta_{10-35}$(D23-K28) relative to the WT [169].

Another application of all-atom MD in explicit solvent to test stability of five preformed $A\beta_{9-42}$ fibril-like assemblies (monomer through pentamer) showed that the initial conformations were stable for trimers through pentamers [170]. In assembly of three or more peptides, two parallel in-register β-sheets with a connecting turn were preserved. A dimer, however, was shown to undergo a larger conformational change at its C-terminus, and the predominant conformation exhibited an additional anti-parallel β-sheet in one of the subunits. The authors indicated that this conformational rearrangement allowed efficient shielding of hydrophobic residues from the solvent, which is not possible for a dimer in the fibril conformation. In addition to the presence of the hydrogen bonds in the β-sheets, they found that trimers through pentamers were stabilized by the intermolecular D23-K28 salt bridges, whereas the D23-N27 contact was found in the dimer. Horn and Sticht's results suggested that the degree of structural similarity between larger assemblies and a fibril may offer a structural explanation for the experimental finding that trimers and tetramers acted as more potent seeds in fibril formation than dimers [170].

A preformed fibril-like pentamer of the $A\beta_{17-42}$ fragment was assessed for stability using stirred MD combined with the umbrella sampling method [171]. By studying the thermodynamics of peptide dissociation from the core of this fibril-like $A\beta_{17-42}$ pentamer at physiological temperature, a finite level of hydration around the D23–K28 salt bridge was found to be crucial for its stability, whereas the mutation F19G had no effect on the binding free energy of the terminal peptide. The packing of I32 and the aliphatic portion of the K28 side chain served to regulate the level of hydration in the core of the assembly, resulting in a more rigid D23-K28 salt bridge [171].

Other Fragments In a stability study of the $A\beta_{10-40}$ fragment, the authors used REMD to study the effect of the D23Y mutation on (10–40) fibril growth. The initial system was a preformed fibril-like hexamer of the $A\beta_{10-40}$ where each peptide received this mutation. By computing the free energy landscapes, the distributions of peptide–fibril interactions, and by comparing with the WT $A\beta_{10-40}$ peptide, the authors found that the D23Y mutation had a relatively minor influence on the docking of $A\beta$ peptides to the fibril. However, they found that it had a strong impact on the locking stage due to profound stabilization of the parallel in-registry β-sheets formed by the peptides on the fibril edge. Their results thus suggested that the D23Y mutation would promote fibril growth [172].

In a stability study of small fibril-like oligomers (dimer to hexamer) using all-atom MD in explicit water, the authors found that at room temperature hexamers were the most stable entity. The tested peptides were the $A\beta_{37-42}$, the GIFQINS from lysozyme, and the GVQIVYK from tau-protein, all of which have the propensity to form cross-β-structures. The authors concluded that (1) dimers were not thermodynamically stable, (2) dissolution of the fibrils was more difficult than aggregation, (3) a tetramer state was the intermediate state, and (4) two transition states correspond to a trimer and a pentamer [173].

In a study of the $A\beta_{25-35}$ fragment using REMD in explicit solvent, the authors investigated energetically favorable dimer conformations adopted by this fragment. Their results showed a diverse ensemble of well-organized dimers with high β-sheet content coexisting with unstructured dimer complexes. In their findings, the structured dimers comprised parallel and anti-parallel extended β-strand, β-hairpin, and V-shaped β-strand conformations. The authors then constructed protofibril models from their already-found extended and V-shaped dimers. Stability analysis on these protofibrils led to the conclusion that they formed stable structures consistent with experimentally available data. Their results suggest that fibril polymorphism may be encoded in the early stages of aggregation for the $A\beta_{25-35}$ peptide [174].

Surfaces Aggregation studies of $A\beta$ fragments have also been performed in the presence of surfaces. In particular, an all-atom MD study with explicit solvent considered two preformed oligomers each composed of a β-sheet of four or five $A\beta_{25-35}$ fragments and separated by a single-walled carbon nanotube (SWNT). The oligomers had a mixture of parallel and anti-parallel orientations of the fragment. Fu et al.'s results showed that the two separated $A\beta_{25-35}$ β-sheets with mixed anti-parallel–parallel strands could assemble into β-barrels wrapping the SWNT. In contrast, a simulation without the SWNT and another with the SWNT but with purely parallel $A\beta_{25-35}$ oligomers in the β-strands led to disordered aggregates. Regarding the dynamics of the formation of the β-barrel, the authors described two relevant steps in its formation: (1) curving of the $A\beta_{25-35}$ β-sheets as a result of strong hydrophobic interactions with the carbon nanotube concomitantly with dehydration of the SWNT–peptide interface and (2) inter-sheet backbone hydrogen bond formation with fluctuating intra-sheet hydrogen bonds. Their results suggested that the β-barrel formation on the SWNT surface resulted from the interplay of dehydration and peptide–SWNT/peptide–peptide interactions [175].

In the presence of a more familiar surface, the adsorption and aggregation of $A\beta$ on the cell membrane may play an important role in the pathogenesis of AD. An all-atom MD study with explicit solvent investigated the interactions of an $A\beta_{17-42}$ pentamer with self-assembled monolayers (SAMs) terminated with either hydrophobic CH3 or hydrophilic OH function groups. Simulation results, shown as snapshots as a function of simulation time in Figure 16.10, showed that regardless of the characteristics of the surface, the hydrophobic C-terminal region of the $A\beta$ fragment was more likely to be adsorbed on the SAM, indicating a preferential orientation and interface for $A\beta$ adsorption. Structural and energetic comparison among six $A\beta$–SAM systems further revealed that $A\beta$ fragment orientation, SAM surface hydrophobicity, and interfacial waters all determined $A\beta$ adsorption behavior on the surface, highlighting the importance of hydrophobic interactions at the interface [176].

16.3.4.2. Fragment Aggregation: Other Diseases
As shown in the previous section, the folding and dynamics of full-length amyloid-forming proteins exhibit many similarities across the diseases discussed in this chapter. This extends also to key fragments of the full-length, in which many of the properties of the full-length are echoed in their fragments and also across disease.

Prions The implicated neurotoxic amyloid-forming protein in prion diseases is the yeast prion Sup35 protein. The heptapeptide fragment GNNQQNY from the N-terminal prion determining domain of the same protein (residues 7–13) displays the same amyloid properties as the full-length Sup35, including cooperative kinetics of aggregation, fibril formation, binding of the dye Congo red, and the cross-β X-ray diffraction pattern [177].

Initial aggregation studies using all-atom MD in implicit solvent considered the aggregation of GNNQQNY [178]. Results from this work obtained an in-register parallel packing of GNNQQNY β-strands, consistent with X-ray diffraction and Fourier transform infrared data. The parallel β-sheet arrangement was found to be favored over the anti-parallel one because of side-chain contacts, in particular, stacking interactions of the tyrosine rings and hydrogen bonds between amide groups. Results considering the mutant sequence SQNGNQQRG yielded no ordered

(A) CH₃-C

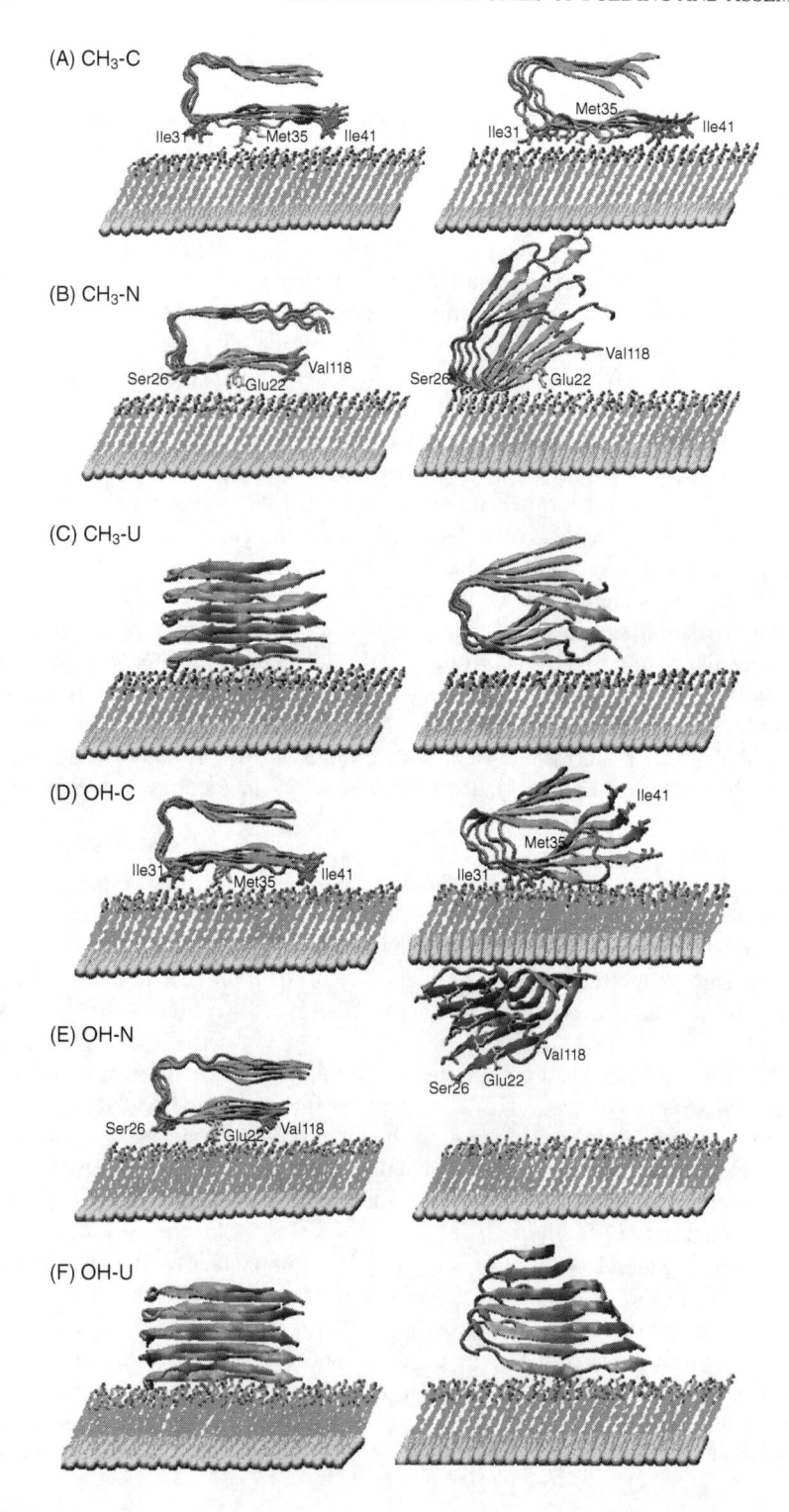

(B) CH₃-N

(C) CH₃-U

(D) OH-C

(E) OH-N

(F) OH-U

Figure 16.10. Different orientation of Aβ on the self-assembled monolayers obtained at 0 and 30 ns of simulations, see Ref. [176]. Three types of Aβ orientations relative to the SAMs were considered in the study: the hydrophobic C-terminal β-strand region facing to the CH3-SAM or the OH-SAM (shown as CH3-C and OH-C in (A) and (D)); the hydrophilic N-terminal β-strand region facing to the CH3-SAM or the OH-SAM (shown as CH3-N and OH-N in (B) and (E)); and both β-strands of the hairpin in contact with the CH3-SAM or the OH-SAM (shown as CH3-U and OH-U in (C) and (F)). Adapted with permission from *Langmuir* (2010) 26, 3308–3316. Copyright (2009) American Chemical Society.

aggregation, in accord with experimental data. The authors pointed out that the statistically preferred aggregation pathway did not correspond to a purely downhill profile of the energy surface because of the presence of enthalpic barriers that originated from out-of-register interactions.

In a similar setup, but using REMD in implicit solvent, a conformational analysis study considering monomers and dimers of the GNNQQNY sequence was investigated [179]. In partial agreement with the study in the previous paragraph, it was found that GNNQQNY dimers formed three stable sheet structures consisting of an in-register parallel, off-register parallel, and anti-parallel. The anti-parallel dimer was stabilized by strong EIS resulting from interpeptide hydrogen bonds. The in-register parallel dimer, which had a structural similarity to the amyloid β-sheet, had fewer interpeptide hydrogen bonds, making hydrophobic interactions more important and increasing the conformational entropy compared with the anti-parallel sheet. The estimated two-state rate constants in this study indicated that the formation of dimers from monomers was fast and that the dimers were kinetically stable against dissociation at room temperature.

In a related study, analysis of the disaggregation dynamics were used to posit plausible mechanisms of aggregation [180]. This and other studies, using REMD on the same peptide fragment GNNQQNY, were carried out focusing on the disaggregation of a hexamer and a 12-mer at high temperature. The results indicated that it was likely that tetramers acted as the transition state in both the hexamer and the 12-mer simulations. In addition, the 12-mer simulations showed that the initial aggregation nucleus had eight peptides. Furthermore, the landscape was rather flat from 8-mers to 12-mers, indicating the absence of major barriers once the initial aggregation nucleus formed. Thus, the likely aggregation pathway could be from monomers to the initial nucleus of 8-mers with tetramers as the transition state.

To study how monomers aggregated to an already-formed amyloid fibril, a study using all-atom MD in explicit solvent added random coil GNNQQNY monomer fragments from the yeast prion Sup35 onto a preformed fibril [87]. These authors' results showed that the random coil → β-strand transition for the Sup35 fragment occurred abruptly over a very narrow time interval, in contrast to similar trajectories using the $Aβ_{37-42}$ of the Aβ. Their results also showed that expulsion of water, resulting in the formation of a dry interface between two adjacent sheets of the Sup35 fibril, occurred in two stages. Ejection of a small number of discrete water molecules in the second stage followed a rapid decrease in the number of water molecules in the first stage. The hydrogen bond network was involved in stabilizing the new aggregated monomer. The importance of the network of hydrogen bonds involving backbone and side chains was further illustrated by considering mutations, where results showed that substitution of the N and Q residues to A compromised the Sup35 fibril stability.

Another important fragment in prion disease is the SNQNNF peptide (human PrP fragment 170–175). This peptide has been experimentally found to form various types of "steric zipper" aggregates [160], a cross-β-spine structure with opposing β-sheets tightly packed against each other in a dry interface [159] that is a representation of the common basic unit of amyloid-like aggregates. In an all-atom MD stability study of these aggregates, De Simone et al. suggested that SNQNNF assemblies were very good candidates to be involved in the structure of PrP fibrils [181]. Their results showed that steric zipper interfaces were able to stabilize assemblies composed of four fragment strands per sheet.

Diabetes In work pertaining oligomer stability, a study on the NFGAIL peptide, derived from the human islet amyloid polypeptide (residues 22–27), performed simulations using all-atom MD in explicit solvent that considered different possible strand/sheet organizations, from dimers to nonamers [100]. These small aggregates were all started from different β-sheet conformations with a host of parallel and anti-parallel configurations. Zanuy et al.'s results showed that the most stable conformation was an anti-parallel strand orientation within the sheets and parallel between sheets. Their results also showed that the mutated NAGAIL oligomer was unstable and disintegrated very quickly after the beginning of the simulation. They attributed this effect to the difference that the aromatic amino acid had on stabilizing the structure of the aggregate and not to the added hydrophobicity introduced by the alanine.

In an all-atom MD with explicit solvent study of the 20–29 segment of the human islet amyloid polypeptide, aggregation of two separated (preformed) β-sheets composed of seven anti-parallel peptides was studied [182]. These authors' results showed that the two-sheet aggregation could occur in both the lateral and the longitudinal direction, indicating an assembly pathway to the early stage of fibril elongation. This mechanism was proposed in addition to the mechanism of one monomer addition onto the preformed fibril nucleus.

Huntington's Disease The hallmark of Huntington's disease are polyglutamine (polyQ) β-stranded aggregates where the number of Q residues exceeds 36 to 40, the "disease threshold" at which point the disease is said to be fully penetrant. In an all-atom MD with explicit solvent study, the stability of polyQ helical structures of different shapes and oligomeric states was studied [183]. The results suggested that the stability of the aggregates increased with the number of monomers, while it was rather insensitive to the number of Q amino acids in each monomer. On the other hand, the authors found that the stability of the single monomer did depend on the number of side-chain intramolecular hydrogen bonds, and therefore on the number of Q amino acids. If such number was lower than that of the disease threshold, the β-stranded monomers were unstable and hence could aggregate with lower probability, consistent with experimental findings. Their results provided an interpretation of the apparent polyQ length dependent-toxicity, while discarding the so-called structural threshold hypothesis, which supposes a transition from random coil to a β-sheet structure only above the disease threshold.

Amyloid-Forming Model Proteins At this point it is clear that the complexity and size of amyloid-forming proteins present a big drawback to the success of computational studies. In many cases, the complexity of the structure, folding, and aggregation make it almost impossible to assign specific results from simulations to individual structural components of these proteins. However, because of the remarkable common characteristics of these proteins in their aggregate morphology and β-sheet fibril architecture, studies have taken to task the design of artificial amino acid sequences that exhibit these same properties despite the obvious dissimilarities in primary sequence between the natural and artificial proteins. The major drive in these studies is that the simpler sequence composition and behavior in these model proteins are more amenable to unambiguous interpretations of results [184].

In a united-atom REMD with implicit solvent study of the $cc\beta$ peptide, which forms a native-like "coiled-coil" structure under ambient solution conditions but which can be converted into amyloid fibrils by raising the temperature, a study showed an α to β-structure conversion at high temperature [185]. The initial conformational structure was a trimer aggregate where each peptide was in an α-helix conformation. For the structural conversion, two mechanisms were identified. In the first, the process goes through the α to β conversion without dissociation of the trimer when the temperature is below the trimer dissociation temperature. In the second, a dissociation/reassociation pathway was identified. The authors' results indicate that this second pathway could be interpreted as the one observed in experiments as a two-stage dock–lock mechanism for the growth of amyloid fibrils.

In a Langevin dynamics simulation study using an off-lattice three-bead coarse-grained peptide model in implicit solvent, a model peptide with the CHPHPA (H-hydrophobic, P-polar, C-positive, A-negative) sequence was considered [186]. The parameters of this model peptide were tuned to have different degrees of β-sheet formation propensity. The results in this work showed the phase diagram as a function of temperature and β-sheet propensity that revealed a diverse family of supramolecular assemblies. These authors showed that peptides with high β-sheet propensity were seen to assemble predominantly into fibrillar structures. Increasing the flexibility of the peptide (reducing β-sheet propensity) led to a variety of structures, including fibrils, β-barrel structures, and amorphous aggregates. The authors suggest that by tuning the β-sheet propensity, their model was consistent with mutations on real amyloid-forming proteins that decreased β-sheet propensity, thus forming nonfibrillar entities suggested as toxic primary causative agents in amyloid diseases.

Other studies have considered molecules made with soluble repeating β-hairpin motifs to study inter- and intra-sheet stability and interactions. For example, one study used the peptide self-assembly mimic (PSAM) from the outer surface protein A (OspA) [12] that formed highly stable but soluble β-sheet structures similar to those formed by amyloid-forming peptides [187]. In this all-atom MD with explicit solvent study, PSAMs with different-sized β-sheets were simulated under several conditions. Results showed that in WT and mutated PSAMs inter-sheet side chain–side chain interactions along with intra-sheet salt bridges were the major driving forces in stabilizing the structural organization.

16.3.4.3. Assembly: Full-Length Proteins

Computational studies of aggregation at the atomic resolution using all-atom MD of full-length amyloid-forming proteins are by all practical considerations intractable under current computing hardware technology. The shear size and number of atoms to be considered would make computational time unreasonably large given that aggregation is a process that ranges from seconds to days in *in vitro* experimental situations. Simplifications are a necessary step to gain any insight into these processes. These simplifications, however, carry the risk that key microscopic ingredients that embody important phenomena are thrown away. Thus, the models have to balance the loss of detail with the gain in computational speed. In the following, work is presented where the simplification of choice comes from coarse-graining the amino acids from tens to a few atomic interaction sites.

In our first consideration, a one-bead simplified model of proteins, the tube model [188], in conjunction with the generic hypothesis of amyloid formation, was used to investigate the phenomenon of ordered aggregation [189]. In this model a polypeptide chain was represented by a tube in which the position of each residue is specified by the coordinates of its C_α atom. Neighboring atoms were connected in a chain where the bonds were thick hard spherocylinders with a diameter of 4 Å. The parameters and variables in this model provided for the symmetry of a peptide chain, amino acid-independent hydrophobicity, and the possibility of forming backbone hydrogen bonds. The simulations were carried out using Monte Carlo with crankshaft, pivot, reptation, displacement, and rotation moves. Under conditions where oligomer formation was a rare event (the most common conditions for forming amyloid fibrils by experiment), the authors calculated directly the nucleation barriers associated with oligomer formation and conversion into cross-β-structure in order to reveal the nature of these species, determined the critical nuclei, and characterized their dependence on the hydrophobicity of the peptides and the thermodynamic parameters associated with aggregation and amyloid formation. Their results support the idea that amyloid formation arises from the interplay between hydrogen bonding and amino acid-independent hydrophobicity that led to a two-step condensation-ordering mechanism in which disordered oligomeric aggregates were formed first and then reorganized into well-organized amyloid fibrils [190].

In subsequent work, the authors, studying these same tube-represented polypeptide chains that aggregate by forming metastable oligomeric intermediate states prior to converting into fibrillar structures, showed that the formation of ordered arrays of hydrogen bonds drove the formation of β-sheets within these disordered oligomeric aggregates [191]. Their results showed that individual β-sheets initially formed with random orientations and subsequently aligned into protofilaments as their lengths increased. Their results suggest that amyloid aggregation represents an example of the Ostwald step rule of first-order phase transitions [192] by showing that ordered cross-β-structures emerged preferentially from disordered compact dynamical intermediate assemblies.

Other models, in addition to using a coarse-grained framework for the protein, also simplify the dynamics by not using the usual continuum equations of motion but by using stepwise potentials in which particles are in free flight until the interactions provoke collisions only for distances smaller than a given threshold [193–196]. These simplifications, plus the consideration of an implicit solvent, have proven very powerful as simulation times have been sped up by at least seven orders of magnitude while retaining consistency with experimental results [157].

Using DMD with a two-bead coarse-grained Gō model of the AD Aβ_{1-40}, a study considered the aggregation of initially separated peptides. The results of Peng et al. showed that for temperatures above the α-helix melting temperature of a single peptide, the model peptides aggregated into a parallel multilayer β-sheet structure with an interstrand distance of 4.8 Å and an inter-sheet distance of 10 Å, consistent with experiments. Their results showed that hydrogen bond interactions gave rise to the interstrand spacing while the Gō side-chain interactions stabilized the β-strands into layers. The authors suggest that the free edges in their aggregates may allow for further aggregation of model peptides to form elongated fibrils [197].

In an initial application of a four-bead coarse-grained model using DMD to the AD Aβ_{1-42} and Aβ_{1-40}, a study considered the formation and stability of dimers of these two peptides. In their study, the authors first classified all dimers that formed during the simulations. Then, the stability of these dimers was tested by expressing these conformations in their all-atom representations, and, using standard molecular mechanics, their relative themodynamic stability was established. In their results, the authors found 10 different planar β-strand dimer conformations. After estimating their relative free energies in the all-atom molecular mechanics simulations with explicit water, they found that dimer conformations had higher free energies compared with their corresponding monomeric states. In addition, they found that the free energy difference between the Aβ_{1-42} and the corresponding Aβ_{1-40} dimer conformation was not significant. Because of this insignificant difference in free energy between these two dimers, their results suggest that planar β-strand Aβ dimers cannot account for experimentally observed differences in Aβ oligomer formation between the Aβ_{1-40} and the Aβ_{1-42} alloforms opening the possibility that different pathways may happen at a higher oligomer formation step [198].

In a subsequent study aimed at understanding oligomer formation in AD, a DMD four-bead coarse-grained protein model was implemented to consider Aβ_{1-40} and Aβ_{1-42} aggregation. Similar to the two-bead model, the four-bead model incorporated backbone hydrogen bond interactions but added specificity of amino acids by implementing an interaction matrix that took into account the hydropathy of amino acids. The authors' results showed that peptides of Aβ_{1-40} formed significantly more dimers than Aβ_{1-42}, whereas pentamers were significantly more abundant in the

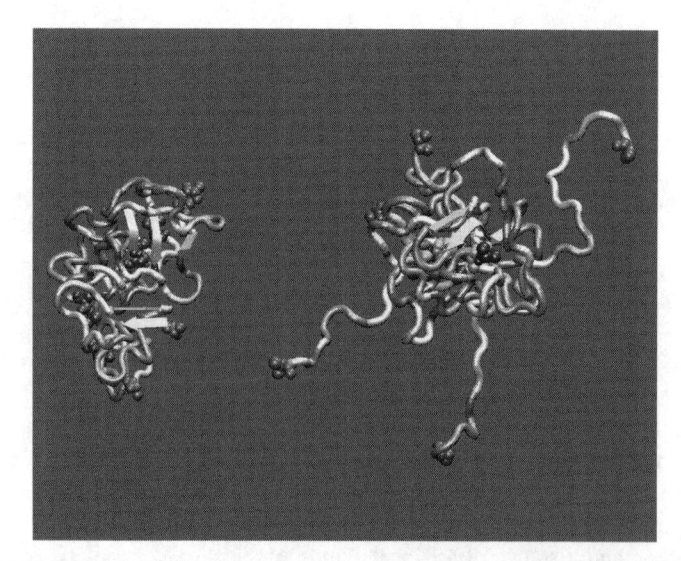

Figure 16.11. Structural features of pentamers of $A\beta_{40}$ (left) and $A\beta_{42}$ (right) obtained from DMD simulations, see Ref. [140]. The secondary structure of pentamers is shown as a silver tube (random coil-like structure), light-blue tube (turn), and yellow ribbon (β-strand). Red spheres in both left and right structures represent the N-terminal D1. (A) The C-terminal amino acids V39 and V40 are shown in purple. (B) The C-terminal amino acid I41 is shown in green, and A42 is shown in blue. Taken with permission from Urbanc, B., Cruz, L., Yun, S., Buldyrev, S. V., Bitan, G., Teplow, D. B., and Stanley, H. E (2004) In silico study of amyloid beta-protein folding and oligomerization, *Proc. Natl. Acad. Sci. U. S. A. 101*, 17345–17350. Copyright (2004) National Academy of Sciences, U.S.A. See color insert.

$A\beta_{1-42}$ relative to $A\beta_{1-40}$. Upon examining the structure of these oligomers, the authors found a turn centered at G37–G38, which was present in a folded $A\beta_{1-42}$ monomer but not in the folded $A\beta_{1-40}$ monomer and that was associated with the first contacts that form during monomer folding. Furthermore, the study showed that pentamers had a globular structure comprising hydrophobic residues within the pentamer's core and hydrophilic N-terminal residues at the surface of the pentamer. In addition, the N termini of the $A\beta_{1-40}$ pentamers were more spatially restricted than for the $A\beta_{1-42}$ pentamers (see Fig. 16.11). Also, the $A\beta_{1-40}$ pentamers formed a β-strand structure involving A2–F4, which was absent in $A\beta_{1-42}$ pentamers. Overall, their results suggest a different degree of hydrophobic core exposure between pentamers of the two alloforms, with the hydrophobic core of the $A\beta_{1-42}$ pentamer being more exposed and thus more prone to form larger oligomers [140].

Even when the previous approach involved substantial simplifications, it nonetheless accounted for experimentally observed differences between the $A\beta_{1-40}$ and $A\beta_{1-42}$. Nonetheless, the model was later expanded to explicitly account for EIs between pairs of charged amino acids. In this extended study, a DMD four-bead protein model was used to study the effect of the strength of EIs on oligomer formation from the $A\beta_{1-40}$ and $A\beta_{1-42}$, which were shown *in vitro* to oligomerize differently.

The results of Yun et al. indicate that EIs promoted the formation of larger oligomers in both $A\beta_{1-40}$ and $A\beta1-42$. Both alloforms displayed a peak at trimers/tetramers, but $A\beta_{1-42}$ displayed additional peaks at nonamers and tetradecamers. Their results suggest that the strength of EIs shifted the oligomer size distributions to larger oligomers but still maintaining the $A\beta_{1-40}$ size distribution unimodal while having the $A\beta_{1-42}$ distribution trimodal, as observed experimentally. Their results preserved previously obtained structural differences such as a turn in the C-terminus of the $A\beta_{1-42}$ that is lacking in the $A\beta_{1-40}$. From their finding that the C-terminal region also has the strongest intermolecular contacts, their results suggest that this C-terminal region played a key role in the formation of $A\beta_{1-42}$ oligomers and that inhibitors targeting this C-terminal region may be able to prevent or alter oligomer formation to reduce their toxicity [199].

A natural extension to this line of work came later with the application of Urbanc et al.'s four-bead model using DMD with amino acid-specific hydropaty and charge plus the possibility of backbone hydrogen bonds to consider the Arctic mutant (E22G). In this and previous work, the authors showed that unlike the $A\beta_{1-40}$, the $A\beta_{1-42}$ had a higher propensity to form paranuclei (pentameric or hexameric) structures that could self-associate into higher-order oligomers. In addition, they found that neither of the Arctic mutants ([E22G] $A\beta_{1-40}$ nor $A\beta_{1-42}$) formed higher order oligomers. However, their (E22G) $A\beta_{1-40}$ formed paranuclei with a similar propensity to that of $A\beta_{1-42}$. Structural analysis revealed that the C-terminal region played a dominant role in $A\beta_{1-42}$ oligomer formation, whereas $A\beta_{1-40}$ oligomerization was primarily driven by intermolecular interactions among the central hydrophobic regions. Also, they found that the N-terminal region A2-F4 played a prominent role in $A\beta_{1-40}$ oligomerization but did not contribute to the oligomerization of $A\beta_{1-42}$ or the Arctic mutants. Of special significance, however, was their result that the oligomer structure of both Arctic peptides resembled $A\beta_{1-42}$ more than $A\beta_{1-40}$, consistent with their potentially more toxic nature [200].

16.3.5. Assembly of Full-Length Proteins in the Presence of Inhibitors and Drugs

In a combined experimental and theoretical study, Fradinger et al. studied how a series of N-terminally truncated $A\beta$-derived peptides, $A\beta_{X-42}$, where X varied from 29 to 39, affected $A\beta_{1-42}$ assembly and cell culture toxicity [126]. Snapshots showing the interaction for four time frames between $A\beta42$ and CTFs during oligomerization are shown in Figure 16.12. All CTFs were shown to significantly inhibit $A\beta_{1-42}$ toxicity. However, the degree of toxicity inhibition was CTF-specific, with $A\beta_{31-42}$ and $A\beta_{39-42}$ demonstrating the highest degrees of toxicity inhibition. Computer simulations using the DMD approach with a four-bead protein model in implicit solvent demonstrated that CTFs co-assembled with $A\beta_{1-42}$ to produce non-toxic hetero-oligomers. This result was consistent with data derived by dynamic light scattering.

Wu et al. studied the structures of selected CTFs ($A\beta_{X-42}$; $X = 29$–31, 39) using replica exchange MD simulations and ion mobility mass spectrometry and showed that the CTFs adopted a metastable β-structure: a β-hairpin for $A\beta_{X-42}$ ($X = 29$–31)

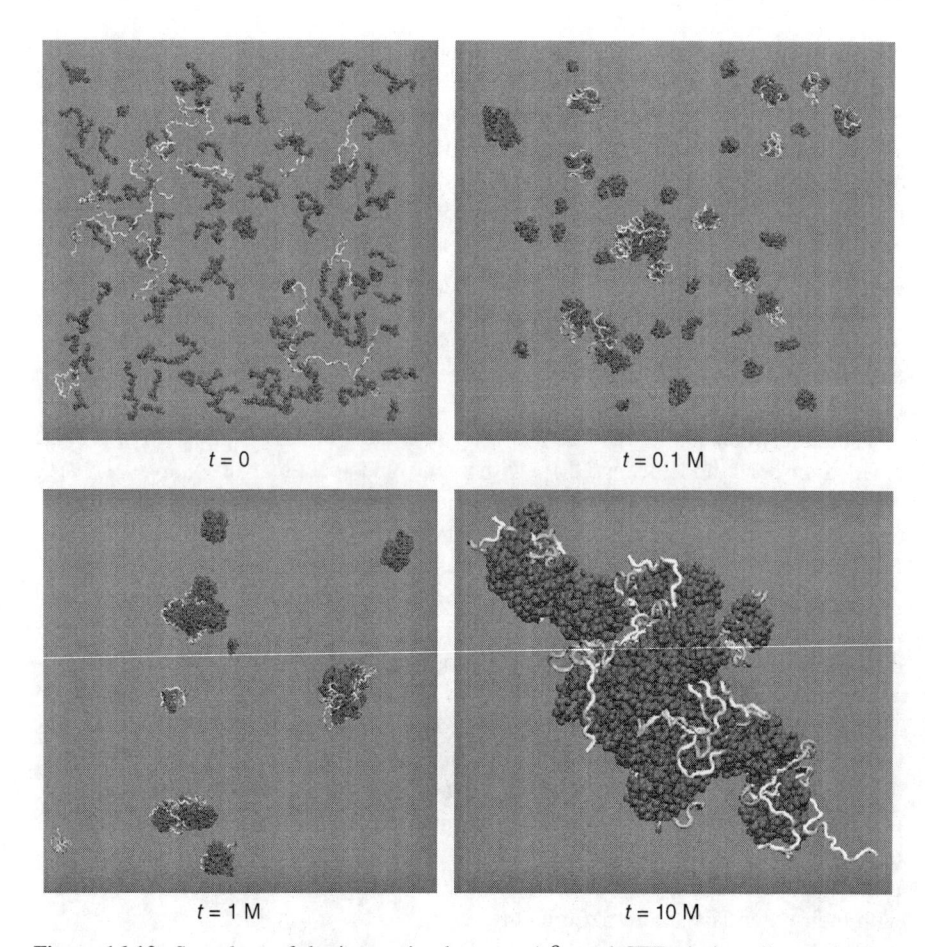

Figure 16.12. Snapshots of the interaction between $A\beta_{42}$ and CTFs during oligomerization. Configurations of 16 $A\beta_{42}$, and 128 $A\beta_{31-42}$ molecules at different time frames measured at given t simulation steps, see Ref. [126]. The dark blue is for CTFs, and $A\beta_{42}$ molecules are indicated by their secondary structure: yellow ribbons, β-strands; blue tubes, turns; silver tubes, random coil. Taken with permission from Fradinger, E. A., Monien, B. H., Urbanc, B., Lomakin, A., Tan, M., Li, H., Spring, S. M., Condron, M. M., Cruz, L., Xie, C. W., Benedek, G. B., and Bitan, G. (2008) C-terminal peptides coassemble into Abeta 42 oligomers and protect neurons against Abeta 42-induced neurotoxicity, *Proc. Natl. Acad. Sci. U.S.A. 105*, 14175–14180. Copyright (2008) National Academy of Sciences, U.S.A. See color insert.

and an extended β-strand for $A\beta_{39-42}$ [127]. The β-hairpin of $A\beta_{30-42}$ was converted into a turn-coil conformation when the last two hydrophobic residues were removed, suggesting that I41 and A42 were critical in stabilizing the β-hairpin in $A\beta_{42}$-derived CTFs. A comparison between structures with solvent and structures without solvent revealed that hydrophobic interactions were critical for the formation of the β-hairpin.

COMPUTER SIMULATIONS

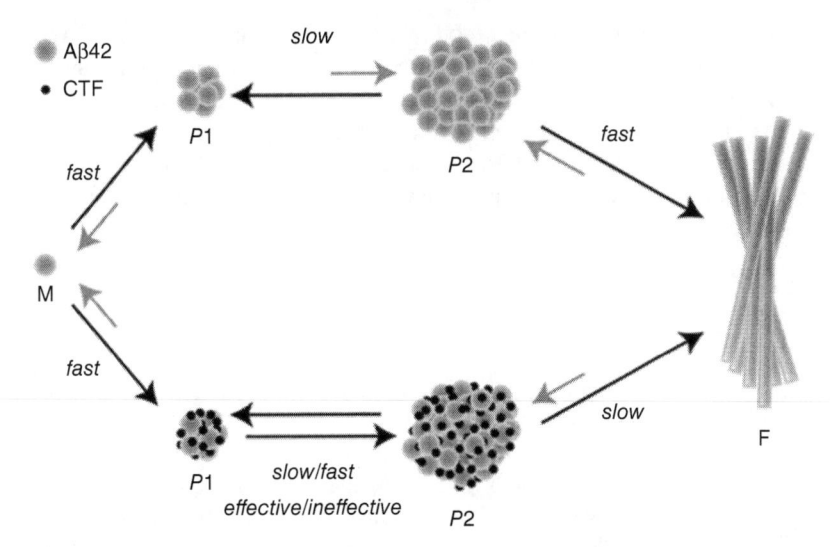

Figure 16.13. Pictorial representation showing how CTFs affect $A\beta_{42}$ assembly, see Ref. [202]. Monomer assembly is a rapid process irrespective of CFTs. The P1→P2 conversion may be promoted by the CFTs, but effective inhibitors of $A\beta_{42}$-induced toxicity induce slower acceleration than ineffective ones, shifting the population toward P1. All CTFs slow the maturation of P2 assemblies into fibrils (F). Adapted with permission from *Biochemistry* (2010) 49, 6358–6364. Copyright (2010) American Chemical Society.

Follow-up studies by Li et al. [201, 202] further explored the biophysical properties of CTFs and their effect on $A\beta_{1-42}$ assembly. Li et al. added to the tested CTFs the $A\beta_{30-40}$, which inhibited $A\beta_{1-42}$ toxicity, and the control peptide $A\beta_{21-30}$, which did not affect $A\beta_{1-42}$ toxicity in cell cultures. In Figure 16.13 a schematic representation of a putative mechanism by which CTFs affect $A\beta_{42}$ assembly is shown. The three most potent inhibitors of $A\beta_{1-42}$ toxicity, $A\beta_{30-40}$, $A\beta_{31-42}$, and $A\beta_{39-42}$, were then chosen for a DMD study of $A\beta_{1-42}$ assembly in the presence of inhibitors and the control peptide $A\beta_{21-30}$ [203]. The toxicity inhibitors were shown to reduce the average amount of the β-strand structure, whereas the control peptide increased the average amount of the β-strand structure in $A\beta_{1-42}$. In addition, solvent exposure of the peptide region D1-D7 was demonstrated to be significantly reduced by toxicity inhibitors, whereas the control peptide increased it, implicating the N-terminal region D1–D7 of $A\beta_{1-42}$ in mediating toxicity [203].

Using REMD with implicit solvent, the impact of ibuprofen on the growth of $A\beta$ fibrils was studied [204]. It was found in this study that binding of ibuprofen to $A\beta$ destabilized the interactions between incoming peptides and the fibril, resulting in a reduced free energy gain upon $A\beta$ peptide binding to the fibril. Furthermore, ibuprofen interactions shifted the thermodynamic equilibrium from fibril-like locked states to disordered docked states. Ibuprofen's anti-aggregation effect was attributed to its competition with incoming $A\beta$ peptides for the same binding site at the fibril edge [204]. In a follow-up study, the nonsteroidal anti-inflammatory drugs naproxen and ibuprofen were both examined with respect to their interactions with Aβ fibril [205]. Naproxen demonstrated significantly higher binding affinity to Aβ fibrils than ibuprofen. The key factor, which was found to enhance naproxen binding, was strong interactions between ligands bound to the surface of the fibril, facilitated by the naphthalene ring in naproxen. These simulations provided plausible microscopic explanation for the differing binding affinities of naproxen and ibuprofen observed experimentally [205].

16.4. SUMMARY

It is evident from the works reviewed in this chapter that computer simulations are helping to elucidate mechanisms and pathways of assembly, both currently experimentally unavailable. New algorithms and computational methodologies can achieve simulation times that are approaching those available from experiments, and ever-increasing computer hardware is making possible computer simulations at increased resolution for longer times.

Studies on monomer folding clearly shed light on the first steps of the chain of events of assembly, from monomer to oligomers and fibrils. Studies using various fragments with the same properties of the full-length amyloid-forming peptide have shown the importance of the initial turns, bends, and even β-hairpins, β-bends, and so on, and how all of these can be related to subsequent aggregation. Salt bridges are important stabilizing agents of these bends as well as the composition of the solvent. Differences in the rate of formation and different structure of mutations make it possible to compare specific substitutions and their pathways.

Moving a step further, studies on the formation of small aggregates and on the stability of preformed oligomers have proven that stable oligomers can exist in both disorganized and protofibrillar structures. A conversion from one to the other is usually proposed, with the first type possibly being the disordered state, which later forms backbone hydrogen bonding leading to β-sheets and finally to the fibrils.

Along the way, we have reviewed surfaces and their importance, as studies indicate that assembly can happen and in some cases be encouraged depending on the nature of the bounding surface. Some of these studies, in particular, address the important question as to whether the membrane affects the initial formation of low-order oligomers and whether it enhances subsequent growth of the fibril.

Finally, we have reviewed some important and promising work on inhibitors of assembly, especially those inhibitors composed of C-terminus fragments of the larger protein, and drugs such as naproxen and ibuprofen. In those studies it was found that the length of the CTFs is a tunable parameter that can affect the degree of inhibition of toxicity. Naproxen and ibuprofen, both anti-inflammatory drugs, showed radically different interactions with already-formed fibrils with intriguing possibilities as agents of aggregation disruption.

The main findings of monomer folding, assembly, and stability have proven common to a range of amyloid diseases, some exemplified in the present chapter. As of now, the answer to the key questions posed in this chapter regarding folding and assembly remains elusive. However, it is clear that continued work on the elucidation of the precise mechanisms responsible for the initial folding and assembly will be key to unraveling the detrimental effects of these devastating diseases.

REFERENCES

1. Lamour, Y. (1994) Alzheimer's disease—A review of recent findings, *Biomed Pharmacother 48*, 312–318.
2. Selkoe, D. J. (2001) Alzheimer's disease: Genes, proteins, and therapy, *Physiol Rev 81*, 741–766.
3. Selkoe, D. J. (2002) Alzheimer's disease is a synaptic failure, *Science 298*, 789–791.
4. Dobson, C. M. (2004) Experimental investigation of protein folding and misfolding, *Methods 34*, 4–14.
5. Lashuel, H. A. and Lansbury, P. T. (2006) Are amyloid diseases caused by protein aggregates that mimic bacterial pore-forming toxins? *Q Rev Biophys 39*, 167–201.
6. Sunde, M. and Blake, C. C. F. (1998) From the globular to the fibrous state: Protein structure and structural conversion in amyloid formation, *Q Rev Biophys 31*, 1–39.
7. Stefani, M. and Dobson, C. M. (2003) Protein aggregation and aggregate toxicity: New insights into protein folding, misfolding diseases and biological evolution, *J Mol Med 81*, 678–699.
8. Sungur, C. I. (2003) Molecular mechanisms of amyloidosis, *N Engl J Med 349*, 1872–1872.
9. van der Hilst, J. C. H., Simon, A., and Drenth, J. P. H. (2003) Molecular mechanisms of amyloidosis, *N Engl J Med 349*, 1872–1873.
10. Merlini, G. and Bellotti, V. (2003) Molecular mechanisms of amyloidosis, *N Engl J Med 349*, 583–596.
11. Nilsson, M. R. (2004) Techniques to study amyloid fibril formation in vitro, *Methods 34*, 151–160.
12. Makabe, K., McElheny, D., Tereshko, V., Hilyard, A., Gawlak, G., Yan, S., Koide, A., and Koide, S. (2006) Atomic structures of peptide self-assembly mimics, *Proc Natl Acad Sci U S A 103*, 17753–17758.
13. Benseny-Cases, N., Cocera, M., and Cladera, J. (2007) Conversion of non-fibrillar beta-sheet oligomers into amyloid fibrils in Alzheimer's disease amyloid peptide aggregation, *Biochem Biophys Res Commun 361*, 916–921.
14. Fandrich, M., Zandomeneghi, G., Krebs, M. R. H., Kittler, M., Buder, K., Rossner, A., Heinemann, S. H., Dobson, C. M., and Diekmann, S. (2006) Apomyoglobin reveals a random-nucleation mechanism in amyloid protofibril formation, *Acta Histochem 108*, 215–219.
15. Makin, O. S., Sikorski, P., and Serpell, L. C. (2006) Diffraction to study protein and peptide assemblies, *Curr Opin Chem Biol 10*, 417–422.
16. Tycko, R. (2006) Molecular structure of amyloid fibrils: Insights from solid-state NMR, *Q Rev Biophys 39*, 1–55.
17. Teplow, D. B., Lazo, N. D., Bitan, G., Bernstein, S., Wyttenbach, T., Bowers, M. T., Baumketner, A., Shea, J. E., Urbanc, B., Cruz, L., Borreguero, J., and Stanley, H. E.

(2006) Elucidating amyloid beta-protein folding and assembly: A multidisciplinary approach, *Acc Chem Res 39*, 635–645.

18 Paravastu, A. K., Leapman, R. D., Yau, W. M., and Tycko, R. (2008) Molecular structural basis for polymorphism in Alzheimer's beta-amyloid fibrils, *Proc Natl Acad Sci U S A 105*, 18349–18354.

19 Das, S., Nikolaidis, N., Klein, J., and Nei, M. (2008) Evolutionary redefinition of immunoglobulin light chain isotypes in tetrapods using molecular markers, *Proc Natl Acad Sci U S A 105*, 16647–16652.

20 Malle, E. and deBeer, F. C. (1996) Human serum amyloid A (SAA) protein: A prominent acute-phase reactant for clinical practice, *Eur J Clin Invest 26*, 427–435.

21 Saraiva, M. J. M. (2002) Hereditary transthyretin amyloidosis: Molecular basis and therapeutical strategies, *Expert Rev Mol Med 4*, 1–11.

22 Kiuru-Enari, S., Somer, H., Seppalainen, A. M., Notkola, I. L., and Haltia, M. (2002) Neuromuscular pathology in hereditary gelsolin amyloidosis, *J Neuropathol Exp Neurol 61*, 565–571.

23 Price, D. L., Sisodia, S. S., and Gandy, S. E. (1995) Amyloid-beta amyloidosis in Alzheimer's disease, *Curr Opin Neurol 8*, 268–274.

24 Mazanec, K., McClure, J., Bartley, C. J., Newbould, M. J., and Ackrill, P. (1992) Systemic amyloidosis of beta-2 microglobulin type, *J Clin Pathol 45*, 832–833.

25 Eulitz, M., Weiss, D. T., and Solomon, A. (1990) Immunoglobulin heavy-chain-associated amyloidosis, *Proc Natl Acad Sci U S A 87*, 6542–6546.

26 Prusiner, S. B. (1987) Prions and neurodegenerative diseases, *N Engl J Med 317*, 1571–1581.

27 Der-Sarkissian, A., Jao, C. C., Chen, J., and Langen, R. (2003) Structural organization of alpha-synuclein fibrils studied by site-directed spin labeling, *J Biol Chem 278*, 37530–37535.

28 Torok, M., Milton, S., Kayed, R., Wu, P., McIntire, T., Glabe, C. G., and Langen, R. (2002) Structural and dynamic features of Alzheimer's A beta peptide in amyloid fibrils studied by site-directed spin labeling, *J Biol Chem 277*, 40810–40815.

29 Dobson, C. M. (2004) Principles of protein folding, misfolding and aggregation, *Semin Cell Dev Biol 15*, 3–16.

30 Dobson, C. M. (2003) Protein folding and misfolding, *Nature 426*, 884–890.

31 Dobson, C. M. (2000) Protein folding, misfolding, and disease, *Abstr Pap Am Chem Soc 219*, U277–U277.

32 Ferreira, S. T., De Felice, F. G., and Chapeaurouge, A. (2006) Metastable, partially folded states in the productive folding and in the misfolding and amyloid aggregation of proteins, *Cell Biochem Biophys 44*, 539–548.

33 Bucciantini, M., Giannoni, E., Chiti, F., Baroni, F., Formigli, L., Zurdo, J. S., Taddei, N., Ramponi, G., Dobson, C. M., and Stefani, M. (2002) Inherent toxicity of aggregates implies a common mechanism for protein misfolding diseases, *Nature 416*, 507–511.

34 Hardy, J. and Selkoe, D. J. (2002) Medicine—The amyloid hypothesis of Alzheimer's disease: Progress and problems on the road to therapeutics, *Science 297*, 353–356.

35 Hardy, J. (2003) Alzheimer's disease: Genetic evidence points to a single pathogenesis, *Ann Neurol 54*, 143–144.

36 Roychaudhuri, R., Yang, M., Hoshi, M. M., and Teplow, D. B. (2009) Amyloid beta-protein assembly and Alzheimer disease, *J Biol Chem 284*, 4749–4753.

REFERENCES

37. Citron, M. (2002) Alzheimer's disease: Treatments in discovery and development, *Nat Neurosci 5*, 1055–1057.
38. Benzinger, T. L. S., Gregory, D. M., Burkoth, T. S., Miller-Auer, H., Lynn, D. G., Botto, R. E., and Meredith, S. C. (1998) Propagating structure of Alzheimer's beta-amyloid((10–35)) is parallel beta-sheet with residues in exact register, *Proc Natl Acad Sci U S A 95*, 13407–13412.
39. Tycko, R. (2003) Insights into the amyloid folding problem from solid-state NMR, *Biochemistry 42*, 3151–3159.
40. Hendriks, L., Vanduijn, C. M., Cras, P., Cruts, M., Vanhul, W., Vanharskamp, F., Warren, A., McInnis, M. G., Antonarakis, S. E., Martin, J. J., Hofman, A., and Vanbroeckhoven, C. (1992) Presenile-dementia and cerebral-hemorrhage linked to a mutation at codon-692 of the beta-amyloid precursor protein gene, *Nat Genet 1*, 218–221.
41. Grabowski, T. J., Cho, H. S., Vonsattel, J. P. G., Rebeck, G. W., and Greenberg, S. M. (2001) Novel amyloid precursor protein mutation in an Iowa family with dementia and severe cerebral amyloid angiopathy, *Ann Neurol 49*, 697–705.
42. Kamino, K., Orr, H. T., Payami, H., Wijsman, E. M., Alonso, M. E., Pulst, S. M., Anderson, L., Odahl, S., Nemens, E., White, J. A., Sadovnick, A. D., Ball, M. J., Kaye, J., Warren, A., McInnis, M., Antonarakis, S. E., Korenberg, J. R., Sharma, V., Kukull, W., Larson, E., Heston, L. L., Martin, G. M., Bird, T. D., and Schellenberg, G. D. (1992) Linkage and mutational analysis of familial Alzheimer's disease kindreds for the app gene region, *Am J Hum Genet 51*, 998–1014.
43. Nilsberth, C., Westlind-Danielsson, A., Eckman, C. B., Condron, M. M., Axelman, K., Forsell, C., Stenh, C., Luthman, J., Teplow, D. B., Younkin, S. G., Naslund, J., and Lannfelt, L. (2001) The "Arctic" APP mutation (E693G) causes Alzheimer's disease by enhanced A beta protofibril formation, *Nat Neurosci 4*, 887–893.
44. Levy, E., Carman, M. D., Fernandezmadrid, I. J., Power, M. D., Lieberburg, I., Vanduinen, S. G., Bots, G., Luyendijk, W., and Frangione, B. (1990) Mutation of the Alzheimer's disease amyloid gene in hereditary cerebral-hemorrhage, dutch type, *Science 248*, 1124–1126.
45. Hardy, J. and Allsop, D. (1991) Amyloid deposition as the central event in the etiology of Alzheimer's disease, *Trends Pharmacol Sci 12*, 383–388.
46. Pike, C. J., Walencewicz, A. J., Glabe, C. G., and Cotman, C. W. (1991) In vitro aging of beta-amyloid protein causes peptide aggregation and neurotoxicity, *Brain Res 563*, 311–314.
47. Hardy, J. A. and Higgins, G. A. (1992) Alzheimer's disease—The amyloid cascade hypothesis, *Science 256*, 184–185.
48. Kirkitadze, M. D., Bitan, G., and Teplow, D. B. (2002) Paradigm shifts in Alzheimer's disease and other neurodegenerative disorders: The emerging role of oligomeric assemblies, *J Neurosci Res 69*, 567–577.
49. Lambert, M. P., Barlow, A. K., Chromy, B. A., Edwards, C., Freed, R., Liosatos, M., Morgan, T. E., Rozovsky, I., Trommer, B., Viola, K. L., Wals, P., Zhang, C., Finch, C. E., Krafft, G. A., and Klein, W. L. (1998) Diffusible, nonfibrillar ligands derived from Abeta(1–42) are potent central nervous system neurotoxins, *Proc Natl Acad Sci U S A 95*, 6448–6453.
50. Walsh, D. M., Klyubin, I., Fadeeva, J. V., Cullen, W. K., Anwyl, R., Wolfe, M. S., Rowan, M. J., and Selkoe, D. J. (2002) Naturally secreted oligomers of amyloid beta

protein potently inhibit hippocampal long-term potentiation in vivo, *Nature 416*, 535–539.

51 Walsh, D. M., Klyubin, I., Shankar, G. M., Townsend, M., Fadeeva, J. V., Betts, V., Podlisny, M. B., Cleary, J. P., Ashe, K. H., Rowan, M. J., and Selkoe, D. J. (2005) The role of cell-derived oligomers of Abeta in Alzheimer's disease and avenues for therapeutic intervention, *Biochem Soc Trans 33*, 1087–1090.

52 Nimmrich, V., Grimm, C., Draguhn, A., Barghorn, S., Lehmann, A., Schoemaker, H., Hillen, H., Gross, G., Ebert, U., and Bruehl, C. (2008) Amyloid beta oligomers (Abeta(1-42) globulomer) suppress spontaneous synaptic activity by inhibition of P/Q-type calcium currents, *J Neurosci 28*, 788–797.

53 Bitan, G., Kirkitadze, M. D., Lomakin, A., Vollers, S. S., Benedek, G. B., and Teplow, D. B. (2003) Amyloid beta-protein (Abeta) assembly: Abeta 40 and Abeta 42 oligomerize through distinct pathways, *Proc Natl Acad Sci U S A 100*, 330–335.

54 Li, X. J., Orr, A. L., and Li, S. H. (2010) Impaired mitochondrial trafficking in Huntington's disease, *Biochim Biophys Acta 1802*, 62–65.

55 Walker, F. O. (2007) Huntington's disease, *Lancet 369*, 218–228.

56 Scherzinger, E., Sittler, A., Schweiger, K., Heiser, V., Lurz, R., Hasenbank, R., Bates, G. P., Lehrach, H., and Wanker, E. E. (1999) Self-assembly of polyglutamine-containing huntingtin fragments into amyloid-like fibrils: Implications for Huntington's disease pathology, *Proc Natl Acad Sci U S A 96*, 4604–4609.

57 Kuemmerle, S., Gutekunst, C. A., Klein, A. M., Li, X. J., Li, S. H., Beal, M. F., Hersch, S. M., and Ferrante, R. J. (1999) Huntingtin aggregates may not predict neuronal death in Huntington's disease, *Ann Neurol 46*, 842–849.

58 Becher, M. W., Kotzuk, J. A., Sharp, A. H., Davies, S. W., Bates, G. P., Price, D. L., and Ross, C. A. (1998) Intranuclear neuronal inclusions in Huntington's disease and dentatorubral and pallidoluysian atrophy: Correlation between the density of inclusions and IT15 CAG triplet repeat length, *Neurobiol Dis 4*, 387–397.

59 Prusiner, S. B. (2001) Shattuck lecture—Neurodegenerative diseases and prions., *N Engl J Med 344*, 1516–1526.

60 Hope, J. (2000) Prions and neurodegenerative diseases, *Curr Opin Genet Dev 10*, 568–574.

61 Garcia, A. D., Albert, E. M., and Sanmarti, L. S. (1998) Prions and transmissible neurodegenerative diseases, *Med Clin (Barc) 110*, 751–757.

62 Wickner, R. B., Masison, D. C., Edskes, H. K., and Maddelein, M. L. (1996) Prions of yeast, [PSI] and [URE3], as models for neurodegenerative diseases, *Cold Spring Harb Symp Quant Biol 61*, 541–550.

63 Prusiner, S. B. (1992) Molecular-biology and genetics of neurodegenerative diseases caused by prions, *Adv Virus Res 41*, 241–280.

64 Prusiner, S. B. (1993) Prions and neurodegenerative diseases, *J Acquir Immune Defic Syndr 6*, 742–742.

65 Prusiner, S. B. (1993) Prions and neurodegenerative diseases, *FASEB J 7*, A1263–A1263.

66 Wickner, R. B., Edskes, H. K., Roberts, B. T., Baxa, U., Pierce, M. M., Ross, E. D., and Brachmann, A. (2004) Prions: Proteins as genes and infectious entities, *Genes Dev 18*, 470–485.

67 Prusiner, S. B. (1998) Prions, *Proc Natl Acad Sci U S A 95*, 13363–13383.

REFERENCES

68. Forno, L. S. (1996) Neuropathology of Parkinson's disease, *J Neuropathol Exp Neurol* 55, 259–272.
69. Uversky, V. N. and Eliezer, D. (2009) Biophysics of Parkinson's disease: Structure and aggregation of alpha-synuclein, *Curr Protein Pept Sci 10*, 483–499.
70. Forno, L. S. (1986) Lewy bodies, *N Engl J Med 314*, 122–122.
71. Lashuel, H. A. (2009) The role of post-translational modification (phosphorylation) in modulating alpha-synuclein, structure, aggregation and toxicity, *J Neurochem 110*, 61–61.
72. Beyer, K. (2006) alpha-Synuclein structure, posttranslational modification and alternative splicing as aggregation enhancers, *Acta Neuropathol 112*, 237–251.
73. Bisaglia, M., Trolio, A., Bellanda, M., Bergantino, E., Bubacco, L., and Mammi, S. (2006) Structure and topology of the non-amyloid-beta component fragment of human alpha-synuclein bound to micelles: Implications for the aggregation process, *Protein Sci 15*, 1408–1416.
74. Uversky, V. N., Yamin, G., Munishkina, L. A., Karymov, M. A., Millett, I. S., Doniach, S., Lyubchenko, Y. L., and Fink, A. L. (2005) Effects of nitration on the structure and aggregation of alpha-synuclein, *Brain Res Mol Brain Res 134*, 84–102.
75. Cleveland, D. W. and Rothstein, J. D. (2001) From Charcot to Lou Gehrig: Deciphering selective motor neuron death in ALS, *Nat Rev Neurosci 2*, 806–819.
76. Rogelj, B. (2009) Tdp-43 mutations associated with amytrophic lateral sclerosis and insights into molecular mechanisms of disease, *J Neurochem 110*, 134–135.
77. Vasil'ev, A. V., Verkhovskaya, L. V., Shmarov, M. M., Tutykhina, I. L., Vorob'eva, A. A., Naroditskii, B. S., and Zakharova, M. N. (2008) The role of vascular endothelial growth factor in the progression of amytrophic lateral sclerosis, *Neurochem J 2*, 297–300.
78. Tortarolo, M., Grignaschi, G., Calvaresi, N., Zennaro, E., Spaltro, G., Colovic, M., Fracasso, C., Guiso, G., Elger, B., Schneider, H., Seilheimer, B., Caccia, S., and Bendotti, C. (2006) Glutamate AMPA receptors change in motor neurons of SOD1(G93A) transgenic mice and their inhibition by a noncompetitive antagonist ameliorates the progression of amytrophic lateral sclerosis-like disease, *J Neurosci Res 83*, 134–146.
79. Aisen, M. L., Sevilla, D., Edelstein, L., and Blass, J. (1996) A double-blind placebo-controlled study of 3,4-diaminopyridine in amyotrophic lateral sclerosis patients on a rehabilitation unit, *J Neurol Sci 138*, 93–96.
80. Unger, M., Wettergren, A., and Clausen, J. (1985) Characterization of DNA and RNA in circulating immunocomplexes in multiple-sclerosis, amytrophic lateral sclerosis and normal controls, *Acta Neurol Scand 72*, 392–396.
81. Bruijn, L. I., Houseweart, M. K., Kato, S., Anderson, K. L., Anderson, S. D., Ohama, E., Reaume, A. G., Scott, R. W., and Cleveland, D. W. (1998) Aggregation and motor neuron toxicity of an ALS-linked SOD1 mutant independent from wild-type SOD1, *Science 281*, 1851–1854.
82. Booth, D. R., Sunde, M., Bellotti, V., Robinson, C. V., Hutchinson, W. L., Fraser, P. E., Hawkins, P. N., Dobson, C. M., Radford, S. E., Blake, C. C. F., and Pepys, M. B. (1997) Instability, unfolding and aggregation of human lysozyme variants underlying amyloid fibrillogenesis, *Nature 385*, 787–793.
83. Ramirez-Alvarado, M., Merkel, J. S., and Regan, L. (2000) A systematic exploration of the influence of the protein stability on amyloid fibril formation in vitro, *Proc Natl Acad Sci U S A 97*, 8979–8984.

84. Kelly, J. W. (1996) Alternative conformations of amyloidogenic proteins govern their behavior, *Curr Opin Struct Biol 6*, 11–17.
85. Ikeda, S. and Morris, V. J. (2002) Fine-stranded and particulate aggregates of heat-denatured whey proteins visualized by atomic force microscopy, *Biomacromolecules 3*, 382–389.
86. Lashuel, H. A. and Wall, J. S. (2005) Molecular electron microscopy approaches to elucidating the mechanisms of protein fibrillogenesis, *Methods Mol Biol 299*, 81–101.
87. Reddy, G., Straubb, J. E., and Thirumalai, D. (2009) Dynamics of locking of peptides onto growing amyloid fibrils, *Proc Natl Acad Sci U S A 106*, 11948–11953.
88. Caughey, B. and Lansbury, P. T. (2003) Protofibrils, pores, fibrils, and neurodegeneration: Separating the responsible protein aggregates from the innocent bystanders, *Annu Rev Neurosci 26*, 267–298.
89. Oda, T., Wals, P., Osterburg, H. H., Johnson, S. A., Pasinetti, G. M., Morgan, T. E., Rozovsky, I., Stine, W. B., Snyder, S. W., Holzman, T. F., Krafft, G. A., and Finch, C. E. (1995) Clusterin (apoj) alters the aggregation of amyloid beta-peptide (α-beta(1–42)) and forms slowly sedimenting α-beta complexes that cause oxidative stress, *Exp Neurol 136*, 22–31.
90. Wang, J., Dickson, D. W., Trojanowski, J. Q., and Lee, V. M. Y. (1999) The levels of soluble versus insoluble brain Abeta distinguish Alzheimer's disease from normal and pathologic aging, *Exp Neurol 158*, 328–337.
91. Klein, W. L., Krafft, G. A., and Finch, C. E. (2001) Targeting small Abeta oligomers: The solution to an Alzheimer's disease conundrum? *Trends Neurosci 24*, 219–224.
92. Klein, W. L., Stine, W. B., and Teplow, D. B. (2004) Small assemblies of unmodified amyloid beta-protein are the proximate neurotoxin in Alzheimer's disease, *Neurobiol Aging 25*, 569–580.
93. Bitan, G., Vollers, S. S., and Teplow, D. B. (2003) Elucidation of primary structure elements controlling early amyloid beta-protein oligomerization, *J Biol Chem 278*, 34882–34889.
94. Gnanakaran, S., Nussinov, R., and Garcia, A. E. (2006) Atomic-level description of amyloid beta-dimer formation, *J Am Chem Soc 128*, 2158–2159.
95. Lee, J. P., Stimson, E. R., Ghilardi, J. R., Mantyh, P. W., Lu, Y. A., Felix, A. M., Llanos, W., Behbin, A., Cummings, M., Vancriekinge, M., Timms, W., and Maggio, J. E. (1995) H-1-nmr of A-beta amyloid peptide congeners in water solution—Conformational-changes correlate with plaque competence, *Biochemistry 34*, 5191–5200.
96. Pike, C. J., Walencewiczwasserman, A. J., Kosmoski, J., Cribbs, D. H., Glabe, C. G., and Cotman, C. W. (1995) Structure-activity analyses of beta-amyloid peptides—Contributions of the beta-25-35 region to aggregation and neurotoxicity, *J Neurochem 64*, 253–265.
97. Teplow, D. B. (1998) Structural and kinetic features of amyloid beta-protein fibrillogenesis, *Amyloid 5*, 121–142.
98. Pan, K. M., Baldwin, M., Nguyen, J., Gasset, M., Serban, A., Groth, D., Mehlhorn, I., Huang, Z. W., Fletterick, R. J., Cohen, F. E., and Prusiner, S. B. (1993) Conversion of alpha-helices into beta-sheets features in the formation of the scrapie prion proteins, *Proc Natl Acad Sci U S A 90*, 10962–10966.
99. Conway, K. A., Harper, J. D., and Lansbury, P. T. (1998) Accelerated in vitro fibril formation by a mutant alpha-synuclein linked to early-onset Parkinson disease, *Nat Med 4*, 1318–1320.

REFERENCES

100. Zanuy, D., Ma, B. Y., and Nussinov, R. (2003) Short peptide amyloid organization: Stabilities and conformations of the islet amyloid peptide NFGAIL, *Biophys J 84*, 1884–1894.

101. Zhang, S., Iwata, K., Lachenmann, M. J., Peng, J. W., Li, S., Stimson, E. R., Lu, Y., Felix, A. M., Maggio, J. E., and Lee, J. P. (2000) The Alzheimer's peptide Abeta adopts a collapsed coil structure in water, *J Struct Biol 130*, 130–141.

102. Straub, J. E., Guevara, J., Huo, S. H., and Lee, J. P. (2002) Long time dynamic simulations: Exploring the folding pathways of an Alzheimer's amyloid Abeta-peptide, *Acc Chem Res 35*, 473–481.

103. Massi, F., Peng, J. W., Lee, J. P., and Straub, J. E. (2001) Simulation study of the structure and dynamics of the Alzheimer's amyloid peptide congener in solution, *Biophys J 80*, 31–44.

104. Murakami, K., Irie, K., Morimoto, A., Ohigashi, H., Shindo, M., Nagao, M., Shimizu, T., and Shirasawa, T. (2003) Neurotoxicity and physicochemical properties of A beta mutant peptides from cerebral amyloid angiopathy—Implication for the pathogenesis of cerebral amyloid angiopathy and Alzheimer's disease, *J Biol Chem 278*, 46179–46187.

105. Massi, F. and Straub, J. E. (2001) Probing the origins of increased activity of the E22Q "Dutch" mutant Alzheimer's beta-amyloid peptide, *Biophys J 81*, 697–709.

106. Massi, F. and Straub, J. E. (2003) Structural and dynamical analysis of the hydration of the Alzheimer's beta-amyloid peptide, *J Comput Chem 24*, 143–153.

107. Watson, D. J., Lander, A. D., and Selkoe, D. J. (1997) Heparin-binding properties of the amyloidogenic peptides Abeta and amylin—Dependence on aggregation state and inhibition by Congo red, *J Biol Chem 272*, 31617–31624.

108. Massi, F., Klimov, D., Thirumalai, D., and Straub, J. E. (2002) Charge states rather than propensity for beta-structure determine enhanced fibrillogenesis in wild-type Alzheimer's beta-amyloid peptide compared to E22Q Dutch mutant, *Protein Sci 11*, 1639–1647.

109. Han, W. and Wu, Y. D. (2005) A strand-loop-strand structure is a possible intermediate in fibril elongation: Long time simulations of amylold-beta peptide (10–35), *J Am Chem Soc 127*, 15408–15416.

110. Baumketner, A. and Shea, J. E. (2007) The structure of the Alzheimer amyloid beta 10–35 peptide probed through replica-exchange molecular dynamics simulations in explicit solvent, *J Mol Biol 366*, 275–285.

111. Tarus, B., Straub, J. E., and Thirumalai, D. (2006) Dynamics of Asp23-Lys28 salt-bridge formation in Abeta(10-35) monomers, *J Am Chem Soc 128*, 16159–16168.

112. Lazo, N. D., Grant, M. A., Condron, M. C., Rigby, A. C., and Teplow, D. B. (2005) On the nucleation of amyloid beta-protein monomer folding, *Protein Sci 14*, 1581–1596.

113. Borreguero, J. M., Urbanc, B., Lazo, N. D., Buldyrev, S. V., Teplow, D. B., and Stanley, H. E. (2005) Folding events in the 21–30 region of amyloid-beta-protein (A beta) studied in silico, *Proc Natl Acad Sci U S A 102*, 6015–6020.

114. Cruz, L., Urbanc, B., Borreguero, J. M., Lazo, N. D., Teplow, D. B., and Stanley, H. E. (2005) Solvent and mutation effects on the nucleation of amyloid beta-protein folding, *Proc Natl Acad Sci U S A 102*, 18258–18263.

115. Chen, W., Mousseau, N., and Derreumaux, P. (2006) The conformations of the amyloid-beta (21–30) fragment can be described by three families in solution, *J Chem Phys 125*, 084911.

116 Baumketner, A., Bernstein, S. L., Wyttenbach, T., Lazo, N. D., Teplow, D. B., Bowers, M. T., and Shea, J. E. (2006) Structure of the 21–30 fragment of amyloid beta-protein, *Protein Sci 15*, 1239–1247.

117 Krone, M. G., Baumketner, A., Bernstein, S. L., Wyttenbach, T., Lazo, N. D., Teplow, D. B., Bowers, M. T., and Shea, J. E. (2008) Effects of familial Alzheimer's disease mutations on the folding nucleation of the amyloid beta-protein, *J Mol Biol 381*, 221–228.

118 Murray, M. M., Krone, M. G., Bernstein, S. L., Baumketner, A., Condron, M. M., Lazo, N. D., Teplow, D. B., Wyttenbach, T., Shea, J. E., and Bowers, M. T. (2009) Amyloid beta-protein: Experiment and theory on the 21–30 fragment, *J Phys Chem B 113*, 6041–6046.

119 Fawzi, N. L., Phillips, A. H., Ruscio, J. Z., Doucleff, M., Wemmer, D. E., and Head-Gordon, T. (2008) Structure and dynamics of the Ass(21–30) peptide from the interplay of NMR experiments and molecular simulations, *J Am Chem Soc 130*, 6145–6158.

120 Balbach, J. J., Ishii, Y., Antzutkin, O. N., Leapman, R. D., Rizzo, N. W., Dyda, F., Reed, J., and Tycko, R. (2000) Amyloid fibril formation by Abeta(16–22), a seven-residue fragment of the Alzheimer's beta-amyloid peptide, and structural characterization by solid state NMR, *Biochemistry 39*, 13748–13759.

121 Klimov, D. K., Straub, J. E., and Thirumalai, D. (2004) Aqueous urea solution destabilizes Abeta(16–22) oligomers, *Proc Natl Acad Sci U S A 101*, 14760–14765.

122 Daidone, I., Simona, F., Roccatano, D., Broglia, R. A., Tiana, G., Colombo, G., and Di Nola, A. (2004) beta-Hairpin conformation of fibrillogenic peptides: Structure and alpha-beta transition mechanism revealed by molecular dynamics simulations, *Proteins 57*, 198–204.

123 Baumketner, A., Krone, M. G., and Shea, J. E. (2008) Role of the familial Dutch mutation E22Q in the folding and aggregation of the 15–28 fragment of the Alzheimer amyloid-beta protein, *Proc Natl Acad Sci U S A 105*, 6027–6032.

124 Baumketner, A. and Shea, J. E. (2006) Folding landscapes of the Alzheimer amyloid-beta(12–28) peptide, *J Mol Biol 362*, 567–579.

125 Wei, G. H. and Shea, J. E. (2006) Effects of solvent on the structure of the Alzheimer amyloid-beta(25–35) peptide, *Biophys J 91*, 1638–1647.

126 Fradinger, E. A., Monien, B. H., Urbanc, B., Lomakin, A., Tan, M., Li, H., Spring, S. M., Condron, M. M., Cruz, L., Xie, C. W., Benedek, G. B., and Bitan, G. (2008) C-terminal peptides coassemble into Abeta 42 oligomers and protect neurons against Abeta 42-induced neurotoxicity, *Proc Natl Acad Sci U S A 105*, 14175–14180.

127 Wu, C., Murray, M. M., Bernstein, S. L., Condron, M. M., Bitan, G., Shea, J. E., and Bowers, M. T. (2009) The structure of Abeta 42 C-terminal fragments probed by a combined experimental and theoretical study, *J Mol Biol 387*, 492–501.

128 Yamamoto, N. and Kuwata, K. (2009) Regulating the conformation of prion protein through ligand binding, *J Phys Chem B 113*, 12853–12856.

129 Chebaro, Y. and Derreumaux, P. (2009) The conversion of helix H2 to beta-sheet is accelerated in the monomer and dimer of the prion protein upon T183A mutation, *J Phys Chem B 113*, 6942–6948.

130 Sticht, H., Bayer, P., Willbold, D., Dames, S., Hilbich, C., Beyreuther, K., Frank, R. W., and Rosch, P. (1995) Structure of amyloid A4-(1–40)-peptide of Alzheimer's disease, *Eur J Biochem 233*, 293–298.

131. Coles, M., Bicknell, W., Watson, A. A., Fairlie, D. P., and Craik, D. J. (1998) Solution structure of amyloid beta-peptide(1–40) in a water-micelle environment. Is the membrane-spanning domain where we think it is? *Biochemistry 37*, 11064–11077.
132. Shao, H. Y., Jao, S. C., Ma, K., and Zagorski, M. G. (1999) Solution structures of micelle-bound amyloid beta-(1–40) and beta-(1–42) peptides of Alzheimer's disease, *J Mol Biol 285*, 755–773.
133. Crescenzi, O., Tomaselli, S., Guerrini, R., Salvadori, S., D'Ursi, A. M., Temussi, P. A., and Picone, D. (2002) Solution structure of the Alzheimer amyloid beta-peptide (1–42) in an apolar microenvironment—Similarity with a virus fusion domain, *Eur J Biochem 269*, 5642–5648.
134. Barrow, C. J., Yasuda, A., Kenny, P. T. M., and Zagorski, M. G. (1992) Solution conformations and aggregational properties of synthetic amyloid beta-peptides of Alzheimer's disease—Analysis of circular-dichroism spectra, *J Mol Biol 225*, 1075–1093.
135. Gursky, O. and Aleshkov, S. (2000) Temperature-dependent beta-sheet formation in beta-amyloid Abeta(1–40) peptide in water: Uncoupling beta-structure folding from aggregation, *Biochim Biophys Acta 1476*, 93–102.
136. Petkova, A. T., Ishii, Y., Balbach, J. J., Antzutkin, O. N., Leapman, R. D., Delaglio, F., and Tycko, R. (2002) A structural model for Alzheimer's beta-amyloid fibrils based on experimental constraints from solid state NMR, *Proc Natl Acad Sci U S A 99*, 16742–16747.
137. Luhrs, T., Ritter, C., Adrian, M., Riek-Loher, D., Bohrmann, B., Doeli, H., Schubert, D., and Riek, R. (2005) 3D structure of Alzheimer's amyloid-beta(1–42) fibrils, *Proc Natl Acad Sci U S A 102*, 17342–17347.
138. Petkova, A. T., Leapman, R. D., Guo, Z. H., Yau, W. M., Mattson, M. P., and Tycko, R. (2005) Self-propagating, molecular-level polymorphism in Alzheimer's beta-amyloid fibrils, *Science 307*, 262–265.
139. Paravastu, A. K., Qahwash, I., Leapman, R. D., Meredith, S. C., and Tycko, R. (2009) Seeded growth of beta-amyloid fibrils from Alzheimer's brain-derived fibrils produces a distinct fibril structure, *Proc Natl Acad Sci U S A 106*, 7443–7448.
140. Urbanc, B., Cruz, L., Yun, S., Buldyrev, S. V., Bitan, G., Teplow, D. B., and Stanley, H. E. (2004) In silico study of amyloid beta-protein folding and oligomerization, *Proc Natl Acad Sci U S A 101*, 17345–17350.
141. Murakami, K., Irie, K., Ohigashi, H., Hara, H., Nagao, M., Shimizu, T., and Shirasawa, T. (2005) Formation and stabilization model of the 42-mer Abeta radical: Implications for the long-lasting oxidative stress in Alzheimer's disease, *J Am Chem Soc 127*, 15168–15174.
142. Yan, Y. L. and Wang, C. Y. (2006) Abeta 42 is more rigid than Abeta 40 at the C terminus: Implications for Abeta aggregation and toxicity, *J Mol Biol 364*, 853–862.
143. Sgourakis, N. G., Yan, Y. L., McCallum, S. A., Wang, C. Y., and Garcia, A. E. (2007) The Alzheimer's peptides Abeta 40 and 42 adopt distinct conformations in water: A combined MD/NMR study, *J Mol Biol 368*, 1448–1457.
144. Xu, Y. C., Shen, J. J., Luo, X. M., Zhu, W. L., Chen, K. X., Ma, J. P., and Jiang, H. L. (2005) Conformational transition of amyloid beta-peptide, *Proc Natl Acad Sci U S A 102*, 5403–5407.
145. Baumketner, A., Bernstein, S. L., Wyttenbach, T., Bitan, G., Teplow, D. B., Bowers, M. T., and Shea, J. E. (2006) Amyloid beta-protein monomer structure: A computational and experimental study, *Protein Sci 15*, 420–428.

146 Shen, L., Ji, H. F., and Zhang, H. Y. (2008) Why is the C-terminus of Abeta(1–42) more unfolded than that of Abeta(1–40)? Clues from hydrophobic interaction, *J Phys Chem B 112*, 3164–3167.

147 Flock, D., Colacino, S., Colombo, G., and Di Nola, A. (2006) Misfolding of the amyloid beta-protein: A molecular dynamics study, *Proteins 62*, 183–192.

148 Yang, M. F. and Teplow, D. B. (2008) Amyloid beta-protein monomer folding: Free-energy surfaces reveal alloform-specific differences, *J Mol Biol 384*, 450–464.

149 Mitternacht, S., Staneva, I., Hard, T., and Irback, A. (2010) Comparing the folding free-energy landscapes of Abeta 42 variants with different aggregation properties, *Proteins 78*, 2600–2608.

150 Anand, P., Nandel, F. S., and Hansmann, U. H. E. (2008) The Alzheimer's beta amyloid (Abeta(1–39)) monomer in an implicit solvent, *J Chem Phys 128*, 165102.

151 Lam, A. R., Teplow, D. B., Stanley, H. E., and Urbanc, B. (2008) Effects of the arctic (E-22→G) mutation on amyloid beta-protein folding: Discrete molecular dynamics study, *J Am Chem Soc 130*, 17413–17422.

152 Lim, K. H., Collver, H. H., Le, Y. T. H., Nagchowdhuri, P., and Kenney, J. M. (2007) Characterizations of distinct amyloidogenic conformations of the Abeta (1–40) and (1–42) peptides, *Biochem Biophys Res Commun 353*, 443–449.

153 Lemkul, J. A. and Bevan, D. R. (2008) A comparative molecular dynamics analysis of the amyloid beta-peptide in a lipid bilayer, *Arch Biochem Biophys 470*, 54–63.

154 Reddy, A. S., Izmitli, A., and de Pablo, J. J. (2009) Effect of trehalose on amyloid beta (29–40)-membrane interaction, *J Chem Phys 131*, 085101.

155 Davis, C. H. and Berkowitz, M. L. (2009) Structure of the amyloid-beta (1–42) monomer absorbed to model phospholipid bilayers: A molecular dynamics study, *J Phys Chem B 113*, 14480–14486.

156 Davis, C. H. and Berkowitz, M. L. (2009) Interaction between amyloid-beta (1–42) peptide and phospholipid bilayers: A molecular dynamics study, *Biophys J 96*, 785–797.

157 Urbanc, B., Borreguero, J. M., Cruz, L., and Stanley, H. E. (2006) Ab initio discrete molecular dynamics approach to protein folding and aggregation. In: Kheterpal, L. and Wetzel, R. (eds.), *Amyloid, prions, and other protein aggregates, Pt B*. Elsevier Academic Press Inc, San Diego, pp. 314–338.

158 Sunde, M., Serpell, L. C., Bartlam, M., Fraser, P. E., Pepys, M. B., and Blake, C. C. F. (1997) Common core structure of amyloid fibrils by synchrotron X-ray diffraction, *J Mol Biol 273*, 729–739.

159 Nelson, R., Sawaya, M. R., Balbirnie, M., Madsen, A. O., Riekel, C., Grothe, R., and Eisenberg, D. (2005) Structure of the cross-beta spine of amyloid-like fibrils, *Nature 435*, 773–778.

160 Sawaya, M. R., Sambashivan, S., Nelson, R., Ivanova, M. I., Sievers, S. A., Apostol, M. I., Thompson, M. J., Balbirnie, M., Wiltzius, J. J. W., McFarlane, H. T., Madsen, A. O., Riekel, C., and Eisenberg, D. (2007) Atomic structures of amyloid cross-beta spines reveal varied steric zippers, *Nature 447*, 453–457.

161 Kayed, R., Head, E., Thompson, J. L., McIntire, T. M., Milton, S. C., Cotman, C. W., and Glabe, C. G. (2003) Common structure of soluble amyloid oligomers implies common mechanism of pathogenesis, *Science 300*, 486–489.

162 Thirumalai, D., Klimov, D. K., and Dima, R. I. (2003) Emerging ideas on the molecular basis of protein and peptide aggregation, *Curr Opin Struct Biol 13*, 146–159.

163　Ma, B. Y. and Nussinov, R. (2002) Stabilities and conformations of Alzheimer's beta-amyloid peptide oligomers (Abeta(16–22') Abeta(16–35') and Abeta(10–35)): Sequence effects, *Proc Natl Acad Sci U S A 99*, 14126–14131.

164　Klimov, D. K. and Thirumalai, D. (2003) Dissecting the assembly of Abeta(16–22) amyloid peptides into antiparallel beta sheets, *Structure 11*, 295–307.

165　Hwang, W., Zhang, S. G., Kamm, R. D., and Karplus, M. (2004) Kinetic control of dimer structure formation in amyloid fibrillogenesis, *Proc Natl Acad Sci U S A 101*, 12916–12921.

166　Liu, F. F., Ji, L., Dong, X. Y., and Sun, Y. (2009) Molecular insight into the inhibition effect of trehalose on the nucleation and elongation of amyloid beta-peptide oligomers, *J Phys Chem B 113*, 11320–11329.

167　Han, W. and Wu, Y. D. (2007) Molecular dynamics studies of hexamers of amyloid-beta peptide (16–35) and its mutants: Influence of charge states on amyloid formation, *Proteins 66*, 575–587.

168　Sciarretta, K. L., Gordon, D. J., Petkova, A. T., Tycko, R., and Meredith, S. C. (2005) Abeta 40-Lactam(D23/K28) models a conformation highly favorable for nucleation of amyloid, *Biochemistry 44*, 6003–6014.

169　Reddy, G., Straub, J. E., and Thirumalai, D. (2009) Influence of preformed Asp23-Lys28 salt bridge on the conformational fluctuations of monomers and dimers of A beta peptides with implications for rates of fibril formation, *J Phys Chem B 113*, 1162–1172.

170　Horn, A. H. C. and Sticht, H. (2010) Amyloid-beta 42 oligomer structures from fibrils: A systematic molecular dynamics study, *J Phys Chem B 114*, 2219–2226.

171　Lemkul, J. A. and Bevan, D. R. (2010) Assessing the stability of Alzheimer's amyloid protofibrils using molecular dynamics, *J Phys Chem B 114*, 1652–1660.

172　Takeda, T. and Klimov, D. K. (2009) Side chain interactions can impede amyloid fibril growth: Replica exchange simulations of Abeta peptide mutant, *J Phys Chem B 113*, 11848–11857.

173　Chen, H. F. (2009) Aggregation mechanism investigation of the GIFQINS cross-beta amyloid fibril, *Comput Biol Chem 33*, 41–45.

174　Wei, G. H., Jewett, A. I., and Shea, J. E. (2010) Structural diversity of dimers of the Alzheimer amyloid-beta(25–35) peptide and polymorphism of the resulting fibrils, *Phys Chem Chem Phys 12*, 3622–3629.

175　Fu, Z. M., Luo, Y., Derreumaux, P., and Wei, G. H. (2009) Induced beta-barrel formation of the Alzheimer's Abeta 25–35 oligomers on carbon nanotube surfaces: Implication for amyloid fibril inhibition, *Biophys J 97*, 1795–1803.

176　Wang, Q. M., Zhao, C., Zhao, J., Wang, J. D., Yang, J. C., Yu, X., and Zhen, J. (2010) Comparative molecular dynamics study of Abeta adsorption on the self-assembled monolayers, *Langmuir 26*, 3308–3316.

177　Balbirnie, M., Grothe, R., and Eisenberg, D. S. (2001) An amyloid-forming peptide from the yeast prion Sup35 reveals a dehydrated beta-sheet structure for amyloid, *Proc Natl Acad Sci U S A 98*, 2375–2380.

178　Gsponer, J., Haberthur, U., and Caflisch, A. (2003) The role of side-chain interactions in the early steps of aggregation: Molecular dynamics simulations of an amyloid-forming peptide from the yeast prion Sup35, *Proc Natl Acad Sci U S A 100*, 5154–5159.

179　Strodel, B., Whittleston, C. S., and Wales, D. J. (2007) Thermodynamics and kinetics of aggregation for the GNNQQNY peptide, *J Am Chem Soc 129*, 16005–16014.

180 Wang, J., Tan, C. H., Chen, H. F., and Luo, R. (2008) All-atom computer simulations of amyloid fibrils disaggregation, *Biophys J 95*, 5037–5047.

181 De Simone, A., Pedone, C., and Vitagliano, L. (2008) Structure, dynamics, and stability of assemblies of the human prion fragment SNQNNF, *Biochem Biophys Res Commun 366*, 800–806.

182 Xu, W. X., Ping, J., Li, W. F., and Mu, Y. G. (2009) Assembly dynamics of two-beta sheets revealed by molecular dynamics simulations, *J Chem Phys 130*, 164709.

183 Rossetti, G., Magistrato, A., Pastore, A., Persichetti, F., and Carloni, P. (2008) Structural properties of polyglutamine aggregates investigated via molecular dynamics simulations, *J Phys Chem B 112*, 16843–16850.

184 Kammerer, R. A., Kostrewa, D., Zurdo, J., Detken, A., Garcia-Echeverria, C., Green, J. D., Muller, S. A., Meier, B. H., Winkler, F. K., Dobson, C. M., and Steinmetz, M. O. (2004) Exploring amyloid formation by a de novo design, *Proc Natl Acad Sci U S A 101*, 4435–4440.

185 Strodel, B., Fitzpatrick, A. W., Vendruscolo, M., Dobson, C. M., and Wales, D. J. (2008) Characterizing the first steps of amyloid formation for the cc beta peptide, *J Phys Chem B 112*, 9998–10004.

186 Bellesia, G. and Shea, J. E. (2009) Effect of beta-sheet propensity on peptide aggregation, *J Chem Phys 130*, 145103.

187 Yu, X., Wang, J. D., Yang, J. C., Wang, Q. M., Cheng, S. Z. D., Nussinov, R., and Zheng, J. (2010) Atomic-scale simulations confirm that soluble beta-sheet-rich peptide self-assemblies provide amyloid mimics presenting similar conformational properties, *Biophys J 98*, 27–36.

188 Hoang, T. X., Trovato, A., Seno, F., Banavar, J. R., and Maritan, A. (2004) Geometry and symmetry presculpt the free-energy landscape of proteins, *Proc Natl Acad Sci U S A 101*, 7960–7964.

189 Dobson, C. M. and Karplus, M. (1999) The fundamentals of protein folding: Bringing together theory and experiment, *Curr Opin Struct Biol 9*, 92–101.

190 Auer, S., Dobson, C. M., and Vendruscolo, M. (2007) Characterization of the nucleation barriers for protein aggregation and amyloid formation, *HFSP J 1*, 137–146.

191 Auer, S., Meersman, F., Dobson, C. M., and Vendruscolo, M. (2008) A generic mechanism of emergence of amyloid protofilaments from disordered oligomeric aggregates, *PLoS Comput Biol 4* (Issue 11), e1000222.

192 Auer, S. and Frenkel, D. (2004) Quantitative prediction of crystal-nucleation rates for spherical colloids: A computational approach, *Annu Rev Phys Chem 55*, 333–361.

193 Smith, S. W., Hall, C. K., and Freeman, B. D. (1997) Molecular dynamics for polymeric fluids using discontinuous potentials, *J Comput Phys 134*, 16–30.

194 Dokholyan, N. V., Buldyrev, S. V., Stanley, H. E., and Shakhnovich, E. I. (1998) Discrete molecular dynamics studies of the folding of a protein-like model, *Fold Des 3*, 577–587.

195 Dokholyan, N. V., Buldyrev, S. V., Stanley, H. E., and Shakhnovich, E. I. (2000) Identifying the protein folding nucleus using molecular dynamics, *J Mol Biol 296*, 1183–1188.

196 Smith, A. V. and Hall, C. K. (2001) alpha-Helix formation: Discontinuous molecular dynamics on an intermediate-resolution protein model, *Proteins 44*, 344–360.

REFERENCES

197 Peng, S., Ding, F., Urbanc, B., Buldyrev, S. V., Cruz, L., Stanley, H. E., and Dokholyan, N. V. (2004) Discrete molecular dynamics simulations of peptide aggregation, *Phys Rev E 69*, 7.

198 Urbanc, B., Cruz, L., Ding, F., Sammond, D., Khare, S., Buldyrev, S. V., Stanley, H. E., and Dokholyan, N. V. (2004) Molecular dynamics simulation of amyloid beta dimer formation, *Biophys J 87*, 2310–2321.

199 Yun, S. J., Urbanc, B., Cruz, L., Bitan, G., Teplow, D. B., and Stanley, H. E. (2007) Role of electrostatic interactions in amyloid beta-protein (Abeta) oligomer formation: A discrete molecular dynamics study, *Biophys J 92*, 4064–4077.

200 Urbanc, B., Betnel, M., Cruz, L., Bitan, G., and Teplow, D. B. (2010) Elucidation of amyloid beta-protein oligomerization mechanisms: Discrete molecular dynamics study, *J Am Chem Soc 132*, 4266–4280.

201 Li, H. Y., Monien, B. H., Fradinger, E. A., Urbanc, B., and Bitan, G. (2010) Biophysical characterization of A beta 42 C-terminal fragments: Inhibitors of A beta 42 neurotoxicity, *Biochemistry 49*, 1259–1267.

202 Li, H. Y., Monien, B. H., Lomakin, A., Zemel, R., Fradinger, E. A., Tan, M. A., Spring, S. M., Urbanc, B., Xie, C. W., Benedek, G. B., and Bitan, G. (2010) Mechanistic investigation of the inhibition of Abeta 42 assembly and neurotoxicity by Abeta 42 C-terminal fragments, *Biochemistry 49*, 6358–6364.

203 Urbanc, B., Betnel, M., Cruz, L., Li, H., Fradinger, E. A., Monien, B. H., and Bitan, G. (2011) Structural basis of $A\beta_{1-42}$ toxicity inhibition by $A\beta$ C-terminal fragments: Discrete molecular dynamics study, *J. Mol. Biol 410*, 316–328.

204 Chang, W. L. E., Takeda, T., Raman, E. P., and Klimov, D. K. (2010) Molecular dynamics simulations of anti-aggregation effect of ibuprofen, *Biophys J 98*, 2662–2670.

205 Takeda, T., Chang, W. L. E., Raman, E. P., and Klimov, D. K. (2010) Binding of non-steroidal anti-inflammatory drugs to Abeta fibril, *Proteins 78*, 2849–2860.

INDEX

Note: Page numbers in **bold** refer to figures.

1SS-α–lac 449
1,2-glucosidase inhibitor miglustat 462, 474
1,2-naphthoquione 458
14-3-3 protein(s)
  domain structures 257, 261
  family 256
  interactions 256
  isoforms/isomers α, β, γ, δ, ε, η, σ, τ, and ζ 256, 262
2,2-dimethyl-2-silapentane-5-sulfonic acid (DSS) 225
2,3-diazabicyclo(2.2.2)oct-2-ene (DBO) 101
2,3,4-trihydroxybenzophenone (THBP) 456, 457
3_{10} helix 90, 135, 137
30S ribosome 21
4-hydroxyindole 458
4-phenylbutyric acid (PBA) 445
4-(2-aminoethyl)benzenesulfonyl fluoride (AEBSF) 224
4E binding protein 1 (4EBP1) 243–244
5-hydroxytryptamine receptor agonists and antagonists 458
5,6-dihydroxyindole 458
53BP1 259, 260
53BP2 259, 260 8-anilinonaphthalene-1-sulfonate, ANS 9

α-aminoisobutyric acid (Aib) 459
α-D-mannosylglyceramide 448
α-D-mannosylglycerate 448
α-helical conformations 24, 210–211, 227, 230, 245, 264, 289, 314, 323, 326, 336, 366, 375–378, 451, 485, 488, 492, 497, 507, 509
α-helical frequencies 92, 176
α-helices 14, 19, 20, 24, 61–63, 69, 92, 132, 141, 160, 164, 170, 175–177, 191, 210–211, 213, 227, 245–246, 230, 264, 289, 313, 314, 323, 326, 357–358, 366, 372–373, 375–378, 451, 461, 485, 488, 492, 495–497, 500, 507
  forming molecular recognition features (α-MoRFs) 20, 23, 36–37, 244
  -like polypeptide stretches 357, 378
  propensity 170, 175–6, 244, 495
α-PrP 449
α-S2C casein (ACas) 354
α-syn oligomerization 464
α-synuclein (α-syn) xv, 5, 12–14, 18, 19, 23, 34, 144, 264, 449, 454, 456–9, 464–465, 482, 485
  fibrillization 449, 456, 458, 464, 482
α-tocopherol 456
$α_1$-antitrypsin deficiency 461

β-alanine 445, 446
β basin preferences for amino acids 92
β-branched amino acids 132, 141, 162, 179
β-carotene 456
β-conformation 397
β-dimethylsulfoniopropionate 445, 446

Protein and Peptide Folding, Misfolding, and Non-Folding, First Edition. Edited by Reinhard Schweitzer-Stenner.
© 2012 John Wiley & Sons, Inc. Published 2012 by John Wiley & Sons, Inc.

β-hairpin 24, 245, 323, 326, 391, 402, 403, 491, 492, 493, 496, 502, 511–512
 conformation 326, 391, 492, 502, 512, 514
β-lactoglobulin 450,
β-pleated sheet secondary structure 481,
β-sheet 7, 14, 24, 69, 73, 92, 93, 131, 132, 138, 141–142, 144–145, 148, 160, 175–177, 191, 210–211, 213, 245, 311–321, 326–333, 336, **337**, 355–356, 358, 375–378, 451, 455, 459, 461, 479–480, 483, 485, 487–488, 491–492, 494–496, 499–509
 conformation 24, 73, 92, 138, 160, 210–211, 315, 321, 323, 326, 329, 337, 375–378, 451, 459, 485, 487, 488, 491–492, 495, 501, 506
 breaker peptides 459
 propensity 7, 92, 93, 145, 147, 148, 488, 507
 -rich aggregates 313, 315, 316, 319
 -rich fibril-like precursor structures 325
 -rich (fibrillar) aggregates/structures 315–316, 319, 325, 338, 455, 501
 -rich macroscopic membrane 332
 -rich structures 325, 338, 501
 sheet/fibril formation 320
 (secondary) structure 131, 132, 142, 320, 321, 327, 328, 330, 331, 336, 338
 tape 311
β-strand 20, 24, 88, 132, 134, 137, 138, 141–145, 227, 245, 310, 314, 319, 377, 403–404, 480, 492–496, 500, 503, 505, 506, 507, 509, 510, **510**, 512, **512**, 513
 Aβ dimers 509
β-synuclein 459, 485
β-turn 396, 491, 492
β$_2$-microglobulin (A$_2$βM) 144, 353, 356, 481–482

γ-synuclein 14, 485

△ASA (accessible surface area) 260, **260**

π-helix conformation 489
π-stacking 366, 374
π-π aromatic interactions 358

φ-preferences of amino acids 188
φ, ψ, and χ$_1$ preferences of amino acids 188
φ,ψ distribution 83, 88

χ$_1$ preferences of amino acids 188

ψ-dependent coupling constants 210
ψ-preferences of amino acids 188

AAIndex Database 175, 176–178, 180
 correlations 180
 scales 177

AANF 353–354
AApoAI 353–354
AApoAII 353–354
AApoAIV 353
ab initio calculations 87, 134–135, 137, 139, 187
ABri 353–354
ABriPP 353
Ac-(AKAAE)$_3$A-NH$_2$ 313
ACal 353
Ac-Ala$_5$-NHMe (Ala5) 61
Ac-A$_4$XA$_4$-amide 141
Ac-KA$_{14}$K-NH$_2$ (KAK) 320
AcN-(AEAEKAKA)$_2$-CNH$_2$ 331
Ac-PPPXPPPGY-NH$_2$ 161
acquired immunodeficiency syndrome (AIDS, HIV-1) 281–282, **283**
AcX$_2$A$_7$O$_2$NH$_2$ 163
adenosine diphosphate ribose (ADP)-ribosylation 28
affibody Z$_{Aβ3}$ 391
AIAPP 353–354
AK-16 336–340, **337**
alanine (Ala) xvii, 6, 15, 17, 60–68, **61**, 68, 80, **80**, 84, 87, 90, 93, 111, 132–141, **133**, 144–149, 160–165, 168–175, 196, 207–215, 250, 309–340, 355, 405, 445–446, 459, 462, 495, 506
 -based (poly)peptides xvii, 162–163, 313, 321, 328, 331, 332, 336, 337, 340
 -based systems 328
 deca- 313, 326–327
 dipeptide 134–139, 160
 hepta- 64, 68
 -rich palindrome sequence 314
 -rich peptides (self-assembling) 309–340
ALB algorithm 247
alpha retroviruses (ASLV) 282
alternative splicing (AS) xvi, 26, 30
Alzheimer's disease (AD) xv, 5, 30, 34, 73, 310, 313, 319, 353, 377, 389, 406, 455, 460, **460**, 479, 481–483, 493, 494, 515
 pathogenesis 403, 463
 pathology 456, 458
ameloblast-associated protein 353
amino acid (AA)
 frequencies 7, 92, 165, 227, 275
 preference xvi, 80, 84, 85, 86, 91, 92, 132, 138, 144, 187, 188, 190, 191, 212,
 property scales 7, 161, 175–180,
 sequence/composition xv, xvi, 5, 7, 8, 16, 83, 85, 89, 91, **91**, 93, 101, 125, 221, 226, **226**, 245, 253, 259, 311, 312, 315, 330, 336, 340, 354, 355, 357, 359, 372, 454, 459, 479, 480, 483, 496, 507, **510**
aminocyclitols 463

amphipathic helix 36, 374
amphiphilic characteristics 452
amphiphilicity 340
amplitude-weighted lifetime 103, 109
amyloid (β)-peptide/protein (Aβ) xvii, 5, 34,
   74–5, 353–356, 365, 377–378, 389–406,
   399, 400, 444, 448–452, 456–459, 463–465,
   481–484, 487–505, 509–514, **510, 512**
 A protein 352, 481–482
 Aβ_{1-29} 392
 Aβ_{1-40} (Aβ40) 377–378, 390–405, **399**,
   444–452, 457–458, 483, 484, 493–497,
   500–501, 509, 510, 511, **512**
 Aβ_{1-42} 448–463, 492–513
 Aβ_{1-42} M35 oxidized 395, 397
 Aβ_{12-28} 397, 492, 493
 Aβ_{21-30} 396, 489–491, 496–497, 513
 Aβ_{30-40} 392, **510, 512**, 513
 affibody complex 396
 aggregation 389, 390, 393, 396, 397, 398, 401,
   458, 459, 463, 500
 aggregation pathway 393
 amyloid aggregation xvii, 34, 73–74, 264, 310
 amyloid diseases 317, 340, 354, 454–455,
   492, 507, 514
 amyloid-forming peptides/proteins 144, 311,
   372, 442, 444, 454, 479–482
 amyloid-induced cytotoxicity 357, 378, 456
 amyloid-like aggregates 506
 amyloid-like fibril formation 310, 312, 313,
   316, 317
 amyloid-like fibrils 264, 310, 311, 314, 315,
   315, 321, 329, 465
 amyloid-like plaques 310
 amyloid plaques 310, 398, 456, 482–483
 amyloid-specific dyes 319, 321
 amyloidogenecity 398
 amyloidogenic diseases 456, 461–462
 amyloidogenic partially folded conformation
   19, 34, 230, 264, 463
 amyloidogenic peptides/proteins 19, 164,
   311–321, 340, 351–358, 375–378, 448–452,
   455–465, 479–481, 486, 500
 amyloidogenic protein fibrillizations 144, 444,
   448–454, 458–459, 464–465, 495
 angiopathy 482
 -derived diffusible ligand (ADDL) 403, 483
 dimer 394, 403, 509
 Dutch, Arctic, Italian mutations of amyloid β
   483, 488–492, 496, 500, 511dynamics 398
 in E. coli 390, 391, 392
 fragments 394, 395, 396, 405, 463, 491, 499, 502
 globulomers 404
 pentamers 403, 501, 509, 510, **510**
 protein precursor (ARPP) 353–354

recombinant 390–392, 406
tetramers 22/23, 394, 395, 403, 502, 511
toxic oligomers 452
toxicity 389, 403
amyotrophic lateral sclerosis (ALS) 463, 479,
   482, 485
analytical ultracentrifugation (AUC) 394
ANCHOR 253–254, **254**
anorexia 358
antagonists 441, 458, 462
anthraquinone 458
anti-cancer properties 457
anti-inflammatories 458, 513–514
anti-retroviral therapy (HAART) 281
AP180 248
apolipoprotein
 AI 353–354, 482
 AII 353–354
 AIV 353
 C-II 375, 454
 E 352
apomorphine 456
apomyoglobin 13, 85, 449
apoptosis 15, 36–37, 256, 289, 290
apo-serum AA 353, 354
aprotinin 224
AR-12 337
Arabidopsis thaliana 8, 223
archaea 8
arginine (Arg) 6, 7, 32, 106, 117, 120, 140, 171,
   250, 264, 334, 336
 -rich IDPs 17
 -rich RNA binding domain 264
argyrophilic grain disease 5, 34
aromatic interactions 316–317, 358, 373
aromatic stacking 367–368, 369, 374
ASemI. See semenogelin I
asparagine (Asn) 6, 7, 91, **91**, 105, 125, 179,
   367, 369
aspartic acid (Asp) 7, 84, 86, 140, 250, 334, 366,
   377, 405, 490
atomic force microscopy (AFM) 311, 315, **315**,
   405, 464, 486
atrial amyloidosis 482
atrial natriuretic factor 353
autosomal dominant disorder oculopharyngeal
   muscular dystrophy 318
auxilin 248
avian leukemia viruses (ALV) 281
axoospermia 30

backbone conformation xvi, 140, 144
backbone conformational preferences xvi, 16, 80,
   84, 92, 81–83, 85–87, 92–93, 132–144,
   147–148, 187–191, 243, 247, 262, 331

backbone dihedral (Ramachandran) angle 4, 50, 61, 64, 70, 73–75, 79, 83, 87, 108, 135, 166, 246
backbone dipole moments 102, 134, 138, 443
backbone dynamics 227, 228, 233, 397
backbone electrostatic interactions 135–142, 147
backbone hydration/solvation 33, 96, 136–147
backbone hydrogen bonds 20, 29, 33, 132, 139, 148, 257, 326, 339, 367, 450, 459, 492, 405, 500, 503, 508, 509, 511, 514
backbone rotamer library 87
baicalein 456
basic-helix-loop-helix-leucine zipper domain (bHLHZip) 38
BAX 256
benzo[c]quinolizinium (MPB) 462
benzoicacid 456
bestatin 224
betaine 445, 448, 452
B-factor values 29
Biacalein 457
bicelles 200–202, 213, 214
binding competent state/conformation 164, 165, **165**
binding incompetent state/conformations 165, **165**
binding promiscuity 241–266
binding-coupled folding 247
biofunctionality 326, 328
BioMagResBank database 146
BLAST searches 250
bond vectors 85, 190, 201
bone crisi 462
bovine pancreatic trypsin inhibitor (BPTI) 360
bovine serum albumin (BSA) 335, 447, 451, 454
aggregation 447
bovine spongiform encephalopathy 455
bovine viral diarrhea virus (BVDV) 283, **283**, 288, 289
Bunyaviridae family 291
butylated hydroxylanisole (BHA) 456
butylated hydroxytoluene (BHT) 456

^{13}C glucose 390
C7$_{eq}$ 135, 137
Caenorhabditis elegans 8, 160
CAG triplet repeats 34, 484
calcitonin xvii, 353, 357–358
hCT (human calcitonin) xvii, 353, 354, 356–378, **364**, **370**, 457, 482
hCT fibril maturation 374–375
hCT fibrillation 358, 361–365, 375, 377
hCT helical oligomers 367, 369, 371
hCT hormone 357, 359, 361, 372, 378

hCT prefibrillar state 358–359, 365
sCT (salmon calcitonin) 357–361, 366–367, 372–375, 378
calcium regulation xvii, 357
caldesmon 636–771
fragment 12–14, 18–19
calpastatin 248
cAMP-dependent protein kinase inhibitor 14
cAMP-regulated chloride 461
cancer 30, 35–37, 256, 281, 455, 457
-associated proteins 35, 221
cannabinoid receptors and agonists 458
Cap-snatching mechanism 292
catechin 456–457
cation–π interactions 316–317
cationic gemini surfactant 452
CBP 23, 256, 259–260
CBP/p300 256
CBr cleavage 392
C-cell thyroid tumors xvii, 35, 353, 357
CCHC zinc finger motif (ZnF) 282, 288
ccβ peptide 507
cDNA 224, 281, 284, 287
cell cycle progression 36, 256, 290
cell death 314, 456, 460, **461**
cell differentiation 31, 332
cell lysate 224, 391
cell malignant transformation 256, 358, 375
cell proliferation 15, 335
cell signaling 8, 34–35, 38, 160
cell toxicity 318, 332, 403, 433, 460–461, **461**, 479–480, 486, 511, 513–514
cell-to-cell transmission 281
cellular ALIX 284
neurotoxicity 463, 512, **512**
proteinaceous deposits 310, 482
restriction factor APOBEC 3G 284
retinoic acid-binding protein (CRABP) 447, 449
central hydrophobic core/cluster (CHC) 397–9, 405, 487, 511
centrifugal filtration 390
ceramide 462
CFTRcor-325 462
chameleon (behavior) 3, 19–20, 23, 264–265
chaotropes 443–444
chaperone xvii, 23–24, 280–284, **283**, 287–293, 351, 378, 414, 441, 444–447, 459–462, **460**, 494
CHAPSO bicelles 213–214
charge-hydropathy (CH) (plot method) 6
Charmm22 force field 368
chemical chaperone hypothesis 445–447, 494
chimeric forms 144

INDEX   533

choline-O-sulfate 445–446
Chou and Fasman (C-F) helical propensities 92–93, 358
CH-plot 6, 15
chromatin 17
chromatogranins A and B 27
chronic fatigue syndrome 281
chymotrypsin 248
circular dichroism (CD) spectroscopy 5, 6, 159, 248, 289, 290
cis-imidazolines (Nutlins) 36–37
cis-trans isomerization 114–116, 125
citrulline 449
class I (RXXPXXP) motifs 246
class II (PXXPXR) motifs 246
class II ligand 246
classical binding paradigm 25
CLUSTALW 247
c-Myc 37–38
  transcription 37
coarse-grain phase behavior 415
coarse-grained models 323, 495, 509
CoCl$_3$ 464
coenzyme Q (CoQ) 457
coil library (Ramachandran probability) xvi, 80, 85–86, **90**, 92–93, 141, 149
coil residues 132, 149
coil-to-globule transition 17
collagen 160, 309
colloidal assembly 443
colloidal dispersions 414, 431
colloidal particles 414–415, 419, 430, 431
collodial scenario 415
colloids 414, 427, 429–432
confocal single-molecule fluorescence spectroscopy 15
conformation-associated diseases 455
conformational analysis (of unfolded peptides) xvi, 505
conformational biases 80, 88, 159–164, 168, 170, 174–175, 178–180
conformational changes 4, 34, 38, 79, 256, 265, 314–315, 356, 375, 377, 425, 485–486, 501
conformational diseases 34, 454–455, 459
conformational ensembles xvi, 3, 160, 168, 170, 188, 213, 228
conformational entropy 100, 139, 173–174, 373, 505
conformational exchange process 230–233
conformational free energies of SH3 ($\Delta G_{con,SH3}$) 164–165
conformational heterogeneity 113,

conformational preferences xvi, 80, 84–85, 87, 92–93, 132, 134, 136, 138–142, 144, 187–191, 243, 247, 331
conformational proclivities 101
conformational relaxation 429
conformational stability 18, 355
conformational transitions 62, 101, 136, 139, 210, 315, 320–323, 377, 488, 495
Congo red 310, 319, 321, 352, 455, 503
connectivity map 62
connectors 20, 23
consecutive predictions of disorder 8
consensus ligand peptides 246
consensus sequence 250
controlled release 331, 335
cooperative conformational changes 4
cooperative formation of local structures 134, 136
cooperativity 18, 132, 138, 142, 210, 421, 423
coplanar pentamer aggregate 73
copolymers 16, 415, 417
copolypeptides 16, 17
copy-choice recombination 290
coronaviruses 287, 290–291
coupled allostery model 291
creatine kinase enzyme 447, 451
CREB 247
Creutzfeldt–Jakob disease (CJD) 314, 455, 485, 494
critical concentration 326, 332, 417
critical core residues 366, 373
critical nucleus 358–361, 363–364, **364**, 366
cross-correlated relaxation 142, 190, 198–200, 398
cross-linked trimer 365
cross-β arrangement 321
cross-β core 310
cross-β spine 455, 506
crystallography 3, 6, 8, 20, 480
CSD1 249
CSN5/Jab1 256
CspTm 15
C-terminal domain (CTD) 256–257, 283, **283**, 291
C-terminal fragments (CTFs) 23, 492, 511–514, **512**
C-terminal hydrophobic tail 495
C-terminal intein fusion 392
C-terminal tetramerization 256
curcumin 357, 462
cyclic side chain 179
cyclin A 254, 260, **12**
cyclin A2 260
cyclohexane 326–327

cystatin C 353, 482
cysteine (Cys) 6, 7, 171, 192, 233, 246, 353, 395, 440, 455, 460–461, **460**, 482
cystic fibrosis (CF) 455, 461–462
cytoplasm(ic) 292, 454, 461
cytoplasmic-domain degradation 462
cytosol composition 464
cytosolic endoproteases 461

dancing proteins 3
database, of physico-chemical property scales 161, 175–178, 337
database propensities model (DPM) 134, 142, 148–149
date hubs 242–243, 262, 280
de novo design of proteins 188, 313, 320, 326, 328, 331–332, 326, 328, 340, 355
dehydration 454, 503
DelPhi algorithm 136
dementia 5, 34, 353, 389, 456, 482
denaturant-induced unfolding 15, 17, 18, 124, 132, 142, 179, 189, 331, 445, 449, 450, 452
denaturants 15–18, 101, 132, 142, 179, 189, 331, 445, 449–450, 452
denaturation 33, 81–82, 282, 289, 359, 361, 391, 443, 445, 493–494
denatured state 81–85, 144, 159, 161, 164, 168, 175, 178, 180, 207, 212, 222
denaturing conditions 16, 81, 82, 131, 132, 212
dengue virus (DENV) 287–288
dentatorubral pallidoluysian 35
Derjaguin–Landau–Verwey–Overbeek (DLVO) theory 430, 433
deubiquitinating enzyme 392
deuterium exchange 403, 450
Dexter mechanism 122
diffusion limited aggregation (DLA) 431–433
diffusion limited cluster growth 429
diffusion ordered spectroscopy (DOSY) 190–191
dihedral angle 16, 59–64, 66, 70, 73–75, 79–80, 87–88, 100, 140, 160, 166, 173–174, 190, 203–205, 246, 367, 404
  principal component analysis (dPCA) 59–75, **61, 70**
DILIMOT algorithm 250
dimeric aggregates 374
dimeric RNA genome 280–281
dimeric structure of hormones 350, 361
dimerization 38, 223, 281, 287–288, 326, 365, 500
dimethylsulfoxide (DMSO) 390, 403, 451
dipole–dipole cross-correlated relaxation rates
dipole–dipole interactions 122, 141, 146, 148, 398

disaggregation 338, 394, 505
discontinuous/discrete molecular dynamics (DMD) 323–325, 486, 489–490, 495, 496, 509–511, **510**, 513
discriminating order parameter 17, 73, 230, 367, 398–399, **19**
DisoPred2 31,
disorder associated binding sites 27, 262, 264
disorder prediction/predictors 8, 27, 31, 33, 37, 244–245, 247, 257, 259, 262, 265
disorder promoting amino acids 6–7, 29
disorder to order transition 20, 25, 28, 37, 39, 243–245, 247, 280
disordered proteins/peptides/regions xv, xvii, 3–10, 14–17, 20–40, **22–23**, 79, **90**, 142–144, 161, 166, 221, 242–266, 282–284, 290–293, 310, 313–314, 354–355, 402, 414, 429, 485–486, 498, 503, 508–509, 513–514
dispersion medium 431
DisProt 4, 251, 283
distance-dependent fluorescence 108
distance distributions 15, 100–101, 104–105, 110–103, 116–119, 122–125
distribution analysis 105, 111, 113, 116, 119–120
disulfide bridge 142, 357
DNA 15, 20–24, 27, 36–37, 222–224, 256, 259–260, 279–288, 331–332, 444, 460, **460**, 484
  binding domain (DBD) 222–223, 256, 259
  binding protein 27, 37, 222–223, 259, 288, 331
  repair 36, 223, 256, 284, 332
  replication 223, 256, 281–282, 289–292
  sequence CAG 484
dobutamine 456
dopamine-producing brain cells 485
dopaminergic neurons 464
dopaminochrome 458
double-kinetics experiments 100, 116, 125
double-labeled peptide 103–108, 109–111, 113–114, 122
double nucleation model 358–364
double wavelength plot 12–13
doughnuts 20, 264
Down's syndrome 5, 34
dPCA. *See under* dihedral angle
Drosophila melanogaster 8, 282
drug delivery systems 326, 331, 338
drug-induced cytotoxicity 357
drug targets xvi, 35–38
ducrose 448
dynamic light scattering (DLS) 311, 365, 433, 511
dynamic switch-like elements 250

INDEX   535

dynamically mobile conformational ensembles 3
dysfunctional heterotypic interactions 414
dysfunctional homotypic interactions 414
dysfunctional protein folding 30

E1-ubiquitin activating enzyme 455, 460, **460**, 461
E2-ubiquitin conjugating enzyme **460**, 461
E3-ubiquitin ligase (MDM2) 36–37, 248, 248, 255–256, 259–260, 391, 393, 460–461, **460**
EAK-16 327, 331–333
Ectoine 448–449
EIF4F 292
elastin 309
electron microscopy 5, 311, 352, 355, 361, 405, 451, 464, 481, 486, 494
electrostatic screening
  model (ESM) 134–147
  in polypeptides 137, 140, 142
electrostatic solvation free energy ESF 136–145, 148
ellagic acid 456
ELM predictiors 250–252, 255, 266
elongated structure 200, 210, 325, 486, 509
EMK16-II 327, 332
emphysema 455
empirical database correlations 190, 206–207
encephalitis virus complex (TBEV) 288
encephalomyocarditis virus 292
endomorphines 459
endoplasmic reticulum (ER) 289, 461–463
entropic brush 16,
enzymatic modification 250
enzymatic phosphorylation 101, 117, 125
enzyme activity 30, 256
epicatechin 456–457
epicatechin gallate 457
epitope mapping 5
epsin1 248
Escherichia coli 23, 32, 33, 224, 390, 391, 392
  BL21:DE3 224
  protein trigger factor (TF) 392
esophagus 36, 256
essential dynamics method 59
ethylammonium nitrate 444
ethylendiamin-tetraacetate (EDTA) 224
Euclidean distance 60,
eugenol 456
eukaryotic cells 8, 29
eukaryotic initiation factor (EIF) 244
eukaryotic linear motif (ELM) 250
eukaryotic proteins 28
eukaryotic proteomes 413–414
eukaryotic translation initiation factor 4E 243

evolutionarily active domain 221
evolutionary rate measurements 26, 223
excluded volume
  constraints 86
  effect 111, 453, 500
  interactions 84–85, 168
  models 168, 169, 170, 171–172, 179
  scales 170
exifone 456–457
exocytosis 248
exogenous 2° structure prediction or homology information 88
exoribonuclease activity 290
experimental propensity scale 92, 169
extracellular amyloid 352, 482
extracellular deposition, of amyloid fibrils 480
extracellular deposits, of transthyretin 482
extracellular lesions 456
extracellular plaques 483
extracellular matrix proteins 309
extremophiles 442

F508del-CFTR protein 462
F-actin 108, 420
factor X_a 392
FAD mutants 396, 483
familial myloidotic poly neuropathy I 482
familial amyloidotic polyneuropathy II 482
familial dementia 353
familial forms (40–44) 483, 485, 501
fatty acid acylation 28–29
fatty acid binding protein (FABP) 392,
fatty acid methylation 29,
fatty liver (steatosis) 289,
feline immunodeficiency virus (FIV) 281
Felix software 225
FF99SB–force field 401
FG-nucleoporins 15, 16, 17
fibril axis 310, 314, 480–483, 487, 494
fibril forming polypeptides and proteins 15, 310–317, 320–325, 340, 352, 357–358, 366, 374–378, 390, 404, 427, 444, 448–458, 464, 480, 482, 486, 488, 492, 495, 498–503
fibril growth kinetics 17, 359, 363, 364, 426–428, 488, 502
fibril inhibitors 312,
fibril-like network 331
fibril solutions 316, 321
fibrillar aggregates 310–311, 313, 315, 359, 366, 429, 483, 499
fibrillar Htn 482
fibrillar structure 455, 315, 321, 359, 363, 364, **364**, 420, 481, 507, 508, 514

fibrillar tangles 5, 34, 456, 482, 483
fibrillization 444, 448–454, 458–459, 464–465
 propensity 144
fibrillogenesis 311, 355, 359, 365, 372, 378, 448, 455, 465
fibrillogenic biopolimers 357
fibrils 20, 24, 34, 73, 144, 264, 310–329, **315**, 352, 356–363, 372–378, 389–404, 420, 424, 427–430, 444, 448, 451–455, 461, 464, 479–488, 491, 494, 499–502, 506–509, 513–514
 cross-β structures 329, 480–481, 502, 508–509
 deposition in tissues 482
fibrinogen α-chain (AFib) 353
fibrous proteins 3
filamentous amyloid-β deposits 5, 34
filamentous structure architecture 310–331
filamentous protein aggregates 34
Finnish hereditary amyloidosis 482
flavanone 456
Flaviviridae family of viruses 288
flavivirus 16, 279, 287–289
flexible worm-like structures 452
FlgM 247
fluorescence correlation spectroscopy (FCS) 335
fluorescence decay 100, 103, 104, 110, **110**, 117, 118, 124, 125
fluorescence lifetime 103, 105, 109–110, 117, 123, 125
fluorescence probes 9
fluroescence resonance energy transfer (FRET) xvi, 15, 99–126
fluorescence spectroscopy 15
fluorine labeled-amino acid 116
fly-casting model 291
folding
 landscape 180, 493
 (mis-)folding disease 310, 351
 pathway xvi, 9, 61–62, 69–73, **70**, 75, 80, 86–90, 374
 transition 18, 79
foldons 87–88
force fields 57, 60–62, 84, 114, 134, 139, 368, 401–402, 489, 491
 empirical 134
 semi-empirical 139, 368, 401–402
forced copy choice recombinations 281
Förster radius 101–103, 107, 110, 111, 113, 123–125
Förster theory 106
Fourier transform infrared (FTIR) spectroscopy 5, 6, 14, 187, 193, 195, 311, 316, 389, 503

fractal dimension 432–433
fractal-shaped aggregates 451
fractional volume occupancy 375
free energy disconnectivity graph 66–67
free energy landscape xvi, 58–75, **61**, 415, 489, 502
free energy surface 61–62, 65, 67, 88, 90, 377
fronto-temporal dementia 353, 482
fructose 448–449
fungus IULD 233
fusion protein 23, 392
fuzzy globule 284
fuzzy proteins 38

GAC trinucleotide repeats coding 34, 313, 317
gag oligomerization 282
gag polyprotein in infected cells 281–282
gag structural polyprotein precursor 282
GagNC (NC domain of Gag) 281
galactose 448–449
gamm-aretroviruses (MoMuLV) 282–283, **283**, 288
Gaucher's disease 455, 462
gel-electrophoresis 5, 224
gel-filtration 5, 359–360, 405
gels 5, 16, 85, 200, 202, 223–224, 317, 326–328, 331–332, 335–340, **337**, 359–360, 405
gelsolin 353, 481–482
genome duplication event 223
genomeNet Japan 175–178
genomic PBS 281–284, 287
genomic RNA (gRNA) 280–292
geometric clustering method 66
GGXGG 162-3, 166–179
globule-forming polar/charged IDPs 17
globulomer-specific antibody 403
globulomers 403–404, 483
glucocerebrosidase (β-Glu) 462–463
glucocorticoid receptor 27
glucose 390, 448, 449, 462
glucosylceramide 462
glutamic acid (Glu) 6, 7, 140, 171 321, 333, 336, 462–463, 490
glutamine (Gln) 6, 7, 159, 162, 171, 179, 318, 366
glutathione-S-transferase (GST) 392
glycerol suppressed aggregation of WT 447
glycine (Gly) 6, 7, 15, 17, 83, 84, 86, **90**, 91, 101, 120, 122, 124, 145, 165–174, 227, 233, 250, 293, 329, 330, 445–446, 495
 -rich host-guest system 171, 174
 -serine block copolypeptides (polyGS) 17, 101, 122, 124, 125
glycoinositol phospholipid anchor 485

INDEX 537

glycolipid substrate 463
glycosaminoglycans 352
glycosylation 28, 461
glycylcistylglycine (GCG) 192
GN8 493–494
GNNQQNY 503, 505
Gō model 509
GOR algorithm 247
gossypetin 456–457
grabbers 20, 23
green fluorescent protein (GRP) 319
GROMOS–force field 61–62, 114, 401
growth media 32
GSK3β 256
GVQIVYK 502
Gypsy retrotransposon 282

HAART. See anti-retroviral therapy
Hallervorden–Spatz disease 5, 34
Hamaker constant 431
hantavirus cardiopulmonary syndrome (HCPS) 291–292
hard sphere collisions 166
hCT. See under calcitonin
H/D exchange 190, 397, 403–405, 450
heat shock protein 15, 23, 461
helical basin, preferences for amino acids 92
helical coil transition 138, 140, 313
helical conformation in micelles 395–397
helical formation 160, 164, 177, 497
helical initiation parameter 177
helical nucleation 62, 373
helical or turn-like segment 373
hemodialysis-related amyloidosis 353, 482
hemoglobin S (HbS) polymerization 352
hemopoietic tissues 36, 256
hemorrhagic fever with renal syndrome (HFRS) 291
hepacivirus 288
hepatic cirrhosis 288
hepatitis C virus (HCV) 287–288, 292–293, 283–289, **283**
hepatocellular carcinoma 288–289
hepatomegaly 462
hereditary forms of amyloidoses 355
hereditary neurodegeneration 34
heterodimerization 38
heterogeneous nucleation mechanism 352, 361
heteropolymers 417
hexafluoroisopropanol (HFIP) 394, 397, 451, 492
hexamers 24, 314, 365–367, 372, 374, 501–502, 505, 511
hGcn5 256
hierarchical self-assembly 311, 329

higher-order assemblies 359
hippocampal neurons 390
His ring stacking 367, 371
histidine (His) 7, 27, 140, 233, 321, 367–368, 371, 373–374, 391–392
 -containing analogue 321
 -rich protein II 27
 tag 391–392
 -tagged ubiquitin 392
histological staining reactions 352
histone H1 27
histone H3 261
HIV-1 integrase 15
HMG-14 27, 458
HMG-17 27
hnRNP A1 (heterogeneous nuclear ribonucleoprotein A1) 293
Hofmeister's series 442–444
homogeneous nucleation 352, 363–365, **364**, 376, 422
homonuclear NMR 225
homo-oligomerization 289, 291
homoserine trans-succinylase 447
homotypic intermolecular interactions 414–415
host–guest system/study 139, 161–175
HP-35. See villin headpiece subdomain
heat shock proteins (hsp) 461
hsl homologues 226
hsp70 461
hsp90 461
HTLV. See human T cell leukemia virus
hub proteins (hubs) 16, 241–243, 256–257, 262, 266
huggers 20–23
human apolipoprotein C-II 352–354, 375, 454, 482
human cystic fibrosis transmembrane conductance regulator (CFTR) gene 461
human interferon gamma (rhIFNγ) 450
human pathologies xiii, 34–35, 280–281, 309, 310, 317, 354, 356, 372, 389, 442, 456, 458, 465, 479, 481, 483, 488, 493–494
human prion protein (hPrP) 465
human proteome 188, 256
human prothymosin α 12–15, 19, 27
human signaling cancer-associated proteins 221
human T cell leukemia virus (HTLV) 281
Huntington protein 484
Huntington disease 35, 310, 313, 317–318, 461, 479, 482–484, 506
hydrodynamic dimensions 6, 9, 15, 16, 222
hydrodynamic radius 9, 199, 335
hydrodynamic techniques/methods 14, 81
hydrodynamic volumes 10, 39, 448

hydrogel 16, 317, 326–328, 331–332, 335–340, **337**
hydrogen exchange rate 87, 148
hydrogen bonding networks 491
hydrogen bonds
  inter-molecular 367, 371, 374, 443, 459, 480, 486, 501
  intra-molecular 33, 160, 323, 331, 443, 451, 506
hydrolysis 105
hydropathy 6–7, 495, 509, 511
hydrophilic pattern 245, 247
hydrophobic cluster 132, 142, 230, 397, 398, 487
hydrophobic collapse model 100, 320
hydrophobic compound 21
hydrophobic core 6, 314, 321, 369–371, 374, 405, 441, 510
hydrophobic interactions 19, 115, 116, 132, 139, 309, 314, 321, 323–326, 332, 334, 358–359, 361, 367, 396, 441, 451, 489, 493, 500, 501, 503, 505, 512
hydrophobic pattern 314,
hydrophobic regions 125, 399, 495, 511
hydrostatic stress 445
hydroxy-3-methyl-glutaryl-CoA reductase (HMG-CoA reductase) 458
hydroxyectoine 448
hydroxylindole 458
hypericin 456–457
hyperphosphorylated filaments 456

iatrogenic syndrome 353
ibuprofen 513–514
IGFBP3 256
immunoglobulin 132, 353–354, 444, 448–450, 481
immunoglobulin G (IgG) 335
imidazole 224, 369
imidazolium cations 444
imidazolium chlorides 444
iminosugars 463
immune response/defense 160, 281, 332, 357
immunodeficiency viruses 264
immunoglobulin 132, 353–354, 444, 448–449, 450, 481
in silico aggregation 314
in silico studies 311
in vitro fibril formation 311, 372, 499
indole 106, 458
indole 3 carbinol 458
indole 3-hydroxyindole 458
indole 4-hydroxyindole 458
indole 5,6-quinone 458
indole dopamine derivatives 458

induced-fit mechanism 242
inhibitors xv, xvii, 14, 23, 35–36, 38, 224, 257, 312, 335, 356–358, 378, 441, 447, 455–459, 462, 463, 465, 487, 492, 511, 513–514
inositol 357, 445, 446
insolubility 414, 485
insoluble aggregates
insoluble fibrils 34, 372, 480–481
insoluble filamentous structures 310
insulin 23, 289, 316, 353, 354, 356, 359–360, 448, 449, 450, 451, 452, 457, 482
insulin B-chain 359–360
insulin fibril 316, 448, 451
insulin fibrillization 450
insulin-related amyloidosis 482
insulin resistance 289
insulinoma 353–354
integrase (IN) 1, 280
intein cleavage 392
interaction potential 253, 256, 430, 431
intercellular amyloid-like deposits 482
intercellular neurofibrillary tangles 5, 34, 456, 482–483
intercellular proteins 28
intercellular quality-control system 455
interhelix aromatic pattern 367
intermediate-resolution model 323–324
interstrand recognition 374
intramolecular packing 11, 377
intraneuronal inclusions 482
intrinsic conformational preferences/properties of amino acids 84–85, 188–189, 243
intrinsic disorder xvi, 7–8, 25, 26–33, 38, 242–243, 246–247, 256–258, 262, 283, **283**
intrinsically disordered aggregation xvii, 414, 418, 420–423, 427, 433
intrinsically disordered binding regions 37, 253–255, 265–266
intrinsically disordered proteins (IDP) xvi, xvii, 20, 23, 31–32, 38, 221–223, 227, 242, 248, 255, 310, 414–415, 418, 420, 427, 429, 433
  degradation 31–33, 414
  expression levels 414
  transcription 5, 32, 414
intrinsically disordered (ID) regions 3, 38, 242, 265, 284, 291
intrinsically folded structural units, IFSUs 248
intrinsically unstructured linker domain (IULD) xvi, 222–233, **226**
ion mobility coupled mass spectroscopy 398
ionic-complementary oligopeptides 332
ionic liquids xvii, 444
internal ribosome entry site (IRES) 292

INDEX   539

islet amyloid polypeptide (IAPP or amylin)
    353–354, 372, 456–457, 459, 465, 482,
    506
islets of Langerhans 353–354
isodesmic aggregation/polymerization 420–423
isoflavone 456
isolated-pair hypothesis 83, 148–149
isoleucine (Ile) 6, 7, 144, 148, 162, 171, 175,
    179, 355
isothermal titration calorimetry (ITC) 161, 164,
    165, 464
isotope enrichment 189, 192
isotope labeling 191–192, 198, 203
ItFix 88–90
ItFix protocol 88
ItFix-SPEED 88, 89
IULD homologues 226–227, 230, 233, **226**
IUPred 8, 251–253
IUPred server 253

Jembrana disease virus (JDV) Tat protein 264

k2d neural network algorithm 373
K382 260
kaempferol 457
Karplus equation/relation 132, 140, 163, 190,
    202–204
Karplus parameters 203–206, 210
Kennedy's disease 35
kerato-epithelin 353
keyword-associated sets 27
KID domain 247
kinase 13–14, 29, 32, 116–117, 250, 447, 451
kinetic fibrillogenesis theory 365
kinetic trapping 500
KLD-12 327
k-means algorithm 66, 71
knowledge-based approach 87
kosmotropes 443–444, 447
Kratky plot 11–12
Kyte–Doolittle hydrophobicity index 142

lactadherin 353
lactoferrin 353
lag-phase 352, 361, 365–366
Langevin algorithm 68
Langevin approach 68
Langevin dynamics 507
Langevin equation 60, 67, 68
Langevin vector 68
Larmor frequency 199
lattice model simulations 245
lavonoids 456
LB variant, of Alzheimer's disease 5, 34

leucine (Leu) 6, 7, 23, 38, 84, 87, 90, 91, 144,
    148, 162, 166, 171, 179, 233, 250, 255, 355,
    367, 372, 374–375, 377
leucine zipper domain/peptide (bHLHZip) 23,
    38, 374
leukemia in rodents and human (Friend MuLV
    and HTLV-1, respectively) 281–283, **283**,
    288
leupeptin 224
Levinthal's paradox/search problem 79–80, 85,
    87–88, 93, 100, 159
Lewy body 5, 34, 310, 456, 482, 485
    disease 482
LiDS 397
Lifson–Roig helix-coil theory 138
ligand-induced folding 27
light scattering 5, 311, 365, 432, 433, 511
linear metric coordinate space 60
linear motif (LM) 250–252, 255
    flanking segments (magenta) 251
linker domain 16, 222, 254, **254**
liver disease 288, 455
local residue packing 4
local structural order 131–134
lock-and-key mechanism 242
long terminal repeat 280
low-dimensional simulation 67
low-dimensional subspace 58
low-molecular-mass inhibitors 356
LPPLP motif 246
Lumry–Eyring nucleated polymerization (LENP)
    model 425
Lyapunov analysis 69
lysine (Lys) 6, 7, 84, 86, 140, 159, 162, 179,
    284, 287, 315–316, 321, 333, 337, 353, 366,
    405, 450, 482, 490
    tripeptides 179
lysosomal storage disease 462
lysozyme 144, 316, 335, 353, 444, 450–454,
    458, 482, 502
    amyloid formation 454
    amyloidosis 482
    fibril formation 454

macromolecular crowding 375
macromolecule stability 445
macrophage 200
macroscopic hydrogel 331–332, 336
mad cow disease 221, 313
magnetic anisotropy 148
major disorder-promoting residues 6–7, 29
major spidroin I and II 328–330
malaria 392
malignancy-caused hypercalcemia 357

maltose 392, 449
  binding protein 392
mammalian (and plant) IULD 227, 231–233
mammalian attachment 331–332
mammalian homologues 232–233
mammalian linkers 228, 231
mannose 448–449
many-to-one signaling 243, 265–266
MAP2 248
MAP2c 248–249
MAPC2 248
Markov chain 68
Markov model 68
Markov dynamics 68
Markov state 60, 67
mass spectrometry 29, 511
Max protein 27
MaxFlux algorithm 488
MAβ1–29 392
MAβ42 390–391, 403
MDM2. *See* E3-ubiquitin ligase
measles virus nucleoprotein 247
melatonin 458
melezitose 449
memoquin 458
mesoglobules 418, 429, 430
metal ion binding 27, 34, 464
metastable aggregates xvii, 365, 432
metastable (conformational) states xvi, 58–60, 62, 64, 65–72, 75, 414, 417–418, 483
metastable oligomeric forms 363–365, 429, 480, 486, 508, **512**
methionine (Met) 7, 332, 355, 390–392
methylamines 445–446, 448
methylation xvi, 26, 28–29, 459
methylene blue 337, 465
methylmannoside 448
methylsulfonium 445–446
Metropolis Monte Carlo method 168
microstate partitioning models (MPMs) 418–420, 424, 427, 430, 433
microtubule-associated protein tau 456
mimetics 357
Mini-Protein MODeler (MPMOD) 172–175, **173**
mixed globular/disordered state 354
modifiers of protein aggregation xvii, 442
molecular anvil 256
molecular crowding 189, 284, 287, 375, 453–454, 489
molecular dynamics (MD) simulation xvi–xvii, 57, 59–62, **61**, 64, 67–73, **70**, 84, 101, 112, 114–115, 119, 126, 170–171, 187, 207–210, 213, 215, 314, 323, 366–368, 374, 375, 390,
399–401, **399**, **400**, 447, 450, 486–489, 492–499, 511
molecular mechanics (MM) simulations 509
molecular recognition 16, 20, 27–28, 37, 57, 242–244, 253, 255, 266, 312
  features (MoRFs) 20, 23, 36–37, 244–245, 254–255, 264, 266
molecular ruler 100
MOLMOL program 368
molten globule 4, 9–18, **11**, 28, 145, 222
molten-globular configurations 17
MoMuLV Gag 282–283, **283**, 288
Monte Carlo simulated annealing (MCSA) 87, 168, 496, 598
MoRFs 20, 23, 36–37, 244–245, 254–257, 264–266
Morin 457
motor neuron disease 5, 34
motor neurons 485
mouse prion fibrillization 444
MPB-07 462
MPB-91 462
MRNAs 30–33, 38, 279–280, 287, 289–292
mSos-derived proline-rich peptide 160
multicellular eukaryotes 30
multicellular organisms 31–32, 38
multiple system atrophy 5, 34
multiplet splitting 191–194, 197–198
multistate conformational equilibria 189
multistranded fibril-ribbons 359
murine leukemia virus (MuLV) 281–283, **283**, 288
muscle contraction 117, 485
musculoskeletal pain 357
mutagenesis 101, 139
mutant rabbit muscle creatine kinase 447
myeloma-associated AH 353
myo-inositol 446
myotonic dystrophy 5, 30, 34
myricetin 456–457

N-alkylaminocyclitols 463
N-alkylated deoxynorjirimycins 463
N'-alkyl-methylimidazolium chlorides 444
N-methylation of peptide inhibitors xvi, 26, 28, 29, 459
N-terminal structured domain (NTD) 291
N-terminal translational activation domain (TAD) 248, 252, 256
N'-(ω-hydroxyalkyl)-methylimidazolium chlorides 444
nanofiber scaffold 327, 332, 334–335, 340
nanofibrillar materials 326
nanostructured biomaterials 326

INDEX 541

NANP$_{19}$ fusion 392
naphthalene 514
naphthoquinones 458
naproxen 513–514
napthylalanine (Naph) 111
native-like folding pattern 9, 149, 356, 461, 499, 507
NC/core proteins 280
NC domain of Gag (GagNC) 281
NC-NC interactions 282
NCp9 282–283, **283**
NC-RT interaction 283–284
NDGA 457
nearest neighbor effects (NN) xvi, 84, 86, 132, 134, 136–141, 148–149, 188
nephrogenic diabetes insipidus 455
network theory 60
neurite growth 332
neurodegeneration 34, 464, 483
neurodegenerative diseases xv, 34, 188, 310, 313, 317, 460, **460**, 479–484, 498
neurodegenerative disorders 5, 34, 35, 264, 352, 456, 463
neurofibrillary tangles 5, 34, 456, 482–483
neurological diseases 281, 485. *See also* neurodegenerative diseases; neurodegenerative disorders
neuromelanin 464
neuronal cytoplasm crowding 454
neuronal loss 389, 483
neuronal survival 34
Neurospora crassa 223
neurotoxicity 463, 512, **512**
neurotoxin 23, 483
neurotransmitter vesicles 34
neutron scattering 486
NFGAIL peptide 506
nicotine 357
Nidovirales order 290
Niemann–Pick disease type C 5, 34
Ni-NTA agarose 224
nonamyloidogenic disease 461–462
nonisotropic small-amplitude motions 4
nonlinear sampling 193
nonradiative pathway occurs 102
nonrandom clusters 188
nonrandom native structure 81
nonrandom non-native structure 81
nonredundant protein chains 6, 245
nonrestricted positions 250
nontoxic monomeric proteins 511
nonfilamentous amyloid-β deposit 5, 34
nonlinear deterministic model of dynamics 60, 67

nonlinear time series analysis/theory 68–69
nonuniform structural properties 228
normal-order sequence alignment 247
Nsp1 17
nuclear magnetic resonance (NMR) spectroscopy xvii, 108, 139, 142, 161, 189–193, 213, 223, 227, 248, 289, 291–292, 373, 376, 377, 496
  ^{13}C chemical shifts 396, 170
  ^{13}C relaxation rates 396, 398
  ^{15}N relaxation rates 397
  ^{15}N-^{1}H (HSQC) NMR spectra 165, 393, 395, 165, 391, 394
  chemical shift index 145, 395
  chemical shift of unfolded proteins 131, 145–147, 395
  heteronuclear 5, 14, 190–192, 225,
  heteronuclear relaxation rates 190–191
  heteronuclear single quantum correlation spectroscopy (HSQC) 165, 198, 202, 213, 391, 394–5, 403
  hydrogen-deuterium exchange studies 450
  J-coupling analysis 163, 213
  J-coupling constants 16, 131, 140, 141, 142, 163, 170, 170, 190–3, 197, 202–5, 210, 401
  nuclear Overhauser effect (NOE) 142–143, 163, 199–200, 210, 222, 395–398, 400, **400**
  nuclear Overhauser effect spectroscopy (NOESY) 199–200
  proton–proton dipolar coupling 202
  pseudo contact shifts (PCS) 190, 248, 266
  pulse field gradient measurement 395, 405
  solid state 329, 376–377, 389, 403, 481, 494, 499
  spin–lattice relaxation rate 250, 230–232, **231**, **232**, 234
  spin relaxation 163, 225, 395, 405
  spin–spin relaxation rates 225, 230–234, **231**, **232**
  TROSY effect 199
nuclear pore complex (NPC) 15–17,
nucleated conformational conversion 311–312, 430
nucleated polymerization 311–312, 316, 321, 425
nucleation 62, 187, 311, 315, 352, 358–365, **364**, 373, 375, 376, 418, 422, 424–428, 448, 454, 465, 508
  elongation mechanism 422, 425, 427
  phase 311
  propagation models 373
  sites 187
  time 352
nucleic acid binding 27, 279
nucleic acid binding proteins (NABP) 279
nucleic acid chaperone proteins 279–293

nucleocapsid (NC) 280–284, **283**, 287–292
  /core proteins 280
  formation 289–290
  -NC interactions 282
  protein sequences 291–292
  -RT interaction 283–284
nucleocytoplasmic transport in eukaryotes 15
nucleoporins (Nups) 14–17
nucleoprotein 247, 281
nucleotide excision-repair activity in vitro 284
nutlins 36–7

octodon degus 354
odontogenic ameloblast-associated protein (AOaap) 353
odontogenic tumors 353
off-pathway aggregation 321, 420
oligomeric conformations 19, 34, 264, 314, 375, 485
oligomeric intermediates 480
oligomeric nucleus 314
oligomeric state of proteins 190, 243, 394
oligomerization 282, 289, 291, 319, 357, 365, 367, 373–374, 452, 458, 460, **460**, 464, 484, 510–512, **510**, **512**
  of Aβ 264
oligomers 16, 19–24, 34, 73, 190, 243, 264, 282, 289, 291, 311, 314, 319, 323–324, 357, 365–375, 389–406, 418, 427–433, 452, 458, **460**, 461, 464, 480–487, 496–503, 506–514, **510**, **512**
oncogene 35
on-column digestion 391
one-dimensional random flight chain 211–212
one-dimensional self-assembly model of chiral molecules 311
one-to-many signaling 243, 262, 264, 265, 266
Oosawa's model 421–424
OPLS–force field 401–402
optical activity 108, 163, 164, 176
optical rotational dispersion (ORD) spectroscopy 5
optical spectroscopy 57, 163, 164
orange projections 367
order-disorder predictions 8, 27, 256–257, 262
orientational disorder 73
Oryza sativa 222
osmolyte xvii, 419, 444–450
osmoprotectants 445
osmoregulation 445
osteocalcine 27
osteonectine 27
osteoporosis 347
Ostwald step rule 509

p21 247, 256
p21Cip1/p27[Kip2] 247
P39A tetra-Cys CRABP 449
p53 23–24, 35–37, 247–248, 255–266, **258**, **259**, **260**, 455
  C-terminal binding region 24
  Mdm2 interaction 36–37, 248, 255, 256, 259–260
  N-terminal domain 255, 257
  -related cancer 35–37, 256, 455
  sequence 257–260
  Taz2 complex 255
  transactivation function 257
Paget's disease 357,
pair-wise alignment 262
palindrome prion fragment 314
palindromic sequences 246, 287–288
parallel β-sheets 377, 483, 494, 499, 50,
paramagnetic relaxation enhancement (PRE) 190–1, 248, 395
paranuclei structures 483–484, 51,
Parkinson's disease xv, 30, 34, 73, 264, 310, 313, 319, 455–456, 461, 463–465, 479, 482, 485
parkinsonian substantia nigra 463–464
PARP-1 256
PARSE 136
party hubs 242–243
pathogenesis 39, 221, 280, 289, 317, 403, 463, 481, 498, 503
pathogenic RNA viruses 16, 279–280
pathological amyloid formation 34, 354, 356, 372, 389, 442
pathway-based protein folding 93
pathway-directed search 88
p-benzoquinone 457
PBS and PAS sequences 281–284, 287
PC12 cells 403,
Pearson correlation 142, 175–8, 401
pentahydroxybenzophenon 456, 457
peptide
  backbone 59, 141, 144, 148, 168, 170, 173–175, 179, 336, 450, 492, 500
  coarse-grained model 484, 494–495, 509
  oligo 139, 320, 326–328, 331–332, 336, 338, 340
  oligo hydrogels 328, 331–332, 340
  –peptide recognition 326
  self-assembly mimic (PSAM) 508
  –solvent interactions 179
  structure xiii, 117
peptidomimetic small molecules 459
persistence length 104, 111–112, 116, 124, 168
pestiviruses 288–289

INDEX
543

pET15b plasmid 224
phage display-derived peptide (R18) 261
phase diagram approach 415–419, **416**, 433, 507
phase diagram for polymer solutions 416–419, **416**
phase-separation boundary 417
phase-separation state 414–5, 418, 419
phenolsulfonphthaleine 456–457
phenylacetic acid 456
phenylalanine (Phe) 6–7, 15, 17, 175, 250, 355, 367, 372–373, 462
 -glycine repeats 15–17
phosgene 105
phosphodiesterase inhibitors 458,
phosphodiesterase γ-subunit 12–13, 18–19
phosphopeptides 262
phosphorylation xvi, 26, 28–29, 32, 101, 117–121, 125, 256, 257
phosphoserine (pS) 117–120, 123
phosphothreonine (pT) 117–123
 residue 117, 120
photochemical cross-linking 365, 367, 375
photoinduced cross-linking 398
physico-chemical property scales in AAIndex 161, 175–178, 337, 351
PKI alpha 247
plant cell wall proteins 160,
plastocyanin 144
pleiotropy 242
PMG 13
polar homo-polypeptides 16
poliovirus 292
polyacrylamide gel (PAA) 202
 electrophoresis 5, 224
polyadenine-binding protein nuclear-1 (PABPN1) 319, 321
polyalanine 60, 62, 207, 210, 213, 313, 317, 318, 319, 321, 323, 325, 326, 328, 329, 330, 331
 diseases 313, 317, 318, 319
 peptides 207, 210, 213, 313, 319, 321, 323, 326
 stretches 313, 317, 319, 328–329
 "tracts" 313, 318
polyampholytes 17
polyampholytic nature 17
poly(ethylene glycol) (PEG) 330, 454
polyglutamine 35, 318, 426, 427, 484, 506
 diseases 318
 repeat expansion 484
 β-stranded aggregates 506
poly-L-lysine aggregation 451
poly-L-proline 107
polymer(s) 16, 416
 aggregation xvii, 415, 418

phase diagram xvii, 415–419, **416**, 433, 507
 -rich phase 418, 420
 theory 15
polymerization (DP) 284, 311–313, 316, 321, 352, 419–425, 428
polypeptide hormones 353, 456
polyphenol 456–457
polyproline II (PII pr PPII) xvi, 80, 83, 85, 86–89, 92, 107–108, 110, 113, 116, 132, 133, **133**, 138–143, 147–148, 159–180, **165**, **173**, 208–210, 213, 245–250, 315, 316
 -based binding motifs 246
 conformation 133, 138–143, 147–148, 160–168, **165**, 170–171, 174–175, 175, 177, 179, 180, 246, 248
 distance distributions 111–113
 helical motif 107, 246
 propensity 16, 161–179, **165**
 propensity scales 16, 139, 161, 166–172, 175–176, 178, 179
 region 83, 85, 89, 92, 173–175, **173**
 stretches 163
polytropic tropism 281
polyvalent ordered complexes 20,
PONDR VL-XT 8, 37, 243–244, 257, 258, **258**
PONDR-FIT 247
PONDR-RIBS 247
population frequencies 170
post-mortem senile plaques 398
posttranslational modification xvi, 26, 257–260, **258**, 290, 414
potential of mean force (PMF) 326–327, 430
power-law behavior 432
PPXPP 162, 167, 168, 169, 170, 172, 176–177, 178
PRE effects 39
prediction algorithms 88, 138, 244, 257
pre-equilibrated nucleus model for nucleation-elongation 425–428
prefibrillar aggregates 359, 365–366, 374
prefibrillar state 358–359, 365
preferential exclusion 447, 501, 503
pre-molten globules 4, 9–18, **11**, 28, 39, 222
pre-mRNA 30, 38, 279
pre-osteoblast cell proliferation 335
primary contact sites (PCSs) 248, 266
primary systemic amyloidosis 482
principal component analysis (PCA) xvi, 59–75, **61, 70**
prion diseases 34, 313–314, 319, 455–456, 479, 484, 485, 493, 494, 503, 506
prion protein (PrP) 34, 221, 284, 313–315, 353, 449, 456–457, 465, 481–482, 485, 493–494, 506

prion Sup35 23, 444, 503, 505–506
prion prokaryotes 8
PROF algorithm 247
prolactinomas 353
proline (Pro) xv, 6–7, 83, 86, 92, 107–108,
    113–116, **114**, 146, 159–175, 230, 233, 246,
    248, 250, 264, 320, 353, 357, 445, 449, 459
  -repeat sequences 108
  residues 108, 113–115, **114**, 230, 357, 459
  -rich (peptide) host 162, 164, 170
  -rich loop region 160
propensity scales 16, 92, 138–41, 161–162,
    166–172, 175–179
propylene glycol 105–123, **110**, **112**, **114**
prostate cancer 281
protamines 17, 27
protease 14, 29, 224, 248, 282–283, 320, 351,
    392, 396, 414, 460–463, **460**, 485, 489
  -resistant core 396
proteasomal degradation 462
protein conformational abnormalities 455
protein crystallization 389, 443
Protein Data Bank (PDB) 8, 28–9, 80, 83–85,
    87–88, **90**, 91, 141, 244–247, 254, **254**,
    258–261, **258**
protein(s)
  aggregation-associated diseases 310, 454
  -based recognition 243–255
  chameleon 3, 19–20, 23, 256–261, 264
  denaturation 33, 81–82, 282, 289, 359, 361,
    391, 443–445, 493–494
  dephosphorylation 28,
  folding; *see* folding
  intermediate-resolution (PRIME) model
    323–325
  intrinsic disorder xvi, 7, 8, 25–33, 38,
    242–243, 246–247, 256–258, 262, 265, 283,
    283
  kinase 13–14, 29, 32, 116–117, 250, 447, 451
  ligation 392
  ligand complexes 26
  modifications (PTMs) 26, 28, 29, 31–32
  modifiers xvii, 441–465
  –nucleic acid complex 414
  phosphorylation; *see* phosphorylation
  precipitation 429, 441, 443, 463
  –protein interaction (PPI) 16, 30–31, 35–8
    241–242, 245–246, 260, 442, 455, 498
    quartet model 4, 28
  sequence alignment 88–89, 222–223, 226,
    245–247, 261
  solubility 233, 391, 413–414, 443–445, 485
  stability 6, 18, 66, 72, 92, 124, 144, 179, 190,
    255, 290, 338, 355, 366, 368, 372, 374, 378,
    396, 413–417, 432, 442–445, 448, 451, 486,

    488, 494, 498, 499, 500, 501–502, 506, 509,
    514
  structure prediction 86–89, 138, 244
  structure-function relationship/paradigm xiii,
    4, 26, 38, 222
  trinity 4, 28
protein/peptide aggregation. *See also* protein/
    peptide aggregation mechanism
  inhibition 459, 500
  kinetics 389
  rates 355, 430–432, 454, 488
  suppression 444, 447, 450
  thermodynamics 389
protein/peptide aggregation mechanism 313, 320,
    324, 351, 352, 358, 390, 393, 398, 400, 401,
    418, 419, 420, 421, 427, 429, 431, 432, 433,
    455, 441, 465, 492, 497, 505, 508
  models 311–313, 320, 323, 324, 355, 413,
    414, 415, 418 420, 422–433
  pathways 321, 389, 393, 403, 404, 420,
    452–453, 505–506, 514
  rate constant 431–432
  resistant analogue 357, 359, 366, 378
  suppressing properties 444
protein/peptide anti-aggregation activity/ability
    458, 513
proteinase digestion 289
proteinase K 248
proteolysis 9, 248, 414, 489
proteolytic cleavages 483
proteolytic mapping 5
proteome 4, 32, 188
prothymosin 12–15, 19, 27
protofibril/protofibrillar 357, 358, 359, 363, 364,
    364, 480, 484, 486, 502, 514
protofibril growth 359, 363, 364
protofibril ribbon 359
protofibrillar intermediate 480, 486
protofilament 310–311, 509
proviral DNA synthesis 281
psi packaging sequence 282
purpurogallin 456–457
putative protein binding sites 255

quality control mechanisms 351, 454
quasiharmonic analysis 59
quasi-spherical oligomers 486
quercetin 457
quinone 457–458
  -bearing polyamines 458
Q_{WT} wild-type peptide 165–166

RADA-16 327, 331, 332, 334, 335
radius of gyration 58, 73, 81–82, 101, 163, 191,
    432

INDEX
545

raffinose 449
Ramachandran basin 85–88
  assignments 85, 87
  frequencies for each amino acid 85
Ramachandran distribution
  of backbone ϕ, χ dihedral angles 79
  of coil library 85
Ramachandran map/plot/sampling distribution
  xv, 83, 88–91, **90**, 168, 170, 173, **173**, 174
Ramachandran preferences for individual amino acid type 85
Ramachandran search space 86
Raman optical activity (ROA) spectroscopy 163–164
Raman spectroscopy 311, 389
random coil
  behavior 82, 168, 188, 207
  conformation 138, 165, 397, 491
  definition 207
  -like structure 223, 510, **510**
  model xv, xvi, 80, 82–83, 132, 188
  peptides 188–189, 213
  polymer 81
random flight-chain 211–212
random fuzziness 25
random hydrophobic collapse 100
random screening of chemical libraries 455
random unbiased search 80
Ras pathway 160
Ras signaling cascade 160
rational design
  of short peptide sequences 455
  strategies 356
Rattus norvegicus 223, 230
RBD, RNA binding domain 283, **283**, 290–293
RdRp, RNA dependent RNA polymerase 291–292
reaction-limited aggregation (RLA) 432–433
recombinant overexpression 390
reconciliation problem 82
reduced-dimensionality representation 58, 64, 66
reduced spectral density mapping 233
regular doublet split 193
regulation xvi–xvii, 4, 5, 8, 15, 26–39, 265, 279, 357, 414, 445
regulatory diversity 38
regulatory network 38
regulatory domain 256–257
regulatory processes 256
relaxation
  measurements 222, 230
  rate 142, 190–191, 198–199, 225–226, 230–234, **231**, **232**, 395–398, 405–407

replica exchange molecular dynamics (REMD) 326, 338, 339, 396, 489–498, 502, 505, 507, 513
replication
  primer tRNA 281–284, 287, 292
  protein A, heterotrimer 222–223
  protein A, RPA70 16, 222–223, 226–228, 233, 248, 259–260
representative ribbon representation 261, 368, 481, 494
residual dipolar couplings (RDCs) 85–86, 190, 200–202, 211–215
reticuloendothelial tissues 36, 356
retro human metallothionein-2 α domain 246
retro-proteins 245–246
retroviral Gag–NC–genomic RNA interaction 282, 284
retroviral nucleocapsid (NC) proteins 280–283
retrovirus 16, 279, 280–284
  replication 16, 280
reverse-phase HPLC 320, 390, 392
reverse transcriptase (RT) 93, 166, 256, 280–284, 287–288
  enzyme (RT) 280–284, 288
reversed fragment rF of protein B 247
reversible transition from a disordered to a ordered conformation 79
rheomorphic 3
ribbon diagram 399–400, **399**, **400**, 404
ribonucleoparticle (RNP) 279, 281, 292, 293
ribozyme activation 282
right-handed coiled coils 21, 86, 402
RNA 15–6, 20–23, 27, 30–33, 38, 226, 264, 279–293, **283**, 313, 460, **460**
  chaperone activity 283, **283**, 291–292
  -dependent RNA polymerase (RdRp) 291
  polymerization 284, 313
  –RNA interactions 279, 291
rnl homologues 226
RNP (ribonucleoprotein) 279–281, 292–293
rosmarinic acid 457
rotameric conformations/rotamers 78, 87, 173
rotational isomeric state model 82, 84
rotational tumbling rate 190, 193, 200, 230, 397,
RPA70 (linkers) 222–223, 226–228, 233, 248, 250–260
Rpeptide 392–393
RT enzyme 281, 283

S100B($\beta\beta$) 256
Saccharomyces cerevisiae 8, 223
sarcosine 449
SARS-coronavirus 283, **283**, 286–287, 290–291
saturation concentration 419–424, 427

saturation transfer 390, 404–405
  (difference) experiment 390, 404–405
Saupe matrix 201,
scale free (network) architecture 241–242
scaling law (function) 81, 394–395
scanning calorimetry 5
Schizosaccharomyces pombe 32
SDRD protein 27
SDS micelle 397
secondary structure
  prediction algorithm 138
  propensity scales 161, 175–177
secondary systemic amyloidosis 482
sedimentation 5
segmental labeling 392
segmental motion 228–231
self-aggregation xvii, 313
self-assembly 309, 313, 317, 326–332, 335–337, 340, 352, 355–357, 374, 444, 450, 455, 458
self-complementary oligopeptides 332
self-recognition elements (SREs) 311–312
Sem-5 SH3 domain 160–161, 164
semenogelin I 353
semi-flexible biopolymers 124
semi-stiff chain 104
senile plaques 398, 482–483
senile systemic 353, 482
sequential stabilization 86, 88
serine (Ser) 6–7, 17, 101, 117–122, 124, 166, 171, 330, 447
serotonin N-acetyltransferase (AANAT) 261
serum amyloid A protein 481
serum amyloid P-component 352
sesamol 456
severe acute respiratory syndrome (SARS) 290
sex muscle-5 (Sem-5) 160–161, 164
SH3 domain 160–161, 164–165, **165**, 246, 250
SH3:Sos system 161, 166
short angle neutron scattering (SANS) 5
short linear motif (SLiM) 250–251
sickle cell hemoglobin 420
side chain
  bulk 7, 122, 141, 148
  functionalities 311
  rotamers 87, 173
  steric blocking effect 148
signal transduction 117, 242, 256–257, 289
silk-mimicking biomaterials 328, 330–331
sin nombre hantavirus (SNV) 291
single-distribution analysis 113
single-molecule fluorescence techniques 15, 464
sirtuin 23, 259–260
site-directed mutagenesis 101, 139
SLiMDisc 250

small angle X-ray scattering (SAXS) 5, 11–12, 163, 213, 291
SNase 12, 27
S-nuclei 191
sodium dodecyl sulfate polyacrylamide (SDS) 320, 367, 397, 403, 452
solvation free energy 136, 137
solvent exclusion 413
solvent–solvent interactions 415
solvophilic surfaces 429
solvophobic cores 429
sorbitol 445, 446, 449
Sos ligand 161
Sos mutant 164–165
Sos peptide 161, 164–165, 173–174
spectral density mapping 233, 398
spermatogenesis 17,
sphere-to-rod transition 430
spider silk 328–330
spidroin proteins 328–330
spinocerebellar atrophy-1, -2, -3, -6, -7, and -17 35
spinodal /stability boundary 417–419
splenomegaly 462,
splice boundaries 30
spongiform encephalopathies 313, 353, 482, 484
Src-homology 3 (SH3) 160–161, 164–166, **165**, 174, 246, 250–252
stackers/β-arcs 20, 23, 24
staphylococcal nuclease 82
staphylococcal protein 245
static light scattering 5, 432
statistical-coil model 81, 188
statistical libraries and energies 85
Steinberg's equation 100, 105, 124
steric collisions 173, 179
steric hinderance 455
steric restrictions 26
steric violations 166
steric zippers 355, 506
sterically allowed conformations xv, 81, 160
sterically driven baseline propensity 174
sterics bias 179
stochastic driving 68
Stokes–Einstein relation 190
stopped-flow mixing 100–101
strand-coil transition model 137–138
strand-loop-strand (SLS) structure 488
structural plasticity 5, 18, 20, 31, 34, 250, 257
structural threshold hypothesis 507
structure
  -based drug discovery 403
  prediction 86–89, 138, 244
  -promoting effects 373

INDEX

subacute sclerosing panencephalitis, 5, 34
subgenomic mRNAs (sgmRNAs) 287, 290–291
subtilisin 248
sucrose 449
sulfobetaines 452
SUMOylation 26
supramolecular assemblies 358, 507
supramolecular structures 309, 311, 340
Swiss Protein database 7–8, 27, 28, 31, 35
Swiss-Prot sequences 8
SWNT-peptide 502–503
synthetic food additives 456
synucleinopathies 39, 264
Syrian hamster prion protein (ShPrP) 313–314
systemic amyloidosis 310, 482

TAF1 256
tannic acid 456–457
tau protein 5, 14, 34, 353–354, 456, 458, 482, 504
taurine 445–446, 449
taxonomic group 226, 233
Taz2 255
template assembly (TA) 311–312
tendon implants 329–330
tenosynovium AA 353
Tetra-Cys (Htt53/proline) 449
tetrafluoroehtylene (TFE) 320, 397, 451, 492
tetramethylsilane 197
TEV protease 392
Tfb1 260
TFIID 256
TFIIH 256
tGcn5 260
thaliana 8, 223, 230
THBP 456–457
thermal denaturation 81–83, 445
*T*hermus thermophilus chaperone ClpB 23–24
theta solvent 81, 83
theta temperature 416, **416**
thioflavin T (ThT) 310, 319, 321, 361–2, 366, 390, 400, **400**, 420, 455
thioflavin binding assay 361
three-helix bundle 391,
threonine (Thr, T) 117–120
thrombin 391
  recognition site 391
thymol 456
thymosin β4 247
time-correlated single photon counting 109
time-resolved fluorescence decays 110, **110**, 124
time-resolved FRET spectroscopy 104
tissue engineering 328, 331–332
  scaffolds 331–332

tissue transplantation 332
tissue-specific modulation of protein function 38
tissue-specific signaling 38
toxic species 315, 403, 480, 487
toxicity 318, 332, 351, 354, 357, 378, 389–403, 433, 456, 460–463, **460**, 479–480, 486, 492, 507, 511, **512**, 514
  of recombinant 390
toy model 365
trafficking-defective mutant proteins 462
trajectory exchange 326
transactivating response element (TAR) RNA sites 264
transactivation domain/region/segment 36, 38, 256–257
transcription
  factors 20, 24, 37, 248, 257, 281, 284, 318
  regulating sequences (TRSs) 287, 290
transcriptional regulation 5, 15
transcriptome organization 33
transient helical segment 227–228
transient structural elements 243–255
translational diffusion 394–397, 405
transmissible gastroenteritis virus (TGEV, a group 1 coronavirus) 290
transmissible spongiform encephalopathies (TSEs) 313, 353, 484–485
transmission electron microscopy (TEM) 358, 486
transthyretin 353, 356, 481–482
  variants and fragments 482
trehalose 445–449, 497, 500–501
trifluoroethanol (TFE) 179, 320, 397, 451, 492
trimethylamine-N-oxide (TMAO) 27, 445–449
trinucleotide
  expansions 317
  repeat sequences 34, 313
tRNA anti-codon loop ACUUUUAA 284, 287
Trp/Dbo FRET pair 101, 106–111, 116, 125
TRS. *See* transcription, regulatory sequences
trypsin 248, 335, 461
  inhibitor 335
tryptophan (Trp) 101, 103, 105, 106, 109–110, 114–115, 117, 120, 124, 162, 458–459
  emission 109
  -naphthoquinonone hybrid 458
tumbling 190, 193, 200, 276, 397
tumor-suppressor transcription factor 248
turbidity measurements 361
turn or bend like structure 396
turn-like polypeptide stretches 357, 378
tweezers 20, 23–24
two-pathway protein structure-function paradigm 26

TY3, yeast retrotransposon 3 282–283, **283**
type II diabetes 310, 455, 457, 479, 482
type II' β-turn 326, 492
tyrosine (Tyr) 6, 7, 106, 117, 162, 171, 175, 250, 316, 320, 368–374, 377, 464, 503

ubiquitin 29, 36, 88, 134, 142–143, 248, 351, 414, 455, 460–462, **460**
  enzyme (E1)-related cancer 455
  proteasome system 351, 414, 460, **460**, 462
ubiquitination 26–29, 461
UCSF Chimera package 368
unfolded peptides xvi, 3–40, 131–149, 160, 180, 201, 202, 206, 211, 212, 241–266, 315, 378,
unfolded proteins xvi–xvii, 3–40, **11**, 59, 62, 71, 74, 79–93, 100, 101, 108, 131–149, 160, 180, 187–191, 201–202, 206, 211–215, 230, 241–266, 313–316, 356, 378, 413, 444, 445, 461, 489, 496
unforced copy choice recombinations 281
unicellular organism 32
united-atom protein model 489
unstructured peptides xvi–xvii, 3, 26, 27, 131, 188, 215, 221, 242, 259, 262, 280, 282, 289–290, 293, 354, 367, 395–396, 403, 491, 502
untranslated regions (UTR) 280, 287
upper critical solution temperature (UCST) 416
urea 10–18, **11**, 81, 82, 131, 142–144, 180, 189, 212, 225, 248, 390, 392, 445, 450, 452, 491–494, 500
  solubilization 392
Usp2-c 392

valine (Val) 6, 7, 84, 87, 90, 139, 140, 141, 144, 148, 149, 162, 175, 179, 355, 405, 490, 504
van der Waals contact 119
van der Waals distance 119, 125
van der Waals interactions 139, 326, 430
van der Waals parameters 148
van der Waals radius 160, 173
vesicula seminalis 353
vibrational circular dichroism (VCD) spectroscopy 311, 316
vibrational spectroscopy 311
villin headpiece 60, 69–70, **70**, 75
  subdomain (HP-35 NleNle) 69–72, **70**, 75

viral DNA
  envelope (XMRV) 281
  infectivity factor Vif 284
  integrase (IN) 280–282
  mRNAs 280, 287, 292
  nucleocapsid shell 280
  RNA 279–280, 290–291
  RNA-dependent RNA polymerase (RdRp) 291–292
  synthesis 281–284, 287–288
  transactivator Tat 264, 281, 284
  VPR 284
virion genomic RNA 280
virus replication xvii, 280–3, 287, 289, 290, 292
viscoelastic properties of peptide-based hydrogels 327
viscogene 123
viscometry 5
viscous solvents 100, 104–105, 113
vitamins 456
VSL2P 258, App 13

water/air interfaces 452
water/solid interfaces 452
West Nile virus (WNV) 283, **283**, 286–289
wormlike-chain
  distribution 111–112, **112**
  model 104, 124, 212
wrappers 20, 23

X-ray crystallography 6, 8, 20, 36, 480
X-ray diffraction 108, 329, 352, 481, 503
X-ray structure 138, 205, 213
XAO peptide 163, 170
xenotropic tropism 281

yeast kinase-substrate network 32
yeast protein interaction network 31, 241, 243
yeast retrotransposon TY3 282–283
yeast ubiquitin hydrolase-1 (YUH-1) digestion 392
yellow fever virus (YFV) 288,
YM-3 centriprep 224

zinc finger (ZnF) (motif/protein) 20–1, 144–145, 282–284, 288
Zn^{2+} coordination 15, 282, 392, 463, 465
zuotin yeast (DNA binding protein) 331